$46.04
MW
1-31-84

Biochemistry of Exercise

Proceedings of the Fifth International Symposium
on the
Biochemistry of Exercise

June 1–5, 1982
Boston, Massachusetts
U.S.A.

International Series
on Sport Sciences, Volume 13

Biochemistry of Exercise

Scientific Editors

Howard G. Knuttgen, Ph.D.
Department of Health Sciences
Boston University
Boston Massachusetts, U.S.A.

James. A. Vogel, Ph.D.
U.S. Army Research Institute
of Environmental Medicine
Natick, Massachusetts, U.S.A.

Jacques Poortmans, Ph.D.
Institut Superieur
d'Education Physique
University Libre de Bruxelles
Bruxelles, Belgium

Human Kinetics Publishers, Inc.
Champaign, IL 61820

Publications Director: Richard D. Howell

Production Director: Margery Brandfon

Editorial Staff: John Sauget

Cover Design: Jack Davis

Series Editors:
Richard C. Nelson, Ph.D., and Chauncey A. Morehouse, Ph.D.
The Pennsylvania State University
University Park, Pennsylvania
U.S.A.

Volumes 1–11 in this series were published by University Park Press. Volumes 2, 4, 5, 7, 8, 9, 10, 11A, and 11B are available; Volumes 1, 3, and 6 are out of print.

Library of Congress Catalog Card Number: 82-84696

ISBN: 0-931250-41-2

Published by Human Kinetics Publishers, Inc.,
Box 5076, Champaign, Illinois 61820.

CONTENTS

Lecturers and Primary Authors   xvii
Symposium Organization   xxv
Acknowledgments   xxvii
Glossary of Exercise Terminology   xxix
Preface   xxxi

KEYNOTE LECTURE   1

Biochemical Bases of Fatigue in Exercise Performance:
Catastrophe Theory of Muscular Fatigue, *Richard H.T. Edwards*   3

INVITED LECTURERS   29

Muscle Fiber Activation and Recruitment, *V. Reggie Edgerton,
R.R. Roy, R.J. Gregor, C.L. Hager and T. Wickiewicz*   31

Pattern of Chemical Energetics in Fast- and Slow-Twitch
Mammalian Muscles, *Martin J. Kushmerick*   50

Substrate Availability, *Eric Hultman and H. Sjöholm*   63

Exercise Metabolism and Endocrine Function, *Atko Viru*   76

Exercise Training in Coronary Heart Disease, *John O. Holloszy,
A.A. Ehsani and J.M. Hagberg*   87

Metabolic Diseases and Exercise Performance, *Michael Berger*   97

PANEL: THE INTRACELLULAR ENVIRONMENT   111

The Intracellular Environment in Peripheral Fatigue,
Jacques R. Poortmans   113

Phosphorus Metabolites and the Control of Glycolysis Studied
by Nuclear Magnetic Resonance, *M. Joan Dawson*   116

Effect of Acidosis on Maximum Force Generation of Peeled
Mammalian Skeletal Muscle Fibers, *Sue K.B. Donaldson*   126

Effects of Temperature on Muscle Metabolism, *Ethan R. Nadel*   134

Control of Metabolism and the Integration of Fuel Supply
for the Marathon Runner, *Eric A. Newsholme*   144

Effect of Acidosis on Energy Metabolism and Force
Generation in Skeletal Muscle, *Kent Sahlin*   151

Morphological Studies of Muscle Fatigue, *Michael Sjöström
and J. Fridén*   161

**THE ELECTROMYOGRAPHIC SIGNAL IN
MEASURING AND UNDERSTANDING FATIGUE**   173

Some Properties of the Median Frequency of the Myoelectric
Signal during Localized Muscular Fatigue, *Carlo J. De Luca,
M.A. Sabbahi, F.B. Stulen and G. Bilotto*   175

Electromyography in Muscle Fatigue Studies: Power Spectrum
Analysis and Signal Theory Aspects, *Lars Lindström
and I. Petersén*   187

Electromyographic, Mechanical, and Metabolic Changes
during Static and Dynamic Fatigue, *Paavo V. Komi*   197

GENERAL METABOLISM   217

Familial Aggregation of PWC_{150}, Blood Lipid Components
and Serum Uric Acid in Adopted and Biological Siblings,
C. Allard, C. Leblanc, J. Talbot, S. Leclerc and C. Bouchard   219

Muscle Metabolism during High Intensity Eccentric
Exercise, *W.J. Evans, J.F. Patton, E.C. Fisher
and H.G. Knuttgen*   225

Effect of Exercise on Metabolism of Energy Substrates
in Pregnant Mother and Her Fetus in the Rat, *J. Gorski*   229

Changes in Muscle ATP, CP, Glycogen, and Lactate after
Performance of the Wingate Anaerobic Test, *I. Jacobs,
O. Bar-Or, R. Dotan, J. Karlsson and P. Tesch*   234

A Computer Simulation Model of Energy Output in Relation
to Metabolic Rate and Internal Environment, *A. Mader,
H. Heck and W. Hollman*   239

Substrate Sparing Effect of Atenolol Compared with
Propanolol during Intensive Short-Term Exercise,
A.A. McLeod, J.E. Brown, R.S. Williams and D.G. Shand   252

Physical Performance and Muscle Metabolic Characteristics,
P.A. Tesch, J.E. Wright, W.L. Daniels and B. Sjödin   258

Metabolic Pattern and Blood Flow of the Contracting Rat
Calf Muscle, *R. Wilke, D. Angersbach, and P. Ochlich*   264

The Influence of Recovery Duration on Repeated
Maximal Sprints, *S.A. Wootton and C. Williams* 269

Metabolic Responses of Single Rat Skeletal Muscle
Fibers to Tetanization and Ischemia, *D.A. Young
and O.H. Lowry* 274

CARBOHYDRATE METABOLISM **279**

Muscle and Liver Glycogen Resynthesis Following Oral
Glucose and Fructose Feedings in Rats, *D.L. Costill,
B. Craig, W.J. Fink and A. Katz* 281

Mitochondrial Substrate Oxidation, Muscle Composition
and Plasma Metabolite Levels in Marathon Runners, *K. Gohil,
D.A. Jones, G.G. Corbucci, S. Krywawych, G. McPhail,
J.M. Round, G. Montanari and R.H.T. Edwards* 286

Glycogen Synthesis: Effect of Diet and Training, *J.L. Ivy,
W.M. Sherman, W. Miller, S. Farrell and B. Frishberg* 291

Liver Glycogen Store and Hypoglycemia during Prolonged
Exercise in Humans, *J.-M. Lavoie, D. Cousineau, F. Peronnet
and P.J. Provencher* 297

Effect of Chronic Exercise on Glycogen Concentration
in Muscle Grafts of Rats, *F.S.F. Mong and J.L. Poland* 302

Plasma FFA Influence on Muscle Glycogen, *J.L. Poland,
J.W. Poland and J.M. Kennedy* 306

The Marathon: Recovery from Acute Biochemical Alterations,
*W.M. Sherman, D.L. Costill, W.J. Fink, L.E. Armstrong,
F.C. Hagerman and T.M. Murry* 312

LIPID METABOLISM **319**

Interactions Between Exercise Training and Sucrose Intake
on Adipocyte Lipolysis in Rat Parametrial
Adipose Tissue, *L.J. Bukowiecki, J. Lupien, G. Côté,
E. Cardinal and A. Vallerand* 321

Effects of Sex, Fatness and Training Status on Human
Fat Cell Lipolysis, *J.P. Després, C. Bouchard,
L. Bukowiecki, R. Savard and J. Lupien* 326

Lipoprotein Lipase Hydrolyzes Intramuscular Triacylglycerols
in Muscle of Exercised Rats, *L.B. Oscai* 331

Alterations in Composition of Venous Plasma FFA Pool during
Prolonged and Sprint Exercises in the Horse, *D.H. Snow,
L.M. Fixter, M.G. Kerr and C.M.M. Cutmore* 336

PROTEIN METABOLISM   343

Blood Ammonia and Glutamine Accumulation and Leucine
Oxidation during Exercise, *P. Babij, S.M. Matthews,
S.E. Wolman, D. Halliday, D.J. Millward, D.E. Matthews,
and M.J. Rennie*   345

Potential Biochemical Basis of Muscle Atrophy during
Prolonged Weightlessness, *N.M. Cintrón-Treviño,
C.S. Leach and P.C. Rambaut*   351

Acute Phase Proteins and Immune Complexes during Several
Days of Severe Physical Exercise, *B. Dufaux,
K. Höffken, and W. Hollman*   356

Alterations of the Content of Free Amino Acids in Skeletal
Muscle during Prolonged Exercise, *A.K. Eller and A.A. Viru*   363

Effect of Intense Prolonged Running on Protein Catabolism,
P.W.R. Lemon, D.G. Dolny and B.A. Sherman   367

Body Composition among College Wrestlers and Sedentary
Students with Emphasis on 3-Methylhistidine Excretion
and Calculation of Muscle Mass, *J. Mendez, W. Vollrath
and M. Druckemiller*   373

Cytochrome C Synthesis Rate is Decreased in the 6th Hour
of Hindlimb Immobilization in the Rat, *P.A. Watson,
A. Srivastava and F.W. Booth*   378

Alanine Formation during Maximal Short-term Exercise,
*H. Weicker, H. Bert, A. Rettenmeier, U. Oettinger,
H. Hägele and U. Keilholz*   385

LACTATE METABOLISM   395

Effects of Training on Oxidation of Injected [U-^{14}C]–
Lactate in Rats during Exercise, *G.A. Brooks and
L. Divine-Spurgeon*   397

Blood Pressure Response in Relation to Blood Lactate
during Exercise, *J. Karlsson, R. Dlin, F. Wahlberg,
R. Sannerstedt and C. Kaijser*   404

Maximal Blood Lactic Acid Concentration and its Recovery
Course after Exhaustive Graded Treadmill Exercise in
Young Men, *J.-S. Lai and I.-N. Lien*   411

Oxygen Uptake Kinetics and Lactate Accumulation in
Heavy Submaximal Exercise with Normal and High
Inspired Oxygen Fractions, *P.K. Pedersen*   415

Oxygen Consumption and Lactate Production in Varanid
and Iguanid Lizards: A Mammalian Relationship,
H.J. Seeherman, R. Dmi'el and T.T. Gleeson   421

Mathematical Approach to Lactate Kinetics in Short
Strenuous Exercise, *X. Sturbois and P. Jacqmin*   428

Physical Training in Humans: A Central or Peripheral Effect,
K. Young and R.J. Maughan   433

ENZYMES   439

Exercise-Induced Loss of Muscle Enzymes, *F.J. Cerny and
G. Haralambie*   441

Enzyme Profiles in Type I, IIA, and IIB Fiber Populations
of Human Skeletal Muscle, *B. Essén-Gustavsson and
J. Henriksson*   447

Cyclic Adenosine Monophosphate in Skeletal Muscle in
Response to Exercise, *A.H. Goldfarb and J.F. Bruno*   453

Kinetic Potential of Succinate Dehydrogenase during
Localized Muscular Exercise and Training, *S. Govindappa,
P. Reddanna and C.V.N. Murthy*   460

The Role of Superoxide Dismutase and Catalase in
Muscle Fatigue, *R.R. Jenkins*   467

Alterations in Substrate Supply and Circulatory Responses
to Exercise in Myophosphorylase Deficiency, *S.F. Lewis,
R.G. Haller, J.D. Cook and C.G. Blomqvist*   472

Effect of Physical Exercise on Erythrocyte Zinc and Carbonic
Anhydrase Isozyme in Men, *H. Ohno, F. Hirata, K. Terayama,
T. Kawarabayashi, R. Doi, T. Kondo and N. Taniguchi*   479

NUTRITION   485

Some Effects of the Quality of Protein, Food Restriction,
and Physical Exercise on Liver Development: Hepatic
Total Lipids, *M.C.R. Belda and S.M. Zucas*   487

Changes in Skeletal Muscle Metabolism Induced by a
Eucaloric Ketogenic Diet, *E.C. Fisher, W.J.Evans,
S.D. Phinney, G.L. Blackburn, B.R. Bistrian and V.R. Young*   497

Effect of Fat and Carbohydrate Supply on Myocardial
Substrate Utilization during Prolonged Exercise,
L. Kaijser, M.L. Wahlqvist, S. Rössner and L.A. Carlson   502

Lipoprotein Lipase Activity and Intramuscular Triglyceride
Stores in Conditioned Men: Effects of High Fat and
Low Fat Diets, *B. Kiens, B. Essén, P. Gad and H. Lithell*   508

Influence of Caffeine on Serum Substrate Changes during
Running in Trained and Untrained Individuals, *J.J. Knapik,
B.H. Jones, M.M. Toner, W.L. Daniels and W.J. Evans*   514

Glycogen Overloading in Rats: Effects of Various
Sugar Intakes, *M. Ledoux, L. Voghel, L. Brassard,
G. Brisson and F. Peronnet*   520

Effects of Exercise Training and Sucrose Intake on Cellular
Proliferation in Rat Parametrial Adipose Tissue, *J. Lupien,
G. Côté, E. Cardinal, A.L. Vallerand and L. Bukowiecki*   524

Effects of Physical Training and High Energy Diet on
Glucose Homeostasis in Male Rats, *D. Richard, A. Labrie,
A. Tremblay and J. Leblanc*   530

Substrate Utilization during Normal and Loading Diet
Treadmill Marathons, *S.E.H. Hall, J.T. Braaten, T. Bolton,
M. Vranic and J. Thoden*   536

ELECTROLYTES   **543**

Calcium Regulation of Myofibril ATPase Activity at
Exhaustion and Recovery, *A.N. Belcastro,
M.M. Sopper and M.P. Low*   545

The Changes in the Concentration and Total Amount of
the Electrolytes in the Blood Serum at Various
Muscular Exercises, *O. Imelik*   550

Muscle Electrolyte Changes in Young Exercised Rats,
*M.M. Jaweed, R.C. DeGroof, G.J. Herbison,
J.F. Ditunno, Jr. and C.P. Bianchi*   557

Sweat Electrolyte Losses during Prolonged Exercise in
the Horse, *M.G. Kerr and D.H. Snow*   564

Electron Microprobe Analysis of Fatigued Fast- and
Slow-twitch Muscle, *W.L. Sembrowich, D. Johnson,
E. Wang and T.E. Hutchinson*   571

ACID-BASE BALANCE   **577**

Leg Muscle pH following Sprint Running, *A. Katz, A. Barnett,
D.L. Costill, W.J. Fink and R.L. Sharp*   579

Skeletal Muscle Buffering Capacity in Elite Athletes,
*D.C. McKenzie, W.S. Parkhouse, E.C. Rhodes,
P.W. Hochochka, W.K. Ovalle, T.P. Mommsen
and S.L. Shinn*   584

The Relationship between Carnosine Levels, Buffering
Capacity, Fiber Type and Anaerobic Capacity in
Elite Athletes, *W.S. Parkhouse, D.C. McKenzie,
P.W. Hochochka, T.P. Mommsen, W.K. Ovalle, S.L. Shinn
and E.C. Rhodes*   590

Buffer Capacity of Blood in Trained and Untrained Males,
R.L. Sharp, L.E. Armstrong, D.S. King and D.L. Costill   595

Metabolic and Acid-base Exchange during Repetitive
Twitch Contractions of in Situ Dog Skeletal Muscle,
W.N. Stainsby and R.W. Barbee   600

TEMPERATURE   605

Economy of Isometric Exercise vs Temperature in the Cat
Soleus: An Isolated, Perfused Muscle Preparation,
A.A. Biewener, R. Karas, G. Goldspink and C.R. Taylor   607

Velocity Dependent Effect of Muscle Temperature on Short
Term Power Output in Humans, *A.J. Sargeant*   612

Exercise and Heat-induced Sweat, *T. Verde, R.J. Shephard,*
P. Corey and R. Moore   618

HORMONES-GENERAL   623

Effect of Exercise and Testosterone on the Active Form
of Glycogen Synthase in Human Skeletal Muscle, *K. Allenberg,*
N. Holmquist, S.G. Johnsen, P. Bennett, J. Nielsen,
H. Galbo and N.H. Secher   625

Dysadrenarche as a Possible Explanation for Delayed Onset
of Menarche in Gymnasts, *G.R. Brisson, M. Ledoux,*
S. Dulac and F. Peronnet   631

Plasma Leucine Enkephalin-like Radioreceptor Activity
and Tension-Anxiety before and after Competitive
Running, *P.A. Farrell, W.K. Gates, W.P. Morgan*
and C.B. Pert   637

Influence of the Training Level on the Dynamics of Plasma
Androgens at Rest and during Exercise in Male Dog,
J. Gagnon, D. DeCarufel, G.R. Brisson and R.R. Tremblay   645

Changes in Plasma Gut Hormones during Exercise in the Horse,
G.M. Hall, J.N. Lucke, T.E. Adrian and S.R. Bloom   652

Effects of Exhaustive Exercise on Prostaglandin
Metabolism in SHR Rat Kidney, *I. Hashimoto, M. Higuchi*
and K. Yamakawa   657

Exercise-induced Changes in Plasma Prostaglandin E and
Fα, Renin and Catecholamines in Hypertensive
Patients, *P. Lijnen, R. Fagard, J. Staessen and A. Amery*   662

The Effects of an Acute Exercise Bout on the Serum Level
of Testosterone and Luteinizing Hormone in Male Subjects
above Forty Years of Age, *G. Métivier*   667

HORMONES-INSULIN   673

The Influence of Physical Training on Glucose
Turnover and Hormonal Responses in Insulin-induced
Hypoglycemia, *H. Galbo, M. Kjaer, K.J. Mikines,
N.J. Christensen, B. Tronier, B. Sonne, J. Hilsted
and E.A. Richter*   675

Enhanced Insulin Sensitivity of Skeletal Muscle
following Exercise, *L.P. Garetto, E.A. Richter,
M.N. Goodman and N.B. Ruderman*   681

Endocrine Function of the Pancreas during Exercise,
*V.V. Menshikov, E.P. Gitel, T.D. Bolshakova,
V.G. Cukes and O.B. Dobrovolsky*   688

Regulation of Insulin Receptor Affinity during Exercise,
G. Michel, T. Vocke and W. Bieger   694

The Influence of Physical Training and High Energy
Intake on Glucose Tolerance and Insulin in
Human Subjects, *A. Tremblay, A. Nadeau, D. Richard
and J. Leblanc*   702

Effect of Exercise on Glucose Homeostasis in Humans
with Insulin and Glucagon Clamped, *R.R. Wolfe,
E.R. Nadel and J.H. Shaw*   707

HORMONES-CATECHOLAMINES   713

Effect of Physical Activity on β-Receptor Activity,
W.P. Bieger and R. Zittel   715

Regulation of Glycogenolysis in Human Muscle during
Epinephrine Infusion and during Exercise, *D. Chasiotis,
K. Sahlin and E. Hultman*   723

Perceived Exertion and Muscle Lactate Accumulation
during Exercise following β-Adrenergic Blockade,
P. Kaiser and P.A. Tesch   728

The Role of Catecholamines and Triiodothyronine on the
Calorigenic Response to Norepinephrine in Cold-
Adapted and Exercise-Trained Rats, *J. Leblanc, A. Labrie,
D. Lupien and D. Richard*   733

Time and Intensity Dependent Catecholamine Responses
during Graduated Exercise as an Indicator of Fatigue
and Exhaustion, *M. Lehmann, R. Kapp, M. Himmelsbach
and J. Keul*   738

Catecholamine Response to Maximal Anaerobic Exercise,
*I.A. Macdonald, S.A. Wootton, B. Muñoz, P.H. Fentem
and C. Williams*   749

The Influence of β-Adrenoceptor Blockade on the
Hemodynamic Response and Physical Performance of
Middle and Long Distance Runners, *D.D.J. Malan,
J.T. Fritz, A.C. Dreyer and G.L. Strydom* 755

Endurance and Metabolic Adjustments to Exercise in
Sympathectomized (6-OHDA) Rats, *F. Peronnet and
A. Imbach* 762

Alpha and Beta Adrenergic Effects on Muscle Metabolism
in Contracting, Perfused Muscle, *E.A. Richter,
N.B. Ruderman and H. Galbo* 766

NERVE AND MUSCLE 773

Differential Inter- and Intra-Muscular Responses to
Exercise: Considerations in Use of the Biopsy
Technique, *R.B. Armstrong, M.H. Laughlin,
J.S. Schwane, and C.R. Taylor* 775

Gas Tensions (O_2, CO_2, Ar and N_2) in Human Muscle during
Static Exercise and Occlusion, *F. Bonde-Petersen
and J.S. Lundsgaard* 781

The Effects of Two Isokinetic Training Regimens on
Muscle Strength and Fiber Composition, *V.M. Ciriello,
W.L. Holden and W.J. Evans* 787

Oxygen Consumption, Work and Efficiency as a Function
of Contraction Velocity in Isolated Frog Sartorius Muscle,
N.C. Heglund and G.A. Cavagna 794

Autonomic Nerve System and Metabolic Adaptations to
Long-Term Exercise, *M. Krotkiewski, P. Björntorp,
U. Smith, V. Marks, T. William-Olsson, A. Wirth
and K. Mandroukas* 799

The Effect of Training on Resting Muscle Membrane
Potentials, *R.F. Moss, P.B. Raven, J.P. Knochel,
J.R. Peckham and J.D. Blachley* 806

MYOGLOBIN 813

Myoglobin Function in Exercising Skeletal Muscle
in Hypoxia, *R.P. Cole* 815

Myoglobin Concentration and Training in Humans,
E. Jansson, C. Sylvén and B. Sjödin 821

Myoglobin Content of Normal and Trained Human
Muscle Fibers, *P.M. Nemeth, M.M.-L. Chi, C.S. Hintz
and O.H. Lowry* 826

CLINICAL EXERCISE   833

Rated Effort Angina, Perceived Leg Fatigue and
Blood Lactate during Graded Exercise, *H. Åström,
A. Holmgren, J. Karlsson and E. Orinius*   835

Blood Lactate Threshold in Trained Ischemic Heart
Disease Patients, *E.F. Coyle, W.H. Martin, A.A. Ehsani,
J.M. Hagberg and J.O. Holloszy*   840

Peripheral Responses and Adaptation to Treadmill
Exercise in Patients with Intermittent Claudication,
U. Maass and K. Alexander   846

The Effect of Long-term Physical Training on the
Relationship of Muscle Morphology to Metabolic
State and Insulin Sensitivity in Normal and
Hyperglycemic Obese and Diabetic Subjects,
*K. Mandroukas, M. Krotkiewski, P. Björntorp,
G. Holm, H. Lithell, U. Smith, P. Lönnroth and
G. Strömblad*   852

FATIGUE   857

Fatigue and Metabolic Patterns of Overloaded Fast-
twitch Rodent Skeletal Muscle Contracting in Situ,
K.M. Baldwin, S.L. Hillman and V. Valdez   859

Does a Reduction in Motor Drive Necessarily Result in
Force Loss during Fatigue? *B. Bigland-Ritchie,
R. Johansson and J.J. Woods*   864

Effects of Varied Dosages of Caffeine on Endurance
Exercise to Fatigue, *B.S. Cadarette, L. Levine,
C.L. Berube, B.M. Posner and W.J. Evans*   871

Membrane Permeability Changes as a Fatigue Factor in
Marathon Runners, *H.M. Gunderson, J.A. Parliman,
J.A. Parker and G. Bell*   877

Relationships between Fiber Type, Enzyme Activities,
Anaerobic Capacity and Human Muscle Fatigue,
*G. Lortie, C. Bouchard, J.A. Simoneau, C. Leblanc
and G. Thériault*   882

Delay of Fatigue Effects during Exhaustive Exercise by
Aldosterone, *W. Skipka and U. Schramm*   886

CLOSING SESSION   893

Fatigue in Retrospect and Prospect: ^{31}P NMR Studies
of Exercise Performance, *Britton Chance, A. Sapega,
D. Sokolow, S. Eleff, J.S. Leigh, T. Graham,
J. Armstrong and R. Warnell*   895

Fatigue in Retrospect and Prospect: Heritage, Present
Status and Future, *Philip D. Gollnick*   909

Author Index   922

Lecturers and Primary Authors

Claude Allard
Laval University, PEPS
Quebec, P.Q. Canada, GIK 7P4

K. Allenberg
August Krogh Institute
Universitetsparken 13
DK-2100, Denmark

Robert B. Armstrong
Department of Physiology
Oral Roberts University
7777 South Lewis
Tulsa, Oklahoma 74171, U.S.A.

Hans Åström
Departments of Thoracic Clinics
Karolinska Hospital
S-104 01 Stockholm, Sweden

Philip Babij
Department of Human Metabolism
Rayne Institute
University College London Medical
School
University Street
London, England WC1E 7HT

Kenneth M. Baldwin
Department of Physiology and
Biophysics
University of California at Irvine
Irvine, California 92717, U.S.A.

Angelo N. Belcastro
Department of Physical Education
University of Alberta
Edmonton, Alberta, Canada T6G 2H9

Cristina R. Belda
Department of Food and Nutrition
Faculty of Pharmaceutical Sciences
of Araraquara
Rua Humaitá 1680
São Paulo, Brazil

Michael Berger
Medizinischche Klinik E
Universitat Düsseldorf
Moorenstrasse 5, D-4000
Düsseldorf 1, Federal Republic
of Germany

W. Bieger
Department of Pathophysiology
University Heidelberg
Heidelberg, Federal Republic
of Germany

Andrew A. Biewener
Concord Field Station
Old Causeway Road
Bedford, Massachusetts 01730, U.S.A.

B. Bigland-Ritchie
John B. Pierce Foundation
290 Congress Avenue
New Haven, Connecticut 06519,
U.S.A.

F. Bonde-Petersen
Kompagnistraede 8
1208 Kobenhavn K, Denmark

Guy R. Brisson
Laboratoire d'Endocrinologie
Université de Québec
Trois-Rivières, P.Q.,
Canada G9A 5H7

George A. Brooks
Department of Physical Education
103 Harmon Gymnasium
University of California
Berkeley, California 94720, U.S.A.

Louis J. Bukowiecki
Laboratory of Endocrinology
Department of Physiology
Medical School
Laval University
Québec, G1K 7P4, Canada

Bruce S. Cadarette
U.S. Army Research Institute of
Environmental Medicine
Natick, Massachusetts 01760, U.S.A.

Frank J. Cerny
Children's Lung Center
219 Bryant Street
Buffalo, New York 14222, U.S.A.

Britton Chance
Johnson Research Foundation
University of Pennsylvania
School of Medicine
Philadelphia, Pennsylvania 19104,
U.S.A.

Dimitrios Chasiotis
Department of Clinical Chemistry II
Huddinge University Hospital
S-141 86 Huddinge, Sweden

Nitza M. Cintrón-Treviño
National Aeronautics and Space
Administration
Johnson Space Center
Houston, Texas 77058, U.S.A.

Vincent M. Ciriello
Liberty Mutual Research Center
Hopkinton, Massachusetts 01748,
U.S.A.

Randolph P. Cole
Department of Medicine
Columbia University
630 West 168th Street
New York, New York 10032, U.S.A.

David L. Costill
Human Performance Laboratory
Ball State University
Muncie, Indiana 47306, U.S.A.

Edward F. Coyle
Department of Preventive Medicine
Washington University Medical
School
509 South Euclid Avenue
St. Louis, Missouri 63110, U.S.A.

M. Joan Dawson
Department of Physiology
University College London
Gower Street
London WC1E 6BT, England

Carlo J. De Luca
Children's Hospital Medical Center
300 Longwood Avenue
Boston, Massachusetts 02115,
U.S.A.

Jean-Pierre Després
Physical Activity Sciences Laboratory
Laval University, Ste-Foy
Québec, G1K 7P4, Canada

Sue K.B. Donaldson
Department of Physiology
College of Health Sciences
Rush University
1753 West Congress Parkway
Chicago, Illinois 60612, U.S.A.

B. Dufaux
Institut für Kreislauff
Carl Diem Weg
5 Köln 41, Federal Republic
of Germany

V. Reggie Edgerton
Department of Kinesiology
University of California
Los Angeles, California 90024, U.S.A.

Richard H.T. Edwards
Department of Medicine
School of Medicine
University College London

University Street
London WC1E 6 JJ, England

A.K. Eller
Department of Sports Physiology
Tartu State University
Ylikooli 18, Tartu 202400
Estonian S.S.R., U.S.S.R.

Birgitta Essén-Gustavsson
Department of Medicine I
College of Veterinary Medicine
Swedish University of Agricultural
Sciences
Uppsala, Sweden

William J. Evans
Physiology Laboratory
U.S.D.A. Human Nutrition Research
Center on Aging
Tufts University
711 Washington Street
Boston, Massachusetts 02111, U.S.A.

Peter A. Farrell
Department of Human Kinetics—
423 Enderis Hall
University of Wisconsin-Milwaukee
Milwaukee, Wisconsin 53201, U.S.A.

Elizabeth C. Fisher
Department of Health Sciences
Boston University
36 Cummington Street
Boston, Massachusetts 02215, U.S.A.

Jacques Gagnon
Laboratoire d'Endocrinologie-
Métabolisme
Le Centre Hospitalier de
l'Université Laval
2705, Boulevard Laurier
Ste-Foy, Québec, Canada G1V 4G2

Henrik Galbo
Department of Medical Physiology,
B. Panum Institute, 3C, Blegdamsvej
DK-2200 Copenhagen N, Denmark

Lawrence P. Garetto
Division of Diabetes, E-211
University Hospital

75 East Newton Street
Boston, Massachusetts 02118, U.S.A.

K. Gohil
Department of Medicine
Faculty of Clinical Sciences
The Rayne Institute
University College London
University Street
London, England WC1 6JJ

Alan H. Goldfarb
Exercise Science Lab
University of Maryland
College Park, Maryland 20742, U.S.A.

Philip D. Gollnick
Department of Physical Education
for Men
Washington State University
Pullman, Washington 99164, U.S.A.

Jan Gorski
Department of Physiology
Upstate Medical Center
766 Irving Avenue
Syracuse, New York 13210, U.S.A.

Sepur Govindappa
Exercise Physiology Division
Department of Zoology
Sri Venkateswara University
Tirupati—517 502, A.P., India

Hans M. Gunderson
Departments of Chemistry and
Mathematics
Northern Arizona University
Flagstaff, Arizona 86011, U.S.A.

George M. Hall
Department of Anaesthetics
Hammersmith Hospital
Royal Postgraduate Medical
London W12 OHS, England

Susan E. Hall
Division of Metabolism and
Endocrinology
Ottawa Civic Hospital
Curling Avenue
Ottawa Ontario, Canada

Isao Hashimoto
Exercise Physiology Laboratory
National Institute of Nutrition
1-23-1 Toyama
Shinjuku-ku, Tokyo 162, Japan

Norman Heglund
Concord Field Station
Old Causeway Road
Bedford, Massachusetts 01730, U.S.A.

John O. Holloszy
Washington University
Department of Preventive Medicine
4566 Scott Avenue
St. Louis, Missouri 63110, U.S.A.

Eric Hultman
Department of Clinical Chemistry II
Karolinska Institutet
Huddinge sjukhuset
S-141 86 Huddinge, Sweden

O. Imelik
Department of Sports Physiology
Tartu State University
Ylikooli 18, Tartu 202400
Estonian S.S.R., U.S.S.R.

John L. Ivy
Department of Physical Education
and Pharmacology
University of South Carolina
Columbia, South Carolina 29208,
U.S.A.

Ira Jacobs
Department of Clinical Physiology
Karolinska Hospital
S-104 01 Stockholm, Sweden

Eva Jannson
Department of Clinical Physiology
Karolinska sjukhuset
S-104 01 Stockholm, Sweden

M. Mazher Jaweed
Department of Rehabilitation
Medicine
Thomas Jefferson University
Tenth and Walnut Street

Philadelphia, Pennsylvania 19107,
U.S.A.

Robert R. Jenkins
Biology Department
Ithaca College
Ithaca, New York 14850, U.S.A.

Lennart Kaijser
Department of Clinical Physiology
Karolinska sjukhuset
S-104 01 Stockholm, Sweden

Peter Kaiser
Department of Clinical Physiology
Karolinska sjukhuset
S-104 01 Stockholm, Sweden

Jan Karlsson
Laboratory for Human Performance
Department of Clinical Physiology
Karolinska sjukhuset
S-104 01 Stockholm, Sweden

Abram Katz
Human Performance Laboratory
Ball State University
Muncie, Indiana 47306, U.S.A.

Morag G. Kerr
Department of Veterinary Medicine
Royal Veterinary College
Hawkshead House, Hawkshead Lane
North Mymms, Hatfield
Hertfordshire, England

Bente Kiens
August Krogh Institute
University of Copenhagen
13, Universitetsparken
2100 Copenhagen Ø, Denmark

Joseph J. Knapik
Exercise Physiology Division
U.S. Army Research Institute of
Environmental Medicine
Natick, Massachusetts 01760, U.S.A.

Paavo V. Komi
Kinesiology Laboratory
University of Jyväskylä
40 100 Jyväskylä 10, Finland

Marcin Krotkiewski
Departments of Rehabilitation
Medicine and Internal Medicine
Sahlgren's Hospital
University of Göteborg
Göteborg, Sweden

Martin J. Kushmerick
Department of Physiology and
Biophysics
Harvard Medical School
25 Shattuck Street
Boston, Massachusetts 02115, U.S.A.

Jin-Shin Lai
Department of Rehabilitation
National Taiwan University Hospital
Taipei, Taiwan 100
Republic of China

Jean-Marc Lavoie
Departement d'éducation physique
Université de Montréal
Montréal, Quebéc, Canada H3C 3J7

Jacques LeBlanc
School of Medicine (Physiology)
Laval University
Quebéc City, Canada

Mariellie Ledoux
Department of Nutrition
Université de Montréal
C.P. 6128
Montréal, Quebéc, Canada H3C 3J7

M. Lehmann
Department of Internal Medicine
Division of Performance and Sports
Medicine
University of Freiburg
Freiburg, Federal Republic
of Germany

Lars Lindström
Department of Clinical
Neurophysiology
University of Göteborg
Sahlgrenska sjukhuset
S-413, 45 Göteborg, Sweden

P.W.R. Lemon
Applied Physiology Laboratory
Kent State University
Kent, Ohio 44242, U.S.A.

Steven F. Lewis
Pauline and Adolph Weinberger
Laboratory of Cardiopulmonary
Research
Department of Internal Medicine
Physiology and Neurology
University of Texas Science Center
Dallas, Texas 75235, U.S.A.

P. Lijnen
Hypertension Unit
Camput Gasthuisberg
Herestraat 49
B-3000 Leuven, Belgium

Gilles Lortie
Physical Activity Sciences Laboratory
Laval University
Québec, Canada G1K 7P4

J. Lupien
Laboratory of Endocrinology
Department of Physiology
Medical School
Laval University Québec
Province of Québec, Canada G1K 7P4

Ulrich Maass
Department of Internal Medicine
Division of Angiology
D-3000 Hannover 61, Federal
Republic of Germany

Ian A. Macdonald
Department of Physiology and
Pharmacology
University of Nottingham Medical
School
Nottingham NG7 2 UH,
United Kingdom

Alois Mader
Institut fur Kreislaufforschung
Carl Diem Weg
5000 Köln 41, Federal Republic
of Germany

xxii   Lecturers and Primary Authors

D.D.J. Malan
Johannes van der Walt Institute
for Biokinetics
Department of Physical Education
University of Potchefstroom
for C.H.E.
Potchefstroom, South Africa 2520

K. Mandroukas
Institute of Rehabilitation Medicine
University of Göteborg
Övre Husargatan 36
S413 14 Göteborg, Sweden

Donald C. McKenzie
Departments of Sports Science,
Sports Medicine, Anatomy
and Zoology
The University of British Columbia
Vancouver, B.C., Canada V6T 1W5

Andrew A. McLeod
Division of Clinical Pharmacology
Box 3813
Duke University Medical Center
Durham, North Carolina 27710,
U.S.A.

Jose Mendez
The Pennsylvania State University
101 Lab for Human Performance
University Park, Pennsylvania 16802,
U.S.A.

Vadim V. Menshikov
State Central Institute of Physical
Education
Syrenevyi bulv., 4
Moscow, 105483, U.S.S.R.

Guy Métivier
School of Human Kinetics
University of Ottawa
Ottawa, Ontario, Canada

G. Michel
Medizin Poliklinik-Medizin VII
Hospitalstrasse 3
D-6900 Heidelberg, Federal Republic
of Germany

Franz S. Mong
Department of Anatomy and
Physiology
Medical College of Virginia
Richmond, Virginia 23298, U.S.A.

Raymond F. Moss
Department of Rehabilitation/Sports
Medicine
1501 Merrimac Circle
Fort Worth, Texas 76107, U.S.A.

Ethan R. Nadel
John B. Pierce Foundation
Laboratories
290 Congress Street
New Haven, Connecticut 06519,
U.S.A.

Patti M. Nemeth
Departments of Neurology,
Neurosurgery and Pharmacology
Washington University School of
Medicine
St. Louis, Missouri 63110, U.S.A.

Eric A. Newsholme
Department of Biochemistry
University of Oxford
South Parks Road
Oxford OX1 3QU, England

Hideki Ohno
Department of Hygiene and
Preventive Medicine
Asahikawa Medical College
Asahikawa 078-11, Japan

Lawrence B. Oscai
Department of Physical Education
University of Illinois at Chicago
Chicago, Illinois 60680, U.S.A.

Wade S. Parkhouse
Department of Sports Science
The University of British Columbia
Vancouver, B.C, Canada V6T 1W5

Preben K. Pedersen
Institute of Physical Education
Odense University

Campusvej 55
DK 5230 Odense M, Denmark

Francois Peronnet
Départment D'Éducation Physique
Université de Montréal, CP-6128-A
Montréal, Québec, Canada H3C 3J7

James L. Poland
Department of Physiology
Medical College of Virginia
Richmond, Virginia 23298, U.S.A.

Jacques R. Poortmans
Chimie Physiologique
Institut Supérieur D'Éducation
Physique et de Kinésithérapie
Université Libre de Bruxelles
B-1050 Bruxelles, Belgium

Denis Richard
Department of Physiology
Faculty of Medicine
Laval University
Québec, Canada G1K 7P4

Eric A. Richter
Division of Diabetes and Metabolism
University Hospital
75 East Newton Street
Boston, Massachusetts 02118, U.S.A.

Kent Sahlin
Department of Clinical Chemistry II
Huddinge University Hospital
S-141 86 Huddinge, Sweden

Anthony J. Sargeant
Human Physiology Laboratory
Polytechnic of North London
London, NW5 3LB, England

Howard J. Seeherman
Concord Field Station
Old Causeway Road
Bedford, Massachusetts 01730, U.S.A.

W.L. Sembrowich
Cardiac Pacemakers, Inc.
Physiology Research Department
4100 North Hamline Avenue

P.O. Box 43079
St. Paul, Minnesota 55112, U.S.A.

Rick L. Sharp
Human Performance Laboratory
Ball State University
Muncie, Indiana 47306, U.S.A.

William M. Sherman
Department of Physical Education
Exercise Biochemistry Laboratory
University of Texas at Austin
Austin, Texas 78712, U.S.A.

Michael Sjöström
Department of Anatomy
University of Umeå
S-901 87 Umeå, Sweden

Werner Skipka
Institute of Physiology
Deutsche Sporthochschule Köln
Köln, D-5000, Federal Republic
of Germany

David H. Snow
Veterinary School
Bearsden Road
Bearsden, Glasgow, Scotland

Wendell N. Stainsby
Department of Physiology
College of Medicine
University of Florida
Gainesville, Florida 32610, U.S.A.

Xavier Sturbois
Laboratory of Effort
Unit EDPH—UCL—1
Place de Coubertain bte
B1348 Louvain-la Neuve, Belgium

Per A. Tesch
Department of Environmental
Medicine
Karolinska Institutet
S-104 01 Stockholm, Sweden

Angelo Tremblay
Department of Physiology
Laval University
Québec, Canada G1K 7P4

T. Verde
320 Huron Street
Toronto, Ontario M5S 1A1, Canada

Atko Viru
Department of Sports Physiology
Tartu State University
Ylikooli 18, Tartu 202400
Estonian S.S.R., U.S.S.R.

Peter A. Watson
Department of Physiology
University of Texas Medical School
at Houston
Houston, Texas 77030, U.S.A.

Helmut Weicker
Abteilung Pathophysiologie und
Sportmedizin
Universität Heidelberg
Hospitalstrasse 3
D-6900 Heidelberg, Federal Republic
of Germany

Rolfe Wilke
Beecham-Wülfing GmbH and
Company KG

Bethelner Landstrasse 15, D-3212
Gronau/Leine, Federal Republic
of Germany

Robert R. Wolfe
51 Blossom Street
Boston, Massachusetts 02114, U.S.A.

Stephen A. Wooton
Department of Physical Education
and Sports Science
University of Technology
Loughborough, Leicestershire, LE11
3TU, England

Douglas A. Young
Pharmacology Department
Washington University Medical
School
St. Louis, Missouri 63110, U.S.A.

Kevin Young
Department of Physiology and
Pharmacology
University Medical School
Queen's Medical Centre
Nottingham, England

Symposium Organization

Research Group on the Biochemistry of Exercise

International Council of Sports and Physical Education

Howard G. Knuttgen
Organizing Committee Chairman

Jacques Poortmans
Research Group Chairman
Symposium President

Organizing Committee
C. Richard Taylor
Harvard University
James A. Vogel
U.S. Army Research Institute
of Environmental Medicine
Vernon R. Young
Massachusetts Institute
of Technology
William J. Evans
Boston University
Howard G. Knuttgen (Chairman)
Boston University

Research Group on the Biochemistry of Exercise
Pietro diPrampero
Geneva, Switzerland
George Haralambie
Freiburg, West Germany
Lars Hermansen
Oslo, Norway
John O. Holloszy
St. Louis, U.S.A.
Hans Howald
Magglingen, Switzerland
Nikolai Jakovlev
Leningrad, U.S.S.R.
Joseph Keul
Freiburg, West Germany
Howard G. Knuttgen
Boston, U.S.A.
Guy Métivier
Ottawa, Canada
Eric Newsholme
Oxford, England
Bengt Saltin
Copenhagen, Denmark
Jacques Poortmans (Chairman)
Bruxelles, Belgium

Acknowledgments

Benefactors

Major financial support for the Symposium was provided by:

- Department of the Air Force, Office of Scientific Research
- Department of the Army, Medical Research and Development Command
- Department of the Navy, Office of Naval Research
- National Aeronautics and Space Administration

The Organizing Committee also wishes to acknowledge the generous contributions of the following:

- Johnson & Johnson
- Pripps Pluss
- National Dairy Council
- Ross Laboratories
- Stokely-Van Camp, Inc.

Editorial Staff

Technical Editors:
Cheryl L. Riegger and Nancy E. O'Hare
Department of Health Sciences
Boston University
Boston, Massachusetts
U.S.A.

Editorial Assistants:
Nancy J. Hochman
Jonathan S. Herland
Joseph J. Knapik
Patricia I. Fitzgerald
Jorge Pinto Ribeiro
Roger A. Fielding

Glossary of Exercise Terminology

To promote consistency and clarity of communication and avoid ambiguity, the following terms were defined as indicated for all symposium presentations:

Exercise: Any and all activity involving generation of force by the activated muscle(s). The activity may involve maintenance of posture, movement of the vertebrate skeleton, movement of fluids, etc. In dynamic exercise, the muscle may perform shortening (concentric) contractions or be overcome by external resistance and perform lengthening (eccentric) contractions. When muscle force results in no movement, the contraction should be termed static or isometric.

Work: Force expressed through a distance but with no limitation on time (unit: joule). Quantities of energy and heat expressed independently of time should also be presented in joules. The term "work" will *not* be used as synonymous with muscular exercise.

Power: Force expressed through a distance per unit of time (unit: watt). Other related processes such as energy release and heat transfer should, when expressed per unit of time, be quantified and presented in watts.

Strength: The maximal force that can be exerted in a specific movement and, therefore, described according to type of contraction (concentric, isometric, eccentric), velocity for dynamic contractions, linear displacement or angular change, etc. Quantification should be expressed in terms of force (unit: newton) or torque (unit: newton-meter). A voluntary strength assessment of a human subject should be identified as involving a maximal voluntary contraction (MVC).

Endurance: The time limit of a person's ability to maintain either a specific isometric force or a specific power level involving combinations of concentric or eccentric muscular contractions.

Fatigue: The inability of a physiological process to continue functioning at a particular level and/or the inability of the total organism to maintain a predetermined exercise intensity.

Exhaustion: Complete inability of a muscle or organism to maintain exercise.

Exercise Intensity: A specific level of maintenance of muscular activity that can be quantified in terms of power (energy expenditure or work performed per unit of time) or isometric force sustained.

Preface

This symposium is the fifth in a series of international meetings dealing with the biochemistry of muscular exercise as organized by the Research Group on the Biochemistry of Exercise (of the International Council of Sports and Physical Education). Earlier meetings with a similar format were held at Bruxelles, Belgium (1968), Magglinen, Switzerland (1973), Quebec City, Canada (1976), and Bruxelles, Belgium (1979). The Research Group was formed with the objectives of furthering the knowledge of muscular exercise to better understand the basic biological mechanisms involved and to provide information of practical value to clinicians, scientists, and educators interested in the role of exercise in health maintenance, disease prevention, and rehabilitation.

Held in Boston, Massachusetts, this is the first time that the United States has been privileged to host such a meeting. With lectures and scientific sessions taking place on the campus of Boston University, the organization of the Fifth International Symposium was accomplished by personnel from Boston University, the US Army Research Institute of Environmental Medicine, Harvard University, and the Massachusetts Institute of Technology.

The format of the 5-day meeting included the presentation of invited lectures, two panel sections, and free communication sessions which included both oral and poster presentations. The manuscripts from all of the investigators making the various presentations have been organized and presented in this publication according to these groupings. The invited lectures all address themselves to various aspects of the symposium theme, "The Biochemical Bases of Fatigue," as well as the two panel sections which dealt with the specific topics of the intracellular environment and the electromyographic signal in measuring and understanding fatigue.

Over 100 free communications deal with a wide range of topics within the general area of the biochemistry of exercise. Presentations were made as 10-minute lectures with slides (followed by 5 minutes of audience discussion) or poster presentations which were on view for periods of 5 hours each day (with the primary author present for questions and discussion during the final hour). Manuscripts from both the oral and poster presentations have been arranged in sections for this publication according to general topic areas subsequently designated. It should be understood by the reader that considerable overlap exists among manuscripts appearing in many of the sections and careful scrutiny of the contents will identify particular areas of interest.

During the last few years, we have witnessed an extremely rapid advancement of our knowledge of physical exercise, particularly at the cellular level, as a result of an increased interest in exercise research and an increased sophistication of investigatory techniques. One of the general objectives of this particular symposium was the dissemination of information relative to the present state of knowledge and understanding of the biochemistry of exercise. Presentation of this publication is made as an additional step in the accomplishment of this objective.

Howard G. Knuttgen
Boston, Massachusetts, U.S.A.

James A. Vogel
Natick, Massachusetts, U.S.A.

Jacques Poortmans
Bruxelles, Belgium

Keynote Lecture

Biochemical Bases of Fatigue in Exercise Performance: Catastrophe Theory of Muscular Fatigue

Richard H.T. Edwards
University College London and
University College Hospital London, London, England

I thank the organizers of this Symposium for the honor of being invited to present this Keynote Address. The invitation requested that I "set the stage" for the following days of meetings by addressing the topic of "fatigue" in exercise in its broadest context:

> "—*for large muscle activity as well as specific muscle groups*
> —*for patients, everyday exercisers, and elite athletes*
> —*for low intensity aerobic exercise, maximal aerobic power and high-intensity anaerobic exercise*
> —*etc.*"

The definition of fatigue proposed by the organizers of the Symposium is stated as "the inability of a physiological process to continue functioning at a particular level and/or the inability of the total organism to maintain a predetermined exercise intensity." I have hitherto defined fatigue as the "failure to maintain a required or expected force" (Edwards, 1981). To show that I learned from participating in the Ciba Foundation Symposium (No. 82, 1981) I am extending my definition to include "failure to maintain the required or expected power output." As such, my revised definition of fatigue, including both force and power output, is thus not far from that recommended by the organizers of this Symposium.

The published proceedings of the Ciba Foundation Symposium provides an up-to-date account of current ideas on muscle fatigue. I do not therefore intend to go over the same ground in this Keynote Address. Rather, I would hope that my remarks here will be a direct development of ideas expressed at that Symposium and will point to questions for further research. Since I have, with co-authors, published detailed reviews of muscle biochemistry and physiology (Edwards & Wiles, 1981; Rennie & Edwards, 1981) I hope I may be forgiven if I concentrate on the last of the categories, the "et ceteras," to which I was asked to address my remarks. Others here are better qualified to discuss the other categories and will do so during the Symposium.

My interest in exercise was one which arose not so much from athletics or sport but from my enduring interest in mountains. In mountaineering, muscles are used in three main ways: "climbing up, climbing down and hanging on" (Edwards, 1975). These functions of muscle correspond to dynamic exercise with concentric contractions (positive work), eccentric contractions (negative work) and isometric contractions, respectively. I shall be discussing fatigue in these different forms of contraction in due course. In particular, I shall attempt to describe the function and biochemistry of human muscle in vivo in the same terms as those which have been applied to the study of isolated preparations in vitro. As such, this presentation owes much in inspiration to the marvelous account of the function of human muscle, including fatigue, given at the Royal Institution Christmas lectures for young people in 1926 by the late Professor A.V. Hill under the title "Living Machinery" (Hill, 1927). The biochemistry of muscular contraction in its historical development is reviewed in that scholarly volume entitled "Machina Carnis" (Needham, 1971).

I have long been an admirer of the physiologists of the first part of this century who mapped out the broad territory of exercise physiology. It is salutary to read the extent of understanding of physiological processes given in Bainbridge's "Physiology of Muscular Exercise" (3rd Edition, Bock & Dill, 1931). Muscle fatigue was the subject of an extensive work by Angelo Mosso (Mosso, 1915). In this book Mosso describes his ergograph which he used to measure fatigue curves on different subjects before and after varied forms of physical and mental stress. He describes fatigue curves before and after a professorial inaugural lecture and after the high altitude ascent to his laboratory near the summit of Monte Rosa. Perhaps most of all I have been influenced by the writings of Sir Joseph Barcroft in his essay entitled "Features in the Architecture of Physiological Function" (Barcroft, 1934). In this he describes exercise as an example of his aphorism "every adaptation is an integration." The idea of "integration" has intrigued several workers since that time (e.g., Edwards et al., 1969; Jones et al., 1975; Wasserman, van Kessel & Burton, 1967). The approach has proved to be of value in analyzing exercise performance in patients as well as explaining superior performance of elite athletes.

In Figure 1a a comparison of the factors contributing to the ventilatory drives in exercise is shown for a patient with obstructive lung disease, a

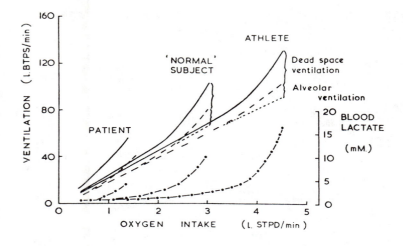

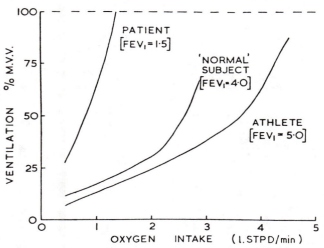

Figure 1—Illustrative examples of ventilatory response to exercise in a patient with obstructive lung disease, "normal subject" and a more athletic subject. Curves are based on typical figures found in published exercise studies (Jones, Campbell, Edwards & Robson, 1975). Ventilation is expressed as % maximum voluntary ventilation (MVV) as defined by Freedman, 1970.

"normal" subject (of fairly sedentary habitual activity) and an athlete. This schematic representation summarizes our understanding of the differences in exercise performance and the factors limiting exercise capacity. It underlines an important concept that in comparing individuals it is necessary to recognize that responses must be related to the individual's

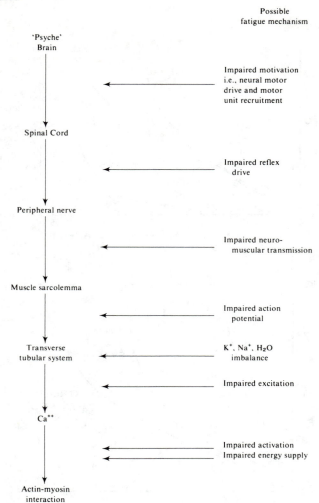

Figure 2—Chain of command for muscular contraction and the possible mechansims underlying fatigue.

capacity. However, biochemical and energetic considerations require energy expenditure to be analyzed in absolute terms. Thus the patient, who is less "fit" than the normal subject or athlete, is for this reason more likely to accumulate lactic acid. Nevertheless the total accumulation and therefore the blood concentration by the time the patient stops exercising is still lower than in the other two groups because the absolute work rate is so much lower. The patient is shown as having been stopped because ventilation has reached the ventilatory capacity. Note that the greater motivation on the part of the athlete can allow a higher proportion of the ventilatory capacity to be achieved (Edwards et al., 1969) and also a greater lactate accumulation to result than in the normal subject. The point of "fatigue" (taken as the end point of the progressive exercise test) cannot thus be simply related to the achievement of any critical lactate concentration. Indeed, recognition of the limiting factor for exercise performance is frequently confused, particularly in the less well trained, by the constellation of symptoms which include muscular fatigue, pain (in muscles and elsewhere), perception of excessive effort and breathlessness. In whole-body exercise there are ways of distinguishing the dominant limiting factor in patients (Jones et al., 1975). The identification of more specifically muscular limiting factors has been the subject of my own particular interest and that of my collaborators over the past 12 years (Edwards, 1978, 1979, 1981). In those 12 years there has been an enormous increase in the interest in exercise physiology and in characterizing the differences in exercise performance between individuals, ranging from patients with cardio-respiratory or metabolic disturbances to elite athletes specializing in different events. Since these will be the subject of presentations and discussions during this Symposium, I crave forgiveness for not making due reference to this type of study in my present review. Suffice it to say that it is now possible to prepare a scheme indicating the chain of command for muscular contraction (Figure 2) and the possible mechanisms underlying fatigue where this might be due to a failure at one or more of the links of that chain. Since this scheme owes much to studies on isolated muscle preparations it is necessary to remember in the extrapolation to the function of muscle groups or man as a whole that only when parallel fatigue, i.e., dropping out of parallel-acting, force-producing elements (motor units or individual cells), has been eliminated can contractile fatigue, i.e., failure of force maintenance by the individual elements, be recognized. Much of what follows in this Symposium is based on the assumption that muscle as a whole reflects the metabolic changes in the individual muscle cell. This practical assumption may be far from justified in fatigued or diseased muscle. A practical classification of fatigue is that based on the response to electrical stimulation at different frequencies (Table 1).

Table 1

Physiological Classification of Fatigue
(after Edwards, 1979)

Type of fatigue	Definition	Possible mechanism
A. Central	Force or heat generated by voluntary effort less than that by electrical stimulation	Failure to sustain recruitment and/or frequency of motor units
B. Peripheral	Same force loss or heat generation with voluntary and stimulated contractions	
a) High frequency fatigue	Selective loss of force at high stimulation frequencies (with impaired E.M.G.)	Impaired neuromuscular transmission and/or propagation of muscle action potential
b) Low frequency fatigue	Selective loss of force at low stimulation frequencies (without E.M.G. impairment)	Impaired excitation-contraction coupling

A "Law of Fatigue"

It is an everyday experience that high-intensity exercise of large contraction forces can be sustained for only short periods of time. This "law" has been variously expressed since first reported in the case of racing animals and man (Kennelly, 1906; Wilkie, 1980). A similar "law" has been found to apply to the respiratory muscles (Freedman, 1970) and to isometric contractions (Rohmert, 1960). Figure 3 is a theoretical consideration of Rohmert's curve which was based on over several thousand observations in more than one group of muscles in normal subjects. Some years ago I was aware of variation in the observed endurance time and was impressed to discover that possible metabolic determinants of endurance were likely to be of far less significance in the statistical description of the curve than was the accuracy of determination of the maximum voluntary contraction force. Here, then, is a simple clue that central mechanisms may in fact play a more important role in determining endurance than do biochemical changes in muscle. A further clue occurred to me when I attempted to simulate breath-holding experiments during "leg-holding" (isometric contractions of the quadriceps sustained to fatigue). In the breath-holding experiments it had been found that the end point of a breath-hold could not be identified with particular alveolar air composition and that further

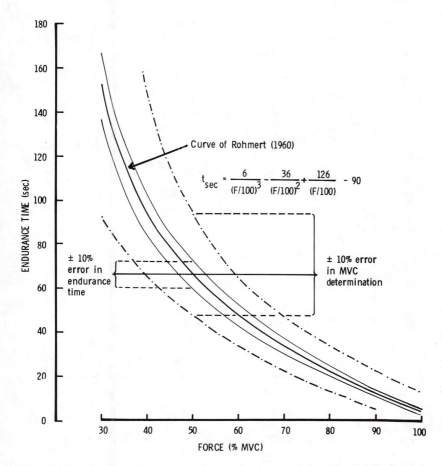

Figure 3—Relation between endurance time and force expressed as % maximum voluntary contraction (MVC) for isometric contractions (curve based on Rohmert, 1960). Limits are shown for 10% error in endurance time determination or in determination of MVC. The equation is the re-writing of Rohmert's equation to allow endurance to be expressed in s.

breath holds could be made despite continuing hypoxia and/or hyper-capnia. The analogous "leg-holding" experiment is illustrated in Figure 4. Since ischemia was continued throughout the experimental period no aerobic recovery could take place following fatigue and yet a further two contractions could be made thus showing that the end point, i.e., point of fatigue, could not be identified with a particular combination of metabolic factors in muscle. It has since become evident that this type of experimental protocol is associated with excitation failure at the point of fatigue but with substantial anaerobic recovery. This experiment indicated the difficulty in

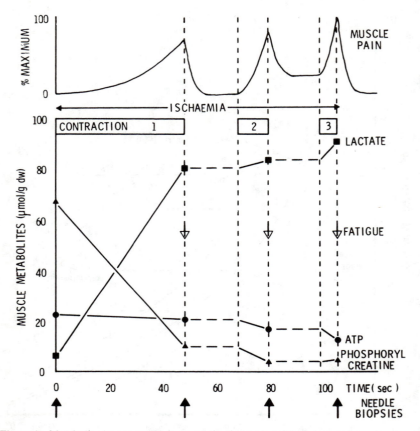

Figure 4—Metabolic changes and pain perception in a series of three isometric contractions made at two-thirds MVC with continuous ischemia (based on Edwards, Nordesjö, Koh, Harris & Hultman, 1971).

distinguishing muscle pain from fatigue and also demonstrated striking reduction in pain intensity with release of muscle tension during the anaerobic recovery intervals. Only when the circulation was restored was complete relief of pain and recovery of function possible. This experiment also illustrates that both pain and fatigue may simply reflect "hard times" in the muscle and the extent to which these cause a limit to performance will depend on central factors, including motivation.

Relation between Energy Exchange and Mechanical Properties of Muscle during Contraction

Measurements of metabolic heat production using a thermistor or thermocouple probe in contractions in which needle biopsies were taken at

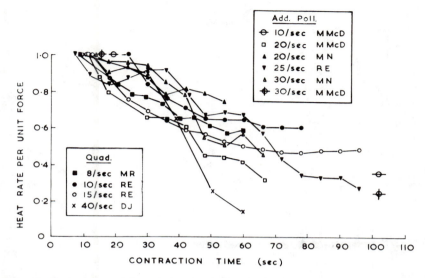

Figure 5—Heat rate per unit force during electrically-stimulated isometric contractions of the adductor pollicis and quadriceps muscles made with ischemia. Note a progressive improvement in economy, i.e., less heat produced for a given force as contractions are sustained, independent of stimulation frequency. (This extends the data previously published by Edwards & Hill, 1975).

the start and at fatigue showed that all the metabolic heat could be accounted for by the measured metabolic changes (Edwards, Hill & Jones, 1975a). In these and other studies it was clearly evident that the heat rate per unit force decreased with continuing contraction in isometric contraction sustained with ischemia (Figure 5). The explanation for this improvement in economy would be sought in the changes in the maximum relaxation rate of the muscle in response to electrical stimulation (Figure 6). Here, contractions were sustained with brief interruptions for the measurement of relaxation rate (Wiles, Young, Jones & Edwards, 1979) while ischemia continued for a minute's anaerobic recovery. Note that in both muscle groups there was no recovery of the relaxation rate until the circulation had been restored. Similar studies involving metabolic heat measurements showed clearly that there was no recovery of heat production until the circulation had been restored (Wiles & Edwards, 1982). This evidence suggests that relaxation rate is determined by metabolic factors as demonstrated in fatigued mouse muscle (Edwards, Hill & Jones, 1975b) and in frog muscle studied by ^{31}P nuclear magnetic resonance (Dawson, Gadian & Wilkie, 1980). The slowing of relaxation during sustained isometric contraction can contribute to force maintenance, thus counteracting the tendency to fatigue (Jones, 1981). This tendency for altered mechanical properties to compensate for fatigue is further demonstrated

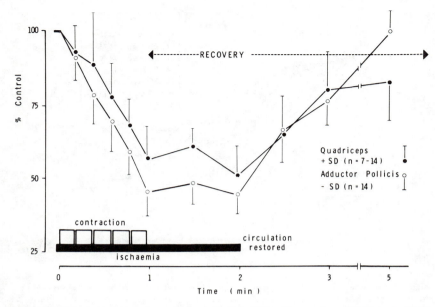

Figure 6—Maximum relaxation rate (as defined by Wiles, Young, Jones & Edwards, 1979) decreasing during sustained isometric contractions in the quadriceps and adductor pollicis muscles. Note that there is no recovery of relaxation rate until the circulation is restored.

by the effects of chronic low frequency stimulation which has been found recently to reduce fatigability by potentiating the forces generated at low frequencies compared with those at high (Dubowitz, Hyde, Scott & Vrbova, 1982; Edwards, Jones & Newham, 1982), possibly as a result of an alteration in fiber type characteristics as indicated from animal studies (e.g., Edgerton, Goslow, Rasmussen & Spector, 1980).

Relation between Mechanical Properties of Muscle and "Central Fatigue"

Though it was shown by Merton that the site of fatigue can be peripheral (Merton, 1954) this presupposes that there has been no more central failure in the chain of motor command. Well-motivated individuals are capable of activating their muscles maximally (Merton, 1954), but the extent to which this is possible varies between subjects (Bigland-Ritchie, Jones, Hosking & Edwards, 1978) and between muscle groups (Belanger & Mc Comas, 1981). For muscle force to be optimally sustained it appears to be necessary for the muscle action-potential frequency to fall (Bigland-Ritchie, Johansson,

Lippold & Woods, 1982; Jones, Bigland-Ritchie & Edwards, 1979; Marsden, Meadows & Merton, 1976). What then could be the factor which controls this graded reduction in firing frequency which matches this frequency to the mechanical properties of the muscle so as to minimize fatigue? Firing frequency is low and diminishes rapidly both under conditions of complete curarization (Freyschuss & Knutsson, 1971) and when there is partial denervation (Miller & Sherratt, 1978). It is an interesting observation by Bigland-Ritchie and colleagues (Bigland-Ritchie, Goto, Johansson & Woods, 1982) that the contractile speed and frequency of innervation decline in parallel during maximum voluntary contractions. Metabolic changes in the muscle could thus influence the firing frequency by reducing the relaxation rate and thence the degree of oscillation of unfused tetani with a resulting decrease in facilitatory afferent neural drive which otherwise helps to sustain firing frequency.

Relation between Energy Exchanges and Excitation as Causes of Peripheral Fatigue

Much has been learned by careful studies of patients with selected defects of metabolism in muscle (Figure 7). Hypothyroid patients sustain force for longer periods than normal at less ATP cost (Wiles, Young, Jones & Edwards, 1979). Contributing to this prolonged endurance rate, however, is a better than normal preservation of excitation with prolonged electrical stimulation at 20 Hz with local ischemia (Figure 7). Conversely, patients with impaired glycolysis due to myophosphorylase or phosphofructo-kinase deficiency show a striking early fatigue due to failure of excitation (Edwards & Wiles, 1981). It is significant in this connection that there is no anaerobic recovery of the action potential in myophosphorylase deficiency whereas this was seen to a substantial degree in a control normal subject (Edwards, Wiles, Gohil, Krywawych & Jones, 1982). Energy exchanges are thus capable of influencing excitation and alterations in sodium and potassium pumping by the transverse tubular system, which may also contribute to fatigue (Bezanilla, Caputo, Gonzalez-Serratos & Venosa, 1972; Bigland-Ritchie, Jones & Woods, 1979). The very rapid recovery of excitation with aerobic recovery from ischemic fatigue suggests a critical readjustment of electrolyte balance which is energy-dependent insofar that it may fail to occur when energy supply is limited, as in the case of the failure of anaerobic recovery of the action potential in myophosphorylase deficiency cited above. Though not studied in skeletal muscle, it is of interest that (in vascular smooth muscle) Na^+-K^+ transport processes appear to depend on ATP supply from aerobic glycolysis (Paul, Bauer & Pease, 1979).

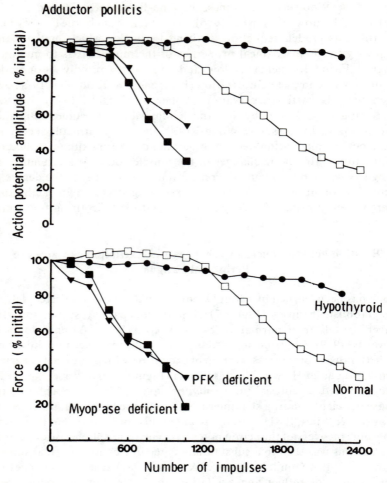

Figure 7—Force, fatigue and failure of excitation during electrical stimulation at 20 Hz with local ischemia in a normal subject and in patients with altered energy supply in muscle.

Activation for a given excitation appears to be linked to metabolic changes in that lactate accumulation and the associated intra-cellular acidosis appear to compete with calcium for actomyosin binding sites (Bolitho-Donaldson & Hermansen, 1978). In heavy dynamic exercise, therefore, a failure of power output may be brought about by the intra-cellular acidosis (Hermansen, 1981). The slow recovery of low frequency fatigue after exercise seems not to be related to resolution of acidosis or restoration of energy stores (Edwards, Hill, Jones & Merton, 1977).

Relation between Mechanical Factors and Activation

Length dependence of activation is a well-known phenomenon in cardiac muscle (Jewell, 1977). In skeletal muscle length dependence is a significant factor at low stimulation frequency (Rack & Westbury, 1969) when it may simulate and add to any pre-existing low frequency fatigue and as such could be of significance in diaphragmatic fatigue (Edwards, 1979). The recent observation that low frequency fatigue is greater in the leg undergoing eccentric contractions of the quadriceps compared with the contralateral leg which carried out concentric contractions (Edwards, Mills & Newham, 1981) shows that the low frequency fatigue tends to be related to damage to the muscle fibers. Such damage, evident in electron-micrograph of needle biopsy samples from the leg undergoing eccentric contractions (Newham, Mills, McPhail & Edwards, 1982), could not be attributed to extensive metabolic changes since the oxygen/ATP cost is much lower in the eccentric contracting compared with the concentrically contracting muscle, whereas the force sustained per fiber is greater (Bigland-Ritchie & Woods, 1976).

Metabolic Limits to Exercise Performance

It has long been suggested that the increase in hydrogen ion concentration with anaerobic glycolysis results in feedback inhibition to phosphofructo-kinase with a resulting reduction in glycolytic flux and ATP supply. This has been found to be a possible factor in isometric contractions and in dynamic exercise (Bergström, Harris, Hultman & Hordesjö, 1971). It is exacerbated in isometric contraction by increasing the metabolic rate by passively warming the muscle (Edwards, Harris, Hultman, Kaijser, Koh & Nordesjö, 1972) and in dynamic exercise by the ingestion of ammonium chloride to induce a metabolic acidosis (Jones, Sutton, Taylor & Toews, 1979; Sutton, Jones & Toews, 1981).

Topical magnetic resonance offers a non-invasive technique to investigate impaired energy supply as a factor underlying muscle pain or fatigue. The first application of this new technique was the examination of a case of McArdle's syndrome (Ross, Radda, Gadian, Rocker, Esiri & Falconer-Smith, 1981). In Figure 8 is shown a similar investigation on a patient of mine with phosphofructokinase deficiency (Edwards, Dawson, Wilkie, Gordon & Shaw, 1982). Notice that there has been virtually no change in muscle pH during the period of exercise and ischemia, though a significant fall was observed in the control normal subject. More interesting, however, is the striking accumulation of sugar phosphates which was evident not

16   Edwards

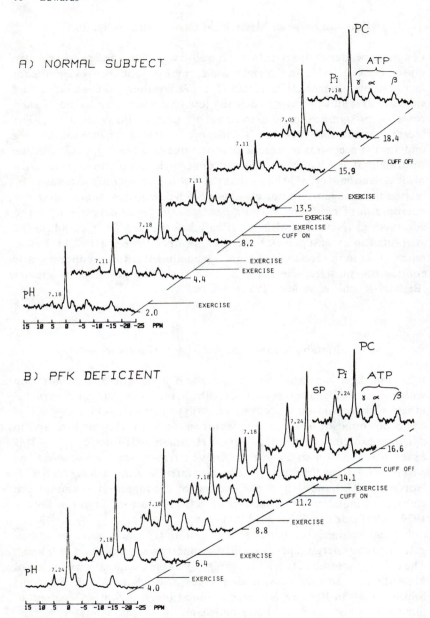

Figure 8—^{31}P spectra and muscle pH values obtained in forearm muscles of a patient with phosphofructokinase deficiency and in a control normal subject. (Reproduced with permission from Edwards et al., 1982.)

only with ischemia but even during the preceding period of aerobic exercise. Phosphofructokinase deficiency is a more troublesome condition than myophosphorylase deficiency. Because of the level of the enzyme block it is not possible to utilize blood-borne glucose in the way that makes possible the "second wind" in patients with myophosphorylase deficiency (Pernos, Havel & Jennings, 1967), though, of course, they can metabolize blood-borne fatty acids.

Alterations in mitochondrial metabolism are also present with impaired exercise tolerance (Edwards, Wiles, Gohil, Krywawych & Jones, 1982; Land & Clark, 1979). A patient with a mitochondrial myopathy was investigated by ^{31}P nuclear magnetic resonance with the finding of excessive muscle acidosis and with evidence of impaired oxidative phosphorylation (Gadian, Radda, Ross, Hockaday, Bore, Taylor & Styles, 1981).

The term "mitochondrial myopathy" covers a heterogeneous group of disorders in which mitochondrial function is impaired. A new technique has been developed which allows parts of the electron transport chain in mitochondria to be investigated in needle biopsy samples (Gohil, Jones & Edwards, 1981). As is well recognized from other workers, mitochondrial function can be improved by exercise training (Henriksson & Reitman, 1977). There are also striking differences in mitochondrial function between elite marathon runners and sedentary controls (Gohil, Jones, Krywawych, McPhail, Round, Corbucci, Montanari & Edwards, 1982). A number of patients presenting at my clinic with muscle pain appear to have low values for one or more of the component activities of the mitochondrial electron transport chain (Gohil, Jones, Mills & Edwards, 1982). Whether this represents a tissue correlate of "unfitness" as a consequence of inactivity or whether there are more specific, though potentially reversible, deficiencies in mitochondrial function remains to be seen. An exercise test is a logical and necessary investigation in patients with muscle pain and fatigue (Brooke, Carroll & Davis, 1977). In the future it is likely that topical nuclear magnetic resonance could play a valuable role in view of the cellular metabolic changes which can be revealed by this non-invasive technique. Patients with impaired mitochondrial fat metabolism may also suffer from muscle pain and/or muscle cell damage with resultant myoglobinuria (reviewed by DiMauro, 1979). It is, however, of considerable interest that though there are certain similarities in the exercise responses of patients with impaired energy supply due to defects in glycolysis, compared with those with defects in fat metabolism, there is a striking absence of accounts describing contracture (electrically silent contractions lasting sometimes for several hours) in the latter group, whereas they are characteristic of the former.

Fatigue: Performance Limit, Catastrophe and/or Protection

In athletics it is clear that training can and does, through skill and metabolic adaptations, extend the frontiers of performance capacity. In striving for "ultimate" performance it may be that physiological mechanisms which are designed to be protective are overridden. (Conversely, they appear to be "underridden" in patients with "Effort Syndromes" who cannot or will not activate their muscles fully or extend themselves to the limits of their performance potential [reviewed in Newham & Edwards, 1979].)

To return to the mountaineering analogy it is evident that if an isometric contraction sustained while hanging on is prematurely terminated by fatigue then there may be risk to life by falling. Apart from this real life catastrophe, there are very interesting possibilities to be considered by applying the new mathematical theory of catastrophes (Zeeman, 1976; Poston & Stewart, 1978) to muscle fatigue. Figure 9 illustrates the possible pathways by which fatigue might come about as a result of energy loss, excitation or activation failure or a combination of both. The axes are drawn on the basis of experimental observations of a number of workers. Relating impairment of energy supply to force loss is a linear function based on data presented by Dawson, Gadian & Wilkie (1978). Excitation has a curvilinear relation to force (Edwards, Young, Hosking & Jones, 1977) as does the relation between ionic calcium concentration and force (Bolitho-Donaldson & Hermansen, 1978). Four illustrative paths are indicated. Path number 1 represents pure energy loss without any impairment of excitation. This, if continued indefinitely, could result in the muscle ATP concentration falling to zero and rigor conditions with resulting damage to the contractile mechanism. In practice it is impossible to drive a contracting muscle into rigor unless it has been given a metabolic poison (though it may be that an analogous situation occurs when a patient with a glycolytic disorder such as myophosphorylase or phosphofructokinase deficiency exercises to contracture formation). The rigor tension is indicated, though not quantitively. Pathway number 4 indicates pure excitation or activation limitation with no loss of energy. In practice, fatigue may demonstrate a combination of energy loss and some impairment in excitation or activation. In dynamic exercise the brief rhythmic innervation provides protection from the development of high frequency fatigue, thus enabling muscles to utilize a large part of their stored energy. It is possible that in such exercise the vector for fatigue may follow some if not all of the pathway indicated by number 3. In isometric contractions, particularly under ischemic conditions, the metabolic demand may be greater than can be met by energy supply mechanisms and a concomitant impairment of excitation or activation may ensue. This is presented in

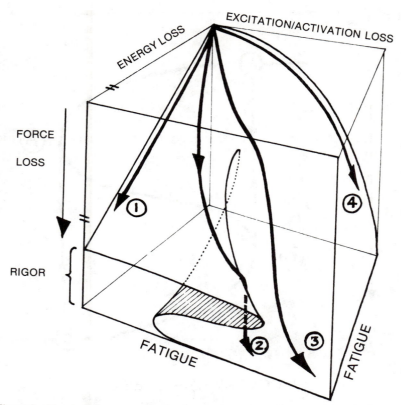

Figure 9—Muscle fatigue: catastrophe theory model. Theoretical pathways for different forms of muscular fatigue.

1. Pure energy loss in absence of excitation failure, which if continued would lead to rigor.
2. Mixed fatigue with catastrophe—sudden force loss with failure of excitation/activation as (but not only) on steep part of curve 4.
3. Mixed energy loss and excitation failure but without "catastrophe."
4. Pure excitation/activation failure thus preventing energy loss.

Figure 3 in which there is a sudden fall over the catastrophe cusp ("the straw that breaks the camel's back") with a failure of excitation and a saving of any further depletion of energy stores, thus preventing the muscle going into rigor. If a section is cut through the point on the catastrophe cusp at which pathway 2 crosses the cusp it is evident (Figure 10) that a hysteresis is predicted in the recovery process. During fatigue the pathway goes from A to B and then suddenly there is a collapse of excitation and a concomitant failure of force or power output (B, C). In the initial phase of

Excitation/Activation Loss

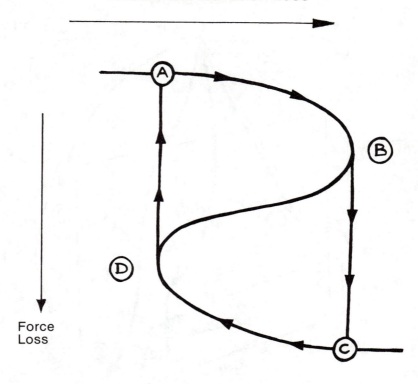

Force
Loss

RECOVERY HYSTERESIS

Figure 10—Section through "catastrophe" in fatigue pathway (2) shown in Figure 8. A, B is pathway to edge of cusp; B, C is "catastrophe"; C, D represents recovery of excitation with little or no increase in force generation and D, A shows the eventual improvement in force generation (corresponding to recovery of excitation-contraction coupling).

recovery a restitution of the action potential with little or no force generation is predicted, as indeed demonstrated in single frog muscle fibers (Grabowski, Lobsiger & Luttgau, 1972) and as thought to occur with low frequency fatigue (Edwards, Hill, Jones & Merton, 1977). With recovery of excitation-contraction coupling the hysteresis loop is completed (D, A).

The application of catastrophe theory to muscle fatigue is speculative but it nevertheless emphasizes the complex interrelation of energy supply for the contractile mechanism with that necessary for excitation/activation processes. It also serves to indicate that it may not be an appropriate question to ask, "What is the cause of fatigue?" The fact is that there are probably many different types of fatigue; each may occur with a particular form of muscular activity. At a cellular level there are built-in protective mechanisms which cause a muscle to cease functioning before it has been irreparably damaged (Wilkie, 1981). At a higher level there is evidence of reflex fatigue and this may serve as suggested by Sherrington as a means to protect individual motor units from excessive and possibly damaging activity (Sherrington, 1947). At higher levels still in the nervous system the limiting factor may not only be a conscious or unconscious need to cease the bombardment from afferent signals indicating movement or other kinesthetic signals but also possibly "hard times" (adverse local chemistry) in the working muscles (Kniffki, Mense & Schmidt, 1981). The idea that central fatigue serves a protective function is an old one (Mosso, 1915; Waller, 1891) but one which in my opinion deserves further attention and investigation. However desirable it is to extend the frontiers of human exercise performance we can but marvel at the integration (Barcroft, 1934) of complex mechanisms which makes this performance possible. Fatigue is thus a manifestation of one or more of the "fail-safe" mechanisms in the organism which call for temperance before damage occurs.

Conclusion

This review has not been, and could not have been, comprehensive nor can it provide an explanation for fatigue in every type of muscular activity. I have emphasized that fatigue may be due to a breaking of any one of the links of the chain of command (Figure 1) for muscular contraction. Catastrophe Theory is a mathematical concept for explaining real or apparent discontinuities in functional relationships. The Catastrophe figure shown (Figure 9) is the simplest, to illustrate the principle, while others with a more complex form could accomodate intermediate fatigue states with impaired energy supply and excitation.

The catastrophe theory of fatigue may validly describe the final common path in cellular function leading to impaired performance despite central or other factors, e.g., increase in muscle temperature (leading to increased ATP turnover) (Edwards et al., 1972) or dehydration (leading to impaired oxygen and substrate supply) which appear to be critical. An important prediction from a catastrophe theory of fatigue is that it may prove to be very difficult if not impossible to recognize *the* limiting factor, e.g., at the end of a marathon, if the factor is at or near the edge of a catastrophe cusp. A marginally deficient rate of ATP supply at one power output with consequent failure of excitation/activation may be adequate and capable of full excitation/activation at an only slightly lower power output in aerobic exercise. In isometric contraction making substantial demands on

ATP is the fuel for contraction:

$$H_2O + ATP \longrightarrow ADP + P_i + energy$$

actomyosin
ATPase

ATP is regenerated by:

(1) Creatine kinase reaction

$$ADP + PCr \rightleftharpoons ATP + Cr$$

(2) Anaerobic glycolysis

$$1/n \, (C_6H_{10}O_5)_n + 3ADP + 3P_i \longrightarrow 2C_3H_6O_3 + 2H_2O + 3ATP$$

(glycogen)   (lactic acid)

(3) Oxidative phosphorylation

$$C_3H_4O_3 + 2\tfrac{1}{2}O_2 + 18ADP + 18P_i \longrightarrow 3CO_2 + 20H_2O + 18ATP$$

(pyruvic acid)

ATP: adenosine tri phosphate
ADP: adenosine di phosphate
P_i: inorganic phosphate
PCr: phosphoryl creatine
Cr: creatine

Figure 11—Energy exchange in muscle.

anaerobic metabolism the instant when contraction stops and aerobic recovery occurs may allow rapid ATP recovery (and recovery of any excitatory loss) simply because aerobic glycolysis has a considerably greater ATP yield than anaerobic glycolysis (Figure 11). A rapid failure of force may occur with high frequency stimulation with an almost instantaneous improvement in force generation when the frequency is suddenly reduced (Bigland-Ritchie, Jones & Woods, 1979) thus illustrating the "critical" state of excitation independent of immediate energy supply. The cause of fatigue may thus be transitory and beyond our measurement, though fortunately not beyond our imagination.

I have in my possession a monograph entitled "Mechanical work and heat production during muscular activity" by Adolf Fick (1882). The monograph was given to A.V. Hill by Otto Meyerhof in 1921 and was passed on to me in 1975 by my friend and collaborator Professor D.K. Hill. This book shows that the relations between metabolism, work and mechanical properties of (frog) muscle were already well recognized. In the century which has followed its publication we have come to recognize the individual links in the chain of command governing muscular activity (Figure 2). Physical training may bring about a "strengthening" of the weaker links and so improve force generation. It can also provide a means by which the choice of substrate (fat) for oxidative metabolism may, as a consequence of increased oxidative capacity of muscle cells (see Gollnick & Saltin, 1982), favor improved endurance in prolonged exercise. The "chain" remains successful, however, despite the adaptive changes. Catastrophe theory may be the best means yet available for describing events immediately influencing the failure of one of its links leading to fatigue.

Acknowledgments

I am particularly grateful to the collaborators I have cited. Support is also gratefully acknowledged from the Wellcome Trust and the Muscular Dystrophy Group of Great Britain, the Muscular Dystrophy Associations (of America).

References

BARCROFT, J., Features in the architecture of physiological function. Cambridge University Press, 1934.

BELANGER, A.Y. & A.J. McComas. Extent of motor unit activation during effort. J. Appl. Physiol. 51:1131-1135, 1981.

BERGSTRÖM, J., R.C. Harris, E. Hultman & L.O. Nordesjö. Energy rich phospagens in dynamic and static work. In B. Pernow & B. Saltin (Eds.), Muscle Metabolism during Exercise. New York: Plenum Press, 1971, pp. 341-355.

BEZANILLA, F., C. Caputo, H. Gonzalez-Serratos & R.A. Venosa. Sodium dependence of the inward spread of activation in isolated twitch muscle fibres of the frog. J. Physiol. 223:507-523, 1972.

BIGLAND-RITCHIE, B., Y Goto., R. Johansson & J.J. Woods. Contractile speed and EMG in fatigue of human voluntary contractions. J. Physiol. (In Press) 1982.

BIGLAND-RITCHIE, B., R. Johansson, O.C.J. Lippold & J.J. Woods. Changes of single motor unit firing rates during sustained maximal voluntary contractions. J. Physiol. (In Press) 1982.

BIGLAND-RITCHIE, B., D.A. Jones, G.P. Hosking & R.H.T. Edwards. Central and peripheral fatigue in sustained maximum voluntary contractions of human quadriceps muscle. Clin. Sci. Mol. Med. 54:609-614, 1978.

BIGLAND-RITCHIE, B., D.A. Jones & J.J. Woods. Excitation frequency and muscle fatigue: electrical responses during human voluntary and stimulated contractions. Experim. Neurol. 64:414-427, 1979.

BIGLAND-RITCHIE, B. & J.J. Woods. Integrated EMG and O_2 uptake during positive and negative work. J. Physiol. 260:267-277, 1976.

BOCK, A.V. & D.B. Dill. The physiology of muscular exercise. F.A. Bainbridge (Ed.), 3rd. Ed. London: Longmans, pp. 225-235, 1931.

BOLITHO-DONALDSON, S.K. & L. Hermansen. Differential, direct effects of H^+ on Ca^{2+}-activated force of skinned fibers from the soleus, cardiac and adductor magnus muscles of rabbits. Pflügers Arch. 376:55-65, 1978.

BROOKE, M.H., J.E. Carroll & J.E. Davis. The prolonged exercise test in patients with muscle pain and fatigue. Transactions of the American Neurological Association, 102, 1-3, 1977.

CIBA Foundation Symposium No. 82. Human Muscle Fatigue: Physiological Mechanism. London: Pitman Medical, 1981.

DAWSON, J.J., D.G. Gadian & D.R. Wilkie. Muscular fatigue investigated by phosphorus nuclear magnetic resonance. Nature 274:861-866, 1978.

DAWSON, J.J., D.G. Gadian & D.R. Wilkie. Mechanical relaxation rate and metabolism studied in fatiguing muscle by phosphorus nuclear magnetic resonance. J. Physiol. 299:465-484, 1980.

DIMAURO, S., Metabolic myopathies. In P.J. Vinker & G.W. Bruyn (Eds.), Handbook of Clinical Neurology, Vol. 41, Diseases of Muscle, Part II. S.P. Ringel (Ed.) Amsterdam: North Holland, 1979, pp. 175-234.

DUBOWITZ, V., S.A. Hyde, O.M. Scott & G. Vrbova. Effect of long-term electrical stimulation on the fatigue of human muscle. J. Physiol. (In Press) 1982.

EDGERTON, V.R., G.E. Goslow, Jr., S.A. Rasmussen & S.A. Spector. Is resistance of a muscle to fatigue controlled by its motor neurones? Nature 285:589-590, 1980.

EDWARDS, R.H.T., Physiology of fitness and fatigue. In C. Clarke, M. Ward & E. Williams, Proceedings of Symposium on Mountain Medicine and Physiology. London: Alpine Club, 1975, pp. 107-110.

EDWARDS, R.H.T., Physiological analysis of skeletal muscle weakness and fatigue. Clin. Sci. Mol. Med. 54:463-470, 1978.

EDWARDS, R.H.T., Physiological and metabolic studies of the contractile machinery of human muscle in health and disease. Phys. Med. Biol. 24:237-249.

EDWARDS, R.H.T., The diaphragm as a muscle: mechanisms underlying fatigue. Amer. Rev. Resp. Dis. 119:81-84, 1979.

EDWARDS, R.H.T., Muscle weakness and fatigue. I.C.I. monograph series on symptoms. London: Gower Medical Publishing, 1981, pp. 1-25.

EDWARDS, R.H.T., Human muscle function and fatigue. In R. Porter & J. Whelan (Eds.), Human Muscle Fatigue: Physiological Mechanisms. Ciba Foundation Symposium, No. 82. London: Pitman Medical, 1981, pp. 1-18.

EDWARDS, R.H.T., M.J. Dawson, D.R. Wilkie, R.E. Gordon & D. Shaw. Clinical use of nuclear magnetic resonance in the investigation of myopathy. Lancet i:725-731, 1982.

EDWARDS, R.H.T. & J.A. Faulkner. Properties of the Respiratory Muscles In Thorax-Vital Pump, Ed. C. Roussos and P.T. Macklem (In Press).

EDWARDS, R.H.T., R.C. Harris, E. Hultman, L. Kaijser, D. Koh and L.O. Nordesjö. Effect of temperature on muscle energy metabolism and endurance during successive isometric contractions, sustained to fatigue, of the quadriceps muscle in man. J. Physiol. 220, 335-352, 1972.

EDWARDS, R.H.T. & D.K. Hill. "Economy" of force maintenance during electrically stimulated contractions of human muscle. J. Physiol. 250:13-14P, 1975.

EDWARDS, R.H.T., D.K. Hill & D.A. Jones. Heat production and chemical changes during isometric contractions of the human quadriceps muscle. J. Physiol. 251:303-315, 1975a.

EDWARDS, R.H.T., D.K. Hill & D.A. Jones. Metabolic changes associated with the showing of relaxation in fatigued mouse muscle. J. Physiol. 251:287-301, 1975b.

EDWARDS, R.H.T., D.K. Hill, D.A. Jones & P.A. Merton. Fatigue of long duration in human skeletal muscle after exercise. J. Physiol. 272:769-778, 1977.

EDWARDS, R.H.T., D.A. Jones & D.J. Newham. Low frequency stimulation and changes in human muscle contractile properties. J. Physiol. (In Press) 1982.

EDWARDS, R.H.T., N.L. Jones, E.A. Oppenheimer, R.L. Hughes & R.P. Knill-Jones. Inter-relation of responses during progressive exercise in trained and untrained subjects. Quart. J. Exp. Physiol. 54:394-403, 1969.

EDWARDS, R.H.T., K.R. Mills & D.J. Newham. Greater low frequency fatigue produced by eccentric than concentric muscle contraction. J. Physiol. 317: 17P, 1981.

EDWARDS, R.H.T., L.O. Nordesjö, D. Koh, R.C. Harris & E. Hultman. Isometric exercies—factors influencing endurance and fatigue. In B. Bernow & B. Saltin (Eds.), Muscle Metabolism During Exercise. New York: Plenum Press, 1971, pp. 357-368.

EDWARDS, R.H.T., C.M. Wiles, K. Gohil, S. Krywawych & D.A. Jones. Energy metabolism in human myopathy. In D.L. Schotland (Ed.), Disorders of the Motor Unit. New York: John Wiley, 1982, pp. 715-726.

EDWARDS, R.H.T., A. Young, G.P. Hosking & D.A. Jones. Human skeletal muscle function: description of tests and normal values. Clin. Sci. Mol. Med. 52:283-290, 1977.

FICK, A., Mechanische Arbeit und Wärmeentwickelung bei der Muskelthätigkeit. Leipzig: F.A. Brockhaus, pp. 1-234, 1882.

FREEDMAN, S., Sustained maximum voluntary ventilation. Respiration Physiology 8:230-244, 1970.

FREYSCHUSS, V. & E. Knutsson. Discharge patterns in motor nerve fibres during voluntary effort in man. Acta. Physiol. Scand. 83:278-279, 1971.

GADIAN, D., G. Radda, B. Ross, J. Hockaday, P. Bore, D. Taylor & P. Styles. Examination of a myopathy by phosphorus nuclear magnetic resonance. Lancet ii:774-775, 1981.

GOHIL, K., D.A. Jones & R.H.T. Edwards. Analysis of muscle mitochondrial function with techniques applicable to needle biopsy samples. Clin. Physiol. 1:195-207, 1981.

GOHIL, K., D.A. Jones, S. Krywawych, G. McPhail, J. Round, G.G. Corbucci, G. Montanari & R.H.T. Edwards. Mitochondrial substrate oxidations, muscle composition and metabolite levels in marathon runners. Proceedings of the 5th International Symposium on Biochemistry of Exercise, 1982, pp. ?

GOLLNICK, P.D. & B. Saltin. Significance of skeletal muscle oxidative enzyme enhancement with endurance training. Clin. Physiol. 2:1-12. 1982.

GRABOWSKI, W., E.A. Lobsiger & H.C. Luttgau. The effect of repetitive stimulation at low frequencies upon the electrical and mechanical activity of single muscle fibres. Pflugers Arch. 334:222-239, 1972.

HENRICKSSON, J. & J.S. Reitman. Time course of changes in human skeletal muscle succinate dehydrogenase and cytochrome oxidase activities and maximal oxygen uptake with physical activity and inactivity. Acta. Physiol. Scand. 99:91-97, 1977.

HERMANSEN, L., Effort of metabolic changes on force generation in skeletal muscle during maximal exercise. In R. Porter & J. Whelan (Eds.), Human Muscle Fatigue: Physiological Mechanisms. Ciba Foundation Symposium, No. 82. London: Pitman Medical, 1981, pp. 75-88.

HILL, A.V., Living machinery. (Royal Institution Christmas Lectures, 1926) London: G. Bell, 1927, pp. 1-250.

JEWELL, B.R., A re-examination of the influence of muscle length on myocardial performance. Circ. Res. 40:221-230, 1977.

JONES, D.A., B. Bigland-Ritchie & R.H.T. Edwards. Excitation frequency and muscle fatigue: mechanical responses during voluntary and stimulated contractions. Exper. Neurol. 64:401-413, 1979.

JONES, D.A., Muscle fatigue due to changes beyond the neuromuscular junction. In R. Porter & J. Whelan (Eds.), Human Muscle Fatigue: Physiological Mechanisms. Ciba Foundation Symposium, No. 82. London: Pitman Medical, 1981, pp. 178-196.

JONES, N.L., E.J.M. Campbell, R.H.T. Edwards & D.G. Robertson. Clinical Exercise Testing. Philadelphia: Saunders, 1975.

JONES, N.L., J.R. Sutton, R. Taylor & C.J. Toews. Effect of pH on cardiorespiratory and metabolic responses to exercise. J. Appl. Physiol. Respirat. Exercise Physiol. 43(6):959-964, 1979.

KENNELLY, A.E., An approximate law of fatigue in the speeds of racing animals. Proceedings of American Academy of Arts and Sciences 42:275-331, 1906.

KNIFFKI, K.-D., S. Mense & R.F. Schmidt. Muscle receptors with five afferent fibres which may evoke circulatory reflexes. Circ. Res. 48(Supp 1):25-31, 1981.

LAND, J.M. & J.B. Clark. Mitochondrial myopathies. Biochemical Review 7:231-245, 1979.

MARSDEN, C.D., J.C. Meadows & P.A. Merton. Fatigue in human muscle in relation to the number and frequency of motor impulses. J. Physiol. 25:94P-95P, 1976.

MERTON, P.A., Voluntary strength and fatigue. J. Physiol. 123:553-564, 1954.

MILLER, R.G. & M. Sherratt. Firing rates of human motor units in partially denervated muscle. Neurology 28:1241-1248, 1978.

MILLS, K.R. & R.H.T. Edwards. Differential diagnosis in muscle pain: review of 109 cases. Eur. J. Clin. Invest. (In Press) 1982.

MOSSO, A., Fatigue. (Translated by M. Drummond & W.B. Drummond). London: Allen & Unwin, 1915.

NEEDHAM, D.M., Machina Carnis: The Biochemistry of Muscular Contraction in its Historical Development. Cambridge: University Press, 1971.

NEWHAM, D. & R.H.T. Edwards. Effort syndromes. Physiotherapy 65:52-56, 1979.

NEWHAM, D.J., K.R. Mills, G. McPhail & R.H.T. Edwards. Muscle damage in response to exercise. Eur. J. Clin. Invest. (In Press) 1982.

PAUL, R.J., M. Bauer & W. Pease. Vascular smooth muscle: Aerobic glycolysis linked to sodium and potassium transport processes. Science 206:1414-1416, 1979.

PERNOW, B.B., R.J. Havel & D.B. Jennings. The second wind in McArdle's syndrome. Acta Med. Scand. 472(suppl):294-307, 1967.

POSTON, T. & I. Stewart. Catastrophe Theory and its Applications. London: Pitman, 1978, pp. 1-430.

RACK, P.M.H. & D.R. Westbury. The effects of length and stimulus rate or tension in the isometric cat soleus muscle. J. Physiol. 204:443-460, 1969.

RENNIE, M.J. & R.H.T. Edwards. Carbohydrate metabolism of skeletal muscle and its disorders. In P.J. Randle, D.F. Steiner & W.J. Whelan (Eds.), Carbohydrate Metabolism and its Disorders. London: Academic Press, 1981, pp. 1-118.

ROHMERT, W., Emittlung von erhulungspausen für statische arbeit des menschen. Intern. A. ange. Physiol. 18:123-164, 1960.

ROSS, B.D., G.K. Radda, D.G. Gadian, G. Rocker, M. Esiri & J. Falconer-Smith. Examination of a cause of suspected McArdle's Syndrome by ^{31}P nuclear magnetic resonance. New Engl. Med. 304:1338-1342, 1981.

SHERRINGTON, C., The Integrative Action of the Nervous System. Cambridge: University Press, 1947, pp. 215-224.

SUTTON, J.R., N.L. Jones & C.J. Toews. Effect of pH on muscle glycolysis during exercise. Clin. Sci. 61:331-338, 1981.

WALLER, A.D., The sense of effort: an objective study. Brain 14:179-247, 1891.

WASSERMAN, K., A.L. van Kessal & G.C. Burton. Interaction of physiological mechanisms during exercise. J. Appl. Physiol. 22:71-85, 1967.

WILES, C.M. & R.H.T. Edwards. Metabolic heat production in isometric ischaemic contractions of human adductor pollicis. Clin. Physiol. (In Press) 1982.

WILES, C.M., A. Young, D.A. Jones & R.H.T. Edwards. Relaxation rate of constituent muscle fibre types in human quadriceps. Clin. Sci. 56:47-52, 1979.

WILES, C.M., A. Young, D.A. Jones & R.H.T. Edwards. Muscle relaxation rate, fibre-type composition and energy turnover in hyper and hypothyroid patients. Clin. Sci. 57:375-384, 1979.

WILKIE, D.R., Equations describing power input by humans as a function of duration of exercise. In P. Cerretelli & B.J. Whipp (Eds.), Exercise Bioenergetics and Gas Exchange. Amsterdam: Elsevier, 1980, pp. 75-80.

WILKIE, D.R., Shortage of chemical fuel as a cause of fatigue: studies by nuclear magnetic resonance and bicycle ergometry. In R. Porter & J. Whelan (Eds.), Human Muscle Fatigue: Physiological Mechanisms. Ciba Foundation Symposium, No. 82. London: Pitman Medical, 1981, pp. 102-119.

ZEEMAN, E.C., Catastrophe theory. Proceedings of the Royal Institution of Great Britain 49:77-92, 1976.

Invited Lectures

Muscle Fiber Activation and Recruitment

V. Reggie Edgerton, R.R. Roy, R.J. Gregor,
C.L. Hager and T. Wickiewicz
University of California, Los Angeles, California, U.S.A.

A large body of information has accumulated over the last fifteen years which describes, in rather extensive detail, acute and chronic biochemical responses to treadmill exercise. The central point of this paper is to emphasize the need to be more precise in the use of the term "exercise" so that data derived from these studies can be interpreted in a more mechanistic manner. Although in a general sense everyone in this audience is aware of the meaning of "exercise," there is little detailed information regarding the specific events in individual cells that are probably responsible for inducing the acute and chronic biochemical responses that have been so well characterized. For example, it is known that muscles are recruited with respect to the size and type of motor units and that some combination of force and velocity is produced as a result of that recruitment (Henneman & Mendell, 1981; Walmsley et al., 1978). It is also clear that there are electrical events that are associated with that recruitment, both in the spinal cord and the higher nervous system as well as in the musculature. Also, there is rather impressive evidence that there is some neurotrophic phenomenon through which the central nervous system can affect the biochemistry of the muscle (Guth, 1968). An attempt will be made in this paper to approach some of these variables in a more quantitative way so that perhaps the specific events that fall under the encompassing term "exercise" can be associated with more specific acute and chronic biochemical responses.

In the evaluation of most exercises, as it might affect acute and chronic biochemical responses, muscle force or torque has received some attention. More recently, velocity has become a frequently discussed topic primarily as a result of developments in technology which permit us to study both force and velocity in vivo as well as in situ (Spector et al., 1980; Walmsley et al., 1978). Under controlled, constant conditions in situ, there is a very specific force-velocity relationship (Figure 1). Note that as velocity increases, force decreases in a shortening contraction. It is also obvious that higher forces can be developed in a lengthening contraction than in a shortening contraction and that these forces are more constant over a range of negative velocities than is true of a similar range of velocities when the muscle is actually shortening. This characteristic of muscle, when stimulated maximally in an in situ preparation, seems to be a characteristic of all skeletal muscle (Close, 1972). Based on these observations, if force is to be studied as a potential inducer of a particular biochemical response, then it is evident that velocity should be evaluated as well.

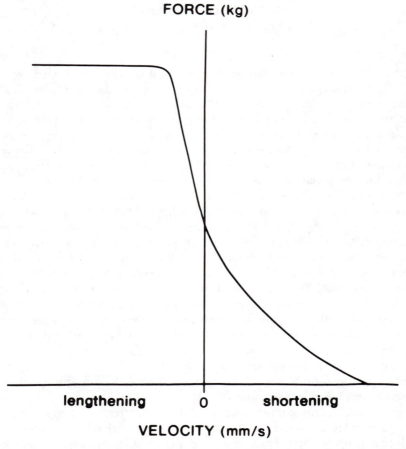

Figure 1—A theoretical relationship of force and velocity. Note the relative independence of force to negative compared to positive velocities.

A second general point which will be discussed is the electrical events that are involved when a motoneuron pool is activated. Impressive work by a combination of authors such as Henneman and his co-workers (1981) and Burke and his co-workers (1977) has demonstrated a predictable relationship between the percentage of a motoneuron pool for a muscle that is recruited and the tension that is produced by those motoneurons (Figure 2). It is also evident that in most cases, the order of recruitment of motoneurons is relatively constant. There is considerable controversy, however, over the commonality of altered order in normal movements as well as the degree of alteration that might occur in performing a specific movement (Desmedt & Godaux, 1978; Grimby & Hannerz, 1977). This point of controversy will not be addressed in this paper. It will be assumed that in most events the order of recruitment is fixed and whatever alterations in the recruitment order that might occur do not impose a dominating factor in determining either the acute or chronic biochemical response to exercise.

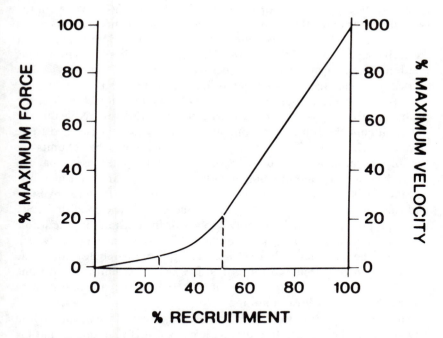

Figure 2—Relationship of the percentage of motor units recruited and the cumulative force produced by motor units up to a given percent recruitment. Also, if it is assumed that as recruitment increases and force is held constant, then velocity would increase as a function of percent recruitment. The percent recruitment and cumulative force is based on the medial gastrocnemius of the cat (Walmsley et al., 1978).

Variables that Affect Force-velocity Properties

The relationship of force and velocity in in situ muscle preparations that have been characterized by a number of authors (most well-known of whom is A.V. Hill [1970]) has been in almost all cases a maximal level of activation. It is also known, however, that the characteristic force-velocity relationship also exists when a muscle is submaximally activated (Phillips & Petrofsky, 1981). As long as the activation level remains constant, the general hyperbolic relationship of force and velocity will be similar at submaximal and maximal levels of stimulation. Although the specific curvature of the force-velocity relationships may vary slightly by activating varying percentages of the motoneuron pool, the fundamental hyperbolic relationship remains whether the muscle is stimulated at 25, 50 or 100% of the maximum output of the muscle (Figure 3). In an intact animal, this force-velocity relationship would also be expected. There is considerable evidence, however, that this may not be the case (Perrine & Edgerton, 1978). But be that as it may, there is every reason to believe at the present time that the level of recruitment is associated in a general way with the types of motor units; that is to say, there is a predominance of slow-twitch oxidative units (SO) recruited initially followed by a predominance of fast oxidative glycolytic (FOG) and finally, fast glycolytic fibers (FG). This relationship between force and velocity and percent recruitment is also related to percentage of fiber types as illustrated (Figure 3).

The potential contributions of SO, FOG and FG motor units in the medial gastrocnemius (MG) of a cat are estimated (Figure 4). Note that the potential contribution to a maximum velocity (Vmax) is assumed to be similar in FG and FOG motor units. On the other hand, the SO units are capable of producing only about 40% of that velocity. The tension that can be produced by each motor unit type within the cat MG varies considerably. The most tension can be produced by the FG motor units followed by FOG and least of all, by the SO units. In fact, under isometric conditions, SO units can probably contribute only 3 to 4% of the tension that the whole MG is capable of producing (Burke et al., 1973). It can also be easily seen that when the velocities of shortening exceed approximately 125 mm/s, the SO units can contribute no force. The FG and FOG units at that same speed can still contribute approximately 60% to the maximum force potential under isometric conditions (Po).

Another factor to consider in estimating the force-velocity properties in an exercise is the contribution of muscle design. It is well known that the length of the muscle fiber as well as other biochemical variables affects the velocity and force potential. A comparison of the force-velocity relationships of the soleus (SOL) and MG of the cat shows that the actual shortening velocity in mm/s for the SOL is 176 (Figure 5). However, when

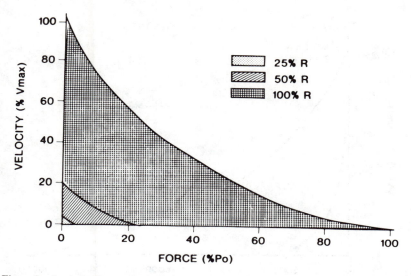

Figure 3—A relative force-velocity plot of a muscle as a function of percent recruitment and motor unit type. The percentages of the forces produced for each unit type are based on data for the cat medial gastrocnemius (Burke et al., 1973).

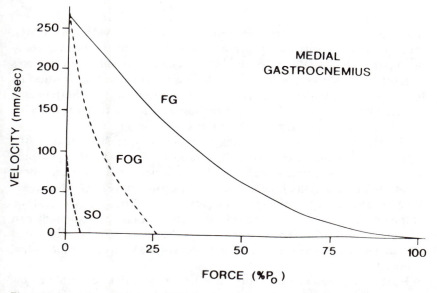

Figure 4—Given the intrinsic velocities and architectural features of the medial gastrocnemius, the relative contribution of each motor unit type to force and velocity of the whole cat medial gastrocnemius muscle.

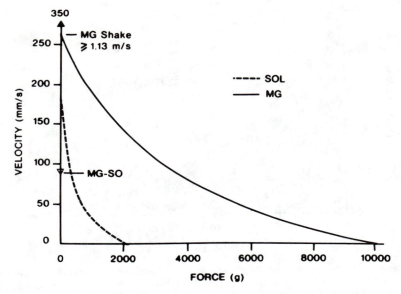

Figure 5—A comparison of the force velocity properties of the whole soleus (slow) and medial gastrocnemius (fast) muscles and the theoretical contribution of the slow motor units in the medial gastrocnemius.

the longer fibers of the soleus are normalized to that equivalent to the MG, the shortening velocity is only 90 mm/s. Thus, even though the shortening maximum velocity (Vmax) of the MG sarcomeres is almost three times faster than the Vmax of soleus, the SOL compensates for this slowness by having more sarcomeres *in series.* Consequently, during shortening a greater displacement occurs within the same period of time for the SOL than would be predicted by the Vmax at the sarcomere level.

The effect of sarcomeres arranged *in series* is most evident in the semitendinosus muscle (Figure 6). This muscle is divided into a proximal and a distal compartment separated by a connective tissue band about one-third of the length of the muscle from its proximal end (Bodine et al., 1982). Each compartment has its own nerve trunk. When the proximal end of the muscle is stimulated independently, its Vmax is approximately half of that when the distal end is stimulated. When both ends are stimulated simultaneously, the Vmax is equal to the sum of the Vmax of the individual compartments.

Human leg musculature represents a third and final example of the relationship of muscle architecture and physiology (Figure 7). Although the torque-velocity relationships are similar in shape for flexors and extensors of the knee and ankle, at the higher velocities they differ markedly with respect to their absolute forces and velocities (Wickiewicz,

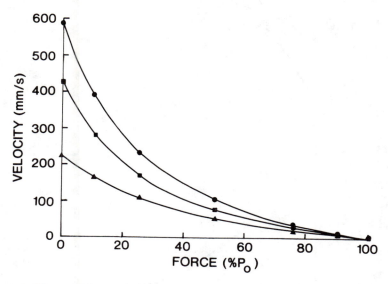

Figure 6—The actual force-velocity properties of the cat semitendinosus muscle as it can function under three conditions—as a distal, proximal, and total muscle (Bodine et al., 1982). From Journal of Neurophysiology, 48(1):192–201. Copyright by the American Physiological Society.

1982). Although part of this difference between muscle groups can be explained by the differences in the mechanical nature of attachments of the muscles to the bones and the consequential lever arms, it has been demonstrated by Wickiewicz and his colleagues (1982) that muscle architecture can account for a large component of the apparent difference in absolute torques and velocities of each of these four muscle groups.

Although the intrinsic properties of muscle at known levels of activation are understood, the effect of the biomechanical attachments on the interaction of the variables during normal movements is virtually unknown. In order to obtain force-velocity relationships during normal movements, Gregor and colleagues (1982) have performed a series of experiments on cats in which the MG tendon and the SOL tendon have been implanted with a force transducer. In addition, these cats have been implanted with wire electrodes which permit a recording of the EMG patterns within those particular muscles. And finally, movement during treadmill locomotion and other selected movements has been studied using high-speed cinematography. A study of these parameters during normal movements permits us to determine the force-velocity relationships as well as the relative level of activation of these particular muscles during movement of a normal cat which is unrestrained. The force-velocity relationship within a single step of a cat on a treadmill belt traveling at a fast speed (65 m/min) is

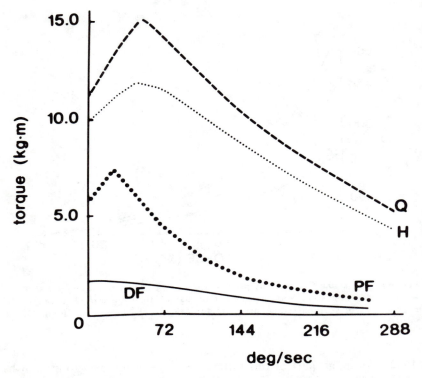

Figure 7—Torque velocity measurements of muscle groups in the human. Note the wide range in values among the muscle groups in spite of relatively similar biochemical properties as reflected by histochemistry (from Wickiewicz et al., 1982).

illustrated (Figure 8). The higher forces in the MG usually occur at a time when the muscle is lengthening (Gregor et al., 1981). A third point is that during the rapid extension phase of the step cycle during stance, the force decreases dramatically in spite of the fact that the muscle continues to receive considerable excitation as suggested by the EMG (Gregor et al., 1981). This rapid drop in force must occur when the velocity increases, even assuming that the level of activation remains relatively constant. Peak forces from both the SOL and the MG occur at a negative velocity, or approximately at the time the muscle is contracting under zero velocity conditions. During the shortening phase, the SOL force decreases more slowly than the MG even at the higher speed. Considering the intrinsic properties of the slow and fast muscle alone, one would expect that the tension would decrease more slowly in the slow than in the fast muscle. Under circumstances when plantarflexion occurs more rapidly, this is not the case because of the force-velocity relationships under which both

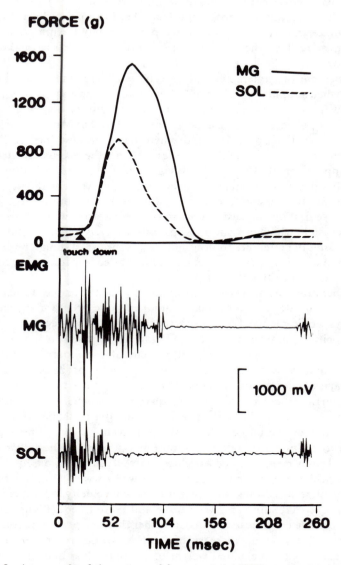

Figure 8—An example of the pattern of forces and the EMG produced by the medial gastrocnemius and soleus muscles of the cat while walking on a treadmill at about 60 m/ min (preliminary data by Gregor et al. [1981]).

muscles must function during normal movement as synergists. Clearly, as the velocity of movement around the ankle increases, the SOL is at a much greater disadvantage for producing force because of its slower Vmax. When ankle extension velocity exceeds approximately 176 mm/s, no force can be produced by the SOL even though it appears to be maximally activated.

Another curious interaction of recruitment, forces and velocities is evident when a rapid oscillation of the leg of a cat is induced by placing the foot in water or by placing tape around the paw. Immediately after this stimulus is applied, there is an apparent attempt to shake the tape or water from the paw. This specialized movement has been described earlier by Smith and co-workers (1980). Note (Figure 9) that the MG is activated at a rate of approximately 10 to 12 Hz. Based on the EMG, the SOL appears to be inactive during this rather violently appearing movement and in spite of the potential for passive forces, there was no fluctuation in force in the SOL. These findings suggest that there are mechanisms within the central nervous system which provide a means for selectively activating fast units in one muscle in preference to slow units in a synergist. It should be pointed out that in practically every other event that the cat performs, the SOL is always active when the MG is active.

Another rather basic relationship of force, velocity and recruitment that should be considered and evaluated in an exercise program expected to induce a specific chemical response is the relationship between recruitment and velocity of movement. Initially Figure 2 was used to illustrate the previously reported relationship of percent of the motoneuronal pool activated and the cumulative tension that would be produced by a given percentage of motoneurons. This, of course, assumes that velocity was zero. If, on the other hand, as is the case in most movements, a negative and positive velocity does occur, then variations in force could be a function of velocity and/or recruitment level. If the percent of recruitment of the motoneuronal pool is modified, the output as a result of the activation of the muscle may be manifested as velocity as well as force. Consequently, if force is held constant and velocity is allowed to vary as a function of recruitment, this relationship of recruitment and velocity theoretically will be similar to that shown between force and recruitment (Figure 2). For example, if one chooses to run faster, rather than the percent recruitment being manifested as an increase in force, it would be in velocity. As recruitment varies, the actual weight of the individual does not change. Consequently, the major variable to control is velocity as a function of recruitment. The fact that this indeed is the case to some extent is suggested by preliminary data from our laboratory and also by Walmsley et al. (1978). An increase in the running rate on a treadmill belt causes a very small variation in peak force during each step, particularly in the SOL, whereas peak velocity must increase substantially.

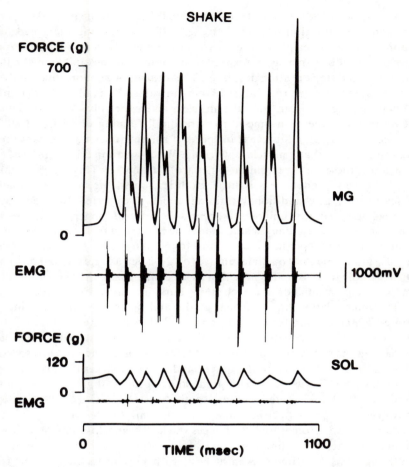

Figure 9—Forces and EMG activity during a rapid shake of the cat hindlimb after placing tape on the paw. Note that the forces in the medial gastrocnemius approach that seen during normal locomotion (Figure 8), while essentially no force fluctuation or EMG is particularly evident from the soleus muscles (preliminary data by Gregor et al. [1981]).

In considering ways in which one can modify the biochemical response of skeletal muscle in response to exercise, duration of exercise is known to be of some importance although there clearly is not a linear relationship between the duration of the exercise and the biochemical adaptation (Terjung, 1976). A factor that almost certainly deserves further consideration is the pattern of activity of muscles while running at various speeds and at varying grades. For example, when an animal runs at 13.4 m/min compared to 40.2 m/min, the duration of on-time of the SOL and lateral gastrocnemius (LG) muscles differs considerably (Gardiner et al., 1982). In this case, two variables to be considered are the number of steps taken per min and the duration of each step. The effect of tripling the speed of running for a guinea pig on these two variables is shown in Figure 10. As the treadmill belt increases in speed, the number of steps taken per min increases and step duration decreases. The total amount of on-time for the ankle extensors when running at 13.4 m/min at a 30% grade is 7.7 s/min. While running at 40.2 m/min, the muscle is active for only 4.8 s/min. This represents a percent on-time during treadmill running of 12.8 and 8%, respectively. Thus, an increase in speed of running does not increase the amount of time that a muscle is working. When an animal is running at a 0% grade, these same percent on-times are 10.6 and 7.3. This suggests that by increasing grade, there is little modification in the actual on-time. Therefore, if any additional training effect is expected to occur by increasing speed, it is unlikely to occur as a result of longer durations of on-time for the muscle as a whole. It may be of interest further to consider, that if this activity per min is continued for 60 min at 13.4 and 40.2 m/min at 30% grade, the total on-time relative to a 24-h period represents 0.53 and 0.34% of the time. But also note that as speed increases, the percent of the motor pool recruited during each EMG burst increases (Figure 10).

The inter-relationship of speed, grade, and percent of maximal integrated EMG per min for the SOL and the LG is illustrated in Figure 11. The change in the activity of the SOL varies little with speed, while the percent grade results in approximately a 20% increase in activity at a speed of 40.2 m/min. The LG on the other hand, shows a larger increase in the integrated EMG with respect to grade and speed. At 0 grade, speed has minimal effect on the total amount of activity in the muscle. At 30% grade, however, the speed effect becomes more evident. At the higher speed, a change in grade from 0 to 30% induces the largest increase in EMG. One underlying factor that undoubtedly is an important parameter to monitor (although this is difficult, even at best, with present technology), is to identify the population of motor units that is activated under varying speeds and grades. Although running at a faster speed reduces the total on-time of the muscle, there is a net increase in total integrated EMG. Consequently, this elevated EMG must be a function of a greater percentage of motoneurons being recruited or an increase in firing frequency of the motor units already recruited. Thus, low threshold units are activated for a shorter duration for each step per min at the higher speeds, than at the lower speeds. At the higher speeds, the higher threshold motor units are activated more often because at a lower speed those motor units probably are not recruited at all. The total integrated EMG provides only a partial picture of the important activation features of the muscle with little information on individual motor units.

The relationship of fatigue and the percent recruitment should also be considered. For example, if motor units function at their maximum force-

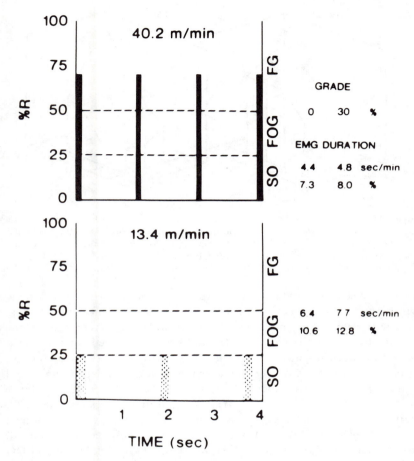

Figure 10—Data derived from the paper of Gardiner et al. (1982) showing the relative frequency and duration of stepping in guinea pigs while running on a treadmill at 13.4 and 40.2 m/min at 0 and 30% grade. The total time that EMG occurred each min in the lateral gastrocnemius is shown on the right. The height of the bars represents an estimated percentage of the motoneuronal pool activated at each speed at a 30% grade.

velocity potential, only a matter of seconds can expire before the subject is unable to continue at that particular force-velocity mode. If the subject recruits approximately 50% of the motoneuron pool, and thereby operates predominantly at half the maximum force-velocity capability, then the time to exhaustion will increase in a manner which will be hyperbolically related to force and velocity. If as little as 5 to 10% of the motoneuron pool is recruited, then the time to exhaustion extends undoubtedly for several hours. The percent recruitment, which is dependent also on the type of motoneurons that are recruited, will determine the time to exhaustion. If

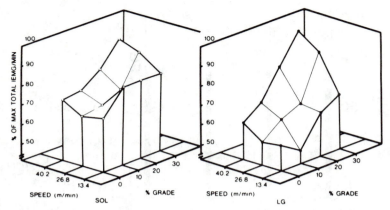

Figure 11—Data from Gardiner et al. (1982) showing the integrated EMG per min of the lateral gastrocnemius and soleus muscle of the guinea pig while running on the treadmill at varying speed and grades. Note that a greater increase in EMG occurs in response to elevation than to speed. From J. Appl. Physiol.: Respirat. Environ. Exer. Physiol. 52(2): 451–457, 1982. Copyright 1982 from The American Physiological Society.

only the lower threshold units are recruited for a long period, little fatigue will be experienced because those particular motor units are the most fatigue-resistant. On the other hand, as a larger percentage of the motoneuronal pool is recruited, the units are less suited for prolonged activation and consequently the time to exhaustion must decrease. These fatigue patterns have been made apparent when stimulating individual motor units in rat (Kugelberg, 1973), cat (Burke et al., 1973), and skunk (Frederick et al., 1978) models. If an FG motor unit is stimulated with trains of impulses lasting 330 ms with individual bursts occurring at a frequency of 40 Hz and each train occurs every second, the motor unit will lose as much as 80 to 100% of its tension within two min. Although FOG units tend to lose some tension, it is usually only around 10%. SO units are the most fatigue-resistant and frequently demonstrate no drop in tension after two min and usually for up to several h.

Using the same rationale and data that has been reported on human performance, it appears almost certain that the percentage of slow-twitch fibers as identified histochemically is a reasonably accurate predictor of the fatigability of the muscle fibers. Although this may be in part due to the fact that a muscle is characterized as slow- or fast-twitch, contraction time is not the variable directly responsible for the fatigue resistance. Because the slow-twitch fibers as identified by a pH sensitive ATPase stain also have a relatively high oxidative capacity in humans, the resistance to fatigue is also high in "slow-twitch fibers." The relationship between the average velocity in m/s that one can run in a 2-h period vs the percentage of slow-twitch fibers can be estimated from information on world record-holders running at a variety of speeds (Riegel, 1981), and data obtained from biopsies taken of elite athletes as reported by Fink et al. (1977). The average running speed for the world record-holder in an event that lasts approximately 2 h is 4.80 m/s. If it is assumed that for an individual to last for 2 h, only the high oxidative fibers can perform for that period of time, and assuming a maximum velocity of running of 10 m/s which requires 100% of the motor units, the percentage of the motoneuron pool that must be active to run 4.8 m/s for 2 h is estimated to be approximately 80% (Figure 12). Interestingly, elite track athletes in these endurance events have about 80% high oxidative fibers, most of which are probably identified as "slow-twitch." Based on the data obtained by Fink et al. (1977), these elite athletes should be able to run 4.95 m/s. It is clear from animal and human studies that the percentage of muscle fibers that have a high oxidative capacity has a significant effect on the ability of that muscle to sustain repetitive activity.

Some Experimental Models Examined with Respect to Level of Activity

A range of models has been used in animal and human experimentation with respect to varying levels of use or disuse. Immobilization has been a common model of disuse. This terminology is at best confusing in that it is

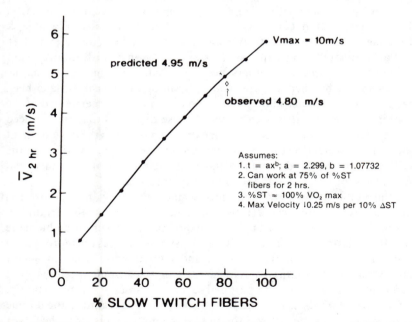

Figure 12—An estimate of the effect of varying percentages of "slow- and fast-twitch" muscle fibers in humans on the maximum, average speed that can be run in a race lasting approximately 2 h. See text for a description of the assumptions that serve as the basis for this graph.

clear that immobilization is not disuse. There is little or no evidence to demonstrate that immobilization indeed induces a marked loss of activity in predominantly fast muscle. To determine to what degree immobilization might be disuse, the relative EMG activity for a 24 h period before and after immobilization of the ankle joint in either an extended, neutral, or shortened position for both a fast and slow muscle was studied (Fournier, 1981). These results show that immobilization is not disuse. At least, in the lengthened and in the neutral position, only a slight, if any, decrement is suggested. Although atrophy is a well-known product of the immobilization, it apparently is not a product of disuse in this particular model. Then, one must ask what is responsible for inducing atrophy and the histochemical and contractile properties that have been reported to have occurred following several weeks of immobilization (Fournier et al., 1981).

A second model that we have been investigating is a spinal cord transection. It also has been employed as a disuse model. Indeed, atrophy need not occur as a result of spinal cord transection. Results from the model suggest a wide range in individual responses to spinal cord transection with respect to EMG activity and atrophy and these two variables do not seem to be tightly coupled. The spinal level of transection as well as the age at transection can be important variables which affect muscle properties and activity-related phenomena. Spinal cord transection plus deafferentation can essentially silence motoneurons (Eldridge & Mommaerts, 1980), but transection alone is not a model of disuse (Smith et al., 1981). In studies over the last 3 yr in which some cordotomized cats have been exercised and others not exercised, a range of contractile, histochemical, biochemical and neurophysiological measures have been made on the neuromuscular system 3 mo to 1 yr after surgery. Two of the more interesting findings with respect to "use-disuse" is that any reduced activity that might occur does not appear to be a factor in the development of a shorter contraction time in the slow muscle. Secondly, regardless of the activity level, the fundamental fatigue characteristics of the individual types of motor units remain the same after cordotomy.

A third model that we have recently been investigating in collaboration with Goslow and co-workers (Edgerton et al., 1980) is cross-reinnervation. Again, the question that should be asked is: can the assumption be made that after cross-reinnervation of slow and fast muscle, does the motoneuron maintain its normal activity pattern and consequently transmit a new activity pattern to the newly reinnervated muscle? Some attempts have been made to assess this with minimal success. In spite of this uncertainty, one important implication from these experiments in regard to the fatigue properties can be stated. It was found that the motor units from the flexor hallucis longus (fast and fatigable) that innervated the SOL muscles reduced the contraction time as would be expected, but did not reduce its resistance to fatigue. These results were surprising in light of the findings which have shown a close relationship between fatigability and the mitochondrial content of muscle fibers and the responses of this system to chronic exercise training (Terjung, 1977). It was expected, for example, that as a result of the SOL muscle being reinnervated by high threshold motoneurons which presumably would then be activated less often would result in a decrease in the mitochondrial content and an increase in the fatigability. This clearly was not the case, however, in that 32 out of 33 motor units innervated by the flexor hallucis longus motoneurons in the SOL muscle maintained their resistance to fatigue as was typical of a normal or a self-reinnervated SOL muscle. These results suggest that there is some inherent genetic program which has determined a minimal level of oxidative capacity and its accommodating vasculature that is independent of the neuromuscular activity level. The contrasting experiment in which muscle fibers of the flexor hallucis longus which are normally fatigable are innervated by high fatigue-resistant motoneurons was not obtained in this experiment. More recent experiments by G.E. Goslow and his co-workers, however, have demonstrated that the fatigue-resistance indeed increases in a normally fatigable muscle as a result of being reinnervated by a SOL motoneuron (personal communication). This is what one would expect since an increase in activity can result in an enhanced mitochondrial content of muscle fibers. These two sets of experiments together suggest 1) that there is a minimal level of mitochondria that is determined in a muscle fiber during development and this minimal level is independent of neuromuscular activity and 2) that this minimal number of mitochondria can be enhanced by a greater than normal level of activity.

There are many other "use-disuse" models that would be appropriate to discuss in relation to force, velocity, percent recruitment and time to exhaustion. The models of immobilization, spinal cord transections and cross-reinnervation, however, provide vivid examples of why it is desirable to examine carefully the specific aspects of an "exercise" program as related to the expected and the observed biochemical responses. Knowledge of the force-velocity-recruitment parameters that characterize a specific model may be as important as the biochemical responses identified if one is interested in identifying regulators of protein metabolism. These "inducers" are as critical as the inducees. Perhaps an appropriate analogy is that, one needs to know more than just the weight of the food that is ingested in order to predict the biological consequence of a given diet.

References

BODINE, S., R.R. Roy, D.A. Meadows, R.F. Zernicke, R. Sacks, M. Fournier & V.R. Edgerton. Architectural, histochemical and contractile characteristics of a unique biarticular muscle: The cat semitendinosus. J. Neurophysiol. 48:192–201, 1982.

BURKE, R. & P. Rudomin. Spinal neurons and synapses. In E. Kandell (Ed.), Handbook of Physiology. Sect. 1. The Nervous System. Bethesda, Md: American Physiological Society, 1977.

BURKE, R.E., D.N. Levine, P. Tsairis & F.E.Zajac, III. Physiological types and histochemical profiles in motor units of the cat gastrocnemius. J. Physiol. (Lond.) 234:723–748, 1973.

CLOSE, R., Dynamic properties of mammalian skeletal muscles. Physiol. Rev. 52:129–197, 1972.

DESMEDT, J.E. & E. Godaux. Ballistic contractions in fast or slow human muscles: discharge patterns of single motor units. J. Physiol. (Lond.) 185:185–196, 1978.

EDGERTON, V.R., G.E. Goslow, Jr., S.A. Rasmussen & S.A. Spector. Is resistance to fatigue controlled by its motoneurons? Nature 285:589–590, 1980.

ELDRIDGE, L. & W. Mommaerts. Ability of electrically silent nerves to specify fast and slow muscle characteristics. In D. Pette (Ed.), Plasticity of Muscle, pp. 325–337. New York: Walter de Gruyter, 1980.

FINK, W.J., D.L. Costill & M.L. Pollock. Submaximal and maximal working capacity of elite distance runners. Part II: Muscle fiber composition and enzyme activities. Ann. N.Y. Acad. Sci. 301:323–327, 1977.

FOURNIER, M., R.R. Roy, H. Perham & V.R. Edgerton. Electromyographic response of fast and slow extensors immobilized at different lengths. Soc. Neurosci. Abst. 7:769, 1981.

FREDERICK, E.C., M.F. Hamant, S.A. Rasmussen, A.K. Chan & G.E. Goslow, Jr. Correlation of histochemical and physiological properties of muscle units in the striped skunk. Experientia 34:372–373, 1978.

GARDINER, K.R., P.F. Gardiner & V.R. Edgerton. Guinea pig soleus and gastrocnemius electromyograms at varying speeds, grades and loads. J. Appl. Physiol.: Respirat. Environ. Exercise Physiol. 52:451–457, 1982.

GREGOR, R.J., C.L. Hager & R.R. Roy. "In vivo" muscle forces during unrestrained locomotion. J. Biomechanics 14:489, 1981.

GRIMBY, L. & J. Hannerz. Firing rate and recruitment order of toe extensor motor unit in different modes of voluntary contraction. J. Physiol. (Lond.) 264:865–879, 1977.

GUTH, L., "Trophic" influences of nerve on muscle. Physiol. Rev. 48:645–687, 1968.

HENNEMAN, E. & L.M. Mendell. Functional organization of motoneuron pool and its input. In J.M. Brookhart and V.B. Mountcastle (Eds.), Handbook of Physiology. The Nervous System II, pp. 423–507. Bethesda, Md: American Physiological Society, 1981.

HILL, A.V., First and Last Experiments in Muscle Mechanics. London: Cambridge University Press, 1970.

HOFFER, J.A., M.J. O'Donovan, C.A. Pratt & G.E. Loeb. Discharge patterns of hindlimb motoneurons during normal cat locomotion. Science 213:466–468, 1981.

KATZ, B., The relation between force and speed in muscular contraction. J. Physiol. (Lond.) 96:45–64, 1939.

KUGELBERG, E., Histochemical composition, contraction speed and fatigability of rat soleus motor units. J. Neurol. Sci. 20:177–198, 1973.

PERRINE, J.J. & V.R. Edgerton. Muscle force-velocity and power-velocity relationships under isokinetic loading. Med. Sci. Sports 10:159–166, 1978.

PHILLIPS, C.A. & J.S. Petrofsky. Velocity of contraction of skeletal muscle as a function of activation and fiber composition: A mathematical model. J. Biomechanics 13:549–558, 1980.

RIEGEL, P.S., Athletic records and human endurance. American Scientist 69:285–290, 1981.

SMITH, J.L., B. Betts, V.R. Edgerton & R.F. Zernicke. Rapid ankle extension during paw shakes. Selective recruitment of fast ankle extensors. J. Neurophysiol. 43:612–620, 1980.

SMITH, J.L., L.A. Smith, R.F. Zernicke & M.G. Hoy. Locomotion in exercised and non-exercised cats cordotomized at two or twelve weeks of age. Exp. Neurol. 76:393–413, 1982.

SPECTOR, S.A., P.F. Gardiner, R.F. Zernicke, R.R. Roy & V.R. Edgerton. Muscle architecture and force-velocity characteristics of the cat soleus and medial gastrocnemius: Implications for motor control. J. Neurophysiol. 44:951–960, 1980.

TERJUNG, R.L., Muscle fiber involvement during training of different intensities and durations. Amer. J. Physiol. 230:946–950, 1976.

WALMSLEY, B., J.A. Hodgson & R.E. Burke. Forces produced by medial gastrocnemius and soleus muscles during locomotion in freely moving cats. J. Neurophysiol. 41:1203–1216, 1978.

WHITING, W.C., R.J. Gregor, R.R. Roy & C.L. Hager. Muscle force production during paw shakes in the cat hindlimb. Med. Sci. Sports Exercise 13:128, 1981.

WICKIEWICZ, T.L., V.R. Edgerton, R.R. Roy, J. Perrine & P. Powell. Human leg muscle architecture and force-velocity properties. Med. Sci. Sports Exercise 14:144, 1982.

Pattern of Chemical Energetics in Fast- and Slow-Twitch Mammalian Muscles

Martin J. Kushmerick
Harvard Medical School, Boston, Massachusetts, U.S.A.

Since the time of Ranvier (1874), it has been known that mammalian musculature is composed of both red and pale cells. We now know that the individual cells in an anatomical muscle range from slow-twitch, highly oxidative fibers containing many mitochondria and much myoglobin (SO fibers), through fast-twitch fibers which are both highly oxidative and glycolytic (FOG fibers) to fast-twitch glycolytic fibers which have a very limited oxidative metabolic capacity (FG fibers) (Barnard et al., 1971). With increased excitation to a motor neuron pool, the cells are excited sequentially in the same order just given (Henneman, Somjen & Carpenter, 1965), so that on average the smaller slow-twitch fibers are recruited more often during normal activity than the larger fast-twitch glycolytic fibers. Thus, information concerning energetics of these different types of cells is crucial to understanding the energetics of exercising muscles and animals. In this discussion, we will focus on the pattern of energy costs for different muscle types insofar as the data allows us, and on their quantitative differences. The issues facing students of muscle energetics are several. First, for the same mechnical output, is the energy cost the same for the different fiber types? Is there a temporal pattern to these energy costs? Is there any evidence that any of the basic energetics are under physiological control? These and many other questions are at the heart of our discussions, but, unfortunately, only some of the questions have been experimentally addressed.

Classical Pattern of Muscle Energetics

Partly due to the ease of experimentation with frog sartorius muscles, but mostly due to the experimental skills and enthusiasm of A.V. Hill, the

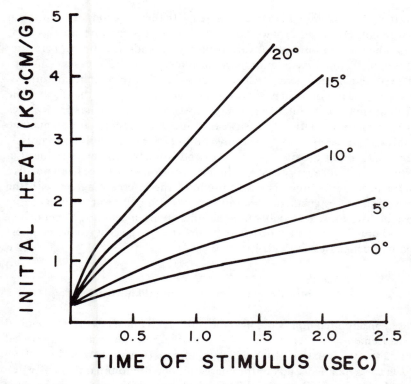

Figure 1—Time course of heat production during a continuous isometric tetanus of frog muscle at different temperatures. Notice that extrapolation of the curves to the y-axis yields a similar value (0.6 to 0.8 kg-cm/g) for all temperatures, whereas the terminal slopes of the curves increase markedly with temperature. Data are redrawn from Hartree and Hill (1921) with the heat output given in mechanical equivalents as reported originally because of an uncertainty in the absolute calibration.

pattern of muscle energetics with which we are most familiar is derived from the study of heat output in amphibian skeletal muscle. Figure 1 illustrates the basic pattern of energy output during an isometric tetanus (Hartree & Hill, 1921). A maximal tetanus is commonly used in these kinds of experiments to achieve a definable and reproducible steady-state. The pattern observed is that of an initial and saturable energy requiring process at the beginning of a contraction superimposed on a steady rate of energy utilization proportional to the maintained isometric force. This temporal distinction gives rise to the notion that two separate mechanisms of energy consumption are responsible for the observed pattern (Curtain & Woledge, 1978; Homsher & Kean, 1978). The steady energy consumption is thought to be due to cross-bridge turnover in the steady-state of force generation. The initial process, which corresponds to an activation energy cost or a "start-up" cost is thought to be associated with the energetics of calcium ion movements (Rall, 1982). Two kinds of experiments corroborate this idea. First, the total amount of activation energy depends little on temperature, although, of course its kinetics speed up as temperature increases. This is to be expected if the energy cost were associated with a finite amount of calcium release or other process required to turn on the contractile mechanism. The maintenance energy cost, on the other hand, increases with temperature with a Q_{10} between 2 and 3 as expected for an enzyme catalyzed process, namely actomyosin ATPase. Secondly, for muscle stretched so that the filament overlap and isometric force are reduced, a graph of the total energy cost as a function of isometric force has a positive intercept at zero force. The magnitude of the intercept agrees with the activation energy cost. Another important feature of the activation energy cost in frog muscle is that once the underlying mechanism operates, a period of repriming must transpire before the energy cost can be manifest again (Kushmerick & Paul, 1977).

Pattern of Energy Cost in Mammalian Muscles

Based on a limited number of heat measurements and a few direct chemical measurements, the energetic pattern in mammalian muscles appeared basically similar to that observed in amphibians (Curtin & Woledge, 1978). Recently, a detailed study of the energetics of a predominantly fast-twitch muscle, the mouse extensor digitorum longus, EDL, and of a predominantly slow-twitch muscle, the mouse soleus, was completed (Crow & Kushmerick, 1982). Some important qualitative and quantitative differences were found. The energetics of the EDL differed from that of the soleus in two major ways as the data in Figure 2 show. First, the energy cost normalized to the isometric force per cross-section area was independent of the tetanus duration in soleus. The longest of these contractions was sufficient to deplete most of the high energy phosphate pool. In the EDL, there was a time-dependent decrease in the energy cost for force maintenance, a feature to which I return later. The second major difference was in the magnitude of the energy cost per unit force per cross-section area. For brief tetani, the energy cost in the fast-twitch EDL was 2.9 times that of the slow-twitch soleus. This was expected from the greater actomyosin ATPase activity from EDL versus soleus muscles. After some 10 s of tetanus, however, the force normalized energy cost in the EDL was reduced so that it was only 1.5 times that of the soleus. We now have evidence that the decreased energy cost in the EDL is due to a down-regulation of actomyosin ATPase activity, as discussed in the fourth section of this paper.

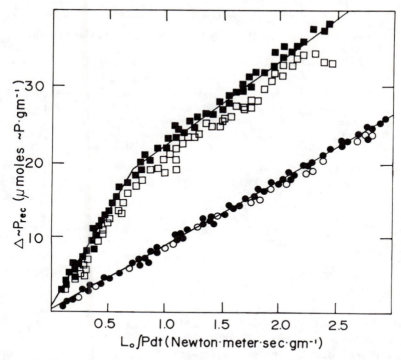

Figure 2—The relationships between total high energy phosphate utilization ($\Delta \tilde{}P_{rec}$) and the tension-time integral for isometric tetani of mouse EDL (squares) and soleus (circles) at 20°. $\Delta \sim P$ is expressed in μmol $\sim P/g$ and is the amount of high energy phosphate resynthesized as calculated from recovery oxygen consumption alone (open symbols) or from recovery oxygen consumption plus recovery lactate production (closed symbols). The lines drawn through the data are linear functions fitted by least squares:

soleus: $\Delta \tilde{}P_{rec} = 0.34 \pm 0.71 + 8.73 \pm 0.51\ L_o \int Pdt$

EDL: $\Delta \tilde{}P_{rec} = 0.68 \pm 0.13 + 32.4 \pm 3.6\ L_o \int Pdt$
   for tetani up to 4 s duration, $0 < L_o \int Pdt < 0.75$

$\Delta \tilde{}P_{rec} = 11.3 \pm 1.5 + 11.1 \pm 1.2\ L_o \int Pdt$
   for tetani longer than 8 s $1.0 < L_o \int Pdt < 2.5$

Data from Crow and Kushmerick (1982a) should be consulted for details.

Stretch experiments indicate the existence in both soleus and EDL of a different kind of activation energy cost or, more properly, a force-independent energy cost (Crow & Kushmerick, 1982c). The magnitude is approximately twice as large in the EDL as in the soleus. The data given in Table 1 were obtained from a series of experiments stretched to various degrees of overlap of the thick and thin filaments and stimulated for 3 or 9 s. It is clear that the reduction in the total ATPase activity observed in EDL

Table 1

Fractionation of the Total Energy Cost Associated with Contraction

Muscle	Tetanus duration (s)	Total[a] energy consumption μmol/gm s	Tension[a] dependent energy consumption μmol/gm s	Tension-[a] independent energy consumption μmol/gm s	Tension-independent cost / Total cost
SOL	3	1.46 ($\pm$0.10)	1.00 ($\pm$0.07)	0.44 ($\pm$0.04)	0.31 ($\pm$0.06)
	9	1.46 ($\pm$0.16)	1.01 ($\pm$0.09)	0.45 ($\pm$0.02)	0.31 ($\pm$0.04)
EDL	3	4.08 ($\pm$0.38)	2.91 ($\pm$0.05)	1.21 ($\pm$0.11)	0.29 ($\pm$0.04)
	9	2.82 ($\pm$0.15)	1.74 ($\pm$0.06)	1.12 ($\pm$0.15)	0.39 ($\pm$0.02)

[a]Mean values $\pm$ SD

is due to a reduction in the force-associated energy cost, not to the force-independent or activation energy cost. These data also show clearly that the force-independent energy cost increases in magnitude as stimulation continues. Thus, the pattern of the force-independent or activation energy costs in mammalian muscles differs from that in the amphibian ones. Therefore, the maintenance energy cost, as usually defined, is not solely attributable to cross-bridge ATP splitting in mammalian muscles.

Furthermore, depending on the duration of the contraction, the relative energy cost for force maintenance in fast-twitch vs slow-twitch muscles differs from 3 to 1 (EDL compared to soleus) for brief tetani to 3 to 2 for longer contractions.

^{31}P NMR Measurements

The ability of modern nuclear magnetic resonance (NMR) specrometers to measure the tissue content of the phosphate compounds relevant to muscle energetics (ATP, PCR, and Pi) is established (Hoult et al., 1974). Sufficiently large magnets exist to monitor these intracellular metabolites in intact human limbs (Ross et al., 1981). One relevant observation is that the inorganic phosphate content in muscles of normal human limbs and in well perfused fast-twitch muscles (Meyer, Kushmerick & Brown, 1982) is substantially lower (approximately 1 μmol/g) than usually reported from analyses of tissue extracts (5 to 7 μmol/g). It is likely that at least some of the discrepancy is due to artifactual breakdown of creatine phosphate, a much more labile compound than ATP, during the extraction or more likely during the freezing procedures. The interesting possibility remains, however, that some form of inorganic phosphate is not detected by NMR and so may be an interesting chemical feature of high energy phosphate metabolism hitherto undetected.

Two recent experimental results from our laboratory are noteworthy. It is well established that intercellular pH can be measured from the chemical shift (the spectral position) of inorganic phosphate (Moon & Richards, 1973). This is due to the fact that the protonated species has a chemical shift different from HPO_4^{2-} and that the measured chemical shift is proportional to the number average of each species with the midpoint at the pK_a. During stimulation and subsequent recovery, the intercellular pH changes are characteristically different in perfused cat biceps, a fast-twitch muscle, and in the perfused cat soleus, an almost pure slow-twitch muscle. During a 15 min period of steady-state twitches, the soleus undergoes a marked intracellular alkalinization from a control value of pH 7.1 to pH 7.4 (Figure 3). In these experiments, the extracellular pH is maintained with a bicarbonate/CO_2 buffer at pH 7. The mechanism for the alkalinization is the proton uptake during net PCr breakdown. In the recovery period, there is a moderate acidification to pH 6.9. By contrast, in the fast-twitch biceps muscle, little alkalinization was detected, even though the PCr breakdown was more rapid. This is most likely due to a greater intracellular buffer capacity and to concomitant lactate production. During the recovery period, the acidification was quite marked, reaching values as low as pH 6.2. The biceps is a mixed muscle with approximately equal proportions of FOG and FG fibers. Interestingly, the shape of the inorganic phosphate peak broadened considerably during the recovery period in the biceps but not in the soleus. This observation indicates a heterogeneity in fiber to fiber intracellular pH not present in the homogeneous soleus muscle. Thus, there are detectable differences in the kinetics of metabolic responses in FOG vs FG fibers. Unfortunately, there are no data on the possible differences in energy costs between FOG and FG fibers, a problem on which we and others are currently working.

A second feature of our NMR results is relevant to the mechanism of muscle fatigue. A well recognized hypothesis is that fatigue is caused by a decrease in intracellular pH. Comparison of the time course of changes in twitch tension, tetanic tension, intercellular pH, PCr concentration and oxygen consumption allowed us to test that idea. All cells were given a 15 min period of twitches at a rate of 30 to 60 per min during which time, twitch and tetanic force decreased as did the PCr level. The recovery of both twitch and tetanic force in the recovery period was much faster than

CHEMICAL SHIFT OF Pi DURING STIMULATION-RECOVERY

A

soleus
60 twitches/min
2 min scans

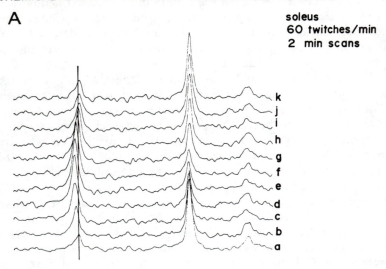

Figure 3—Portions of ^{31}P NMR spectra which show the chemical shift of the inorganic phosphate peak as a measure of intracellular pH during a stimulation-recovery cycle. Cat soleus (panel A) and biceps (panel B) were perfused with an erythrocyte containing physiological saline (Meyer et al., 1982) at 25° and mounted in a Bruker HX-270 spectrometer. The spectra were acquired in the Fourier transform mode using 90° pulses at intervals of 15 s at a frequency of 109.3 MHz. The three peaks shown are, left to right, inorganic phosphate, phosphorylcreatine and ATP (γ phosphorus). Each set of spectra is aligned vertically at the frequency of PCr which is independent of pH under the conditions of this experiment. The vertical line through the Pi region indicates an intracellular pH of 7.1. A shift to the left indicates an alkalinization and a shift to the right indicates an intracellular acidification.

Panel A—soleus: Sequential spectra takes every 2 min, 8 scans averaged.
  a.   unstimulated control
  b–f. sequential spectra during a 10 min period of maximal isometric twitches at 60 per min
  g–h. sequential spectra during the next 10 min period of recovery
Panel B—biceps: sequential spectra takes over 5 min intervals, 20 scans averages.
  a.   unstimulated control
  b–d. sequential spectra during a 15 min period of maximal isometric twitches at 30 per min
  e–j. sequential spectra during the next 30 min recovery period
Note there is little detectable alkalinization during stimulation of the biceps and a marked acidification during recovery. Notice also that in both muscles, the PCr level was reduced to similar levels; data not shown established that a similar steady-state oxygen consumption was reached during the stimulation period of both muscles.

B

biceps
30 twitches / min
5 min scans

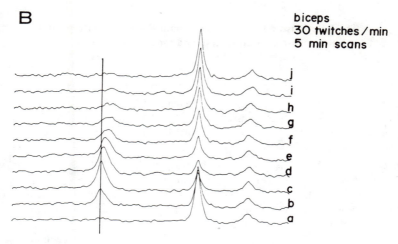

Figure 3 (continued)

the restoration of intracellular p H to control values. Recovery of isometric force paralleled recovery of PCr to control values, therefore suggesting that isometric tension correlates with intracellular chemical potential of the high energy phosphates.

Regulation of Chemomechanics

The decrease in energy cost for force maintenance in the mouse fast-twitch EDL muscle has been described above. Calcium ion acts as a switch on the thin filament by removing an inhibition to actomyosin ATPase. As such, there is no mechanism for calcium ion modification of the kinetics of actomyosin interaction despite some experimental evidence to the contrary (Julian, 1971; Levy et al., 1976). Nonetheless, there was a decrease in the energy cost in the EDL which we attribute to a decreased rate of actomyosin turnover as the following data and arguments show. First, the decrease in the total energy cost was found to be entirely in the tension-dependent fraction as already described (Table 1). Secondly, the maximal velocity of unloaded shortening, which is an excellent measure of cross-bridge turnover rate, was reduced to 56% of controls in muscles given a prior prolonged tetanus (Table 2). Thirdly, there was a specific and reversible covalent modification of the 18,000 dalton light chain of fast-twitch myosin, sometimes called the regulatory light chain by analogy with the function of a comparable protein in scallop muscle. This modification is the phosphorylation of the fast-twitch light chain achieved by a calcium-

Table 2

The Relationship between Light Chain Phosphorylation and the Energy Cost for Contraction and the Maximum Velocity of Shortening in the Mouse Soleus and EDL

Tetanus Duration (Seconds)	$P_o{}^a$ N	$\Delta{\sim}P_{init}{}^b$ $\Delta{\sim}P_{rec}{}^b$ $\mu mol{\sim}P.N.^{-1}m^{-1}s^{-1}$	Maximal velocity[c] of shortening Fiber lengths s^{-1}	$\dfrac{LC2f - p^d}{LC2f - P + LC2f}$	
Soleus					
3	0.12 (±.013)	9.1 (±1.8)	8.9 (±0.4)	1.88 (±0.05)	0.11 (±0.05)
9	0.11 (±.011)	8.9 (±1.9)	8.7 (±0.3)	1.89 (±0.04)	0.12 (±0.06)
15	0.11 (±.006)	8.8 (±2.5)	8.8 (±0.3)	1.88 (±0.03)	0.10 (±0.05)
EDL					
3	0.17 (±0.008)	24.1 (±3.6)	22.1 (±1.1)	5.75 (±0.13)	0.22 (±0.07)
9	0.16 (±.012)	14.8 (±3.9)	14.5 (±1.9)	4.07 (±0.06)	0.45 (±0.06)
15	0.13 (±.015)	10.0 (±1.2)	11.4 (±1.4)	3.21 (±0.10)	0.51 (±0.05)

All measurements are given as the mean ± one standard deviation.
[a] Isometric tetanic force prior to measurement of maximal velocity of shortening or at end of stimulation for the chemical measurements.
[b] $\Delta{\sim}P_{init}$ and $\Delta{\sim}P_{rec}$ designate the extents of high energy phosphate utilization for the isometric tetanus by direct measurements of breakdown of PCr and ATP ($\Delta{\sim}P_{init}$) or by total recovery metabolism ($\Delta{\sim}P_{rec}$). Data from Crow and Kushmerick (1982a,c) are normalized per unit of force per cross-section area per s.
[c] The maximal velocity of shortening is expressed in units of muscle cell length per s and was measured by a quick release to slack length.
[d] The fractional extent of phosphorylation of the 18,000 dalton light chain of fast-twitch myosin, LC2f, is expressed as the fraction of total light chain present. The light chain content was measured by densitometric scanning two dimensional gel electrophoretograms of the muscle proteins. Details are given in Crow and Kushmerick (1982b).

calmodulin-myosin light chain kinase complex. Fourthly, none of the chemical, mechanical or phosphorylation changes were noted in the slow-twitch soleus muscles. Figure 4 summarizes all of these experimental results. Not only is this thick filament regulation of great interest for the mechanism of actomyosin ATPase, but also it establishes the important notion that quantitative aspects of the energy costs of muscle contraction are amenable to physiological control. This physiological down-regulation

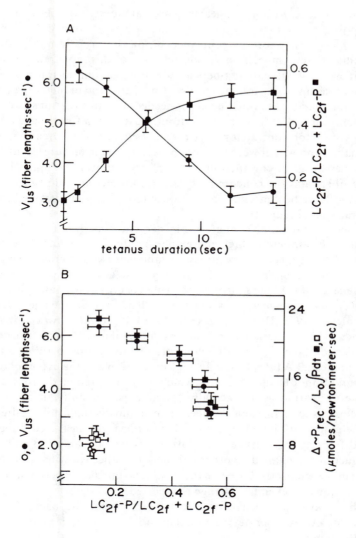

Figure 4—The relationships between maximal velocity of shortening, isometric energy cost and the extent of myosin light chain phosphorylation.

Panel A—The fractional extent of phosphorylation of the 18,000 dalton fast-twitch myosin light chain, LC2f, (closed squares) and velocity of shortening (closed circles) are plotted as a function of isometric tetanus duration. Average of at least 6 muscles were used; error bars give one standard deviation.

Panel B—The rate of normalized energy utilization (squares) and the maximal velocity of shortening (circles) in EDL (closed symbols) and in soleus (open symbols) are given as a function of the fractional extent of LC2f phosphorylation. Data from Crow and Kushmerick (1982b).

of cross-bridge activity appears to operate only the EDL, a muscle composed predominantly in fast-twitch fibers (Crow & Kushmerick, 1982b). The functional significance of phosphorylation in vertebrate skeletal muscle myosin may be related to this apparent fiber-type specificity. Fast glycolytic fibers are the ones optimized for large power outputs because of their large cross-sectional area and of their organization into motor units. Maximal power is needed only in situations of extreme acceleration and of large power output. Fast-twitch fibers have a higher energy cost for contraction that do slow energy cost. Great speed and mechanical power, however, are not necessary and in fact are not useful for sustained forceful contractions. Phosphorylation of the light chains in these fibers and the concomitant reduction in actomyosin turnover rate may therefore be a mechanism to reduce fatigability in sustained forceful contractions (Figure 4).

A second potential regulation of energetics involves the relationship between the amount of high energy phosphate splitting and the extent of oxidative and glycolytic recovery metabolism (Kushmerick, 1978). The problem of the energy imbalance in frog muscles with respect to heat and high energy phosphate splitting and to high energy phosphate splitting and recovery metabolism is well known and will not be described here (Curtain & Woledge, 1978; Homsher & Kean, 1978). It is clear, however, from our studies of mouse muscles that there is no evidence for any biochemical imbalance. That is, the stoichiometric relationships between creatine phosphate and ATP splitting during contraction and the extent of oxygen consumption during recovery which can be predicted from known metabolic pathways are actually found experimentally in mammalian muscles as shown in Table 2 where there are no distinguishible differences in the extent of high energy phosphate utilization measured directly by rapid freezing methods or by recovery oxygen consumption and lactate production. A detailed myothermal energy balance study has not yet been done in these muscles. Based on these data, no imbalance between enthalpy production and chemical change is expected.

Thus, possible mechanisms wasteful of metabolic energy, such as substrate cycles, operate at an undetectably low level, if at all, in mouse hindlimb muscles.

Acknowledgment

The efforts by Howard Seeherman made these experiments possible by developing the arterially perfused cat bicep muscle. Michael Crow worked energetically and persistently during the initial phases of all of this work and performed the experiments with the mouse muscles. Ronald Meyer developed the perfused soleus muscle technique and continues to work productively on all aspects of the NMR experiments. Truman Brown is a valuable colleague and critic whose help with the NMR experiments is invaluable.

References

BARNARD, R.J., V.R. Edgerton, T. Furukawa & J.B. Peter. Histochemical, biochemical and contractile properties of red, white, and intermediate fibers. Am. J. Physiol. 220:410-414, 1971.

CROW, M. & M.J. Kushmerick. Chemical energetics of slow- and fast-twitch muscles of the mouse. J. Gen. Physiol. 79:147-166, 1982a.

CROW, M. & M.J. Kushmerick. Myosin light chain phosphorylation is associated with a decrease in the energy cost for contraction in fast twitch mouse muscle. J. Biol. Chem. 257:2121-2124, 1982b.

CROW, M. & M.J. Kushmerick. The effect of previous contractile activity on the mechanical performance of mouse fast-twitch and slow-twitch muscles. J. Gen. Physiol. In Review, 1982c.

CURTIN, N.A. & R. Woledge. Energy changes and muscle contraction. Physiol. Rev. 58:690-761, 1978.

HARTREE, W. & A.V. Hill. The regulation of the supply of energy in muscular contraction. J. Physiol. (London) 55:133-158, 1921.

HENNEMAN, E., G. Somjen & D.O. Carpenter. Functional significance of cell age in spinal motoneurons. J. Neurophysical. 28:560-580, 1965.

HOMSHER, E. & C.J. Kean. Skeletal muscle energetics and metabolism. Annu. Rev. Physiol. 40:93-131, 1978.

HOULT, D.I., S.J.W. Busby, D.G. Gadian, G.K. Radda, R.E. Richards & P.J. Seeley. Observation of tissue metabolites using ^{31}P nuclear magnetic resonance. Nature (London) 252:285-287, 1974.

JULIAN, F.J., The effect of calcium on the force velocity relation of briefly glycerinated frog muscle fibers. J. Physiol. (London) 218:117-145, 1971.

KUSHMERICK, M.J. Energy balance in muscle contraction: a biochemical approach. Curr. Top. Bioenerg. 6:1-37, 1978.

KUSHMERICK, M.J. & R.J. Paul. Chemical energetics in repeated contractions of frog sartorius muscles at 0° C. J. Physiol. 267:249-260, 1977.

LEVY, R.M., Y. Umazume & M.J. Kushmerick. Ca^{++} dependence of tension and ADP production in segments of chemically skinned muscle fibers. Biochem. Biophys. Acta 430:352-365, 1976.

MEYER, R.A., M.J. Kushmerick & T.R. Brown. Application of ^{31}P-NMR spectroscopy to the study of striated muscle metabolism. Am. J. Physiol. 242:C1–C11, 1982.

MOON, R.B. & J.H. Richards. Determination of intracellular pH by ^{31}P nuclear magnetic resonance. J. Biol. Chem. 248:7276–7278, 1973.

RALL, J.A., Energetics of Ca^{2+} cycling during skeletal muscle contraction. Fed. Proc. 41:155–160, 1982.

RANVIER, L., De quelque faits relatifs a l'histologie et a la physiologie des muscles stries. Arch. Physiol. Nom. Pathol. 6:1–15, 1874.

ROSS, B.D., G.K. Radda, D.G. Gadion, G. Rocker, M. Esiri & J. Falconer-Smith. Examination of a case of suspended McArdle's syndrome by ^{31}P nuclear magnetic resonance. New Eng. J. Med. 304:1338–1342, 1981.

Substrate Availability

Eric Hultman and H. Sjöholm
Huddinge University Hospital, Huddinge, Sweden

The skeletal muscle system is unique in its ability to adjust energy production to large changes in energy demand. Energy expenditure from rest to maximum power output of short duration can increase by as much as 1000-fold, and this can be achieved within fractions of a second. Increased energy output by muscle can under certain conditions be sustained for several hours at a time without rest. The different rates of energy production in response to changes in environmental demands are met by the muscle through the utilization of a spectrum of different substrates.

Time Course of Substrate Availability

The immediately available substrates for energy production are those stored in muscle tissue itself, such as high energy phosphates and glycogen. These are in close contact with the contractile machinery within the muscle cells and can thus be utilized without delay when the command for increased energy output arrives to the muscle. Both may be used anaerobically, enabling metabolism to begin without prior adjustment of the circulation. In addition to high energy phosphate and glycogen, endogenous hexose phosphates and free glucose are also available for immediate, anaerobic utilization. The intracellular contents of these substrates at rest, however, are small, and the penetration by glucose of the plasma membrane is a slow process compared to the rate of formation of hexose phosphate from locally stored glycogen (about 1 mmol $kg^{-1}d$ m min^{-1} compared to > 100).

At the start of contraction the store of oxygen available to the muscle cells in the form of oxymyoglobin and in the capillaries as oxyhemoglobin is in terrestrial mammals extremely small. Delays in the activation of mitochondrial respiration by ADP and inorganic phosphate will further reduce the likely importance of this during the first of increased energy demand. Availability of lipid in the form of blood-borne free fatty acids (FFA) is low at rest but will increase during exercise. The time scale for the increase is, however, not seconds but rather minutes or hours. The peak level of FFA during exercise is reached after 3 to 4 h of continuous exercise. Triacylglycerol stored in muscle or transported to the muscle via the blood may also be used; however, relatively little is known regarding the mechanism by which this substrate is utilized and the rate at which this occurs during exercise. The time course of substrate utilization is consequently the following:

1. $ATP \rightarrow ADP + P_i + energy$
2. $ADP + PCr \rightarrow ATP + Cr$
3. $ADP + glucose\ (glycogen) \rightarrow ATP + lactate$
4. $ADP + O_2 + substrate \rightarrow ATP + CO_2$

Power Output and Substrate Choice

Variation in maximal energy production (sic ATP production) will further influence the choice of substrate(s) during exercise. The rate of energy demand by the muscle is determined by the rate of ATP degradation through the activity of ATPase in the contracting muscle. It is important that a large decrease in ATP does not occur as this would significantly decrease the free energy change during continued ATP degradation, and ultimately inhibit further contraction by the muscle. A rapid loss of ATP during contraction would thus be selflimiting. The increase in ADP when ATP is hydrolyzed stimulates the utilization of PCr, whilst the combined effects of increased Ca^{2+} and inorganic phosphate will be the signal to increased glycogen breakdown through transformation and activation of phosphorylase. PCr and glycogen as substrates for ATP resynthesis are utilized at the start of exercise and will continue to be used if the energy demand is higher than can be met by oxidative metabolism (i.e., during exercise with work loads higher than maximal aerobic power or $\dot{V}o_2max$). At exercise intensities lower than $\dot{V}o_2max$ the relative importance of pyruvate and lipid oxidation will be increased; the lower the exercise intensity the greater the contribution from oxidative metabolism. A discussion of the maximum power output from different fuels for exercise was presented by McGilvery (1975).

It should be pointed out that the values given in this paper concerning contents of substances in muscle are referred to dry muscle tissue and not as, in many other papers, to wet tissue. The reason for this is that the muscle samples are freeze-dried and powdered before being analyzed, and further that the dry tissue is a more accurate base of reference than wet tissue during exercise when water shifts can be expected to occur. Recalculation of dry to wet tissue content can be done by dividing values by 4.3. To calculate the concentration in intracellular water the value in dry muscle should be divided by 3, i.e., 1 kg of dry tissue is normally associated with 3 l of intracellular water and 0.3 l of extracellular water.

Energy Rich Phosphagens and Anaerobic Glycolysis

The levels of the principal energy rich phosphagens in human muscle determined in acid extracts of biopsy samples are 24.0 ± 2.6 mmol ATP and 75.5 ± 7.6 mmol PCr per kg dry muscle. The total creatine and total adenine contents are 124.4 ± 11.2 and 27.4 ± 2.5 mmol per kg dry muscle, respectively (Harris et al., 1974). The glycogen content of the muscle is of the order of 350 mmol glucosyl units per kg dry muscle.

Maximum power output during short bursts of exercise has been determined under various conditions such as climbing stairs at top speed (Margaria et al., 1966) pedaling on a cycle ergometer (Ikuta & Ikai, 1972) and standing high jump (Davies, 1971; di Prampero & Mognoni, 1981). Estimates of the maximum rate of ATP utilization in these studies vary from 7 mmol active phosphate ($\sim$P) kg^{-1} d m s^{-1} (Margaria et al., 1964, 1966) to 25 to 30 mmol in the studies of the standing high jump. Undoubtedly much of the difference between estimates can be attributed to the exercise model used. In the study of Margaria et al. the exercise period was of the order of 3 s and contained both contraction and relaxation while maximum work output in the other two studies consisted of only one maximum contraction lasting 0.2 s.

In order to study the utilization of the different substrates during contractions of short duration we have used electrical stimulation of the intact quadriceps femoris muscle via surface electrodes. The electrical stimulation produced an isometric contraction and the power output by the knee extensors was measured by means of a strain gauge. It is possible using this technique to stimulate the muscle at different frequencies and for a constant time period. In Figure 1 is shown two experiments, the first employing 50 Hz stimulation during 1.28 s (64 impulses) and during 2.56 s (128 impulses). The blood flow was occluded by means of a tourniquet round the upper part of the thigh. In a second experiment 20 Hz stimulation was utilized and the contraction time was 3.2 s (64 impulses). This stimulation was repeated after 60 s rest with continued occlusion of the circulation. Biopsies were taken before and after each contraction

ATP-turnover mmol·kg^{-1}·s^{-1}

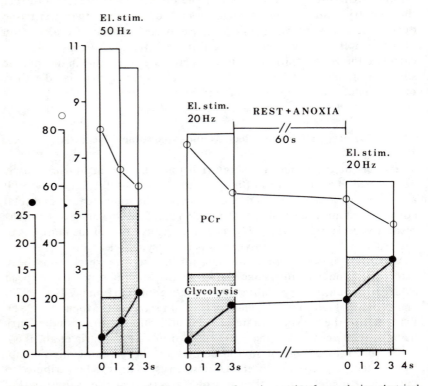

Figure 1—ATP turnover rate in the quadriceps femoris muscle of man during electrical stimulation at frequencies of 50 and 20 Hz. Results shown in the lefthand part of the figure are from an experiment in which the quadriceps was stimulated at 50 Hz for 1.28 and 2.56 s. Results shown in the right are from a second experiment in which the muscle was stimulated for two periods at 20 Hz with an interval of rest of 60 s. Throughout the experiment the local circulation was occluded by means of a tourniquet located around the upper thigh.

Columns show the rate of ATP production per s per kg dry muscle from degradation of PCr and from glycolysis. Symbols denote the muscle contents of PCr (o) and lactate (•) in mmol per kg dry muscle.

period making it possible to determine the substrate utilization during contractions at 50 Hz and at 20 Hz, and also to investigate the metabolic change in the rest period without circulation.

During the 1.26 s period of stimulation with 50 Hz, which gives a near maximum contraction force, the total ATP turnover was 11 mmol kg^{-1} d m s^{-1}, 80% of this being derived from breakdown of PCr and approximately 20% from degradation of glycogen to lactate. Clearly, formation of lactate begins very early following the start of contraction. The longer stimulation at 50 Hz gave a slightly lower energy outut per s and showed increased contribution of glycolysis to total energy production. During the period 1.28 to 2.56 s glycolysis accounted for 50% of the ATP resynthesis. At 20 Hz stimulation (using the same voltage as the 50 Hz stimulation) the power output is in the order of 70% of the maximum (Hultman et al., 1981). The energy expenditure was 7.8 mmol ATP kg^{-1} s^{-1}. About 30% of the energy was derived from glycolysis during the first 3 s of contraction while repeated contraction resulted in further increase in glycolytic rate, providing more than 50% of the utilized energy.

The total power output was only slightly decreased (5%) during the second 3 s contraction period, but the measured energy output was decreased by 23%. The reason for this is not known. Possibly some of the stored intracellular oxygen is utilized during the second period when ADP and P_i content are increased both in the cytoplasm and in the mitochondria. This can also be true for the 60 s rest period without circulation during which time the measured ATP turnover was of the order of 0.06 mmol kg^{-1} s^{-1}; this value is however uncertain as the lactate and PCr values at the start of the second contraction are not significantly different from those at the end of the first (see Figure 1). Prolonged isometric contraction was studied during 30 s electrical stimulation at 20 Hz and biopsy specimens were obtained before and after 10, 20 and 30 s. Contraction force increased during the first 10 s and thereafter remained relatively stable. The utilization of the different substrates, PCr and glycogen, is shown in Figure 2 (normal leg). Total ATP turnover decreased only slightly during contraction and was close to the value determined by Margaria et al. (1964). PCr utilization was highest during the first 10 s but declined as the PCr store was progressively emptied. In contrast ATP resynthesis from glycolysis increased successively and during the last period accounted for most of the energy production.

Glycogen Store and Isometric Exercise

In order to study the effect of low glycogen content upon metabolism during isometric contraction the following experiment was performed. A group of subjects worked on a cycle ergometer with one leg each to exhaustion and for the remainder of that day were given a carbohydrate-low diet. The exercise was continued in the morning of the following day in order to deplete as far as possible the glycogen store in the quadriceps femoris muscle of the exercise leg; the glycogen content of the other leg of each subject was maintained at a normal level. Mean glycogen contents were 58 ± 5 and 250 ± 27 mmol glycosyl units per kg d m in the two legs after the exercise periods. The quadriceps femoris in either leg was then stimulated electrically using the same voltage and frequency, for equal duration, and the metabolic intermediates measured in biopsy samples.

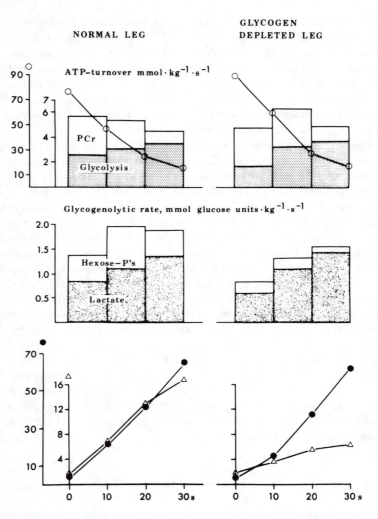

Figure 2—The effect of altered glycogen level upon ATP turnover rate and glycogenolytic rate in the quadriceps femoris of man during elecrical stimulation. Stimulation of either leg of each subject (5 in all) was in each case for 30 s and at a frequency of 20 Hz. Prior to stimulation one leg of each subject was subjected to continuous bicycle exercise in order to decrease the muscle glycogen content. At the start of contraction the mean glycogen level in the "normal leg" was 250 ± 27 mmol glucosyl units per kg dry muscle and in the "glycogen depleted leg" 58 ± 5 mmol glucosyl units per kg dry muscle.

The upper part of the figure shows ATP production from PCr degradation and glycolysis. The middle part, the rate of glycogenolysis calculated from the accumulation of hexose phosphates and lactate.

The symbols denote the muscle contents of PCr (o), lactate (●) and glucose 6-P (△) in mmol per kg dry muscle.

Results are given in Figure 2. Total energy production from PCr degradation and from glycolysis was practically the same in the two legs. Glycogen degradation, as calculated from accumulation of hexose phosphates and lactate was, however, reduced in the leg with the low glycogen content. The difference was dependent on a much lower accumulation of hexose phosphates; while the lactate formation was of the same order of magnitude in both the legs. Excess glycogen degradation with accumulation of hexose phosphates may thus be of less importance to the maintenance of adequate phosphofructokinase activity in vivo than has been previously suggested. The lower levels of glucose- and fructose-6-phosphates observed were clearly adequate in each case for stimulating PFK activity. It can be concluded that even after prolonged glycogen depleting exercise with apparent exhaustion, there is no impairment of the muscle's ability to perform short lasting bouts of isometric exercise, and that the glycolytic rate can be kept high even when the glycogenolytic rate is decreased due to a shortage of available substrate.

Muscle Glycogen Store and Dynamic Exercise

During dynamic exercise with exercise intensities above the subject's $\dot{V}O_2max$, degradation of PCr and glycogen proceeds rapidly as described above. Exercise intensities between 90 and 100% of $\dot{V}O_2max$ result in a rapid utilization of locally stored glycogen. The amount available is normally about 350 mmol glucosyl units kg^{-1} d m (or 80 mmol kg^{-1} wet tissue). Degradation proceeds at a rate of approximately 10 to 12 mmol glucosyl units kg^{-1} d m min^{-1} (Saltin & Karlson, 1971) but at this high exercise intensity the total store of glycogen is never fully utilized. The high rate of glycogen degradation results in a rapid accumulation of lactate in the muscle to values inhibiting further energy production at the same rate. At exercise intensities demanding 70 to 80% of the subject's $\dot{V}O_2max$ the whole glycogen store in the muscle can be utilized and in this situation the amount of substrate available is a determinant for exercise capacity. It has been shown that the glycogen store in muscle can be varied by the food intake, i.e., a carbohydrate-rich diet increases and a carbohydrate-poor decreases the glycogen store (Hultman & Bergström, 1967). It has been further shown that resynthesis of glycogen after depleting exercise is very sensitive to the type of food taken during the recovery period. Thus a carbohydrate-rich diet taken after depletion of the glycogen store will increase the amount of glycogen in the depleted muscle groups to normal values within 1 day, and may increase the store further to 2 to 3 times the normal value when the diet is continued for another 2 to 3 days (Bergström & Hultman, 1966). This increase in glycogen store will increase exercise performance capacity to the same degree, provided that an exercise intensity of 70 to 80 % of $\dot{V}o_2$ max is employed (Bergström et al., 1967). During exercise with sub-maximal intensities blood glucose and lipids are also utilized.

Blood Glucose and Liver Glycogen

Blood glucose per se constitutes a very small store which would be rapidly emptied if it was not continuously renewed by the liver. Two sources are available for glucose production by the liver; the glycogen store and the utilization of gluconeogenic substrates. Gluconeogenic substrates are used at rest to an amount equal to 0.30 mmol glucose per min. The rate of utilization can increase during exercise up to 1.7 mmol per min (Hultman, 1978). Gluconeogenesis is dependent mainly upon lactate, derived from the exercising muscles and to a lesser extent upon alanine and glycerol. At the same time the glycogen store in the liver is used at rest at a rate corresponding to 0.5 mmol glucosyl units per min and during exercise up to 5 mmol per min (Hultman, 1978). The total amount of glycogen in the liver is normally of the order of 500 mmol glucosyl units, though this is known to vary with the diet (Nilsson & Hultman, 1973). One day of carbohydrate-free diet will decrease the liver glycogen store to as low as 60 to 70 mmol in the whole liver, while 1 day of carbohydrate-rich diet increases the store to 800 to 900 mmol (Nilsson & Hultman, 1973). From these values it can be seen that hard exercise will deplete the glycogen store in the liver within a short period if the diet during the previous day was low in carbohydrate. The result of depletion of the liver glycogen store is a decrease in blood glucose level resulting in pronounced central nervous effects such as dizziness, and ultimately incapacity to continue to exercise (Bergström et al., 1967). It can, however, also be seen from the values given above that previous feeding with a carbohydrate rich diet will insure the availability of liver glycogen for long periods of hard exercise. The precise mechanism by which glycogen stored in the liver is released is still not fully understood, though it is known that a series of hormonal changes occur during exercise resulting in increased blood levels of glucagon and adrenalin and decreased levels of insulin. These changes will tend to increase the release of glucose from the liver. Direct nervous control of the liver may also be important in regulating glycogenolysis.

Fat Utilization

During prolonged exercise with work loads of 70% of $\dot{V}o_2$max and lower, the importance of fat as an energy source is greatly increased. According to

McGilvery (1975), maximal energy output from fat combustion is lower than from oxidation of glucose, being 1 mmol ~P per kg dry muscle per second for fat compared to 2 mmol for glucose. Actual rates, however, will vary from one individual to another as a result of variation in enzyme composition in the exercising muscles. Lipids which are available as energy substrates are local triacylglycerols in muscle, and triacylglycerols and FFA in blood. The amount of triacylglycerols stored in muscle is of the order of 48 mmol per kg d m corresponding to 24 000 mmol of ~P if fully utilized. Relatively little is known with regard to the mechanisms regulating the use of the local fat store, though it has been shown by Fröberg et al. (1971) and Essén et al. (1977) that prolonged exercise results in a reduction of local triacylglycerols. More important as an energy substrate is blood-borne fat in the form of triacylglycerol and FFA. The uptake of circulating triacylglycerol has not been studied in detail but it is known that the activity of lipoproteinlipase in muscle, the enzyme responsible for the primary degradation of the lipid, is increased in long distance runners but not in sprinters (Nikkilä et al., 1978). Blood-borne FFA are released from adipose tissue and this release is enhanced during exercise. It is known that uptake of FFA by the muscle is directly related to the concentration of FFA in blood. It was shown by Havel et al. (1963) using ^{14}C labelled long chain fatty acids that the turnover rate of plasma FFA increased during prolonged exercise and that a considerable fraction of the energy used during such exercise was derived from the oxidation of FFA from adipose tissue. The increase in blood FFA during exercise is a relatively slow process; peak levels are generally not reached until the third or fourth h of exercise. At this time FFA oxidation can account for 65% of the total energy utilized. It has been shown that the uptake of FFA by working muscle generally exceeds its rate of oxidation (Gollnick, 1977) but in spite of this, additional increase in the availability of FFA will further increase its utilization during exercise.

Increased FFA Availability

The effect of increased FFA availability is decreased utilization of both muscle glycogen and blood glucose with a prolongation of the exercise period before exhaustion is reached (Costill et al., 1977; Hickson et al., 1977; Rennie et al., 1976). Increase in plasma FFA in these studies was brought about by infusion of heparin with or without peroral administration of additional fat. Plasma FFA may also be increased by ingestion of caffeine for 1 h before exercise (Costill et al., 1978; Ivy et al., 1979). Again this may lead to increased performance time during exercise as a result of increased lipid metabolism and lowered rate of carbohydrate utilization.

Effect of Inhibition of FFA Release

Decreased FFA availability results in a shift in metabolism to higher rates of carbohydrate utilization. This has been shown in studies of Bergström et al. (1969) and Pernow & Saltin (1971) in which nicotinic acid was used to block FFA release from adipose tissue. In both studies utilization of muscle glycogen increased and the blood sugar level decreased when FFA release was blocked. A decreased level of FFA in blood may also occur following administration of β receptor blocking agents, resulting in increased rate of carbohydrate degradation (Brzezinska & Nazar, 1970; Galbo et al., 1976; Nazar et al., 1971). As previously discussed administration of glucose during exercise will also decrease FFA utilization. In this situation glucose has a double effect. Firstly there is a direct effect upon blood FFA levels as a result of increased re-esterification in adipose tissue due to increased availability of glycerol 1-phosphate when the glucose content is raised. The second effect of glucose is to increase the release of pancreatic insulin which in turn will inhibit lipolysis in adipose tissue at the enzyme level. The effect of glucose ingestion can be particularly important during prolonged exercise when it can, paradoxically, give rise to hypoglycemia. This results from the rapid depletion of the carbohydrate store in the liver due to decreased FFA availability (and thus lowered contribution to overall energy production) from levels of insulin in the blood.

Summary

1. ATP is the primary substrate for energy production and its constant renewal is a prerequisite for continuation of muscle contraction. Decreases in the level of ATP in muscle, in the order of 20 to 40%, are seen only at the end of very hard exercise.

2. Phosphocreatine (PCr) serves as an immediate store for energy production. Although the total store may be utilized, this never occurs before activation of glycolysis has begun. Its main function appears as a buffer to sudden changes in energy demand. A surprisingly constant relationship between PCr and lactate contents is found in human muscle, when sampled by the needle biopsy technique, during both isometric and dynamic exercise.

3. Glycogen is the largest locally available store of energy to the muscle. Utilization of this has been observed within the first second of the start of exercise.

At moderate to low exercise intensities the whole store is available but, at high intensities, total utilization may be impeded by the accumulation of H^+ and/or early attainment of fatigue.

As judged by the marked accumulation of hexose monophosphate during isometric exercise, degradation of glycogen generally seems to occur at a rate in excess of the glycolytic rate. The amount of glycogen stored in the muscle is greatly influenced by diet received and exercise performed during the previous days. At moderate exercise intensities, this can be a determinant for exercise capacity.

4. The availability of blood glucose as an energy source is dependent mainly upon the size of the liver glycogen store, and to a lesser extent upon synthesis in the liver from gluconeogenic substrates. Compared to the muscle glycogen store, that in the liver is considerably more labile and dependent upon the previous day's diet.

Utilization of blood glucose is of greatest importance during prolonged exercise.

5. Fat may only be used aerobically and, as an energy store, constitutes the body's largest reserve. The total amount available is always in excess of the need imposed by exercise. The mechanisms governing the utilization of the different types of blood-borne and locally stored lipids are, even today, still relatively unknown.

During prolonged exercise, free fatty acids (FFA) from adipose tissue may account for up to 60% of the total energy needs, but the rate of utilization is dependent upon the blood level which increases slowly during exercise.

During exercise the rate of FFA uptake is generally higher than the rate at which it is used. In spite of this, additional increase in the availability of FFA will further increase its utilization.

Acknowledgment

This work was supported by grants from the Swedish Medical Research Council (No. 03X-02647).

References

AHLBORG, G. & P. Felig. Influence of glucose ingestion on fuel-hormone response during prolonged exercise. J. Appl. Physiol. 41:683–187, 1976.

BERGSTRÖM, J., L. Hermansen, E. Hultman & B. Saltin. Diet, muscle glycogen and physical performance. Acta Physiol. Scand. 71:140–150, 1967.

BERGSTRÖM, J. & E. Hultman. Muscle glycogen synthesis after exercise: an enhancing factor localized to the muscle cells in man. Nature 210:309–310, 1966.

BERGSTRÖM, J., E. Hultman, L. Jorfeldt, B. Pernow & J. Wahren. The effect of nicotinic acid on physical working capacity and metabolism of muscle glycogen in man. J. Appl. Physiol. 26:170–176, 1969.

BRZEZINSKA, Z. & K. Nazar. Effect of beta-adrenergic blockade on exercise metabolism in dogs. Arch. Int. Physiol. Biochem. 78:883, 1970.

COSTILL, D.L., E. Coyle, G. Dalsky, W. Evans, W. Fink & D. Hoopes. Effects of elevated plasma FFA and insulin on muscle glycogen usage during exercise. J. Appl. Physiol. Respirat. Environ. Exercise Physiol. 43:695–699, 1977.

COSTILL, D.L., G.P. Dalsky & J.W. Fink. Effects of caffeine ingestion on metabolism and exercise performance. Med. Sci. Sports 10:155–158, 1978.

DAVIES, C.T.M., Human power output in exercise of short duration in relation to body size and composition. Ergonomics 14:245–256, 1971.

DiPRAMPERO, P.E. & P. Mognoni. Maximal anaerobic power in man. Medicine and Sport 13:38–44, 1981.

ESSÉN, B., L. Hagenfeldt & L. Kaijser. Utilization of blood-borne and intramuscular substrates during continuous and intermittent exercise in man. J. Physiol. (London) 265:489–506, 1977.

FELIG, P. & J. Wahren. Role of insulin and glucagon in the regulation of hepatic glucose production during exercise. Diabetes 28(suppl):71–75, 1979.

FOSTER, C., D.L. Costill & W.J. Fink. Effects of pre-exercise feedings on endurance performance. Med. Sci. Sports 11:1–5, 1979.

FRÖBERG, S.O., L.A. Carlsson & L.G. Ekelund. Local lipid stores and exercise. Adv. Exp. Med. Biol. 11:307–314, 1971.

GALBO, H., J.J. Holst, N.J. Christensen & J. Hilsted. Glucagon and plasma catecholamines during beta-receptor blockade in exercising man. J. Appl. Physiol. 40:855–863, 1976.

GOLLNICK, P.D., Free fatty acid turnover and the availability of substrates as a limiting factor in prolonged exercise. Ann. New York Acad. Sci. 301:64–71, 1977.

HARRIS, R.C., E. Hultman & L.-O. Nordesjö. Glycogen, glycolytic intermediates and high-energy phosphates determined in biopsy samples of musculus quadriceps femoris of man at rest. Methods and variance of values. Scand. J. Clin. Lab. Invest. 33:109–120, 1974.

HAVEL, R.J., A. Naimark & C.F. Borchgrevink. Turnover rate and oxidation of free fatty acids of blood plasma in man during exercise: Studies during continuous infusion of palmitate-1-C^{14}. J. Clin. Invest. 42:1054–1063, 1963.

HICKSON, R.C., M.J. Rennie, R.K. Conlee, W.W. Winder & J.O. Holloszy. Effects of increased plasma fatty acids on glycogen utilization and endurance. J. Appl. Physiol. 43:829–833, 1977.

HULTMAN, E., Muscle glycogen store and prolonged exercise. In E.J. Shephard (Ed.), Frontiers of Fitness, p. 30. Springfield, Ill.: C.G. Thomas Publisher, 1971.

HULTMAN, E., Dietary manipulations as an aid to preparation for competition. XXth World Congress in Sports Medicine. Melbourne, 1974. Congress Proceedings, 1976, pp. 239–265.

HULTMAN, E. & L. H:son Nilsson. Liver glycogen in man, effect of different diets and muscular exercise. In B. Pernow & B. Saltin (Eds), Adv. Exp. Med. Biol. II, pp. 143–151. New York-London: Plenum Press, 1971.

IKUTA, K. & M. Ikai. Study on the development of maximal anaerobic power in man with bicycle ergometer. Res. J. Physiol. 17:151–157, 1972.

IVY, J.L., D.L. Costill, W.J. Fink & R.W. Lower. Influence of caffeine and carbohydrate feedings on endurance performance. Med. Sci. Sports 11:6–11, 1979.

MARGARIA, R., P. Aghemo & E. Rovelli. Measurement of muscular power (anaerobic) in man. J. Appl. Physiol. 21:1662–1664, 1966.

MARGARIA, R., P. Cerretelli & F. Mangeli. Balance and kinetics of anaerobic energy release during strenuous exercise in man. J. Appl. Physiol. 19:623–628, 1964.

McGILVERY, R.W., The use of fuels for muscular work. In H. Howald & J.R. Poortmans (Eds.), Metabolic Adaptation to Prolonged Physical Exercise. Basel: Birkhäuser Verlag, 1975.

NAZAR, K., Z. Brzezinska & W. Kowalski. Mechanism of impaired capacity for prolonged muscular work following beta-adrenergic blockade in dogs. Pflugers Arch. 336:72, 1972.

NIKKILÄ, E.A., M.R. Taskinen, S. Rehnnen & M. Härkönen. Lipoprotein lipase activity in adipose tissue and skeletal muscle of runners: relation to serum lipoproteins. Metabolism 27:1661–1671, 1978.

NILSSON, L. H:son & E. Hultman. Liver glycogen in man—the effect of total starvation or a carbohydrate-poor diet followed by carbohydrate refeeding. Scand. J. Clin. Lab. Invest. 32:325–330, 1973.

PERNOW, B. & B. Saltin. Availability of substrates and capacity for prolonged heavy exercise in man. J. Appl. Physiol. 31:416–422, 1971.

RENNIE, M.J., W.W. Winder & J.O. Holloszy. A sparing effect of increased plasma fatty acids on muscle and liver glycogen content in the exercising rat. Biochem. J. 156:647–655, 1976.

SALTIN, B. & J. Karlsson. Muscle glycogen utilization during work of different intensities. In B. Pernow & B. Saltin (Eds.), Adv. Exp. Med. Biol. 11:289–300, 1971.

Exercise Metabolism and Endocrine Function

Atko Viru
Tartu State University, Tartu,
Estonian S.S.R., U.S.S.R.

Hormonal Ensemble during Muscular Exercises

Endocrine systems take part in specific homeostatic reactions as well as in the mechanism of general adaptation. Consequently, various alterations in the body's hormonal ensemble are required for effective muscular activity.

The responses of blood level of hormones on exercise can be divided into fast responses, responses of modest rate and responses with a lag period (Figure 1). Examples of the fast responses are a rapid increase in the concentrations of catecholamines (Peqingnot et al., 1979) and cortisol (Lehnert et al., 1968) in blood plasma within the first few min of an exercise. The corticotropin response (Few et al., 1975; Viru et al., 1981a) must also be included in this group. The responses of modest rate are the elevation of aldosterone (Sundsford et al., 1975), thyroxine (Kirkeby et al., 1977; Terjung & Tipton, 1971) and probably also of vasopressin (Melin et al., 1980) levels. The responses of their humoral regulators are usually more rapid. It is demonstrated in the relationship of renin vs aldosterone (Aldercreuz et al., 1976; Sundsford et al., 1975) The dynamics of blood thyrotropin level (Galbo et al., 1977; Viru et al., 1981a) allow the suggestion of the same in the relationship of thyrotropin vs thyroxine. A lag period is characterized by the elevation of somatotropin levels (Buckler, 1972; Hartog et al., 1967; Sutton & Lazarus, 1976), glucagon (Felig et al., 1972; Luyckx et al., 1978) and calcitonin (Drževetskaya & Limanski, 1978) and by the reduction of insulin level (Hunter & Sukkar, 1968; Kuyckx et al., 1978).

Obviously, there exist kinds of mechanisms for activating endocrine glands during exercise, mechanisms for rapid and for delayed activation. It is reasonable to suggest an important role of neural factors in the first. The second is dependent on some effects of exercise which are cumulative. In

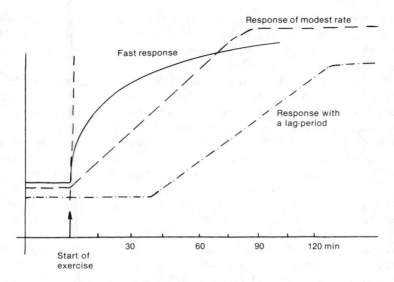

Figure 1—Blood hormone levels in response to exercise.

fast responses, the mechanism of rapid activation prevails, while in the responses of modest rate the mechanism of delayed activation seems to play a decisive role. The response with a lag-period is, probably, due to the lack of activity of mechanisms of rapid activation.

Most hormonal responses are dependent on the intensity and duration of exercise. The mechanism of rapid activation is more sensitive to the intensity and the mechanism of delayed activation to the duration of exercise. That is the reason why the threshold load eliciting the response of the blood level of hormones is more uniformly found for fast than for other responses. The threshold load for blood catecholamines (Galbo et al., 1975; Harggendahl et al., 1970; Lehmann et al., 1981) and cortisol (Davies & Few, 1973) responses are between 50 and 75% of maximal oxygen uptake. A very long duration of exercise, however, can cause an elevation of the blood glucocorticoid level even at the intensity below the threshold load (Keibel, 1974).

The mechanism of rapid activation seems to be sensitive to the inhibitory action of high hormone level prior to exercise (Few et al., 1980; Viru et al., 1981).

Mobilization of the Body's Energy Reserve

During the supramaximal exercise the high power output correlates with the high blood level of noradrenaline (Häaggendahl et al., 1970). It suggests that the maximal anaerobic capacity may be related to the possible magnitude of blood catecholamine response to exercise. The action of catecholamines on muscle glucogenolysis via the system of cycle AMP-phosphorylase can be considered as an explanation of the effect of catecholamines on the anaerobic processes.

In experiments with the isolated, perfused rat hindquarter, it has been found that contractions per se stimulate glycogenolysis only for a brief period and that a direct effect of adrenaline is needed for continued glycogenolysis during exercise. The effect of adrenaline on muscular net glycogen breakdown involved the production of cyclic AMP, activation of glycogen phosphorylase, and inactivation of glycogen synthetase (Galbo, 1981). Additional evidence for the role of catecholamines in the glyco-genolysis of exercising muscles was obtained in experiments by excluding the sympatho-adrenal activity. In dogs, adrenergic blockade diminished the lactate' production and the magnitude of muscle glycogen usage (Issekutz, 1978). During swimming, in the muscular tissue of adrenal demedullated rats the phosphorylase (a + b) activity decreased and the depletion of muscle glycogen was less pronounced than in intacts. Disagreeing results were obtained by Gollnick and Ianuzzo (1975).

On exercising rats, the adrenaline promotes hepatic glycogen depletion as well as gluconeogenesis (Galbo, 1981). Adrenaldemedulation dimin-ished the hepatic glycogenolysis (Richter et al., 1980) and caused hypo-glycemia (Malig et al., 1980). Gollnick and Ianuzzo (1975), however, did not establish any changes in liver glycogen depletion during exercise when the actions of sympatho-adrenal systems and pituitary hormones were excluded. A potent stimulus for hepatic glycogenolysis arises from the increased blood level of glucagon. From experiments, where the glucagon deficiency was induced, it was concluded that the basal level of glucagon normally accounts for a certain hepatic glucose production at rest and during exercise. The basal glucagon level, however, need not be essential for the exercise-induced rise in hepatic glucose output. Moreover, the effect of exercise-induced hyperglucagonemia on glucose production is only transient (Galbo, 1981). Wahren and Björkman (1981) assume that the hyperglucagonemia as well as hypoinsulinemia have a permissive rather than a regulatory influence on exercise-induced changes in hepatic glucose output.

A lot of evidence has been obtained for the role of hormones in lipolysis during exercise. There are pieces of evidence for the stimulatory action of sympatho-adrenal hormones (Gollnick & Ianuzzo, 1975; Issekutz, 1978; Malig et al., 1966), somatotropin (Hunter et al., 1965) and thyroxine (Kaciuba-Uscilko et al., 1979), and for the inhibitory action of insulin (Issekutz et al., 1967). As was suggested by Hunter and Sukker (1968), hypoinsulinemia, during prolonged exercise, diminished the usage of glucose and promoted the usage of free fatty acids for oxidative substrates in muscles. As a result the blood glucose will be reserved for fuel of nerve cells.

Hormonal Mechanisms of Plastic Attainment of Muscular Activity

The plastic attainment of function consists in the renewal of related structural proteins, the additional synthesis of enzyme proteins and the supply of these syntheses by "building materials." Among the inducers of these protein syntheses, many hormones have their role. The mobilization of the amino acid pool is also connected with the induction of syntheses of enzyme proteins. The main role here belongs to glucocorticoids. The action on the systhesis of gluconeogenic enzymes extends the role of the enzymes to energy attainment. The usage of amino acids is affected by the influence of adrenaline, somatotropin and insulin on their transport.

During exercise, the amino acids are used first of all in the liver. In rats after 3 h of swimming the content of free amino acids decreased in the liver together with the elevation of blood corticosterone concentration and augmentation of alanine-aminotransferase activity in hepatic tissue (Figure 2). The adrenalectomy excluded, and the administration of corticosterone restored these responses (Viru & Eller, 1976). The adrenalectomy suppressed and the previous administration of cortisol elevated the tryptophan oxidase activity in the liver after prolonged swimming. If the RNA synthesis were blocked by the treatment with actinomycin D, the cortisol would not elevate the enzyme activity. It was also found in muscle tissue that the removal of the action of glucocorticoids excludes the changes in transaminase activity during exercise (Critz & Withrow, 1965).

After exercise, the intensity of glycogen synthesis was quite low in the skeletal and heart muscles of adrenalectomized rats. The administration of dexamethasone restored the rate of glycogen synthesis. As the blockade of protein synthesis by cyclohexamide excluded the effect of dexamethasone; this action of glucocorticoid was also mediated by the synthesis of regulatory protein(s) (Kôrge et al., 1982).

Protein synthesis can be considered as a main tool in the action of glucocorticoids on the working capacity. Our experiments showed that the restoration of working capacity measured by the duration of swimming with an additional load of 3% of body weight did not take place if during the previous 5 days, adrenalectomized rats were treated with cortisol and actinomycin D or with cortisol and a high dose of progesterone, competing glucocorticoids for specific glucocorticoid receptors in cellular cytoplasm. There are, of course, a number of various proteins (enzymes of amino acid and of gluconeogenesis, etc.) whose enhanced synthesis due to the action of glucocorticoids gives a cumulative effect on physical working capacity.

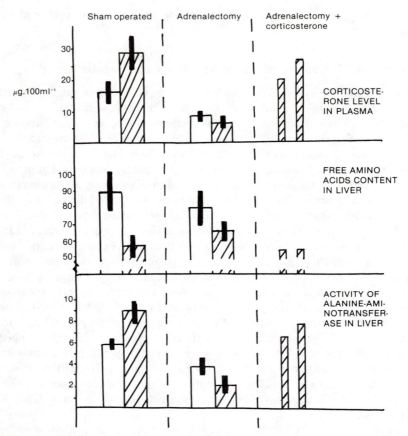

Figure 2—Plasma corticosterone, liver amino acids, and liver alanine-aminotransferase after 3 h swimming exercise.

Among them are Na, K-ATPase of the sarcolemma of myocardial cells. The activity of this decreases in the states of adrenal insufficiency. The physical working is in good correlation with the activity of the enzyme (Kôrge, 1976).

In persons of varying fitness, we blocked the glucocorticoid response to exercise for the assessment of maximal oxygen uptake by treating with dexamethasone during 3 previous days or enhanced the response by administering corticotropin. The levels of maximal oxygen uptake, PWC_{170}, or maximal rate of pedaling, did not change (Viru & Smirnova, 1982). Obviously, the duration of exercise (7 to 9 min) was too short to the role of glucocorticoid response. Naturally, if the role of glucocorticoids in the induction of protein syntheses, the results of glucocorticoid response exercise will appear only after a lag period.

Hormonal Ensemble in Fatigue

During long-lasting exercises, the blood levels of glucocorticoids (Viru, 1977) and catecholamines (Matlina, 1976b) may decrease to values below that of initial ones. Usually this tendency is preceded by the increased level of both glucocorticoids and catecholamines. This kind of dynamics is common for a long-lasting action of various stressors. Consequently the stress-response of the pituitary-adrenocortical system is divided into (1) the phase of preliminary activation, (2) the phase of subnormal activity, and (3) the phase of secondary activation (Viru, 1979). In the sympathoadrenal response to stressors the following phases are distinguished: (1) the phase of fast activation, (2) the phase of steady continuous activation, and (3) the depletion phase (Matlina, 1976).

During prolonged exercise a tendency to decrease towards the initial level is observed also in the concentration of somatotropin after its elevation (Hartog et al., 1967; Hunter et al., 1965).

The subnormal phase of the pituitary-adrenocortical system associates with metabolic disturbances. After a very long period of swimming (12 to 18 h duration) in untrained rats the decrease of blood corticosterone level associates with the reduction of Na, K-ATPase activity in the microsomal fraction of skeletal and heart muscle tissue and with the intracellular accumulation of sodium and water in these tissues (Kôrge & Viru, 1971; Kôrge et al., 1974; Kôrge, 1976). In the subnormal phase of adrenocortical activity the elevated activity of hepatic tryptophan oxidase was decreased (Figure 3.). The level of liver glucogen decreased promptly at the beginning of exercise which was repeated in the subnormal phase of adrenocortical activity (Viru et al., 1981). In humans the subnormal phase was accompanied by decreased blood pressure as a response to exercise (Viru, 1977).

The appearance of the phase of subnormal activity depends on the state of adrenals and on fitness. In rats the hypotrophy of z. fasciculata of the adrenal cortex due to treatment with dexamethasone for 3 wk promoted the decrease of blood corticosterone level during prolonged exercise in association with reduced working capacity (Smirnova & Viru, 1977). On the other hand, training elevates the functional stability of the pituitary-adrenocortical system and it makes it difficult to obtain the phase of subnormal activity (Kôrge, 1976; Seene et al., 1978; Viru, 1977) (Figure 2). This can be considered as a reason why after marathon races or other long lasting sport events increased cortisol level is detected in blood instead of a decreased level (Dessypris et al., 1976; Keul et al., 1981; Maron et al., 1975; Sundsford et al., 1975). The effect of training on the level of adrenal cortex (Frenkl et al., 1975; Tharp & Buuck, 1974), by including an elevated content of cytochrome aa_3, increased the number of mitochondria, vesicular christae in them, elements of the cytoplasmic reticulum and

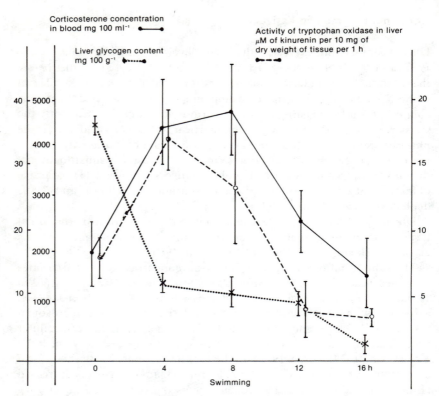

Figure 3—Plasma corticosterone, liver glycogen, and liver tryptophan oxidase activity during prolonged swimming in water 32–33° C by untrained rats.

polysomes in the cells of the zona fasciculata (Seene et al., 1978), can be considered a foundation for the elevated functional stability of the pituitary-adrenocortical system.

The decreased activity of the sympatho-adrenal system may also cause a reduction of physical working capacity as it was shown in experiments with adrenergic blockade or a lesion of the adrenal medulla (Malig et al., 1966). The training increases the functional stability of the sympatho-adrenal system and improves the possibilities of catecholamine synthesis (Matlina et al., 1976).

During prolonged exercise, the subnormal adrenocortical activity can be withdrawn or avoided by the administration of corticotropin (Viru & Äkke, 1969; Viru, 1977). Thus, the subnormal adrenocortical activity is not due to the exhaustion of adrenocorticocytes. After a hippocampal lesion the phase of subnormal activity was not observed when it occurred in sham operated rats during prolonged swimming (Viru, 1975). Consequently, the subnormal adrenocortical activity is the result of a regulatory mechanism. It excludes the possibility of the exhaustion of the adrenocorticocytes and hypothalamic neurosecretory cells. It limits also the duration of the action of the glucocorticoids on the body tissues.

The depletion phase of the sympatho-adrenal system is directly connected with the loss of possibilities of adrenomedullary cells for adrenaline synthesis (Matlina, 1976a, b). This disturbance occurred in connection with the subnormal phase of adrenocortical activity. It was excluded by the administration of glucocorticoids (Matlina et al., 1978).

Consequently, during prolonged exercise in the state of fatigue the defense reaction is introduced by increasing the activity of hippocampal centers (Viru, 1975). Due to the decreased activity of the pituitary-adrenocortical system the fatal depletion of the body's plastic resources is avoided and the synthesis of catecholamines is blocked. The latter excludes the activity of the mechanism of mobilization of the energy reserve.

References

ADLERCREUTZ, H., M. Harkonen, K. Kuopposalmi, K. Kosunen, H. Naveri & S. Rehunen. Physical activity and hormones. Adv. Cardiol. 18:144–157, 1976.

BUCLER, J.M., Exercise as a screening test for growth hormone release. Acta Endocrin. 69:219–229, 1972.

CRITZ, J.B. & T.J. Withrow. Adrenocortical blockade and the transaminase response to exercise. Steroids 5:719–728, 1965.

DAVIES, C.T.M. & J.D. Few. Effect of exercise on adrenocortical function. J. Appl. Physiol. 35:888–891, 1973.

DESSYPRIS, A., K. Kuoppasalmi & H. Adlercreutz. Plasma cortisol, testosterone, androstenedione and luteinizing hormone (LH) in a non-competitive marathon run. J. Steroid Biochem. 7:33–37, 1976.

DRŽEVETSKAYA, I.A. & N.N. Limanski. Thyrocalcitonin activity and calcium level in plasma during muscular activity. Sechenov Physiol Journal of USSR 64:1498–1500, 1978.

FELIG, P., J. Wahren, R. Hendler & G. Ahlborg. Plasma glucogen levels in exercising man. New England J. Med. 287:184–185, 1972.

FEW, J.D., F.J. Imnas & J.S. Weiner. Pituitary-adrenal response to static exercise in man. Clin. Sci. 49:201–206, 1975.

FEW, J.D., G.C. Cashmore & G. Turton. Adrenocortical response to one-leg and two-leg exercise on a bicycle ergometer. Eur. J. Appl. Physiol. 44:167–174, 1980.

FRENKL, R., L. Csalay & G. Csákváry. Further experimental results concerning the relationship of muscular exercise and adrenal function. Endokrinologie 66:285–291, 1980.

GALBO, H., Endocrinology and metabolism in exercise. Int. J. Sports Med. 2:203–211, 1981.

GALBO, H., J.J. Holst & N.J. Christensen. Glucagon and plasma catecholamine responses to graded and prolonged exercise in man. J. Appl. Physiol. 38:70–76, 1975.

GALBO, H., L. Hummer, I.B. Petersen, N.J. Christensen & W. Bie. Thyroid and testicular hormonal responses to graded and prolonged exercise in man. Eur. J. Appl. Physiol. 36:101–106, 1977.

GOLLNICK, P.D. & C.D. Ianuzzo. Acute and chronic adaptation to exercise in hormone deficient rats. Med. Sci. Sports 7:12–19, 1975.

HÄGGENDAL, J., L.H. Hartley & B. Saltin. Arterial noradrenaline concentration during exercise in relation to the relative work levels. Scand. J. Clin. Lab. Invest. 26:337–342, 1970.

HARTOG, M., R.J. Havel, G. Copinshi, J.M. Earli & B.C. Ritchie. The relationship between changes in serum levels of growth hormone and mobilization of fat during exercise in man. Quart. J. Exp. Physiol. 52:86–96, 1967.

HUNTER, W.M. & M.Y. Sukkar. Changes in plasma insulin levels during muscular exercise. J. Physiol. 196:110P–112P, 1968.

ISSEKUTZ, B., Role of beta-adrenergic receptors in mobilization of energy sources in exercising dog. J. Appl. Physiol. 44:869–876, 1978.

KACIUBA-USCILKO, H., Z. Brezezinska & A. Kobryn. Metabolic and temperature responses to physical exercise in thyroidectomized dogs. Europ. J. Appl. Physiol. 40:219–226, 1979.

KEIBEL, D., Nebennierenrinden-Hormone and sportliche Leistung. Med. u. Sport. 14:65–76, 1974.

KEUL, J., B. Kohler, G. von Glutz, U. Lüthi, A. Berg & H. Howald. Biochemical changes in a 100 km run: carbohydrates, lipids, and hormones in serum. Eur. J. Appl. Physiol. 47:181–189, 1981.

KIRKEBY, K., S.B. Strömme, I. Bjerkedal, L. Hertzenberg & H.E. Refsum. Effects of prolonged strenuous exercise on lipids and thyroxine in serum. Acta Med. Scand. 202:463–467, 1977.

KÔRGE, P. & A. Viru. Water and electrolyte metabolism in skeletal muscle of exercising rats. J. Appl. Physiol. 31:1–4, 1971.

KÔRGE, P., A. Eller, S. Timpmann & E. Seppet. Role of glucocorticoids in regulation of glycogen resynthesis after exercise and mechanism of their action. Sechenov Physiol. J. USSR. (in press) 1982.

KÔRGE, P.K., Myocardial sodium-potassium pump and its adrenocortical regulation as factors limiting cardiac adaptation to severe physical exertion. Kardiologia (Moscow) 9:15–21, 1976.

KÔRGE, P., S. Roosson & M. Oks. Heart adaptation to physical exertion in relation to work duration. Acta Cardiol. 29:303–320, 1974.

LEHMANN, M., J. Keul, G. Huber & M. Da Prada. Plasma catecholamines in trained and untrained volunteers during graduated exercise. Int. J. Sports Med. 2:143–147, 1981.

LEHNERT, G., H. Leiber & K.H. Scheller. Plasmacortisol und Plasmacorticosterone im Anpassungstadium der dosierten köperlichen Arbeit. Endokrinologie 52:402–405, 1968.

LUYCKX, A.S., F. Pirnay & P.J. Lefebore. Effect of glucose on plasma glucagon and free fatty acids during prolonged exercise. Eur. J. Appl. Physiol. 39:53–61, 1978.

MALIG, H., D.Stern, P. Altland, B. Highman & B. Brodie. The physiologic role of the sympathetic nervous system in exercise. J. Pharmacol. Exp. Ther. 154:35–45, 1966.

MARON, M.B., S.M. Horvath & J.E. Wilkerson. Acute blood biochemical alteration in response to marathon running. Eur. J. Appl. Physiol. 34:173–181, 1975.

MATLINA, E., Main phases of catecholamine metabolism under stress. In E. Usdin, R. Kvetnansky & I.J. Kopin (Eds.), Catecholamines and Stress. Oxford: Pergamon Press, 1976a.

MATLINA, E., The influence of muscular loadings on the catecholamines in men and animals. Acta et Commentationes Universitatis Tartuensis 381:3–49, 1976b.

MATLINA, E.S., G.S. Pukhova, S.D. Galimov, S.N. Almae & A.I. Galentchik. The catecholamines metabolism during adaptation to muscular activity. Sechenov Physiol. J. USSR 62:431–437, 1976.

MATLINA, E., G. Schreiberg, M. Voinova & L. Dunaeva. The interrelationships between catecholamines and corticosteroids in the course of muscular fatigue. Sechenov Physiol. J. USSR 64:171–176, 1978.

PEQUIGNOT, J.M., L. Peyrin, R. Favier & R. Flandrois. Adrenergic response to intense muscular work in sedentary man in relation to emotivity and physical training. Eur. J. Appl. Physiol. 40:117–135, 1979.

RICHTER, E.A., H. Galbo, B. Sonne, J.J. Holst & N.J. Christensen. Adrenal medullary control of muscular and hepatic glycogenolysis and of pancreatic hormonal secretion in exercising rats. Acta Physiol. Scand. 108:235–242, 1980.

SEENE, T., R. Masso, M. Oks, A. Viru & E. Seppet. Changes in the adrenal cortex during adaptation to different regimes of physical activity. Sechenov Physiol. J. USSR 64:1444–1450, 1978.

SMIRNOVA, T. & A. Viru. Dependence of physical working capacity on the state of adrenal cortex. Acta et Commentationes Universitatis Tartuensis 419:130-133, 1977.

SUNDSJORD, J.A., S.B. Stromme & A. Aakwaag. Plasma aldosterone (PA), plasma renin activity (PRA) and cortisol (PF) during exercise. In H. Howald & J.P. Poortmans (Eds.), Metabolic Adaptation to Prolonged Physical Exercise. Basel: Birkhäuser Verlag, 1975.

SUTTON, J.R. & L. Lazarus. Growth hormone in exercise: comparison of physiologic and pharmacologic stimuli. J. Appl. Physiol. 41:523-527, 1976.

TERJUNG, R.L. & C.M. Tipton. Plasma thyroxine and thyroid stimulating hormone levels during submaximal exercise in humans. Am. J. Physiol. 220:1840-1845, 1971.

THARP, G. & R. Buuck. Adrenal adaptation to chronic exercise. J. Appl. Physiol. 37:720-722, 1974.

VIRU, A., Defense reaction theory of fatigue. Schweiz. Z. Sportmed. 23:171-178, 1975.

VIRU, A., Functions of the Adrenal Cortex in Muscular Activity. Moscow: Medizina, 1977 (in Russian).

VIRU, A., Dynamics of response of hypothalamo-pituitary-adrenocortical system in stress. Uspekhi Sovr. Biol. (Moscow) 87:271-286, 1979.

VIRU, A. & H. Äkke. Effect of muscular work on cortisol and corticosterone content in the blood and adrenals of guinea pigs. Acta Endocrin. 62:385-390, 1969.

VIRU, A. & A. Eller. Adrenocortical regulation of protein metabolism during prolonged physical exertions. Bull. Eksp. Biol. Med. (Moscow) 82:1439, 1976.

VIRU, A. & T. Smirnova. Independence of physical working capacity from increased glucocorticoid level during short-term exercise. Int. J. Sports Med. (In Press) 1982.

VIRU, A., T. Smirnova, K. Tomson & T. Matsin. Dynamics of blood levels of pituitary trophic hormones during prolonged exercise. In J. Poortmans & G. Niset (Eds.), Biochemistry of Exercise IVB. Baltimore: Univ. Park Press, 1981a.

VIRU, A., T. Smirnova & S. Roosson. Dynamics of adrenocortical activity and glycogen in muscles and liver during prolonged muscular work. Acta et Commentationes Universitatis Tartuensis 562:47-53, 1981b.

WADE, C.E. & J.R. Claybaugh. Plasma renin activity, vasopressin concentration and urinary excretory responses to exercise in men. J. Appl. Physiol. 49:930-936, 1980.

WAHREN, J. & Björkman. Hormones, exercise and regulation of splanchnic glucose output in normal man. In J. Poortmans & G. Niset (Eds.), Biochemistry of Exercise IVB. Baltimore: Univ. Park Press, 1981.

Exercise Training in Coronary Heart Disease

John O. Holloszy, A.A. Ehsani and J.M. Hagberg
Washington University School of Medicine,
St. Louis, Missouri, U.S.A.

Exercise is widely used in the treatment of patients with coronary artery disease (CAD). The use of exercise in the rehabilitation of coronary patients is relatively recent in the U.S.A.; it first gained acceptance in the 1960's, largely due to the pioneering work of Hellerstein and coworkers (Hellerstein et al., 1965; Hellerstein, 1968). At that time it was still customary to keep patients on bed rest for weeks following a myocardial infarction; this resulted in severe deterioration of physical condition. The major purpose of the early rehabilitation programs was physical reconditioning with the goal of enabling patients to resume their normal activities and return to work (Hellerstein, 1979).

It soon became evident that some coronary patients responded to exercise-training with an improvement in cardiovascular function (Hellerstein et al., 1963; Holloszy et al., 1964a; Naughton et al., 1969; Varnauskas et al., 1966). This observation, together with accumulating evidence that men in physically active jobs have only 30 to 50% as high an incidence of clinical CAD as sedentary men (cf. Fox et al., 1971; Holloszy, 1963) led to a change in perspective, and to investigation of the possibility that exercise might have a beneficial effect on the heart and its blood supply in CAD. More recent epidemiological studies have provided evidence that leisure time physical activity also has a protective effect against CAD (Froelicher & Brown, 1981; Morris et al., 1980; Paffenberger et al., 1978). A further impetus to the investigation of the effect of training on the heart in patients with CAD has come from studies of the effects of exercise on the myocardial vasculature in animals.

Studies on Experimental Animals

Studies on rats (Leon & Bloor, 1968; Stevenson et al., 1964; Tepperman & Pearlman, 1961) and dogs (Wyatt & Mitchell, 1978) have shown that exercise can induce an enlargement of the coronary arteries. Increases in capillary density have also been reported (Leon & Bloor, 1968; McElroy et al., 1978). Studies on rats (Scheuer et al., 1974; Spear et al., 1978) and dogs (Laughlin et al., 1978) suggest that these anatomical adaptations result in increased maximum coronary blood flow per unit mass of heart. Perhaps of more relevance to CAD are studies on animals with experimental coronary occlusion. Eckstein (1957) measured retrograde coronary flow in dogs with chronic single vessel occlusions and found a higher index of coronary collateral function in trained than in sedentary dogs. Heaton et al. (1978) measured regional myocardial blood flow using radioactive microspheres in dogs with chronic, multivessel occlusions; blood flow to underperfused regions of myocardium during exercise was 39% greater after a period of training than prior to training.

Training also brings about changes in cardiac function in experimental animals. These adaptations are of two types. One, probably mediated by adaptations in the autonomic nervous system and skeletal muscles, results in increased efficiency with a reduction in energy requirement of the heart during submaximal exercise; the other, which involves the myocardium itself, results in enhanced maximum contractile function (Dowell et al., 1977; Scheuer et al., 1974; Stone, 1977). For a detailed review see Scheuer and Tipton (1977) and Stone (1980).

Studies on experimental animals have also shown that exercise-training can protect against the development of atherosclerosis. This was demonstrated most convincingly in a recent study by Kramsch et al. (1981) on monkeys fed an atherogenic diet. The sedentary monkeys developed marked coronary atherosclerosis and stenosis. The exercised animals had substantially reduced atherosclerotic involvement; they also had wider coronary arteries, further minimizing luminal narrowing. The sedentary and training monkeys had similar marked elevations of total serum cholesterol. Total triglyceride and LDL and VLDL triglyceride levels, however, were much lower, while HDL cholesterol was higher in the exercise group. Exercise induces similar changes in triglycerides (Gyntelberg et al., 1977; Holloszy et al., 1964b) and HDL cholesterol (Wood & Haskell, 1979) in man. These favorable changes in lipid and lipoprotein levels could play a role in protecting against atherosclerosis.

Studies on Patients

Two types of study of the effects of training on CAD have been conducted. One approach has been to evaluate the effectiveness of relatively long-term exercise in the secondary prevention of CAD. The other has been to study the adaptations of patients with CAD to relatively short periods of training.

Studies on Patients with CAD

Two types of study of the effects of training on CAD have been conducted. One approach has been to evaluate the effectiveness of relatively long-term exercise in the secondary prevention of CAD. The other has been to study the adaptations of patients with CAD to relatively short periods of training.

Exercise in the Secondary Prevention of CAD

The results of four randomized trials of the effectiveness of exercise in reducing mortality from CAD in patients with one myocardial infarction have been published. In a fifth study, exercise was one of the interventions in a multifactorial intervention program (Kallio et al., 1979). The final results of a sixth study, the Ontario Exercise Heart Collaborative Study, had not been published at the time this review was written. The mortality data from the four exercise studies are summarized in Table 1. Although all four studies showed a positive trend favoring the exercise group, none showed a statistically significant difference in mortality. This is hardly surprising, as the prospective randomized trial is an inefficient study design for evaluating the effects of training.

Randomly assigning a patient to an exercise program does not assure that he will exercise. There is also no assurance the controls will refrain from exercise. Thus the question addressed by the statistical analysis of a prospective randomized trial is: "Does assignment of a patient to an exercise group protect against progression of CAD?" rather than "Does exercise protect against CAD?" The general experience has been that the drop-out rate (for nonmedical reasons) from exercise programs for patients with CAD is between 30 to 60% in the first 6 to 12 mo. Even if exercise has a powerful protective effect against the progression of CAD, as appears likely (cf. Kramsch et al., 1981; Morris et al., 1980; Paffenberger et al., 1978), it is apparent that a considerably larger group of patients than was included in these studies (Table 1) would be needed to overcome the confounding effects of the large number of nonexercising and inadequately exercising patients in the "exercise" group and of exercising (on their own initiative) patients in the "control" group. Since it seems unlikely, both for ethical and financial reasons, that a large randomized trial of the effect of exercise will be conducted, it might be worthwhile to examine the validity of combining data from the studies that have been completed for statistical

Table 1

Randomized Trials of Exercise in the Secondary Prevention of CAD

Trial	Months duration	Assigned to control group			Assigned to exercise group		
		Number	Deaths	% Mortality	Number	Deaths	% Mortality
a	48	157	35	22.3	158	28	17.7
b	12	81	8	9.9	77	5	6.5
c	29	200	28	14.0	180	18	10.0
d	36	328	24	7.3	323	15	4.6

[a]Sanne et al. (1973) and Wilhelmsen et al. (1975)
[b]Kentala (1972)
[c]Platsi (1976
[d]Shaw (1981), The National Exercise and Heart Disease Project

analysis. (When the mortality data from the four studies in Table 1 are combined, without consideration of differences in study design, duration, etc., the average total percent mortality for the exercising groups is significantly lower than that for the control groups; 8.9% vs 12.4%, $P < 0.05$.)

Cardiovascular Adaptations to Exercise-Training of Patients with CAD

The literature on the physiological and anatomical adaptations of the cardiovascular system to exercise in patients with CAD has been exhaustively reviewed (Clausen, 1976; Mitchell, 1975), and will be referred to only selectively here to provide the background for more recent studies.

Exercise-training increases the minimum exercise intensity required to induce ST-segment depression or angina in some patients (Detry & Bruce, 1971; Redwood et al., 1972; Sim & Neill, 1974). A decrease in the magnitude of ST displacement at a given submaximal level of exercise has also been reported (Detry & Bruce, 1971). Previous studies, however, in which objective end points were used did not show an improvement in myocardial blood supply or a decrease in myocardial ischemia in response to training. In three studies in which coronary arteriograms were performed before and after training no changes in coronary collateral circulation were detected (Ferguson et al., 1974; Nolewajka et al., 1979; Sim & Neill, 1974). Ferguson et al. (1978) measured coronary sinus blood flow and left ventricular O_2 consumption at various exercise intensities in

CAD patients before and after 6 mo of training; coronary sinus blood flow and left ventricular (LV) O_2 uptake at the same double product (i.e., heart rate $\times$ systolic BP) were unchanged, and maximum coronary sinus blood flow and LV O_2 consumption were unaffected by the training.

Nolewajka et al. (1979) evaluated the effects of an exercise program on intercoronary collaterals in patients with CAD using angiography and infusion of labeled microspheres. No changes in the extent of collateralization or myocardial perfusion were evident after training. Detry and Bruce (1971) found that the double product and the magnitude of ST-segment depression at the same submaximal work rate were less after training, but the relationship between the magnitude of the ST-segment depression and the double product was unchanged. Maximum ST depression was actually greater after training, probably because a higher double product was attained. The double product is a useful clinical indicator of myocardial O_2 requirement. A number of studies have shown that coronary blood flow and myocardial O_2 consumption closely parallel the double product (Ferguson et al., 1978; Gobel et al., 1978).

These negative results provide evidence that training did not improve myocardial blood supply or ischemia at a given rate of cardiac energy utilization and O_2 need. Instead, it appears that the beneficial effects of training on exercise capacity and exertional angina threshold in these studies were mediated by the adaptations in the autonomic nervous system and skeletal muscles which result in a slower heart rate and lower blood pressure and, therefore, in a reduced myocardial O_2 requirement during submaximal exercise (cf. Clausen, 1976).

It seemed possible that the apparent failure of myocardial oxygenation to improve in response to training may have been due to an insufficient exercise stimulus. In the studies on animals, reviewed above, in which training appeared to improve myocardial blood supply, moderately strenuous exercise lasting 1 to 3 h/day, 5 days/wk was used. In contrast the exercise used in the studies in which training did not appear to improve myocardial ischemia in CAD patients was generally mild and brief. In this context, it seemed possible that the lack of improvement may have been due to termination of training before a sufficient adaptive stimulus was attained.

To test this possibility, we are evaluating the effectiveness of a 12 mo-long program of exercise of progressively increasing intensity, duration and frequency. During the last 3 mo of this program the patients are generally running 4 to 5 miles/day, 4 to 5 days/wk, or doing an equivalent amount of exercise on a bicycle ergometer (Ehsani et al., 1981). One indicator of myocardial ischemia that is being used to evaluate the effectiveness of this training program is the relationship between the double product and ischemic ST-segment depression during exercise. In the first 10 patients with ST depression to complete 12 mo of training, VO_2 max increased from 26 to 35 ml/kg/min. The double product at which ST depression (0.1 mV) first appeared was 22% greater after training (Ehsani et al., 1981). The extent of ST-segment displacement at the same double product was significantly less after training. Despite a 20% increase in maximum double product during exercise, the maximum degree of ST depression after training was less for the group. An additional five patients have now shown similar improvements (unpublished).

The development of noninvasive methods for evaluating the effects of training in CAD. Exercise acutely induces an increase in myocardial contractility with an increase in ejection fraction in normal individuals. In contrast the majority of patients with CAD have an impaired capacity to increase contractile function and show no increase or a decrease in ejection fraction in response to dynamic or isometric exercise (Mitchell & Blomqvist, 1979). The abnormal response of ejection fraction and other indicators of contractile function to exercise reflects the development of regional abnormalities of myocardial contraction, i.e., wall motion abnormalities not present at rest. These are due to an inadequate increase in blood flow to meet the exercise-induced increase in O_2 requirement of involved regions of myocardium, resulting in development of myocardial ischemia.

Jensen et al. (1980) have used radionuclide ventriculography to evaluate the response of left ventricular ejection fraction to bicycle exercise in CAD patients before and after a 6 mo training program of moderate intensity. Ejection fraction did not increase in response to maximal bicycle exercise either before or after training. There was also no improvement in ejection fraction during maximal exercise in response to training. At equivalent submaximal work rates, however, at which similar double products were reached, a significantly greater mean ejection fraction was attained after training (55% pre vs 59% post).

Radionuclide ventriculography is also being used to evaluate the effect of our intense 12 mo-long program of training mentioned earlier. In the first eight patients who have been studied before and after adapting to this more frequent, intense and prolonged training, the maximal exercise intensity, maximal systolic blood pressure and maximal double product attained during a bicycle exercise test were all significantly higher after, as compared to before, training. Despite the higher double product, ejection fraction at maximal exercise was higher after training (48 ' 15%, pre vs 57 ' 16%, post; $P < 0.001$); the ejection fraction increased in response to maximal exercise after, but not before, training (Ehsani et al., 1982a). Similar improvement occurred in the ejection fraction response to isometric exercise.

Another noninvasive procedure that has been used to evaluate the effect of the 12 mo-long training program on the response of myocardial contractile function to acute exercise is echocardiography (Ehsani et al., 1982b). Eight patients with an abnormal response to isometric exercise on whom it has been possible to obtain adequate echocardiograms have now completed 12 mo of training. Their VO_2 max increased from 26^{r}1 to 37^{r}2 ml/kg/min in response to the training. Left ventricular fractional shortening and mean velocity of circumferential fiber shortening (mVcf) decreased progressively in response to graded isometric hand grip exercise before, but not after, training. At comparable levels of blood pressure during static exercise, m Vcf was significantly higher after training; 0.76^{r}0.04 d/s before vs 0.98^{r}0.07 d/s after, P < 0.01 (Ehsani et al., 1982b).

The results of these three studies on the effects of intense, prolonged exercise-training should be considered preliminary, as the number of CAD patients who have completed the training program is still small. The results thus far are encouraging, however, because they provide evidence that exercise-training, if sufficiently frequent, intense and prolonged, can result in a reduction in myocardial ischemia at the same or a higher double product, with an associated improvement in cardiac function.

Acknowledgment

This research was supported by NIH Research Grant HL-22215.

References

CLAUSEN, J.P., Circulatory adjustments to dynamic exercise and effect of physical training in normal subjects and in patients with coronary artery disease. Prog. Cardiovasc. Dis. 18:459–495, 1976.

DETRY, J.-M. & R.A. Bruce. Effects of physical training on exertional ST segment depression in coronary heart disease. Circulation 44:390–396, 1971.

DOWELL, R.T., H.L. Stone, L.A. Sordahl & G.K. Asimakis. Contractile function and myofibrillar ATPase activity in the exercise-trained dog heart. J. Appl. Physiol. 43:977–982, 1977.

ECKSTEIN, R.W., Effect of exercise and coronary artery narrowing on coronary collateral circulation. Circulation Res. 5:230–235, 1957.

EHSANI, A.A., D.R. Biello, S.A. Bloomfield & J.O. Holloszy. Exercise training improves intrinsic left ventricular performance in ischemic heart disease. Clin. Res. 30:480A, 1982a.

EHSANI, A.A., G.W. Heath, J.M. Hagberg, B.E. Sobel & J.O. Holloszy. Effects of twelve months of intense exercise training on ischemic ST-segment depression in patients with coronary artery disease. Circulation 64:1116–1124, 1981.

EHSANI, A.A., W.H. Martin, G.W. Heath & E.F. Coyle. Cardiac effects of prolonged and intense exercise training in patients with coronary artery disease. Am. J. Cardiol. 49 (In Press), 1982b.

FERGUSON, R.J., P. Cote, P. Gauthier & M.G. Bourassa. Changes in exercise cornary sinus blood flow with training in patients with angina pectoris. Circulation 58:41–47, 1978.

FERGUSON, R.J., R. Petitclerc, G. Choquette, L. Chaniotes, P. Gauthier, R.F. Huot, C. Allard, L. Jankowski & L. Campeau. Effect of physical training on treadmill exercise capacity, collateral circulation and progression of coronary disease. Am. J. Cardiol. 34:764–770, 1974.

FOX, S.M., J. Naughton & W.L. Haskell. Physical activity and the prevention of coronary heart disease. Ann. Clin. Res. 3:404–432, 1971.

FROELICHER, V.F. & P. Brown. Exercise and coronary heart disease. J. Cardiac. Rehab. 1:277–288, 1981.

GOBEL, R.L., L.A. Nordstrom, R.R. Nelson, C.R. Jorgensen & Y. Yang. The rate-pressure product as an index of myocardial oxygen consumption during exercise in patients with angina pectoris. Circulation 57:549–556, 1978.

GYNTELBERG, F., R. Brennan, J.O. Holloszy, G. Schonfeld, M.J. Rennie & S.W. Weidman. Plasma triglyceride lowering by exercise despite increased food intake in patients with Type IV hyperlipoproteinemia. Am. J. Clin. Nutrition 30:716–720, 1977.

HEATON, W.H., K.C. Marr, N.L. Capurro, R.E. Goldstein & S.E. Epstein. Beneficial effect of physical training on blood flow to myocardium perfused by chronic collaterals in the exercising dog. Circulation 57:575–581, 1978.

HELLERSTEIN, H.D., Exercise therapy in coronary disease. Bull. NY Acad. Med. 44:1028, 1968.

HELLERSTEIN, H.D., Cardiac rehabilitation: a retrospective view. In M.L. Pollack & D.H. Schmidt (Eds.), Heart Disease and Rehabilitation. Boston: Houghton Mifflin, 1979.

HELLERSTEIN, H.D., E.Z. Hirsch, W.W. Cumler, L. Allen, S. Polster & N. Zucker. Reconditioning of the coronary patient: a preliminary report. In W. Likoff & L.H. Moyer (Eds.), Coronary Heart Disease, pp. 448–454. New York: Grune & Stratton, 1963.

HOLLOSZY, J.O., The epidemiology of coronary heart disease. National differences and the role of physical activity. J. Am. Geriatrics Soc. 11:718–725, 1963.

HOLLOSZY, J.O., J.S. Skinner, A.J. Barry & T.K. Cureton. Effect of physical conditioning on cardiovascular function. A ballistocardiographic study. Am. J. Cardiol. 14:761–770, 1964a.

HOLLOSZY, J.O., J.S. Skinner, G. Toro & T.K. Cureton. Effects of a six month program of endurance exercise on the serum lipids of middle-aged men. Am. J. Cardiol. 14:753–760, 1964b.

JENSEN, D., J.E. Atwood, V. Froelicher, M.D. McKirnan, A. Battler, W. Ashburn & J. Ross, Jr. Improvement in ventricular function during exercise studied with radionuclide ventriculography after cardiac rehabilitation. Am. J. Cardiol. 46:770–777, 1980.

KENTALA, E., Physical fitness and feasibility of physical rehabilitation after myocardial infarction in men of working age. Ann. Clin. Res. 4(Suppl. 9), 1972.

KRAMSCH, D.M., A.J. Aspen, B.M. Abramowitz, T. Kreimendahl & W.B. Hood, Jr. Reduction of coronary atherosclerosis by moderate conditioning exercise in monkeys on an atherogenic diet. New Eng. J. Med. 305:1483–1489, 1981.

LAUGHLIN, H.H., J.N. Diana & C.M. Tipton. Effects of exercise-training on coronary reactive hyperemia and blood flow in the dog. J. Appl. Physiol. 45:604–610, 1978.

LEON, A.S. & C.M. Bloor. Effects of exercise and its cessation on the heart and its blood supply. J. Appl. Physiol. 24:485–490, 1968.

MCELROY, C.L., S. A. Gissen & M.C. Fishbein. Exercise-induced reduction in myocardial infarct size after coronary occlusion in the rat. Circulation 57:958–962, 1978.

MORRIS, J.N., R. Pollard, M.G. Everitt, S.P.W. Chave & A.M. Semmence. Vigorous exercise in leisure time: Protection against coronary heart disease. Lancet 2:1207–1210, 1980.

MITCHELL, J.H., Exercise training in the treatment of coronary heart disease. Adv. Intern. Med. 20:249–272, 1975.

MITCHELL, J.H. & C.G. Blomqvist. Responses of patients with heart disease to dynamic and static exercise. In M.L. Pollock & D.H. Schmidt (Eds.), Heart Disease and Rehabilitation, p. 86–96. Boston: Houghton Mifflin, 1979.

NAUGHTON, J., J. Bruhn, M.T. Lategola & T. Whitsett. Rehabilitation following myocardial infarction. Am. J. Med. 46:725–734, 1969.

NOLEWAJKA, A.J., W.L. Kostuk, P.A. Rechnitzer & D.A. Cunningham. Exercise and human collateralization: an angiographic and scintigraphic assessment. Circulation 60:114–121, 1979.

PAFFENBERGER, R.S., A.L. Wing & R.T. Hyde. Physical activity as an index of heart attack risk in college alumni. Am. J. Epidemiol. 108:161–175, 1978.

PALATSI, I., Feasibility of physical training after myocardial infarction and its effect on return to work, morbidity and mortality. Acta Med. Scand. Suppl. 599, 1976.

REDWOOD, D.R., D.R. Rosing & S.E. Epstein. Circulatory and symptomatic effects of physical training in patients with coronary artery disease and angina pectoris. New Engl. J. Med. 286:959–965, 1972.

SANNE, H., Exercise tolerance and physical training of non-selected patients after myocardial infarction. Acta Med. Scand. Suppl. 551, 1973.

SCHEUER, J., S. Penpargkul & A.K. Bhan. Experimental observations on the effects of physical training upon intrinsic cardiac physiology and biochemistry. Am. J. Cardiol. 33:744–751, 1974.

SCHEUER, J. & C.M. Tipton. Cardiovascular adaptations to physical training. Ann. Rev. Physiol. 39:221–251, 1977.

SHAW, L.W., Effects of a prescribed supervised exercise program on mortality and cardiovascular morbidity in patients after a myocardial infarction. The National Exercise and Heart Disease Project. Am. J. Cardiol. 48:39–46, 1981.

SIM, D.N. & W.A. Neill. Investigation of the physiological basis for increased exercise threshold for angina pectoris after physical conditioning. J. Clin. Invest. 54:763–770, 1974.

SPEAR, K.L., J.E. Koerner & R.L. Terjung. Coronary blood flow in physically trained rats. Cardiovasc. Res. 12:135–143, 1978.

STEVENSON, J.A.F., V. Feleki, P. Rechnitzer & J.R. Beaton. Effect of exercise on coronary tree size in the rat. Circ. Res. 15:265–269, 1964.

STONE, H.L., Cardiac function and exercise training in conscious dogs. J. Appl. Physiol. 42:824–832, 1977.

STONE, H.L., The heart and exercise training. In Hearts and Heart-Like Organs, Vol. 2, pp. 389–418. New York: Academic Press, 1980.

TEPPERMAN, J. & D. Pearlman. Effects of exercise and anemia on coronary arteries of small animals as revealed by the corrosion-cast technique. Circ. Res. 9:576–584, 1961.

VARNAUSKAS, E., H. Bergman, P. Houk & P. Bjorntorp. Haemodynamic effects of physical training in coronary patients. Lancet 2:8–12, 1966.

WILHELMSEN, L., H. Sanne, D. Elmfeldt, G. Grimby, G. Tibblin & H. Wedel. A controlled trial of physical training after myocardial infarction. Prev. Med. 4:491–508, 1975.

WOOD, P.D. & W.L. Haskell. The effect of exercise on plasma high density lipoproteins. Lipids 14:417–427, 1979.

WYATT, H.L. & J. Mitchell. Influences of physical conditioning and deconditioning on coronary vasculature in dogs. J. Appl. Physiol. 45:619–625, 1978.

Metabolic Diseases and Exercise Performance

Michael Berger
Universität Düsseldorf,
Federal Republic of Germany

This short review focuses upon limitations of exercise performance as related to classical metabolic diseases. Thus the large—and ever increasing—number of well defined specific disorders of muscle energy metabolism (Engel, 1981) will not be discussed in detail. Rather, the pathophysiology of fatigue states precipitated by metabolic diseases and complications, such as the various forms of diabetes mellitus or in the course of caloric deprivation will be reviewed. In addition, the consequences of repeatedly performed exercise, i.e., training, as to the various classical disorders of a number of metabolic diseases will be delineated. Thus this chapter is primarily focused upon the reciprocal relationship between classical metabolic diseases and exercise performance.

Metabolic Diseases Interfering with Exercise Capacity

A number of metabolic disorders are associated with performance incapacitites and premature fatigue of the skeletal musculature. In fact, the most severe limitation of any physical exertion is frequently the key symptom of these diseases. The pathophysiological and pathobiochemical bases of these diseases will be more specifically discussed elsewhere during this symposium. Recently, Wiles et al. (1981) have put forward a most helpful systematization of the various metabolic (and other) myopathies interfering with exercise capacities (Table 1). Such a scheme should assist—at least in most cases—to classify the ever growing number of disorders as they continue to be published, mostly on very limited numbers of patients. In many of these rare disorders, the diagnosis can only be made using elaborate techniques, at best available to medical units specializing in

Table 1

Human Metabolic Myopathies:
Sites of Defects[a]

---

1. Neuromuscular transmission, e.g., myasthenia gravis

2. Sarcolemmal excitation, e.g., myotonia

3. Excitation-contraction coupling, e.g., acidosis

4. Energy economy

    (a) hypermetabolic myopathy, e.g., Luft's disease
    (b) thyro-toxikosis

5. Energy-generation

    (a) deficiencies of myophosphorylase, phosphofructokinase or other enzymes of the glycogenolytic/glycolytic pathways, e.g., McArdle's disease, various types of glycogenoses, etc.
    (b) mitochondrial abnormalities, such as defects of substrate utilization or transport, respiratory chain disorders, etc.

6. Contractile processes

---

[a]Adapted from Wiles et al., 1981.

these areas (Edwards, 1982). In others, such as the various types of glycogenesis or in the case of the McArdle's syndrome(s), the diagnosis may be made most likely merely on the basis of the clinical symptoms and a number of characteristic abnormalities in circulating metabolite levels, i.e., lactate, amino acids and ammonia in the case of McArdle's disease (Berger et al., 1978; Rumpf et al., 1981; Wahren et al., 1973). Other disorders, especially those involving mitochondrial defects, may be more complex, involving overlap of a number of abnormalities (Engel, 1981); very recently new non-invasive diagnostic techniques have been introduced which may prove most helpful in defining the particular locus of the biochemical defects in these rare cases of metabolic myopathies (Edwards et al., 1982).

At least on a quantitative basis, clinical medicine is more often confronted with performance incapacities due to the classical metabolic diseases, diabetes and obesity.

At least on a quantitative basis, clinical medicine is more often confronted with performance incapacities due to the classical metabolic diseases, diabetes and obesity.

As to the diabetic state, the deranged metabolism may, in principle, interfere with exercise performance via three different ways:

(1) due to chronic hypoinsulinemia resulting in protein catabolism and muscle wasting,

(2) due to acute insulin deficiency resulting in hyperglycemia/hyperosmolarity and keto-acidosis,

(3) due to iatrogen hyperinsulinemia resulting in hypoglycemia and a block of fuel supply for the exercising muscles.

Muscle Wasting of Chronic Hypoinsulinemia/Negative Nitrogen Balance

"Muscle weakness and fatigue" in response to inappropriate challenges is a common symptom of almost any disease state, inasmuch as diseases are usually associated with protein catabolism (and negative nitrogen balances). More specifically, "weakness and fatigue" are the leading symptoms of the manifestation phase of diabetes mellitus: close to 100% of all patients complain of a decrease in working capacity at the onset of the disease; this subjective symptom is obviously most striking in patients with particularly high physical activities, be they at work or during leisure time. The muscle wasting associated with chronic hypoinsulinemia ("The melting down of the flesh into the urine") was first properly documented by Benedict and Joslin (1912). Historically it might be of interest that Minkowski who had observed the excessive nitrogen excretion in his pancreatectomized dogs as early as 1889 never really believed that this massive protein wasting was part of the metabolic consequences of the pancreatectomy: in 1893, he suggested that the high excretion of nitrogen in his pancreatectomized dogs was due instead to the malabsorption syndrome of these animals (Minkowski, 1893). In any case, the dramatic protein catabolism and muscle wasting of severe diabetes was well known to the physicians in the first two decades of this century and Benedict and Joslin (1912) were even able to demonstrate clear-cut correlations between the quantitative excretion rates of glucose, ketoacids and nitrogen in the urine of diabetic patients of various severities. The causal role of insulin in the development of diabetic protein catabolism became immediately apparent when the muscle wasting and the excessive nitrogen excretion were blocked along with glucosuria and ketonuria as insulin treatment was performed for the first time in a diabetic patient, almost to the month 60 yr ago (Figure 1) (Burrow et al., 1982).

Apart from the diabetic state, obviously, protein catabolism leading to muscle wasting and fatigue can be induced by (semi-)starvation or malnourishment as documented by Benedict (1915), Cahill et al. (1972), Keys et al. (1950), and Winick (1979). In a more recent study in obese patients on hypocaloric diets, a striking decrease in the individual's capacity for endurance exercise at approximately 75% of the $\dot{V}_{O_2max}$ was observed in association with a particularly low carbohydrate intake (Bogardus et al., 1981); in fact, clear-cut relationships were found between the (almost total) restriction of carbohydrate intake, a substantial decrease

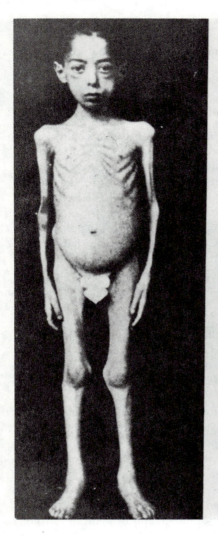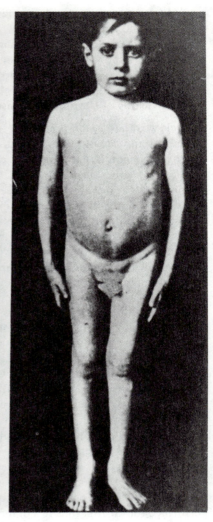

Figure 1—Kachexia and apparent muscle wasting in a juvenile patient with type-I-diabetes mellitus before and after institution of insulin therapy. (Courtesy of Dr. J.P. Assal, Geneva)

of skeletal muscle glycogen contents and the inability to perform strenuous endurance exercise. Albeit the levels of glycemia were distinctly lower during the carbohydrate restriction diet, in the light of newer findings reported by Felig et al. (1982) these differences in blood glucose concentrations do not appear to offer a plausible explanation for the diet-induced inhibition of exercise performance. Rather, the decrease of skeletal muscle glycogen content and/or the negative nitrogen balance most likely induced by the severe carbohydrate restriction (Zimmermann-Telschow & Müller-Wecker, 1976) are the metabolic bases for the fatigue state under these circumstances.

Finally, muscle weakness and the susceptibility to fatigue can be secondary to a state of negative nitrogen balance induced by endocrine disorders, as reviewed by Dr. Viru during this symposium. It should be added at this point that the glucagonoma syndrome often referred to under the heading of "metabolic disease" induces strikingly negative nitrogen balances in association with hypoaminoacidemia and muscle weakness (Berger et al., 1980; Müller et al., 1982; Teuscher et al., 1979).

Acute Insulin Deficiency Resulting in Hyperglycemia/Hyperosmolarity and Ketoacidosis

As a consequence of acute insulin deficiency, a rapid deterioration of metabolic homeostasis develops—the principle features being hyperglycemia, hyperosmolarity and keto(acido)sis. The resulting circulatory problems, in addition to the adverse effects the systemic acidosis might have on skeletal muscle function (Hermansen, 1981), will obviously cause an equally rapid decrease of exercise capacity. Finally, the full-blown clinical syndrome of diabetic ketoacidosis represents a metabolic catastrophe most often associated with a circulatory shock syndrome (Chantelau et al., 1982) and a deterioration of cerebral function—a state of immediate danger for the life of the patient. But even in less dramatic, earlier phases of insulin deficiency, often without any impairment of the patient's well-being, performance of physical exercise may precipitate a further, and sometimes dramatic worsening of the metabolic disequilibrium: in accordance with earlier clinical recommendations not to exercise severely diabetic patients (Allen, et al. 1919; Errerbro-Knudsen, 1948) we have demonstrated in insulin-deficient type-I diabetic patients that a bicycle ergometer challenge of relatively minor intensity induces an additional rise of blood glucose and free fatty acid levels (Berger et al., 1977); in fact, these adverse effects were associated with a rapid aggravation of ketosis (i.e., an augmentation of blood ketone levels from approximately 2 to 4 mmol/1 within 3 h). It follows from these clinical investigations that in patients with insulin-deficient-states physical activity can precipitate ketoacidosis within surprisingly short intervals—a conclusion which is unfortunately not too rarely substantiated by practical experiences with diabetic patients. The pathophysiological basis of these malign effects of physical activity in insulin-deficient patients has been discussed elsewhere (Berger & Berchtold, 1981; Berger et al., 1982; Vranic & Berger 1979; Wahren et al., 1978).

In addition, in some studies (albeit not confirmed by others) relatively mild muscular exercise in diabetic patients both in insulin-deficiency and in moderate, but not optimal, metabolic control was associated with an inappropriately high increase in circulating lactate levels or increased lactate release from the contracting muscles (Berger et al., 1977; Wahren et al., 1975) (Figure 2). Ultimately, such a hyperlactatemia might limit the capacity of diabetic patients for endurance exercise unless they are in optimal metabolic control.

In fact, nowadays an increasing number of diabetic patients participate in sports competitions. We have learned from these patients that optimal exercise performance depends on strictly optimized metabolic control.

Persistance of Hyperinsulinemia during Exercise Resulting in Hypoglycemia and a Block of Fuel Supply for Exercising Muscles

Of even greater clinical relevance is the impairment of exercise performance in insulin requiring diabetic patients due to iatrogen hyperinsulinemia, resulting in hypoglycemia, i.e., cerebral glycopenia. As early as 1926, Lawrence has most precisely documented the potentiation of the hypoglycemic effect of subcutaneously injected insulin during physical activity (Lawrence, 1926) (Figure 3). Actually, exercise-induced hypoglycemia has become one of the principal acute complications of the current insulin replacement therapy in patients with diabetes mellitus and, on the other hand, it represents the major obstacle for exercise performance in insulin-treated diabetic patients. Only quite recently, various hypotheses have been proposed as to the pathophysiological basis of exercise-induced hypoglycemia in insulin-treated diabetics and respective recommendations as to its prevention have been put forward (Berger et al., 1982; Richter et al., 1981; Vranic, Wahren & Horvath, 1979). As it stands, exercise-induced hypoglycemia in insulin-dependent diabetes is due to the imbalance of peripheral glucose utilization as increased by muscular contractions and the inappropriate glucose production by the liver being inhibited by persisting (or even increasing) levels of circulating exogenous insulin. In order to ensure normoglycemia during exercise, insulin dosages before (and following) physical activity have to be reduced and/or additional carbohydrate intake is required. These preventive measures depend largely on the duration and intensity of the physical activity, the individual's degree of physical fitness and his/her prenutrition as well as the actual degree of metabolic control. In a recent study, we have observed that well-controlled insulin-dependent diabetic patients need to reduce their morning insulin dose by more than 2/3 in order to be able to bicycle for 3 h at a relatively mild intensity (Kemmer et al., 1982). Insulin-requiring diabetic patients running the marathon are used to reducing their morning insulin dose by 90% (D. Costill, personal communication).

In addition to preventing hypoglycemia, as a central-nervous-system

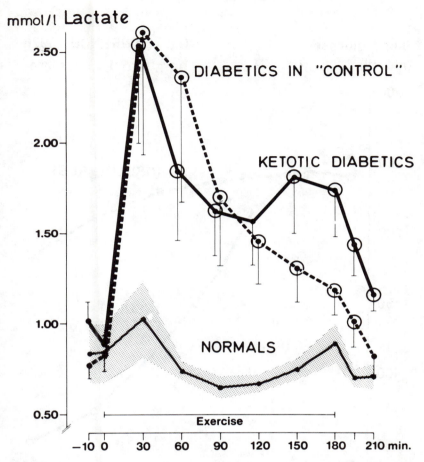

Figure 2—Effect of prolonged exercise on blood lactate in healthy control subjects (NORMALS), diabetic patients in "moderate control" (DIABETICS IN "CONTROL") and Ketotic diabetic patients (KETOTIC DIABETICS). Encircled values are significantly different from corresponding values of the control group at $p < 0.05$. The shaded area indicates the SEM of the mean values of the normal controls. Adapted from Berger et al. (1977).

related reason for the inability to perform exercise, an inappropriate fuel supply for the periphery may be another problem induced by the persisting hyperinsulinemia in insulin-dependent diabetic patients. Thus, we have observed in moderately well controlled patients with type-I diabetes mellitus that approximately 30 min following the injection of regular insulin a fall of circulating free fatty acid levels during a bicycle ergometer test of 40 min duration (Kemmer et al., 1979) (Figure 4): due to high circulating insulin levels the mobilization of free fatty acids from adipose

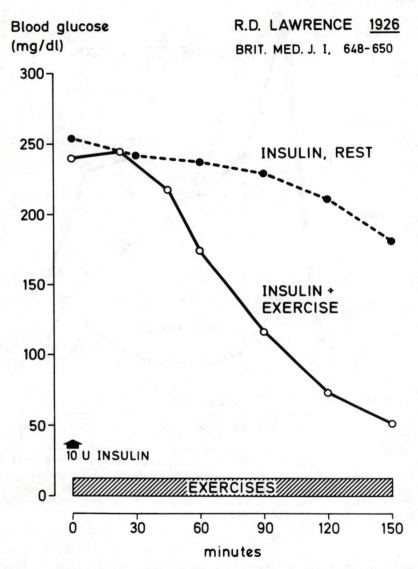

Figure 3—Effect of exercise on the hypoglycemic effect of 10 units subcutaneously-injected regular insulin in a young diabetic patient. Adapted from Lawrence (1926).

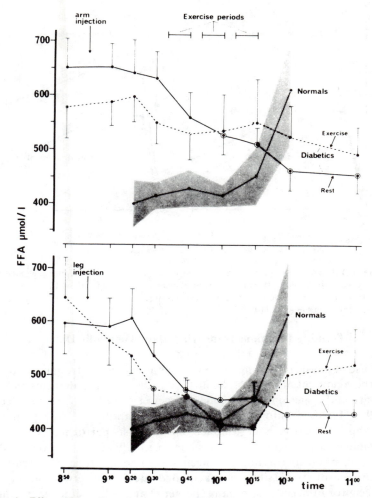

Figure 4—Effects of exercise and subcutaneously-injected regular insulin on serum levels of free fatty acids in nine diabetics and three normal controls (shaded area). Circled values are significantly different from values obtained before the insulin injection in the diabetics and from values obtained before the onset of exercise in the normals (p < 0.05). Adapted from Kemmer et al. (1979).

tissue depots was impaired resulting in a limitation of fuel supply to the contracting muscles in the periphery. It is obvious that such impairments of the complex regulation of fuel homeostasis during exercise in insulin-dependent diabetic patients will ultimately limit the exercise performance capacities unless a more physiological and flexible mode of the insulin replacement therapy is found.

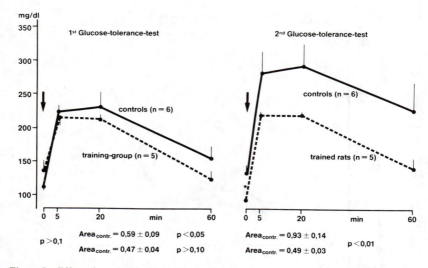

Figure 5—Effect of treadmill training (9 wk) on intravenous glucose tolerance (1 g glucose per kg body wt) in 7-wk-old fatty Zucker rats. Statistical evaluations for the differences between the various areas under the glucose concentration time curves are presented. Adapted from Becker-Zimmermann et al. (1982).

Training Programs Improving upon Metabolic Diseases

The use of physical training programs has always been strongly advocated for the treatment of the classical metabolic diseases, i.e., diabetes mellitus, obesity and hyperlipoproteinemia (Berger et al., 1982; Vranic & Berger, 1979). Qualified information as to the quantitative aspects of any metabolic improvement induced by exercise performance and the possible persistence of such effects is surprisingly limited, however. More recently, physical training has been shown to improve the organism's insulin sensitivity possibly via a facilitated interaction between insulin and its receptor in peripheral organs (Berger et al., 1982; Berger & Berchtold 1982; Richter et al., 1981). In the obese Zucker rat, which resembles metabolically in many ways the obese maturity-onset type diabetic patient, a physical training program has been shown to improve glucose tolerance, hyperlipoproteinemia and insulin resistance—and, if started in younger animals, to partially prevent the development of insulin-resistance/-glucose-intolerance (Figure 5) (Becker-Zimmermann et al., 1982). These investigations should encourage the initiation of detailed studies of the potential benefits of physical training programs in humans with type-II diabetes mellitus (Berger & Berchtold, 1981). Such beneficial consequences of physical training programs in humans would be particularly relevant in patients with type-II-diabetes since the insulin resistance—very often associated with obesity and hyperlipoproteinemia—appears to be of principal importance as to the etiology of this metabolic syndrome—more recently often referred to as "diabesity." At the same time, physical training should be instrumental in any attempt to treat obesity (Franklin & Rubinfire, 1980) as well as various disorders of lipoprotein-metabolism albeit—again—detailed studies substantiating these often proposed benefits of exercise performance are rather limited and respectively lacking.

Conclusions

Impairments of exercise performance due to metabolic diseases may, in essence, be summarized by relating them to the consequences of (a) chronic hypoinsulinemia/negative nitrogen balance and muscle wasting, (b) effects of acute insulin deficiency resulting in hyperglycemia/hyperosmolarity and ketoacidosis and (c) effects of persisting hyperinsulinemia in insulin-treated diabetic patients resulting in hypoglycemia, i.e., cerebral glycopenia, and various inhibitions of the peripheral fuel supply during physical activity. On the other hand, there is a growing—albeit at present still limited—body of evidence pointing to the benefits of physical training programs in the therapy of metabolic diseases and syndromes, such as the almost endemic combination of type-II-diabetes mellitus, obesity and hyperlipoproteinemia (plus hypertension). It is on this basis that the reciprocal interrelationship between exercise performance and metabolic diseases appears to become increasingly important for clinical medicine in the near future.

References

ALLEN, F.M., E. Stillman & R. Fitz. Total dietary regulation in the treatment of diabetes. Chapter V: Exercise. New York: The Rockefeller Institute of Medical Research, 1919, pp. 468–499.

BECKER-ZIMMERMANN, K., M. Berger, P. Berchtold, F.A. Gries, L. Herberg, & M. Schwenen. Treadmill training improves intravenous glucose tolerance and insulin sensitivity in fatty Zucker rats. Diabetologia 22, 1982. (In Press)

BENEDICT, F.G. & E.P. Joslin. A Study of Metabolism in Severe Diabetes. Carnegie Institute of Washington, Publ. No. 176, 1912.

BENEDICT, F.G., A Study on Prolonged Fasting. Carnegie Institute of Washington, Publication No. 203, 1915.

BERGER, M., P. Berchtold, H.J. Cüppers, H. Drost, H.K. Kley, W.A. Müller, W.

Wiegelmann, H. Zimmermann-Telschow, F.A. Gries, H.L. Krüskemper & H. Zimmermann. Metabolic and hormonal effects of muscular exercise in juvenile type diabetics. Diabetologia 13:355–365, 1977.

BERGER, M. & P. Berchtold. Diabetes and muscular exercise in man. In J. Poortmans & G. Niset (Eds.), Biochemistry of Exercise IV-A, International Series of Sport Sciences, Vol. II-A, pp. 239–259. Baltimore: University Park Press, 1981.

BERGER, M. & P. Berchtold. Effects of exercise in obese diabetics. In P. Björntorp, M. Cairella & A.N. Howard (Eds.), Recent Advances in Obesity Research III, pp. 341–347. London: J. Libbey, 1981.

BERGER, M. & P. Berchtold. Insulin Action. In A. Marble (Ed.), Joslin's Diabetes Mellitus. Philadelphia: Lee & Febiger, 1982 (in press).

BERGER, M., P. Christacopoulos & J. Wahren. Diabetes and Exercise. Bern, Switzerland: Huber, 1982.

BERGER, M., E. Gibbels, B. Leven & D. Seiler. Das McArdlesche Syndrom. Fortschr. Neurol. Psychiat. 46:312–326, 1978.

BERGER, M., A. Teuscher, P.A. Halban, E. Trimble, P.P. Studer, C.B. Wollheim, H. Zimmermann-Telschow & W.A. Müller. In vitro and in vivo studies on glucagonoma tissue. Hormon. Metb. Res. 12:144–150, 1980.

BOGARDUS, C., B.M. LaGrange, E.S. Horton & E.A.H. Sims. Comparison of carbohydrate-containing and carbohydrate restricted hypocaloric diets in the treatment of obesity. Endurance and metabolic fuel homeostasis during strenuous exercise. J. Clin. Invest. 68:399–404, 1981.

BURROW, G.N., B.E. Hazlett & M.J. Phillips. A case of diabetes mellitus. New Engl. J. Med. 306:340–343, 1982.

CAHILL, G.F., T.T. Aoki, E.B. Marliss. Insulin and Muscle Protein. In D.F. Steiner & N. Freinkel (Eds.), Handbook of Physiology, Section VII, Vol. I.: The Endocrine Pancreas, pp. 563–575. Washington: The American Physiological Society, 1972.

CHANTELAU, E.A., G.E. Sonnenberg & M. Berger. Kreislaufinsuffizienz bei Coma diabeticum. Dtsch. Med. Wochenschr. 107:203–204, 1982.

EDWARDS, R.H.T., Keynote Address. This symposium.

EDWARDS, R.H.T., J. Dawson, D.R. Wilkie, R.E. Gordon, & D. Shaw. Clinical use of nuclear magnetic resonance in the investigation of myopathy. Lancet 1:725–731, 1982.

ENGEL, A.B., Metabolic and endocrine myopathies. In Sir J. Walton (Ed.), Disorders of Voluntary Muscle. 4th Edition, pp. 664–711. Edinburgh: Churchill Livingstone, 1981.

ERREBRO-KNUDSEN, E.O., Diabetes Mellitus and Exercise. Copenhagen: G.E. Gad, 1948.

FELIG, P., A. Cherif, A. Minagawa & J. Wahren. Hypoglycaemia during prolonged exercise in normal man. New Engl. J. Med. 306:895–900, 1982.

FRANKLIN, B.A. & M. Rubinfire. Losing weight through exercise. J.Am. Med. Ass. 244:377-378, 1980.

HERMANSEN, L., Effect of metabolic changes on force generation in skeletal muscle during maximal exercise. In R. Porter & J. Whelan (Eds.), Human Muscle Fatigue: Physiological Mechanisms, pp. 75-88. London: Pitman Medical (Ciba Foundation Symposium No. 82), 1981.

KEMMER, F.W., P. Berchtold, M. Berger, A. Starke, H.J. Cüppers, F.A. Gries & H. Zimmermann. Exercise-induced fall of blood glucose in insulin-dependent diabetics unrelated to alterations of insulin mobilization. Diabetes 28:1131-1137, 1979.

KEMMER, F.W., H.J. Cüppers, P. Berchtold, A. Woll, V. Jörgens, A. Starke & M. Berger. Decreased risk of exercise induced hypoglycaemia by reduced insulin dose or increased carbohydrate intake in insulin-dependent diabetics. In M. Berger, P. Christacopoulos & J. Wahren (Eds.), Diabetes and Exercise, p. 89. Bern, Switzerland: Huber, 1982.

KEYS, A., J. Brozeck, A. Henschel, O. Mickelsen & H.L. Taylor. The Biology of Human Starvation. Volumes I and II. Minneapolis: The University of Minnesota Press, 1950.

LAWRENCE, R.D., The effect of exercise on insulin action in diabetes. Brit. Med. J. 1:648-652, 1926.

MINKOWSKI, O., Untersuchungen über den Diabetes Mellitus nach Exstirpation des Pankreas. Leipzig: F.C.W. Vogel, 1893.

MÜLLER, W.A., H.J. Cüppers, H. Zimmermann-Telschow, H. Micheli, H. Wyss, A.E. Renold & M. Berger. Amino acids and lipoproteins in plasma of duodeno-pancreatectomized patients. Effects of glucagon in physiological amounts. Europ. J. Clin. Invest. (In Press), 1982.

RICHTER, E.A., N.B. Ruderman & S.S. Schneider. Diabetes and exercise. Am. J. Med. 70:201-209, 1981.

RUMPF, K.W., H. Wanger, H. Kaiser, H.-M. Meinck, H.H. Goebel & F. Scheler. Increased ammonia production during forearm ischemic work test in McArdle's disease. Klin. Wochenschr. 59:1319-1320, 1981.

TEUSCHER, A., P.P. Studer, A. Krebs, M. Berger & P. Aeberhard. Diabetes-verlauf und klinisches Bild bei Glukagonom. Schweiz. Med. Wschft. 109:1273-1280, 1979.

VRANIC, M. & M. Berger. Exercise and diabetes mellitus. Diabetes 28:147-167, 1979.

VRANIC, M., J. Wahren & S. Horvath (Eds.), Diabetes and Exercise. Diabetes 28 (Suppl. 1), 1979.

WAHREN, J., P. Felig, R.J. Havel, L. Jorfeldt, B. Pernow & B. Saltin. Amino acid metabolism in McArdle's syndrome. New Engl. J. Med. 288:774-777, 1973.

WAHREN, J., P. Felig & L. Hagenfeldt. Physical exercise and fuel homeostasis in diabetes mellitus. Diabetologia 14:213-222, 1978.

WAHREN, J., L. Hagenfeldt & P. Felig. Splanchnic and leg exchange of glucose, amino acids and free fatty acids during exercise in diabetes mellitus. J. Clin. Invest. 55:1303–1314, 1975.

WILES, C.M., D.A. Jones & R.H.T. Edwards. Fatigue in human metabolic myopathy. In R. Porter & J. Whelan (Eds.), Human Muscle Fatigue: Physiological Mechanisms. London: Pitman Medical (Ciba Foundation Symposium 82), 1981, pp. 264–282.

WINICK, M. (Ed.), Hunger Disease. Studies by the Jewish Physicians in the Warsaw Ghetto. Current Concepts in Nutrition, Vol. 7., New York: John Wiley & Sons, 1979.

ZIMMERMANN-TELSCHOW, H. & H. Müller-Wecker. Beziehungen zwischen Stickstoff-Bilanz und Aminosäuren, Glukose und Insulin im Blut in verschiedenen Stoffwechselsituationen des menschlichen Organismus. Hoppe-Seyler's Z. Physiol. Chem. 357:695–706, 1976.

Panel:
The Intracellular
Environment

The Intracellular Environment in Peripheral Fatigue

Jacques R. Poortmans
Chimie Physiologique, Institute Supérieur d'Education
Physique et de Kinésithérapie, Universite Libre de Bruxelles,
Bruxelles, Belgium

As suggested by Bigland-Ritchie et al. (1978), when one investigates the failure to generate force during sustained or repeated muscle contractions it is necessary to consider both central fatigue and peripheral fatigue. Among the factors causing peripheral fatigue are the impairment of: excitation-contraction coupling, energy exchange, and fuel supply. These basic mechanisms depend on cellular integrity which may be disturbed by different parameters such as changes in water and electrolytes, temperature, pH, substrate depletion and local depots of metabolites.

Peripheral fatigue, on the biochemical side, has been investigated by in vitro and in vivo techniques. Among the latter ones, invasive methods have been the choice for years. Arterio-venous differences measured for the whole muscle are the oxygen uptake, the flux of substrates, and the release of metabolites (Keul et al., 1967; Poortmans et al., 1979; Rowell et al., 1965; Sanders et al., 1964; Wahren, 1966). The muscle-biopsy needle technique introduced by Bergstrom and Hultman (1966) made possible the study of intracellular stores, substrate depletion, and metabolic profiles of the different muscle fibers.

Nowadays, non-invasive probes of tissue metabolism using physical methods are being developed (Cohen, 1982). Nuclear magnetic resonance (NMR) methods are used for intracellular pH measurements, phosphorus monitoring, and studies of gluconeogenesis. Fluorometric measurements for NADH and NADPH levels in the cytoplasm and the mitochondria in intact tissue are becoming common techniques in several laboratories. Mass spectrometry (MS) and positron computed tomography (PCT) are used for transcutaneous blood gas monitoring and measurement of the local cerebral metabolic rate for glucose. Application of these techniques will lead to some new insights into the question of muscular fatigue. The

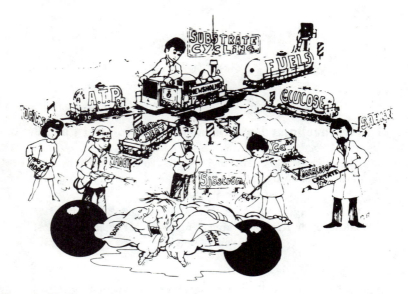

Figure 1—The interrelationships among the panel members as they seek an understanding of peripheral fatigue.

speakers on this panel give a survey of what is known of the factors that perturb the intracellular environment when fatigue appears (Figure 1).

Together with David Wilkie, Joan Dawson has applied the phosphorus NMR to living muscle to monitor the levels and rates of utilization of ATP, ADP, phosphorylcreatine and phosphorus simultaneously, directly or indirectly and to relate changes in them to concurrent changes in the mechanical performance of muscles.

Sue Donaldson is working on the intracellular changes accompanying fatigue with emphasis on the effects of a decrease in pH on Ca^{2+}-activation of force generation. The introduction of the skinned fiber technique has allowed her to study force generation in single muscle fibers.

The effects of muscle temperature on contractile properties and body temperature have been analyzed by Ethan Nadel. He has demonstrated the importance of elevated body temperatures on the ability to deliver oxygen to muscle during exercise.

Eric Newsholme is well known for his studies on the regulation of metabolism (Newsholme & Start, 1973). The control of the rate of individual reactions in a pathway and hence the flux of the pathway as a whole is of primary importance for fuel supply in exercising muscle. Why and how does a muscle change its energy production from carbohydrates to lipids? What are the limitations in the use of these fuels, especially when physical exhaustion is concerned? The analysis of substrate cycles and flux-generating steps could help in understanding the metabolic basis of fatigue in endurance exercise.

During these last years Kent Sahlin has been involved in intracellular pH of human muscle during exercise. The hydrogen ions produced in excess during severe exertion are taken up to some extent by the buffer systems of muscle and blood. Meanwhile, there is still a pronounced fall in muscle pH during brief intense work. This will affect the intracellular energy metabolism and consequently physical performance.

It has been observed that severe muscular discomfort may appear after unusually heavy physical exercise. This delayed muscle soreness has been studied on the morphological side by Michael Sjoström's group. Indeed it is important to know if tissue injury is a common situation that appears after overloading the muscle fibers.

From these various contributions, it is clearly established that peripheral fatigue is a complex phenomenon. The identification of a single or several factors leading to the impairment of performance depends upon the conditions under which exercise is conducted.

References

BERGSTRÖM, I. & E. Hultman. Muscle glycogen synthesis after exercise. An enhancing factor localized to the muscle cells in man. Nature 210:309–310, 1966.

BIGLAND-RITCHIE, B., D.A. Jones, G.P. Hosking & R.H.T. Edwards. Central and peripheral fatigue in sustained maximum voluntary contractions of human quadriceps muscle. Clin. Sci. Molec. Med. 54:609–614, 1978.

COHEN, J.S. Noninvasive Probe of Tissue Metabolism. New York: J. Wiley and Sons, 1982.

KEUL, J., E. Doll & Keppler. The substrate supply of the human skeletal muscle at rest, during and after work. Experientia 23:974–979, 1967.

NEWSHOLME, E.A. & C. Start. Regulation in Metabolism. New York: J. Wiley and Sons, 1973.

POORTMANS, J., I. Delescaille-Vandenbosch & R. Leclereq. Lactate uptake by inactive forearm during progressive leg exercise. J. Appl. Physiol. 45:835–839, 1979.

ROWELL, L.B., E.J. Masoro & M.J. Spencer. Splanchnic metabolism in exercising man. J. Appl. Physiol. 20:1032–1037, 1965.

SANDERS, C.A., G.E. Levinson, W.H. Abelmann & N. Freindel. Effect of exercise on the peripheral utilisation of glucose in man. New Engl. J. Med. 271:220–225, 1964.

WAHREN, J., Quantitative aspects of blood flow and oxygen uptake in the forearm during rhythmic exercise. Acta Physiol. Scand. 67(Suppl. 269): 1–93, 1966.

Phosphorus Metabolites and the Control of Glycolysis Studied by Nuclear Magnetic Resonance

M. Joan Dawson
University College London, London, England

We have been using nuclear magnetic resonance (NMR) in conjunction with other techniques to study the biochemistry and physiology of isolated frog muscle and, more recently, the limb muscles of normal human subjects and patients with muscular disorders. In most of our studies we have related mechanical performance to energetically important phosphorus metabolites (P-metabolites) and intracellular pH, which are observed directly in a ^{31}P NMR spectrum, and to calculated quantities derived from the spectrum, including $dG/d\xi$ (free-energy change) for ATP hydrolysis, rate of glycolysis and rate of ATP utilization. In this paper I will stress the following results related to the intracellular environment of skeletal muscle:

1. NMR is suitable for quantitative analysis of [P-metabolites] and intracellular pH in human skeletal muscle as well as in other tissues. The results are in some cases more accurate than those obtained by needle biopsy techniques.

2. [ADP] and [AMP] which are free to take part in enzymatic reactions in vivo are far lower than those measured by chemical analysis. Contrary to information presented in popular textbooks, changes in [ADP], [AMP], "phosphate potential" or "energy charge" do not activate glycolysis during contractions of isolated frog skeletal muscles or limb muscles of normal human subjects.

The studies of isolated frog muscle were done in collaboration with D.G. Gadian and D.R. Wilkie; those on human subjects are in collaboration with R.H.T. Edwards, R.E. Gordon, D. Shaw and D.R. Wilkie.

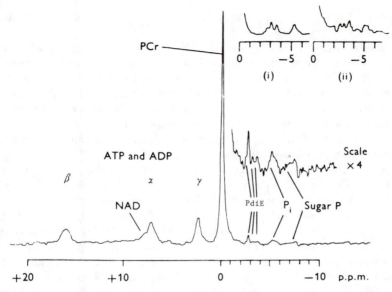

Figure 1—[31]P spectrum from 4 resting frog sartorius muscles perfused with oxygenated Ringer's solution at 4° C. May 19, 1976. The x-axis is frequency in ppm (parts per million deviation from reference frequency); the y-axis is signal intensity. The accumulation time was nearly 6 h (10,000 pulses of 40 μs duration at 2 s intervals). The short length of record on the right-hand side of the PCr peak has been enlarged 4 x vertically to show fine detail. The insert Figures (i) and (ii) are from different experiments and show how the P-diester peaks vary in size, though not in position. From Dawson, Gadian & Wilkie, 1977. Note that the sign convention for the x-axis has altered since this spectrum was obtained (see Figure 2).

Methods

[31]P (the naturally occurring phosphorus) possesses the quantum mechanical characteristic of "spin" and thus can give rise to NMR signals. The ^{31}P spectrum of intact tissues is relatively simple, with peaks arising from only a few energetically important P-metabolites that are free in solution; this characteristic, together with the fact that it is totally nondestructive, makes ^{31}P NMR a highly satisfactory tool for scientific and clinical investigations. The NMR technique requires that the object of interest be placed in a powerful and uniform magnetic field (B_0). At suitable intervals a brief pulse of radiofrequency magnetic field is applied at right angles to B_0. If this field (B_1) is equal to the resonance frequency of a particular atomic nucleus, the nucleus absorbs energy and subsequently emits a weak radiofrequency signal for a few ms. The precise frequency of resonance (ν) is proportional to the field (B) experienced by the nucleus: $\nu = kB$. B is shifted slightly from B_0 by the chemical environment of the various nuclei, including their circulating electrons. This leads to the "chemical shift" shown along the

x-axis of Figure 1, which makes it possible to identify the various compounds present.

In studies of isolated frog muscle, ^{31}P spectra were recorded at 129.2 MHz on a spectrometer constructed at the Biochemistry Department, University of Oxford. The muscles were contained within a 7 mm diameter NMR sample tube at 4°C, and were either perfused with oxygenated Ringer's solution, or in some cases poisoned with sodium cyanide to ensure uniformly anaerobic status. The experimental chamber was equipped with stimulating electrodes and an isometric force transducer so that mechanical and metabolic measurements could be made simultaneously.

Recent technical advances have made it possible to obtain high quality NMR spectra from isolated areas in intact human subjects; this variation of NMR has been given the name "topical magnetic resonance"(TMR), after the Greek word "τoπoσ" meaning "a place" (Gordon et al., 1980). TMR differs from conventional NMR in that the magnetic field is shaped so that it is uniform within a localized sensitive volume from which intelligible NMR signals are obtained, while outside this volume the field changes very rapidly, giving rise to a broad spectral hump which is mathematically removed. The TMR spectrometer was designed and built by Oxford Research Systems (Oxford, England, Product Note 111). The 1.9 tesla horizontal magnet has a bore size of 20 cm, large enough to admit an adult human arm or leg. The physiological problem is to locate the limb comfortably and securely in relation to the sensitive volume in such a way that no pressure is applied to nerves, arteries or veins. The subject gripped a force transducer mounted on a shaped and padded limb support. The spectra were obtained on a sensitive volume of 22 cm^3, located 6 cm distal to the ulnar tuberosity at the muscular parts of the flexor carpi radialis and palmaris longis.

Those interested in further explanation of the NMR technique and its application to metabolic studies on intact tissue are referred to a recent book on the subject (Gadian, 1982).

Results

Figure 1 shows a resting ^{31}P NMR spectrum of well-oxygenated frog sartorius muscles maintained at 4°C. This is to be compared with Figure 2 which is a ^{31}P spectrum obtained from the resting forearm of a normal human subject. Clearly resolved peaks can be observed, indicating the presence of Pi, PCr and ATP. NAD (total NAD and NADH) appears as a highly significant shoulder to the right-hand side of the α-adenosine peak. The hexose and triose phosphates, which resonate together at + 7.5 ppm at physiological pH are too low in concentration to be observed. Both spectra show significant peaks at + 3.5 ppm; this is the phosphodiester region

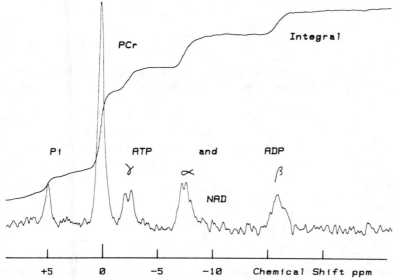

Figure 2—[31]P spectrum of human forearm muscle obtained at 32 M Hz using a surface coil. The spectrum was obtained in approximately 27 min (800 pulses of 80 μs duration at 2 s intervals), and was enhanced by 10 Hz line broadening and convolution differencing. October 8, 1980. Subject M.J.D. From Cresshaw et al., 1981.

(labeled PdiE in Figure 1 and Table 2), which may have special interest for the study of muscle disease (Chalovich et al., 1979). The position of the Pi peak is sensitive to pH and after appropriate calibration can be used to determine intracellular pH; in resting human muscle we find $pH = 7.08 \pm 0.04$, a value which agrees well with determinations on muscle homogenates (Sahlin, Harris & Hultman, 1975).

Under suitable conditions the areas of the resonances obtained in NMR spectra are directly proportional to the concentrations of compounds from which they are derived. This is not usually the case in biological experiments, however, and quantitative interpretation of the NMR peak integrals must take account of a number of variables related to the methodology and to characteristics of the tissue being observed. We have set out the relevant problems in other publications (see Dawson, Gadian & Wilkie, 1977; Edwards et al., 1982; Wilkie et al., 1983). In this short presentation I will confine myself to a comparison of the NMR results with those obtained by chemical analysis.

Undoubtedly the most accurate estimates of [P-metabolites] by chemical methods have been obtained on frog sartorius and semitendinosis muscles maintained at 0° C and subjected to very fast freezing to terminate chemical reactions. It is now well known that freezing must be accomplished in less than 100 ms if no artifactual breakdown of PCr is to occur (Kretzschmar &

Table 1

Relative Concentration of P Compounds in Resting Frog Sartorius as Determined by NMR and by Chemical Analysis

		NMR	Chemical	P
PCr	$\bar{x}$	6.74	8.14	
βATP	SE	0.309	0.744	n.s.
	n	6	18	
PCr	$\bar{x}$	16.02	13.10	
Pi	SE	1.58	3.03	n.s.
	n	6	17	

Rough estimates of actual concentrations may be made by assuming that the resting PCr content is 27 mmol kg^{-1} (see Dawson, Gadian & Wilkie, 1977); n.s. = not significant.

Table 2

Composition of Resting Human Muscle

[P-Metabolite]	Biopsy mean ± SE(n) mmol kg^{-1} wet	^{31}P TMR [ATP] constant at 5.5 mmol kg^{-1} mean ± SE(n = 7) mmol kg^{-1} wet	^{31}P TMR Integral constant at 49.5 mmol kg^{-1} mean ± SE(n = 7) mmol kg^{-1} wet
ATP	5.5 ± 0.07(81) ≡	5.5	5.1 ± 0.10
PCr	17.4 ± 0.19(81)	29.0 ± 0.69	27.4 ± 0.23
Pi	10 ± ? (3)	4.4 ± 0.33	4.3 ± 0.27
PCr + Cr	28.6 ± 0.28(81)	—	—
PCr + Pi	31.6 ± 3.27(11)	33.4 ± 0.77	31.7 ± 0.29
P di E	—	0.7 ± 0.09	0.8 ± 0.09
NAD + NADH	0.7	1.0 ± 0.09	0.9 ± 0.07

From Wilkie et al., 1983. P di E = phosphodiesters (see text). [NAD + NADH] in column 1 is not from human biopsies but from fast glycolyzing porcine muscle.

Wilkie, 1969; Kretzschmar, 1970). Table 1 relates the ratios of [P-metabolites] in aerobic frog sartorius obtained by NMR to those determined by ultra-fast freezing followed by chemical analysis. The two techniques are subject to totally different kinds of experimental artifacts, and so the good agreement shown in Table 1 confirms the accuracy of both. In order to convert the ratios determined by NMR into absolute concentrations, the peak integrals must be calibrated with reference to the results obtained by chemical analysis. In our original studies we set the PCr peak integral in oxygenated resting frog muscle equal to the best estimate available from chemical analysis, 27 mmol kg^{-1}. This procedure is exactly equivalent to the calibration of chemical measurements with reference to a standard solution.

We used similar calibration methods in our TMR measurements of forearm muscle of normal human subjects. In experiments that are described elsewhere (Wilkie et al., 1983) we found that skin, fat, blood and bone make no detectable contribution to our ^{31}P spectra, which were therefore calibrated as shown in Table 2. The results of chemical analyses of needle and open biopsies are shown in column 1. In column 2 the TMR results are calibrated in relation to the relatively stable [ATP] determined by chemical analysis, and in column 3 they are related to the analytically determined total mobile P (i.e., 3 × [ATP] + [PCr + Pi] + 2x[NAD + NADH] = 49.5 mmol kg^{-1}). The general agreement between these two essentially independent calibration procedures attests to the reliability of the results. Unlike the case of frog muscle, however, there are differences in the results obtained by NMR and by chemical analysis of open and needle biopsy samples. The TMR results show a higher [PCr] and a lower [Pi] than is obtained by biopsy, while the total [PCr] + [Pi] is not significantly different by the two methods. This result tends to confirm long-held suspicions that PCr is hydrolyzed during the minimum of 6 s required to freeze the needle biopsy samples.

Accurate assessment of [ADP] and [AMP] is particularly important because of the key role these substances are thought to play in regulation of metabolism. [ADP] determined from chemical analysis of extracts of frog and human skeletal muscles is about 0.5 mmol kg^{-1}, a value which is large enough to be detected by NMR. Analysis of large numbers of frog and human ^{31}P spectra, however, show that the integrals of the γ and β-adenosine peaks do not differ significantly, meaning that the free [ADP], which lacks the β-resonance, is too small to be detected. The answer to this discrepancy is that the commonly employed extraction procedures strip ADP from its in vivo binding sites where it is inaccessible to NMR. The known binding sites for ADP on actin and myosin (see Ferenzi et al., 1978) together account for virtually all of the ADP present; indeed, the functional integrity of the muscle is dependent upon such binding.

For this reason, the [ADP] determined by chemical analysis is not directly relevant to metabolic studies. At present, the [ADP] that is free in solution in vivo can only be determined by calculation from ^{31}P NMR spectra and knowledge about the creatine kinase reaction (Dawson, Gadian & Wilkie, 1978; Gadian et al., 1981; Wilkie, 1981a). This yields a value of approximately 20 μmol l^{-1} in well-maintained resting frog skeletal muscles. The fact that [ADP] which is free in solution is much less than that determined by chemical analysis has profound consequences: it can be shown quantitatively that suggestions of compartmentation of the creatine kinase reaction in skeletal muscle (see Wilkie, 1981a) and heart (Dawson, 1983) arise from failure to take this difference into account. Widespread failure to appreciate the extent and significance of ADP binding in muscles has led to generally incorrect assumptions about the in vivo environment by biochemists studying enzyme control systems in vitro (Wilkie, 1981b; Wilkie et al., 1983).

AMP is in equilibrium with ADP through the myokinase reaction, so the [AMP] that is free to take part in metabolic reactions is proportional to $[ADP]^2$. It amounts to about 0.2μmol^{-1} in resting frog muscle.

Control of Glycolysis

The increase in rate of glycolysis as a result of muscle contraction is a case in point. This depends on the activation of at least two enzymes, phosphorylase and phosphofructokinase, and is widely believed to be caused by increases in the products of ATP hydrolysis, notably [ADP] or [AMP] (Newsholme & Start, 1973) or quantities derived from them such as "adenylate charge" (Atkinson, 1977) or "phosphate potential." The NMR results on frog and human muscle show that normally glycolysis is activated by something else. The rate of lactic acid (LA) formation can be determined from ^{31}P NMR spectra of anaerobic muscles on the basis of the change in pH combined with information about the internal buffers of frog muscle (Dawson, Gadian & Wilkie, 1978) or human muscle (Wilkie et al., 1983). In repetitively stimulated anaerobic frog gastrocnemii, we have found that glycolysis is activated during each contraction, but quickly comes to a halt despite the fact that [PCr], [ADP] and [AMP] have not been restored to resting values. As stimulation is continued [ADP] and [AMP] build to higher and higher levels—but glycolysis is activated only during and shortly after each contraction (see Dawson, Gadian & Wilkie, 1980) (Figures 8 & 9).

The results are the same in humans. If the human forearm is made ischemic by application of a sphygomomanometer cuff for periods as long as 1 h, the rate of lactic acid build-up is extremely low, about 3 mmol kg$\|^1$ h^{-1} (Cresshull et al., 1981; Wilkie et al., 1983) despite a decline in [PCr] to as low as 1/2 its resting level, and a large rise in [ADP], inferred from the

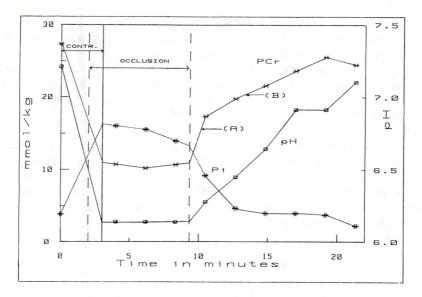

Figure 3—Changes in PCr (*), Pi, (#) and pH (o) as a result of maximum voluntary isometric contraction. After 2 min of contraction a sphygmomanometer cuff was inflated; contraction was continued for 1 additional min and ischemic conditions were maintained for a further 6 min.

Changes in P-metabolites were calculated by assuming the total metabolite phosphorus remains unchanged during the course of the experiment (see text). Spectra were averaged in 2 min bins, beginning when the contraction was terminated. There was no apparent change in [ATP] during the course of this experiment, but there was an approximately three-fold increase in [SP] by the end of the ischemic period, which did not return appreciably toward normal by the end of the experiment.

When occlusion was terminated recovery of force development was studied by measuring peak force roughly even 20 s in a brief (1 to 2 s) test contraction; 50% of the initial force was developed at point A and more than 90% at point B. From Wilkie et al., 1983. Subject M.J.D.

creatine kinase equilibrium. Figure 3 shows that the situation is dramatically different during a maximal voluntary contraction maintained for 3 min. PCr dropped to 1/2 of its initial level and pH fell by a full pH unit, representing an increase in [LA] of approximately 34 mmol kg^{-1}. Neither pH nor PCr recovered measurably during the post-contractile ischemic period. Thus, although the changes in [P-metabolites] are roughly similar, contraction produces a several hundred-fold greater rise in glycolytic rate than does ischemia.

Conclusions

The results we have recently obtained on forearm muscles of normal human subjects are complementary to those on isolated frog muscles and they lead us to the same conclusions. We are satisfied that our ^{31}P NMR measurements of [ATP], [PCr] and [Pi] are reliable, as are the results obtained by chemical analysis of frog muscles frozen within 100 ms. The relatively slow freezing times for needle biopsy samples from human muscles presents problems since approximately 20% of the total PCr is artifactually broken down during the sampling procedure; it remains to be seen whether measurements by this method of *changes* in [PCr] as a result of experimental intervention are seriously in error.

One of the important contributions of NMR to the study of muscular mechanisms has been its ability to estimate [ADP] and [AMP] that are free to take part in metabolic reactions, and how these quantities vary as a result of experimental interventions. Our results in both frog and human muscle show that during normal contractions, glycolysis is not activated by changes in the concentrations of these metabolites.

References

ATKINSON, D.E., Cellular Energy Metabolism and its Regulation. New York: Academic Press, 1977.

CHALOVICH, J.M., C.T. Burt, M.J. Danon, T. Glonek & M. Bárány. Phospho-diesters in muscular dystrophies. Ann. N.Y. Acad. Sci. 317:649–668, 1979.

CRESSHULL, I., M.J. Dawson, R.H.T. Edwards, D.G. Gadian, R.E. Gordon, G.K. Radda, D. Shaw & D.R. Wilkie. Human muscle analysed by ^{31}P nuclear magnetic resonance in intact subjects. J. Physiol. 317:18P, 1981.

DAWSON, M.J., Nuclear Magnetic Resonance. In A.J. Drake-Holland & M.I.M. Noble (Eds.), Cardiac Metabolism. Wiley and Sons Ltd, 1983 (In Press).

DAWSON, M.J., D.G. Gadian & D.R. Wilkie. Contraction and recovery of living muscle studied by ^{31}P nuclear magnetic resonance. J. Physiol. 267:703–35, 1977.

DAWSON, M.J., D.G. Gadian & D.R. Wilkie. Muscular fatigue investigated by phosphorus nuclear magnetic resonance. Nature 274:861–866, 1978.

DAWSON, M.J., D.G. Gadian & D.R. Wilkie. Studies of the biochemistry of contracting and relaxing muscle by the use of ^{31}P NMR in conjunction with other techniques. Phil Trans. R. Soc. Lond. B289:445–455, 1980.

EDWARDS, R.H.T., M.J. Dawson, D.R. Wilkie, R.E. Gordon & D. Shaw. Clinical use of nuclear magnetic resonance in the investigation of myopathy. Lancet 1:725–731, 1982.

FERENCZI, M.A., E. Homsher, R.M. Simmons & D.R. Trentham. Reaction mechanism of the magnesium ion-dependent adenosine triphosphatase of frog muscle myosin and subfragment 1. Biochem. J. 171:165–175, 1978.

GADIAN, D.G., Nuclear Magnetic Resonance and its Applications to Living Systems. Oxford: Clarendon Press, 1982.

GADIAN, D.G., G.K. Radda, T.R. Brown, E.M. Chance, M.J. Dawson & D.R. Wilkie. The activity of creatine kinase in frog skeletal muscle studied by saturation transfer nuclear magnetic resonance. Biochem. J. 194:215–228, 1981.

GORDON, R.E., P.E. Hanley, D. Shaw, D.G. Gadian, G.K. Radda, P.E. Styles, P.J. Bore & L. Chan. Localisation of metabolites in animals using ^{31}P topical magnetic resonance. Nature 287:736–738, 1980.

KRETZSCHMAR, K.M., Energy production and chemical change during muscular contraction. Ph.D. Thesis, University of London. p. 126 and Table 3, 1970.

KRETZSCHMAR, K.M. & D.R. Wilkie. A new approach to freezing tissues rapidly. J. Physiol. 202:66–67P, 1969.

NEWSHOLME, E.A. & C. Start. Regulation in Metabolism. London: Wiley, 1973.

SAHLIN, K., R.C. Harris & E. Hultman. Creatinine kinase equilibrium and lactate content compared with muscle pH in tissue samples obtained after isometric exercise. Biochem. J. 152:173–180, 1975.

WILKIE, D.R., Shortage of chemical fuel as a cause of fatigue: studies by nuclear magnetic resonance and bicycle ergometry. Ciba Foundation Symposium 82. London: Pitman Medical, 1981a.

WILKIE, D.R., Discussion at Royal Society Meeting on the enzymes of glycolysis: structure, activity and evolution. Phil. Trans. R. Soc. Lond. B293:40–41, 1981b.

WILKIE, D.R., M.J. Dawson, R.H.T. Edwards, R.E. Gordon & D. Shaw. ^{31}P NMR Studies of resting muscle in normal human subjects. In G. Pollack & H. Sugi (Eds.), Cross-bridge Mechanisms in Muscle Contraction, proceedings of the 2nd International Symposium, Seattle, WA, June 1982. (In Press), 1983.

Effect of Acidosis on Maximum Force Generation of Peeled Mammalian Skeletal Muscle Fibers

Sue K.B. Donaldson

Rush University, Chicago, Illinois, U.S.A.

Intracellular acidosis is one of the factors associated with decline of maximum force generation, or fatigue, of skeletal muscle fibers (Hermansen, 1981). Decrease in force generation was found by Dawson and co-workers (1978) to be linearly related to the increase in intracellular H^+ and free ADP concentrations during direct stimulation of unpoisoned frog muscles. In contrast, ATP levels were not significantly lowered during the development of fatigue. Given that $MgATP^{2-}$ is the fuel for actomyosin force generation, the finding that fatigue occurs without depletion of ATP makes the observed correlation between depletion of creatine phosphate and decline of force generation (Spande & Schotelius, 1970) difficult to interpret. Furthermore, force is not always proportional to phosphocreatine level (Dawson et al., 1978) during fatigue, making depletion of fuel an unlikely causative or contributing factor in the decline of force generation. The loss of force generation during fatigue, however, is associated with a decrease in ATP utilization, suggesting a decrease in actomyosin cross-bridge cycling (Crow & Kushmerick, 1982; Dawson et al., 1978; Edwards, 1981). Failure of Ca^{2+}-activation (Fabiato & Fabiato, 1978) or product inhibition of the actomyosin ATPase by H^+, ADP, or Pi (Dawson et al., 1978; Trentham et al., 1976) might contribute to a decline in cross-bridge cycling and force generation.

   The purpose of this paper is to review some recent results related to the effects of acidosis on force generation of the functionally isolated skeletal muscle contractile apparatus. These data indicate that even if acidosis is not a causative factor in fatigue, it may contribute to the difference in fatigability among the various mammalian skeletal fiber types.

Mammalian Peeled Fibers

The contractile apparatus force generation can be studied directly by using mammalian fibers which have been skinned by peeling off of their sarcolemma. Natori (1954) first introduced this important preparation using frog fibers; Dr. Hermansen and I modified the technique for mammalian skeletal fibers (Donaldson & Hermansen, 1978.) I will refer to these fibers as "peeled" rather than the more general term "skinned" to avoid confusion with other methods of disrupting, rather than removing, the sarcolemma.

Although peeled fibers have the disadvantage of losing all of their soluble constituents and they cannot be activated via physiological mechanisms, the peeled fibers do allow precise control of intracellular environment for the study of Ca^{2+}-activated isometric force generation under known conditions.

The major bathing solution variables in our studies were Ca^{2+} concentration and pH; a variety of saturating and subsaturating Ca^{2+} levels was used at both pH 7.0 and 6.5, since these pHs are known to be achieved intracellularly in humans at rest and during vigorous exercise to exhaustion, respectively (Hermansen & Osnes, 1972; Sahlin, 1978.) Thus, using the peeled fiber preparation, the effect of acidosis on contractile protein Ca^{2+}-activated force generation was studied as an isolated variable.

Fibers were taken from rabbit soleus, adductor magnus, and tibialis muscles. Typical isometric force records for the various rabbit fiber types are shown in Figure 1. These mammalian peeled fibers are quite resilient and allow repeated contractures and relaxations according to variations in bathing solution Ca^{2+} concentration without loss of force generating ability. In each of the traces, contracture is activated at standard subsaturating and saturating Ca^{2+} levels, alternating between pH 6.5 and 7.0. The steady isometric force levels at pH 7.0 illustrate that the slow-twitch (SO, type I oxidative) fibers are the most sensitive to Ca^{2+} at pH 7.0 and are not greatly affected by acidosis at any Ca^{2+} level. An interesting result is that the fast twitch oxidative (FOG, type II oxidative) fibers are quite distinct in force generating characteristics from the fast twitch glycolytic (FG, type II nonoxidative) fibers. Fast-twitch glycolytic fibers are the least sensitive to Ca^{2+} at pH 7.0 and their force generation at subsaturating and saturating Ca^{2+} levels is greatly depressed by acidosis. The fast-twitch oxidative fibers have intermediate force generating properties relative to the other fiber types at both pHs. All effects of acidosis were completely reversible (Donaldson, 1982). Dr. Hermansen and I have obtained similar data using peeled human fibers obtained via needle biopsy of vastus lateralis (unpublished).

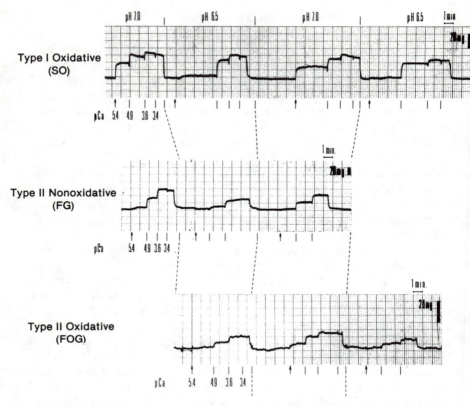

Figure 1—Force records from three peeled fibers of the same animal, identified as to histochemical type. (Stained pieces of the same fibers are shown in Figure 2.) For each trace, ordinate is isometric force and abscissa is time. Changes in bathing solution pH are indicated at the top of the figure with portions of the record at identical pH marked off with dashed lines. The first four time marks below each force trace, beginning with the arrow, signal the indicated changes in bathing solution pCa. This pattern of Ca changes is repeated (starting at each arrow) in each trace at alternating pH's.

Each mammalian fiber was identified histochemically from the pattern of staining of its pieces for myofibrillar ATPase and NADH tetrazolium reductase. Stained pieces of the fibers illustrated in Figure 1 are shown in Figure 2. Type I, slow-twitch, fibers stained darkest for myofibrillar ATPase following acid preincubation; whereas, type II, fast-twitch, fibers stained most darkly for myofibrillar ATPase following alkaline preincubation. Type II fibers were further subcategorized as to oxidative or nonoxidative (glycolytic) according to the NADH tetrazolium reductase stain (Donaldson, 1982).

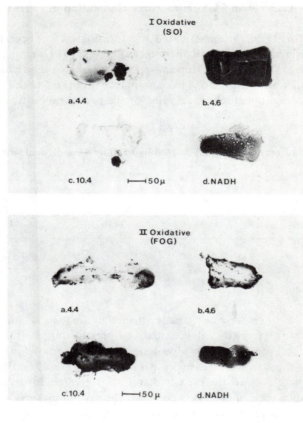

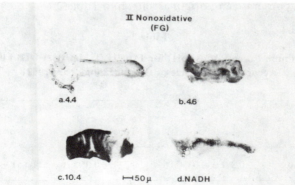

Figure 2—Collages of histochemically stained pieces of the fibers from Figure 1. The histochemical stains for the pieces of each fiber are a), b) and c): myofibrillar ATPase (pH 9.4) following pre-incubation at the designated pH's and d): NADH-tetrazolium reductase.

Effect of Acidosis on Maximum Force Generation of Peeled Fibers

Maximum force generation activated at saturating Ca^{2+} concentrations is depressed by acidosis in frog (Robertson & Kerrick, 1979) and mammalian skeletal fibers (Donaldson & Hermansen, 1978; Fabiato & Fabiato, 1977). As shown in Table 1, the ratio of maximum force at pH 6.5 to that at pH 7.0 varies according to histochemical type for the peeled rabbit fibers.

Table 1

Maximum Force Ratios (pH) According to Fiber Type (Donaldson, 1982)

Maximum force ratio	Type I oxidative (SO)	Type II oxidative (FOG)	Type II non-oxidative (FG)
pH 6.5/pH 7.0	0.880 ± 0.013(47)*	0.745 ± 0.030(18)	0.659 ± 0.013(39)

*Mean ± SEM(n)

Slow-twitch fibers were not greatly affected by acidosis, losing only 12% of maximum force. Fast twitch oxidative fibers lost approximately 25% maximum force and fast twitch glycolytic fibers declined 44% in maximum force with the same decrease in pH. The SO and FG fibers were indistinguishable as to their particular Ca^{2+}-activated force generation regardless of muscle of origin. In contrast, the maximum force generating characteristics of the FOG fibers although intermediate to the other types, depended upon muscle of origin as shown in Table 2.

Table 2

Maximum Force Ratios (pH) for Type II Oxidative (FOG) Fibers According to Muscle of Origin (Donaldson, 1982)

Maximum force ratio	Soleus	Tibialis anterior	Adductor magnus
pH 6.5/7.0	0.881 ± 0.035(5)*	0.778 ± 0.050(4)	0.651 ± 0.031(7)

*Mean ± SEM(n)

Type II oxidative fibers from soleus showed the acid resistance typical of the slow-twitch fibers that are most characteristic of this muscle. Similarly, type II oxidative fibers derived from adductor magnus muscles displayed the extreme sensitivity to pH characteristic of FG fibers. Fast-twitch oxidative fibers from tibialis anterior, a mixed fiber type muscle, were overall intermediate in terms of acidotic depression of maximum force.

These data are particularly relevant to the mechanism of fatigue in two respects. First, the relative sensitivity of the various fiber types to acidotic depression of force generation correlates with their relative fatigability; (Burke et al., 1973; Close, 1972). Second, these data demonstrate that the major histochemical fiber types constitute functionally distinct categories. The contractile protein basis of these functional differences is not known (Donaldson, 1982).

Mechanisms of Acidotic Depression of Maximum Force Generation

In considering the mechanism of acidotic depression of force generation, it is perhaps easiest to eliminate from consideration certain mechanisms of action of H^+ on the contractile apparatus. The acidotic depression of maximum force generation cannot be explained as a competition of H^+ for activating Ca^{2+} sites, since saturating Ca^+ concentrations were used. The decrease in Ca^{2+} sensitivity, evidenced at subsaturating Ca^{2+} concentrations, however, may be due to H^+ competition with Ca^{2+} and would be manifested as additional loss of force if Ca^{2+} release were impaired in fatigue (Donaldson & Hermansen, 1978; Fabiato & Fabiato, 1977). The effect of H^+ is also not likely to be due to shielding of negative charges on the thin filaments with concomitant alteration of lattice distances, since Dr. Hermansen and I (1978) found that effect of acidosis on maximum force generation was essentially the same at both 1mM and 10mM free Mg^{2+} concentrations. Furthermore, the change in Mg^{2+} concentration itself did not alter maximum force generation at pH 7.0. This substantial change in a divalent cation concentration should have altered thin filament negative charges and thus modified any charge shielding by H^+.

Nor is the effect of H^+ via a decrease in $MgATP^{2-}$ concentration, since we solved the complex equilibrium for each bathing solution taking into account all known ionic complexes. In our studies $MgATP^{2-} = 2\,mM$, CP = 15 mM and CPK = 15 Units/ml at each pH, Mg^{2+} and Ca^{2+} level.

The most likely explanation for the acidotic depression of maximum force of the peeled fibers is an alteration at the level of actomyosin cross-bridges. Acidosis might alter the energy available for work per ATP hydrolyzed and thus affect force (Dawson et al., 1978); but the magnitude and direction of this effect cannot be calculated for our fibers, since we did not measure concentrations of phosphocreatine, creatine, Pi and pH within the fibers during contracture. Change in energy available for work per ATP hydrolyzed due to acidosis per se should affect all fibers equally, however, and thus would not explain the variation in acidotic depression among the fiber types.

The most likely explanation is that the increase in H^+ concentration causes a product inhibition of the actomyosin ATPase, thus lowering force generation (Trentham et al., 1976.) This would also be expected for increased free ADP and Pi concentrations, although we have not yet tested the effect of these variables on peeled fiber force generation. The linear relationship of decline of force and increase in free ADP and H^+ concentrations observed by Dawson and coworkers (1978) in fatiguing frog fibers and the decline in ATP utilization that occurs during fatigue (Crow & Kushmerick, 1982; Dawson et al., 1978; Edwards, 1981) are consistent with product inhibition of myofibrillar ATPase. The differences in the contractile apparatus behavior we observed for the various mammalian fiber types may very well be due in part to differences in degree of product inhibition at the cross-bridge level.

Summary

Acidotic depression of contractile force generation should contribute to skeletal fiber fatigue even if it is not a primary causative factor. Data from peeled fibers demonstrate that mammalian skeletal fibers display type-specific acidotic depression of maximum force generation that correlates with their variable fatigability.

Acknowledgments

This research was supported by grants from USPHS, NIH (HL 23128) and Muscular Dystrophy Association. The assistance of D. Huetteman and S. Bickham was invaluable.

References

BURKE, R., D. Levine, P. Tsairis & F. Zajac. Physiological types and histochemical profiles in motor units of the cat gastrocnemius. J. Physiol. (Lond.) 234:723–748, 1973.

CLOSE, R., Dynamic properties of mammalian muscles. Physiol. Rev. 52:129–197, 1972.

CROW, M. & M.J. Kushmerick. Chemical energetics of slow- and fast-twitch muscle of the mouse. J. Gen. Physiol. 79:147–166, 1982.

DAWSON, M.J., D.G. Gadian & D.R. Wilkie. Muscular fatigue investigated by phosphorus nuclear magnetic resonance. Nature 274:861–866, 1978.

DONALDSON, S.K.B., Rabbit peeled skeletal muscle fibres: Ca-activated force generating properties according to histochemical type. J. Physiol. (Lond.) (under review), 1982.

DONALDSON, S.K.B. & L. Hermansen. Differential direct effects of H^+ on Ca^{2+}-activated force of skinned fibers from the soleus, cardiac, adductor magnus muscles of rabbits. Pflugers Arch. 376:55–65, 1978.

EDWARDS, R.H.T., Human muscle function and fatigue. In R. Porter & J. Whelan (Eds.), Human Muscle Fatigue: Physiological Mechanisms. London: Pitman Medical Ltd., 1981.

FABIATO, A. & F. Fabiato. Effects of pH on the myofilaments and the sarcoplasmic reticulum of skinned cells from cardiac and skeletal muscles. J. Physiol. 276:233–235, 1978.

HERMANSEN, L., Effect of metabolic changes on force generation in skeletal muscle during contraction. In R. Porter & J. Whelan (Eds.,), Human Muscle Fatigue: Physiological Mechanisms. London: Pitman Medical Ltd., 1981.

HERMANSEN, L. & J.B. Osnes. Blood and muscle pH after maximal exercise. J. Appl. Physiol. 32:304–308, 1972.

NATORI, R., The property and contraction process of isolated myofibrils. Jikeikai. Med. J. 1:119, 1954.

ROBERTSON, S. & W.G.L. Kerrick. The effects of pH on Ca^{2+}-activated force in frog skeletal muscle fibers. Pflugers Arch. 380:41–45, 1979.

SAHLIN, K., Intracellular pH and energy metabolism in skeletal muscle of man. Acta. Physiol. Scand. Suppl. 455:1–56, 1978.

SPANDE, J.I. & B.A. Schottelius. Chemical basis of fatigue in isolated mouse soleus fibers. Am. J. Physiol. 219:1490–1495, 1970.

TRENTHAM, D.R., J.F. Eccleston & C.R. Bagshaw. Kinetic analysis of ATPase Mechanisms. Quart. Rev. Biophys. 9:217–281, 1976.

Effects of Temperature on Muscle Metabolism

Ethan R. Nadel
John B. Pierce Foundation Laboratory and
Yale University School of Medicine,
New Haven, Connecticut, U.S.A.

The effects of temperature on muscle metabolic processes are diverse and seemingly paradoxical, depending upon the experimental approach and the variable of interest. For instance, Asmussen and Böje (1945) demonstrated that humans had an "improved capacity" for dynamic exercise as a result of prior heating of muscle. In a similar vein, Bergh and Ekblom (1979) found that peak oxygen uptake in humans performing dynamic exercise fell linearly as muscle temperature was reduced by prior cooling. It is well established, however, that whole body heating reduces maximal aerobic power (Rowell, 1974) and the duration of an isometric contraction sustained to fatigue is greater when muscle temperature is 27 to 32 ° C than when higher (Clarke et al., 1958; Edwards et al., 1972). The purpose of the present paper is first to describe the pattern of thermal energy exchange within the muscle during dynamic exercise, then to discuss whether muscle temperature can limit dynamic exercise and finally to consider the mechanisms by which whole body heating affects muscle metabolism.

Determinants of Skeletal Muscle Temperature

In simplest analysis, the temperature of any tissue is a function of its rate of energy production and the rate of blood flow through that tissue. Resting skeletal muscle has a relatively low metabolic rate with respect to many other body tissues, averaging around 2 ml O_2/min/kg of muscle, and is also relatively underperfused, with blood flow averaging around 25 ml/min/kg. Because of the constancy of both metabolic rate and blood flow,

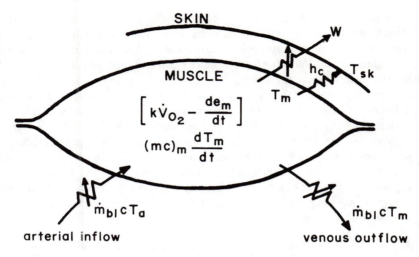

Figure 1—A schematic diagram of the energy flows in muscle. See text for definition of terms. Modified from Mitchell (1977).

temperature in most muscle groups at rest tends to be stable and predictable, between 33° and 35°C.

A more detailed and quantitative examination of the energy exchange in muscle is shown in Figure 1, as modified from Mitchell (1977). In steady state conditions of rest (or of exercise), the change of chemical energy stores (e_m) and of muscle temperature (T_m) with time are by definition set to zero and the muscle temperature is a function of the energy in and energy out, as solved by the following relation:

$$\dot{m}_{bl}c(T_a - T_m) + k\dot{V}_{O_2} = h_c(T_m - T_{sk}) \pm W \tag{1}$$

where

$$\dot{m}_{bl} = \text{mass flow rate of blood in ml/min}$$

$$c = \text{specific heat of blood in kJ/ml/°C}$$

$$T_a = \text{arterial blood temperature in °C}$$

$$T_m = \text{muscle temperature in °C}$$

$$\dot{V}_{O_2} = \text{oxygen uptake in ml } O_2/\text{min}$$

$$k = \text{energy equivalent in kJ/ml } O_2$$

$$h_c = \text{conductive energy transfer coefficient in kJ/min/°C}$$

$$T_{sk} = \text{skin temperature in °C}$$

In healthy people performing dynamic exercise involving concentric muscle contractions, muscle metabolic rate can increase around 100 times more than during rest and this elevated metabolism can be sustained for extended periods. Nearly all of the increase in metabolism is accomplished by oxidation of fatty acids and glucose and requires the appropriate organ system adjustments that provide for maintenance of fuel and oxygen delivery. Conversion of one form of energy to another in a biological system releases a certain amount of heat and ultimately all of the energy appears as heat. Thus, at the onset of dynamic muscular exercise, the rate of thermal energy (heat) production is accordingly elevated to a level proportional to the intensity of exercise. In this case, the term $k\dot{V}O_2$ in Equation 1 becomes large, the term representing the storage of thermal energy is no longer equal to zero and the energy balance equation for muscle must be rewritten as follows:

$$\dot{m}_{b1}c(T_a - T_m) + [k\dot{V}O_2 - \frac{de_m}{dt}] = h_c(T_m - T_{sk}) \pm W + (mc)_m \frac{dT_m}{dt} \quad (2)$$

where

$(mc)_m =$ the mass specific heat product for muscle

$\dfrac{dT_m}{dt} =$ the change in muscle temperature with time (thermal storage)

$\dfrac{de_m}{dt} =$ the change chemical energy equivalents with time (this term is small).

During the initial transient of exercise, the thermal storage term is proportional to the metabolic rate term and therefore it is evident that the muscle temperature increases as a function of the exercise intensity. This is the case because, during the initial transient, the new rate of heat production greatly exceeds the combined rates of heat loss. However, during dynamic exercise several factors occur that promote increases in the rate of heat dissipation from muscle, thereby progressively attenuating the rate of rise in muscle temperature until, eventually, a new steady state is reached. As muscle temperature increases, the temperature gradient from muscle to surrounding tissue also increases and heat flow by conduction is accordingly enhanced. Similarly, storage of thermal energy in muscle reverses the temperature gradient between arterial blood, which enters muscle at body core temperature, and muscle. Heat is then transferred from muscle to blood during the period of its residence in muscle capillaries. Most importantly, however, the rate of muscle blood flow is

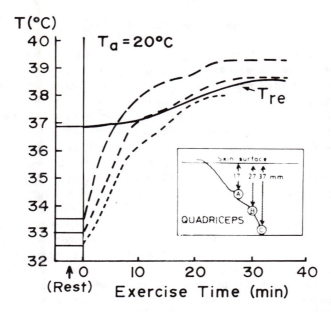

Figure 2—Pattern of temperature change in quadriceps muscle at three depths during cycle ergometer exercise at 76% of maximal aerobic power. Modified from Saltin et al. (1968).

able to increase at least 30-fold above resting conditions to around 750 ml/min/kg of muscle during heavy exercise. Since the rate of heat transfer is the product of the blood flow rate, specific heat and temperature gradient, it is easily seen that the marked elevation in blood flow to muscle during dynamic exercise is the primary factor accounting for the dissipation of the heat produced. Thus, most of the heat produced in muscle during dynamic exercise is carried away from the muscle in blood by convection, is distributed throughout the body, which serves as a heat sink, and is ultimately dissipated to the environment by physical processes as mandated by the body's temperature regulatory system. Since the rate of muscle blood flow is to a great extent determined by the muscle metabolic rate during dynamic exercise, at least up to moderately high intensities (c.f. Mitchell et al., 1972), the rates of heat production in muscle and heat loss from muscle become relatively well matched after a period of adjustment and a new, elevated steady state muscle temperature is achieved for the duration of the exercise period.

Figure 2 illustrates the pattern of muscle and esophageal temperature changes during 35 min of cycle ergometer exercise at 76% of maximal aerobic power. These data support the theoretical description given in the previous paragraphs.

Muscle Temperature and Fatigue

One approach to the question of whether high muscle temperatures can limit dynamic exercise has involved the study of subjects performing extended bouts of eccentric exercise, in which the muscles resist elongation due to external force, resulting in the addition of heat to the body above the metabolic heat rather than the converse, as occurs with concentric exercise. Because of the extra input of heat to the muscles during eccentric contractions, and the relatively low metabolic rate (Knuttgen et al., 1971) and therefore low muscle blood flow (Bonde-Petersen et al., 1970) associated with eccentric exercise, it would follow that at any rate of total energy production muscle temperature should be higher during eccentric than during concentric exercise. This is indeed the case (Nadel et al., 1972). Figure 3 illustrates the steady state muscle, esophageal and mean skin temperatures measured between 30 and 40 min of eccentric and concentric cycle ergometer exercise production. At any rate of total energy production, H (equal to metabolic energy production, M, $\pm$ work, W), the muscle temperature was greater by about 1.2° C during eccentric exercise than during concentric exercise. At the highest exercise intensity neither of the two subjects who began the bout were able to complete the 40 min, one stopping at 35 min, when $T_m = 41.1°$ C. Muscle temperatures exceeding 40° C were commonly observed.

The inability of muscle temperature to rise above 40° C during concentric exercise at high intensities has caused the suggestion that a muscle temperature of 40° C may be among the factors limiting exercise. However, the ability of subjects to continue exercise with eccentric contractions when muscle temperatures exceeded 40.5° C indicates that 40° C is not limiting. It is not clear that muscle temperature becomes limiting above 41° C, although muscle biopsy samples have shown that even at such heavy intensities there is little indication of anaerobiosis (Bonde-Petersen et al, 1972).

A novel approach to the question of whether high muscle temperature could limit exercise was recently attempted by Knuttgen et al. (1982). Recognizing that naive subjects cannot prolong heavy exercise with eccentric contractions for extended periods, but following a period of training the same subjects can continue exercise at the same intensity seemingly indefinitely, they asked the question whether the visco-elastic properties of muscle might be reduced by high muscle temperature and whether training was associated with lower muscle temperatures and an increased ability to resist fatigue. They found, however, that training for five wk, while allowing the subjects a greatly increased capacity to tolerate eccentric exercise at heavy intensities, did not result in any significant change in quadriceps muscle temperature. They concluded that the fatigue

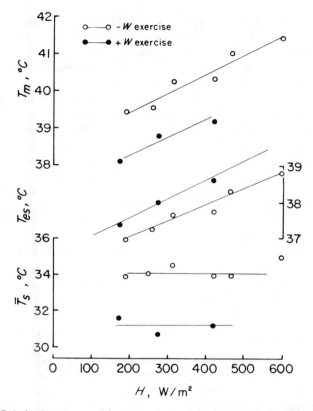

Figure 3—Relation between quadriceps muscle, esophageal and mean skin temperatures and rate of total heat production (H) after 40 min of eccentric (-W) and concentric (+) cycle ergometer exercise. From Nadel et al. (1972) with permission of the American Physiological Society.

in the naive condition, leading to a voluntary cessation of exercise, could not be attributed to the high muscle temperatures that developed during exercise with eccentric muscle contractions. Thus, here again there is no evidence that muscle temperature around 40° C in and of itself contributes the fatigue state.

On the other hand the attainment of a sufficiently high muscle temperature may be essential for achieving a maximal aerobic exchange within the muscle. Bergh and Ekblom (1979) studied eight well trained subjects in conditions in which they had undergone various degrees of whole body cooling induced by swimming in cold (13 to 15° C) water. Following pre-cooling, subjects performed cycle ergometer exercise to exhaustion. The peak attainable oxygen uptake fell linearly with decreasing body temperatures below a quadriceps muscle temperature of 38.6° C and an esophageal temperature of 37.7° C. The amount of decrement was on the order of 5 to 6% of peak oxygen uptake per degree for both muscle and body core temperatures. While these data underscore the importance of the attainment of sufficiently high body temperatures in order to maximize the efficiency of the oxygen transport and uptake systems and minimize the onset of fatigue (time to exhaustion was also progressively shorter below the critical body temperatures), they do not allow the conclusion that muscle temperature is the primary limiting factor. Maximal attainable heart rate is a function of body core temperature, so it could well be that the peak oxygen uptake by muscle is not limited by any reduction in temperature dependent muscle energetics but rather by a reduction in oxygen delivery rate due to a reduced cardiac output. Because muscle temperature and body core temperature are linked by the convective properties of the bloodstream, it would be difficult to separate the two to determine which of these is the preponderant factor. While it is clear that the optimal muscle temperature for force development over short periods (up to 80 s) is less than 32° C (Clarke et al., 1958; Edwards et al., 1972), the optimal muscle temperature for power output, oxygen supply notwithstanding, is presently not known.

Impairment of Oxygen Delivery with Body Heating

The body is able to employ several strategies to resist muscle anaerobiosis in conditions of simultaneous elevations in metabolic and thermoregulatory demands, such as occur during exercise in the heat. The problem is that the heart must deliver adequate blood flow to both muscle and skin in the face of reductions in cardiac filling due to filtration of fluid out of the vascular volume, increases in cutaneous venous volume accompanying the increased flow and the progressive decrease in the total body fluid content associated with the high evaporative rate. If the combined demand is not excessive, e.g., during mild exercise in the heat or heavier exercise in cooler conditions, the heart is able to increase its output to meet the needs of both muscle and skin (Nadel et al., 1979). During moderately heavy exercise in the heat, the body calls upon three reflexes that serve to maintain adequate cardiac filling and blood flow to contracting muscle groups. The first of these is a relative cutaneous vasoconstriction superimposed upon the high vasodilator drive called for by the thermoregulatory mechanism (Nadel et al., 1979). A second reflex is a certain increase in cutaneous venomotor tone at the onset of exercise and an increasing degree of tone as exercise progresses (Nadel et al., 1980a). Finally, as blood volume becomes diminished there is an inhibition of thermoregulatory activity manifested as an upward shift in the internal temperature threshold for cutaneous vasodilation (Nadel et al., 1980b). All of these reflexes have the consequence of reducing heat transfer from core to skin, thereby insuring blood flow and therefore oxygen delivery to muscle at the expense of the thermoregulatory system.

We recently attempted to place a greater burden on the oxygen transport systems than in the previous studies by asking four subjects to exercise to exhaustion in the heat (35° C) at intensities designed to produce 80 and 100% of maximal aerobic power (Brown et al., 1982). Although subjects reached the same oxygen uptake in the heat as they did in cool conditions (20° C), they were incapable of exercising for as long a duration in the heat. At 80% of maximal aerobic power the mean time to exhaustion was only 10.5 ± 1.4 min at 35° C, compared with 18.7 ± 3.5 min at 20° C. At 100% of maximal aerobic power, the pattern was identical, with subjects becoming exhausted at 3.3 ± 0.2 min at 35° and at 4.7 ± 0.7 min at 20° C.

Arterialized blood samples were taken at frequent intervals during exercise and recovery and analyzed for a number of variables, including lactic acid as an index of anaerobiosis. In the heat lactate appearance time was shorter than in the cooler condition at both intensities of exercise. Furthermore, circulating lactate was greater in the heat at any given time of exercise (Figure 4). Peak lactate values were practically identical in the four conditions, implying that a given level of muscle anaerobiosis was associated with an inability to continue exercise. The more rapid appearance and build-up of lactate in blood during heavy exercise in the heat indicated to us that the rate of oxygen delivery (hence, blood flow) to muscle was compromised with respect to that in a cooler environment. This is presumably attributable to the effects of the lower central circulating blood volume that accompany heat exposure. Thus, while able to be compensated for during moderate exercise by reflexes described in previous paragraphs, the effects of high ambient temperature on the distribution of the blood volume cannot be compensated for during heavy exercise and oxygen delivery to muscle is compromised. Presumably, although only inferred from our previous studies and the data of Rowell (1974), the compromise results from an inability to achieve the same high cardiac output that is possible in cooler conditions, this the result of a falling cardiac stroke volume that cannot be compensated for by a heart rate that is already near-maximal.

In conclusion, the effects of high ambient temperature and of low muscle temperature appear to be the most important in the induction of fatigue during dynamic exercise. The inherent difficulties associated with teasing apart experimentally the body temperatures from one another in an intact human, in order to investigate the effect of muscle temperature per se on muscle energetics in situ, make this a particularly challenging area of study.

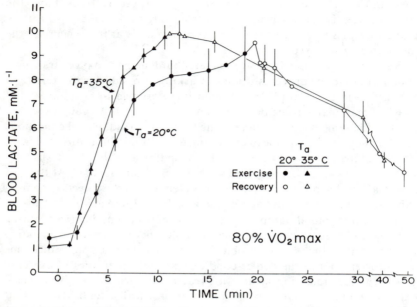

Figure 4—Pattern of arterialized blood lactate change during cycle ergometer exercise (80% maximal aerobic power) at 20° and 35°. Data are means ± SE of four subjects.

Acknowledgment

Studies reported herein were supported by N.I.H. Grants HL-20634 and HL-17732.

References

ASMUSSEN, E. & O. Bøje. Body temperature and capacity for work. Acta Physiol. Scand. 10:1–22, 1945.

BERGH, U. & B. Ekblom. Physical performance and peak aerobic power at different body temperatures. J. Appl. Physiol. 46:885–889, 1979.

BONDE-PETERSEN, F., H.G. Knuttgen & J. Henriksson. Muscle metabolism during exercise with concentric and eccentric contractions. J. Appl. Physiol. 33:792–795, 1972.

BONDE-PETERSEN, F., B. Nielsen, S. Levin Nielsen & L. Vanggård. [133]Xe clearance from musculus quadriceps femoris during concentric and eccentric bicycle exercise at different temperatures and loads. Acta Physiol. Scand. 79:10A, 1970.

BROWN, N.J., L.A. Stephenson, G. Lister & E.R. Nadel. Relative anaerobiosis during heavy exercise in the heat. Fed. Proc. 41:1677, 1982.

CLARKE, R.S.J., R.F. Hellon & A.R. Lind. The duration of sustained contractions of the human forearm at different muscle temperatures. J. Physiol. 143:454–473, 1958.

EDWARDS, R.H.T., R.C. Harris, E. Hultman, L. Kaijser, D. Koh & L.-O. Nordesjö. Effect of temperature on muscle energy metabolism and endurance during successive isometric contractions, sustained to fatigue, of the quadriceps muscle in man. J. Physiol. 220:335–352, 1972.

KNUTTGEN, H.G., F. Bonde-Petersen & K. Klausen. Oxygen uptake and heart rate responses to exercise performed with concentric and eccentric muscle contractions. Med. Sci. Sports 3:1–5, 1971.

KNUTTGEN, H.G., E.R. Nadel, K.B. Pandolf & J.F. Patton. Effects of training with eccentric muscle contractions on exercise performance, energy expenditure and body temperature. Int. J. Sports Med. 1982 (in press).

MITCHELL, J.W., Energy exchanges during exercise. In E.R. Nadel (Ed.), Problems with Temperature Regulation during Exercise. New York: Academic Press, 1977.

MITCHELL, J.W., J.A.J. Stolwijk & E.R. Nadel. Model simulation of blood flow and oxygen uptake during exercise. Biophys. J. 12:1452–1466, 1972.

NADEL, E.R., U. Bergh & B. Saltin. Body temperatures during negative work exercise. J. Appl. Physiol. 33:553–558, 1972.

NADEL, E.R., E. Cafarelli, M.F. Roberts & C.B. Wenger. Circulatory regulation during exercise in different ambient temperatures. J. Appl. Physiol. 46:430–437, 1979.

NADEL, E.R., S.M. Fortney & C.B. Wenger. Circulatory adjustments during heat stress. In P. Cerretelli & B.J. Whipp (Eds.), Exercise Bioenergetics and Gas Exchange. New York: Elsevier/North Holland, 1980a.

NADEL, E.R., S.M. Fortney & C.B. Wenger. Effect of hydration state on circulatory and thermal regulations. J. Appl. Physiol. 49:715–721, 1980b.

ROWELL, L.B., Human cardiovascular adjustments to exercise and heat stress. Physiol. Rev. 54:75–159, 1974.

SALTIN, B., A.P. Gagge & J.A.J. Stowlijk. Muscle temperature during submaximal temperature in man. J. Appl. Physiol. 25:679–688, 1968.

Control of Metabolism and the Integration of Fuel Supply for the Marathon Runner

Eric A. Newsholme

University of Oxford, Oxford, England

The distance of the marathon race is 26 mi, 385 yd (42.2 km) which is completed by top runners in about 130 min with a total energy expenditure of almost 12,000 kJ. Many non-elite runners are able to complete the distance in about 180 min with a similar total energy expenditure. One of the problems in completing the distance in such times is the provision of sufficient fuel to satisfy the rate of energy expenditure. The fuel reserves of the body are stored in two forms, triacylglycerol and glycogen, which are oxidized to release the energy that is used to power the contraction of the muscles. Although physiological factors such as dehydration and hyperthermia may limit performance in a marathon run, there is increasing evidence that provision of fuel for the muscle is a major limiting factor and that satisfactory control and selection of fuels for oxidation by the muscle is of considerable importance in delaying the onset of fatigue.

Fuels and Energy Demand by the Marathon Runner

Triacylglycerol is stored in adipose tissue, which is distributed diffusely throughout the body (for example, under the skin, around the major organs, in the abdominal cavity). The amount of triacylglycerol stored in the body is very large and is sufficient to last even the top class runner for 5 days and 5 nights of marathon running (Table 1). This energy store is released into the bloodstream in the form of long chain fatty acids (e.g., palmitic, oleic and stearic acids). Since these fatty acids have a low solubility in the aqueous medium of the blood, they have to be transported in association with the plasma protein, albumin. This protein possesses two high affinity binding sites for fatty acids and, at physiological concentra-

Table 1

Fuel Reserves and Rates of Utilization under Different Conditions in Humans

Tissue of store	Approximate total fuel reserve		Estimated period for which fuel store would provide energy		
	g	kj	Days of starvation	Days of walking	minutes of marathon running
Adipose tissue triacylglycerol	16 000	600 000	60	19.2	7 143
Liver glycogen	90	1 500	0.15	0.05	18
Muscle glycogen	350	6 000	0.6	0.20	71
Blood and extra	20	320	0.03	0.01	4

Note: For references and calculations, see Newsholme & Leech (1983).

tions of albumin, saturation of these sites produces an upper limit of about 2 mM fatty acids (Spector & Fletcher, 1978). Nonetheless, it is the *free* concentration of fatty acids in the blood, which is probably only 0.1% of the total concentration, that governs the rate of diffusion into the muscle. This low concentration means that only a small proportion of the fatty acid passing through the muscle is extracted and this may represent a major limitation in the ability of muscle to oxidize fatty acids and this poses a problem for the marathon runner. In other words, it is suggested that the *rate* of uptake of fatty acid from the blood into the muscle cannot provide sufficient fuel to meet the energy *demands* of the marathon runner, even if all this fatty acid is oxidized. There are five lines of evidence in support of this view:

1. If the carbohydrate stores of the body are depleted (e.g., by restriction of dietary carbohydrate intake for 2 to 3 days) a given level of exercise produces exhaustion more quickly than for subjects on a normal diet and especially in comparison to those on a high carbohydrate diet (Christensen & Hansen, 1939).

2. If the carbohydrate store in the muscle is increased prior to exercise (by eating a diet high in carbohydrate) a given level of exercise can be maintained for a longer period of time (Berström et al., 1967; Costill & Miller, 1980).

3. If the fatty acid concentration in the blood is artificially elevated prior to exercise, a given level of exercise can be maintained for a longer period of time (Costill et al., 1977; Hickson et al., 1977). This manipulation ensures that hydrolysis of some of the stores of triacylglycerol occurs prior to the exercise so that the blood fatty acid concentration is elevated and fatty acids are available for oxidation as soon as the exercise begins rather than after 30 or more min later.

4. In elite ultradistance runners, running for 24 h, the power output declined towards 50% of $\dot{V}O_2$max during the race while the respiratory exchange ratio decreased to about 0.7 indicating mainly oxidation of fatty acids (Davies & Thompson, 1979). It is unlikely that carbohydrate stores could be obtained from the oxidation of fatty acids by the muscle.

5. The $\dot{V}O_2$max of a patient suffering from a deficiency of 6-phosphofructokinase in muscle has been measured (P. Cerretelli, personal communication). Since these patients cannot carry out glycolysis in muscle, fatty acids must be the major fuel for this tissue during exercise. Dr. Cerretelli has observed that the $\dot{V}O_2$max is only about 60% of what he would expect from other physiological characteristics of the patient.

These findings suggest that a very high power output (e.g., > 60 to 70% $\dot{V}O_2$max) cannot be maintained solely from fatty acid oxidation. If the limitation is the rate of diffusion from the bloodstream into the muscle the question arises why this rate could not be increased by raising the blood fatty acid concentration above 2 mM (i.e., beyond the capacity of the high-affinity binding sites on albumin). Undoubtedly when total concentration does increase above 2 mM the free concentration increases markedly and this would be expected to increase the rate of diffusion markedly. This is unlikely to occur, however, since a high concentration of free fatty acids produces a micellar solution which can be dangerous; for example, micellar solutions can damage cell membranes, increase the stickiness of platelets (which can result in thrombus formation) and interfere in electrical conductivity in the heart (which can result in arrthymias) (see Cowan & Vaughan-Williams, 1977; Newsholme & Leech, 1983). That 2 mM fatty acids in the blood may be an upper safe limit is supported by reports of the blood concentrations of fatty acids under various extreme conditions (Table 2). In no situation does the concentration exceed 2 mM.

In order to maintain a high power output during the marathon run (and elite runners can maintain power output greater than 85% of their $\dot{V}O_2$max for the duration of the race, Costill, 1979), both carbohydrate and fatty acids must be oxidized by the muscle. Glycogen is stored in the liver, from where it is made available to muscle via the bloodstream as glucose, and in muscle where it is broken down and used directly via the process of glycolysis. Unlike fatty acids, however, the rate of glucose utilization and oxidation by muscle may be able to provide sufficient energy to satisfy the

Table 2

Plasma Fatty Acid Concentrations in Humans under Different Conditions

Conditions	Plasma fatty acid concentration (mM)
Normal fed	0.30–0.60
Stress (racing driver)	1.72
Starvation (8 days)	1.88
Prolonged exercise (240 min)	1.83
Immediate pre-surgery	0.90
Three hour post-surgery	0.80
Diabetic coma	1.60
Glycogen storage disease (Type III) (after 12 h starvation)	1.90
Severely burned patients	1.30
Severely injured patients	0.75

For references and further details see Newsholme & Leech (1983).

energy demands of the marathon runner in the absence of other fuels. Two lines of evidence support this view. First, the respiratory exchange ratio during the early stages of running in man when the power output is very high, can be close to unity, indicating oxidation of carbohydrate. Secondly, the activity of hexokinase in muscle of humans is sufficient, on the basis of the energy that would be released due to complete oxidation of the glucose 6-phosphate, to account for the energy demands of the marathon runner and to account for the oxygen consumption during a marathon run (Newsholme & Leech, 1973). Nonetheless, there is a limitation in the use of carbohydrate as a fuel for the marathon runner and this is the small reserve of carbohydrate in the body. The total glycogen in the liver and that in the muscles which will be used during running cannot last for longer than about 90 min of marathon running (at an energy expenditure of 84 kJ each min—Table I). Thus *optimal* running performance for a prolonged period of time is achieved only if both glucose and fatty acids are oxidized by the muscle. In other words, since the runner cannot maintain a sufficiently high power output from fatty acid oxidation alone, and since reliance solely on carbohydrate would deplete the total carbohydrate stores well before the end of the race, both fuels must be used simultaneously. Hence, as much fatty acid as possible must be oxidized to allow the limited carbohydrate reserves to be used for the duration of the marathon run. This conservation of carbohydrate at the expense of fatty acid oxidation is facilitated by a specific intracellular control mechanism.

Control of the Rate of Glucose Utilization by Fatty Acid Oxidation

Probably all tissues of body utilize glucose as the normal fuel for energy production in the fed state; indeed for some tissues (e.g., red and white cells of the blood, brain and nervous tissue) it may be the *only* fuel that they can use for energy generation. Hence, in order for survival of these tissues, the normal concentration of glucose in the blood must be maintained under all conditions, even sustained exercise. If it is reduced seriously below the normal concentration, the brain fails to function and coma can result. Indeed the demand for fuel is so large in elite marathon runners that the total amount of glucose in the blood will be used in less than 2 min. Such a decrease is normally prevented by the degradation of the liver store of glycogen to glucose which is then released into the bloodstream and, when this is depleted, by gluconeogenesis in the liver. In addition to this, however, it is known that fatty acid oxidation in the muscle reduces the rate of glucose utilization and oxidation by that tissue. This ensures that, if fatty acids and glucose are available in the blood, fatty acids will be oxidized and this will reduce glucose utilization and help to maintain the normal or near normal blood glucose concentrations.

The mechanism of this regulation by fatty acids is as follows. Fatty acid oxidation in muscle raises the intracellular concentrations of the important allosteric regulators of glycolysis and pyruvate oxidation, acetyl-CoA, citrate and glucose 6-phosphate. An increase in the acetyl-CoA/CoA concentration ratio will inhibit pyruvate dehydrogenase and hence markedly reduce carbohydrate oxidation; an increase in the concentration of citrate will inhibit 6-phosphofructokinase and this will result in a decreased rate of glycogenolysis and glucose utilization since inhibition of 6-phospho-fructokinase will raise the concentration of glucose 6-phosphate, which inhibits both hexokinase and glycogen phosphorylase (Newsholme & Start, 1973). If the rate of fatty acid oxidation were reduced, exercise would demand a greater rate of utilization of the glucose and hence the carbohydrate stores would be used more quickly and this would eventually deplete these stores and result in hypoglycemia; this has been observed when fatty acid mobilization is reduced by nicotinic acid (Carlson et al., 1963). Indeed, with the widespread clinical use of β-blockers in humans, it has been noted that this treatment leads to the early onset of fatigue (Simpson, 1977). Since β-blockers inhibit the release of fatty acids from adipose tissue, the lower blood concentration of fatty acids and hence their lower rate of oxidation would demand a greater rate of utilization of carbohydrate and this might provide an explanation of the fatigue.

It should be noted that, although fatty acid oxidation reduces the rate of glycolysis and pyruvate oxidation, this is not a fixed reduction and the rates of these processes will be higher than at rest. The rates of glycolysis and pyruvate oxidation, however, will be less than they would be if there were no fatty acid oxidation. The mechanism of regulation of glycolysis is such that if the intensity of exercise is increased, and there is no compensatory change in the rate of fatty acid oxidation, the rate of glycolysis will increase.

The important point that is derived from the above discussion is that fatty acid not only provides a fuel for exercising muscle but also regulates the rate of glucose utilization and oxidation so that the latter provides for only the deficit in energy which would occur if only fatty acids were oxidized. Hence this extends the period during which both carbohydrate and fat can be oxidized so that a high power output can be maintained for a longer period. Ideally, the competitive marathon runners should finish the race with almost no glycogen in their muscles: if they finish with some glycogen left they could have run faster; if they used all the glycogen prior to the end of the race, so that the muscles could only function on fatty acid oxidation, the running pace would have to be reduced by perhaps 50% (that is, exhaustion). It is possible that elite marathon runners have a higher-than-normal capacity to oxidize fatty acids so that their power output can be high and yet glycogen stores can last for the duration of the race. The notorious "wall" experienced by many non-elite runners may be the point when the muscle glycogen reserve becomes very low and fatty acid oxidation is the major means for providing energy. This discussion emphasizes that one important effect of training is to increase the ability of the muscle fibers to remove fatty acids from the extracellular fluid and oxidize them. It remains to be seen how important the intracellular fatty acid binding protein is in muscle in increasing the rate of fatty acid diffusion into muscle just as myoglobin is considered to improve the diffusion of oxygen. If this is an important factor, changes in its concentration may provide an easy-to-measure index of the best training regime not only for elite athletes but joggers intent on improving their physical fitness.

References

BERGSTRÖM, J., L. Hermansen, E. Hultman & B. Saltin. Diet, muscle glycogen and physical performance. Acta Physiol. Scand. 71:140–150, 1967.

CARLSON, L.A., R.J. Hovel & L.G. Ekelund. Effect of nicotinic acid on the turnover rate and oxidation of the free fatty acids of plasma in man during exercise. Metabolism 12:837–845, 1963.

CHRISTENSEN, E.H. & O. Hansen. Arbeitsfähigket und Ehrnährung. Skand. Arch. Physiol. 81:160–175, 1939.

COSTILL, D.L., A Scientific Approach to Distance Running, 1979.

COSTILL, D.L. & J.M. Miller. Nutrition for endurance sport: carbohydrate and fluid balance. Int. J. Sports Med. 1:2–14, 1980.

COSTILL, D.L., E. Coyle & G. Dulsky. Effect of elevated plasma FFA and insulin on muscle glycogen usage during exercise. J. Appl. Physiol. 43:695, 1977.

COWAN, J.C. & E.M. Vaughn-Williams. The effects of palmitate on intracellular potentials recorded from Langendorff-perfused guinea pig hearts in roimoxia and hypoxia and during perfusion at a reduced rate of flow. J. Molec. Cellul. Cardiology. 9:327–342, 1974.

DAVIES, C.T.M. & M.W. Thompson. Aerobic performance of female marathon and male ultramarathon athletes. Eur. J. Appl. Physiol. 41:233–245, 1979.

HICKSON, R.C., M.J. Rennie, R.K. Conlee et al. Effects of increased plasma fatty acids on glycogen utilization and endurance. J. Appl. Physiol. 43:829–833, 1977.

NEWSHOLME, E.A. & A.R. Leech. Metabolism in Medicine. London, Sydney and Toronto: Wiley & Sons, 1983.

NEWSHOLME, E.A. & C. Start. Regulation in Metabolism. London, Sydney and Toronto: Wiley & Sons, 1973.

SIMPSON, W.T., Nature and incidence of unwanted effects with atenolol. Postgrad. Med. J. 53(Suppl 3): 162–167, 1977.

SPECTOR, A.A. & S.E. Fletcher. Transport of fatty acid in the circulation. In S.M. Dietschy, A.M. Gotto & S.A. Ontko (Eds.), Disturbances in Lipid and Lipoprotein Metabolism. Bethesda: American Physiological Society, 1978.

Effect of Acidosis on Energy Metabolism and Force Generation in Skeletal Muscle

Kent Sahlin

Huddinge Hospital, Huddinge, Sweden

An adequate muscle function is critically dependent upon a fairly constant electrolyte composition in the external and internal milieu. During intense muscular activity the high energy demand requires an accelerated anaerobic energy production and lactic acid will be formed and accumulated in the contracting muscle. Lactic acid is at physiological pH almost completely dissociated and hydrogen ions are formed in an equivalent amount to lactate. The resulting acidification will influence many of the processes involved in the transformation of chemical energy into mechanical energy. I will in this paper focus on the effect of acidosis on production of ATP and on the generation of force.

Effect of Exercise on Intramuscular pH in Humans

During intense dynamic and static exercise lactate content in muscle increases about 30-fold (Table 1). In addition to lactic acid also other acidic compounds such as glucose 6-P, glycerol 1-P, pyruvate, citrate and malate will accumulate. Lactic acid contributes, however, to more than 85% of liberated H^+ions. Most of the formed H^+ions will be buffered within the tissue and only a small fraction ($< 0.001\%$) appears as free ions. Muscle pH is decreased with about 0.5 pH units at fatigue compared to resting condition (Table 1) and a close relationship exists between lactate accumulation and decrease in muscle pH (Figure 1). Muscle pH during bicycle exercise tends to be slightly higher for a certain lactate than during isometric contraction. This might be due to an exchange of H^+ions with the blood during bicycle exercise (due to escape of CO_2) but not during isometric contraction where no circulation occurs in the contracting muscle.

Table 1

Changes in Muscle pH and Lactate of Human Skeletal Muscle after Exercise

Conditions	Muscle pH	Lactate (mmol/kg dry wt)
Rest; n = 12	7.08 ± 0.03	4.1 ± 2.2
Bicycle exercise at a constant load (200 to 360 W) to fatigue (6 to 11 min); n = 7	6.60 ± 0.16	114 ± 19
Isometric contraction to fatigue (45 s); n = 10	6.56 ± 0.07	94 ± 17

Results are expressed as means ± SD and are from Sahlin et al. (1975 & 1976).

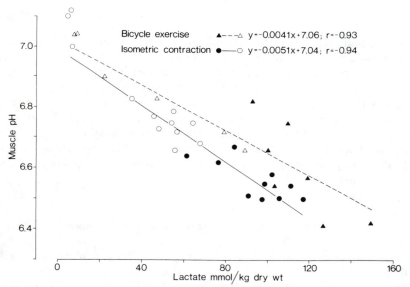

Figure 1—Relation between lactate content and muscle pH in muscle biopsies taken from human quadriceps muscle after exercise. Isometric contraction was sustained at a force corresponding to 68% of the subject's maximal voluntary contraction force. Bicycle exercise was performed at about 50 to 100% of $\dot{V}_{O_2}$max for 5 to 11 min. Filled symbols denote samples taken at fatigue. Results are from Sahlin et al. (1975 & 1976).

Accumulation of lactic acid will increase the osmotic pressure and result in a water shift. Water content in muscle has been found to increase about 10 to 20% during exercise (Sahlin et al., 1978) and might exert a restricting effect on the local circulation.

Effects of Acidosis on Energy Metabolism

Changes in the Ionic Form of ATP, ADP and P_i

ATP and ADP in the cell are composed of different ionic species involving Mg^{2+}, K^+ and H^+. Both ATP and ADP have pKa values in the physiological pH-range and their ionic composition will be changed by H^+ion accumulation. The concentration of the Mg-complexes of ATP and ADP decrease during acidosis (Figure 2). In most enzymatic reactions involving ATP and ADP the active ionic species are the Mg-complexes and changes in their concentrations are thus of physiological importance. The protonated form of ATP ($HATP^{3-}$) is known to be a very potent inhibitor of phosphofructokinase (PFK). The large relative increase in $HATP^{3-}$ during acidosis will form the basis of the extreme pH sensitivity of PFK.

Inorganic phosphate and glycogen are substrates for phosphorylase. It has been shown by Kasvinsky and Meyer (1977) for rabbit muscle and by

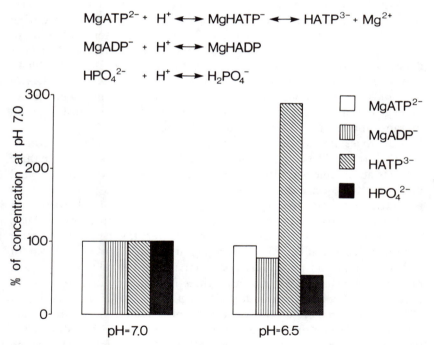

$$MgATP^{2-} + H^+ \longleftrightarrow MgHATP^- \longleftrightarrow HATP^{3-} + Mg^{2+}$$

$$MgADP^- + H^+ \longleftrightarrow MgHADP$$

$$HPO_4^{2-} + H^+ \longleftrightarrow H_2PO_4^-$$

□ $MgATP^{2-}$

▥ $MgADP^-$

▨ $HATP^{3-}$

■ HPO_4^{2-}

% of concentration at pH 7.0

pH=7.0   pH=6.5

Figure 2—Effect of hydrogen ion accumulation upon the relative concentration of the ionic forms of ATP, ADP and HPO_4^{2-}. Calculations were performed by utilizing the previously derived equations (Sahlin, 1978) at 1 mmol/1 Mg^{2+} and 160 mmol/1 K^+. At a pH of 7.0, $MgATP^{2-}$ and $HATP^{3-}$ amount to 92% and 2% of total ATP, respectively, and MgADP amounts to 52% of the total free ADP. A pKa of 6.8 was used for $H_2PO_4^-$.

Chasiotis for human muscle (personal communication) that HPO_2^{2-} is the true substrate for phosphorylase. A decrease in pH from 7.0 to 6.5 transfers about 50% of the available substrate (HPO_2^{2-}) to the unreactive form ($H_2PO_4^-$). In contracting muscle where lactic acid is formed the store of creatine phosphate is utilized and P_i is liberated. It has been shown recently that the activity of phosphorylase is critically dependent upon the concentration of P_i (Chasiotis et al., 1982). The liberation of P_i from creatine phosphate and the effect of H^+ accumulation on the active form will thus be important for the regulation of glycogenolytic activity.

Inhibition of Glycolysis and Glycogenolysis

A continuous high rate of glycolysis during exercise would result in a severe acidosis in muscle and in the whole body and could affect the homeostasis and the function of acid-labile components in muscle and other more sensitive and critical tissues (i.e., heart and brain). Prevention against excessive lactic acid production during exercise is therefore necessary. Several investigations have shown that glycolysis is enhanced during alkalosis and inhibited during acidosis. The main regulatory step seems to be phosphofructokinase (PFK) which activity shows a marked pH dependence and is almost completely inactive at a pH of 6.4 (Danforth, 1964).

Accumulation of fructose 6-P which is a substrate for PFK can overcome the pH inhibition of PFK and necessitates a control point higher up in glycolysis. It has been demonstrated in vitro that the transformation of phosphorylase b to a is slowed down during acidosis (Danforth, 1964). In order to test if this effect also is present in vivo the following experiment was performed (Chasiotis, Hultman & Sahlin, unpublished data). In three subjects adrenaline was infused intravenously (0.15 μg kg^{-1} body weight min^{-1}) on two occasions. In one case the infusion was preceded by an isometric contraction to fatigue resulting in a muscle pH of 6.60, whereas in the other case no exercise was performed (i.e., normal muscle pH of 7.08). Biopsies from the quadriceps femoris muscle were taken prior to the start of infusion and after 0.5 min and 2 min of infusion and were analyzed for cAMP content and phosphorylase a activity. In the non-exercised leg, muscle content of cAMP increased three-fold already after 0.5 min of adrenaline infusion, whereas in the "low pH leg" cAMP increased only 1.7-fold. An even more striking pattern was observed for the phosphorylase b to a transformation. In the nonexercised leg a rapid transformation occurred (phosphorylase a increased 4-fold after 0.5 min), whereas no transformation occurred in the muscle with low pH (a decrease in phosphorylase was actually observed). After 2 min of infusion of adrenaline phosphorylase a increased also in the muscle with low pH but was still lower than for the non-exercised leg. These results demonstrate in vivo that during acidosis cAMP formation is inhibited and transformation of phosphorylase *b* to *a* in response to adrenaline is slowed down. The latter effect is at least partly due to the lower cAMP content in muscle but the total absence of any increase in phosphorylase *a* after 0.5 min suggest also that phosphorylase *b* kinase is affected by acidosis.

Inhibition of Lipolysis

It has been shown by Fredholm and Hjemdahl (1976) that acidosis resulted in less formation of cAMP and an inhibition of lipolysis when isolated fat cells were exposed to noradrenaline. Jones et al. (1977) found that during exercise after induced acidosis, plasma content of glycerol and free fatty acids was lower than normal. Evidence thus exists that lipolysis is inhibited also in vivo.

Effect of Acidosis on Muscular Function

Relaxation Time and Tension Development in Isolated Rat Muscle

It is well known that when muscle fatigues the rate of relaxation is slowed down. This will impair the performance in dynamic exercise where a well-timed rhythm of contraction and relaxation of several muscle groups is required. It was shown by Dawson et al. (1980) that slowing of relaxation was related to many of the biochemical changes which occur in fatiguing muscle (i.e., PCr depletion, accumulation of H^+, ADP or P_i and to the free energy change for ATP hydrolysis). To differentiate between these factors it is necessary to change one factor independent of the others. We have compared isolated rat muscles poisoned with iodoacetate, IAA, (which inhibits lactic acid formation) with unpoisoned muscles. By this technique the effect of H^+ ion accumulation can be separated from other metabolic changes in fatigued muscle. Stimulation of unpoisoned muscle resulted in lactic acid formation and decrease of muscle pH simultaneously with an increase of the relaxation time (Figure 3). In contrast stimulation of IAA poisoned muscles resulted in unchanged muscle pH and relaxation time.

To investigate the effect of H^+ ion accumulation on the capacity to generate force we have incubated isolated rat muscle in an atmosphere of 30% CO_2 in oxygen. By this procedure muscle pH decreased to about the same level as after electrical stimulation to fatigue (50% of initial tension). Acidosis was induced by two totally independent ways but the resulting mechanical response was similar (Figure 4), i.e., decreased tension and increased relaxation time. The experiment provides a strong evidence that H^+ ion accumulation in muscle causes these effects in a direct or indirect way.

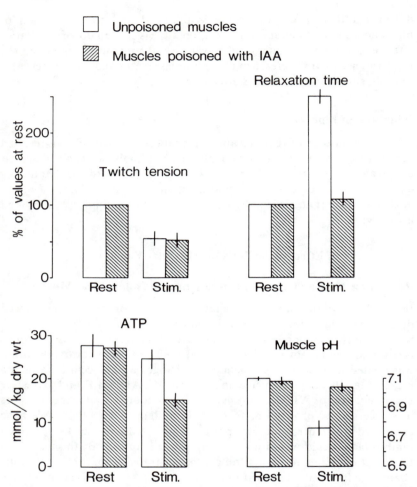

Figure 3—Effect of electrical stimulation on isolated rat muscle. Stimulation was performed at 2 Hz under anaerobic conditions until tension decreased to about 50% of the initial. Unpoisoned muscles required about 3.1 min stimulation whereas muscles poisoned with iodoacetic acid (IAA) required only 1 min stimulation. Relaxation time was defined as the time elapsed when tension decreased to 50% of peak tension. Results are from Sahlin et al. (1981).

Shortage of Chemical Fuel

I have above pointed out that several of the enzymatic systems involved in the generations of ATP are inhibited during acidosis. Although acidosis also has some positive effects on the oxygen transport into the working

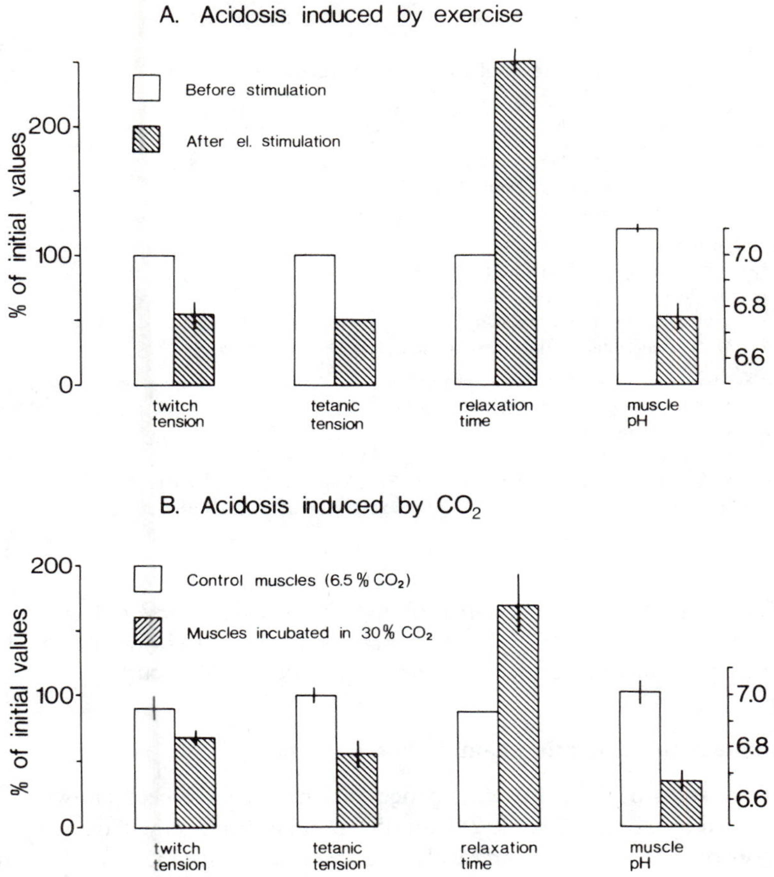

A. Acidosis induced by exercise

B. Acidosis induced by CO$_2$

Figure 4—Comparison between acidosis induced by muscle contraction (A) and CO$_2$ (B) on the contraction characteristics of isolated rat muscle. Contraction (Figure 4A) was performed by electrical stimulation at 2Hz during 3.1 min under anaerobic conditions. Muscles in Figure 4B were incubated for 90 min in 6.5% CO$_2$ (controls) or 30% CO$_2$ in oxygen. They were only stimulated for 1.2 s before and at the end of the incubation period. Muscle pH was in A obtained from measurements in muscle homogenate and in B by the CO$_2$-technique. For definition of relaxation time see Figure 3. Results for Figure 4A are form Sahlin et al. (1981) and Figure 4B from Sahlin, Edström and Sjöholm (unpublished data).

muscle (Table 2) it seems possible that the generation of ATP might be inadequate to meet the demands. Muscle content of ATP decreases, however, only about 20% at fatigue (Sahlin et al., 1975) and experiments with isolated rat muscles (Figure 3) demonstrate that about half of the store can be utilized during muscular contraction. The ATP content as such does thus not seem to be limiting for the contractile machinery. The increases in

Table 2

Effects of Lactic Acid Accumulation

A. Positive effects on ATP generation
  1. Increased O_2 delivery (Bohr effect)
  2. Local vasodilation

B. Negative effects on ATP generation
  1. Inhibition of glycolysis and glycogenolysis
  2. Inhibition of lipolysis in adipose tissue
  3. Increased water content of muscle
  4. Decreased lactate efflux from muscle (Hirche et al., 1975)

C. Negative effects on the contraction process
  1. Increased requirement for Ca^{2+} (Donaldson et al., 1978)
  2. Decreased maximal tension (Donaldson et al., 1978)
  3. Decreased myosine ATP-ase activity (Schädler, 1967)
  4. Increased protein binding of Ca^{2+} in SR (Nakamaru & Schwarts, 1972)
  5. Increased concentration of K^+ in the extracellular space (Saltin et al., 1981)

ADP, P_i and H^+ which occur in fatigued muscle will, however, decrease the energy yield for the ATP splitting process (Sahlin, 1978) and a too low energy yield in some critical step utilizing ATP might occur.

Fatigue in the Excitation-Contraction Process

Several steps in the contractile process in muscle have been shown to be impaired by acidosis (Table 2). Inhibition of contraction will decrease the energy output of muscle and results in decreased requirement for ATP and further glycolysis. Acidosis might be regarded as a safety mechanism by which the muscle cell is protected from total ATP depletion and from destructive lactic acid accumulation.

An increase of extracellular potassium has been observed during isometric contraction (Saltin et al., 1981). Activation of muscle is critically dependent upon the extracellular electrolyte composition and increases of potassium can interfere with the excitation process. Increases of intra-cellular H^+ ions might be involved in the transfer of K^+ as these ions are known to have a reciprocal relation.

Conclusions

Intensive short-term exercise results in pronounced changes in the pH of working muscles. This will markedly change the intracellular environment and affect the ionic composition and the activity of several enzymes involved in the generation of ATP. From experiments with isolated rat muscle, evidence has been put forward that slowing of relaxation and impairment of tension development is caused by accumulation of H^+ions.

References

CHASIOTIS, D., K. Sahlin & E. Hultman. Regulation of glycogenolysis in human muscle at rest and during exercise. J. Appl. Physiol. (accepted for publication).

DANFORTH, W.H., Activation of glycolytic pathway in muscle. In B. Chance & R.W. Estabrook (Eds.), Control of Energy Metabolism. New York: Academic Press, 1965.

DAWSON, M.J., D.G. Gadian & D.R. Wilkie. Mechanical relaxation rate and metabolism studied in fatiguing muscle by phosphorous nuclear magnetic resonance. J. Physiol. 299:465–484, 1980.

DONALDSON, S.K.B., L. Hermansen & L. Bolles. Differential direct effects of H^+ on Ca^{2+}-activated force of skinned fibers from the soleus, cardiac and adductor magnus muscles of rabbits. Pflügers Arch. 376:55–65, 1978.

FREDHOLM, B. & P. Hjemdahl. Inhibition by acidosis of adenosine 3'-5'-cyclic monophosphate accumulation and lipolysis in isolated rat fat cells. Acta Physiol. Scand. 96:160–169, 1976.

HIRCHE, H., V. Hombach, H.D. Langohr, U. Wacker & J. Busse. Lactic acid permeation rate in working gastrocnemii of dogs during metabolic alkalosis and acidosis. Pflügers Arch. 356:209–222, 1975.

JONES, N.L., J.R. Sutton, R. Taylor & C.J. Toews. Effect of pH on cardio-respiratory and metabolic responses to exercise. J. Appl. Physiol. 43:959–964, 1977.

KAVINSKI, P.J. & W.L. Meyer. The effect of pH and temperature on the kinetics of native and altered glycogen phosphorylase. Arch Biochem. Biophys. 181:616–631, 1977.

NAKAMARU, Y. & A. Schwarts. The influence of hydrogen ion concentration on calcium binding and release by skeletal muscle sacroplasmic reticulum. J. Gen. Physiol. 59:22–32, 1972.

SAHLIN, K., Intracellular pH and energy metabolism in skeletal muscle of man, with special reference to exercise. Acta Physiol. Scand. (Suppl. 455):7–45, 1978.

SAHLIN, K., A. Alvestrand, R. Brandt & E. Hultman. Intracellular pH and bicarbonate concentration in human muscle during recovery from exercise. J. Appl. Physiol. 45:474–480, 1978.

SAHLIN, K., L. Eström, H. Sjöholm & E. Hultman. Effects of lactic acid accumulation and ATP decrease on muscle tension and relaxation. Am. J. Physiol. 240:C121–C126, 1981.

SAHLIN, K., R.C. Harris & E. Hultman. Creatine kinase equilibrium and lactate content compared with muscle pH in tissue samples obtained after isometric contraction. Biochem. J. 152:173–180, 1975.

SAHLIN, K., R.C. Harris, B. Nylind & E. Hultman. Lactate content and pH in muscle samples obtained after dynamic exercise. Pflügers Arch. 367:143–149, 1976.

SALTIN, B., G. Sjogaard, F.A. Gaffney & B. Rowell. Potassium, lactate and water fluxes in human quadriceps muscle during static contractions. Circ. Res. 48:118–124, 1981.

SCHÄDLER, M. Proportionale Aktivierung von ATPase-Aktivität und Kontraktionsspannung durch Calciumionen in isolierten contractilen Strukturen verschiedener Muskelarten. Pflügers Arch. 296:70–90, 1967.

Morphological Studies of Muscle Fatigue

Michael Sjöström and J. Fridén
University of Umeå, Umeå, Sweden

"Muscle fatigue" means that muscle fibers are unable to continue functioning at a particular level. The functional inability as such cannot be visualized by the use of morphological techniques. The "fatigued" subcellular structures involved in energy metabolism and the contractile machinery can be examined, however. The phenomenon of fatigue can thereby be indirectly studied and, hopefully, be better understood.

The morphological methodology permits the study of different fiber types. The human skeletal muscles are characteristically composed of fibers with different structural and functional properties. Fiber recruitment varies with the type of muscular exercise and the ability of fibers to respond to various functional demands may also vary. One classical example, which illustrates the usefulness of microscopic techniques, is the fiber type specific glycogen depletion after different types of muscular exercise (e.g., Costill et al., 1973).

Up-to-date methodologies include specimen examination both at the light and the electron microscopic level. Among the LM-techniques are immuno- and enzyme histochemistry in combination with enzyme determination in microdissected fibers (Lowry et al., 1978). EM-techniques may include ultra-thin sectioning of rapidly frozen tissue (cryoultramicrotomy) followed either by studies of the macromolecular structure in negatively stained sections using image analysis or by examination in the electron microscope of freeze-dried sections using X-ray microanalysis (Sjöström & Squire, 1977; Somlyo et al., 1978). The latter procedure may be used to analyze the quantities of several ions in one single thin-sectioned mitochondria, or a lateral sac belonging to the sarcoplasmic reticulum, from a normal, fatigued or exhausted muscle fiber (Somlyo et al., 1981).

Unfortunately, many of these advanced techniques are available only to a limited number of laboratories. Morphologists in general, as well as physiologists and biochemists, still have to rely on conventional techniques, such as standardized enzyme histochemistry at light microscopic level and plastic embedding for electron microscopy, in combination with morphometric methods. Using these techniques a description of the overall morphology is possible. A number of cellular and subcellular structures are available for qualitative and/or quantitative analysis. Among these structures are a) fiber type occurrence and size, b) mitochondrial volume and appearance, c) lipid droplet distribution and volume fraction, d) myofibrillar composition and architecture, and e) glycogen amount and particle distribution. Ultrastructural criteria for classification of the fibers into different fiber types are now also available (Sjöström et al., 1982). But to what extent can the nature and degree of the pathophysiology behind muscle fatigue be clarified by the conventional morphological approach? Are there signs at the cellular level of fiber necrosis or of fiber rupture? Or are the abnormalities eventually found only seen at subcellular level? This report attempts to answer some of these questions.

We report results from a series of morphological studies of fibers from fatigued human skeletal muscles. The muscle tissue demonstrated in all cases an inability to maintain a predetermined exercise intensity. The reason for development of the fatigue varied, however. One group of individuals had subjected their leg muscles to repeated relative ischemia (individuals with intermittent claudication). Another group had performed a long-term sub-maximal muscular exercise (running 30 km), while a third group had performed short-term high-load muscular exercise in the form of maximal eccentric contractions.

Materials and Methods

Individuals Demonstrating Ischemic Fatigue

This study group consisted of 20 men (mean age 59 ± 6 yr) admitted to our surgical department because of unilateral intermittent claudication. All the individuals had a history of peripheral arterial insufficiency (without rest pain in one leg only) for 6 mo or more. The walking distance covered by the subjects before the pain occurred (initial walking tolerance) and maximum walking tolerance were determined in a passage way in the hospital (1 m/s, $0°$) and/or on a treadmill (1 m/s, $0°$). Mean initial walking distance, when tested in the hospital passage, was 185 m (range 45 to 495). Further physical, clinical and physiological data on the individuals are given elsewhere (Sjöström et al., 1980). Muscle biopsies were obtained from the tibialis anterior muscle of both the symptomatic and the asymptomatic leg.

Individuals Performing Long-Term Submaximal Muscular Exercise

This group consisted of six men (mean age 28 ± 7 yr), who participated in a long distance cross-country run (Lidingöloppet 30 km, Stockholm). The race was run on good tracks through forests and across fields and included several small hills with a total elevation of 450 m. The individuals finished the race on average 43% slower than the winner (range 19 to 65%). After the finish, the individuals were transported to the laboratory where a muscle biopsy was taken without any delay (30 to 40 min after the race finish) from the vastus lateralis muscle.

Individuals Performing Short-Term High-Load Muscular Exercise

The individuals in this group (nine men, mean age 26 ± 8 yr) performed a well defined heavy muscular exercise in the form of repeated eccentric contractions. The load was individually adjusted so that an exercise intensity demanding 80 to 100% of maximal oxygen uptake was obtained. Duration of exercise was 30 min, although two individuals failed to fulfill the program. The perceived exertion was for all individuals, according to the Borg-scale, very, very hard. Muscle biopsies were obtained from the vastus lateralis muscle 2 h after the exercise as well as 1, 2, 3, 4 or 6 days postexercise. At the same time strength measurements at preset angular velocities were performed with a Cybex II standard equipment. The individuals experienced severe muscular discomfort 1 to 3 days after the exercise. The major symptoms were muscular stiffness, tenderness, and pain, especially when making active movements.

Specimen Preparation

Muscle biopsies were surgically obtained from well defined portions of the muscle. Each biopsy was initially divided into halves, one of which was prepared for enzyme histochemistry, the other for electron microscopy. Detailed descriptions of the preparative procedures, fiber terminology at light- and electron microscopic level, sampling for morphometry, collection of morphometric data and statistical methods are given in detail elsewhere (Ängquist & Sjöström, 1980; Sjöström et al., 1980; Sjöström et al., 1982). Glycogen particles were identified at electron microscopic level by the method of Thiery (Thornell et al., 1977).

Results

Morphology in Muscles Repeatedly Subjected to Ischemic Fatigue

Two different patterns of morphological deviations were seen in the symptomatic fatigued muscles (Table 1). One suggested primary muscle

Table 1

Summary of Morphological Findings in Biopsies from the Tibialis Anterior Muscle of 20 Male Patients with Unilateral Intermittent Claudication

Frequency	Internal nuclei[1]	De- and re-generation[2]	Angular fibers[2]	Type grouping[3]
Symptomatic leg				
—	3	2	13	10
X	1	4	2	3
XX	11	4	—	—
XXX	5	8	2	5
XXXX	—	2	3	2

[1] X = 1 to 3% of the fibers have internal nuclei; XX = 3 to 10%; XXX = 10 to 30%; XXXX = > 30%.
[2] X = 1 to 3 fibers in 1 to 3 fascicles; XX = > 3 fibers in 1 to 3 fascicles; XXX = 1 to 3 fibers in > 3 fascicles; XXXX = > 3 fibers in > 3 fascicles.
[3] X = 1 to 3 areas with small type grouping; XX = > 3; XXX = 1 to 3 whole fascicles of single fiber type; XXXX = > 3 whole fascicles of single fiber type.

The numerals represent the number of subjects.

fiber damage (internal nuclei and signs of de- and regeneration) and the other indicated denervation (angular fibers and fiber type grouping). Significant correlations were found between the initial walking tolerance and the relative number of type 1 fibers as well as the diameters of both type 1 and type 2 fibers (Table 2). Data on mitochondrial volumes, obtained from structurally intact fibers, often correlated significantly with the results of the functional tests (Table 2). Above all, the volume of mitochondria in type 2A fibers showed a strong correlation with the initial walking tolerance.

Muscle Morphology after Long-Term Submaximal Muscular Exercise (Figure 1)

The enzyme histochemically treated cross-sections showed well-preserved fascicles with tightly packed polygonal fibers. There was no evidence of any frequent focal or diffuse morphological abnormality in any of the sections at light as well as electron microscopical level. Thus, mitochondria showed a normal appearance. Numerous glycogen depleted fibers, which were exclusively of type 1, were evident at light microscopic level in PAS-stained sections. Some of these fibers were practically completely

Table 2

Correlations between Morphological (Tibialis Anterior Muscle) and Physiological Variables in 20 Male Patients with Unilateral Intermittent Claudication

| | Hospital passage (16) | | Treadmill (11) | |
	S	A	S	A
	Initial walking tolerance			
Relative number of				
type 1 fibers	NS	NS	0.72**	0.65
type 2A fibers	NS	NS	−0.83***	−0.83***
Diameters of type 1 fibers	0.51*	NS	NS	NS
type 2 fibers	0.49*	NS	NS	NS
Relative cross-sectional area				
of type 1 fibers	NS	NS	0.43	0.46
	Max walking tolerance			
Mitochondrial volumes				
total	NS	0.41*	NS	0.63*
type 1 fibers	NS	0.40*	NS	0.66*
type 2 fibers	0.48*	0.34	NS	0.48
type 2A fibers	0.64***	NS	NS	0.51
type 2B fibers	NS	0.46*	NS	NS

Pearson correlation coefficients with $p < 0.1$ are given; NS = not significant; *$p < 0.05$, **$p < 0.001$; the number of X/Y variables in parentheses; S = symptomatic leg, A = asymptomatic leg.

depleted of glycogen, while others appeared to have a low glycogen content. Ultrathin sections, treated for visualization of glycogen, clearly revealed quantitative differences in the particle content between the fibers. Type 1 fibers showed a low or negligible particle content. They showed no subsarcolemmal accumulation and practically no intermyofibrillar particles. Shorter rows of particles located in the I-band were relatively frequent, however.

Morphology of Muscle Fibers after Short-Term High-Load Muscular Exercise (Figure 2)

There were no signs at the cellular level of either ischemic fiber necrosis or fiber rupture in these muscles. At subcellular level abnormalities were seen, however. Frequent focal disturbances of the characteristic cross-striated

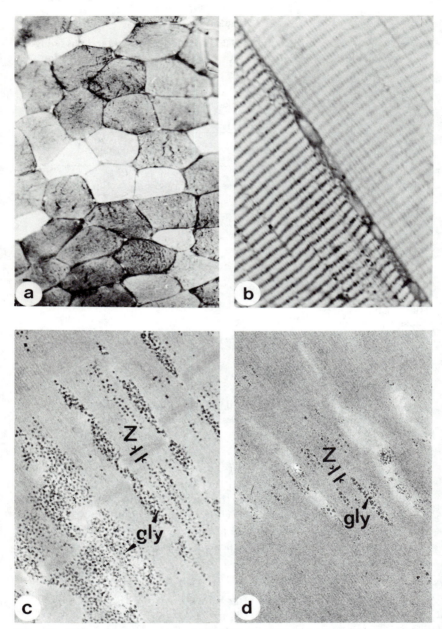

Figure 1—PAS-stained cross and longitudinal cryo sections from a muscle biopsy obtained immediately after prolonged exercise (*a–b*). Electron micrographs stained with PA-TSC-SP (*c* = control, *d* = after prolonged running). The "glycogen depleted" fibers are of type 1. Z = Z-band, gly = glycogen particles. Magnification × 375 (*a*), × 1.500 (*b*) and × 18.000 (*c–d*).

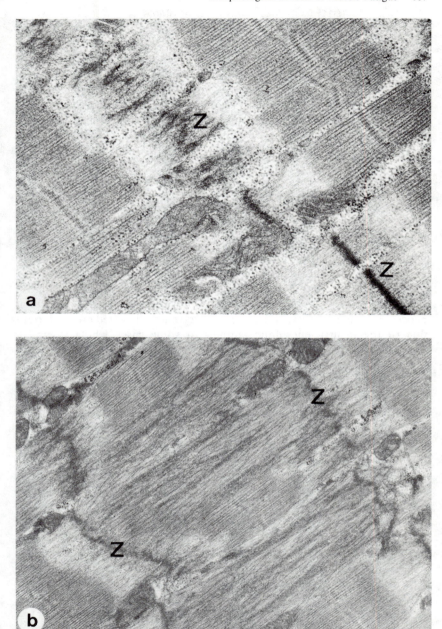

Figure 2—Electron micrographs showing distorted contractile material in fibers subjected to exhaustive eccentric exercise. Z = Z-band. Magnification × 30.000 (*a*) and × 20.500 (*b*).

contractile band pattern were revealed. The disturbances were found to originate from the myofibrillar Z-band which showed a marked broadening, streaming and, at places, total disruption. In some cases only one single Z-band of one myofibril was affected, while there was also evidence for involvement of several sarcomeres and myofibrils. The disturbances were found mainly in type 2 fibers. A significant parallel decline of the torque at all different angular velocities was found. The structural changes in the muscle fibers and the muscular strength both returned to normal 6 days after the eccentric exercise.

Discussion

Ischemic Fatigue

Obvious morphological abnormalities were seen in the individuals with peripheral arterial insufficiency. These muscles repeatedly had demonstrated fatigue when they were subjected to an ischemia which was relative to the functional demands. The findings indicated that subcellular processes in many fibers, and nerves, are seriously distorted during fatigue due to the insufficient blood supply. In some cases whole fibers, or segments of fibers, simply died and regenerative processes were initiated.

Perhaps more interesting were the relationships, demonstrated between the morphometric and physiological data, which suggested several fiber type specific adaptive processes in the fatigued muscle. An example was the relationship between fiber diameters for both fiber types in the symptomatic muscles and the walking tolerance. These results supported the hypothesis that stretch, or the physical conditions for muscular contraction, controls the muscular size, i.e., the synthesis or degradation or both of myofibrillar proteins. This conclusion could be drawn on the basis of the following. The muscular pain induced by the relative ischemia prevented work above a certain level of intensity and duration and this caused the fatigue and limited the physical performance of each subject. Thus, the functional demands on the contractile apparatus were defined and could be measured. An optimal relationship between fiber diameters and walking capacity could, thus, be demonstrated. In this regard, therefore, human subjects with insufficient blood supply seem to constitute a unique experimental model.

Is the ischemic fatigue that limits the physical performance relative to a certain fiber type? The walking distance increased with a higher relative amount of type 1 fibers and decreased with an increasing number of type 2 (type 2A) fibers. The answer should be, so far, that the different fibers are selectively used and, eventually, adapted to the insufficient blood supply. Both the maximum and the initial pain walking tolerance correlated significantly with mitochondrial data from the asymptomatic leg. In relation to this leg the initial walking tolerance may be taken as a measurement of the ordinary activity demands. Thus, a subcellular adaptation to this relatively low working load, albeit due to the disease of the opposite leg, seems to have occurred.

In the symptomatic leg, on the other hand, such a relation did not exist in that only data on mitochondrial volumes in type 2A fibers were correlated to the results of the functional test; for example, the walking distance covered by the subjects before pain occurred. Data on mitochondrial volumes in type 1 fibers did not show such a covariation although the parameter apparently was increased when compared with data on mitochondrial volumes in type 1 fibers in the asymptomatic leg. The difference, however, between data on mitochondrial volumes in type 1 fibers of the two legs was correlated to walking tolerance (especially maximum walking tolerance—$p < 0.01$). These results emphasize the importance of the mitochondrial adaptation in relation to blood flow and development of pain and fatigue. The question arises whether the volume fraction in type 1 or type 2A fibers, or both in combination, serves as a limiting factor within the muscle fiber or whether the results only reflect the characteristics of the muscle fiber environment, such as degree of capillarization of recruitment pattern.

Fatigue Due to Long-Term Submaximal Muscular Exercise

In this case, no obvious muscle fiber damage was seen, probably because of a sufficient supply of oxygen and substrates to the muscle fibers. A selective metabolic activity in relation to the functional demands could be indirectly demonstrated however, in that only certain fibers were depleted of glycogen. It was evident that the fiber type that had the lowest glycogen content was that of type 1. The impression one gets from PAS-stained sections of almost completely empty fibers was, however, misleading since there was obviously an abundance of glycogen particles even in empty fibers. It was remarkable that these particles were mainly situated in the myofibrils, i.e., between the myofilaments. This may be because they are probably lying inaccesible to and at the same time a distance from the mitochondria which is why they were not readily utilized in glycolysis. It was, furthermore, interesting that the remaining intermyofibrillar particles were small in size and seemed almost rudimentary. This could be because a gradual breakdown of individual particles takes place as the substrate need increases.

Fatigue Due to Short-Term High-Load Muscular Exercise

Morphological changes were here seen only at the subcellular level in the form of disturbances of the myofibrillar structure, especially of type 2 fibers. The immediate interpretation of the findings was that the high myofibrillar tension developed during activation of the contractile material, for example, the interdigitating arrays of thin and thick myofilaments, had resulted in some mechanical disruption of the Z-bands. These Z-bands connect adjacent sarcomeres to each other. The findings therefore indicate that the Z-bands constitute a weak link in the myofibrillar contractile chain. Similar conclusions have been made previously (Fridén et al., 1981).

Eccentric muscular exercise is an important component in most exercises. This type of exercise causes greater tension per active motor unit than corresponding concentric exercise and therefore increases the risk for mechanical damage to the myofibrillar material. The present results are not, on the other hand, necessarily direct proofs for the theory of mechanical Z-band disruption. The structural disturbances may also be secondary resulting from an activation of lysosomal enzymes, bringing about a concomitant inflammation. It does appear, however, that the overloaded, fatigued type 2 muscle fibers seem to have their contractile machinery partially distorted.

Conclusion

In this report, a few examples have been given which show how the morphological appoach can be designed to study the phenomenon of muscle fatigue. Pathophysiological events accompanying different forms of muscle fatigue have been visualized. Fiber type selective recruitment and specific metabolic adaptation have been demonstrated. Thus, although progress in the development of the morphological methodology is continuously reported and new detailed information is possible to obtain, there is still a great deal of information to be found in tissue sections prepared by the use of already established and easily available light and electron microscopic techniques. Objective morphometric information, then, makes many studies still more meaningful. The delicate structural de-and rearrangement during and after fatigue and the adaption to various functional demands may otherwise be difficult to detect.

References

ÄNGQUIST, K.A. & M. Sjöström. Intermittent claudication and muscle fiber fine structure. Morphometric data on mitochondrial volumes. Ultrastruct. Pathol. 1:461–470, 1980.

COSTILL, D.L., P.D. Gollnick, E.D. Jansson, B. Saltin & E.M. Stein. Glycogen depletion pattern in human muscle fibers during distance running. Acta Physiol. Scand. 89:374–383, 1973.

FRIDÉN, J., M. Sjöström & B. Ekblom. A morphological study of delayed muscle soreness. Experientia 37:506–507, 1981.

LOWRY, C.V., J.S. Kimmey, S. Felder, M.M.Y. Chi, K.K. Kaiser, P.N. Passonneaus, K.A. Kirk & O.H. Lowry. Enzyme patterns in single human muscle fibers. J. Biol. Chem. 253:8269, 8277, 1978.

SJÖSTRÖM, M., K.A. Ängquist & O. Rais. Intermittent claudication and muscle fiber fine structure: Correlation between clinical and morphological data. Ultrastruct. Pathol. 1:309–326, 1980.

SJÖSTRÖM, M., S. Kidman, K. Henriksson-Larsén & K.A. Ångquist. Z- and M-band appearance in different histochemically defined types of human skeletal muscle fibers. J. Histochem. Cytochem. 30:1–11, 1982.

SJÖSTRÖM, M. & J. Squire. Cryo-ultramicrotomy and myofibrillar fine structure. A review. J. Microsc. 111:239–278, 1977.

SOMLYO, A.P., A.V. Somlyo, H. Shuman, B. Sloane & A. Scorpa. Electron probe analysis of calcium compartments in cryosections of smooth and striated muscles. Ann. N.Y. Acad. Sci. 307:523–544, 1978.

SOMLYO, A.V., H. Gonzalez-Serratos, H. Schuman, G. McClellan & A.P. Solyo. Calcium release and ionic changes in the sarcoplasmic reticulum of tetanized muscle: an electron probe study. J. Cell Biol. 90:577–594, 1981.

THORNELL, L.E., M. Sjöström, U. Karlsson & E. Cedergren. Variable opacity of glycogen in routine electron micrographs. J. Histochem. Cytochem. 25:1069–1073, 1977.

The Electromyographic Signal in Measuring and Understanding Fatigue

Some Properties of the Median Frequency of the Myoelectric Signal during Localized Muscular Fatigue

Carlo J. De Luca, M.A. Sabbahi, F.B. Stulen and G. Bilotto
Children's Hospital Medical Center,
Boston, Massachusetts, U.S.A.,
and Liberty Mutual Research Center,
Hopkinton, Massachusetts, U.S.A.

During the past 5 yr, various works related to the development of a technique and device for objectively measuring localized muscular fatigue have been performed in our NeuroMuscular Research Laboratory. The following text is intended to provide the reader with a quick overview of our past work. For additional details on some of the following projects, the reader is referred to recent publications from our laboratory.

Background

The involvement of various physiological and psychological phenomena has made the issue of fatigue in humans extremely complex. By limiting our interests to localized muscular fatigue in a group of muscles, it may be possible to devise a technique for its objective evaluation (Stulen & De Luca, 1979). The technique is based on the observable fact that the power density spectrum of the myoelectric (ME) signal detected on the surface of the skin undergoes a compression (towards the lower frequencies) as a function of time during a sustained muscle contraction. It has been suggested by others (Lindstrom, 1970) as well as by our group (Stulen & De Luca, 1981) that this phenomenon is directly related to the conduction velocity of the muscle fibers and various time-dependent (fatiguing) processes within a muscle. In fact, data to be discussed later will provide evidence of a direct relationship between the frequency compression of the ME signal and metabolic events within the muscle.

In order to exploit this phenomenon, it has become necessary to decide how to measure this compression easily and reliably. A mathematical analysis based on empirical observations was undertaken to model the power density spectrum of the ME signal as a rational (polynomial) function of the conduction velocity of the active muscle fibers producing the ME signal. Four parameters of the power density spectrum were considered: the mean frequency; the mode frequency; the median frequency; and the ratio of the low-frequency root mean square (RMS) to high-frequency RMS (the median frequency being the demarcation value). These parameters were chosen because they had all been used by other investigators as a measure of localized muscular fatigue.

The ratio parameter was found to be the most sensitive to conduction velocity but was the least reliable because it depends on the shape of the power density spectrum. Due to the stochastic nature of the ME signal, the shape of the power density spectrum varies. It is precisely this point that also renders the mode frequency useless because the power density spectrum does not have a well-defined peak value. The mean and median frequencies were found to have similar sensitivity to the conduction velocity. But there was an important distinction; in the presence of noise the estimate of the median frequency was always better than that of the

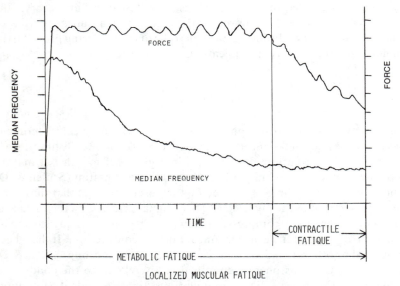

Figure 1—Suggested distinction between metabolic and contractile fatigue in localized muscular fatigue.

mean frequency. Based on our calculations, the median frequency is the preferred parameter for measuring the frequency compression as a function of the conduction velocity of the muscle fibers. For additional details, refer to Stulen & De Luca (1981). Note that the median frequency decrease (a measure of metabolic fatigue) always occurs before contractile fatigue (Figure 1).

To track the spectral compression which accompanies localized muscle fatigue, an analog device was developed to measure the median frequency on-line and in real-time. Figure 2 is a photograph of the actual device, "The Muscle Fatigue Monitor." The ME signal is pre-amplified and bandpass filtered. The preamplified signal is simultaneously passed through high and low pass filters whose cut-off frequencies may be modulated but are constrained to be equal. Each signal then passes through RMS circuits. The high and low RMS voltages are differentially amplified and passed through an integrator. The integrator output modulates the cut-off frequencies of the high and low pass filters. As the integrator adjusts the cut-off frequencies of the filters, it forces the RMS voltages to be equal. At this point, the integrator uniquely determines the median frequency. For additional details, refer to Stulen and De Luca (1982).

General Properties of the Median Frequency

A variety of experiments have been performed to assess the convenience of using the Muscle Fatigue Monitor and the reliability and usefulness of using the median frequency as a measurement of localized muscular fatigue. The experiments to be described were all performed on either the first dorsal interosseous (FDI) or deltoid muscles. A schematic illustration

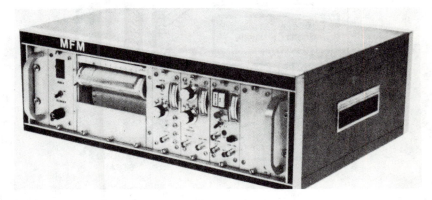

Figure 2—The Muscle Fatigue Monitor.

of a typical experimental arrangement recording the isometric force and the ME signal of the FDI is presented in Figure 3. The thermocouple shown in the figure was only used during the cooling experiments (to be described in sequel). The recording arrangement for the deltoid muscle was essentially similar in concept. Figure 4 presents typical recordings from these types of experiments.

The time-dependent behavior of the median frequency was investigated during sustained, isometric, constant-force contractions at 20%, 50%, and 80% of maximal voluntary contraction (MVC). Figure 5 shows the median frequencies of the ME signal obtained from the FDI and deltoid muscles, respectively, in one subject. The median frequency was observed to decrease as the contraction time progressed. The higher the force of contraction, the greater the rate of decrease. The average starting value of the initial median frequency (defined as the peak frequency recorded during the first 0.5 s of contraction) for the FDI was found to be much greater than that for the deltoid, while their ending values were approximately the same. This difference was determined to be statistically significant ($p < 0.05$). The curves obtained from other subjects have shown a large variation with force in the initial values, while yet other subjects have insignificant variations. The results have shown a smaller variation for the deltoid as compared to the FDI. The final values in both the FDI and the deltoid showed less variation than the initial values as a function of force.

It was observed that the value of the median frequency at the end of the sustained contractions at 50% MVC was slightly less than that obtained at

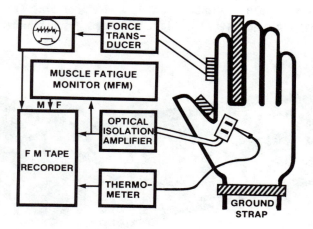

Figure 3—Illustration of experimental arrangement for recording the myoelectric signal and force output of the FDI.

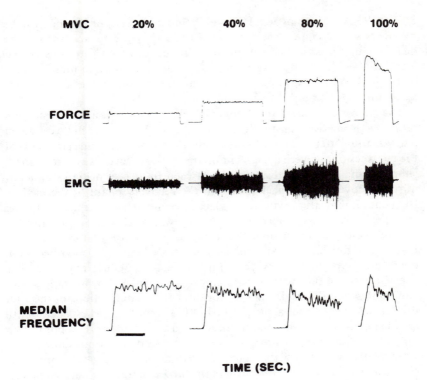

Figure 4—Recording of the force output, myoelectric signal, and median frequency of the FDI at 20%, 40%, 80% and 100% of Maximal Voluntary Contraction (MVC). The horizontal bar represents 3 s of sustained contraction.

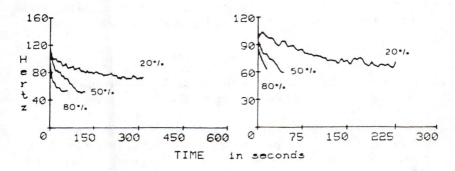

Figure 5—The median frequencies of myoelectric signals obtained from one subject performing sustained, isometric, constant-force contractions of the first dorsal interosseus (FDI) (left graph) or of the deltoid (right graph) at 20%, 50% and 80% of Maximal Voluntary Contraction (MVC). The ordinate is the median frequency in Hertz.

80% MVC. A preliminary analysis would find this result conflicting with the hypothesis that lactic acid accumulation within the muscle causes the decrease in the value of the median frequency. However, Tesch and Karlsson (1977) found maximal lactic acid accumulation in muscle tissue which had contracted at 50% MVC. This coincidence in results requires further investigation.

The recovery of the median frequency to its initial value after a sustained, fatiguing, isometric, constant-force contraction has also been investigated and was found to be essentially complete after 5 min. The time constants of the recovery phase were dependent on the previous level of contraction. A simple rising exponential curve can be used to fit the median frequency during recovery. It is speculated that the recovery phase of the median frequency is indicative of the rate of lactate removal from the muscle tissue.

Repeatability of the median frequency was analyzed by plotting the average value of the median frequency as a function of the force level maintained during each particular contraction. Data were obtained from contractions performed by the same subject on 3 different days. The average value of the median frequency was determined by averaging four values obtained during isometric, constant-force contractions of the FDI or deltoid performed at 10%, 20%, 40% and 60% MVC held until exhaustion. The average coefficient of variation of all the data points was only 2.4% with a range from 1.0% to 5.4%. For any one subject, the curves were relatively close together and followed the same trend with force. The coefficient of variation of the mean of the three average median frequencies was 9.6%. Therefore, the measurement of the median frequency was consistent from day to day, even though the recording electrode was applied and removed each day. Similarly, measurements made without removing the electrode indicated a 5.4% variation among the averaged contractions. The data indicate that measurement of the median frequency provides a reliable parameter for assessing localized muscular fatigue.

Effect of Ischemia and Cooling on the Median Frequency of the Myoelectric Signal

Ischemia was induced by inflating a pneumatic cuff to 180 mmHg for 10 min around the subject's wrist. Cooling was achieved by placing an ice pack on the skin over the FDI. Ten males volunteered for the study. Each subject elicited brief isometric contractions at 20% and 80% MVC at prescribed intervals during and after ischemia or cooling. In this experiment, only the value of the initial median frequency (IMF) of the constant-force contraction was measured. In both cases, the IMF decreased and subsequently recovered; also, the value of the IMF during 80% MFC was always lower than during the corresponding 20% MVC (Figure 6). These results (Sabbahi et al., 1979a,b) indicate that the decrease in the median frequency is related to the accumulation of acidic byproducts which are present in the muscle tissue in greater quantities when the sequence of 80% MVC is performed (Tesch & Karlsson, 1977). During cooling, the decrease was linearly related to the intra-muscular temperature. The post-ischemia and post-cooling recovery was also consistent with the notion that the median frequency behaves in a similar fashion to conduction velocity.

Differences in the Median Frequencies of the Myoelectric Signal due to Handedness

Recently a study was begun for the purpose of investigating the commonality of the median frequency (obtained under similar conditions) of contralateral corresponding muscles (Sabbahi et al., 1981). We wished to determine if handedness (dominance) affects the median frequency. These experiments were limited to the FDI. So far, the FDI muscle in the hand of 10 right-handed males, 10 left-handed males, 10 right-handed females, and 10 left-handed females have been tested. The experimental set-up was similar to that described previously (Figure 3). Each subject was asked to abduct the index finger isometrically at 20%, 40%, 80% and 100% MVC. In this set of experiments, only the value of the IMF at the beginning of the constant-force contractions was measured.

In the right-handed subject, the IMF of the FDI of the non-dominant hand was either equal to or higher than that of the FDI of the dominant

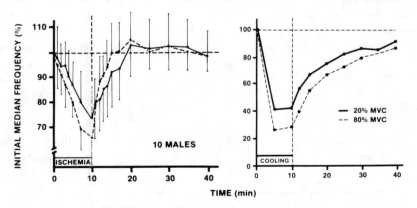

Figure 6—Time course of the average initial median frequency (as percent of control) during and subsequent to local ischemia (left graph) or localized cooling of the FDI (right graph) of a representative subject.

hand at levels greater than 20% MVC. The value of the grand mean of the IMF in all the right-handed subjects was substantially higher in the left hand than in the right hand at every contraction level. In the left-handed subjects, however, no relationship was found between the value of the IMF and the muscle of the hand tested. Similar results were not obtained from female subjects who appeared to have a higher IMF in their FDI than their male counterparts. Further study is needed to effectively evaluate the results obtained in females.

Our current interpretation of the preliminary results favors the concept that the more used muscle (in this case the right FDI of right-handed subjects) should have, over the many years of preferential usage, developed relatively more slow-twitch fibers than its contralateral corresponding muscle. Mild evidence of fast-twitch to slow-twitch fiber changes as a function of exercise has been presented by some investigators. (For a review of this literature, see Salmons & Henriksson, 1981). In the FDI slow-twitch fibers have smaller diameters and slower conduction velocities (Polgar et al., 1973); therefore, one would expect to obtain lower median frequency values from a muscle with relatively more slow-twitch fibers. Admittedly, this argument requires proof.

Possible Muscle Fiber Typing by Analysis of Surface Myoelectric Signals

In this study (Rosenthal et al., 1981), the surface ME signal was detected from the soleus, gastrocnemius, vastus medialis, vastus lateralis, as well as the FDI muscles of six males. These muscles were specifically chosen due to their different fiber-type composition as reported in the literature (Johnson et al., 1973). Because this was the beginning of a pilot study, no muscle biopsies were performed. The subject performed brief isometric constant-force contractions at 20% and 80% MVC. The isometric contractions of the lower limb muscles were measured on an adapted Cybex II dynamometer, and again, only the IMF was considered.

Figure 7 indicates that at higher levels of contraction, the percentage difference of the IMF (between the two contractions) was minimal in the soleus muscle, whereas in the other muscles, it gradually increased proportionally with the increased percentage of fast-twitch fibers within these muscles.

Median Frequency and Muscle pH Measurements Obtained from Normal Human Subjects

To obtain an indication of the correlation between localized muscular fatigue and the change in pH in human subjects, we recorded the median

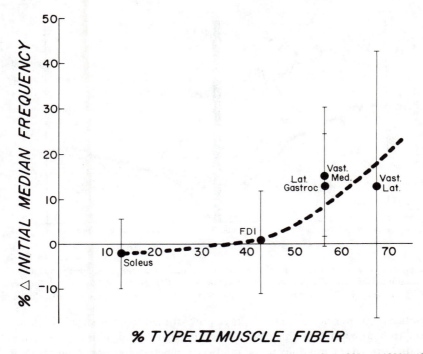

Figure 7—The percentage change of the initial median frequency (from 20% to 100%) of Maximal Voluntary Contraction (MVC) increases exponentially with the percentage composition of Type II muscle fibers in a particular muscle. The intersubject variations are probably due to the difference in the percentage of Type II fiber composition.

frequency as a function of time during a sustained, isometric, constant-force contraction along with the intramuscular pH. Figure 8 shows the median frequency of the ME signal of the FDI muscle as a function of such a contraction, with the corresponding changes in pH levels of the muscle tissue. The pH is expressed as the difference from its value at rest and the value obtained during the contractions. The right side of the graph shows the pH recovery phase while the muscle was relaxed. During the contraction, both the median frequency and the muscle pH decreased from their initial starting levels. In the recovery phase, the pH rose and eventually leveled off and returned to its approximate starting level.

It is difficult to perform a quantitative analysis on this type of experiment due to some of the inaccuracies inherent in the experimental arrangement. Figure 8, however, does show the qualitative behavior of the muscle pH during localized muscle fatigue. The similarity in the trends for both the median frequency and the muscle pH during contraction is strongly indicative of their correlation. This figure represents the results

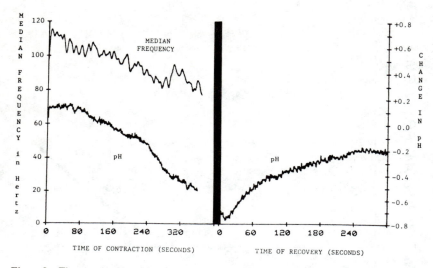

Figure 8—The muscle pH and median frequency obtained from a human subject performing a sustained, isometric, constant-force contraction of the FDI at 50% of Maximal Voluntary Contraction (MVC). The pH recovery after termination of the contraction is also shown.

obtained from one subject. We have now studied five subjects. The same relationships appear to have been present in all of them. This is yet another indication of the possible causal relationship between the accumulation of acidic byproducts within the muscle and the decrease in the median frequency of the myoelectric signal.

Preliminary Results Obtained from the
Gastrocnemius Muscle of the Rabbit

The pH measurements made in the human subjects could not be easily analyzed quantitatively. In order to perform similar measurements in more controlled conditions, an animal preparation was developed In this preparation, we measured the muscle fiber conduction velocity, the pH of the muscle surface, and the median frequency of the ME signal during a sustained stimulation on in situ nerve-muscle preparation.

The preparation consisted of White New Zealand rabbits anesthetized with ethyl carbamate. A hook stimulation electrode was placed on the sciatic nerve; the gastrocnemius muscle was exposed and prepared for recording the ME signal. A glass-bulb type pH electrode was firmly positioned on the medial head of the gastrocnemius. A glass reference electrode was placed between the medial head and skin. Prior to electrode placement, the pH electrode and meter were calibrated using buffered pH solutions. The two electrodes were connected to a pH meter whose output was recorded on an FM tape recorder. To obtain the conduction velocity, a specially constructed electrode was used, consisting of two bipolar electrodes (DISA Model 13K780) with a 10 mm separation to optimize conduction velocity measurements. The placement was parallel to the muscle fiber and away from the motor endplate region.

Only two such experiments have been performed to date. The conduction velocity of the muscle fiber was calculated by measuring the delay in the arrival time of the compound action potential at the two recording contacts of the conduction velocity electrode. The pH and median frequency were obtained as described previously. There was striking interaction between the above parameters, strongly suggesting a causal relationship between muscle pH, conduction velocity, and the power density spectrum of the myoelectric signal.

Acknowledgment

This work was supported in part by Liberty Mutual Insurance Company and by the Department of Education Research Grant GOO-800-3004 (National Institute of Handicapped Research).

References

JOHNSON, M.A., J. Polgar, D. Weightman & D. Appleton. Data on the distribution of fiber types in 36 human muscles: an autopsy study. J. Neurol. Science 18:111–129, 1973.

LINDSTROM, L., On the frequency spectrum of e.m.g. signals. Thesis. RES Lab. Med. Electronics. Chalmers Univ. of Tech., Göteborg, Sweden, 1970.

POLGAR, J., M.A. Johnson, D. Weightman & D. Appleton. Data on fiber size in 36 human muscles: an autopsy study. J. Neurol. Science 19:307–318, 1973.

ROSENTHAL, R.G., M.A. Sabbahi, R. Merletti & C.J. De Luca. Possible fiber typing by analysis of surface e.m.g. signals. Proc. of the 11th Ann. Meeting of the Soc. for Neuroscience, p. 683, November 1981.

SABBAHI, M.A., C.J. De Luca & W.R. Powers. The effect of ischemia, cooling and local anesthesia on the median frequency of the myoelectric signal. Proc. of the 4th Conf. of I.S.E.K., pp. 94–95, August, 1979A.

SABBAHI, M.A., R. Merletti, C.J. De Luca, & R.G. Rosenthal. How handedness, sex and force level affect the median frequency of the myoelectric signal. Proc. of the 4th Ann. Conf. on Rehab. Engrg., pp. 232–234, August, 1981.

SABBAHI, M.A., W.R. Powers, C.J. De Luca & F.B. Stulen. Intramuscular

186   De Luca, Sabbahi, Stulen and Bilotto

conduction velocity during ischemia and cooling. Physical Therapy 59:579, May, 1979B.

SALMONS, S. & J. Henriksson. The adaptive response of skeletal muscle to increased use. Muscle and Nerve 4:94–105, 1981.

STULEN, F.B., A technique to monitor localized muscular fatigue using frequency domain analysis of the myoelectric signal. Ph.D. Thesis. M.I.T., 1980.

STULEN, F.B. & C.J. De Luca. A non-invasive device for monitoring localized muscular fatigue. Proc. of the 14th Ann. Meeting of A.A.M.I., p. 268, May, 1979.

STULEN, F.B. & C.J. De Luca. Frequency parameters of the myoelectric signal as a measure of muscle conduction velocity. IEEE Trans. on Biomed. Engrg. 28:515–523, 1981.

STULEN, F.B. & C.J. De Luca. Muscle Fatigue Monitor: A non-invasive device for observing localized muscular fatigue. IEEE Trans. on Biomed. Engrg., 29:760–768, 1982.

TESCH, P. & J. Karlsson. Lactate in fast and slow twitch skeletal muscle fibres of man during isometric contractions. Acta Physiol. Scand. 99:230–236, 1977.

Electromyography in Muscle Fatigue Studies: Power Spectrum Analysis and Signal Theory Aspects

Lars Lindström and I. Petersén

University of Göteborg Sahlgren's Hospital,
Göteborg, Sweden

One of the first observations made with electromyographic (EMG) methods concerned so-called muscle fatigue. In 1912, Piper found that the dominating wave of the myoelectric signal—later named the Piper rhythm—decreased during forceful contractions from about 50 Hz to 40 or 30 Hz (Piper, 1912). Subsequent research added the finding that the amplitude of the surface derived signal increased during the development of the fatigue. An illustration to these phenomena is shown in Figure 1. Later, with intramuscular electrodes, it was found that the potential duration of the fatigued motor unit increased, whereas the amplitude—surprisingly enough—decreased slightly or remained constant. In order to

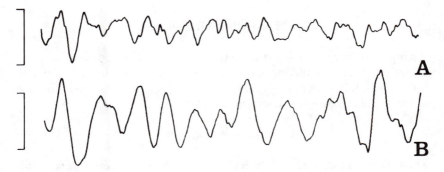

Figure 1—Myoelectric changes during fatigue. Surface EMG from biceps brachii muscle loaded with 8 kg at the wrist. Curve (A) shows the signal at the beginning of an exhausting contraction and (B) at the end. The vertical bars represent 5 mV and the length of each tracing is 250 ms.

explain this contradiction the synchronization concept was commonly adopted. It also seemed to explain the occurrence of the Piper rhythm and its slowing with fatigue.

At increased contraction levels so many motor units are recruited and their repetition rates are so high that the recorded signal seldom allows study of individual potentials. The signal has the character of random noise. A convenient way to analyze such signals is by power spectrum analysis. Early applications of such EMG analyses in muscle fatigue studies are those of Kaiser and Petersén (1962) and Kogi and Hakamada (1962) with the findings that the high frequency power decreased and that of low frequencies increased during forceful contractions. These findings have since then been confirmed by a large number of investigators. The explanations offered have ranged from alterations in the CNS efferent pattern to local muscle phenomena.

Parallel with experimental EMG studies, theoretical models of the myoelectric signal generation have emerged. Two main approaches can be distinguished in such modelling: one focusing on properties of the motor unit impulse trains (e.g., De Luca, 1975; De Luca & Van Dyk, 1975) and the other one on volume conduction phenomena (e.g., Rosenfalck, 1969). In an attempt to make a partial infusion of the two approaches, we have used the volume conduction theory and included the time series aspects of the signal summation within the motor unit (for a review see Lindström & Magnusson, 1977).

In this presentation a few elements from our theoretical calculations on the myoelectric signal, relevant to the study of muscle fatigue phenomena, will be briefly outlined. We will also summarize some of our experiments, including hitherto unpublished ones, which have led to our present standpoint concerning EMG and muscle fatigue.

Theoretical Methods

A simple Fourier transformation of the expression for a source signal (e.g., an action potential) moving with a certain velocity (v) shows (Lindström, Magnusson & Petersén, 1970) that the power spectrum W(ω), as a function of the angular frequency ω, can be written:

$$W(\omega) = v^{-2} \, G(\omega/v) \tag{1}$$

where G(ω/v) includes all filtering effects of the myoelectric signal. This general relation is valid for signals from single fibers, motor units, as well as the whole muscle. It says that any change in the velocity (v) is accompanied by a frequency scaling of the power spectrum together with a change in the total power of the signal.

The modification the signal undergoes when the electrode is moved a distance (h) from the source (a muscle cell of radius, a) is expressed with modified Bessel functions, which are strongly decreasing with increasing argument:

$$\text{Distance effect} = K_0(\omega\, h/v)\; /\; K_0(\omega\, a/v) \tag{2}$$

This signal modifying factor tells us that the high frequency components are strongly attenuated for increased observation distances, the attenuation being velocity dependent. It also tells us that the low frequency components are hardly attenuated at all and therefore can be conducted over considerable distances.

The motor unit signal is obtained by summation of the single fiber contributions having a certain time dispersion caused by the different starting points (standard deviation, s) in the innervation zone. The modification of the spectrum through the summation of N single contributions is:

$$\text{Summation effect} = N^2 \exp(-\omega^2 s^2/v^2) \tag{3}$$

Here the duration of the motor unit signal is:

$$T_{mu} = s/v \tag{4}$$

With the single fiber source signal equal to ω/v we can calculate the motor unit signal power and its RMS (root-mean-square) value:

$$RMS_{mu} \sim N\, s^{-1}\, v^{-1/2} \tag{5}$$

The motor unit signal amplitude is then obtained from Equations 4 and 5:

$$A_{mu} \sim N\, s^{-3/2} \tag{6}$$

The relation (3) demonstrates that the motor unit signal has a strong roll-off on the high frequency side besides that caused by the observation distance.

The final step in obtaining the whole muscle signal is to make a summation of the contributions from all motor units in the muscle. The expression yielded by this summation is quite complex, but it still shows the same interrelation between frequency and velocity as pointed out in

Equation 1. The determinants of the signal power are still easy to obtain. We find for the amplitude (in this case proportional to the RMS value):

$$A_{tot} \sim N^{1/2} \, v^{-1/2} \, s^{-1/2} \tag{7}$$

With respect to muscle fatigue, one of the more interesting phenomena in the power spectrum is caused by the use of bipolar electrodes where the signal first passes under one of the electrode plates and, after a certain time delay, under the other plate. This causes the power spectrum to show so-called dips, determined by the modifying factor:

$$\text{Electrode effect} = \sin(\omega \, d/v) \tag{8}$$

From the position of the dips (f_{dip}), i.e., frequencies where the sin()-function is zero, the propagation velocity of the action potentials can be calculated according to the question:

$$v = 2d \, f_{dip} \tag{9}$$

It should be observed that this relation contains only the velocity, the electrode size and the dip frequency but no other parameter of biological origin.

Experimental Results

Early EMG experiments at our laboratory (Kadefors, Kaiser & Petersén, 1968) confirmed and extended previous findings that the high frequency content decreased and the low frequency content increased during fatigue. The load dependence as well as the rate of recovery after fatiguing contractions was also quantified.

Investigations concerning the underlying mechanisms for the spectral changes were performed in 1970 (Lindström, Magnusson & Petersén, 1970). These investigations were largely guided by the theoretical findings concerning the influence of the propagation velocity on the myoelectric signal. The "dip" method was used to prove that the propagation velocity did decrease during forceful contractions. An approximately exponential time course was found:

$$v(t) \cong v(O) \exp(-t/T_f) \tag{10}$$

Here, t is time of contraction and T_f is a characteristic fatigue time constant. Later (Lindström, Kadefors & Petersén, 1977), the "dip" method was replaced by an indirect determination of the velocity change through the center frequency of the myoelectric spectrum (Equation 1 immediately reveals that the center frequency is proportional to the propagation velocity). Application of this method has shown that there exists a threshold for muscle force below which time dependent changes in the EMG cannot be monitored.

In 1970, studies were made on the effect of anoxia and ischemia on the EMG (Mortimer, Magnusson & Petersén, 1970). It was found that the oxygen lack as such, but with intact perfusion, gave small velocity decrements for "red" muscles and no changes at all for "white" ones. On the other hand alterations in the perfusion flow had very large effects on the propagation velocity of the action potentials for both muscle types.

Although it is well known that the temperature has an influence on the myoelectric signal, few investigations have been made concerning the spectral influence of heat production in fatiguing contractions (e.g., Petrofsky, 1979). Results in this respect from our laboratory indicate that the total velocity change during forceful contractions is caused by two counteracting factors, one varying with temperature, and the other one depending on other phenomena. The latter one is dominating, being approximately five times that of the temperature factor in strength.

In order to apply the EMG methods to fatigue studies of dynamic work, Örtengren (1972) modelled this type of work as a series of forceful contractions with pauses of rest in between, characterizing the work situation by the force, the period time, and the duty cycle (i.e., the work time in relation to the total period time). A strong dependence of the EMG changes on force and duty cycle was found. Calculations on the periodic muscle loading as a series of exponential velocity decreases during the work phase alternating with exponential increases (Broman, 1977) during the rest phase, have shown that the velocity decreases initially and, in contrast to the isometric case, levels out at a plateau (e.g., Örtengren, 1979). The value of this level is determined by the time constants for fatigue (T_f reflects the force) and recovery (T_r) and by the duty cycle κ. The calculations—extensively verified experimentally—show that the relative velocity decrease is:

$$\Delta v / v(O) = [1-(1-\kappa)(T_f/T_r)/\kappa]^{-1} \qquad (11)$$

The interrelation between muscle fatigue, as monitored with EMG, and the subjective experience of the work performed as well as of the fatigue it causes, has been investigated to some extent. The subjective fatigue was assessed with a rating scale of perceived exertion (RPE scale) originally developed for general fatigue tests. A parallel development was found for the velocity decrease and the subjective sensation of exertion, although the latter measure also included the influence of the loading as such. At the end of the (endurance) experiments, the velocity decrease was deeper with high load experiments than with low ones. This finding led us to experiments in which the EMG fatigue time constant was used as a predictor of endurance. A close agreement was found between the endurance time and the EMG fatigue time constant (T_f) raised to the power 2/3 (Lindström & Petersén, 1981).

Discussion

The myoelectric signal has a power spectrum which is peaked. This peaked shape has its origin in filtering effects caused by electrodes, observation distances and signal summation. The shape explains the existence of the Piper rhythm which consists of components mainly from the peak frequency. It also explains why a spectral shift alters the low and high frequency regions.

Apart from the fact that measurements have shown the propagation velocity to decrease during forceful contractions, virtually all observations concerning EMG changes during fatigue are explained by a velocity decrease. Other phenomena may occur but their influence on the myoelectric signal is less dominating. The theoretical investigation on volume conduction and signal summation shows that the frequency always appears in connection with the velocity as a quotient ω/v. Any change in velocity thus gives a shift (c.f. Stulen & De Luca, 1981) of the spectrum along the frequency axis. Also, the earlier mentioned contradiction between the amplitude changes of motor unit and whole muscle signals during fatigue is dissolved if one considers the signal power dependence on the velocity.

A possible chain of events in the development of fatigue, as seen with EMG methods, is indicated in Figure 2. Forceful contractions will impair the blood flow through the muscle (e.g., Barcroft & Millen, 1939) which causes a lack of oxygen and an alteration of the metabolism to the anaerobic type. The accumulated metabolites then lower the propagation velocity of the myoelectric action potentials, probably through a change in excitability of the cell membranes. Intracellular hydrogen ions are since long known as strong depressors of membrane excitability (Tasaki, Singer & Takenaka, 1967), and thus of the propagation velocity. Of interest is that the excitation process is less sensitive to extracellular pH changes than to intracellular ones (Terakawa, Nagano & Watanabe, 1978). The mechanism is also quite well described from a theoretical point of view (Bass & Moore, 1968).

The EMG fatigue experiments indicate the existence of a force threshold (at about 20% of the maximum voluntary contraction) below which fatigue

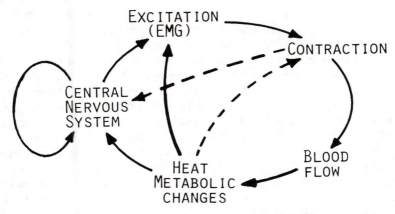

Figure 2—Schematic diagram illustrating some "pathways" involving EMG in the development of muscle fatigue.

is not detected with myoelectric methods. When considering dynamic work this "free" interval is much enlarged due to the repeated possibilities for recovery. In fact, for work with moderate force development very little fatigue will persist as long as the duty cycle is within or below the range 0.3 to 0.5. In most activities of biological significance, the duty cycle remains low (e.g., in walking, running, respiration, chewing) and fatigue is only seen in pathological or otherwise abnormal cases (e.g., diseases and sports activities).

The measurements of temperature development in relation to electromyographic spectral changes indicate that the heat has only a minor influence on the signal. The production and elimination of the heat follows essentially the same physical laws as with the metabolites. The time constants involved are different, however, and will thus give different EMG changes depending on the time relations as expressed by the duty cycle and the period time. For strong contractions and low values of the duty cycle one might even arrive at the situation in which the velocity increase due to temperature rise is larger than the depression caused by accumulated metabolites. Such a balance thus gives a net increase in velocity and consequently also in the spectral center frequency.

The measurements of perceived exertion demonstrate a definite parallelism between the subjective experience and the fatigue development of the myoelectric signal. There are several possible mechanisms that underlie this relation: irradiation in the CNS reticular system, nervous feed-back through free nerve endings or nociceptors, and humoral transmission to the CNS. Our data fit reasonably well with the interpretation that the myoelectric signal reflects intracellular events, whereas a diffusion limited outflow of metabolites is sensed extracellularly.

The application of EMG power spectrum analysis has mainly been used in the following disciplines: basic muscle physiology, clinical neurophysiology, ergonomics, and respiratory physiology. In ergonomics, EMG spectrum analysis has proven valuable, to quantify incidents of fatigue in industrial work (Örtengren, Andersson, Broman, Magnusson & Petersén, 1975) and is in this respect a necessary complement to the methods commonly used in work physiology. The EMG analysis has also been used to spot highly fatigued muscles in complicated interplay with synergists (Kadefors, Petersén & Herberts, 1976). In respiratory physiology the spectrum analysis of diaphragmatic EMG has become an interesting tool in the evaluation of the fatigue state of that muscle (e.g., Grassino & Bellemare, 1978; Gross, Ladd, Riley, Macklem & Grassino, 1980) and in the monitoring of early signs of respiratory failure (Muller, Gulston, Cade, Whitton, Froese, Bryan & Bryan, 1979).

The application of EMG methods to the study and quantification of muscle fatigue seems to have a future. A few areas have been mentioned but many others are obvious. One application is to the measurement of peripheral circulatory disturbances where the EMG spectrum analysis, in contrast to the plethysmographic methods now in use, has an increased sensitivity with decreasing blood flows. Finally we want to point out the possibility of using spectrum analysis of myoelectric signals as an indicator of the intracellular metabolic state.

References

BARCROFT, H. & J.L.E. Millen. The blood flow through muscle during sustained contraction. J. Physiol. 97:17–31, 1939.

BASS, L. & W.J. Moore. A model of nervous excitation based on the Wien dissociation effect. In A. Rich & N. Davidson (Eds.), Structural Chemistry and Molecular Biology. San Francisco: W.H. Freeman & Co., 1968.

BROMAN, H., An investigation on the influence of a sustained contraction on the succession of action potentials from a single motor unit. Electromyogr. Clin. Neurophysiol. 17:341–358, 1977.

DE LUCA, C.J., A model for a motor unit train recorded during constant force isometric contractions. Biol. Cybern. 19:159–167, 1975.

DE LUCA, C.J. & E.J. Van Dyk. Derivation of some parameters of myoelectric signals recorded during sustained constant force isometric contractions. Biophys. J. 15:1167–1180, 1975.

GRASSINO, A. & F. Bellemare. Respiratory muscle fatigue and its effects on the breathing cycle. In C. von Euler & H. Lagrercrantz (Eds.), Central Nervous Control Mechanisms in Breathing. Stockholm: Wenner-Gren Center Int. Symposium Series, 1978.

GROSS, D., H. Ladd, E. Riley, P.T. Macklem & A. Grassino. The effect of training on strength and endurance of the diaphragm in quadriplegia. Amer. J. Med. 68:27–35, 1980.

KADEFORS, R., E. Kaiser & I. Petersén. Dynamic spectrum analysis of myopotentials with special reference to muscle fatigue. Electromyography 8:39–74, 1968.

KADEFORS, R., I. Petersén & P. Herberts. Muscular reaction to welding work: An electromyographic investigation. Ergonomics 19:543–558, 1976.

KAISER, E. & I. Petersén. Frequency analysis of action potentials during tetanic contraction. Electroenceph. Clin. Neurophysiol. 14:955, 1962.

KOGI, K. & T. Hakamada. Slowing of surface electromyogram and muscle strength in muscle fatigue. Rep. Inst. Sc. Lab. 60:27–41, 1962.

LINDSTRÖM, L., R. Kadefors & I. Petersén. An electro-myographic index for localized muscle fatigue. J. Appl. Physiol.: Respirat. Environ. Exercise Physiol. 43:750–754, 1977.

LINDSTRÖM, L. & R.I. Magnusson. Interpretation of myoelectric power spectra: A model and its applications. Proceed. IEEE. 65:653–661, 1977.

LINDSTRÖM, L., R. Magnusson & I. Petersén. Muscular fatigue and action potential conduction velocity changes studied with frequency analysis of EMG signals. Electromyography 10:341–355, 1970.

LINDSTRÖM, L. & I. Petersén. Power spectra of myoelectric signals: motor unit activity and muscle fatigue. In E. Stålberg & R.R. Young (Eds.), Clinical Neurophysiology. London: Butterworths, 1981.

MORTIMER, J.T., R. Magnusson & I. Petersén. Conduction velocity in ischemic muscle: Effect on EMG frequency spectrum. Am. J. Physiol. 219:1324–1329, 1970.

MULLER, N., G. Gulstone, D. Cade, J. Whitton, A.B. Froese, M.H. Bryan & A.C. Bryan. Diaphragmatic muscle fatigue in the newborn. J. Appl. Physiol.: Respirat. Environ. Exercise Physiol. 46:688–695, 1979.

ÖRTENGREN, R., Spectrum analysis of myoelectric signals from the brachial biceps muscle during isometric dynamic loading. The Third Int. Conf. on Med. Physics, incl. Med. Eng., Göteborg, Sweden. 13:5, 1972.

ÖRTENGREN, R., Electromyographic evaluation of localized muscle fatigue during static and dynamic loading. In A. Persson (Ed.), Symposia: Invited Contributions to 6th Intern. Congr. EMG. Stockholm: Huddinge University Hospital, 1979.

ÖRTENGREN, R., G. Andersson, H. Broman, R. Magnusson & I. Petersén. Vocational Electromyography: Studies of Localized Muscle Fatigue at the Assembly Line. Ergonomics 18:157–174, 1975.

PETROFSKY, J.S., Frequency and amplitude analysis of the EMG during exercise on the bicycle ergometer. Eur. J. Appl. Physiol. 41:1–15, 1979.

PIPER, H., Elektrophysiologie menschlicher Muskeln. Berlin: Verlag von Julius Springer, 1912.

ROSENFALCK, P., Intra- and Extracellular Potential Fields of Active Nerve and Muscle Fibers. Kobenhavn: Akademisk forlag, 1969.

STULEN, F.B. & C.J. De Luca. Frequency parameters of the myoelectric signal as a measure of muscle conduction velocity. IEEE Trans. Biomed. Eng. 28:515–523, 1981.

TASAKI, I., I. Singer & T. Takenaka. Effects of internal and external ionic environment on excitability of squid giant axon. J. Gen. Physiol. 48:1095–1123, 1967.

TERAKAWA, S., M. Nagano & A. Watanabe. Intracellular pH and plateau duration of internally perfused squid giant axons. Jpn. J. Physiol. 28:847–862, 1978.

Electromyographic, Mechanical, and Metabolic Changes during Static and Dynamic Fatigue

Paavo V. Komi

University of Jyväskylä, Jyväskylä, Finland

In electromyographic (EMG) experiments of fatigue in human skeletal muscle three major agreeable findings have been observed: 1) In submaximal maintained isometric contractions integrated EMG (IEMG) activity increases and the slope of this increase is dependent on the maintained tension level (e.g., Laurig, 1970; Rau & Vredenbregt, 1970; Viitasalo & Komi, 1978); 2) the relationship between IEMG and force throughout the entire range of isometric forces is shifted to the left when muscle is fatigued (e.g., Komi & Viitasalo, 1977); 3) under maximal contractions while the maximum force decreases the maximum IEMG also decreases (e.g., Bigland-Ritchie, 1981). Although these basic observations do not necessarily explain the mechanism of neuromuscular fatigue, they, however, demonstrate how the input (activation)/output (force production) relationship is changing under maximal and submaximal contractions. The first part of this report discusses this problem of input/output relationship by comparing the fatigue responses of repeated maximal concentric and eccentric contractions.

It is logical to expect that fatigue causes changes not only in the amount of EMG activity but also in the characteristics of the recorded signals. Power spectral density function has often been used to indicate changes in the signal characteristics. An increase of the lower spectral components with a concomitant decrease in the high frequency components has been well documented (e.g., Kadefors et al., 1968; Viitasalo & Komi, 1977). Less is, however, known about the EMG spectrum changes in the skeletal muscles which have different composition of fast-twitch (FT) and slow-twitch (ST) fibers. The second part of this paper presents some evidence that the magnitude of the EMG spectrum changes is related to muscle fiber composition. The different muscle fibers have different metabolic profiles, and therefore it is of interest to examine also if the production of lactic acid in the muscle is related to EMG changes during fatigue contractions. Finally, fatigue is expected to cause changes in the reflex activation of the muscle. Therefore input/output characteristics of the reflexly induced contraction are compared with those of the voluntary contraction, both measured under similar fatigue conditions. Surface EMG recording was employed in all studies which will be reported in more detail in this report.

Fatigue under Maximal Concentric and Eccentric Contractions

Eccentric work is characterized by smaller expenditure of energy (e.g., Asmussen, 1953) and lower motor unit activity (e.g., Bigland & Lippold, 1954) than the concentric work performed with similar muscle tension and at constant contraction velocity. Under maximal contractions performed with relatively slow contraction velocity, integrated EMG activity is similar under both conditions but the maximal muscle tension is much higher in eccentric than in concentric work (Komi & Buskirk, 1972; Komi, 1973).

Our interest in studying the fatigability of muscle during eccentric contraction was stimulated by the observations that muscle becomes easily sore after repeated eccentric contractions (Asmussen, 1956; Komi & Buskirk, 1972). Two experimental studies were performed, one with elbow flexors (Komi & Rusko, 1974) and the other with both leg extensors (Komi & Viitasalo, 1977). In both cases the respective muscles were loaded with 40 repeated maximal contractions, in one session concentrically and in the other session, eccentrically. Special dynamometers were employed to ensure that movement velocity and contraction intervals were constant and the same in both types of work.

As Figure 1 demonstrates, the decrease in relative muscle tension during eccentric fatigue loading with elbow flexors was much greater than in the concentric exercise. The absolute forces in the beginning of fatigue loading were 440 N and 240 N, respectively, in eccentric and concentric contraction. Maximum IEMG of the examined elbow flexor muscles, biceps brachii, brachialis, and brachioradialis, were on the same level in the beginning of both types of exercise, and the fatigue caused IEMG values to decrease in a similar manner in both conditions. As Figure 2 demonstrates the reduction in IEMG activity in the biceps brachii muscle was down to 80% from the control situation. In line with the changes in IEMG the energy expenditure, as measured with $\dot{V}O_2$ calculations from the expired air, was the same in the two conditions. Thus the conclusion was made that the metabolic "loading" of the maximal eccentric and concentric fatigue contractions was similar, but because the force decline was greater in

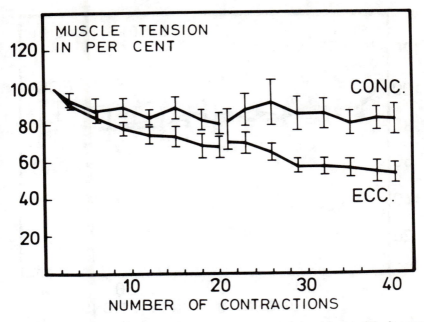

Figure 1—Relative decrease in concentric (conc.) and eccentric (ecc.) forces of the forearm flexors during repeated respective 40 maximal contractions. The absolute maximum force levels in the beginning of the concentric and eccentric exercises were 430 and 240 N, respectively. The force decline was much greater during eccentric fatigue. The total work time in both conditions was 12 min (from Komi & Rusko, 1974).

eccentric exercise and the subjects felt soreness in their exercised muscles only after the eccentric condition, the reason for the greater fatigue (= reduction in force) was termed as mechanical.

In the experiment with leg extensors (Komi & Viitasalo, 1977) the muscle glycogen levels and blood lactate values were determined before, immediately after, and two days after the fatigue loading. No marked differences in these parameters, as well as in the work energy expenditure could be seen between the eccentric and concentric fatigue contractions. EMG was recorded continuously during the work loadings, but also before the work, and in the second day after the test in isometric one leg knee extension employing the following submaximal loads: 150, 250, 400, 550 and 700 N. After the fatigue, the 700 N load was often too heavy, and thus could not always be recorded. In addition the force time curve in maximum isometric two leg extension contraction, was recorded before, immediately after, and two days after the fatigue loading.

While the maximal IEMG during the fatigue contractions did not change significantly differently in the two conditions, the tests in isometric

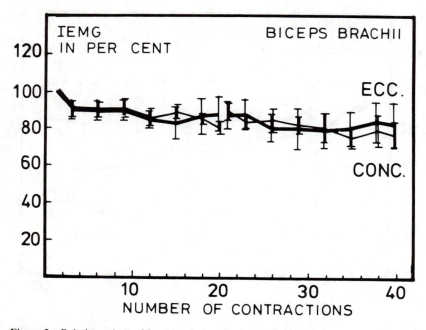

Figure 2—Relative values of integrated electromyographic (IEMG) activity of the biceps brachii muscle during eccentric and concentric fatigue of Figure 1. The absolute maximum IEMG levels were the same in the beginning of exercise in both conditions. IEMG declined in the same manner in concentric and eccentric fatigue.

conditions revealed marked differences. The relationship between IEMG and muscle tension in isometric knee extension showed that more neural energy was needed for the production of a certain muscle tension after than before the fatigue loading. This kind of shift in the regression line to the left (Figure 3, middle) was greater in eccentric than in concentric work. Average motor unit potential (AMUP) computed from the EMG recordings at submaximal isometric load of 250 N, was increased substantially more after eccentric work (Figure 3, lower part). Much greater differences could be seen in the force-time and submaximal EMG parameters when the tests were taken two days after the fatigue loading. The recovery was complete after concentric work but substantially delayed after eccentric work (Figure 3). In fact after eccentric work, no return to the before-test condition occurred either in the force-time or IEMG/muscle tension variables. AMUP curves from the EMG analysis were only slightly different in concentric condition for the three test instances (before, after, and 2nd day), but were substantially separated in eccentric work.

Muscle soreness experienced by all subjects during the few days after eccentric fatigue was undoubtedly associated both with the decreased force

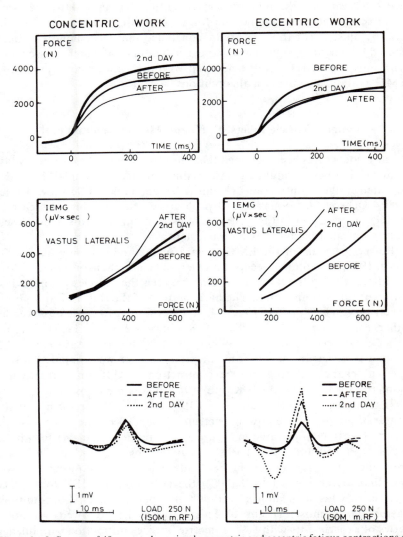

Figure 3—Influence of 40 repeated maximal concentric and eccentric fatigue contractions of the leg extensor muscles on the isometric force-time curve of both leg extension (above), on IEMG/force relationship in unilateral isometric knee extension (middle), and on average motor unit potential (AMUP) recorded from the rectus femoris muscle during isometric knee extension load of 250 N (lower part). The measurements were taken before, immediately after and 2 days after the fatigue loading (data from Komi & Viitasalo, 1977).

capacity (especially in force-time measurements) and with the increased neural activation of the muscle at a given tension. It is natural to expect that more central activation is needed for a sore muscle as compared to the healthy muscle. It can therefore be concluded that "mechanical" fatigue caused by repetitive high tension eccentric loading and which is associated with muscle soreness, can be detected with quantitative and qualitative EMG analysis in submaximal conditions.

Fatigue, Muscle Fibers, EMG and Muscle Lactate Level

Some evidence has been obtained that muscles composed of primarily fast twitch (FT) fibers produce a greater reduction in maximum IEMG under repeated fatigue contractions (Komi & Tesch, 1979; Viitasalo & Komi, 1980). Ochs et al. (1977) have also observed relatively high increase in EMG/tension ratio in the gastrocnemius as compared to soleus muscle during repeated plantar flexion movement.

Power spectrum of the EMG signal (for methods of analysis, see De Luca, 1982, this volume) seems, however, to be more sensitive than IEMG in differentiating the fatigue patterns between the muscles composed of different fiber composition. The basic pattern of the shift in EMG power spectrum can be shown in Figure 4, and it is similar to those observed by others earlier (e.g., Kadefors et al., 1968; Viitasalo & Komi, 1977). Our first observations on the influence of muscle structure on the shift in EMG frequency spectrum came from the study, where we compared power spectral changes during fatigue contraction of different submaximal isometric loads (Viitasalo & Komi, 1978). It was then observed that the change in mean power frequency (MPF), which is the frequency at the center of gravity of the power spectrum, depended on the force in static effort. This is demonstrated in Figure 5, with reductions of MPF with the loads of 30, 50 and 70%. The slopes indicating the shift to the lower frequencies were higher with higher loads. The second major observation was that at the highest load (70%) the reduction of MPF took place significantly faster in subjects with a higher percent of FT fibers in their vastus lateralis muscle (Figure 6). This study was then followed by another series of experiments with 50% isometric fatigue contraction with one leg knee extension (Larsson, Viitasalo & Komi, 1978). The previous finding was confirmed by the result that during the fatigue contraction the relative increase of signals in the lower bandwidth (24 to 48 Hz) increased more, the greater the relative cross-sectional area of FT fibers in the vastus lateralis muscle (r = .73, p < .01).

The major reason for the shift in EMG spectral density function towards lower frequency components during fatigue has recently been questioned

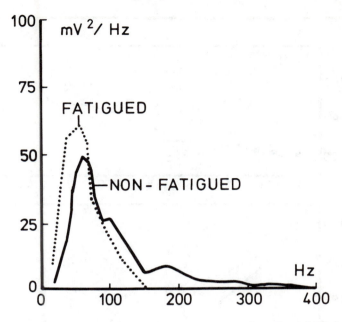

Figure 4—A typical fatigue response in EMG power spectrum (from Komi & Tesch, 1979).

(Edwards, 1981). This shift is thought to be due to the reduction in muscle action potential conduction velocity as suggested by Mortimer, Magnusson & Petersén (1970). Because the latter phenomenon is likely to be related to ischemic conditions in the muscle (see also Stålberg, 1966), and because ischemia increases muscle lactate levels (Karlsson, 1971), we tend to support the concept of Mortimer et al. (1970), and that of Lindström et al. (1970) that EMG spectral shifts indicate changes in muscle action potential conduction velocities (see also Lindström, 1982, this volume).

Accumulation of muscle lactate might be one of the major biochemical events occurring simultaneously with EMG spectral changes. Although the significant interrelationship may not necessarily imply that muscle lactate has a definite causal role, confirmation of such a relationship would be a step for further examination of the detailed mechanisms. To study this association we utilized the experience that muscle lactate formation under maximal dynamic fatigue contractions occurs preferentially in FT fibers (Tesch et al., 1978a & b). The possible association of EMG spectral changes and muscle structure under these dynamic conditions was studied with 11 subjects who performed 100 maximal voluntary concentric knee extensions through a motion range from 100° flexion to a fully extended knee joint at an angular velocity of 3.14 rad x s⁻¹. The passive recovery time between contractions was 0.7 s (for other details see Komi & Tesch, 1979).

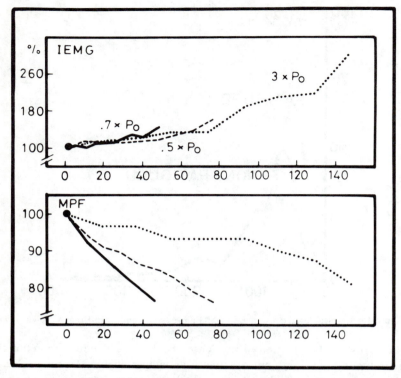

Figure 5—Relative changes in IEMG and in mean power frequency (MPF) of EMG power spectrum at three different levels of isometric contractions (knee extensors) maintained until exhaustion (from Viitasalo & Komi, 1978).

The subjects in the study were divided into two separate groups: group I with percent FT fiber area of 43.2% in the vastus lateralis muscle, and group II (percent FT area 67%). Figure 7 shows that the torque decline was greater in subjects with greater percent FT area. MPF of the EMG power spectrum declined by $12 \pm 4\%$ (n.s.) in group I and $25 \pm 3\%$ (p $<$ .001) in group II. As is shown in Figure 8 the group responses in MPF changes became significant after 25 to 30 contractions. When the changes in two extreme bandwidths of the EMG spectrum were plotted against the number of contractions in the two groups of subjects (Figure 9), the group II had a significantly (p $<$ .01) higher relative content of the lower bandwidth (24 to 48 Hz) and a lower content (p $<$ .05) of the higher bandwidth (136 to 400 Hz) as compared to group I, which had smaller percent FT area in the vastus lateralis muscle. The differences became significant again after 25 to 30 contractions.

Komi and Tesch (1979) did not measure muscle lactate values in the study explained above. They, however, assumed that the marked changes

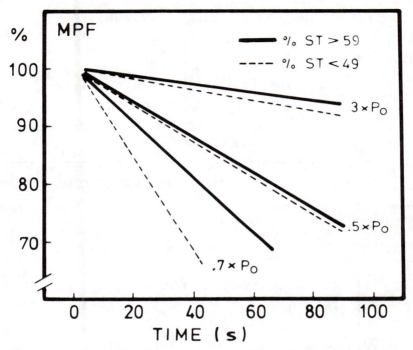

Figure 6—Relative change in mean power frequency (MPF) of EMG spectrum of the rectus femoris, vastus lateralis and vastus medialis muscles under three different isometric loads. Note that at the load of .7 x P_O the decrease in MPF occurred at a significantly ($p < .05$) greater rate in subjects with less than 49% of slow-twitch (ST) fibers in the vastus lateralis muscle (Viitasalo & Komi, 1978).

in force output as well as in MPF were related to greater fatigue of FT fibers of the respective motor units. This assumption was based on the earlier findings of Tesch et al. (1978a & b) with the similar testing protocol that a greater rate of lactate accumulation occurred in FT fibers during the first 25 contractions. In order to obtain direct comparisons between EMG spectral changes and muscle lactate concentration another study (Tesch et al., 1982) was designed. Eight physical education students performed maximal unilateral concentric leg extensions in a speed controlled dynamometer (Komi, 1982). The exercise comprised 120 contractions at an angular velocity of 1 rad x s^{-1}. Passive returns to the starting position (knee angle 90°) intervened the contractions. EMGs from the vastus lateralis, vastus medialis and rectus femoris muscles were recorded with surface electrodes together with the force throughout the total exercise time of 15 min. For measurements of muscle lactate concentrations muscle biopsies (Bergström, 1962) were obtained from the vastus lateralis muscle before and immediately after exercise.

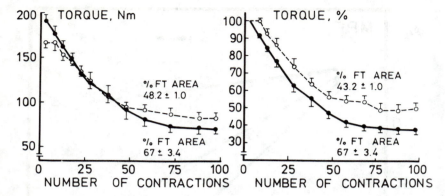

Figure 7—Absolute (left) and relative (right) values of maximal torque of the knee extensor muscles during 100 repeated maximal concentric contractions. Note the greater reduction of the torque in subjects who had predominance of fast-twitch (FT) fibers in their vastus lateralis muscle (Komi & Tesch, 1979).

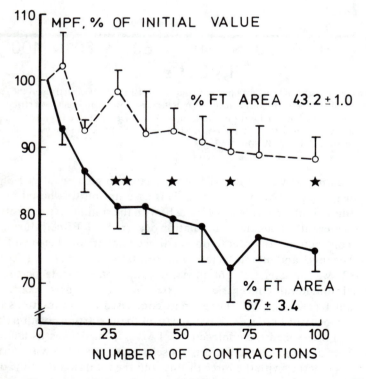

Figure 8—Change in mean power frequency (MPF) under the same conditions as in Figure 6. MPF declined at a greater rate in subjects with greater percent FT area in the vastus lateralis muscle (Komi & Tesch, 1979).

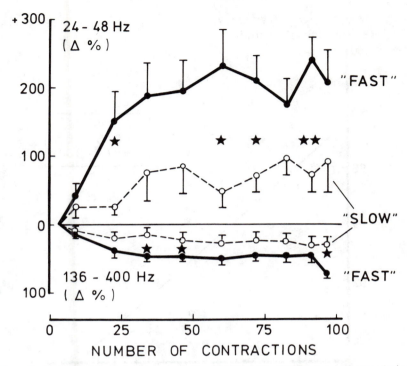

Figure 9—Relative changes in two extreme bandwidths of the EMG power spectrum during the condition of Figures 7 and 8. The terms "fast" and "slow" refer to the respective percent FT areas of the two previous figures (Komi & Tesch, 1979).

The changes in force and EMG patterns were essentially the same as those observed in the earlier study of Komi and Tesch (1979). However, the individual values (range 2.0 to 18.3 mmol x kg^{-1} ww) in lactate concentrate were found to be higher with greater reduction in MPF (Figure 10). Thus evidence has been obtained that metabolic, mechanical and electrical changes in the muscle are interrelated during fatigue contractions. More specifically, the observed results must indicate that mechanical and metabolic changes in force generation are associated with changes in motor unit potentials so that the characteristics of the muscle fiber membrane are changed with a likely reduction in propagation velocity of the action potentials during fatigue.

Influence of Fatigue on Reflex Contractions

Fatigue contractions are expected to have influence not only on the input/output relationship in the voluntary contractions, but also in the

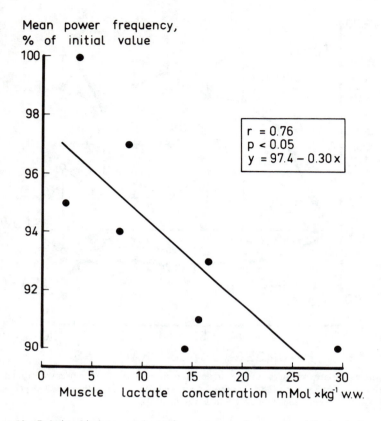

Figure 10—Relationship between change in mean power frequency and the level of muscle lactate during the repeated 120 maximal concentric contractions of the knee extensors (Tesch et al., 1982).

reflexly induced contractions. Häkkinen and Komi (1982) examined this with 14 male subjects who had experience in weight training. Utilizing the dynamometer of Komi (1982) the subjects were instructed to maintain a 50% isometric contraction level by extending the right knee until exhaustion. The required contraction level was shown by an x-y plotter placed in front of the subject. The testing was terminated after an abrupt decrease of the required force level. EMG readings from rectus femoris, vastus medialis and vastus lateralis muscles were recorded during the entire voluntary contraction. Figure 11 presents the results of IEMG and MPF changes, which were, as expected, increased in IEMG and decreased in MPF.

Patellar reflexes of the right leg were measured 1) in a resting condition as well as 2) immediately after the fatigue test and again 3) after 3 min of

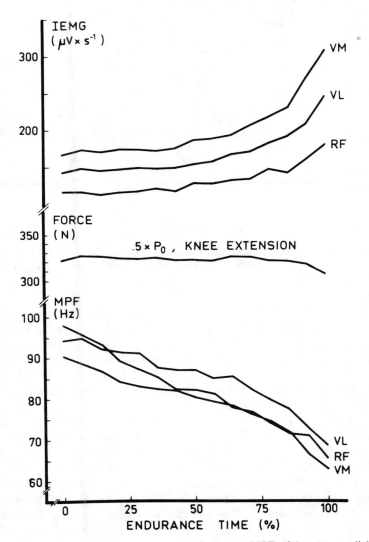

Figure 11—Levels of IEMG and mean power frequency (MPF) of the vastus medialis (VM), vastus lateralis (VL) and rectus femoris (RF) muscles during isometric contraction at a load of .5 x P_O maintained until exhaustion (Häkkinen & Komi, 1982).

recovery period. During the measurements the subjects kept their eyes closed. The reflex hammer was dropped in each condition from an angle of 90° with respect to the patellar tendon. Figure 12 depicts the records of the tendon tap, EMG and the subsequent force response. Two time parameters were measured from the individual records: 1) reflex latency (LAT), which indicates the time of the impulse travel both along the sensory and motor

REFLEX   RESPONSE

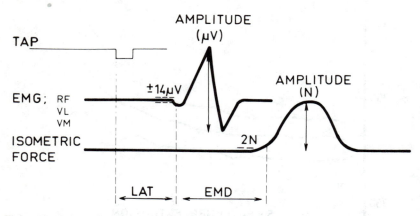

Figure 12—Schematic presentation of the records obtained in patellar reflex measurements performed before and immediately after the fatigue contraction of Figure 11. LAT = reflex latency; EMD = electromechanical delay.

pathways of the spinal reflex, 2) reflex electromechanical delay (EMD) which is the time between the arrival of the electrical signal in the muscle and beginning of the force, which was sensed by strain gauges attached to the cuff around the subject's right ankle. The values of $\pm 14\,\mu$V and $+2$ N denote the threshold levels used for detecting the beginning of the EMG and force signals, respectively. Peak-to-peak amplitudes of EMG and force responses were also calculated.

As is shown in Figure 13 reflex LAT did not change during the three different test conditions. Reflex EMD, on the other hand, increased significantly ($p < .01$) during fatigue from 32.6 ± 6.9 ms to 41.9 ± 6.8 ms and decreased after 3 min of recovery to 33.7 ± 9.2 ms. This increase in reflex EMD during fatigue was greater with greater reduction in MPF of the voluntary contraction (Figure 14). Peak-to-peak amplitude of the reflex EMG increased ($p < .05$) and the corresponding force decreased ($p < .01$) during fatigue. The EMG/force ratio increased therefore considerably during fatigue (Figure 15).

When these results are supplemented with the finding that the changes in IEMG/force ratio between the voluntary and reflex contractions were significantly interrelated ($r = .63$, $p < .05$), the observations point to the following three aspects: 1) Voluntary fatigue contractions induce mechanical and electrical changes both in the voluntary response itself and in the reflex contraction produced with a constant force of the tendon tap; 2) these changes are in many respects interrelated; 3) the marked increase in the amplitude of the reflex EMG potential must indicate that changes occur in the sensitivity of the muscle spindles to the stretch during fatigue.

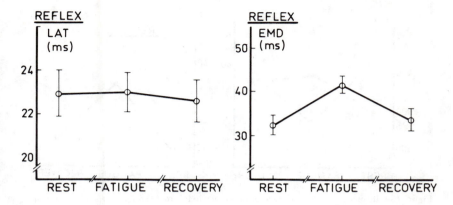

Figure 13—Mean (± SD) values of reflex latency (LAT) and electromechanical delay (EMD) before (rest), immediately after (fatigue) and 3 min after fatigue (recovery) of knee extensor muscles at a load of .5 x P_O as shown in Figure 11.

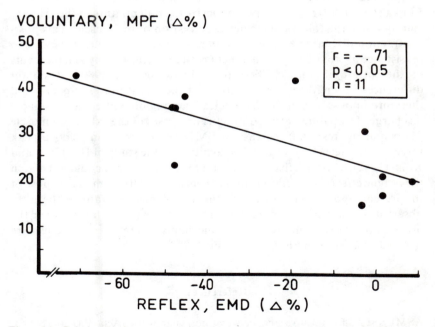

Figure 14—Relationship between change (decrease) during fatigue in mean power frequency (MPF) of EMG during voluntary contraction and change (increase) in electromechanical delay (EMD) of the reflex contractions. For the fatigue condition see Figure 11.

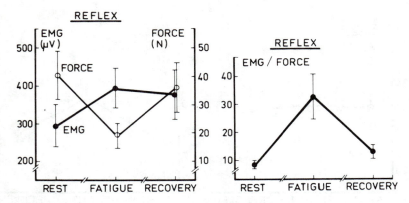

Figure 15—Influence of isometric fatigue (as shown in Figure 11) on peak-to-peak amplitudes of reflex EMG and reflex force (left) and on their ratios (right).

This latter point needs further explanation. Suggestions have been given that the increase in sensitivity of the muscle spindle may take place during fatiguing isometric contraction (Sato & Takemoto, 1972). Moreover, some animal (e.g., Smith et al., 1974; Suzuki & Hutton, 1976) and human (e.g., Enoka et al., 1980) experiments emphasize the potentiation of the primary ending discharge following isometric activation of the muscle. This post-contraction discharge and increase in stretch sensitivity might be due to persistence of cross-bridge attachment between actin and myosin filaments of the intrafusal muscle fibers (e.g., Brown et al., 1980). Resultant differences in length between extra- and intrafusal muscle fibers could therefore impose stretch on the nuclear bag region and enhance spindle discharge. The failure of the contractile processes of the extrafusal muscle fibers and the resultant increase in EMD and decrease in peak-to-peak force amplitude were of so great a magnitude in the study of Häkkinen and Komi (1982), however, that the increased excitatory drive could not match the fatigue effects in the muscle fibers themselves. Although the purpose of the present report was not to discuss the sites of muscular fatigue, the latter observations strongly support the concept of peripheral fatigue over the others. For an excellent review of the mechanism of fatigue the readers are referred to a report by Bigland-Ritchie (1982).

References

ASMUSSEN, E., Positive and negative muscular work. Acta Physiol. Scand. 28:364-382, 1953.

ASMUSSEN, E., Observations on experimental muscle soreness. Acta Rheum. Scand. 2:109–116, 1956.

BERGSTRÖM, J., Muscle electrolytes in man. Scand. J. Clin. Lab. Invest. (Suppl.) 68:1–110, 1962.

BIGLAND-RITCHIE, B., EMG and fatigue of human voluntary and stimulated contractions. In Human Muscle Fatigue: Physiological Mechanisms, pp. 130–156. London: Pitman Medical (Ciba Foundation Symposium), 1982.

BIGLAND, B., & O.C.J. Lippold. The relation between force, velocity and integrated electrical activity in human muscles. J. Physiol. (London) 123:214–224, 1954.

BROWN, M.C., G.M. Goodwin & P.B.C. Matthews. The persistence of stable bonds between actin and myosin filaments on intrafusal muscle fibres following their activation. J. Physiol. (London) 210:9–10P, 1970.

DE LUCA, This volume, 1982.

EDWARDS, R.H.T., Human muscle function and fatigue. In Human Muscle Fatigue: Physiological Mechanisms, pp. 1–18. London: Pitman Medical (Ciba Foundation Symposium), 1981.

ENOKA, R.M., R.S. Hutton & E. Eldred. Changes in excitability of tendon tap and Hoffmann reflexes following voluntary contractions. Electromyogr. Clin. Neurophysiol. 48:664–672, 1980.

HÄKKINEN, K. & P.V. Komi. Electromyographic and mechanical characteristics of human skeletal muscle during fatigue under voluntary and reflex conditions. Submitted for publication, 1982.

KADEFORS, R., E. Kaiser & I. Petersén. Dynamic spectrum analysis of myopotentials with special reference to muscle fatigue. Electromyography 8:39–74, 1968.

KARLSSON, J., Lactate and phosphagen concentrations in working muscle of man. Acta Physiol. Scand. (Suppl.) 358:1–72, 1971.

KOMI, P.V., Relationship between muscle tension, EMG and velocity of contraction under concentric and eccentric work. In J.E. Desmedt (Ed.), New Developments in Electromyography and Clinical Neurophysiology, Vol. 1:596–606. Basel: Karger, 1973.

KOMI, P.V., Dynamometer for measurement of concentric and eccentric forces in knee extension and flexion, 1982. (in preparation)

KOMI, P.V. & E.R. Buskirk. Effect of eccentric and concentric muscle conditioning on tension and electrical activity of human muscle. Ergonomics 15(4):417–434, 1972.

KOMI, P.V. & H. Rusko. Quantitative evaluation of mechanical and electrical changes during fatigue loadings of eccentric and concentric work. Scand. J. Rehab. Med. (Suppl.) 3:121–126, 1974.

KOMI, P.V. & P. Tesch. EMG frequency spectrum, muscle structure and fatigue during dynamic contractions in man. Eur. J. Appl. Physiol. 42:41–50, 1979.

KOMI, P.V. & J.T. Viitasalo. Changes in motor unit activity and metabolism in human skeletal muscle during and after repeated eccentric and concentric contractions. Acta Physiol. Scand. 100:246–254, 1977.

LARSSON, L., J.H.T. Viitasalo & P.V. Komi. Changes in reflex time and EMG signal characteristics in the ageing quadriceps muscle. In L. Larsson, "Morphological and functional characteristics of the ageing skeletal muscle in man. A cross-sectional study." Acta Physiol. Scand. (Suppl.) 457:V:1–V:17, 1978.

LAURIG, W., Electromyographie als arbeitswissenschaftliche Untersuchungs-methode zur Beurteilung von statischer Muskelarbeit (Electromyography as an ergonomic method for evaluation of static muscular work). Berlin-Köln-Frankfurt: Beuth Vertrieb, 1970.

LINDSTRÖM, L., This volume, 1982.

LINDSTRÖM, L., R. Magnusson & I. Petersén. Muscular fatigue and action potential conduction velocity changes studied with frequency analysis of EMG signals. Electromyogr. Clin. Neurophysiol. 10:341–356, 1970.

MORTIMER, J.T., R. Magnusson & I. Petersén. Conduction velocity in ischemic muscle: effect on EMG frequency spectrum. Am. J. Physiol. 219:1324–1329, 1970.

RAU, G. & J. Vredenbregt. The relationship between the EMG activity and the force during voluntary static contractions of the human m. biceps. Instituut voor Perceptie Onderzoek, Insulindelaan 2, Eindhoven, 1970.

SATO, M. & R. Takemoto. Facilitation and suppression of the patellar reflex by muscular exercise. J. Hum. Ergol. 1:67–73, 1972.

SMITH, J.L., R.S. Hutton & E. Eldred. Postcontraction changes in sensitivity of muscle afferents to static and dynamic stretch. Brain Res. 78:193–202, 1974.

STÅLBERG, E., Propagation velocity in human muscle fibers in situ. Acta Physiol. Scand. (Suppl.) 287, 1966.

SUZUKI, S. & R. Hutton. Postcontractile motoneuronal discharge produced by muscle afferent activation. Med. Sci. Sports 8:258–264, 1976.

TESCH, P., P.V. Komi, I. Jacobs, J.T. Viitasalo & J. Karlsson. Muscle lactate, muscle glycogen and EMG power spectrum during dynamic fatigue. In preparation, 1982.

TESCH, P., B. Sjödin, A. Thorstensson & J. Karlsson. Muscle fatigue and its relation to lactate accumulation and LDH activity in man. Acta Physiol. Scand. 103:413–420, 1978a.

TESCH, P., B. Sjödin & J. Karlsson. Relationship between lactate accumulation, LDH activity, LDH isozyme and fibre type distribution in human skeletal muscle. Acta Physiol. Scand. 103:40–46, 1978b.

VIITASALO, J.T. & P.V. Komi. Changes in motor unit activity and metabolism in human skeletal muscle during and after repeated eccentric and concentric contractions. Acta Physiol. Scand. 100:246–254, 1977.

VIITASALO, J.T. & P.V. Komi. Isometric endurance, EMG power spectrum, and fiber composition in human quadriceps muscle. In E. Asmussen & K. Jørgensen (Eds.), Biomechanics VI-A, pp. 244–250. Baltimore: University Park Press, 1978.

VIITASALO, J.T. & P.V. Komi. EMG, reflex and reaction time components, muscle structure, and fatigue during intermittent isometric contractions in man. Int. J. Sports Med. 1:185–190, 1980.

General Metabolism

Familial Aggregation of PWC$_{150}$, Blood Lipid Components and Serum Uric Acid in Adopted and Biological Siblings

C. Allard, C. Leblanc, J. Talbot,
S. Leclerc and C. Bouchard
Laval University, Québec, Canada

Family and twin studies have indicated that some of the population variance in plasma levels of cholesterol, triglyceride, high-density lipoproteins, uric acid, and in physical working capacity is associated with genetic variation. Iselius (1979) has made a path analysis of published reports on lipids and lipoproteins. Genetic studies on uric acid have been reviewed (Ahern et al., 1980; Bouchard et al., 1981), and have discussed the few studies on the heritability of indicators of aerobic capacities and powers. Preliminary data on the PWC$_{150}$ of adopted and biological siblings have been presented by Bouchard et al. (1982).

The purpose of the present study is to report preliminary data on the familial aggregation of serum lipid components, serum uric acid and PWC$_{150}$ for adolescents sharing a common environment. This work is part of a larger research program that proposes to estimate the contribution of genes to several physiological and biochemical traits.

Procedures

From June 1978 to August 1981, 375 volunteer families of the Quebec region comprising a total of 897 children of French descent were investigated in this laboratory. All the siblings formed the sample for the present study. They belong to the following sibships: adopted (n = 48); biological (natural children) (n = 222); dizygotic (DZ) twins (n = 55); and monozygotic (MZ) twins (n = 54). Diagnosis of zygosity was established with the help of several red blood cell antigens, the HLA system from leucocytes, and the red cell phosphoglucomutase isozymes at the PGM_1 locus as described by Chagnon et al. (1981). Fasting blood was collected and treated as described by Chagnon et al. (1981). All the members of a family were investigated on the same day. Total cholesterol (TC), high-density lipoprotein cholesterol (HDL-C), uric acid, and triglyceride (TG) concentrations were determined in sera frozen for up to 3 yr using the Abbott V.P. System (S. Pasadena, CA 91030) and enzymatic assays. PWC_{150}/kg^{-1} was computed from three work loads on a Monark bicycle, each 6-min in duration.

In the analysis of twin data, the mean test values for MZ and DZ twins were compared and the test values by twin types were evaluated for homogeneity as described by Christian et al. (1976). The values for serum triglycerides were normalized by a log_{10} transformation. The statistical analyses of the data were made on residual scores after the removal of the effects of age and sex of subjects ($y = age + sex + age \times sex + age^2 + age^3$).

Table 1

Comparison of Mean Test Values by Sibling Types

| Variable | Relatedness of sibling groups | | | | Signif. diff. in means* |
| | 1 | 2 | 3 | 4 | |
	Adopted	Biological	DZ	MZ	
Age, yr	14 (113)**	15 (545)	14 (120)	15 (119)	1 vs 2
PWC_{150} kg^{-1}	8.7 (110)	8.8 (531)	8.6 (117)	8.0 (110)	2 vs 4
TC	170 (111)	169 (539)	176 (119)	179 (119)	2 vs 4
TG log_{10}	1.83 (111)	1.84 (538)	1.87 (119)	1.90 (119)	NS
HDL-C	56.8 (110)	54.6 (539)	54.8 (118)	53.0 (119)	1 vs 4
Uric acid	4.6 (111)	4.6 (539)	4.6 (119)	4.6 (119)	NS

*$p \leq 0.01$
**Number of subjects

Results

Table 1 compares the mean test values of each group of siblings and indicates that the twinning process itself or the adoption process did not affect the values.

Table 2 shows that the total variance calculated for MZ and DZ twins was similar which permits the conclusion that the test values by twin types were homogeneous. When such variances do not differ, the standard test for quantitative traits is appropriate (Christian et al. 1976), and comparisons of intraclass correlations are used.

Table 3 shows the intraclass correlation coefficients for the various groups of siblings. Genetic variation was suggested for all the test values because the correlations were significantly higher in MZ twins than in DZ twins and biological siblings, and much smaller in adopted siblings.

Table 4 shows the F ratios between siblings' within-pair variance for the test values. The evaluation of the statistical significance of the within-pair variations in the adopted, biological, DZ and MZ siblings suggests that the familial influence was the strongest for HDL-C followed by total cholesterol, uric acid and triglyceride. These data suggest that familial aggregation is lowest for the physical working capacity.

Table 2

F′ Test* for Equality of Total Variance of MZ and DZ Twins

Variable	DZ pairs		MZ pairs		F′ ratio	Signif- icance** p ≤ 0.05
	Total variance⁺	Approx. DF	Total variance	Approx. DF⁺⁺		
PWC_{150} kg⁻¹	6.42	49	5.48	76	1.2	NS
TC	1135	60	1027	78	1.1	NS
TG log₁₀	0.04	54	0.04	78	1.1	NS
HDL-C	156	55	196	69	1.3	NS
Uric acid	1.15	44	1.15	72	1.0	NS

*Use of a two-tailed F test
**Christian et al. (1974) have proposed to perform the F′ test at higher probability level ($\alpha =$ 0.2). For reasons presented by Sistonen et al. (1980), the usual levels of significance have been used.

Table 3

Intraclass Correlations* by Sibling Groupings

Variable	Correlations			
	Adopted	Biological	Twins	
			DZ	MZ
PWC$_{150}$ kg^{-1}	0.01	0.24	0.41	0.60
TC	0.19	0.36	0.41	0.79
TG, log$_{10}$	0.22	0.32	0.47	0.69
HDL-C	0.01	0.40	0.40	0.88
Uric acid	0.03	0.32	0.48	0.74

*Computed on residuals after the removal of effects of age and sex.

Table 4

F Ratios between Siblings' within-pair Variance for the Test Values

Variable	Sibling Groups			
	Adopted	Biological*	DZ*	MZ*
PWC$_{150}$ kg^{-1}	1.02	1.78	2.46	3.99
TC	1.53	2.32	2.41	8.84
TG, log$_{10}$	1.63**	2.11	2.84	5.56
HDL-C	0.98	2.59	2.36	15.41
Uric acid	1.07	2.13	2.91	6.67

*$p \leq 0.01$
**$p \leq 0.05$

Discussion

Theoretically the genetic influence should be insignificant among adoptees, of similar magnitude among natural siblings and DZ twins, and high in MZ twins. Such a situation was apparently met for HDL-C. Evidence for a significant genetic component of variance for apo A-II has been found by

Sistonen et al. (1980). Considering the other lipid components (TC and TG, log$_{10}$), the r$_i$ for MZ twins were high but were dissimilar among biological and DZ twins which indicated that an environmental influence existed.

The conclusion that there is no evidence that genetic factors are of importance for the regulation of serum cholesterol fraction (Christian et al., 1979) has been challenged by Iselius (1979) who indicated that the recalculated estimates of Christian et al. (1976) were in good agreement with other studies. Our data generally agree with those reported by Iselius (1979).

The important genetic influence on the variation of serum uric acid evaluated in the present investigation on young subjects was the same as described by Havlik (1977) for adults.

Bouchard et al. (1980) concluded that genetic studies on aerobic capacities and power have yielded inconsistent estimates mainly because of a lack of control over age and sex, smallness of samples, and inadequate quantitative procedures. The preliminary results presented in this paper show that, among the traits examined, the PWC$_{150}$ was the least influenced by genetic factors. Nevertheless the heritability appears important.

Acknowledgments

Supported by CRSQ (78004), FCAC Québec (EQ-1300 and CE-29) and MLCP Québec (HCSR-7712 and 7912).

References

AHERN, F.M., R.C. Johnson & G.C. Ashton. Family resemblances in serum uric acid level. Behavior. Genet. 10:303–307, 1980.

BOUCHARD, C., C. Leblanc, G. Lortie, J.A. Simoneau, G. Thériault & A. Tremblay. Submaximal physical working capacity in adopted and biological siblings. Med. Sc. Sports. Exerc. 14(2), (in press) 1982.

BOUCHARD, C., M.-C. Thibault & J. Jobin. Advances in selected areas of human work physiology. Yearbook of Phys. Anthropol. 24:1–36, 1981.

CHAGNON, Y.C., C. Bouchard & C. Allard. Isoelectric focusing of red cell phosphoglucomutase (E.C.: 2.7.5.1) at the PGM$_1$ locus in a French-Canadian population. Hum. Genet. 59:36–38, 1981.

CHRISTIAN, J.C., M. Feinleib, S.B. Hulley, W.P. Castelli, R.R. Fabsitz, R.J. Garrison, N.O. Borhani, R.H. Rosenman & J. Wagner. Genetics of plasma cholesterol and triglycerides: a study of adult male twins. Acta Genet. Med. Gemellol. (Roma) 25:145–149, 1976.

CHRISTIAN, J.C., K.W. Kang & J.A. Narton. Choice of an estimate of genetic variance from twin data. Am. J. Hum. Genet. 26:154–161, 1974.

HAVLIK, R., R. Garrison, R. Fabsitz & M. Feinleib. Genetic variability of clinical chemical values. Clin. Chem. 23:659–662, 1977.

ISELIUS, L., Analysis of family resemblance for lipids and lipoproteins. Clin. Genet. 15:300–306, 1979.

SISTONEN, P. & C. Enholm. On the heritability of serum high-density lipoprotein in twins. Am. J. Hum. Genet. 32:1–7, 1980.

Muscle Metabolism during High Intensity Eccentric Exercise

W.J. Evans, J.F. Patton, E.C. Fisher and H.G. Knuttgen
U.S. Army Research Institute of Environmental Medicine,
Natick, Massachusetts, U.S.A., and Boston University,
Boston, Massachusetts, U.S.A.

The muscular effort of concentric exercise produces external work while the muscular effort of eccentric exercise resists the external force and "absorbs" the mechanical energy imposed by the ergometer. Two studies in the early 1950's (Abbot et al., 1952; Asmussen, 1953) investigated the metabolic response of eccentric and concentric contractions. Both investigations presented a general response in which identical power levels for concentric and eccentric exercise produce markedly lower oxygen cost for the eccentric contraction exercise along with lower circulatory and pulmonary responses. Exercise intensities for eccentric and concentric muscle contractions which produce identical force generation by the muscles produce a much lower set of responses for the exercising human. For example, the oxygen cost of eccentric vs concentric exercise can range between 1:3 and 1:10. The integrated electromyogram of muscles producing identical tension in the two types of exercise produces a much lower voltage for the eccentric muscle contractions (Asmussen, 1953).

In addition to the above observations, later investigations using eccentric muscular contractions have reported a training effect that was not previously observed (Bonde-Petersen et al., 1973; Klaussen et al., 1971; Knuttgen et al., 1982). The training effects of concentric exercise on skeletal muscle have been well-studied. These effects indicate a general shift in the carbon source for the Krebs cycle from carbohydrate to fat during exercise and an increase in the activities of key enzymes involved in oxidative metabolism (Holloszy, 1975). Little is known, however, about the specific training effects of eccentric exercise.

The purpose of the present investigation was to determine 1) the recruitment patterns among the different types of extrafusal muscle fibers during high-intensity exercise with eccentric contractions and 2) the effects of eccentric exercise training on both the recruitment patterns and the accompanying metabolic events.

Methods

Seven male subjects (mean age $= 21$ yr, mean wt $= 72.8$ kg, mean $\dot{V}_{O_2 max} = 2.91$ liter min^{-1}) were recruited for participation in the study. All subjects were informed of the potential risks and hazards of their participation in the study and signed an informed consent form. Following an initial period of instruction and practice, the subjects were given a pre-training (Pre-T) test, 4 to 5 wk of training with eccentric contraction exercise, and a post-training (Post-T) test. The pre-T and post-T tests were identical for each subject and consisted of eccentric exercise at 250 W for 45 min. Exercise was performed on a cycle ergometer of special design, the pedal axle of which is driven in a reverse direction form normal cycling by an electric motor (Knuttgen et al., 1982). The pedals are driven at the seated subject who resists the tendency of the ergometer to increase the pedal axle rpm from 60 to 66. The ergometer is so controlled that the subject has to provide a specific resistance power in order to maintain the 60 rpm.

Pulmonary ventilation and respiratory exchange were determined at various periods during the exercise by the Douglas bag collection technique using the Beckman LB-2 and Applied Electrochemistry S3-A for gas analysis.

Muscle biopsies were obtained before and after the exercise periods from the vastus lateralis muscle. Pre-exercise (pre-ex) samples were divided and frozen in liquid N_2 for later determination of glycogen levels and the activities of the following enzymes: HK, GP, MDH and LDH. Post-exercise (post-ex) samples were divided into two portions, one of which was frozen in liquid N_2 for later determination of glycogen levels. The second portion was mounted in OCT, frozen in isopentane, and cooled to the temperature of liquid N_2 for later histochemical myosin ATPase and PAS staining.

The pre-T and post-T exercise tests were separated by a 5 wk eccentric training program (3 days 2 wk^{-1}, 60 min day^{-1}, 250 W) on the eccentric cycle ergometer.

Results

Glycogen use during exercise fell from a mean of 22.4 ± 3.4 mmol kg^{-1} wet wt before training to 13.0 ± 2.3 mmol kg^{-1} wet wt post training, a 42% reduction ($p < .05$).

In the pre-T experiments, $\dot{V}_{O_2}$ during the first 4 min of exercise evidenced a range among the subjects of 0.80 to 1.18 l min^{-1} and then continued to demonstrate increases for the remainder of the exercise periods. The averaged final values represented 130% of the 2 to 4 min values. Heart rates at the end of the exercise periods ranged between 119 to 150 beats min^{-1} (mean = 136 beats min^{-1}).

The training period resulted in a general decrease in $\dot{V}_{O_2}$ and heart rate throughout exercise. The $\dot{V}_{O_2}$ was significantly lower post-T compared to pre-T for all subjects. At 45 min of exercise $\dot{V}_{O_2}$ was reduced by 32% from 1.4 l min^{-1} pre-T to a post-T mean of 0.95 l min^{-1} ($p < .05$). Heart rate at the end of exercise ranged from 98 to 120 beats min^{-1} (mean = 109 beats min^{-1}), and was significantly lower for all subjects post-T compared to pre-T. $\dot{V}_{O_2max}$ as measured in concentric contraction exercise testing did not change as a result of the training program. Both before and after training, exercise resulted in a significant glycogen depletion in Type I compared to Type II fibers. Muscle enzyme activities did not change as a result of the training (Table 1).

Table 1

**Muscle Enzyme Activities Before and After
Eccentric Exercise Training (μmoles g^{-1} min^{-1})**

	HK	GP	MDH	LDH
Pretraining	1.61 ± .16	6.59 ± 1.89	51.57 ± 4.02	145.37 ± 10.75
Post-training	1.10 ± .29	4.39 ± 1.44	36.13 ± 9.30	125.65 ± 28.8

Discussion

The results of this study are in agreement with previous studies indicating a low $\dot{V}_{O_2}$ during high intensity eccentric exercise. It is interesting to note that eccentric exercise training not only caused an expected drop in heart rate at the same intensity of exercise, but also caused a reduction in $\dot{V}_{O_2}$. The mechanism for this reduction in $\dot{V}_{O_2}$ may involve an adaptation of the intrafusal fibers to the forced stretch induced by the eccentric contractions. An exaggerated stretch reflex was noted in all subjects during the pre-T exercise. This response was generally absent in the post-T exercise.

Associated with the reduction in $\dot{V}_{O_2}$ was a reduction in glycogen utilization. This may have been caused also by an adaptation of the intrafusal fibers to the exercise or by the subjects learning the proper amount of force to apply (the "proper" number of motor units to recruit) during the training period. Unfortunately the testing procedures were done only at the beginning and end of the training program, so it is unknown how rapidly this reduction in $\dot{V}O_2$ and glycogen use takes place.

Previous studies of both human and rodent skeletal muscle involved in endurance exercise training have demonstrated increases in the activities of key enzymes of carbohydrate and fat oxidation (Holloszy, 1975). Of the enzymes measured in the present study, there were no changes in activities resulting from the training. This suggests that the amount of force generated per se is not the most important stimulus to induce a training response but rather the percent $\dot{V}O_{2max}$. This is supported by the fact that none of the subjects demonstrated a change in $\dot{V}O_{2max}$. Percent $\dot{V}O_{2max}$ also seems to be the most important factor in muscle fiber recruitment. Using concentric contractions, a preferential type I recruitment is observed at low intensities (Gollnick et al., 1973). In the present study, however, high intensity eccentric exercise uses predominantly type I fibers.

Conclusions

High intensity eccentric exercise training results in a reduction in $\dot{V}O_2$ and glycogen utilization at the same absolute intensity. This adaptation may be a motor learning response since $\dot{V}O_{2max}$ and muscle enzyme activities remained unchanged with training.

References

ABBOTT, B.C., B. Bigland & J.M. Ritchie. The physiological cost of negative work. J. Physiol. (London) 117:380–390, 1952.

ASMUSSEN, E., Positive and negative muscular work. Acta Physiol. Scand. 28:364–382, 1953.

BONDE-PETERSEN, F., J. Henriksson & H.G. Knuttgen. Effect of training with eccentric muscle contractions on skeletal muscle metabolites. Acta Physiol. Scand. 88:564–570, 1973.

GOLLNICK, P.D., R.B. Armstrong, C.W. Saulmert IV, R.E. Shepherd & B. Saltin. Glycogen depletion patterns in human skeletal muscle fibers after exhausting exercise. J. Appl. Physiol. 34:615–618, 1973.

HOLLOSZY, J.O., Adaptation of skeletal muscle to endurance exercise. Medicine and Science in Sports 7:155–164, 1975.

KLAUSEN, K. & H.G. Knuttgen. Effect of training on oxygen consumption in negative muscular work. Acta Physiol. Scand. 83:319–323, 1971.

KNUTTGEN, H.G., G.R. Nodel, K.B. Pandolf & J.F. Patton. Effects of training with eccentric muscle contractions on exercise performance, energy expenditure, and body temperature. Int. J. Sports Medicine 3:13–17, 1982.

Effect of Exercise on Metabolism of Energy Substrates in Pregnant Mother and Her Fetus in the Rat

J. Gorski

Medical School, Bialystok, Poland

Pregnant mothers usually undertake muscular efforts of different duration and intensity. There are a number of reports concerning the cardiorespiratory adjustments to muscular exercise during pregnancy (e.g., Erkkola, 1976). There are also several reports showing that acute exercise by the mother can affect fetal heart rate, blood pH and blood gas tensions (Emmanoulides et al., 1972; Hon & Wohlgemuth, 1961). Pregnancy extensively alters the metabolic and humoral balance in the mother's body. Therefore, it is reasonable to expect that the metabolism of energy fuels in exercising pregnant mothers may differ from that observed in unpregnant females or males. Another problem of muscular exertion during pregnancy is its possible indirect effect on the metabolism of the fetus. One might expect that during a mother's exercise, the supply of nutrients to the fetus could be changed. This in turn could cause subsequent alterations in fetal metabolism. However, there are no data available on the metabolism of energy substrates either in an exercising pregnant mother or in her fetus. Thus, the aim of the present study was to examine this problem.

Methods

The experiments were carried out on female Wistar rats whose body wt at the beginning of the project was 190 to 210 g. They were kept with males and daily vaginal smears were taken and watched for spermatozoa. The day in which the spermatozoa were found was considered as the first day of pregnancy. The rats were accustomed to running on the treadmill (12 m/min, 0° incline) as they were made to run for 10 min daily for 6 days preceding the final exercise bout. Rats that were pregnant for 20 days were divided into two groups: 1) fed and 2) fasted. The rats of the second group were deprived of food for 24 h before they were sacrificed. In each group the determinations were carried out at rest, after 60 min of exercise and after exercise until exhaustion. Exhaustion was defined as the point at which the rats were unable to continue running. The average time of running until exhaustion for the fed group was 143 ± 36 min, while that for the fasted group was 117 ± 37 min ($p > 0.05$). The rats were anesthetized with urethane administered intraperitoneally. Two samples of the mother's liver, placenta and fetal liver were taken. Next, blood from the abdominal aorta was withdrawn. The chemical determinations for plasma glucose and free fatty acids and tissue glycogen and triglycerides were done as described previously (Stankiewicz-Choroszucha & Gorski, 1978). The results obtained were evaluated statistically with the Student-t test for unpaired data. Differences with a P value less than 0.05 were considered significant. Each result presented is the mean $\pm$ standard deviation. The mean values were calculated from the data obtained from 10 rats.

Results

The Mother's Blood

In the fed group only exercise to exhaustion resulted in significant ($p < 0.001$) reduction of the blood glucose level. In the fasted rats 60 min of exercise resulted in profound hypoglycemia ($p < 0.001$) that remained until exhaustion. Plasma free fatty acid (FFA) levels were more than doubled after 60 min of exercise in the fed group and then remained unchanged during exercise until exhaustion. Fasting markedly elevated the plasma FFA level in the resting rats where it remained stable during exercise.

The Mother's Liver Glycogen and Triglycerides

Exercise gradually reduced liver glycogen levels in the fed rats (60 min vs the rest $p < 0.05$, exhaustion vs the 60 min $p < 0.001$). In the fasted rats, resting glycogen level was very low and was further reduced during 60 min of exercise (60 min vs the rest $p < 0.001$). Prolonged exercise to exhaustion resulted in a marked accumulation of triglycerides in the liver of the fed rats ($p < 0.001$). Fasting doubled the level of triglycerides in the rested rats and it remained unchanged during exercise.

The Placenta

The placenta glycogen and triglyceride levels were stable during exercise in both groups. However, fasting resulted in accumulation of triglycerides in the placenta.

Table 1

The Effect of Exercise on Energy Substrate Levels in the Pregnant Mother, Placenta and Fetal Liver in Rats

Substrate	Group	Rest	Exercise 60 min	Exercise Exhaustion
Mother's blood				
Glucose	Fed	5.71 ± 0.89	4.78 ± 0.84	3.89 ± 1.05
	Fasted	4.08 ± 0.88[a]	2.44 ± 0.87[b]	2.56 ± 0.37[a]
Free fatty	Fed	372.1 ± 93.4	906.2 ± 174.4	824.7 ± 54.0
acids	Fasted	849.3 ± 222.2[b]	1064.4 ± 298.1	978.5 ± 275.6
Mother's liver				
Glycogen	Fed	175.8 ± 45.9	135.5 ± 30.9	22.8 ± 13.7
	Fasted	5.0 ± 1.9	1.5 ± 0.5	2.4 ± 0.9
Triglycerides	Fed	15.01 ± 3.05	18.08 ± 3.57	24.86 ± 6.31
	Fasted	29.63 ± 5.88[b]	28.16 ± 3.55[b]	29.82 ± 3.88[a]
Placenta				
Glycogen	Fed	18.4 ± 5.1	16.6 ± 3.7	15.4 ± 4.0
	Fasted	19.3 ± 4.7	16.4 ± 3.3	15.5 ± 3.4
Triglycerides	Fed	4.98 ± 0.98	5.59 ± 1.04	5.66 ± 0.74
	Fasted	8.73 ± 0.44[b]	8.77 ± 1.04[b]	8.48 ± 1.34[b]
Fetal liver				
Glycogen	Fed	169.4 ± 36.3	113.2 ± 13.8	120.0 ± 28.4
	Fasted	118.7 ± 29.8[a]	57.6 ± 15.1[b]	69.6 ± 26.4[b]
Triglycerides	Fed	6.86 ± 1.26	6.75 ± 1.17	7.36 ± 1.05
	Fasted	21.41 ± 4.66[b]	18.43 ± 3.44[b]	19.54 ± 5.86[b]

Fasted vs the corresponding fed value: [a]$p < 0.01$; [b]$p < 0.001$.
Units: glucose—mmol/l, free fatty acids—μmol/l, glycogen—μmol of glucose/g, triglycerides—μmol/g.

The Fetal Liver

Exercise lasting for 60 min reduced the glycogen level in the fetal liver in the both groups ($p < 0.001$) and it remained unchanged during exercise to exhaustion. The exercise had no effect on the levels of triglycerides either in the fed or in the fasted group. Fasting itself reduced the glycogen level and resulted in marked accumulation of triglycerides in the fetal liver.

Discussion

The results show that in late pregnant rats exercise induced mobilization of glycogen and accumulation of triglycerides in the liver, elevation of plasma free fatty acids and hypoglycemia. Thus, in this respect exercise-induced changes in metabolism of the examined energy substrates in pregnant rats are similar to those in unpregnant individuals. The only important difference from the typical exercise response (Stankiewicz-Choroszucha & Gorski, 1978) seems to be the rapid elevation of the plasma free fatty acids in our exercising pregnant mothers. However, this may have an important consequence. Hickson et al. (1977) have shown that elevation of the plasma free fatty acid levels during exercise have a sparing effect on the body carbohydrate stores. Thus, it is reasonable to assume that a marked, rapid rise in plasma free fatty acid levels in exercising pregnant rats would spare glucose and thus save it for her fetuses. This might be even more important for the fasted rats in which the resting plasma free fatty acid levels were already very high at the beginning of exercise. The other important observation of the present study is that 60 min of exercise affected the metabolic homeostasis of the fetus by reducing the glycogen level in the fetal liver. Freud et al. (1980) have shown that a mother's hypoglycemia induces a reduction in the fetal liver glycogen level in the rat. However, the exercise-induced reduction in the fetal liver glycogen level was equal in the fed and the fasted group despite the marked difference in the mother's blood glucose level between these two groups. Besides, the fetal liver glycogen level remained constant when the fetuses were exposed to longer periods of maternal hypoglycemia while the animals continued the exercise to exhaustion. Thus, the mother's hypoglycemia itself probably did not cause the reduction in the fetal liver glycogen level. Muscular exercise was found to decrease the uterine blood flow in women (Morris et al., 1956) and to cause hypoxia of the fetus in ewes (Emmanoulides et al., 1972). If this were true for rats, then the reduction in the fetal liver glycogen level could be caused both by hypoxia and by a decreased amount of glucose reaching the fetus as the consequence of the reduced uterine blood flow during the first period of exercise.

Conclusions

1.   In fed, late pregnant rats, exercise of moderate intensity resulted in hypoglycemia, rapid and profound elevation in the plasma free fatty acid levels, use of liver glycogen and accumulation of triglycerides in the liver. No change in the plasma free fatty acids or the liver triglyceride level occurred during exercise in the fasted-pregnant rats.

2.   Exercise had no effect on the placental glycogen and triglyceride levels.

3.   The mother's exercise resulted in reduction of the glycogen level but no change in triglycerides in the livers of the fetuses.

References

EMMANOULIDES, G.G., C.J. Hobel, K. Yashire & G. Klyman. Fetal responses to maternal exercises in the sheep. Am. J. Obstet. Gynecol. 112:130–137, 1972.

ERKKOLA, R., Physical work capacity and pregnancy. Ann. Univ. Turk(uensis), Series D, Turku, 1976.

FREUD, N., A. Kervran, R. Assan, J.P. Gelos & J. Girard. Fetal metabolism response to phloridzin-induced hypoglycemia in pregnant rats. Biol. Neonate 38:321–327, 1980.

HICKSON, R.C., M.J. Rennie, R.K. Conlee, W.W. Winder & J.O. Holloszy. Effects of increased plasma fatty acids on glycogen utilization and endurance. J. Appl. Physiol. 43:829–833, 1977.

HON, E.H. & R. Wohlgemuth. The electronic evaluation of fetal heart rate: IV. The effect of maternal exercise. Am. J. Obstet. Gynecol. 81:361–371, 1961.

MORRIS, N., S.B. Osborn & H.P. Wright. Effective uterine blood-flow during exercise in normal and pre-eclamptic pregnancies. Lancet 2:481–484, 1956.

STANKIEWICZ-CHOROSZUCHA, B. & J. Gorski. Effect of decreased availability of substrates on intramuscular triglyceride utilization during exercise. Eur. J. Appl. Physiol. 40:27–35, 1978.

Changes in Muscle ATP, CP, Glycogen, and Lactate after Performance of the Wingate Anaerobic Test

I. Jacobs, O. Bar-Or, R. Dotan, J. Karlsson and P. Tesch
Karolinska Hospital, Stockholm, Sweden, and Wingate
Institute, Natanya, Israel

The Wingate Anaerobic Test (WT) is used in a number of exercise laboratories in various countries for the evaluation of the capacity for muscle power generation during short term, exhaustive exercise. The classification of the WT as "anaerobic" has been based on indirect assessments of the contribution of anaerobic energy metabolism to its performance and includes reports of oxygen deficits, oxygen debts (for references see Bar-Or, 1981), blood lactate (Jacobs, 1981) and muscle lactate (Jacobs, 1981) and muscle lactate concentrations (Jacobs & Tesch, 1981). The present study was undertaken to provide more direct evidence of the utilization of intramuscular, so-called "anaerobic" energy stores with performance of the WT.

Methods

Muscle biopsies were taken from the vastus lateralis muscle of 14 female physical education students at rest and immediately following the WT. The WT was performed as described earlier (Bar-Or et al., 1980; Bar-Or, 1981) on a mechanically braked Fleisch cycle ergometer following a warm-up. The muscle tissue samples were analyzed for the concentrations of adenosine triphosphate (ATP), creatine phosphate (CP), lactate and glycogen (Karlsson, 1971). In addition, muscle fiber type composition was determined with established histochemical stainings for the activity of myofibrillar ATPase and expressed as the percentage of fast-twitch (type II, % FT) fibers.

For comparison, an additional group of 22 subjects (15 males, 7 females) performed the WT twice: once with the normal 30 s protocol, and a second time when the test was terminated after 10 s. After both bouts a biopsy was taken and the concentrations of CP and lactate determined.

Results

The mean values for all female subjects for mean power output during the 30 s test, for peak power during the best 5 s period, and for the relative power decrease during the 30 s were 7.2 W x kg^{-1} body wt, 9.1 W x kg^{-1}, and 41%, respectively. The males' values were 118% and 113% (p $<$ 0.001) of the females' mean power x kg^{-1} and peak power x kg^{-1}, respectively, while the relative power decrease was only 82% of the female value (p $<$ 0.01).

The relative changes in the measured metabolites in the 14 females from whom both resting and post exercise biopsies were taken are depicted in Figure 1. The differences between the resting and post exercise metabolite concentrations were all highly significant (p $<$ 0.001) and averaged 7.1, 37.7, 50.5 and 102 mmol x kg^{-1} dry wt for ATP, CP, lactate and glycogen, respectively.

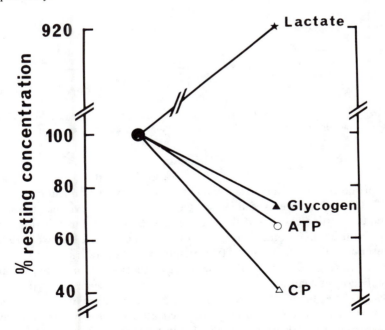

Figure 1—The change in muscle metabolite concentrations in biopsy samples obtained after performance of the Wingate Anaerobic Test, expressed relative to the resting concentrations (100%).

The relationship between the post exercise lactate and CP concentrations is shown in Figure 2 for all subjects. For those subjects who performed both the 10 s and 30 s exercise bouts, the muscle lactate concentrations after the 10 s bout averaged approximately 60% of the 30 s value (36 vs 61 mmol x kg^{-1} dry wt). The mean concentrations for CP after the 10 and 30 s bouts averaged 42.2 and 18.3 mmol x kg^{-1} dry wt, respectively.

The change in the CP concentration from rest to post exercise in the 14 females was inversely related to the % FT fibers in the vastus lateralis muscle (r = –0.80, p < 0.01).

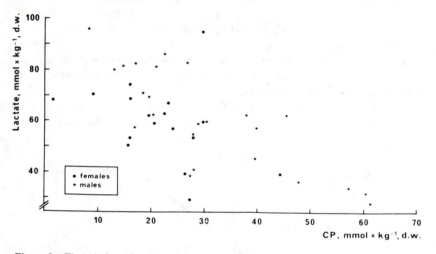

Figure 2—The relationship of lactate concentration to creatine phosphate (CP) in muscle tissue samples obtained after performance of the Wingate Anaerobic Test.

Discussion

The pronounced changes measured in the intramuscular concentrations of ATP, CP and lactate exceed those changes reported previously for short term, exhaustive exercise if the changes are expressed per unit time, i.e., metabolite change x min exercise^{-1} (Hultman et al., 1967; Karlsson, 1971). Consequently, the use of the WT as an indicator of the rate at which muscle power can be generated from the splitting of energy rich phosphagens and glycogenolysis leading to lactate formation would seem to be justified. However, the absolute depletion of CP and accumulation of lactate are not as great as has been previously observed after exhaustive exercise of a longer duration (Hultman et al., 1967; Karlsson, 1971; Tesch, 1980). Therefore the maximal capacity of the endogenous phosphagens and the capacity of glycogenolysis leading to lactate formation is not directly reflected by performance of the WT. Exhaustive exercise of approximately 3 min is probably required to tax the capacity of these energy yielding pathways (Karlsson, 1971).

The relationship between fiber type composition and CP utilization is of interest in light of other studies which suggest that FT muscle fibers may have higher creatine phosphokinase activities than slow-twitch muscle fibers thus enabling a more rapid utilization of CP (Powell, 1982; Thorstensson, 1976). The CP utilization to fiber type relationship in the present study, which employed female subjects, does not support the relationships described above but may offer some insight into the differences between males and females with regard to high intensity exercise (Jacobs & Tesch, 1981).

Another interesting observation concerns the marked lactate accumulation after only 10 s of exercise. The classical theory of a given CP depletion being required prior to significant lactate accumulation during exercise may be questionable. The present data confirm the observations of Saltin et al. (1971) in two subjects who performed high intensity exercise for 10 s, and suggest that such exercise induces glycogenolysis leading to lactate accumulation within a few seconds of the onset of muscular contractions.

In conclusion, the direct measurement of selected intramuscular metabolite concentrations in the present study provides further physiological support for the growing use of the Wingate Anaerobic Test in the assessment of the ability to generate muscular power during short term, high intensity exercise.

References

BAR-OR, O., Le test anaérobie de Wingate. Symbioses 13:157–172, 1981.

BAR-OR, O., R. Dotan, O. Inbar, A. Rotstein, J. Karlsson & P. Tesch. Anaerobic capacity and muscle fiber type distribution in man. Int. J. Sports Med. 1:89–92, 1980.

DOWELL, R.T., Activity of phosphorylcreatine shuttle enzymes in rat cardiac, fast-, and slow-twitch skeletal muscles. Biochem. Biophys. Res. Comm. 104: 740–745, 1982.

HULTMAN, E., J. Bergström & N. McLennan-Anderson. Breakdown and resynthesis of phosphorylcreatine and adenosine triphosphate in connection with muscular work in man. Scand. J. Clin. Lab. Invest. 19:56–66, 1967.

JACOBS, I., The effects of thermal dehydration on performance of the Wingate Anaerobic Test. Int. J. Sports Med. 1:21–24, 1980.

JACOBS, I. & P. Tesch. Short time, maximal muscular performance: relation to muscle lactate and fiber type in females. Med. Sport (Karger) 14:125–132, 1981.

KARLSSON, J., Lactate and phosphagen concentrations in working muscle of man. Acta Physiol. Scand. Suppl. 358, 1971.

SALTIN, B., P.D. Gollnick, B.O. Eriksson & K. Piehl. Metabolic and circulatory adjustments at onset of maximal work. In A. Gilbert & P. Guille (Eds.), Onset of Exercise. Toulouse: University of Toulouse, 1971.

TESCH, P., Muscle fatigue in man. Acta Physiol. Scand. Suppl. 480, 1980.

THORSTENSSON, A., Muscle strength, fiber types and enzyme activities in man. Acta Physiol. Scand. Suppl. 443, 1976.

A Computer Simulation Model of Energy Output in Relation to Metabolic Rate and Internal Environment

A. Mader, H. Heck and W. Hollman

Deutschen Sporthochschule, Köln, Federal Republic of Germany

It is generally accepted that during exercise energy is supplied by three available sources, which are the alactic capacity (usable creatine phosphate, CP), the glycolysis and the oxidative phosphorylation ($\dot{V}_{O_2}$). Assuming that the lactic acid production occurs only when more than 60% of creatine phosphate is broken down and assuming that the only difference between the energy required for mechanical power and the energy supplied by oxygen uptake is supplied by the production of lactic acid, then we can describe the relation between work (Joule) and the metabolic energy supply by the equation:

$$Wg = \int E \, dt = K_{CP} \, CP + K_{La} \, La + K_{O_2} \frac{\dot{V}_{O_2}}{60} \int_{to}^{t} (1 - e^{\frac{(x - to)}{\tau}}) \, dx \quad (1)$$

In Equation 1 K_{CP}, K_{La} and K_{O_2} are transformation-coefficients and CP, La and $\dot{V}_{O_2}$ are the given amounts of creatine phosphate used, lactic acid produced and oxygen uptake (Mader et al., 1981). Equation 1 can be solved if we know the amount of work and the energy contribution of available resources. The solution of Equation 1, however, is only possible if no lactic acid production occurs (aerobic work conditions) or if the mechanical power exceeds the maximal aerobic power to such an extent that the available CP is used within a few s and the speed of oxygen uptake is maximal. In that case the increase of performance could only be energetically supplied by an increase of lactic acid production (Mader et al., 1978; Margaria, 1963). In case of no lactic acid production the relation between the breakdown of creatine phosphate (P_E) and the resulting rephosphorylation from the rise of oxygen uptake (P_Z) to establish a steady state can be described as follows (Stegemann, 1977):

$$\frac{dP}{dt} = K_1 \, P_Z - P_E \qquad (2)$$

with the solution

$$P_{(t)} = Po - \frac{P_E}{K_1} (1 - e^{-k_1 \, t}) \qquad (3)$$

the required breakdown of CP is proportional to the work load and the time constant ($K_1 = 1/T$) of oxygen uptake.

It is well known that the change from an exclusive aerobic energy supply to a partial energy contribution from the production of lactic acid at the so called "aerobic-anaerobic-threshold" is nonlinear and occurs at a level between 60% and 90% $\dot{V}o_{2max}$. It is evident that the relation between mechanical power and $\dot{V}o_2$ near $\dot{V}o_{2max}$ is also nonlinear. Equations 1, 2 and 3 do not explain the mechanisms which relate a given power to a certain rate of ATP production from oxygen uptake and from lactic acid production.

As stated by Newsholme and Start (1973), McGilvery (1973) and many others, it seems to be generally accepted that the phosphorylation state of the system of energy rich phosphates adjusts the rate of oxidative phosphorylation as well as the activity of glycolysis.

The main factor of the regulation of oxidative phosphorylation is the ADP-concentration (McGilvery, 1973). The activity of the glycolytic system is due to the activity of phosphofructokinase (PFK), which is activated by an increase of ADP and AMP and the decrease of ATP (Newsholme & Start, 1973). To establish a causal relation between the work load dependent breakdown of energy rich phosphates and the adjustment of the respiratory chain as well as the rate of glycolysis, we must try to describe the $\dot{V}o_2$ and the lactic acid production as a function of ADP or ADP and AMP concentrations under steady state conditions as well as in the situation of the dynamic change intended to establish a new equilibrium state.

Whereas the rate of oxidative phosphorylation remains uninfluenced by the change of internal pH, the rate of glycolysis decreases with the drop of the intracellular pH and stops the lactic acid production if the internal pH is about 6.3–6.4 (Danforth, 1965).

Based on the "quasi equilibrium assumption" according to Walter (1977) citing Henri (1903), Michaelis and Menten (1913) it is possible to establish a relation between the substrate concentration (S) and the activity of enzymes (Ys), where Ys is proportional to the speed of the reaction (Vs), if the substrate concentration remains constant. According to Hill the equation is:

$$\log \left(\frac{Ys}{1 - Ys}\right) = \log \left(\frac{Vs}{Vmax - Vs}\right) = \log Ks + n \log [S] \qquad (4)$$

respectively

$$Ys = \frac{Ks [S]^n}{1 + Ks [S]^n} = \frac{1}{\left(1 + \frac{1}{Ks [S]}\right)} \qquad (5)$$

In the following, Equation 5 is transformed into

$$Ys = \left(1 + \frac{1}{K_s [S]^n}\right) \qquad (5a)$$

where

$$Ks' = \frac{1}{Ks} \qquad (5b)$$

If we accept that a given constant substrate concentration leads to an equilibrium of the substrate-enzyme-complex (=Ys), then the proportionality of Ys and the reaction speed (Vs/[Vmax–Vs]) includes the "steady state" or "quasi steady state assumption" (Walter, 1977).

In the case of $n = 2$ the dependency between enzyme-activity and substrate-concentration is sigmoidal. As described by Newsholme and Start (1973) and Walter (1977), it was Hill and later Monod et al. (1963) who first showed the importance of a sigmoidal relationship for the possiblity of regulating enzyme catalyzed metabolic reactions or processes.

Between 10% and 90% of the maximal enzyme activity, only a 10-fold increase of the substrate concentration is necessary if a sigmoidal relation exists and in this range the dependency is approximately linear.

Based on a simple model, Newsholme and Start (1973) showed that there are some reasons to assume that the rate of oxidative phosphoration and therefore $\dot{V}_{O_2}$ is dependent on the $[ADP]^2$ concentration to establish a sigmoidal relationship.

Only if this precondition is fulfilled a possible equilibrium state between measured intracellular ADP-concentrations (McGilvery, 1973; Karlsson, 1971) and the activity of the respiratory chain leads to a "steady state" or "quasi steady state" between the ADP-concentration and the speed of rephosphorylation, which can be measured by the $\dot{V}_{O_2}$. In that case the "steady state" between ADP and $\dot{V}_{O_2}$ can be described by reformulating Equation 5:

$$\dot{V}_{O_2} = \frac{\dot{V}_{O_2}max}{(1 + Ks'/[ADP]^2)} \tag{6}$$

According to Newsholme and Start (1973) the "steady state" between ADP and oxidative phosphorylation on the cellular level occurs within 1 s. Therefore it can be accepted that under all conditions a "quasi steady state" exists which can be described by Equation 6.

If we assume, according to Newsholme and Start (1973), that the rate of glycolysis and the resulting ATP-production is dependent on the product of ADP and AMP, the possible dependency under steady state conditions or "quasi steady state conditions" is:

$$\frac{dLa}{dt} = \dot{V}La = \frac{\frac{dLa}{dt}max}{\left(1 + \frac{Ks''}{[ADP][AMP]}\right)} \tag{7}$$

As shown in Figure 1 the simulated relation between $\dot{V}_{La}$ and the product $[ADP] \times [AMP]$ is nearly equivalent $\dot{V}_{La} = f([ADP]^3)$.

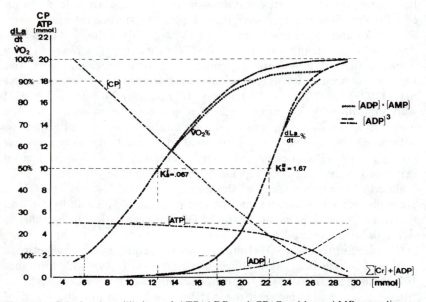

Figure 1—Calculated equilibrium of ATP/ADP and CP/Cr without AMP according to Equation 10. The steady state of $\dot{V}_{O_2}$ and V_{La} (expressed in percent from maximum) is calculated as a function of only HDP (—·—) or ADP and AMP (ooooo). If AMP is included in the calculation a reduced maximum of $\dot{V}_{O_2}$ and $\dot{V}_{La}$ occurs because the ADP-concentration is lower in spite of higher ATP and AMP.

Therefore $\dot{V}_{La}$ can be calculated by:

$$\dot{V}La = \frac{\frac{dLa}{dt}\ max}{(1 + Ks''/[ADP]^3)} \tag{7a}$$

Equations 6 and 7 define the static relation between the ATP-production resulting from oxygen uptake and lactic acid production which are dependent on the phosphorylation state of the ATP, ADP, AMP-system.

To include a description of a pH-dependent activity of glycolysis, particularly the influence on the PFK-activity according to the results of Danforth (1965), more than one solution is possible. If we assume that in the physiological range of intracellular pH (pH_i) the relation beween the change per mmol of added bases or acids and the pH is linear, then we calculate as follows:

$$pH_i = 7.05 - .0195\ La_m + .0195\ \Delta Cr \tag{8}$$

where the pH_i decreases with increasing lactate concentration and increases with a breakdown of creatine phosphate.

There is no correction made for a changing pCO_2. The maximal rate of glycolysis ($\dot{V}_{La'}max$) compatible with pH_i can be expressed by using Equation 5 in the following way:

$$V_{La'} = (1 - \frac{1}{(1 + Ks'''/[H^+]^3})\ V_{La}max \tag{9}$$

With $Ks''' = 10^{-20}$, Equation 9 fits with the results of Danforth (1965) (Figure 2). If we take into account a pH dependent rate of glycolysis, a steady state of the glycolysis as well as a constancy of the phosphorylation in the system of energy rich phosphates occurs only in the case of a constant lactic acid concentration in the muscle which results from an equilibrium of lactic acid production within the muscle itself and its elimination from the muscle. This is only possible in a small range of low lactic acid production when a larger part of the body muscle mass is involved in lactic acid production (see Figure 3).

Hohorst (1963) first, and later McGilvery (1974), calculated in a more accurate and complex way the phosphorylation equilibrium between ATP, ADP, AMP and creatine phosphate (CP) and creatine (= Cr). To avoid complex equations considering complicated conditions, as given in the paper of McGilvery et al. (1974), the equilibrium is calculated by the equation:

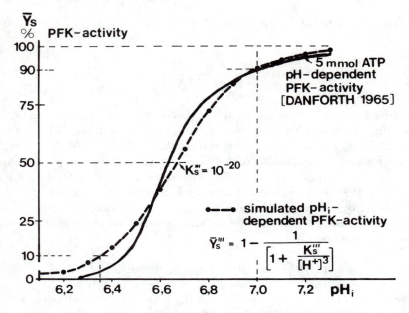

Figure 2—The maximal $\dot{V}_{La}$ due to PFK-activity is calculated as a function of the intracellular H^+- concentration. The simulated PFK-activity (—·—) has been adapted to the experimental observed curve of the pH-dependent PFK-activity of a level of 5 mmol ATP (Danforth, 1965).

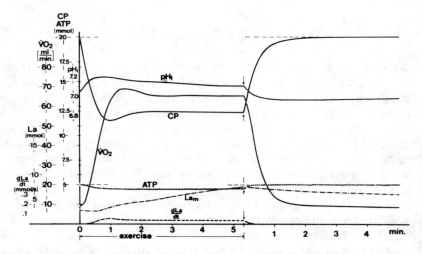

Figure 3—Simulation of time-curves of a power dependent adjustment of a new equilibrium in the ATP/ADP and CP/Cr-system and a new steady state of oxygen uptake in the case of low increasing muscle lactate. The time constant for the oxygen transport and diffusion ($T_1 = 1/K_1$) is 15 s.

$$\frac{[ATP]}{[ADP]} = m' \frac{[CP]}{[Cr]} \tag{10}$$

and the equilibrium between ATP, ADP and AMP by the equation (Newsholme & Start, 1973):

$$m'' = \frac{[ATP]\,[AMP]}{[ADP]^2} \tag{11}$$

The algebraic solution is:

$$[ATP] = \frac{A\,m'\,[CP]}{C + [CP]\,(m' - 1)} \tag{12}$$

$$[ADP] = \frac{A}{\left(\dfrac{m''\cdot(C-[CP])}{m'[CP]} + \dfrac{m'\cdot[CP]}{C-[CP]} + 1\right)} \tag{13}$$

$$[ATP] = \frac{m'\cdot[CP]\cdot[ADP]}{C-[CP]} \tag{13a}$$

$$[AMP] = A - [ATP] - [ADP] \tag{13b}$$

$$A = [ATP] + [ADP] \quad \text{or} \quad A = [ATP] + [ADP] + [AMP]. \tag{14}$$

$$C = [CP] + [Cr] \tag{15}$$

which give the equilibrium concentrations after a small change in creatine phosphate.

The reactions as described in Equations 10 and 11 take place with such a high velocity that we can assume that under all conditions the concentrations are near an equilibrium or that a "quasi equilibrium state" exists.

Using Equations 11 to 15 the steady state relation within the system of energy-rich phosphates can be plotted as a function of the dephosphorylation, i.e., the amount of ADP, AMP and Cr. Using Equations 6 and 7 the steady state rate of $\dot{V}o_2$ and La-production can be plotted against dephosphorylation of the system of energy rich phosphates as shown in Figure 1.

If we assume further that at the molecular level, ADP, AMP concentration and the activity of the respiratory chain or the activation of PFK at all possible speeds of ADP and AMP turnover are near a steady state (quasi steady state assumption), then we can use Equations 7 and 8 to calculate the resulting rephosphorylation speed.

Therefore the delay in oxidative rephosphorylation is dependent mainly on oxygen diffusion and transport processes, which can be described by Equations 2 and 3.

Despite the fact that the relation between (ADP) and the resulting $\dot{V}_{O_2}$ is nonlinear, Equation 3 can be reformulated as follows:

$$\frac{dP}{dt} = K_1 \, P_{\dot{V}_{O_2}} + K_2 \, P_{La} - Pe \tag{16}$$

where $P_{\dot{V}_{O_2}}$ is:

$$P_{\dot{V}_{O_2}} = \frac{K_{O_2}}{K_{CP}} \left(\frac{\dot{V}_{O_2}max}{1 + Ks'/[ADP]^2} \right) \tag{17}$$

and P_{La} is:

$$P_{La} = \frac{K_{La}}{K_{CP}} \left(\frac{\dot{V}_{La'}max}{1 + Ks''/[ADP]^3} \right) \tag{18}$$

Equation 16 can be solved using the Runge Kutta Method under the assumption that in the case of small steps the given relation seemed to be close to linear. To obtain the exact solutions, as shown in Figures 3, 4, 5, and 6, a second time constant for the delay in the oxygen transport system is included.

To describe the lactic acid distribution from the working muscle compartment to the rest of the body, especially the blood, we use a two compartment model including a nonlinear relation between the blood lactate concentration and the speed of lactate diffusion from the muscle according to Freund et al., (1978) and our own results (Mader et al., 1978).

The used system of differential equations is:

$$\frac{dLa}{dt} m = - K_{1,2} \, La_m + K_{2,1} \, La_b - K_{1,0} \, La_m \tag{19}$$

$$\frac{dLa}{dt} b = + K_{1,2} \, La_m + K_{2,1} \, La_b - K_{2,0} \, La_b \tag{20}$$

In this equation the invasion time constants are calculated as follows:

$$K_{2,1} = K_{1,2} = .2 \, La_b^{-1,6} \tag{21}$$

The resulting time curves blood lactate concentration after the end of the exercise fit the real measured blood lactate curves after short maximal running. For more details see the paper of Freund et al. (1978).

Equations 19 and 20 can be solved with a digital computer by using the method of Runge Kutta.

Some results of the model-simulation are shown in Figures 3, 4 and 5 at different speeds of breakdown of creatine phosphate according to the value of P_E and different performance times. The assumed values of CP before starting exercise and ATP are in the range of reported experimental results (Karlsson, 1971).

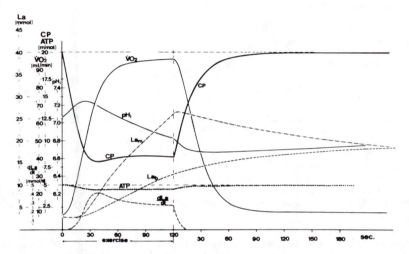

Figure 4—Calculated is the time dependent reaction of phosphorylation state of the system of energy rich phosphate and the time-curves of oxygen uptake ($\dot{V}_{O_2}$), rate of lactate production and lactate accumulation. A relative steady state occurs in the ATP and CP-concentration. A slight increase of ADP causes a continuous slight increase of $\dot{V}_{O_2}$ after the fast adjustment. The continuous increase of muscle lactate (La_m) causes a decrease of pH; and this results in a decrease of $\dot{V}_{La}$. The resulting demand of energy is compensated by the slight increase of $\dot{V}_{O_2}$. The slight increase of $\dot{V}_{O_2}$ shows that a true steady state cannot be established.

The speed of the rephosphorylation of CP is in the range which was reported by Harris et al. (1976). If we plot the resulting $\dot{V}_{O_2}$ and time constants against the load, we see a near linear increase of the oxygen uptake with the load, the expected "leveling off" and a shortening of the time constants with the rising $\dot{V}_{O_2}$. This was shown by Astrand and Saltin (1961) and others (Margaria et al., 1965); see Figure 6.

The relation between $\dot{V}_{O_2}$ and the resulting maximal blood lactate is mainly dependent on the relation between Ks' and Ks'' in a possible physiological range.

A steady state of $\dot{V}_{O_2}$ under the level of $\dot{V}_{O_2 max}$ under the condition of a pH dependent regulation of the glycolysis does not occur if lactate

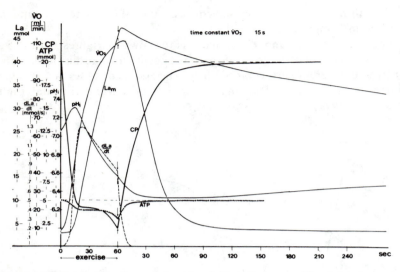

Figure 5—Simulated exhaustive exercise. After the fast adjustment of $\dot{V}_{La}$ and the increasing $\dot{V}_{O_2}$ a relative steady state in the CP and ATP-concentration occurs for a short time. The continuous reduction in the intracellular pH (pH$_i$) leads to decreasing rate of glycolysis (= $\dot{V}_{La}$) which can be compensated by the further increase of $\dot{V}_{O_2}$. Finally the decreasing rate of glycolysis leads to a fast breakdown of CP and ATP, which makes it impossible to continue the work.

concentration increases because the lowering of the rate of glycolysis leads to slightly increasing ADP-concentrations caused by a continual breakdown of CP. This results in further slight increase of the $\dot{V}_{O_2}$ after the fast adjustment, as demonstrated in Figures 4 and 5.

In the case of a higher rate of lactic acid production there occurs a sudden breakdown of CP and ATP when glycolysis slows down. This leads to the breakdown of power or the lowering of power to a level which can be supplied by oxygen uptake with no further lactic acid production as shown in Figure 5.

References

ÅSTRAND, P.O. & B. Saltin. Oxygen uptake during the first minutes of heavy muscular exercise. J. Appl. Physiol. 16(6):971–976, 1961.

DANFORTH, W.H., Activation of glycolytic pathway in muscle. In B. Chance, R.W. Estabrook & J.R. Williamson (Eds.), Control of Energy Metabolism. New York: Academic Press Inc., 1965.

FREUND, H. & P. Gendry. Lactate kinetics after short strenuous exercise in man. Europ. J. Appl. Physiol. 39:123, 1978.

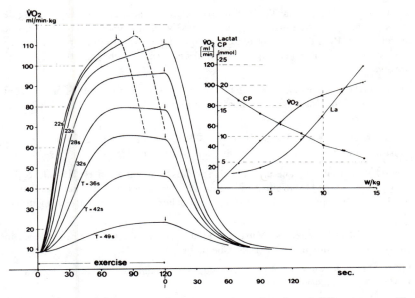

Figure 6—Simulated time-curves of oxygen uptake as function of different power. The increase of power leads to a shortening of the resulting time contant (=T). This is in accordance with the experimental results from Astrand & Saltin (1961) and Margaria et al. (1965). At higher loads leading to lactate production no steady state of oxygen uptake occurs according to the results of Linnarsson (1972) and Whipp & Wassermann (1972).

HARRIS, R.C., T. Edwards, E. Hultmann, L.O. Nordesjö, B. Nylind & K. Sahlin. The time course of phosphorylcreatine resynthesis during recovery of the quadriceps muscle in man. Pflügers Arch. 367:173–142, 1976.

HARRIS, R.C., K. Sahlin & E. Hultmann. Phosphagen and lactate contents of m. quadriceps of man after exercise. J. Appl. Physiol. Respirat. Exercise Physiol. 43(5):852, 1977.

HOHORST, H.J., M. Reim & H. Bartels. Studies on the creatine kinase equilibrium in muscle and the significance of ATP and ADP levels. Biochem. Biophys. Res. Comm. 7:142–146, 1962.

KARLSSON, J., Lactate and phosphagen concentration in working muscles of man. Acta Physiol. Scand.: Suppl. 358, 1971.

LINNARSSON, D., Dynamics of pulmonary gas exchange and heart rate changes at start and end of exercise. Acta Physiol. Scand.: Suppl. 415, 1974.

MADER, A., H. Heck & W. Hollmann. Evaluation of lactic acid anaerobic energy contribution by determination of postexercise lactic acid concentration of ear capillary blood in middle-distance runners and swimmers. In the International Congress of Physical Activity Sciences 1976. Quebec, Canada. Book 4, Exercise Physiology 1978, Miami, Florida.

MADER, A., H. Heck & W. Hollmann: Leistung und Leistungsbegrenzung des menschlichen Organismus, interpretiert am Modell thermodynamisch offener Systeme. Ein Beitrag zur Diskussion biologischer Leistungsgrenzen im Hochleistungssport. In: H. Rieckert, Sport an der Grenze menschlicher Leistungsfähigkeit. Springer, 1981.

MARGARIA, R., P. Cerretelli & F. Mangili. Balance and kinetics of anaerobic energy release during strenuous exercise in man. J. Appl. Physiol. 19(4):623–628, 1964.

MARGARIA, R., F. Mangili, F. Cuttica & P. Cerretelli. The kinetics of the oxygen consumption at the onset of muscular exercise in man. Ergonomics 8:49–54, 1965.

McGILVERY, R.W., The use of fuels for muscular work. In H. Howald & J.R. Poortmans (Eds.), Metabolic Adaptation of Prolonged Physical Exercise. Proc. of the Second International Symposium on Biochemistry of Exercise. Magglingen, 1973.

McGILVERY, R.W. & W. Murray. Calculated equilibrium of phosphocreatine and adenosine phosphates. J. Biol. Chem. 249:4845, 1974.

MONOD, J., J. Wymen & J.-P. Changeux. On the nature of allosteric transitions. J. Mol. Biol. 12:88–118, 1965.

NEWSHOLME, E.A. & C. Start. Regulation in Metabolism. London: Wiley, 1973.

STEGEMANN, J., Leistungsphysiologie. Stuttgart: Thieme, 1977.

WALTER, C., Contributions of enzyme models. In: D.L. Solomon and C. Walter: Mathematical Model in Biological Discovery. Lecture Notes in Biomathematics. Berlin-Heidelberg-New York: Springer, 1977.

WHIPP, B.J. & K. Wasserman. Oxygen uptake kinetics for various intensities of constant-load work. J. Appl. Physiol. 33:351, 1972.

Glossary of Symbols

Wg	amount of work (I)
E	power (W)
t	time in second (minute)
$t_1 = 1/K_1$	time constant V_{O_2} resulting from oxygen transport and diffusion
$T_2 = 1/K_2$	time constant V_{La}
K_{CP}, K_{La}, K_{O_2}	
CP	Creatine phosphate (mmol/kg) muscle
Cr	Creatine
ΔCr	difference between Po and the actual concentration P(t) of creatine phosphate

ATP	Adenosintriphosphate (mmol/kg)
ADP	Adenosindiphosphate (mmol/kg)
AMP	Adenosinmonophosphate (mmol/kg)
m', m''	equilibrium constant
P_E	power-dependent CP-splitting (mmol/s kg)
P_Z	rephosphorylation rate of ATP, CP (mmol/s kg)
P_O	CP-concentration before starting exercise
$P_{(t)}$	actual CP-concentration
$P_{V_{O_2}}$	rephosphorylation rate resulting from the oxygen uptake
P_{La}	rephosphorylation rate resulting from the glycolysis
$\dot{V}_{O_2}$	oxygen uptake per kg/muscle (ml/min kg)
$\dot{V}_{O_2}max$	max. oxygen uptake per kg/muscle (ml/min kg)
La	Lactate concentration mmol
La_m	muscle lactate concentration (mmol/kg)
La_b	blood lactate concentration (mmol/l)
$dLa/dt = \dot{V}_{La}$	rate of lactate production (mmol/s kg)
$\frac{dLa}{dt}max = \dot{V}_{La}max$	maximal rate of lactate (mmol/s kg) production
$V_{La'}max$	pH; dependent maximal rate of lactate production
pH_i	intra-cellular pH
K_S	dissociation constant of enzyme substrate complex
S	substrate concentration (mmol/l)
Ys	relative activity of the enzyme in % of maximum
Vs	equivalent relative speed of reaction compatible with Ys
Vmax	maximal reaction speed
Ks'	constant for ADP-dependent half maximal rate of oxidative rephosphorylation
Ks''	constant for half maximal rate of glycolysis
Ks'''	constant pH-dependent maximal glycolysis ($= V_{La'}max$)
$K_{1,2}$; $K_{2,1}$	Invasion coefficients of lactate distribution to the blood
$K_{1,0}$	lactate elimination coefficient from the muscle
$K_{2,0}$	lactate elimination coefficient from the blood

Substrate Sparing Effect of Atenolol Compared with Propanolol during Intensive Short-term Exercise

A.A. McLeod, J.E. Brown, R.S. Williams and D.G. Shand
Duke University Medical Center, Durham, North Carolina,
U.S.A.

Radioligand binding methods, cyclase activation experiments and other biochemical studies have demonstrated that both alpha and beta adrenoceptors may mediate glycogenolysis within the liver (Aggerbeck et al., 1978; Exton, 1979; Schmelck & Hanoune, 1980). Hepatic glucose production and disposition have been well studied in response to catecholamine infusions, with or without manipulation of insulin or glucagon levels (Best et al., 1982; Rizza et al., 1980; Sacca et al., 1979). The role of adrenergic mechanisms in counterregulation to insulin-induced hypoglycemia has been studied (Abramson & Arky, 1968; Clarke et al., 1979). The adrenergic mediation, and particularly the subtype specificity of the metabolic response to the physiologic stimulus of exercise has been less well characterized, however. The studies reported here utilized the beta-adrenoceptor antagonist propanolol (a non-subtype selective agent) and a newer compound, atenolol, which shows selectivity of antagonism for $beta_1$ adrenoceptors (Harms, 1976).

Preliminary studies in our laboratory with healthy human subjects indicated that during short term exercise (16 min, increasing exercise intensity to 200 W), blood glucose fell after propanolol administration but did not change after a similar dose of atenolol. Propanolol appeared to suppress the lactate response, particularly at intermediate exercise levels. Identical suppression of free fatty acid (FFA) levels was found, indicative of a $beta_1$ adrenoceptor mechanism for lipolysis.

$Beta_2$ adrenoceptor mechanisms have been claimed to mediate muscle glycogenolysis (Arnold et al., 1968), but these may be irrelevant under the stress of exercise. Accordingly, a hypothesis was developed to explain the propanolol-atenolol differences on glucose metabolism by antagonism of intramuscular glycogenolysis by propanolol (but not atenolol), thus necessitating an increased dependence on circulating glucose for energy substate. This hypothesis was tested by administering the drugs after the subject had exercised vigorously to induce muscle glycogen depletion. If the hypothesis was correct, after this maneuver, differences between propranolol and atenolol should be eliminated if the major energy substrates were FFA and circulating blood glucose.

Methods

In preliminary studies, six healthy male subjects aged 21 to 27 each performed three graded exercise tests on a bicycle ergometer. Tests were performed fasting, 1 wk apart, after single oral doses of propranolol 40 mg, atenolol 50 mg, or placebo, given in random order according to a Latin square design. The results of this preliminary study form part of a general hemodynamic, hormonal, and metabolic characterization of differences between non-selective and beta$_1$-selective blockade and will be described in full elsewhere.

Four other subjects aged 27 to 33 performed the same graded exercise test (50, 100, 150, and 200 W, 4 min at each work rate). They then continued to exercise at maximum tolerable workload (125 to 200 W) for a further 30 min. Immediately after this they took one of the treatments described above. Two h later, the subjects repeated the original graded exercise protocol. Blood samples were taken during both exercise periods for measurement of glucose, lactate, and free fatty acid levels.

Glucose was measured using a modified glucose oxidase method. Lactate was measured using a NADH production method. Free fatty acids (FFA) were measured by a gas chromatographic method (Sampson & Hensley, 1975).

Statistical analysis was performed using two-way ANOVA. The study is represented schematically below (Figure 1).

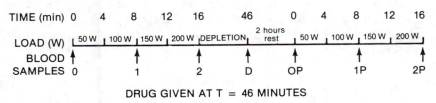

Figure 1—Drug given at T = 46 min.

Results

During 16 min of increasing exercise intensity, FFA levels fell, and then rose again during continued heavy exercise (Table 1). Lactate levels rose to 4.5 ± 0.8 mmol/liter and were sustained at high levels (5.0 ± 0.5 mmol/liter) during the exercise aimed at provoking glycogen depletion. Glucose levels rose slightly at first but essentially showed little change (Table 1).

Table 1

Metabolic Alterations During Graded Exercise Test and 30-min Glycogen Depletion Phase

Exercise stage	FFA (μmol/l)	Lactate (mmol/l)	Glucose (mmol/l)
0 (Rest)	449 ± 77	1.0 ± 0.1	5.02 ± 0.07
1 (8 min)	360 ± 54	1.5 ± 0.1	4.97 ± 0.08
2 (16 min)	287 ± 31	4.5 ± 0.8	5.13 ± 0.08
D (depletion)	379 ± 27	5.0 ± 0.5	5.08 ± 0.07

n = 12; 4 subjects.

When the graded exercise protocol was repeated 2 h later following medication (Table 2), FFA levels behaved in similar fashion, although they were higher at the outset. Propranolol and atenolol both produced similar degrees of antagonism and levels of FFA and at the end of exercise were 47% and 40% of the control respectively ($p = 0.02$ for both drugs against placebo).

All four subjects had lower resting lactate levels on propranolol ($p = 0.02$ by paired t-test). Thereafter, there were no significant differences between treatments for lactate values when tested by analysis of variance.

Major differences were seen in glucose levels. During the second graded exercise test blood glucose fell as exercise progressed irrespective of treatment. However, the fall in glucose on placebo was small. The fall on atenolol was greater but not significantly different. However, a large fall in blood glucose occurred on propranolol ($p = 0.01$ for difference from placebo).

Discussion

The findings in this study show clearly that the differences observed in glucose levels between propranolol and atenolol treatments are accentu-

Table 2

**Metabolic Alterations During Second Graded Exercise Test,
2 H after Drug Administration**

	FFA (μmol/l)		
Stage	Placebo	Propranolol	Atenolol
0 Post drug	529 ± 148	365 ± 51	391 ± 110
1 Post drug	491 ± 111	320 ± 74	337 ± 93
2 Post drug	431 ± 95	203 ± 55	173 ± 41

	Lactate (mmol/l)		
Stage	Placebo	Propranolol	Atenolol
0 Post drug	1.2 ± 0.1	0.9 ± 0.2	1.3 ± 0.2
1 Post drug	1.4 ± 0.1	1.3 ± 0.3	1.7 ± 0.2
2 Post drug	4.2 ± 1.3	4.5 ± 1.6	5.3 ± 1.6

	Glucose (mmol/l)		
Stage	Placebo	Propranolol	Atenolol
0 Post drug	4.88 ± 0.19	4.79 ± 0.13	4.85 ± 0.14
1 Post drug	4.85 ± 0.22	4.56 ± 0.16	4.68 ± 0.18
2 Post drug	4.71 ± 0.16	4.02 ± 0.06	4.57 ± 0.14

ated when subjects have first performed heavy exercise to partially deplete skeletal muscle glycogen. In this situation, exercising muscle becomes more dependent on substrate delivered by the circulation, namely FFA and glucose. In our preliminary study, we showed that the doses used here of propranolol and atenolol produced equal blockade of lipolysis as reflected in FFA levels. Thus the marked reduction in plasma glucose after propranolol administration suggests a potent intrahepatic block of glucose production and can be taken as evidence against our initial hypothesis that beta$_2$ glycogenolysis in muscle is important during exercise. Some experimental evidence already suggests that during exercise, catecholamine mediation of muscle glycogenolysis is unimportant (Stull & Mayer, 1971).

Importantly, differences seen in lactate levels on earlier study were not evident after the glycogen depletion protocol. A slight reduction in resting lactate level was seen after propranolol, possibly implying impaired production at rest, but at peak exercise no differences were found between treatments. Thus there appears to be no difference in production of this important gluconeogenic precursor during exercise. The equal effect of the drugs on lipolysis is also likely to lead to equal reductions of glycerol, another gluconeogenic precursor. Other studies have not demonstrated major alterations in alanine levels after beta blockade. In view of this, we suggest that it is likely that in our subjects, propranolol is producing an important blockade of $beta_2$ receptor mediated hepatic glycogenolysis.

Experimental evidence for a physiologic role for the beta receptor in hepatic glycogenolysis is plentiful, but several authors have claimed to show that alpha receptor mechanisms are of dominant importance in humans (Antonis et al., 1967). It is probable that failure to demonstrate beta receptor mechanisms in such studies was due to concomitant stimulation of glucagon release. During the stress of heavy exercise however, there is a demand for glucose which clearly activates release of catecholamines and glucagon, and the physiological importance of a beta receptor mechanism becomes clear when, as in this study, one of these stimulatory pathways is antagonized by propranolol.

In conclusion, our study supports data that indicate $beta_2$ glycogenolysis in muscle does not play a role in substrate mobilization during vigorous exercise. Laboratory data (Lefkowitz, 1975) suggests that $beta_2$ hepatocyte receptors activate adenylate cyclase to set in train the glycogenolysis needed to provide circulating glucose, and our findings suggest that this mechanism is physiologically important to humans.

References

ABRAMSON, E.A. & R.A. Arky. Role of beta-adrenergic receptors in counter-regulation to insulin-induced hypoglycemia. Diabetes 17:141–146, 1968.

AGGERBECK, M., G. Guellaen & J. Hanoune. Biochemical evidence for the dual action of labetalol on α-and β-adrenoceptors. Brit. J. Pharmacol. 62:543–548, 1978.

ANTONIS, A., M.L. Clark, R.L. Hodge, M. Molony & T.R.E. Pilkingon. Receptor mechanisms in the hyperglycaemic response to adrenaline in man. Lancet 1:1135–1137, 1967.

ARNOLD, A., J.P. McAuliff, D.F. Colella, W.V. O'Connor, & Th.G. Brown, Jr. The β_2 receptor mediated glycogenolytic responses to catecholamines in the dog. Arch. Int. Pharmacodyn. 176:451–457, 1968.

BEST, J.D., G.J. Taborsky, Jr., H.A. Pfeifer, J.B. Halter & D. Porte, Jr. Alpha adrenergic stimulation does not directly increase hepatic glucose production in man. Clin. Res. 30:60A, 1982.

CLARKE, W.L., J.V. Santiago, L. Thomas, E. Ben-Galim, M.W. Haymond & P.E. Cryer. Adrenergic mechanisms in recovery from hypoglycemia in man: adrenergic blockade. Am. J. Physiol. 236:E147–152, 1979.

EXTON, J.H., Mechanisms involved in effects of catecholamines on liver carbohydrate metabolism. Biochem. Pharmacol. 28:2237–2240, 1979.

HARMS, H.H., Isoproterenol antagonism of cardioselective beta adrenergic receptor blocking agents: a comparative study of human and guinea-pig cardiac and bronchial beta adrenergic receptors. J. Pharmacol. Exp. Ther. 199:329–335, 1976.

LEFKOWITZ, R.J., Heterogeneity of adenylate cyclase-coupled β-adrenergic receptors. Biochem. Pharmacol. 24:583–590, 1975.

RIZZA, R.A., P.E. Cryer, M.W. Haymond & J.E. Gerich. Adrenergic mechanisms for the effects of epinephrine on glucose production and clearance in man. J. Clin. Invest. 65:682–689, 1980.

SACCA, L., R. Sherwin & P. Felig. Influence of somatostatin on glucagon- and epinephrine-stimulated hepatic glucose output in the dog. Am. J. Physiol. 236:E113–E117, 1979.

SAMPSON, D. & W.J. Hensley. A rapid gas chromatographic method for the quantitation of underivatised individual free fatty acids in plasma. Clin. Chim. Acta. 61:1–8, 1975.

SCHMELCK, P-H. & J. Hanoune. The hepatic adrenergic receptors. Moll. Cell. Biochem. 33:35–48, 1980.

STULL, J.T. & S.E. Mayer. Regulation of phosphorylase activation in skeletal muscle in vivo. J. Biol. Chem. 246:5716–5723, 1971.

Physical Performance and Muscle Metabolic Characteristics

P.A. Tesch, J.E. Wright, W.L. Daniels and B. Sjödin
Karolinska Institutet, Stockholm, Sweden, U.S. Army
Research Institute of Environmental Medicine, Natick,
Massachusetts, U.S.A., and National Defense Research
Institute, Stockholm, Sweden

Human skeletal muscle is composed of a mixture of slow-twitch (ST or Type I) and fast-twitch (FT or Type II) fibers. These two main fiber types have distinctly different metabolic profiles as indicated by the activity of enzymes involved in oxidative and anaerobic metabolism (Essén et al., 1975), for example, as well as capillarization (Andersen & Henriksson, 1977). Various measures of physical performance have been shown to be related to fiber type composition, enzyme activity and capillary density of the exercising muscle. For example, muscle strength, especially at higher angular velocities, is greater the larger the proportion of FT fibers (Coyle et al., 1979; Thorstensson et al., 1977); maximal oxygen uptake increases with the percentage of ST fibers (Forsberg et al., 1979) and cytochrome oxidase activity (Booth & Narahara, 1974). Moreover, susceptability to muscle fatigue is related to the percentage of FT fibers (Thorstensson & Karlsson, 1976) and lactate dehydrogenase activity (Tesch et al., 1978). The exercise intensity, where significant blood lactate accumulation occurs, is increased the greater the percentage of ST fibers (Farrell et al., 1979; Ivy et al., 1980), capillary supply, and muscle respiratory potential (Ivy et al., 1980; Sjödin & Jacobs, 1981; Sjödin et al., 1981). The present study was undertaken to further clarify the influence of individual variations in muscle metabolic profile on various physical performance characteristics.

Methods and Procedures

On separate days the following experiments were carried out in thirteen healthy and physically active men: 1) *Maximal oxygen uptake* ($\dot{V}O_{2max}$)

was measured during cycling at 60 rpm and defined according to the "leveling off" criterion; 2) *The onset of blood lactate accumulation* (OBLA) was determined during continuous cycling exercise with power output increased every fourth min (Tesch et al., 1981). Venous blood samples were collected before each increment. The power output, which corresponded to a lactate concentration of 4 mmol l^{-1} blood (Mader et al., 1976) was defined as W_{OBLA}. In addition %OBLA, i.e., the fractional utilization of $\dot{V}O_{2max}$ at W_{OBLA}, was calculated; 3) *Maximal peak torque* during knee extensions performed at 30 and 300° s^{-1} was measured using a speed controlled dynamometer (Cybex II®, Lumex Inc, N.Y.); 4) *Muscle fatigue* was calculated from the relative decline in peak torque during 50 consecutive leg extensions (Thorstensson & Karlsson, 1976); 5) *Recovery* was calculated as the maximal peak torque attained in five additional contractions executed 40 s following the muscle fatigue test and defined as the maximal peak torque relative to maximal peak torque during the first 50 contractions.

Muscle biopsies, obtained from the vastus lateralis muscle at rest, were analyzed for fiber type composition expressed as the relative area occupied by fast-twitch fibers (%FT area), and capillary density (capillaries per mm^2 and capillary per fiber). Freeze-dried muscle tissues were analyzed for enzyme activities of phosphofructokinase (PFK), lactate dehydrogenase (LDH), and citrate synthase (CS) using fluorometric techniques.

Results

Maximal oxygen uptake averaged ($\pm$ SD) 51(7) ml kg^{-1} whereas values for W_{OBLA} and %OBLA were 166($\pm$ 38)W and 64($\pm$ 12)%, respectively. Maximal peak torques at 30 and 300° s^{-1} were 3.1($\pm$ 1.3) and 1.6($\pm$ 0.4) Nm kg^{-1} b.w. Peak torque decreased 66($\pm$ 10)% during 50 repeated muscle contractions. Following a 40 s rest period peak torque was 72($\pm$ 12)% of the initial peak torque.

Mean ($\pm$ SD) values for fiber type composition, capillary density and enzyme activities are summarized in Table 1.

Significant correlation coefficients were established between $\dot{V}O_{2max}$ and PFK activity ($r = -0.66$, $p < 0.05$) and LDH CS^{-1} ($r = -0.58$, $p < 0.05$). W_{OBLA} was best correlated with PFK activity ($r = -0.86$, $p < 0.001$) and PFK CS^{-1} ($r = -0.84$, $p < 0.001$). %OBLA was negatively related to %FT area ($r = -0.81$, $p < 0.001$). Maximal peak torque at the high angular velocity was highly correlated with %FT area ($r = 0.82$, $p < 0.001$), whereas a low positive relationship was present at the low angular velocity ($r = 0.56$, $p < 0.05$). Fatigue was positively related to LDH activity and LDH CS^{-1} ($r = 0.79$, $p < 0.001$). Recovery was inversely related to LDH

Table 1

Mean (± SD) Values for Fiber Type Composition, Capillary Density and Enzyme Activities

	$\overline{\times}$	± SD
%FT area	48	20
cap mm^{-1}	284	84
cap fib^{-1}	1.6	0.3
CS, mol g^{-1} min^{-1} 10^{-6}	3.7	0.6
LDH, mol g^{-1} min^{-1} 10^{-4}	0.77	0.30
PFK, mol g^{-1} min^{-1} 10^{-6}	6.3	1.7
LDH CS^{-1}	21.6	9.9
PFK CS^{-1}	17.5	6.3

CS^{-1} ($r = -0.83$, $p < 0.001$), LDH activity ($r = -0.70$, $p < 0.01$), PFK CS^{-1} ($r = -0.71$, $p < 0.01$) and positively related to capillary density ($r = 0.70$, $p < 0.01$).

Discussion

In agreement with numerous recent reports (e.g., Farrell et al., 1977; Sjödin & Jacobs, 1980; Thorstensson & Karlsson, 1976; Thorstensson et al., 1977) the metabolic profile of the exercising muscle was found to influence exercise performance.

The relative distribution of the two main muscle fiber types is essentially established during early childhood (Colling-Saltin, 1980) and seems to be largely genetically determined (Komi & Karlsson, 1979), even though some studies suggest that the proportions of FT and ST fibers may change with intense physical training (Jansson et al., 1978; Tesch et al., 1982). In the present study fiber type composition was expressed as the relative area occupied by FT fibers. Since fiber type selective hypertrophy is possible due to changes in activity pattern (Andersen & Henriksson, 1977; Thorstensson et al., 1976) variability in fiber type composition, as defined here, might be a function of specific individual physical training.

Capillary supply and enzyme activities, however, are factors much more sensitive to modified environmental conditions (Andersen & Henriksson, 1977; Henriksson & Reitman, 1976; Sjödin, 1976).

Although maximal aerobic capacity, as indicated by $\dot{V}_{O_2max}$, exhibited only a modest relationship with the estimated muscle metabolic potential for oxygen supply and utilization as well as lactate reduction, the ability to perform sustained exercise at a high absolute and relative intensity was strongly influenced by these variables. Thus, the activities of phosphofructokinase, and lactate dehydrogenase seemed to be important factors in determining onset of blood lactate accumulation. Muscle strength at a relatively high angular velocity was highly correlated with the percentage of FT fibers, which supports earlier observations (Coyle et al., 1979; Eccles & Sherrington, 1930; Tesch, 1980; Thorstensson et al., 1977).

In experiments where an identical muscle fatigue test was applied, individuals rich in FT fibers (Tesch et al., 1978; Thorstensson & Karlsson, 1976) and high in LDH and/or muscle specific LDH isozymes (Tesch et al., 1978; Tesch 1980) fatigued to a greater extent when compared to individuals rich in ST fibers and having a LDH enzyme and isozyme pattern favoring lactate oxidation (Sjödin, 1976). Differences in intracellular lactate accumulation and associated changes in pH were suggested to be the main cause for the different fatigue response (Tesch, 1980). Likewise, an improved lactate release was implied in subjects with a greater ST fiber frequency. Thus, in agreement with Graham et al. (1978), the muscle/blood lactate gradient increased with increased FT percentage (Tesch, 1980). The present results support the relevance of the vascular bed or lactate release to recovery in light of the relationship of capillary supply to recovery. Moreover, the activity of citrate synthase correlated with recovery emphasizing the relative importance of oxidative metabolism.

In conclusion, we have been able to confirm in one study what has been previously demonstrated in a series of separate studies, i.e., that muscle strength, susceptibility to fatigue, recovery from intense short-term exercise as well as the capacity to perform prolonged exercise are strongly influenced by the metabolic profile of the exercising muscle, as reflected by fiber type composition, capillarization and enzyme activities.

References

ANDERSEN, P. & J. Henriksson. Capillary supply of the quadriceps femoris muscle of man: adaptive response to exercise. J. Physiol. 270:677–690, 1977.

BOOTH, F.W. & K.A. Narahara. Vastus lateralis cytochrome oxidase activity and its relationship to maximal oxygen consumption in man. Pflügers Arch. 349: 319–324, 1974.

COLLING-SALTIN, A.-S., Skeletal muscle development in the human fetus and during childhood. In B.-O. Eriksson (Ed.), Proceedings of the IX International Congress on Pediatric Work Physiology. Baltimore, MD: University Park Press, 1980.

COYLE, E.F., D.L. Costill & G.R. Lesmes. Leg extension power and muscle fiber composition. Med. Sci. in Sports 11:12–15, 1979.

ECCLES, J.C. & Sir C.S. Sherrington. Numbers of contraction-values of individual motor-units examined in some muscles of the limb. Proc. Roy. Soc., Ser. B. 106:326–357, 1930.

ESSÉN, B., E. Jansson, J. Henriksson, A.W. Taylor & B. Saltin. Metabolic characteristics of fiber types in human skeletal muscle. Acta Physiol. Scand. 95:153–165, 1975.

FARRELL, P.A., J.H. Wilmore, E.F. Coyle, J.E. Billing & D.L. Costill. Plasma lactate accumulation and distance running performance. Med. Sci. in Sports 11:338–344, 1979.

FORSBERG, A., P. Tesch, B. Sjödin, A. Thorstensson & J. Karlsson. Skeletal muscle fibers and athletic performance. In P. Komi (Ed.), Biomechanics V-A. International Series on Biomechanics. Baltimore, MD: University Park Press, 1976.

GRAHAM, T., G. Sjøgaard, H. Löllgren & B. Saltin. NAD in muscle of man at rest and during exercise. Pflügers Arch. 376:35–39, 1978.

HENRIKSSON, J. & J.S. Reitman. Quantitative measures of enzyme activities in type I and type II muscle fibers of man after training. Acta Physiol. Scand. 97:392–397, 1976.

IVY, J.L., R.T. Withers, P.J. Van Handel, D.H. Elger & D.L. Costill. Muscle respiratory capacity and fiber type as determinants of the lactate threshold. J. Appl. Physiol. 48:523–527, 1980.

JANSSON, E., B. Bjödin & P. Tesch. Changes in muscle fiber type distribution in man after physical training. A sign of fiber type transformation. Acta Physiol. Scand. 104:235–237, 1978.

KOMI, P.V. & J. Karlsson. Physical performance, skeletal muscle enzyme activities, and fiber types in monozygous and dizygous twins of both sexes. Acta Physiol. Scand. (Suppl) 462, 1979.

MADER, A., H. Heck & W. Hollmann. Evaluation of lactic acid anaerobic energy contribution by determination of post-exercise lactic acid concentration of ear capillary blood in middle-distance runners and swimmers. In F. Landry & W. Orban (Eds.), The International Congress of Physical Activity Sciences. Vol. 4. Exercise Physiology. Miami: Symposia Specialists, 1978.

SJÖDIN, B., Lactate dehydrogenase in human skeletal muscle. Acta Physiol. Scand. (Suppl) 436, 1976.

SJÖDIN, B. & I. Jacobs. Onset of blood lactate accumulation and marathon running performance. Int. J. Sports. Med. 2:23–26, 1981.

SJÖDIN, B., I. Jacobs & J. Karlsson. Onset of blood lactate accumulation and enzyme activities in m. vastus lateralis in man. Int. J. Sports Med. 2:166–170, 1981.

TESCH, P., B. Sjödin, A. Thorstensson & J. Karlsson. Muscle fatigue and its relation to lactate accumulation and LDH activity in man. Acta Physiol. Scand. 103:413–420, 1978.

TESCH, P., Muscle fatigue in man. Acta Physiol. Scand. Suppl. 480, 1980.

TESCH, P.A., D.S. Sharp & W.L. Daniels. Influence of fiber type composition and capillary density on onset of blood lactate accumulation. Int. J. Sports Med. 2:252–255, 1981.

TESCH, P., J. Karlsson & B. Sjödin. Muscle fiber type distribution in trained and untrained muscles of athletes. In P.V. Komi (Ed.), Exercise and Sport Biology. Champaign, IL: Human Kinetics Publishers, 1982.

THORSTENSSON, A. & J. Karlsson. Fatiguability and fiber composition of human skeletal muscle. Acta Physiol. Scand. 98:318–322, 1976.

THORSTENSSON, A., B. Hultén, W. von Döbeln & J. Karlsson. Effect of strength training on enzyme activities and fibre characteristics in human skeletal muscle. Acta Physiol. Scand. 96:392–398, 1976.

THORSTENSSON, A., L. Larsson, P. Tesch & J. Karlsson. Muscle strength and fiber composition in athletes and sedentary men. Med. Sci. in Sports 9:26–30, 1977.

Metabolic Pattern and Blood Flow of the Contracting Rat Calf Muscle

R. Wilke, D. Angersbach, and P. Ochlich

R & D Laboratories, Beecham-Wülfing, Gronau(Leine),
Federal Republic of Germany

Muscular exercise is accompanied by marked metabolic changes and an increase in blood flow. During isometric contractions, muscle blood flow depends on the balance between locally released vasodilating metabolites, the degree of increase in muscular pressure and the increase in perfusion pressure that attends exercise (Shepherd et al., 1981). Contractions at high tensions lead to fatigue, and the maximal duration of exercise decreases. Fatigue is considered to be mainly due to a decrease in the amount of energy supply available for electromechanical processes (Edwards & Wiles, 1981). This is particularly true for glycogen stores (Jacob, 1981). The duration of isometric exercise seems to be a function of fiber composition (Petrofsky et al., 1981; Tesch, 1980) and the rate of energy exchange.

Our experiments were performed to study the changes in muscle blood flow, metabolite levels, pH and pO_2 in relation to the contractile force of rat calf muscle during isometric exercise at submaximal tension.

Methods

Wistar-strain male rats (body wt 280 to 350 g) were used. Under thiopentone anesthesia (100 mg/kg body wt i.p.), the sciatic nerve was severed circa 2 cm proximal to the knee, and the tendon of the calf muscles was cut and connected with an isometric force transducer at a tension of about 1 N. Muscles were directly stimulated with square wave pulses (duration 4 ms; frequency 2.5 Hz; voltage 5 V). For pH- and pO_2 – measurements, muscle contractions were induced by stimulation of the sciatic nerve (0.5 V). The muscles were continuously superfused with 0.9% NaCl-solution (38° C).

Tissue samples were taken at different time points by freeze-clamping the calf muscle. Metabolites were measured using standard methods (Faupel et al., 1972). Samples from the non-contracting hind limb were used as controls.

Muscle blood flow was determined by use of the xenon-133 clearance method. For pH-measurements, we used glass-microelectrodes (tip diameter circa 5 μ), inserted to a depth of 3 to 4 mm into the gastrocnemius muscle. Muscle pO_2 was measured by micro-needle electrodes with a tip diameter of 5 to 10 μm. Owing to their short length (10 to 15 mm), the electrodes were able to follow easily the movements of the contracting muscle. Both pH and pO_2 were recorded simultaneously. Xenon-133 clearance was measured in a separate group of rats. All experiments were conducted in at least five contracting calf muscles. Duration of static contractions lasted for 30 min, followed by a 10 min recovery period.

Training was performed by allowing the rats to run for 30 min/day on a rotating wheel (30 rpm, diameter 9 cm).

Results

The mean resting blood flow of the calf muscle was 8.6 ml/min $\times$ 100 g. With onset of muscular contractions, this increased to a peak of 24.0, followed, 2 min later, by a continuous decline (Figure 1). Muscle contractility force at a tension of about 60 to 30% of initial strength decreased by 50% within about 3 min after onset of exercise, reaching a plateau (about 30% of initial muscle force) after 5 to 10 min of exercise.

Muscle lactate concentrations increased sharply with the onset of contraction. A significant increase was seen after only 10 sec, and a further four- and eight-fold increase was measured after 1 and 4 min, respectively (Figure 2). Thereafter, lactate decreased, almost reaching resting values after 30 min of exercise.

Muscle or, in this case, interstitial pH showed a transient increase immediately after the start of static exercise, followed by a continuous decrease to a pH of 6.86 after 5 min. Thereafter, pH increased continuously, reaching a pH of 7.24 at the end of exercise. After cessation of contractions, a small transient decrease was observed, followed by a return to normal values after the 10 min recovery period. There was a linear inverse relationship between muscle lactate levels and muscle pH (r = –0.74, p < 0.01). Average resting muscle pO_2 was about 29 Torr (n = 15), falling by about 14 Torr within 19 s after the onset of isometric exercise. Thereafter, pO_2 stabilized at different levels, probably depending on the position of the electrode. Generally, muscle pO_2 showed a maximum at a similar time to that of muscle blood flow.

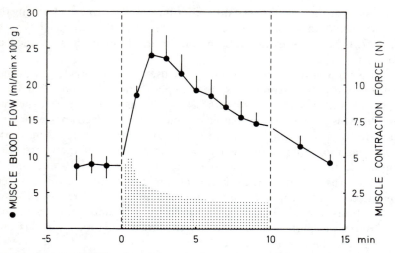

Figure 1—Muscle blood flow (●) at rest, during isometric exercise and at recovery and muscle contractile force (dotted area) in the calf muscle of the rat (Mean values ± SEM).

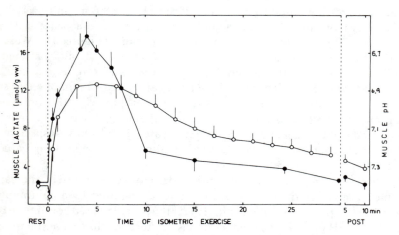

Figure 2—Muscle lactate concentrations (●) and muscle pH (O) at rest, during isometric exercise and at recovery (post) of the calf muscle of the rat (Mean values ± SEM).

Glycogen depletion rate was maximal during the onset of exercise and became slower with further exercise. Glycogen concentration returned to almost 50% of the resting value after the 10 min recovery period (Figure 3). Creatine phosphate (CP) decreased by about 50% in the first 10 s of isometric contractions, and decreased more slowly within the next 15 min. Thereafter, a small increase was observed. Resting values were reached

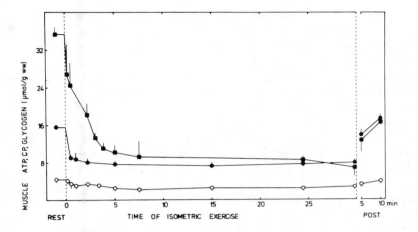

Figure 3—Muscle ATP (o), CP (●) and glycogen (■) concentrations at rest, during isometric exercise and at recovery (post) in the calf muscle of the rat (Mean values ± SEM).

within the 10 min recovery period. A similar, but smoother profile was seen with adenosine triphosphate (ATP).

In trained (30 days) and untrained animals, the time taken to reach 50% of initial muscle contraction force was significantly negatively correlated with lactate concentration ($y = -0.4 + 25.1$, $r = -0.84$, $p < 0.01$ untrained; $y = -0.01 + 14.8$, $r = -0.85$, $p < 0.05$ trained). The different increase in lactate per unit time of exercise is evidenced by the slope. The regression lines are significantly different ($p < 0.05$). No such significant difference was seen between untrained rats and those trained for 8 or 15 days.

Discussion

With isometric exercise, the pattern of muscle metabolites and that of muscle contractile force follows a biphasic course—an initial fatiguing phase is followed by a long-lasting, virtually non-fatiguing phase. Muscle force declines steeply in the initial phase, whereas muscle pO_2, after a transient fall, increases approximately to resting levels. Under such conditions, enough oxygen would be available for the maintenance of aerobic metabolism. Furthermore, anoxic conditions were not found in any of the pO_2 measuring sites. In spite of the adequate oxygen supply, glycogen is rapidly broken down, and lactate accumulates in the contracting muscle. Such a metabolic pattern suggests that, in this early phase, white muscle fibers, with a mainly anaerobic metabolism, contribute substantially to muscle contraction. Owing to the rapid exhaustion of white muscle fibers, contraction force decreases markedly within 5 min. The reason for the increase in muscle blood flow remains obsure.

Proton activity (pH) showed a close correlation with lactate concentrations and could influence blood vessel tone. The transient pH increase with the start of contractions is explained by the rapid hydrolysis of CP (Sahlin, 1978; Steinhagen et al., 1976).

In the first 10 s, muscle CP decreased by almost 50%, and then reached a steady-state level—a typical pattern during continuous work at submaximal workload.

The second, almost non-fatiguing phase, is characterized by virtually constant metabolite levels, indicating that the rate of energy exchange is stabilized. Muscle blood flow is obviously sufficient to meet the oxygen demand of contracting muscle. This mainly aerobic phase indicates that red muscle fibers are preferentially involved, and this is assumed to be favorable for avoiding fatigue (Petrofsky et al., 1981). In contrast to muscle CP and ATP, which show a slight increase at the end of this period, glycogen shows a continuous depletion during isometric exercise.

A significantly prolonged time to decrease muscular contraction force to 50% of initial value was seen in animals trained for 30 days. The slower decrease of muscle force in trained rats seems to be due to an increased aerobic oxidation of metabolic fuels with a decreased intramuscular lactate formation.

References

EDWARDS, R.H.T. & C.M. Wiles. Energy exchange in human skeletal muscle during isometric contraction. Circ. Res. 48 (Suppl. I):11-17, 1981.

FAUPEL, R.P., H.J. Seitz, W. Tarnowski, V. Thiemann & C. Weiss. The problem of tissue sampling from experimental animals with respect to freezing technique, anoxia, stress and narcosis. Arch. Biochem. Biophys. 148: 509-522, 1972.

JACOBS, J., Lactate, muscle glycogen and exercise performance in man. Acta Physiol. Scand., Suppl. 495, 1981.

PETROFSKY, J.C., A.B. Chandler, M.N. Sawaka, D. Hanpeter & D. Stafford. Blood flow and metabolism during isometric contractions in cat skeletal muscle. J. Appl. Physiol. 50(3):493-502, 1981.

SAHLIN, K., Intracellular pH and energy metabolism in skeletal muscle of man. Acta Physiol. Scand., Suppl. 455, 1978.

SHEPHERD, J.T., C.G. Blomqvist, A.R. Lind, J.H. Mitchell & B. Saltin. Static (isometric) exercise. Circ. Res. 48 (Suppl. I):179-188, 1981.

STEINHAGEN, C., H.J. Hirche, H.W. Nestle, U. Bovenkamp & J. Hosselmann. The interstitial pH of working gastrocnemius muscle of the dog. Pflügers Arch. 367:151-156, 1976.

TESCH, P., Muscle fatigue in man. Acta Physiol. Scand., Suppl. 480, 1980.

The Influence of Recovery Duration
on Repeated Maximal Sprints

S.A. Wootton and C. Williams
University of Technology, Loughborough,
Leicestershire, England

During brief, high-intensity exercise, rapid changes in metabolism and muscle function ensue, which ultimately result in an inability to maintain force or exercise intensity. These processes, which collectively contribute to the phenomenom of fatigue, have been principally studied using static isometric exercise while little attention has been directed towards maximal dynamic exercise.

It has been suggested that, even during very heavy exercise on a cycle ergometer (400 W), 10 s bouts of exercise performed every 30 s could be sustained over long periods with no discernible limitation to work output and little increase in blood lactate (Saltin & Essén, 1971). Such studies, however, are performed at exercise intensities which, although equal to or greater than the power output that would elicit maximal aerobic capacity, are considerably lower than those attained during truly maximal "all-out" exercise.

The aim of this study was to further examine the ability to perform repeated bouts of maximal dynamic exercise with different recovery durations.

Methods

Sixteen male athletes, who had given informed consent and were well familiarized with the experimental protocol, performed the following exercise on a modified Monark cycle ergometer.

Five minutes after two warm-up rides (30 s at 120 and 150 W, load 14.7 N) two 25 μl capillary blood samples were obtained from the pre-warmed hand for the determination of pre-exercise blood lactate concentration. The 'warm-up' allowed the subjects to reacquaint themselves with the high pedal frequencies and provide calibration data while causing minimal metabolic disturbance.

The exercise task was randomly assigned and consisted of five 6 s maximal sprint bouts performed at either 30 (30 R) or 60 (60 R) s intervals against a heavy resistance and is similar, in principle, to the Wingate Test (Bar-Or, 1978). Each subject was instructed to attain an initial pedal frequency of 75 rpm against little resistance and the load was employed within 2 s. A timer was activated and the subject then accelerated to a peak maximum pedal frequency and attempted to maintain this power output for 6 s. Passive recovery between bouts was established with the subject remaining seated on the ergometer. The loading was predetermined to ensure that each subject would achieve the maximal power output attainable in this experimental model while pedaling within the range 150 to 160 rpm.

On completion of the fifth bout, the subject remained seated and after 5 min of passive recovery two further blood capillary samples were collected. The blood samples were deproteinized with 250 μl of 2.5% perchloric acid and duplicate assays were performed on the supernatant for blood lactate by a modification of the method of Olsen (1971).

In order to discriminate the rapid changes in pedal frequency a photo-optic counting device was used (Lakomy & Wootton, 1981). The voltage output, proportional to pedal frequency and hence power output, was recorded on a pen-recorder as a fatigue profile. The trace was digitized in order to calculate peak (PPO), end (EPO) and mean (MPO) power outputs for each of the 6 s bouts and statistically analyzed using an analysis of variance with repeated measures on two factors. A student's t-test for correlated means was used to compare blood lactate values.

Results

The capacity to perform repeated 6 s bouts of maximal exercise is markedly influenced by the preceding number of sprint bouts and the recovery duration (Table 1).

The PPO/60 R falls only slightly over the five bouts (3.0% decrease from bout 1 to 5: NS) whereas the ability to sustain this power output with 60 s recovery (EPO) is more noticeably affected and decreases by 12.7% ($p < 0.01$) by the fifth bout.

In contrast, with shorter recovery, the performance characteristics deteriorated more rapidly, affecting both peak and end power outputs. After the initial bout the EPO/30 R has fallen by 6.4% ($p < 0.01$) in sprint bout

Table 1

**Performance Characteristics (Peak, End and Mean Power Outputs:
mean ± sem) for the Two Recovery Protocols**

		No of 6 s Sprint bouts				
	Recovery	1	2	3	4	5
Peak power output (W)	30s	840 ± 31	827 ± 29	774 ± 29	738 ± 29	690 ± 26
	60s	825 ± 28	848 ± 30	838 ± 30**	800 ± 31**	800 ± 30**
End power output (W)	30s	777 ± 26	727 ± 24	629 ± 24	595 ± 24	551 ± 22
	60s	757 ± 28	760 ± 29*	716 ± 28**	676 ± 31**	661 ± 28**
Mean power output (W)	30s	794 ± 28	772 ± 26	710 ± 25	668 ± 27	631 ± 24
	60s	780 ± 25	792 ± 28	774 ± 28**	734 ± 28**	733 ± 30**

*60 s recovery significantly different at $P < 0.05$ from the 30 s recovery protocol.
**60 s recovery significantly different at $P < 0.01$ from the 30 s recovery protocol.

2 and the PPO/30 R attained in the third bout is 7.9% ($p < 0.01$) lower than the initial value. The performance on the final sprint bout reflects the considerable fatigue induced by the preceding exercise: the PPO/30 R has fallen by 17.9% ($p < 0.01$) and the EPO/30 R has decreased by 29.1% ($p < 0.01$). Consequently, the amount of work the subject is able to perform (as demonstrated by the MPO) decreases with each subsequent sprint bout. With 60 s recovery, although the initial three bouts are similar, after five bouts the MPO has reduced by 6.0% ($p < 0.01$). However, the MPO performed on the third and final bouts with 30 s recovery was only 89.4% ($p < 0.01$) and 79.5% ($p < 0.01$), respectively, of the initial bout value.

Furthermore, the blood lactate concentrations increased considerably over the five sprint bouts with both recovery protocols ($p < 0.001$). Blood lactate rose from a pre-exercise level of 1.07 ± 0.06 mM (mean ± SEM) to 11.52 ± 0.38 mM 5 min after five 6 s bouts with 30 s recovery (range 9.11 to 15.52 mM). Although the increase in blood lactate was lower with 60 s recovery ($p < 0.001$) the lactate concentration rose from 1.16 ± 0.08 mM to 10.29 ± 0.46 mM (range 7.17 to 13.93 mM).

Discussion

The fatigue-induced decrements in performance and highly elevated blood lactate concentrations observed during intermittent maximal exercise in this study differ markedly from the findings of previous studies employing very heavy or intense exercise intensities (Essén, 1978; Saltin & Essén, 1971). These differences reflect the magnitude of exercise intensity and, consequently, the energy demands at power outputs which are 2.5 to 3.5 times greater than those which would elicit maximal aerobic capacity.

The rapid replenishment of ATP during high-intensity exercise occurs at the expense of the phosphocreatine (PCr) stores within the muscle. However, with each subsequent sprint bout, inadequate recovery duration would lead to incomplete replenishment of the PCr stores, thereby placing ever increasing demands on glycolysis to maintain the ATP levels within the cell (Harris, Edwards, Hultman & Nordesjö, 1976). Therefore during the 30 s recovery protocol the greater decrements in performance and increases in blood lactate may reflect the earlier and more pronounced reduction in PCr stores than occurs during the 60 s recovery protocol.

Although it appears that depletion of PCr, per se, may not be the limiting factor during high-intensity exercise, the enhanced glycolysis with concomitant acidosis may reduce the ability of the muscle fibers to maintain contractile force by influencing either the production or utilization of ATP (Hermansen, 1981).

Acknowledgment

This work is supported by the Sports Council.

References

BAR-OR, O., A new anaerobic capacity test—Characteristics and applications. Proc 21st World Congress in Sports Medicine, Brasilia, 1978.

ESSÉN, B., Studies on the regulation of metabolism in human skeletal muscle using intermittent exercise as an experimental model. Acta Physiol. Scand. Suppl. 454, 1978.

HARRIS, R.C., R.H.T. Edwards, E. Hultman & L-O Nordesjö. The time course of phophorylcreatine resynthesis during recovery of the quadriceps muscle in man. Pflugers Archiv. 367:137–142, 1976.

HERMANSEN, L., Effect of metabolic changes in force generation in skeletal muscle during maximal exercise. In Porter & Whelan (Eds.), Human Muscle Fatigue: Physiological Mechanisms, pp. 75–88. London: Pitman Medical, 1981.

LAKOMY, H.K. & S.A. Wootton. Discrimination of rapid changes in pedal frequency. J. Physiol. (Lond.) 316, 1P, 1981.

OLSEN, C., An enzymatic fluorometric micromethod for the determination of acetoacetate, β-hydroxybutyrate, pyruvate and lactate. Clin. Chim. Acta. 33: 293–300, 1971.

SALTIN, B. & B. Essén. Muscle glycogen, lactate and PC in intermittent exercise. In Pernow & Saltin (Eds.), Muscle Metabolism during Exercise, pp. 419–424. London: Plenum Press, 1971.

Metabolic Responses of Single Rat Skeletal Muscle Fibers to Tetanization and Ischemia

D.A. Young and O.H. Lowry
Washington University Medical School,
St. Louis, Missouri, U.S.A.

Mammalian skeletal muscle, is a heterogeneous tissue consisting of individual fibers which are histochemically, physiologically and biochemically distinguishable. This heterogeneity has posed a problem for the biochemist who wished to characterize skeletal muscle at the level of the single cell. This obstacle has been largely overcome by the application of sensitive microchemical techniques to the study of single muscle fibers (Essén et al., 1974; Hintz et al., 1982; Lowry et al., 1978). By using this approach it is possible to classify fibers on the basis of enzyme activity alone (Hintz et al., 1980; Spamer & Pette, 1979). Analysis of rat plantaris (fast-twitch) in this manner revealed the presence of two broad groups of fibers with a continuous spectrum of enzyme levels within each group. Fibers which had a large glycogenolytic capacity but contained low levels of oxidative enzymes were designated Type IIB, while fibers with high levels of both oxidative and glycogenolytic enzymes were referred to as Type IIA (Hintz et al., 1980). Recent work has suggested that these fibers exhibit a type specific metabolic response to electrical stimulation (Hintz et al., 1982). We therefore designed an experiment to determine if there was any correlation between enzyme levels and energy utilization in rat plantaris muscle fibers.

Methods

A male Sprague-Dawley rat (380 g) was given pentobarbital (50 mg/kg) and, 30 min later, curare (1.5 mg/kg). The animal was mechanically respired and the plantaris carefully exposed. After the muscle had been allowed to recover for 20 min, it was tetanized directly for 30 s and immediately frozen with copper tongs cooled to liquid N_2 temperature. Stimuli were delivered at a frequency of 50 hz (pulses of 50 v and 0.3 ms duration) which has been shown to maximally tetanize rat plantaris muscle in situ (unpublished observation). Three min before stimulation, the iliac artery and vein were ligated and a thin layer of mineral oil poured over the muscle surface to inhibit diffusion of oxygen. This design results in a closed anaerobic system in which glycogen, ATP, and Phosphocreatine (P-creatine) are almost the exclusive sources of energy. Also, movement of metabolites (i.e., glucose, lactate, etc.) into or out of the muscle is eliminated due to the ischemia. The muscle tissue was then freeze-dried and the fibers dissected. The fibers were classified by their lactate dehydrogenase (LDH) and malate dehydrogenase (MDH) activity. Stimulated fibers were assayed for ATP, P-creatine, lactate, glycogen and glucose-6-P as described (Hintz et al., 1982) and their high energy P expenditure was compared to that of control fibers originating from the contralateral leg. The total high energy P expenditure was estimated by multiplying any changes in ATP, P-creatine and lactate by the appropriate factor. These factors are 1.4, 1.0 and 1.5 for ATP, P-creatine and lactate, respectively, and represent the moles of high energy P anaerobically available/mole substrate. The factor 1.4 is an approximation which takes into account the fact that some of the ATP is converted to ADP and some into AMP as dictated by the equilibrium catalyzed by adenylokinase (Lowry et al., 1964; Nelson et al., 1966).

Results

When LDH was plotted against MDH the fibers fell into two distinct groups which were designated Type IIA (high LDH, high MDH) and Type IIB (high LDH, low-med MDH) (Figure 1). In the unstimulated fibers, glucose-6-P was significantly higher (32%) in the IIA fibers, but none of the other substances measured differed between the two fiber groups (Table 1). Statistical analysis revealed a significant positive correlation between MDH and P-creatine in control IIA fibers ($r = 0.69$, $P < 0.05$) (Figure 2). Following 30 s of stimulation, P-creatine levels fell an average of 79 mmol/kg (dry w) in Type IIB fibers but only 37 mmol/kg in IIA (Table 1). ATP decreased 16 mmol/kg in Type IIB fibers and 3 mmol/kg in Type IIA. After stimulation, there were no overlapping values for either P-creatine or ATP among individual members of the two fiber groups (Figure 2). Lactate increased an averge of 88 mmol/kg in IIB fibers and 69 mmol/kg in IIA fibers. Glucose-6-P increased almost equally in the two groups. Glycogen decreased 40% more in the IIB group, but the difference from group IIA

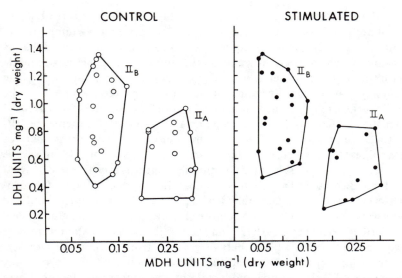

Figure 1—Comparison of malate dehydrogenase (MDH) and lactate dehydrogenase (LDH) on the same fibers selected at random from rat plantaris. Control fibers (open circles) and stimulated fibers (closed circles) were designated IIA or IIB depending upon the activity of these enzymes. Activities are μmol min^{-1} mg^{-1} (Units mg^{-1}).

was not statistically significant. Based on the changes in these metabolites, it was calculated that the consumption of high energy P (on a dry wt basis) was 467 mmole kg^{-1} min^{-1} in IIB fibers and only 288 mmol kg^{-1} min^{-1} in IIA fibers.

Discussion

These data show that under identical conditions of stimulation, rat plantaris Type IIB fibers expend significantly more energy than IIA fibers. There is ample evidence for whole muscle which suggests that with similar stimulation, Type I fibers utilize less energy than Type II fibers (Conlee et al., 1979; Crow and Kushmerick, 1982); however, single fiber studies which distinguish between Type IIA and Type IIB are rare. Hintz et al. (1982) analyzed single fibers after stimulation through the nerve (trains of stimuli 100 ms in duration delivered at 50 Hz at a frequency of 60/min), with intact circulation. Their data are in general agreement with present findings. They demonstrated that there was significantly more ATP and P-creatine lost from tibialis IIB fibers than from Type IIA or Type I fibers.

The method of calculating total high energy P expenditure is not new (Lowry et al., 1964; Nelson et al., 1966) nor is it restricted to this laboratory. Crow and Kushmerick (1982) have recently calculated the extent of high

Table 1

Average Metabolite Values of Rat Plantaris Muscle Fibers

	Control		Stimulated	
	IIA	IIB	IIA	IIB
ATP	31.0 ± 1.9 (9)	31.9 ± 1.5 (15)	27.9 ± 1.5 (8)	15.4 ± 0.8 (16)
P-Creatine	97.5 ± 3.0	100.1 ± 2.2	60.6 ± 3.1	21.4 ± 2.4
Lactate	32.1 ± 2.7	33.4 ± 1.7	99.7 ± 2.7	121.0 ± 2.8
Glucose-6-P	26.0 ± 0.4	19.7 ± 0.5	36.2 ± 1.7	28.6 ± 1.3
Glycogen	182.0 ± 9.0	191.0 ± 8.0	153.0 ± 7.0	150.0 ± 4.0

Metabolites are expressed as mmol kg^{-1}. Values are means ± SE for the number of fibers in parentheses at the top of each column.

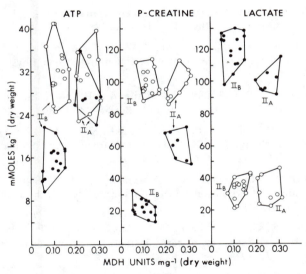

Figure 2—Correlation of MDH and ATP, P-creatine, and lactate. Selected control (open circles) and stimulated fibers (closed circles) are plotted. MDH activity is μmol min^{-1} mg^{-1} (Units mg^{-1}) and metabolites are expressed as mmol kg^{-1}.

energy P splitting in tetanized whole mouse muscle. Their method of calculation is similar, although somewhat more involved. We are presently studying energy expenditure during shorter periods of tetanization. Because the plantaris loses a large amount of tension during 30 s of tetanization (unpublished observations) it is likely that the average rate of high energy phosphate expenditure during this time period may be significantly lower than it is initially. In fact, preliminary data suggest that the rate is almost 3 times higher during the first 5 s of tetanization than the average for a 30 s period.

Acknowledgments

This study was supported in part by a Research Center Grant from the Muscular Dystrophy Association of America, Amer. Canc. Soc. Grant BC-4U and NIH Grants T32-NS07129 and NS-08862.

References

CONLEE, R.K., J.A. McLane, M.J. Rennie, W.W. Winder & J.O. Holloszy. Reversal of phosphorylase activation despite continued contractile activity. Am. J. Physiol. 237:R291-R296, 1979.

CROW, M.T. & M.J. Kushmerick. Chemical energetics of slow- and fast-twitch muscles of the mouse. J. Gen. Physiol. 79:147-166, 1982.

ESSÉN, B., E. Jansson, J. Henriksson, A.W. Taylor & B. Saltin. Metabolic characteristics of fibre types in human skeletal muscle. Acta Physiol. Scand. 95:153-155, 1975.

HINTZ, C.S., M. Chi, R. Fell, J. Ivy, K.K. Kaiser & O.H. Lowry. Enzyme levels in individual rat muscle fibers. Am. J. Physiol. 239:C58–C65, 1980.

HINTZ, C.S., M.M. Chi, R.D. Fell, J.L. Ivy, K.K. Kaiser, C.V. Lowry & O.H. Lowry. Metabolite changes in individual rat muscle fibers during stimulation. Am. J. Physiol. 242 (Cell Physiol. 11):C218-C228, 1982.

LOWRY, O.H., J.V. Passonneau, F.X. Hasselberger & D.W. Schulz. Effect of ischemia on known substrates and cofactors of the glycolytic pathway in brain. J. Biol. Chem. 239:18-30, 1964.

LOWRY, C.V., J.S. Kimmey, S. Felder, M.M. Chi, K.K. Kaiser, P.N. Passonneau, K. Kirk & O.H. Lowry. Enzyme patterns in single human muscle fibers. J. Biol. Chem. 253:8269-8277, 1978.

NELSON, S.R., O.H. Lowry & J.V. Passonneau. Changes in energy reserves in mouse brain associated with compressive head injury. In W. Caveness & E. Walter (Eds.), Head Injury. New York: Lippincott, 1966.

SPAMER, C. & D. Pette. Activities of malate dehydrogenase, 3-hydroxyacyl-CoA dehydrogenase and fructose-1,6-diphosphatase with regard to metabolic sub-populations of fast- and slow-twitch fibers in rabbit muscles. Histochemistry 60:9-19, 1979.

Carbohydrate Metabolism

Muscle and Liver Glycogen Resynthesis Following Oral Glucose and Fructose Feedings in Rats

D.L. Costill, B. Craig, W.J. Fink and A. Katz
Ball State University, Muncie, Indiana, U.S.A.

Earlier studies have determined the effectiveness of glucose and fructose on the resynthesis of muscle and liver glycogen using direct venous infusion (Bergstrom & Hultman, 1967). Since this method of administration does not elicit the same hormonal responses as seen after oral ingestion of these sugars, it is difficult to judge the role of dietary glucose and fructose on glycogen storage. Thus, the intent of this study was to examine the influence of orally administered glucose and fructose on liver and muscle glycogen resynthesis in rats that had been exercised to exhaustion.

Procedures

Forty female rats (Wistar), having an average body weight of 257.1 g were divided into five equal groups. All of the animals were deprived of food for 14 h before the experiment. One group (C1) was sacrificed at rest, whereas a second group (C2) exercised on a treadmill until exhaustion and was immediately sacrificed. A third group of control animals (C3) was exercised to exhaustion, allowed only water, and sacrificed 3 h later. The two experimental groups were exercised to exhaustion and immediately fed 2 ml of a 20% (22.2 mmol) fructose (F3) or glucose (G3) solution via a gastric tube. These animals were sacrificed 3 h after the feeding. The average exercise time to exhaustion was 98.7 min, with no difference between groups.

The sugar solutions contained 1 μCi of U-14C glucose or fructose. The rats were sacrificed with an overdose of nembutal. All tissue specimens were obtained before the animals died. Samples of blood (cardiac puncture), perirenal adipose tissue, liver, and selected muscles (plantaris, soleus, red vastus and white vastus) were obtained from each animal. Blood was analyzed for glucose, triglyceride, and 14C activity. The adipose specimen was used solely to determine 14C activity, while the muscle and liver samples were analyzed for glycogen content, total 14C activity, and 14C in glycogen isolated from a separate tissue sample. In addition to the tissues mentioned above, the stomach and small intestine were flushed with 10 ml of water and the rinse counted for 14C activity.

Results

At exhaustion blood glucose was 2.2 mM (SE ± .3) as compared to the pre-exercise value of 7.4 (SE ± .4) mM (P < .05). After 3 h of recovery from the treadmill run, the C3 animals' blood glucose averaged 5.2 mM, which was significantly different (P < .05) from the pre- and post-exercise levels. The rats that were fed either glucose or fructose had blood glucose concentrations similar to the C1 group and were higher (P < .05) than the C3 animals.

On the average, muscle glycogen was 61% lower (24.1 to 9.5 μmol/g glucose units) after exhaustive exercise. When the control rats (C3) were examined 3 h after the exercise, muscle glycogen was 18.2 μmol/g of tissue, which was higher (P < .05) than immediately after the exercise. Three h after the glucose and fructose feedings, muscle glycogen averaged 25.0 and 24.4 μmol/g of muscle, respectively. Although there was no difference between these means, they were both significantly greater (P < .05) than the C3 value. The only muscle that showed a difference (P < .05) in glycogen concentration was the plantaris, where the glucose-fed animals stored more glycogen (35.9 μmol/g) than the fructose-fed rats (27.7 μmol/g).

Liver glycogen declined from 24.6 (C1) to 8.1 μmol/g (C2) as a result of the treadmill exercise. Despite the fact that the C3 rats were not fed after the exercise, their liver glycogen (11.2 μmol/g) was significantly higher (P < .05) than that measured in the C2 animals. The G3 and F3 rats, on the other hand, showed marked increases in liver glycogen as a result of the feedings. The values for liver glycogen in these groups averaged 41.6 and 43.0 μmol/g, respectively. No difference was found between these two means.

It was observed that relatively more of the carbon label was contained in the liver glycogen from the fructose than glucose feedings (Figure 1). The average 14C activity in muscle of the glucose fed rats (4495 cpm/g) was greater (P < .05) than that observed in the F3 group (3469 cpm/g). The ratio of muscle 14C to liver 14C activities averaged 0.77 (SE ± .08) and 0.29 (SE ± .04) for the G3 and F3 groups, respectively. These data demonstrate

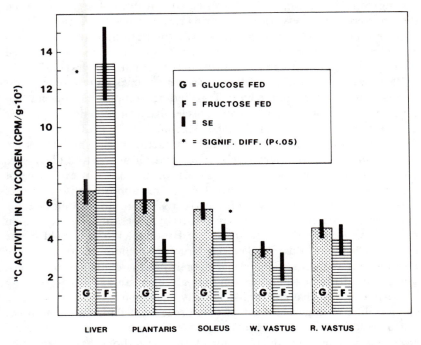

Figure 1—14C activity in the liver and muscles of rats fed 14C-glucose and 14C-fructose immediately after exhaustive exercise.

that the administered U-14C glucose was more evenly distributed between muscle and liver than when the rats were fed the U-14C fructose solution.

Examination of the residue from the stomach and small intestine of these rats revealed that more (P <05) of the carbon label remained in the gut of the F3 animals (66,803 cpm) as compared to those fed glucose (14,390 cpm). These unabsorbed portions of the feedings constituted 4.0% and 0.9% of the total 14C load, respectively.

Discussion

Despite carbohydrate starvation, rats that are exercised to exhaustion show a considerable accumulation of muscle glycogen in the h following the activity (Fell et al., 1980). The present findings demonstrate that after 3 h of recovery (fasting), 60% of the muscle glycogen used during exercise had been restored. Similar patterns of muscle glycogen resynthesis have been observed in exercised human subjects after 10 to 15 h of continued fasting (Maehlum et al., 1978). We observed that 19% of the glycogen used during exercise was resynthesized during recovery despite fasting.

If we assume that the glycogen resynthesis in the samples of the liver was representative of the whole organ (avg wt = 7.9 g) and that the muscles used during running constituted at least 10% of the animals' body wt (257 g × .1 = 25.7 g), then it is estimated that at least nine times more glycogen was stored in muscle than in the liver. This finding supports the concept that hepatic glucose production favors muscle glycogen repletion over liver glycogen resynthesis (Fell et al., 1980; Maehlum et al., 1978).

Both rats and humans absorb fructose intact from the small intestine, thereby increasing the portal vein concentration of fructose (Jeanes & Hodge, 1975). Peripheral circulating levels of fructose, however, are not elevated, which is indicative that hepatic metabolism of fructose is extensive. Studies that have infused fructose were able to elevate the peripheral blood fructose levels and, thereby, assess its effectiveness in muscle glycogen synthesis (Bergstrom & Hultman, 1967). In light of the difference between infused and ingested fructose on blood glucose, fructose and insulin, some questions remain regarding the contribution of oral glucose and fructose to muscle and liver glycogen storage.

Despite the fact that greater 14C activity was found in the liver glycogen, the absolute glycogen content of that tissue was similar after both the glucose and fructose feedings. This suggests that fructose metabolism in the liver results in a preferential retention of fructose carbons than when orally fed glucose is incorporated into liver glycogen. This may be explained by the fact that fructose is phosphorylated in the liver before being released as glucose, whereas a greater proportion of an oral glucose load may escape hepatic retention.

With the exception of the plantaris muscle, the muscle glycogen resynthesis was not appreciably different following the glucose/fructose feedings. At the same time, 14C activity was greater in the muscle glycogen of the plantaris and soleus 3 h after the glucose feeding than after the fructose feeding. It should be noted that the red and white vastus muscles resynthesized only about half as much glycogen during the 3 h period that followed the glucose and fructose feedings, which may explain why no difference was found between the glucose and fructose treatments.

This study demonstrates that glucose and fructose feedings (1.56 mg/g body wt) given immediately after exhaustive exercise result in similar glycogen storage in the liver, although the mechanisms responsible for this resynthesis are different. More of the orally administered glucose than fructose escapes incorporation into liver glycogen and is used to replenish the muscle glycogen depots.

Acknowledgment

This research was supported by a grant from Ross Laboratories, Columbus, Ohio.

References

BERGSTROM, J. & E. Hultman. Synthesis of muscle glycogen in man after glucose and fructose infusion. Acta Med. Scand. 182:93–109, 1967.

FELL, R.D., J.A. McLane, W.W. Winder & J.O. Holloszy. Preferential resynthesis of muscle glycogen in fasting rats after exhausting exercise. Am. J. Physiol. R328–R332, 1980.

JEANES, A. & J. Hodge. Physiological effects of food carbohydrates. Wash., D.C.: Am. Chem. Society, 1975.

MAEHLUM, S., P. Felig & J. Wahren. Splanchnic glucose and muscle glycogen metabolism after glucose feeding during postexercise recovery. Am. J. Physiol. 235:E255–260, 1978.

Mitochondrial Substrate Oxidation, Muscle Composition and Plasma Metabolite Levels in Marathon Runners

K. Gohil, D.A. Jones, G.G. Corbucci, S. Krywawych,
G. McPhail, J.M. Round, G. Montanari and R.H.T. Edwards

University College London, London, England,
and Center of Physiology Applied to Sport, Gubbio (PG) Italy

Endurance atheletes are well known to have very high oxidative capacities (Holloszy & Booth, 1976) in part as a consequence of having muscles largely composed of slow-twitch oxidative fibers (Costill et al., 1976).

We have examined a group of marathon runners and obtained data concerning the fiber type composition of the muscle and the qualitative and quantitative differences in oxidative capacity of mitochondria compared with normals. The results show that overall mitochondrial activities were increased more evenly than would be expected from the fiber type composition of the muscle. There were also differences in morphological features and in relative activities of the components of the oxidative pathways of mitochondria.

The pattern of accumulation of glycolytic and tricarboxylic (TCA) cycle intermediates in plasma suggests that even though there are major adaptive changes in mitochondria it is oxidation that may still limit the utilization of carbohydrates in marathon runners.

Methods

Seven international class marathon runners were studied before and after a 22 km run. Two days before the run plasma samples were obtained and the right quadriceps sampled by needle biopsy (Edwards et al., 1980). Within 2 min of finishing the run further blood samples were taken and the left

286

quadriceps biopsied. Portions of the muscle were taken for quantitative histochemistry (Round et al., 1982) and electron microsopy. The remainder of the muscle sample (20 to 50 mg) was used to assess the capacity of mitochondria to oxidize various substrates. The method involved spectrophotometric measurements of the rates of reduction or oxidation of added cytochrome c (Gohil et al., 1981a). Glucose, lactate, pyruvate and 2-hydroxybutyrate were measured in neutralized perchloric acid extracts of plasma. Organic acids in plasma were measured by Gas Liquid Chromatography (GLC) (Chalmers & Watts, 1972).

Results

Histochemical examination showed an overwhelming predominance of type I fibers in the marathon runners; the proportion of type I fibers is about twice that in normal subjects (Table 1). Mitochondrial staining showed high activity and at the ultrastructural level there were increased numbers of interfibrillary and sub-sarcolemmal mitochondria, many of

Table 1

Fiber Type Composition and Activities of Mitochondrial Substrate Oxidation in Marathon Runners

Subject	% Type I Fibers	Pyruvate plus malate cytochrome c reductase	Succinate cytochrome c reductase	Cytochrome c oxidase
1	73	0.7	1.5	11.8
2	74	4.5	1.8	36.3
3	100	3.1	3.8	24.1
4	88	3.0	2.4	20.4
5	68	1.6	2.0	14.7
6	91	1.20	1.6	20.3
7	81	2.9	1.1	19.2
	35	0.18	1.27	6.21
Untrained Normals	56	0.46	1.39	7.0
	33	0.21	1.3	6.5
	45	0.38	1.35	6.8
	34	0.35	1.29	7.1

Activity: μmol cyt.c/g(ww)/min 22°C ± 1°C.

which were enlarged and contained significant numbers of dense matrix granules. Occasional cristae were seen to be thickened and arranged in parallel bundles.

Quantitative measurements of the rates of mitochondrial substrate oxidation showed these to be increased compared with those of normal subjects (Table 1). It is notable that although the proportion of type I fibers in marathon runners was twice as high compared with normal subjects, pyruvate and cytochrome c oxidations were higher by about three- to six-fold. Not only were the overall mitochondrial activities considerably increased but there was also evidence of altered composition. While the largest increases were seen in pyruvate and cytochrome c oxidations, the increase in succinate oxidation was relatively small. This disparity between activities is evident when they are expressed as ratios. In Table 2 it can be

Table 2

Activity Ratios of Substrate Oxidations

	$\dfrac{(P + M)CR}{SCR} \times 10$	$\dfrac{COx}{SCR}$	$\dfrac{(P + M)CR}{COx} \times 10$
Runners n = 7	13.3 ± 8.9	11.6 ± 5.5	1.1 ± 0.4
Normals n = 5	2.3 ± 0.9	4.9 ± 0.6	0.44 ± 0.2

$(P + M)CR$ = pyruvate + malate cyt. c. reductase
   SCR = succinate cytochrome c reductase
   COx = cytochrome c oxidase

seen that when expressed relative to succinate oxidation, the oxidations of pyruvate and cytochrome c were increased two- to six-fold compared with untrained normals. Of the two, pyruvate oxidation showed a slightly greater increase indicating that different portions of the ETC may change independently. If all activities were uniformly raised then the ratios would be similar in normal subjects and athletes.

The resting and post-exercise levels of plasma metabolites are shown in Table 3. The resting concentrations of all the metabolites were within normal range (S. Krywawych, unpublished observations). After the run the largest increases were seen in levels of malate and succinate six- to ten-fold) with smaller increases in the levels of other TCA cycle intermediates and glucose, 2-hydroxybutyrate, pyruvate and lactate (1.5-5.4 fold).

Table 3

Plasma Metabolite Levels in Marathon Runners

Metabolites	Resting	Post-exercise
Lactate*	1.5 ± 0.3	7.3 ± 4.0
Glucose*	4.1 ± 0.7	12.4 ± 2.4
Pyruvate	43.0 ± 12.0	231.0 ± 65.0
Malate	6.0 ± 2.0	36.0 ± 16.0
Succinate	4.0 ± 1.0	45.0 ± 30.0
Citrate	67.0 ± 4.0	134.0 ± 27.0
2-oxoglutarate	11.0 ± 5.0	18.0 ± 7.0
2-hydroxybutyrate	38.8 ± 10.0	77.4 ± 27.9

Results are given as means ± SD (N = 5) in μmol dm^{-3}.
*mmol dm^{-3}

Discussion

We have found clear evidence for both qualitative and quantitative changes in skeletal muscle mitochondria of marathon runners. The quantitative changes were increases in the oxidative activities which were greater than might be expected for the fiber type composition of muscle and the qualitative changes were the differences in the ratio of the activities of the components of the oxidative pathway.

Although in normal untrained subjects the mitochondrial activity correlated well ($r = 0.89$) with the % type I fibers (Gohil et al., 1981b), in the marathon runners the oxidative activities were greater than that which might be expected for the fiber type composition. It is possible that both the period and the type of training of the individual runners profoundly affect the observed levels of oxidative capacity even when the muscle fiber types are predominantly oxidative. The training stimulus also seems to cause specific increases in oxidations of pyruvate and cytochrome c in contrast to succinate oxidation suggesting that the regulatory portions of the oxidative pathway may be at the start (e.g., PDH activity or complex I of electron transport chain) and/or the end (complex IV) of the oxidative pathway. Because cytochrome c oxidase is an integral membrane protein of mitochondrial cristae, it is tempting to suggest that the thickening of this membrane observed in some mitochondria of the runners is related to the large specific increases in cytochrome c oxidase activity.

The increases in plasma concentrations of glucose (2.5- to 3 fold) and 2-hydroxybutyrate emphasize the importance of liver glycogen and triglyceride mobilization during this type of exercise. The relatively modest increase in lactate further indicates the high oxidative capacity of the marathon runners.

The plasma levels of pyruvate and TCA cycle intermediates show large increases after the run; the pyruvate concentration increases (up to six fold) and there were concomitant increases in malate and citrate. The interpretation of these observations is complicated but it suggests that it is not the supply of pyruvate that is limiting, but its oxidation. If the supply of pyruvate from the glycolytic pathway was limiting, then only small increases in plasma pyruvate and TCA cycle intermediates would be expected, while if transport into mitochondria was limiting, increased plasma pyruvate but not TCA cycle intermediates would be expected. An implication of these results is that in parallel with the increased mitochondrial oxidative capacity there are also changes in the activities of the pathways supplying substrates to mitochondria.

Acknowledgments

We are grateful to Miss V. Patel for her excellent technical assistance in histochemical analysis and to the Muscular Dystrophy Group of Great Britain and to Federazione Italiana Di Atletica for financial support.

References

CHALMERS, R.A. & R.W.E. Watts. Quantitative extraction and gas liquid chromatographic determination of organic acids in urine. Analyst 97:958-967, 1972.

COSTILL, D.L., W.J. Fink & A.L. Pollock. Muscle fibre composition and enzyme activities of elite distance marathon runners. Med. Sci. Sports 8:96-100, 1976.

EDWARDS, R.H.T., A. Young & M. Wiles. Needle biopsy of skeletal muscle in the diagnosis of myopathy and the clinical study of muscle function and repair. New Engl. J. Med. 302:261-271, 1980.

GOHIL, K., D.A. Jones & R.H.T. Edwards. Analysis of muscle mitochondrial function with techniques applicable to needle biopsy samples. Clin. Physiol. 1(2):195-207, 1981a.

GOHIL, K., D.A. Jones, J.M. Round, K.R. Mills & R.H.T. Edwards. Mitochondrial function in patients with muscle pain. Clin. Sci. 61:10P, 1981b.

HOLLOSZY, J.O. & F.W. Booth. Biochemical adaptations to endurance exercise in muscle. Ann. Rev. Physiol. 38:273-291, 1976.

ROUND, J.M., D.A. Jones & R.H.T. Edwards. Flexible microprocessor system for measurement of cell size. J. Clin. Path. 1982. (In press)

Glycogen Synthesis: Effect of Diet and Training

J.L. Ivy, W.M. Sherman, W. Miller, S. Farrell and B. Frishberg
University of South Carolina, Columbia,
South Carolina, U.S.A.

Endurance capacity at work loads between 60 and 85% of maximum oxygen consumption is directly related to the muscle glycogen concentration (Hermansen et al., 1967), thus making muscle glycogen of paramount importance during endurance work. This dependency on muscle glycogen has initiated the practice of raising the muscle glycogen concentration above normal levels before an endurance event, a process referred to as glycogen loading. The classical approach to glycogen loading involves the depletion of glycogen by exercise, the maintenance of low muscle glycogen for 3 days with a high fat-protein diet, and the rapid resynthesis of glycogen above normal levels with a high carbohydrate diet (Bergström et al., 1967). Although this process has been used quite effectively for many years, the mechanism by which the glycogen level is increased above normal is still not fully understood. To determine the mechanisms controlling glycogen storage, we investigated the changes in insulin response to a carbohydrate challenge and activity of selected muscle enzymes during a glycogen loading regimen. It has also been observed that endurance training will enhance glycogen storage (Lamb et al., 1969). Thus a second purpose was to differentiate between the mechanisms controlling glycogen storage during a glycogen loading regimen and those during physical training.

Methods

Subjects

The subjects were seven male volunteers whose average ($\pm$ SD) age, ht and wt were 22.7 yr ($\pm$ 2.1), 176.5 cm ($\pm$ 6.3) and 71.7 kg ($\pm$ 3.1), respectively. The subjects were in good physical health but were not engaged in regular physical training. Prior to experimental testing the subjects were informed as to the nature of the experiment and their informed consent obtained.

Procedures

Muscle biopsies were taken from the vastus lateralis and glucose tolerance tests (GTT) administered following 3 days on a mixed diet (M), following exhaustive exercise and 1 day on a high fat-protein diet (FP), following exhaustive exercise and 1 day on a mixed diet (EM), following exhaustive exercise and 3 days on a high carbohydrate diet (CHO), and following 3 days on a mixed diet after 8 wk of endurance training (T). The M treatment was considered the control treatment. The GTT consisted of ingesting a commercially available drink (Tru Glu, Fisher Scientific Co.) containing 100 g of glucose in 10 fluid oz. following a 12 h overnight fast. Blood samples were drawn from an antecubital vein prior to and at 30, 60, 90, 120 and 150 min after glucose administration.

The caloric contents of the diets were based on 5 day diet recalls completed by each subject. The average calories per day were 3,252, 3,078 and 3,304 for the mixed, high fat-protein and high CHO diets, respectively. The mixed diet consisted of 41% carbohydrates, 30% fats and 29% proteins; and the high carbohydrate diet 75% carbohydrates, 14% fats and 11% protein.

Blood samples were collected in chilled tubes containing EDTA (24 mg/ml, pH 7.4) to prevent clotting and an aprotinin solution (Trasylol, 10,000 KIU/ml) to prevent proteolysis of insulin. Plasma was recovered by centrifugation and assayed for glucose, insulin, free fatty acids and triglycerides as previously described (Ivy et al., 1980). In addition, 1 ml of whole blood was deproteinized in 8% $HClO_4$, centrifuged and the acid extract neutralized and assayed for B-hydroxybutyrate (Dermot et al., 1962).

Muscle glycogen was assayed as described by Lowry and Passoneau (1972). Muscle samples were placed in 2 N HCl and incubated at 100° C for 2 h. After neutralization with 0.66 N NaOH, the liberated glucose units were assayed fluorometrically. For determination of enzyme activities, muscle samples were homogenized at –20° C with 4 vol (w/v) of 25 mM KF, 20 mM EDTA and 60% glycerol (pH 7.0, 23° C) per gram tissue. The homogenization was continued at 4° C after an additional 15 vol of 25 mM KF and 20 mM EDTA (pH 7.0, 23° C) per gram tissue. The homogenate was centrifuged at $8,000 \times g$ for 20 min at 4° C and the supernatant assayed for glycogen synthase in the presence of various concentrations of G-6-P (Kochan et al., 1981) and hexokinase (Lowry & Passoneau, 1972). Proteins in the pellet were precipitated with 5% TCA, resuspended and hydrolyzed in 0.1 N NaOH and total protein determined by the method of Lowry et al. (1951).

Results

The changes in muscle glycogen, plasma insulin levels and the activity of selected muscle enzymes were examined during a glycogen loading regimen and following endurance training. Biopsies from the vastus lateralis revealed that muscle glycogen following the FP and EM treatments were significantly reduced in comparison to glycogen values following the M treatment. Hexokinase activity was unaffected by the FP and EM treatments, but the activity of glycogen synthase was doubled as measured by the activity ratio $(^I/I + D)$. An increased glycogen synthase activity was also indicated by the $A_{0.5}$ and fractional velocity. Fractional velocity is obtained by dividing the velocity of the enzyme at a given concentration of G-6-P by the velocity obtained with saturating amounts of G-6-P (Kochan et al., 1981). Fractional velocity is similar to the $A_{0.5}$ in that it is a measure of the activity of the enzyme in the presence of less than saturating concentrations of G-6-P. Glycogen synthase total activity, although elevated, was not significantly different than M (Table 1).

Table 1

Changes in Muscle Glycogen and Activities of Specific Muscle Enzymes During a Glycogen Loading Regimen and Following Training

	M	FP	EM	CHO	T
Glycogen	79.6	49.1[2]	59.6[24]	124.8[1]	119.8[1]
(mmol/kg ww)	± 4.2	± 2.4	± 2.1	± 11.1	± 10.0
Hexokinase	43.7	49.1	44.9	68.1[1]	63.5[1]
(μmol/mg prot./min)	± 6.5	± 4.2	± 5.5	± 5.3	± 8.1
Glycogen synthase					
Activity ratio	3.1	9.7[1]	8.4[1]	2.7	1.4[2]
(O μM G6P/SAT. G6P) ±	0.6	± 1.1	± 1.7	± 0.1	± 0.1
Fractional velocity	12.4	48.9[13]	44.0[13]	21.2[1]	11.5
(1 mM G6P/SAT. G6P) ±	1.3	± 3.8	± 4.0	± 2.7	± 1.8
$A_{0.5}$	9,627	1,338[2]	2,861[2]	7,252	12,544
(μM G6P)	±2,243	± 589	± 280	±1,561	± 973

Mean ± SEM. M, 3 days on a mixed diet; FP, exhaustive exercise and 1 day on a high fat-protein diet; EM, exhaustive exercise and 1 day on a mixed diet; CHO, three days on a high carbohydrate diet; T, 8 weeks of endurance training followed by 3 days on a mixed diet. 1 indicates mean is greater than M; 2 indicates mean is less than M; 3 indicates mean is greater than CHO; 4 indicates mean is greater than FP ($p < 0.05$). The accuracy of the $A_{0.5}$ values are restricted. The $A_{0.5}$ are overestimated due to the limited number of G-6-P concentrations and large changes in enzyme activity between 1,500 to 10,000 μM G-6-P.

In contrast to the FP and EM treatments, the CHO and T treatments resulted in significant increases in muscle glycogen above normal and a 55.8 and 45.0% increase in hexokinase activity, respectively. The glycogen synthase activity ratios following the CHO and T treatments were not different from the activity ratio for the M treatment, indicating the enzyme was in an inactive state. Fractional velocity, however, remained significantly higher than control after the CHO treatment and this elevation in fractional velocity occurred in spite of a significant rise in glycogen synthase total activity. Fractional velocity was not elevated following the T treatment, but glycogen synthase total activity was increased (Table 1).

Oral glucose tolerance tests (GTTs) were administered following each treatment to examine the effect of the treatments on glucose disposal rate and insulin sensitivity. Mean plasma glucose values were similar during GTTs following the M (98.4 ± 3.9 mM), FP (98.7 ± 4.8 mM), EM (92.5 ± 5.0 mM) and T (90.8 ± 4.0 mM) treatments. Following the CHO treatment (87.0 ± 3.7 mM), however, the glucose response to the GTT was significantly suppressed in comparison to the response following the M treatment (Figure 1a). The insulin responses following the carbohydrate challenge were the same for the M ($\overline{X} = 27.7 \pm 1.9 \mu U/ml$) and FP ($\overline{X} = 27.1 \pm 2.5 \mu U/ml$) treatments, but the insulin responses were lowered by the EM ($\overline{X} = 19.8 \pm 2.3 \mu U/ml$) and T ($\overline{X} = 19.2 \pm 1.8 \mu U/ml$) treatments. Although no significant difference in the mean insulin response was noted for the CHO treatment ($\overline{X} = 34.2 \pm 3.5 \mu U/ml$) when compared to the M treatment, plasma insulin concentrations were significantly elevated at 30, 60 and 90 min during the GTT following the CHO treatment (Figure 1b). Prior to the FP GTT free fatty acids and B-hydroxybutyrate were increased above normal by 67 and 552%, respectively.

Discussion

Exhaustive exercise reduced the muscle glycogen stores and activated glycogen synthase as determined by the activity ratio. With the enzyme in this activated state, glycogen levels can return to normal within 24 h, but only if a diet high in carbohydrates is administered (Bergström et al., 1967). The present study supports this previous finding and also suggests that when a high fat-protein diet is administered, increases in plasma FFA and ketones may also help prevent glycogen synthesis by reducing skeletal muscle insulin sensitivity. This is evidenced by the fact that the insulin response during the GTT following the FP treatment was higher and the muscle glycogen level lower than that observed for the EM treatment.

The elevated glycogen level following the CHO treatment was probably the result of the increased pancreatic response to carbohydrates coupled

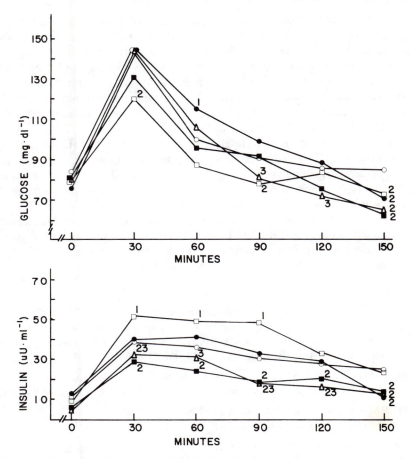

Figure 1a—Illustrates the response of plasma glucose during the glucose tolerance tests for each treatment.

Figure 1b—Illustrates the insulin response during the glucose tolerance test for each treatment, O treatment M; ● treatment FP; △ treatment EM; □ treatment CHO; ■ treatment T. 1 indicates mean is greater than M; 2 indicates mean is less than M; 3 indicates mean is less than FP ($p < 0.05$).

with an increased sensitivity of glycogen synthase to its allosteric activator G-6-P. This increased sensitivity of glycogen synthase to G-6-P (determined by fractional velocity and not detectible by the activity ratio) following a high carbohydrate diet was first noted by Kochan et al. (1981). They hypothesized that this intermediate form of the enzyme was responsible for glycogen synthesis during the supercompensation phase. It has also been demonstrated, however, that insulin must be present for supercompensation to occur (Ivy, 1977). In this regard, we believe it is the hyperinsulinemic response to carbohydrates that is responsible for the increased sensitivity of glycogen synthase to G-6-P. This increased insulin concentration may also serve to increase the rate of glucose transport, thus increasing the availability of glucose to glycogen synthase, as well as the intracellular G-6-P concentration. The increase in hexokinase activity would be functionally important in that it would prevent the rate limiting step in glucose uptake from shifting from transport to glucose phosphorylation as the G-6-P concentration increased.

In the trained state, muscle glycogen levels were comparable to those following the CHO treatment, but unlike the CHO treatment, the elevated glycogen levels following training could not be attributed to an increased pancreatic response to carbohydrates or improved glucose tolerance. It was noted, however, that insulin sensitivity was improved and that the maximum activities of glycogen synthase and hexokinase were significantly increased following training. Thus the elevated muscle glycogen stores following training were probably the result of an increased muscle insulin sensitivity and improved muscle enzyme profile.

References

BERGSTRÖM, J., L. Hermansen, E. Hultman & B. Saltin. Diet, muscle glycogen, and physical performance. Acta Physiol. Scand. 71:140–150, 1967.

DERMOT, H.W., J. Mellonby & H.A. Krebs. Enzymatic determination of D (-) – B – hydroxybutyric acid and acetoacetic acid in blood. Biochem J. 82:90–96, 1962.

HERMANSEN, L., E. Hultman & B. Saltin. Muscle glycogen during prolonged severe exercise. Acta Physiol Scand. 71:129–139, 1967.

IVY, J.L., Role of insulin during exercise-induced glycogenesis in muscle: effect of cyclic AMP. Am. J. Physiol. 233:E509–E513, 1977.

IVY, J.L., D.D. Costill, W.J. Fink & E. Maglischo. Contribution of medium and long chain triglyceride intake to energy metabolism during prolonged exercise. Inter. J. Sports Med. 1:15–20, 1980.

KOCHAN, R.G., D.R. Lamb, E.M. Reimann & K.K. Schlender. Modified assays to detect activation of glycogen synthase following exercise. Am. J. Physiol. 240:E197–E202, 1981.

LAMB, D.R., J.B. Peter, R.N. Jeffrees & H.A. Wallace. Glycogen, hexokinase and glycogen synthetase adaptations to exercise. Amer. J. Physiol. 217:1628–1632, 1969.

LOWRY & Passoneau. Flexible System of Enzymatic Analysis. New York: Academic Press, 1972.

LOWRY, O.H., N.J. Rosebrough, A.L. Farr & R.J. Randall. Protein measurement with the Folin phenol reagent. J. Biol. Chem. 193:265–275, 1951.

Liver Glycogen Store and Hypoglycemia during Prolonged Exercise in Humans

J.-M. Lavoie, D. Cousineau, F. Peronnet and P.J. Provencher

Université de Montréal, Montréal, Quebec, Canada

Evidence has accumulated indicating that depletion of body carbohydrate stores can play an important role in the development of physical exhaustion during prolonged strenuous exercise (Bergstrom et al., 1967; Hermansen, 1967). Under certain experimental conditions, the development of exhaustion during exercise has been attributed to the development of hypoglycemia, which in animals is thought to be secondary to liver glycogen depletion (Clark & Conlee, 1979). The purpose of the present investigation was to evaluate the relative importance of a decrease in human liver glycogen depots on the development of hypoglycemia during prolonged exercise. This was done by studying the effect of a decrease in liver glycogen with and without an increase in muscle glycogen stores. Liver glycogen was depleted by a 60 min period of arm exercise followed by a 24 h low carbohydrate (CHO) diet.

Methods

Two series of experiments were conducted in which muscle and liver glycogen stores were manipulated (Table 1). In the first series of experiments, metabolic and hormonal comparisons between a normal (NLG) and a low liver glycogen (LLG), both combined with a high muscle glycogen stores (HMG) were studied in seven male subjects ($\overline{X} \pm SD$; age $= 23 \pm 4$; wt $= 68 \pm 7$ kg; leg $\dot{V}O_2$max $= 52 \pm 5$ ml kg^{-1} min^{-1}) during a period of prolonged exercise (90 min; 60% leg $\dot{V}O_2$max).

In the second series of experiments, blood glucose response to LLG combined with normal muscle glycogen stores (NMG) was studied in six

Table 1

Description of Muscle and Liver Glycogen Manipulations

First experiment

HMG + NLG = Leg CHO loading

HMG + LLG = Leg CHO loading + [arm exercise* + 24 h low CHO diet]

Second experiment

NMG + LLG = [arm exercise* + 24 h low CHO diet]

HMG: High muscle glycogen
NLG: Normal liver glycogen
LLG: Low liver glycogen
NMG: Normal muscle glycogen

*arm exercise: 60 min; 70% arm $\dot{V}_{O_2}$max

male subjects ($\overline{X} \pm$ S.D.; age = 24 $\pm$ 5; wt = 78 $\pm$ 8 kg; leg $\dot{V}_{O_2}$max = 48 $\pm$ 6 ml kg^{-1} min^{-1}) during an exercise period identical to the first series.

Blood samples during the exercise were collected by means of a polyethylene catheter inserted into an antecubital vein. Free fatty acids (FFA) were determined spectrophotometrically by the method of Pinelli (1973); blood lactate and plasma glucose and glycerol were determined enzymatically using kits of reagents from the Calbiochem-Behring Corporation. Insulin (IRI) and glucagon (IRG) were determined by commercially available radioimmunoassay (Bioria, Montreal and RSL, Carson, CA). Plasma epinephrine (E) and norepinephrine (NE) were determined by a radioenzymatic method (Cat-a-kit, UpJohn Diagnostics, Kalamazoo, MI). Statistical evaluation of the data in both experiments was made by means of t-tests for paired comparisons.

Results

A decrease in liver glycogen combined with an increase in muscle glycogen stores, as studied in the first series of experiments, did not produce any significant alterations in blood glucose concentration during 90 min of leg exercise (Table 2).

LLG as compared to NLG in the first experiment was, however, associated with higher concentrations of FFA, glycerol (Figure 1), and catecholamines, whereas insulin concentration was lower and glucagon concentration unchanged (Figure 2).

Table 2

Blood Glucose Concentrations and Exercise Duration for the Two Experiments

	Glucose, mmol l^{-1}		Duration, min:s
	Rest	End of exercise	
First experiment			
HMG + NLG	4.6 ± 0.2	4.5 + 0.4	90
HMG + LLG	4.7 + 0.1	4.5 + 0.4	90
Second experiment			
NMG + LLG	4.4 + 0.1	3.32 + 0.2*	75:58 + 6:14

Values are means + SE. *indicates significant difference vs rest; $P < 0.01$. Explanations as in Table 1.

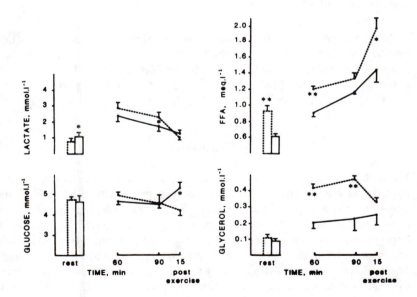

Figure 1—Metabolic responses at rest, during, and after exercise with increased muscle glycogen stores and with a) intact (—), or b) decreased hepatic glycogen depots (---). Values are means ± SE;* indicates significant difference between the two exercise conditions; $P < 0.05$; **$P < 0.01$.

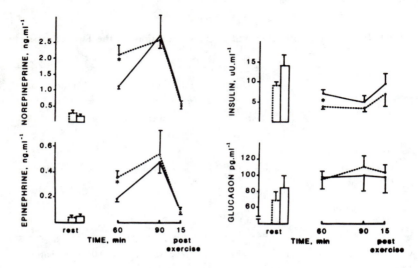

Figure 2—Hormonal responses at rest, during, and after exercise with increased muscle glycogen stores and with a) intact (—), or b) decreased hepatic glycogen depots (---). Explanations as in Figure 1.

When a decrease in liver glycogen was combined with a normal level of muscle glycogen reserves, as studied in the second series of experiments, prolonged exercise resulted in a significant decrease ($p < 0.01$) in blood glucose concentration (Table 2). The exercise duration was then decreased by approximately 15 min compared to the HMG conditions (Table 2).

Discussion

The results of this study show that, in humans, a decrease in liver glycogen content, as seen in the first experiment, did not result in hypoglycemia during a period of prolonged exercise. The second series of experiments demonstrates the importance of the muscle glycogen content in relation to hypoglycemia with a reduction of liver glycogen depots. Hypoglycemia occurs with LLG only when muscle glycogen reserves are not increased. These findings indicate that, contrary to what has been reported in animals, liver glycogen without muscle glycogen depletion is not necessarily associated with hypoglycemia in humans during prolonged moderate exercise (Clark & Conlee, 1979), and that liver glycogen in man is not a more important source of energy than muscle glycogen (Baldwin et al., 1973). The reason for this could be that the absolute quantity of glycogen stored in human tissue is much more important than in rat muscle (Clark & Conlee, 1979).

The protection against hypoglycemia during exercise in the HMG + LLG, as compared to the HMG + NLG condition of the first experiment, was insured, in addition to the elevated muscle glycogen content, by a higher adipose tissue lipolysis, as indicated by higher concentrations of FFA and glycerol, and probably by a higher rate of gluconeogenesis. These metabolic adaptations were probably triggered by higher concentrations of catecholamines and greater hypoinsulinemia. These compensatory mechanisms, helping to spare liver and muscle glycogen (Hickson et al., 1977) and reduce blood glucose uptake (Rennie & Holloszy, 1977), were probably insufficient, combined with a non-elevated muscle glycogen store (NMG + LLG), to avoid blood glucose utilization in a greater proportion than hepatic release, which resulted in a decrease in blood glucose level. This suggests that during prolonged exercise in man, the increase in glucose uptake by working muscles, secondary to a decrease in muscle glycogen, is more likely to cause hypoglycemia than just a decrease in liver glycogen content.

Acknowledgment

This study was supported by a grant from CRSNG, Canada (A7594).

References

BALDWIN, K.M., J.S. Reitman, R.L. Terjung, W.W. Winder & J.O. Holloszy. Substrate depletion in different types of muscle and in liver during prolonged running. Am. J. Physiol. 225:1045–1050, 1973.

BERGSTRÖM, J., L. Hermansen, E. Hultman & B. Saltin. Diet, muscle glycogen and physical performance. Acta Physiol. Scand. 71:140–150, 1967.

CLARK, J.H. & R.K. Conlee. Muscle and liver content: diurnal variation and endurance. J. Appl. Physiol.: Respirat. Environ. Exercise Physiol. 47:425–428, 1979.

HERMANSEN, L., E. Hultman & B. Saltin. Muscle glycogen during prolonged severe exercise. Acta Physiol. Scand. 71:129–139, 1967.

HICKSON, R.C., M.J. Rennie, R.K. Conlee, W.W. Winder & J.O. Holloszy. Effects of increased plasma fatty acids on glycogen utilization and endurance. J. Appl. Physiol.: Respirat. Environ. Exercise Physiol. 43:829–833, 1977.

PINELLI, A. A new calorimetric method for plasma fatty acid analysis. Clin. Chem. Acta 44:385–390, 1973.

RENNIE, M.J. & J.O. Holloszy. Inhibition of glucose uptake and glycogenolysis by availability of oleate in well-oxygenated perfused skeletal muscle. Biochem. J. 168:161–170, 1977.

Effect of Chronic Exercise on Glycogen Concentration in Muscle Grafts of Rats

F.S.F. Mong and J.L. Poland
Medical College of Virginia,
Richmond, Virginia, U.S.A.

Free skeletal muscle transplantation has been performed in animals (Markley et al. 1978; Mufti et al., 1977) and in humans (Hakelius, 1979) with success both morphologically and functionally. Exploration of the metabolic capacity for such transplants to utilize substrates, specifically glycogen, has recently been initiated (Mong & Poland 1982). Since exercise can change a variety of basic characteristics of skeletal muscle (Bagby et al., 1972; Barnard et al., 1970; Binkhorst, 1969; Gollnick & King, 1969), including glycogen concentration (Poland et al., 1980), and since metabolically the transplants behave similarly to normal muscles (Mong & Poland, 1982) it was decided to see whether exercise could modify the glycogen concentration of transplants.

Materials and Methods

Male Sprague-Dawley rats, 100 to 120 g in wt, were anesthetized with chloral hydrate (4 mg/100 g wt). The extensor digitorum longus (EDL) and soleus (SOL) muscles were removed from their origins and insertions. They were then switched to each other's muscle bed and sutured to the proximal and distal tendon stumps left in situ. EDL muscles, therefore, became SOL-transplants and vice versa. No attempt was made to secure neural and vascular anastomoses. The animals were then divided into experimental (exercised) and control (sedentary) groups. For the control group, the animals were returned to their cages. For the experimental group, the animals were trained to run on a treadmill beginning 10 days after the surgery (the time needed for wound healing). The animals were forced to run one mile/h for two 30-min periods daily with at least four h in between. These animals were sacrificed after either 12, 25 or 50 days of exercise. At least 48 h were allowed to elapse after the last bout of exercise before the conditioned rats were sacrificed. Control (sedentary) animals were sacrificed concurrently. The exercised and sedentary transplants, as well as contralateral exercised and sedentary EDL and SOL muscles were quickly removed, weighed and processed for glycogen analysis. This involves digestion of the tissue in 30% KOH and precipitation of the glycogen with ethanol. The precipated glycogen was then re-dissolved in water and its concentration determined with anthrone reagent giving a color reaction in which maximal absorbency was at 620 mμ.

Results

Table 1 shows the results of glycogen analysis. It can be seen that with 12 days of exercise, neither the contralateral muscles nor the transplants showed any significant changes. With 25 days of exercise, the contralateral muscles showed significant increase in glycogen, while the transplants remained unchanged. With 50 days of exercise, the contralateral muscles,

Table 1

Glycogen Levels (mg/g) in Exercised and Sedentary EDL, SOL, EDL-transplants, and SOL-transplants

| | Control | | Transplanted | |
	EDL	SOL	EDL	SOL
Sedentary exercised (12 days)	5.46 ± 0.11 (4)	3.01 ± 0.40 (4)	4.34 ± 0.44 (4)	2.06 ± 0.29 (4)
	5.09 ± 0.22 (5)	3.98 ± 0.31 (5)	4.40 ± 0.31 (5)	1.81 ± 0.49 (4)
Sedentary exercised (25 days)	4.66 ± 0.26 (7)	3.26 ± 0.26 (7)	3.53 ± 0.24 (7)	2.44 ± 0.41 (6)
	5.67 ± 0.10* (4)	5.00 ± 0.38* (4)	3.66 ± 0.71 (4)	2.11 ± 1,39 (3)
Sedentary exercised (50 days)	3.95 ± 0.16 (4)	3.38 ± 0.13 (4)	3.41 ± 0.20 (4)	3.01 ± 0.84 (3)
	4.85 ± 0.30* (4)	5.44 ± 0.40* (4)	4.27 ± 0.34* (4)	3.92 ± 0.61 (4)
	$p < 0.05$	$p < 0.01$	$p < 0.05$	

Values are mean ± SE. Numbers of muscles or transplants analyzed are in the parentheses. Asterisk (*) represents significant difference from control.

as well as EDL-transplants, showed significant increases in glycogen. The glycogen in SOL-transplants also increased, though the increase was not statistically significant.

Discussion

Our results not only confirm other studies (Poland et al., 1980) illustrating that exercise can indeed change glycogen concentration of skeletal muscles, but also show that exercise can modify the glycogen of muscle transplants as well. The SOL- and EDL-transplants also respond with increases in glycogen, though the elevation in SOL-transplants is not statistically significant. The difference between the responses of SOL- and EDL-transplants is difficult to explain, although it could be due to the fact that SOL-transplants do not regenerate as well as EDL-transplants. Our previous studies also show changes of SOL-transplants to be less pronounced than that of EDL-transplants (Mong & Poland, 1982).

Our results seem to suggest that normal muscles change their glycogen in response to exercise sooner (25 days of exercise) than muscle transplants do (50 days of exercise). This conclusion should be avoided, however, because morphological and histochemical maturation of the muscle transplants require 35 to 50 days after surgery. It was during this period that our exercise program was applied and, thus, the developmental process of muscle fibers in the transplants could have complicated the results. Further experiments in which exercise is applied to mature muscle grafts should be performed so that a comparison of responses to exercise between normal muscles and muscle grafts can be evaluated. Nevertheless, our results clearly indicate that muscle grafts are metabolically sound and can change their glycogen in response to exercise as do normal muscles.

References

BAGBY, G.R., W.L. Sembrowick & P.D. Gollnick. Myosin ATPase and fiber composition from trained and untrained rat skeletal muscle. Am. J. Physiol. 223:1414–1417, 1972.

BARNARD, R.J., V.R. Edgerton & J.B. Peter. Effect of exercise on skeletal muscle, I. Biochemical and histochemical properties. J. Appl. Physiol. 28:762–766, 1970.

BINKHORST, R.A., The effect of training on some isometric contraction characteristics of a fast muscle. Pflüg. Arch. 309:193–202, 1969.

GOLLNICK, P.D. & W.D. King. Effect of exercise and training on mitochondria of rat skeletal muscle. Am. J. Physiol. 216:1502–1509, 1969.

HAKELIUS, L., Free muscle grafting. Clin. Plast. Surg. 6:301–316, 1979.

MARKLEY, J.M., J.A. Faulkner & B.M. Carlson. Regeneration of skeletal muscle after grafting in monkey. Plast. Reconstr. Surg. 62:415–422, 1978.

MONG, F.S.F., J.L. Poland & J.W. Poland. Glycogen and histological changes in muscle grafts of rats during fasting or exercise. Canad. J. Physiol. Pharmacol. 1982, (in press).

MUFTI, S.A., B.M. Carlson, L.E. Maxwell & J.A. Faulkner. The free autografting of entire limb muscles in the cat morphology. Anat. Rec. 188:417–430, 1977.

POLAND, J.L., C. Trowbridge & J.W. Poland. Substrate repletion in rat myocardium, liver and skeletal muscle after exercise. Canad. J. Physiol. and Pharmacol. 58:1229–1233, 1980.

Plasma FFA Influence on Muscle Glycogen

J.L. Poland, J.W. Poland and J.M. Kennedy
Medical College of Virginia,
Richmond, Virginia, U.S.A.

It has previously been shown that, in rats, following moderate exercise, glycogen recovery is different in cardiac and skeletal muscles. Myocardial glycogen readily supercompensates, while concurrently skeletal muscle glycogen merely returns to its pre-exercise level (Poland et al., 1980 Terjung et al., 1974). Similarly, injected glucocorticoids differentially affect glycogen of cardiac and skeletal muscles with myocardial glycogen rising faster and peaking sooner (Poland et al., 1982).

Exercise or glucocorticoids might affect muscle glycogen by generating an increase in plasma free fatty acids (FFA) levels which in turn creates an increase in tissue citrate. High citrate concentrations inhibit phosphofructokinase activity and thus reduce glycogenolysis to promote glycogen buildup. Cardiac muscle is more sensitive to FFA induced citrate changes than is skeletal muscle (Adrouny, 1969). The purpose of the present study was to determine if the glycogen changes in cardiac and skeletal muscle following exercise or glucocorticoid administration could be mediated by FFA induced citrate changes.

Methods

Male Sprague-Dawley rats were housed two per cage, fed ad libitum and exposed to a 12 lighting cycle. Some rats were injected intraperitoneally with 400 μg of dexamethasone with control rats receiving only saline. Rats in the exercise experiments were forced to run on a motor driven treadmill at 1 mph. Untrained rats ran for approximately 15 min and trained rats for 30 min. Other rats were made lipemic by administering corn oil intraperitoneally and heparin subcutaneously, to increase plasma lipoprotein lipase activity. Some lipemic rats were sacrificed at various times following the corn oil injection to determine the effect of high plasma FFA levels per se on muscle citrate and glycogen. Other lipemic rats began exercising 30 min after the injection, when FFA were elevated, and were then sacrificed after 2 or 4 h of exercise recovery to determine if high FFA levels would alter the pattern of glycogen recovery.

At the time of sacrifice, each rat was anesthetized with Nembutal, and the vastus lateralis (VL) muscles from both legs and the heart removed and quickly frozen for citrate and glycogen determinations. A blood sample was collected from the blood pooled in the thoracic cavity, centrifuged and the plasma removed and frozen for later FFA analysis.

Glycogen in the muscles was extracted by the procedure of Good et al. (1933) and quantitatively determined by the anthrone method of Seifter et al. (1950). The analytic procedure for citrate (Stern, 1957) involved extraction in 10% trichloroacetic acid during homogenization, oxidizing the citrate to pentabromoacetone which then reacted with a thiourea color reagent. Plasma FFA levels were determined by a modification of the method of Dole and Meinertz (1960) but using TAC [2-(2-thiazolylazo)-p-cresol] as the agent to give a color reaction with copper salts of FFA as suggested by Noma et al. (1973). Statistical analysis of the data was done by the Student's "t" test to determine differences between mean values. A value of $P < 0.05$ was accepted as indicative of a significant difference.

Results

Table 1 compares citrate levels in cardiac and vastus lateralis muscles following dexamethasone injection with previously reported glycogen and plasma FFA changes that occur concurrently (Poland et al, 1982). Injected dexamethasone produced a prolonged elevation of plasma FFA and tissue glycogen, with cardiac glycogen rising faster and peaking sooner than did skeletal muscle glycogen. In contrast to these prolonged effects, dexamethasone produced an elevation of myocardial citrate at only 4 and 6 h and no change in citrate levels of VL muscles.

The effects of exercise on plasma FFA, muscle citrate and muscle glycogen are shown in Table 2. In both untrained and trained rats, exercise produced an increase in plasma FFA. Concurrently myocardial citrate was elevated, though the increase proved to be statistically significant only in the untrained hearts. The moderate intensity of exercise employed did not significantly reduce myocardial glycogen and the elevated levels post exercise were not a statistically significant supercompensation. In the vastus lateralis the citrate levels were also elevated, but less pronounced than those of cardiac muscle. The exercise employed did significantly lower

Table 1

Citrate, Glycogen and FFA Data Showing Changes After Glucocorticoid Administration

	Controls	Hours After Dexamethasone Injection				
		2	4	6	17	26
Glycogen ($\mu g/mg$)						
Heart	4.53 ± .31	6.25 ± .51*	7.78 ± .75*	10.55 ± .37*	9.43 ± .78*	7.35 ± 1.53*
RVL	5.32 ± .58	4.36 ± .23	5.66 ± .52	6.03 ± .21	10.78 ± .91*	6.27 ± 1.16
WVL	5.12 ± .51	4.75 ± .66	5.62 ± .72	5.73 ± .21*	13.04 ± .62*	9.04 ± 1.19*
Citrate ($\mu g/g$)						
Heart	71 ± 4		106 ± 12*	104 ± 5*	67 ± 9	
VL	26 ± 1		24 ± 2	22 ± 2	26 ± 2	
Plasma FFA($\mu Eq/l$)	299 ± 21	488 ± 27	378 ± 13	471 ± 40	504 ± 41	490 ± 42

Values are means ± SEM. An * indicates a significant difference from the control value.

Table 2

Plasma FFA, Muscle Citrate and Muscle Glycogen Changes Before and After Exercise

	Controls	Hours After Exercise			
		0 H	1 H	2 H	4 H
FFA (μEq/l)					
Untrained	239 ± 8	339 ± 46*	374 ± 38*	401 ± 52*	291 ± 24*
Trained	389 ± 23	989 ± 103*	549 ± 60*	428 ± 47	437 ± 47
Cardiac Citrate (μg/g)					
Untrained	54.0 ± 5.7	78.8 ± 12.9*	70.3 ± 9.3	94.9 ± 8.7*	42.8 ± 5.8
Trained	67.1 ± 7.8	93.8 ± 13.8	73.8 ± 5.6	64.5 ± 8.1	61.5 ± 7.2
Cardiac Glycogen (μg/mg)					
Untrained	3.68 ± .31	3.45 ± .34	4.68 ± .44	3.98 ± .28	4.22 ± .35
Trained	4.59 ± .35	3.62 ± .46	5.38 ± .39	5.41 ± .78	5.42 ± .49
VL Citrate (μg/g)					
Untrained	29.6 ± 1.4	41.2 ± 4.1*	29.7 ± 4.5	27.1 ± 4.0	25.7 ± .9*
Trained	22.6 ± 1.8	29.0 ± .6*	24.9 ± 2.3	24.0 ± 2.6	26.5 ± 2.8
VL Glycogen (μg/mg)					
Untrained	5.32 ± .35	1.12 ± .35*	3.93 ± .44*	3.70 ± .26*	4.45 ± .19
Trained	6.06 ± .24	4.99 ± .37*	4.92 ± .30*	5.07 ± .48*	5.44 ± .22

Values are means ± SEM. An * indicates a significant difference from the control value.

VL glycogen with the untrained rats, exhibiting a greater decrease than the trained rats.

Following a single IP injection of corn oil, plasma FFA levels (Controls $= 239 \pm 8 \, \mu Eq/l$) were elevated at 2, 4 and even 17 h post injection (2850 ± 75; 1891 ± 308; $553 \pm 23 \, \mu Eq/l$, respectively). In spite of these sustained high FFA levels, both muscle citrate and muscle glycogen levels were relatively stable. Statistically significant increases occurred with citrate only at 4 h in the VL ($33.5 \pm .2$ vs $29.6 \pm 1.4 \, \mu g/g$) and with glycogen only at 2 h in the heart ($4.66 \pm .27$ vs $3.68 \pm .31 \, \mu g/mg$). At 17 h the citrate levels actually seemed depressed and were significantly lower in the myocardium (32.4 ± 10.9 vs $54.0 \pm 5.7 \, \mu g/mg$). Similarly, elevated plasma FFA levels did not enhance glycogen recovery or citrate levels at either 2 or 4 h after exercise.

Discussion

The data show that glucocorticoids do differentially affect skeletal and cardiac glycogen and citrate. The increase in myocardial citrate occurred early in the response to dexamethasone injection while skeletal muscle citrate concurrently remained stable. This would help explain the faster increase in cardiac glycogen than skeletal muscle glycogen following dexamethasone administration. However, at a later time, the glycogen levels in both cardiac and skeletal muscles are elevated though the citrate levels have returned to normal values. At this later time the dexamethasone-induced glycogen increases must be due to some mechanism other than the citrate inhibition of glycolysis. Since plasma FFA levels were maintained at elevated values during the entire experimental period, glycogen sparing might come into play.

Exercise also produced elevations in plasma FFA and briefer elevations in muscle citrate, predominantly in the heart. The brief increase in citrate would tend to promote glycogenesis during the early period of recovery after exercise. The fact that citrate increases occur predominantly in cardiac rather than skeletal muscle may help account for the differential effect of either exercise or glucocorticoid administration on glycogen levels.

The theory that citrate and glycogen changes observed following glucocorticoid injection or after exercise could be initiated by elevated FFA levels is not supported by the exerimental data. Elevation of plasma FFA levels per se, even to extraordinary heights, did not generate any dramatic or consistent increases in either citrate or muscle glycogen at rest or during recovery after exercise. Thus, if elevated FFA levels are to influence muscle glycogen, they must occur in conjunction with some other regulatory phenomenon.

References

ADROUNY, G.A., Differential patterns of glycogen metabolism in cardiac and skeletal muscles. Am. J Physiol. 217:686–693, 1969.

DOLE, V.P. & H. Meinertz. Microdetermination of long-chain fatty acids in plasma and tissues. J. Biol. chem. 235:2595–2599, 1960.

GOOD, C.A., H. Kramer & K.M. Somogyi. The determination of glycogen. J. Biol. Chem. 100:485–491, 1933.

NOMA, A., H. Okabe & M. Kita. A new colorimetric microdetermination of free fatty acids in serum. Clin. Chim. Acta 43:317–320, 1973.

POLAND, J.L., J.W. Poland & R.N. Honey. Differential response of rat cardiac and skeletal muscle glycogen to glucocorticoids. Can. J. Physiol. and Pharmacol. 1982. (In press)

POLAND, J.L., C. Trowbridge & J.W. Poland. Substrate repletion in rat myocardium, liver and skeletal muscles after exercise. Can. J. Physiol. and Pharmacol. 58:1229–1233, 1980.

SEIFTER, S., S. Dayton, B. Novic & E. Muntwyler. The estimation of glycogen with the anthrone reagent. Arch. Biochem. 25:191–200, 1950.

STERN, J.R., Estimation of citric acid. In Methods in Enzymology. 3:426–428, 1957.

TERJUNG, R.L., K.M. Baldwin, W.W. Winder & J.O. Holloszy. Glycogen depletion in different types of muscle and in liver after exhausting exercise. Am. J. Physiol. 226:1387–1391, 1974.

The Marathon: Recovery from Acute Biochemical Alterations

W.M. Sherman, D.L. Costill, W.J. Fink,
L.E. Armstrong, F.C. Hagerman and T.M. Murry
Ball State University, Muncie, Indiana, U.S.A., and
Ohio University, Athens, Ohio, U.S.A.

It is well documented that training as well as acute exercise affect the activity of various intramuscular enzymes. In particular, these effects have been documented for enzymes mediating the disposal of intracellular glucose, i.e., hexokinase, glycogen synthase, and the hexose mono-phosphate pathway, as well as other enzymes (Beaconsfield & Reading, 1964; Kochan et al., 1981; Peter et al., 1968). It is not known, however, how long the acute effects of exercise persist or if subsequent rest or exercise modify the enzymatic alterations induced by acute exercise. The purpose of this study, therefore, was to determine the effect of rest or exercise during the wk following an endurance exercise task (a marathon) on muscle glycogen and the activities of hexokinase, glycogen synthase, and the hexose monophosphate pathway. Thus, the results of this study describe the recovery of acute alterations in enzyme activity following marathon running.

Methods

Ten well-trained male runners volunteered to participate in this study. Informed consent was obtained and all procedures complied with the Declaration of Helsinki. The mean ($\pm$ SD) age, wt, percent body fat, $\dot{V}O_2$max and percent ST, FTa and FTb fibers were 25.8 ($\pm$ 3.5), 66.0 ($\pm$ 3.8), 9.3 ($\pm$ 1.7), 68.7 ($\pm$ 5.4), 61.8 ($\pm$ 8.5), 32.4 ($\pm$ 8.0) and 5.7 ($\pm$ 3.5), respectively.

Five subjects were randomly assigned to either a postmarathon rest group or exercise group. The rest group performed no physical activity while the exercise group performed exercise of gradually increasing duration during the 7 day recovery period. This exercise consisted of 20, 25, 30, 35, 40 and 45 min runs on a treadmill ergometer at the subject's selected intensity of running. The day following the marathon the subjects consumed 800 g of carbohydrate in 4,000 kcal. During each succeeding day of the recovery wk the subjects consumed 450 g of carbohydrate in 3,000 kcal/day.

Muscle samples were obtained from the lateral head of the gastrocnemius before the marathon (pre), within 15 min following the marathon (post), and 1, 3, 5 and 7 days following the marathon. Muscle glycogen was determined after acid hydrolysis (2.0 N HCl) as the glycosyl units · kg wet tissue^{-1} (Lowry & Passanneau, 1972). Glycogen synthase activity was determined at several concentrations of glucose-6-phosphate as described by Kochan et al. (1981). The glycogen synthase activity ratio (I-form I + D - form $^{-1}$) and the fractional velocity (activity at 250 μM glucose-6- phosphate) were calculated. The activity of hexokinase was determined fluorometrically by following the reduction of NADP$^+$ using glucose as substrate (Lowry & Passanneau, 1972). The activity of glucose-6-phosphate dehydrogenase and 6-phosphogluconate dehydrogenase was determined as the reducing capacity of the hexose monophosphate pathway (HMSRC) by following the reduction of NADP$^+$ using glucose-6-phosphate as substrate (Lowry & Passanneau, 1972). Protein was determined using the Hartree-modified Lowry procedure (Hartree, 1965). To determine treatment, time, and subject effects, a two-way analysis of variance technique (split-plot nested design) with repeated measures was used (Winer, 1962). When the F-ratio indicated significance (p < .05) means were compared with Student's New Multiple Range Test (Winer, 1962).

Results

The mean (± SD) marathon time for the rest and exercise group was 187.7(± 24) and 163.8 (± 11), respectively (p > .05). The pattern of fatigue was the same for both groups (determined as % $\dot{V}O_2$max for split intervals) and followed previously described patterns (Karlsson & Saltin, 1972). The running intensity through 27.4 km was 75% $\dot{V}O_2$max after which it averaged 65% $\dot{V}O_2$max for the remainder of the race. During the recovery wk, the mean (± SD) intensity of exercise selected by the exercise group was 52(± 11), 57(± 12), 56(± 16), 56(± 16), 59(± 10) and 60(± 10), respectively.

Statistical analysis of the data indicated that the treatments, rest or exercise, did not affect muscle glycogen or the enzymes mediating the disposal of intracellular glucose during the recovery period. There was, however, a significant time effect on the measured variables which will be described in the following sections.

Muscle Glycogen

Muscle glycogen was supercompensated prior to the marathon and decreased to the lowest levels reported following running exercise (Table 1). One day following the marathon, muscle glycogen levels had increased by 38% and continued to increase to 64% of premarathon levels through the third day postmarathon. Thereafter, muscle glycogen levels changed very little through the seventh day postmarathon.

Glycogen Synthase Activity

The glycogen synthase activity ratio was significantly elevated following the marathon (Table 1). Thereafter, the activity ratio had returned to normal resting levels during the recovery period. This occurred in spite of continued muscle glycogen synthesis through day 3 postmarathon. The fractional velocity, on the other hand, was significantly increased immediately, postmarathon, and 1 day postmarathon. It is likely that the fractional velocity remained elevated through the second day post marathon and accounts for continued glycogen synthesis through day 3 postmarathon.

Hexokinase Activity

Hexokinase activity was 340 pkat mg protein^{-1} prior to the marathon and increased to 435 pkat mg protein^{-1} after the marathon. Hexokinase activity was still increased 5 days following the marathon (424 pkat mg protein^{-1}).

Hexose Monophosphate Pathway

The activity of the HMSRC was 760 pkat mg protein^{-1} prior to the marathon and decreased to 710 pkat mg protein^{-1} after the marathon. Thereafter, the activity of the HMSRC continued to decrease to 550 pkat mg protein^{-1} the seventh day postmarathon.

Discussion

This study confirms the acute biochemical alterations that accompany prolonged strenuous running and describes the recovery from these acute

Table 1

**Levels of Muscle Glycogen and Activities of the Measured Enzymes
Premarathon and During the Postmarathon Recovery Period**

	Premarathon	Postmarathon	1	3	5	7
Glycogen (mmol glucosyl units kg wet tissue^{-1})	196.1 ($\pm$ 6.3)	25.5* ($\pm$ 3.2)	79.0* ($\pm$ 4.7)	117.1* ($\pm$ 7.4)	131.4* ($\pm$ 5.9)	125.1* ($\pm$ 6.6)
Glycogen synthase activity ratio	5.4 ($\pm$ .86)	18.5* ($\pm$ 1.9)	9.84 ($\pm$ 2.3)	5.12 ($\pm$ .68)	3.48 ($\pm$ .59)	4.44 ($\pm$ .82)
Glycogen synthase fractional velocity	13.1 ($\pm$ 1.1)	63.1* ($\pm$ 2.7)	40.7* ($\pm$ 5.2)	16.4 ($\pm$ 1.3)	15.1 ($\pm$ 1.4)	16.5 ($\pm$ 1.1)
Hexokinase (pkat mg protein^{-1})	340 ($\pm$ 6.9)	446* ($\pm$ 17.0)	418* ($\pm$ 5.3)	436* ($\pm$ 12.1)	424* ($\pm$ 9.2)	392 ($\pm$ 7.3)
HMSRC (pkat mg protein^{-1})	758 ($\pm$ 5.7)	708* ($\pm$ 7.1)	699* ($\pm$ 4.9)	643* ($\pm$ 5.6)	609* ($\pm$4.8)	545* ($\pm$4.1)

Values are mean ($\pm$ SE). * indicates significantly different from premarathon values.

changes during a 7 day recovery period. It is somewhat surprising that exercise during the recovery week did not differentiate the two recovery groups. It is likely, however, that the lowered intensity of running selected by the exercise group and the length of the recovery runs was not sufficient to acutely stress the muscle on a day-to-day basis. It was not until the seventh day postmarathon that the combination of intensity and duration (60% $\dot{V}_{O_2}$max, 45 min) began to differentiate the two groups (data not shown).

Muscle glycogen content was dramatically affected by the marathon which resulted in a net utilization of 165 mmol glucosyl units kg wet tissue^{-1}. Concomitant with the severe reduction in muscle glycogen (to 25 mmol glucosyl units kg wet tissue^{-1}) was the activation of glycogen synthase. The initial resynthesis of muscle glycogen through day 1 postmarathon could be accounted for by the increased activity ratio and fractional velocity while continued synthesis through day 3 postmarathon could be accounted for by the increased fractional velocity. Seven days postmarathon muscle glycogen was still 60% lower than premarathon levels (Table 1). This might be due to the fact that 130 mmol glucosyl units kg wet tissue^{-1} is approximately normal for trained athletes (Sherman et al., 1981) and results from the direct inhibition of muscle glycogen on glycogen synthase (Danforth, 1965). In addition, since insulin has direct hormonal effects on glycogen synthase (Cohen et al., 1976), it is possible that changes in insulin sensitivity affected glycogen synthesis. An increased insulin sensitivity (Ivy et al., this conference), however, should result in enhanced glycogen synthesis, but the 50% carbohydrate diet during the recovery period may not have been sufficient substrate or insulin releasing stimulus (Ivy, 1976) to return muscle glycogen to the high premarathon levels.

Previous investigators have reported both acute and chronic effects of exercise on hexokinase (Barnard & Peters, 1969; Lamb et al., 1969). Hexokinase activity was increased 22% by the marathon and was still 18% elevated 5 days postmarathon. Whether this increased activity is a result of activation of an inactive form of the enzyme, a result of de novo enzyme synthesis, or a combination of both cannot be determined from this investigation. It is also possible that this observation is a consequence of changes in the ratios of the soluble form and insoluble form of this enzyme.

The activity of the HMSRC is generally low in skeletal muscle, but has been shown to be induced by muscle trauma (Beaconsfield & Reading, 1964). In addition HMSRC activity was elevated for 2 to 8 days following eccentric running exercise in rats as a consequence of leucocyte infiltration into the muscle's interstitum (Armstrong et al., 1980). Based on the protracted nature of the marathon, we expected the activity of the HMSRC to increase. Surprisingly, however, the HMSRC was not acutely affected by the marathon and continued to decrease throughout the postmarathon recovery period. Thus, this protracted exercise did not elicit a response to trauma which was observable by measuring this pathway. It is possible that this response might be different in runners that are not as well trained as those participating in the present study.

Acknowledgment

This study was supported, in part, by the National Dairy Council, and Sigma Xi, the Scientific Research Society, 1981.

References

ARMSTRONG, R.B., V. Garshnek & J.A. Schwane. Muscle inflammation: Response to eccentric exercise. Med. Sci. Sports Exercise 12:95, 1980.

ARMSTRONG, R.B., P. Marum, P. Tullson & C.W. Saubert. Acute hypertrophic response of skeletal muscle to synergists. J. Appl. Physiol. 46:835–842, 1979.

BARNARD, R.J. & J.B. Peter. Effect of training and exhaustion on hexokinase activity of skeletal muscle. J. Appl. Physiol. 27:691–695, 1969.

BEACONSFIELD, P. & H.W. Reading. Pathways of glucose metabolism and nucleic acid synthesis. Nature 202:464–466, 1964.

COHEN, P., H.G. Nimmo & C.G. Proud. How does insulin stimulate glycogen synthesis? Biochem. Soc. Symp. 43:69–95, 1979.

DANFORTH, W.H., Glycogen synthetase activity in skeletal muscle. Interconversion of two forms and control of glycogen synthesis. J. Biol. Chem. 240:588–593, 1965.

HARTREE, E.F., Determination of protein: A modification of Lowry method that gives a linear photometric response. Anal. Biochem. 25:486–499, 1968.

IVY, J.L., Role of insulin during exercise induced glycogenesis in rats: Effect on cyclic AMP. Am. J. Physiol. 233:509–513, 1977.

KARLSSON, J. & B. Saltin. Diet, muscle glycogen and endurance performance. J. Appl. Physiol. 31:203–206, 1971.

KOCHAN, R.G., O.R. Lamb, E.M. Reimann & K.K. Schleinder. Modified essays to detect activation of glycogen synthase following exercise. Am. J. Physiol. 240:197–202, 1981.

LOWRY, O.J. & J.V. Passanneau. A Flexible System of Enzymatic Analysis. New York: Academic Press, 1973.

PETER, J.B., R.N. Jeffress & D.R. Lamb. Exercise: Effects on hexokinase activity in red and white skeletal muscle. Science 160:200–201, 1968.

SHERMAN, W.M., D. Costill, W. Fink & J. Miller. Effect of exercise-diet manipulation on muscle glycogen and its subsequent utilization during performance. Int. J. Sports Medicine 2:114–118, 1981.

WINER, J.B., Statistical Principles in Experimental Design. New York: McGraw-Hill, 1971.

Lipid
Metabolism

Interactions Between Exercise Training and Sucrose Intake on Adipocyte Lipolysis in Rat Parametrial Adipose Tissue

**L.J. Bukowiecki, J. Lupien, G. Côté,
E. Cardinal and A. Vallerand**
Laval University, Québec, Canada

The hormone-sensitive triglyceride lipases in white adipose tissue represent the flux-generating step controlling fatty acid metabolism in working muscles, particularly during long-term exercise (Newsholme, 1979). The recent finding that exercise training increases adipocyte responsiveness to lipolytic hormones at metabolic steps distal to stimulus recognition by adrenoreceptors supports the concept that lipases in white adipose tissue play an important regulatory role in adaptation to exercise (Bukowiecki et al., 1980). Because we observed, in the same studies, that food restriction partially mimicked the effects of exercise training, we next investigated the interactions between hyperphagia induced by sucrose consumption and exercise training on adipocyte lipolytic capacity. The experiments were carried out with female rats because it is well known that exercise training increases appetite in females but not in males. The main questions asked in the present studies were whether hyperphagia reduces adipocyte lipolytic response to stimulation by epinephrine in sedentary animals and/or whether it inhibits the enhancing effects of exercise training on adipocyte lipolytic capacity.

Methods

Female Wistar rats (140 to 150 g) were divided into four groups of 10 animals each: sedentary controls receiving Purina chow and water ad libitum (C–P), sedentary controls receiving a 32% sucrose solution in addition to the Purina diet (C-S), and two corresponding groups that were submitted to a training program consisting of 2 h of daily swimming at 36° C during 10 wks (E-P and E-S). The animals were killed 2 days after their last training period. Adipocytes were isolated from parametrial white adipose tissue exactly as previously described (Bukowiecki et al., 1980).

Lipolysis was estimated by measuring the release of extracellular glycerol. The washed adipocytes were first preincubated with gentle shaking in polyethylene vials at 37° C for 15 min in Krebs-Ringer bicarbonate buffer containing 50 mg% glucose and 1% fatty acid-free albumin under an atmosphere of 95% O_2/5% CO_2. After preincubation, the cells were washed twice with warm (37° C) buffer in which the concentration of albumin was raised to 4%. This was done to minimize the accumulation of unbound fatty acids in the extracellular medium during the final incubation (Bukowiecki et al., 1981). Aliquots (0.5 ml) containing 5 x 105 cells were removed and incubated for 15 min in 20 ml polyethylene vials containing 1 ml of buffer under the same conditions as above. At the end of the incubation period, the vials were cooled on melting ice, the floating adipocytes were rapidly discarded and aliquots of the infranatant were sampled in triplicate for glycerol determination by enzymatic methods.

Results

The effects of exercise training and sucrose consumption on rat body weight, food intake, body weight gain efficiency, PWAT weight, adipocyte size and tissue cellularity were reported in the accompanying paper. The same animals were used in the present studies.

Preliminary experiments were carried out to check whether the rate of lipolysis was linear under the present experimental conditions. Indeed, it has been reported that fatty acids, adenosine and possibly other compounds released in the extracellular medium during static incubation of isolated adipocytes might significantly retroinhibit lipolysis if they are allowed to accumulate (Bukowiecki et al., 1981; Fain & Wieser, 1975). It was found that if adipocytes are preincubated, washed and incubated exactly as described under Methods, the rate of lipolysis stimulated by 10 μM epinephrine remained linear during approximately 30 min. Thereafter, lipolysis decreased significantly. An incubation period of 15 min was therefore selected for performing all subsequent experiments.

Although sucrose consumption slightly increased basal lipolysis in sedentary rats, it significantly reduced lipolysis stimulated by 10 μM epinephrine both in sedentary and exercise trained animals (Figure 1). On the other hand, exercise training increased the capacity of adipocytes to respond lipolytically to epinephrine. Moreover, it also reversed the

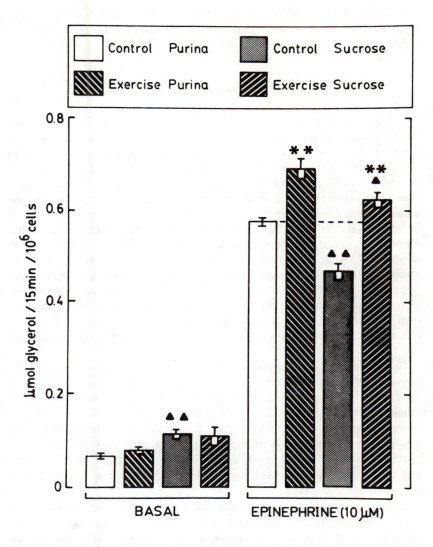

Figure 1—Effects of epinephrine on basal and stimulated lipolysis in adipocytes isolated from C-P, E-P, C-S and E-S groups.

Adipocytes were isolated from parametrial adipose tissue of female rats (175 to 200 g) and incubated in Krebs-Ringer bicarbonate buffer in the presence and absence of epinephrine (10 μM) as described under Methods. The bars represent the means $\pm$ SEM of 8 to 10 experiments performed on separate occasions. The data were analyzed by variance analysis. The asterisks and triangles indicate significant effects of exercise training or sucrose consumption, respectively (one symbol: $p < 0.05$ and two symbols: $p < 0.01$)

inhibitory effects of sucrose consumption on stimulated lipolysis. Thus, exercise training enhances adipocyte responsiveness to epinephrine in controls as well as in hyperphagic animals.

Discussion

The data described in Figure 1 confirm and extend our previous observations (Bukowiecki et al., 1980) by the demonstration that exercise training increases the capacity of adipocytes to respond lipolytically to epinephrine in normal as well as in hyperphagic animals (the detailed data on energy intake are given in Figure 1, paper presented by Lupien et al. at this Symposium). They also show that hyperphagia induced by sucrose consumption inhibits adipocyte responsiveness to epinephrine both in sedentary and in exercising animals. These results demonstrate that adipocyte capacity for responding lipolytically to epinephrine can be modulated by at least two physiological conditions exerting opposite effects on lipolysis, namely exercise training (stimulates lipolysis) and hyperphagia (inhibits lipolysis). The fact that increased energy consumption reduced adipocyte lipolytic capacity agrees with our previous observation that food restriction partially mimicked the enhancing effects of exercise training on lipolysis, both in male and female rats (Bukowiecki et al., 1980).

In female animals, exercise increases appetite and food consumption, particularly if palatable sucrose solutions are offered to the animals (Figure 1 of paper presented by Lupien at this Symposium). Nevertheless, an enhancement of adipocyte lipolytic capacity was still observed in exercising animals that were markedly hyperphagic (the total calorie intake of the E-S groups was nearly double of that of C-P animals). This indicates that, under the present experimental conditions, exercise training was a more potent factor than increased energy consumption for modulating adipocyte lipolytic capacity.

Correlation analysis (Snedecor & Cochran, 1967) revealed that adipocyte capacity for responding lipolytically to norepinephrine was negatively but significantly correlated with adipocyte size, estimated by cellular triglyceride content (Table 1, accompanying paper) ($y = -0.33x + 0.35$; $r = -0.43$; $p < 0.05$). However, no correlation could be demonstrated between lipolysis and adipocyte size within each of the four experimental groups. It is therefore likely that factors other than adipocyte size modulate adipocyte responsiveness to lipolytic hormones. A significant correlation ($y = -31.79x + 34.47$; $n = -0.75$; $p < 0.05$) was found between total food intake and epinephrine-stimulated lipolysis in sedentary (C-P and C-S) animals but not in the exercising groups (E-P and E-S). It therefore appears that exercise training attenuates the inhibitory effects of excessive calorie intake.

Taken as a whole, results from our studies support the conclusion that exercise training per se exerts a predominant role, not only in inhibiting adipocyte proliferation in parametrial white adipose tissue (see accompanying paper), but also in simultaneously increasing adipocyte capacity for mobilizing fatty acids in response to epinephrine stimulation. Whether exercise training exerts similar effects in other adipose tissue depots and whether some of these effects are linked to sexual differences still remains to be investigated.

References

BUKOWIECKI, L., N. Folléa, J. Lupien & A. Paradis. Metabolic relationships between lipolysis and respiration in rat brown adipocytes. The role of long chain fatty acids as regulators of mitochondrial respiration and feedback inhibitors of lipolysis. J. Biol. Chem. 256(24):12840–12848, 1981.

BUKOWIECKI, L., J. Lupien, N. Folléa, A. Paradis, D. Richard & J. LeBlanc. Mechanism of enhanced lipolysis in adipose tissue of exercise-trained rats. Am. J. Physiol. (Endocrinol. Metab. 2): E 422–429, 1980.

FAIN, J. & P.B. Wieser. Effects of adenosine deaminase on cyclic adenosine monophosphate accumulation, lipolysis and glucose metabolism of fat cells. J. Biol. Chem. 250(3):1027–1034, 1975.

SNEDECOR, G.W. & W.G. Cochran. Statistical Methods. New Dehli: Oxford & IBH Publishing Co., 1967.

NEWSHOLME, E.A., The control of fuel utilization by muscle during exercise and starvation. Diabetes 28(Suppl. 1):1–7, 1979.

Effects of Sex, Fatness and Training Status on Human Fat Cell Lipolysis

J.P. Després, C. Bouchard, L. Bukowiecki,
R. Savard and J. Lupien
Laval University, Québec, Canada

Several authors have investigated the relationship between the magnitude of fat deposition and the adipose tissue metabolism in humans (Jacobsson et al., 1976; Ostman et al., 1973). The association between adipose tissue lipolysis and body fatness as well as adipocyte size remains unclear. Thus, some authors have suggested a positive correlation while others have reported a negative correlation between fat cell size and adipocyte lipolysis. Exercise and training are factors that could play a significant role in modulating adipocytes lipolytic activity. Although training has been shown to increase adipocytes lipolytic activity in rats (Askew et al., 1975; Bukowiecki et al., 1980), to our knowledge, the problem has not been investigated in humans. In this study, adipose tissue morphology and lipolytic activity are analyzed with respect to sex of subjects, body fatness indicators and level of physical training.

Methods

Fifty-three subjects (14 women and 39 men), $25.9 \pm 7.6 (\overline{X} \pm SD)$ years of age, weighing 64.8 ± 10.8 kg and with $14.8 \pm 4.7\%$ body fat were submitted to a biopsy of subcutaneous adipose tissue in the suprailiac region. After local anesthesia, between 50 and 150 mg of adipose tissue were removed, using a Steel's biopsy needle. Adipocytes were isolated by a modification of Rodbell's method (1964). Extracellular glycerol was chosen as the indicator of adipocytes lipolytic activity following procedures described by Bukowiecki et al. (1980). Stimulated lipolysis was also determined by adding epinephrine bitartrate to obtain final concentrations of 10^{-5} and 10^{-4} M.

Total body fat was determined by hydrostatic weighing with the Siri equation (1956). Skinfold thicknesses were obtained for women from: biceps, triceps, abdominal, subscapular, suprailiac, thigh, calf; for men: from the above regions and two additional regions, namely the chest and axillary. Analysis of variance (ANOVA) designs were used in the data analysis. Differences between means were tested using the Duncan Multiple Range Test (Kirk, 1968).

Results

Women of the sample have larger adipocytes, higher percent fat ($p \leq 0.001$) and lower basal, sub-maximal and maximal lipolytic activities than men (Figure 1). Moreover, partial correlations between the various fatness indicators and adipocyte lipolytic activity, after statistical control over age and sex effects, reveal that all fatness indicators are negatively correlated with lipolytic activity under all conditions ($-0.29 \leq r \leq -0.45$; $p \leq 0.05$).

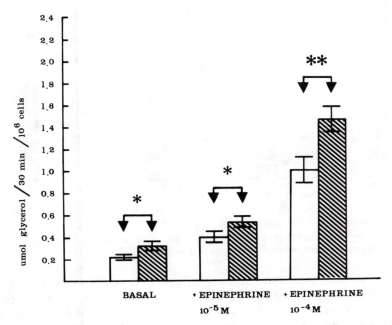

Figure 1—Women (n = 14) □ and men (n = 30) ▨ comparisons of basal and epinephrine stimulated lipolysis in isolated adipocytes. Values are means ± SE. *$p \leq 0.05$; **$p \leq 0.01$.

Table 1

ANOVA-Completely Randomized Factorial Design for Fatness and Training Effects on Basal Lipolysis of Fat Cells

	Sum of squares	Degrees of freedom	Mean squares	F ratio
Source of Variation				
Fatness	0.142	1	0.142	11.94*
Training	0.028	1	0.028	2.33
Fatness X training	0.001	1	0.001	0.06
Within cell	0.286	24	0.012	
Total	0.457	27		

*$P \leq 0.01$ (for $p < 0.01$; 1, 24; $F = 7.82$)

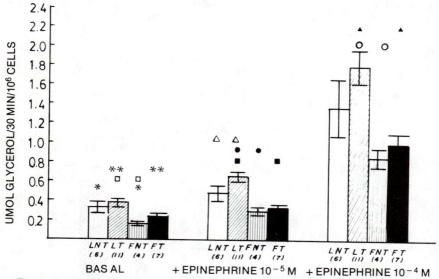

Figure 2—Basal and stimulated lipolytic activity of isolated human adipocytes of lean non-trained (LNT), lean trained (LT), fat non-trained (FNT) and fat-trained subjects (FT). Values are means ± SE; number of subjects is in parentheses.

To dissociate fatness and training effects, an ANOVA factorial design with 28 male subjects classified according to their levels of fatness and training was used. Trained subjects of the study were running more than 100 km/wk, while the non-trained subjects were sedentary. There were no significant differences between trained and non-trained subjects of the study for lipolytic activity. However lean subjects had significantly higher basal, sub-maximal and maximal lipolytic activities than fat subjects ($p < 0.01$). Moreover, results of the ANOVA Completely Randomized Factorial Design analysis showed that there were significant effects of fatness on basal lipolysis (Table 1), as well as on stimulated lipolysis, but no significant effects of training and no significant fatness and training interaction.

Figure 2 further illustrates the dominant fatness effect on fat cells lipolysis under all three conditions.

Discussion

The observation that women have higher percent body fat and mean fat cell size than men is concordant with results of others. Results of the present study indicate, however, that women have also lower basal, sub-maximal and maximal lipolysis.

The negative correlations obtained between mean fat cell size and lipolysis are in agreement with data reported by Zinder and Shapiro (1971) and Bukowiecki et al. (1980), but not with those of Jacobsson et al. (1976) and Osman et al. (1973). Such discrepancies can be caused by several factors: site of tissue sampling, age, sex, diet, hormonal status, exercise, heredity, etc. Thus, with so many factors involved and without direct experimental control over several of them, the observation of a negative correlation between lipolytic activities and mean fat cell size after statistical control over age and sex must be viewed with caution.

There is some evidence that training increases lipolytic activity of adipose tissue in rats (Askew et al., 1975; Bukowiecki et al., 1980). However, to our knowledge, no study has ever been reported about the effect of training on human adipose tissue lipolysis. As a first step toward the understanding of this phenomenon, a completely randomized factorial ANOVA design revealed a significant effect of fatness level on lipolytic activities but a nonsignificant trend for the effect of physical training on the same variables. It is concluded that lipolytic activity is more related to body fatness than to the training status. However, training tends to accentuate lipolytic activity as shown by values found in the lean highly trained subjects of the study.

Acknowledgment

Supported by Ministère de l'éducation du Québec (EQ-1330), FCAC-centre LABSAP du Ministère de l'éducation du Québec (CE-29) and Natural Sciences and Engineering Research Council of Canada (E6227).

References

ASKEW, E.W., R.L. Huston, C.G. Plopper & A.L. Hecker. Adipose tissue cellularity and lipolysis. Response to exercise and corticol treatment. J. Clin. Invest. 56:521–529, 1975.

BUKOWIECKI, L., J. Lupien, N. Follea, A. Paradis, D. Richard & J. Leblanc. Mechanism of enhanced lipolysis in adipose tissue of exercise-trained rats. Am. J. Physiol. 239:422–429, 1980.

JACOBSSON, B., G. Holm, P. Bjorntorp & U. Smith. Influence of cell size on the effects of insulin and noradrenaline on human adipose tissue. Diabetol. 12:69–72, 1976.

KIRK, R.E., Experimental Design: Procedure for the Behavioral Sciences. Belmont: Wadsworth Publishing Company, 1968.

OSTMAN, J., L. Backman & D. Hallberg. Cell size and lipolysis by human subcutaneous adipose tissue. Acta Med. Scand. 193:469–475, 1973.

RODBELL, M., Metabolism of isolated fat cells. J. Biol. Chem. 239:375–380, 1964.

SIRI, W.E., The gross composition of the body. Adv. Biol. Med. Phys. 4:239–280, 1956.

ZINDER, O. & B. Shapiro. Effect of cell size on epinephrine and ACTH-induced fatty acid release from isolated fat cells. J. Lipid Res. 12:91–95, 1971.
Captions to figures

Lipoprotein Lipase Hydrolyzes Intramuscular Triacylglycerols in Muscle of Exercised Rats

L.B. Oscai

University of Illinois at Chicago, Chicago, Illinois, U.S.A.

Intracellular triacylglycerols (TG) can be decreased in heart and skeletal muscle as a result of exercise (Holloszy & Booth, 1976; Oscai et al., 1982). However, the lipase responsible for hydrolyzing endogenous TG in muscle has never been identified (Lech et al., 1977). A possible role for intracellular lipoprotein lipase (LPL) in the regulation of intramuscular TG in rat heart and skeletal muscle is proposed.

Two Distinct Functions for LPL

It is well established that LPL plays a regulatory role in the metabolism of plasma chylomicrons and very low density lipoproteins in the capillary beds of extra-hepatic tissue (Borensztajn, 1979; Robinson, 1970). It is our working hypothesis that LPL also functions in the intracellular regulation of stored TG in heart and skeletal muscle (Oscai, 1979; Oscai et al., 1982; Palmer et al., 1981). In the past, investigators have referred to the intracellular LPL fraction as the non-functional fraction (Borensztajn et al., 1975b). Nevertheless, the evidence indicates that LPL is synthesized in parenchymal cells from where it is secreted and eventually transported to the endothelial surface of capillaries, the site of plasma TG hydrolysis (Borensztajn, 1979). It is possible to wash LPL from the capillaries of heart and skeletal muscle by perfusing the tissue with heparin (Borensztajn & Robinson, 1970; Robinson, 1970). Therefore, tissue used in our work is first perfused with heparin until the perfusates are free from LPL activity and TG.

Since we are in the early stages of this work, care must be taken to insure that we are measuring what we claim and not an artifact. To document more conclusively the intracellular role for LPL, an attempt is currently being made to examine LPL activity in its activated form in myocytes and to purify the enzyme to homogeneity. Although our results provide evidence that LPL may act as a mobilizer and a regulator of intracellular TG in muscle, only with a purified enzyme preparation can the mechanisms of control be identified.

Name Change

Presently, an attempt is being made to change the name of our intracellular enzyme from LPL to type L hormone-sensitive lipase. Justification for the name change is that endogenous LPL activity is modified by epinephrine and glucagon possibly through cyclic AMP and protein kinase (Oscai, 1979; Palmer et al., 1981). Thus, hormone sensitivity has been established. Further justification for the type L classification is that lipoproteins do not exist in parenchymal cells of heart and skeletal muscle. Finally, our intracellular enzyme possesses many of the classical characteristics described for LPL. These common properties include alkaline pH optimum (pH 8.1), serum requirement (LPL is activated by apoprotein C-II present on the surface of chylomicrons and very low density lipoproteins), activation by heparin, inhibition by protamine sulfate, and inhibition by 1 M NaCl (Borensztajn, 1978; Lukens & Borensztajn, 1978).

LPL Hydrolyzes Endogenous TG in Skeletal Muscle of Exercise Trained Rats

A strenuous program of treadmill running increased intracellular LPL activity 2.7-fold in soleus and 2.8-fold in fast-red fibers of the quadriceps (Figure 1). At the same time, intracellular TG decreased 25% in the soleus and 44% in fast-red muscle fibers (Oscai et al., 1982). These results suggest that LPL may be the enzyme responsible for intracellular TG hydrolysis in skeletal muscle and that the capacity to hydrolyze intramuscular TG may be increased with exercise training.

LPL Hydrolyzes Endogenous TG in Heart after a 2 H Swim in Untrained Rats

Figure 1 shows that a 2 h swim in untrained rats increased intracellular LPL activity 64% in the myocardium. With elevated levels of endogenous LPL activity in the heart, TG were reduced 49% (Oscai et al., 1982). In

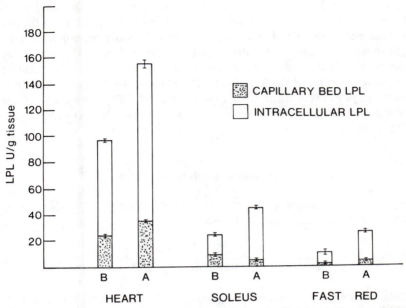

Figure 1—In heart, the effect of a 2 h swim on intracellular lipoprotein lipase activity in untrained rats. In soleus and fast-red fibers of the quadriceps, the effects of regularly performed treadmill running on intracellular lipoprotein lipase activity. B = before exercise. A = after exercise. Rats were sacrificed 24 h after exercise and after an overnight fast. The vertical lines represent twice the Standard Error (SE) from muscle of 6 or 7 rats.

contrast, LPL activity was not elevated in the hearts of rats subjected to a 12 wk long program of treadmill running (Borensztajn et al., 1975a). It now appears that exercise is capable of increasing LPL activity in the heart and that the magnitude of the increase brought about by exercise depends on the intensity of the exercise stimulus. For example, a single bout of high-intensity exercise in untrained rats resulted in a 64% increase in intracellular LPL activity; whereas, a 2 h swim of mild intensity had no effect on enzyme activity in the heart (Oscai et al., 1982)

Sustained Reductions in Heart and Skeletal Muscle
TG Content after Exercise

Endogenous TG content of muscle was measured at least 24 h after exercise in our rats. A prolonged reduction in the TG content of heart and skeletal muscle, that can last about 24 h after exercise, has been reported by other investigators (Carlson & Fröberg, 1969; Fröberg et al., 1972; Scheuer et al., 1970). These results indicate that exercise is associated with a decreased availability of TG to be oxidized by working muscle. The data seem paradoxical to the well-established concepts that the oxidation of fat can provide essentially all of the energy required by working muscle during light to moderate exercise (Holloszy et al., 1978), and that, of the total fatty acid pool utilized during exercise, intramuscular TG stores can contribute a large proportion of the fatty acids (Havel et al., 1964; Issekutz & Paul, 1968). In spite of this, it appears from our data that LPL may be the enzyme responsible for the sustained reductions in intramuscular TG content seen with exercise.

Acknowledgment

This work was supported by National Institute of Health Grants Am-17357, KO4 Am-00216 and HD-10987.

References

BORENSZTAJN, J., Lipoprotein lipase. In A.M. Scanu, R.W. Wissler & G.S. Getz (Eds.), The Biochemistry of Atherosclerosis. New York: Marcek Dekker, Inc., 1979.

BORENSZTAJN, J. & D.S. Robinson. The effect of fasting on the utilization of chylomicron triglyceride fatty acids in relation to clearing factor lipase (lipoprotein lipase) releasable by heparin in the perfused rat heart. J. Lipid Res. 11:111–117, 1970.

BORENSZTAJN, J., M.S. Rone, S.P. Babirak, J.A. McGarr & L.B. Oscai. Effect of exercise on lipoprotein lipase activity in rat heart and skeletal muscle. Am. J. Physiol. 229:394–397, 1975a.

BORENSZTAJN, J., M.S. Rone & T. Sandros. Effects of colchicine and cycloheximide on the functional and non-functional lipoprotein lipase fractions of rat heart. Biochem. Biophys. Acta 398:394–400, 1975b.

CARLSON, L.A. & S.O. Fröberg. Effect of training with exercise on plasma and tissue lipid levels of ageing rats. Gerontologia 15:14–23, 1969.

FRÖBERG, S.O., I. Östman & N.O. Sjöstrand. Effect of training on esterified fatty acids and carnitine in muscle and on lipolysis in adipose tissue in vitro. Acta Physiol. Scand. 86:166–174, 1972.

HAVEL, R.J., L.A. Carlson, L.-G. Ekelund & A. Holmgren. Turnover rate and oxidation of different free fatty acids in man during exercise. J. Appl. Physiol. 19:613–618, 1964.

HOLLOSZY, J.O. & F.W. Booth. Biochemical adaptations to endurance exercise in muscle. Ann. Rev. Physiol. 38:273–291, 1976.

HOLLOSZY, J.O., W.W. Winder, R.H. Fitts, M.J. Rennie, R.C. Hickson & R.K. Conlee. Energy production during exercise. In F. Landry & W.A.R. Orban (Eds.), Regulatory Mechanisms in Metabolism During Exercise. Miami, FL.: Symposia Specialists, 1978.

ISSEKUTZ, B. & P. Paul. Intramuscular energy sources in exercising normal and pancreatectomized dogs. Am. J. Physiol. 215:197–204, 1968.

LECH, J.J., G.J. Jesmok & D.N. Calvert. Effects of drugs and hormones on lipolysis in heart. Federation Proc. 36:2000–2008, 1977.

LUKENS, T.W. & J. Borensztajn. Effects of C apoproteins on the activity of endothelium-bound lipoprotein lipase. Biochem. J. 175:1143–1146, 1978.

OSCAI, L.B., Role of lipoprotein lipase in regulating endogenous triacylglycerols in rat heart. Biochem. Biophys. Res. Commun. 91:227–232, 1979.

OSCAI, L.B., R.A. Caruso & A.C. Wergeles. Lipoprotein lipase hydrolyzes endogenous triacylglycerols in muscle of exercised rats. J. Appl. Physiol. 52:, 1982. (in press)

PALMER, W.K., R.A. Caruso & L.B. Oscai. Possible role of lipoprotein lipase in the regulation of endogenous triacylglycerols in the rat heart. Biochem. J. 198:159–166, 1981.

ROBINSON, D.S., The function of the plasma triglycerides in fatty acid transport. Compr. Biochem. 18:51–116, 1970.

SCHEUER, J., L. Kapner, C.A. Stringfellow, C.L. Armstrong & S. Penpargkul. Glycogen, lipid, and high energy phosphate stores in hearts from conditioned rats. J. Lab. Clin. Med. 75:924–929, 1970.

Alterations in Composition of Venous Plasma FFA Pool during Prolonged and Sprint Exercises in the Horse

D.H. Snow, L.M. Fixter,
M.G. Kerr and C.M.M. Cutmore
University of Glasgow, Glasgow, Scotland

Plasma free fatty acids (FFA) are an important oxidizable substrate during prolonged exercise in all mammalian species examined including the horse while, in very short term, high intensity exercise, it is thought to be of little importance. The purpose of the present study was to investigate changes in plasma FFA in response to prolonged aerobic exercise and very short anaerobic work periods in the horse, a species which has evolved to possess both speed and stamina. In order to avoid making any assumptions about the metabolism of individual FFA, the composition of the plasma pool was determined by Gas Liquid Chromatography (GLC)

Materials and Methods

Animals

Healthy Thoroughbred racehorses were used in all investigations.

Investigations

Study 1. Following a 2 month training period, four horses were exercised over a distance of 80 km or less, if withdrawal because of fatigue was necessary. Full details on this investigation and alterations in blood, sweat, urine and muscle composition have been described (Snow et al., 1982). Blood samples for FFA analysis were taken in their boxes prior to exercise and at 40 and 80 km (or at time of withdrawal).

Study 2. The effects of racing on various blood constituents were determined at race meeting in Great Britain. Blood samples were collected within 2 to 8 min after the completion of the race (mean 5 ± 0.7 SEM). For studying alterations in FFA, blood was collected from eight horses after a 1.2 km race and four horses after a 2.4 km race. Both races were run at a speed of approximately 57 km/hr. As it was impossible to collect blood samples immediately prior to racing, resting samples were collected from a similar group of horses (7) at the same time of day and the same length of time after feeding a similar diet.

Study 3. To overcome the problems of not having immediately preexercise blood samples and the lag time before post exercise sampling, a further investigation was carried out in four horses at a Hong Kong racecourse on a fast workout day. The horses were walked (about 15 min) from their boxes to the racecourse and then galloped at approximately three-quarter pace over a distance of either 800 m (1) or 1,200 m (3), at a speed of approximately 47 km/h. Blood samples were collected from the horses when resting in their boxes prior to being taken out, immediately prior to galloping over the above distances, and within 20 to 30 s of completion of the gallop.

Analytical Methods

Samples for harvesting of plasma were taken into tubes coated with lithium heparin. Prior to centrifugation hematocrit was determined. In study 2 blood samples for lactate were deproteinized with trichloroacetic acid (TCA), and samples were collected anaerobically for blood pH determination.

Samples of plasma for FFA analysis were extracted by the method of Folch, Leco, and Sloan Stanley (1957) after the addition of known amounts of heptadecanedioic acid or an internal standard. FFA were isolated by Thin Layer Chromatography (TLC) transesterified using 2% v/v sulphuric methanol and analyzed by GLC. Glycerol, lactate, glucose and venous pH were determined using conventional techniques.

Results

The effect of prolonged exercise on plasma FFA and some other parameters are shown in Table 1. The total FFA concentration increased three-fold by the finish and there was a significant change in composition which was particularly obvious in the content of myristic, stearic and oleic acids. As there appeared to be no difference in changes in total or pattern of FFA composition after 1.2 or 2.4 km races, the results have been

Table 1

Percent Composition of Major Venous FFAs at Rest, After 40 km and at Finish of Exercise

Fatty Acid		Rest (n = 8)	40 km [+](n = 6)	Finish [+](n = 6)	% difference Rest v fin.
Myristic	(C14:0)	4.1 ± 1.30	3.5 ± 0.2	2.8 ± 0.2	− 32
Myristoleic	(C14:1)	1.8 ± 0.5	1.9 ± 0.3	1.3 ± 0.2	− 28
Palmitic	(C16:0)	26.6 ± 2.1	28.4 ± 0.6	28.7 ± 1.4	+ 8
Stearic	(C18:0)	19.2 ± 1.6	11.53 ± 0.62[xx]	10.2 ± 1.1[xx]	− 47
Oleic	(C18:1)	23.0 ± 2.0	30.0 ± 1.3[xx]	33.8 ± 1.0[xx]	+ 47
Linoleic	(C18:2)	17.1 ± 2.6	16.4 ± 0.5	16.3 ± 0.9	− 5
Linolenic	(C18:3)	6.4 ± 1.6	5.9 ± 0.5	6.3 ± 0.3	− 3
Total FFA	(μEq/l)	336 ± 36	897 ± 88[xxx]	1370 ± 192[xxx]	+ 307
Glycerol	(μmol/l)	30 ± 6	314 ± 18[xxx]	995 ± 83[xxx]	
Hematocrit	(l/l)	0.38 ± 0.01	0.55 ± 0.04[xxx]	0.62 ± 0.01[xxx]	

Mean ± SEM
+ FFA concentrations only determined for those horses that completed more than 64 km.
[xxx] $p < .001$
[xx] $p < .01$

considered together in Table 2. Following exercise there is no significant change in total FFA but there are again large changes in FFA composition with a reduction in percentage of myristic and increased oleic and stearic acids. The results of the third study are shown in Table 3. In this case it should be noted that in three of the four horses an elevation of total plasma FFA from rest to the pre gallop situation, followed by a reduction immediately after the gallop, occurred. There are again changes in composition with increased levels of 18:1 and 18:2.

In all studies, a very marked elevation in plasma glycerol levels occurred with exercise. The marked decrease in plasma glucose and rise in hematocrit are indicative of the severe demands on the horses in the prolonged exercise. The increases in plasma glucose, PCV and blood lactate are indicative of the maximal effort involved in racing.

Discussion

The progressive increases in total plasma FFA and glycerol are in agreement with previous studies on endurance rides in horses (Lucke & Hall, 1978, 1980). The changes in the composition of the plasma FFA pool

Table 2

Percent of Composition of Venous FFAs in Resting Thoroughbred Racehorses (7) and after Racing 1.2 (8) or 2.4 km (4)

		Rest	Post-racing	% Difference
Myristic	(C14:0)	26.9 ± 4.4	5.5 ± 1.5	– 80
Myristoleic	(C14:1)	tr	1.5 ± 0.4	
Palmitic	(C16:0)	36.1 ± 1.9	33.3 ± 1.7	– 8
Stearic	(C18:0)	8.4 ± 1.3	16.6 ± 1.1	+ 98
Oleic	(C18:1)	14.2 ± 2.3	25.9 ± 1.5	+ 82
Linoleic	(C18:2)	13.5 ± 1.4	16.8 ± 1.1	+ 24
Linolenic	(C18:3)	Trace	Trace	—
Total FFA Eq/1		202 ± 25	251 ± 25	+ 24
Glycerol mol/1		28 ± 4	712 ± 73	
Lactate mmol/1		0.2 ± 0.03	25.6 ± 1.3	
pH		—	7.021 ± 0.023	
Glucose mmol/1		5.5 ± 0.1	9.2 ± 0.07	
Hematocrit 1/1		0.43 ± 0.02	0.61 ± 0.01	

Mean ± SEM

are largely due to increased lipolysis, increasing the influx of fatty acids from adipose tissue triglycerides. At intermediate and final stages of the exercise period, plasma FFA composition is similar to that reported in equine adipose tissue (Robb et al., 1972) containing relatively fewer fatty acids shorter than palmitic and a higher ratio of unsaturated to saturated 18C species. Any selectivity in fatty acid oxidation or possible re-esterification will also have affected plasma FFA composition. In humans, Havel et al. (1964) also found a change in the composition of the circulating FFA pool during prolonged exercise. However, in contrast to Havel et al. (1964), an opposite effect was seen for myristic acid, although the decrease was similar for stearic acid. Obviously, fatty acid in adipose tissue and dietary fatty acid composition will play a major role in determining the extent and direction of the changes.

In study 2, in short term maximal exercise, although there was little change in total plasma FFA, glycerol was increased. As there were large changes in the plasma FFA composition, indicating an influx of adipose tissue fatty acids, the increase in glycerol was not a reflection of a futile cycle between FFA and triglyceride (TG) inside adipocytes. The changes in composition and individual absolute concentrations indicate that in this type of exercise a process is operating with selectivity different from the resting state. This is likely to be a result of fatty acid oxidation, as selective

Table 3

Changes in Total Plasma FFA and Percent Composition Following a Short Gallop (Average Speed 47 km/h) in Four Horses

Horse	Distance (m)	Sample	Fatty acid composition (moles %)							Total FFA μEq/l	Glycerol μmol/l	Hematocrit 1/1
			12:0	14:0	14:1	16:0	18:0	18:1	18:2			
1	800	Rest	3.4	5.3	2.6	25.1	16.1	22.9	24.1	117	5	0.47
		Pre	0.5	3.2	0.5	34.6	8.6	27.7	24.8	398	26	0.59
		Post	19.5	3.0	1.1	27.8	5.8	23.8	18.9	287	191	0.59
2	1200	Rest	tr	4.3	0.1	37.9	14.6	21.7	21.3	210	20	0.43
		Pre	tr	2.5	1.2	34.8	5.3	27.2	29.0	230	35	0.51
		Post	0.6	2.6	1.2	40.3	20.4	20.6	14.3	225	161	0.62
3	1200	Rest	tr	1.5	0.9	30.5	13.4	27.6	25.9	215	20	0.38
		Pre	tr	2.4	1.7	31.0	5.8	31.8	27.0	476	53	0.57
		Post	tr	1.5	1.2	34.5	8.9	33.9	19.8	304	250	0.65
4	1200	Rest	tr	3.9	1.1	48.4	29.1	9.6	8.0	314	9	0.40
		Pre	tr	0.7	0.6	32.3	32.8	21.9	21.9	427	53	0.54
		Post	tr	1.0	2.4	25.2	28.2	21.9	21.2	307	232	0.65

fatty acid esterification in both liver and adipose tissue, interacting with dietary input, would be expected to produce the resting plasma FFA composition.

In the third study, the different resting FFA composition may be a reflection of a different dietary regimen to the horses. The mean elevation in three of the four horses, following a short walk to the gallop track, indicates the ready mobilization of fatty acids in the horse. The decrease immediately following exercise, together with changes in composition and marked elevations in plasma glycerol, provides further evidence for fatty acid utilization during short periods at high workloads.

Although the evidence presented for fatty acid oxidation during both prolonged submaximal and highly anaerobic exercise in the horse is indirect, it is relevant to consider differences in response between horses and humans to short term maximal exercise. The horse is capable of increasing its Vo_2max by more than 35-fold (Thomas & Fregin, 1981) which is two times greater than that generally seen in man. This is largely due to the "warm-blooded" equine breeds' capacity to increase circulating red blood cells by 50%, and also possibly due to a greater O_2 extraction (Thomas & Fregin, 1981). In addition it appears that a number of adrenergically mediated responses are very sensitive to increasing concentrations of circulating catecholamines in the Thoroughbred horse. This latter effect is likely to bring about even greater pre-race increases in circulating FFA's, as horses are paraded and exercised prior to the actual race. Therefore the increased capacity for O_2 uptake, together with the possibility of at least a doubling in plasma FFA, argues in favor of increased FFA utilization in short term exercise. Furthermore, extrapolation from Respiratory Quotient (RQ) and oxygen uptake data in treadmill exercise of horses (Thomas & Fregin, 1981) indicates a four-fold increase over resting free fatty acid utilization, although the percentage of fatty acid oxidation has been reduced from 50% to 14% of the total.

Therefore this preliminary study suggests that although high blood lactate levels occur during maximal or near maximal galloping, the prior elevation of free fatty acids due to pre-race warm-ups permit them to be oxidized to a significant degree.

Acknowledgments

This work was financed by the Horserace Betting Levy Board. Studies in Hong Kong were supported by a Carnegie Trust for the Universities of Scotland Travel Grant. M.G. Kerr was a Horserace Betting Levy Board Research Student and C.M.M. Cutmore was a Wellcome Trust Vacation Scholar. Permission of the Jockey Club and Royal Hong Kong Jockey Club and trainers to carry out these studies is gratefully acknowledged.

References

FOLCH, J., H. Lees, S.G.H. Sloan-Stanley. A simple method for the isolation and purification of total lipid from animal tissues. J. Biol. Chem. 226:497–509, 1957.

HAVEL, R.J., L.A. Carlson, L-G. Ekelund, & A. Holmgren. Turnover rate and oxidation of different free fatty acids in man during exercise. J. Appl. Physiol. 19:613–618, 1964.

LUCKE, J.N. & G.M. Hall. Biochemical changes in horses during a 50 mile endurance ride. Vet. Rec. 102:350–358, 1978.

LUCKE, J.N. & G.M. Hall. Long distance exercise in the horse. Golden Horseshoe Ride. Vet. Rec. 106:405–407, 1980.

ROBB, J., R.B. Harper, H.F. Hintz, J.T. Reid, J.E. Lower, H.F. Schryver & M.S.S. Rhee. Chemical composition and energy value of the body, fatty acid composition of adipose tissue and liver and kidney size in the horse. Anim. Prod. 14:25–34, 1972.

SNOW, D.H., M.G. Kerr, M.A. Nimmo & E.M. Abbott. Alterations in blood, sweat, urine and muscle composition during prolonged exercise in the horse. Vet. Rec. 110:377–384, 1982.

THOMAS, D.P. & G.F. Fregin. Cardiorespiratory and metabolic responses to treadmill exercise in the horse. J. Appl. Physiol. 50:864–868, 1981.

Protein
Metabolism

Blood Ammonia and Glutamine Accumulation and Leucine Oxidation during Exercise

P. Babij, S.M. Matthews, S.E. Wolman, D. Halliday, D.J. Millward, D.E. Matthews and M.J. Rennie
University College London Medical School, England, London School of Hygiene and Tropical Medicine, London, England, Northwick Park Hospital, Harrow, England, and Washington University School of Medicine, St. Louis, Missouri, U.S.A.

Exercise causes a marked increase in leucine oxidation in humans (Rennie et al., 1981). It appears that the carbon chain of leucine may be completely oxidized during exercise but we have no information about the fate of the amino group from leucine or other branched chain amino acids during exercise. We have therefore examined the effects of exercise on blood ammonia and glutamine, which are mainly produced by skeletal muscle, and the relationship of the changes observed to leucine oxidation during exercise.

Methods

The studies described here were approved by the Ethical Committee of University College London School of Medicine. Twelve healthy, normal male volunteers were studied after an overnight fast. For measurements of blood amino acids and ammonia, eight of the subjects exercised continuously on a bicycle ergometer at workloads equivalent to 25, 50, 75 and 100% of previously determined $\dot{V}o_2max$. For measurement of leucine oxidation four subjects exercised on a bicycle ergometer at workloads of 25, 50 and 75% of $\dot{V}o_2max$ during the infusion of $1[^{13}C]$-leucine at rates between 4.5 and 7.6 μmol/kg/h. The exercise was carried out after labelling of blood leucine to a plateau value. In both studies blood was sampled from an antecubital vein at rest and during exercise and at intervals for 30 min post exercise. Respiratory measurements were made by conventional techniques. Expired $^{13}CO_2$ was collected from a mixing chamber connected to the output side of the respiratory mouthpiece and samples stored in previously evacuated glass vials. Blood samples were analyzed by standard enzymatic techniques for ammonia and lactate and plasma amino acids were analyzed using a Rank-Hilger amino acid analyzer. Gas chromatographic-mass spectrometric methods of analysis were used to measure the enrichment of blood leucine (Matthews, Ben-Galim & Bier, 1979) and α-ketoisocaproate (αKIC) (Schwarz, Karl & Bier, 1980) and the enrichment of expired CO_2 with ^{13}C was measured using isotope ratio mass spectrometry (Halliday & Read, 1981).

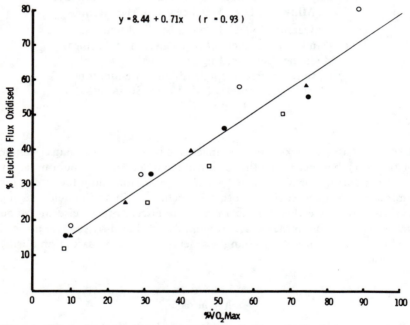

Figure 1—Effect of exercise of increasing intensity on leucine oxidation in four normal men. Each point represents mean of three values taken over 30 min at each workload.

Results

Leucine oxidation appeared to be linearly related to metabolic rate between 8 and 89% of $\dot{V}o_2$max (Figure 1). We have chosen to calculate leucine flux on the basis of the αKIC-labelling since αKIC is produced mainly in muscle and therefore its labelling should give a better indication

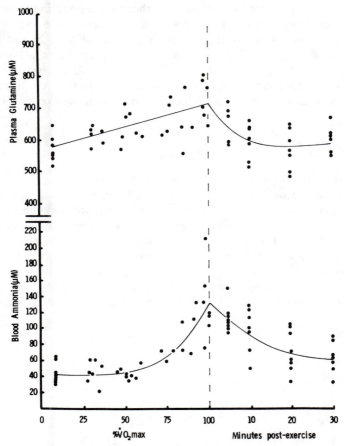

Figure 2—Blood ammonia and glutamine concentrations measured in eight normal subjects during exercise and recovery. Curve for ammonia best approximated an exponential, y = $30.9e^{0.012x}$ ($r^2 = 0.57$) where coefficient of determination (r^2) for individual subjects was 0.71. Relationship between blood glutamine and exercise intensity was linear, y = $564 + 1.45x$ ($r^2 = 0.44$) where r^2 for individual subjects was 0.78.

of whole-body leucine turnover during exercise. The results suggested that at 100% of $\dot{V}_{O_2max}$, leucine oxidation would account for approximately 80% of whole-body leucine flux determined αKIC-labelling. When leucine flux was calculated on the basis of αKIC-labelling, there was no significant change as a result of exercise. However, leucine flux calculated on the basis of leucine labeling showed an apparently artifactual fall which was progressive during exercise of increasing power output because of an increase in the labeling of blood leucine.

During exercise, blood ammonia concentrations did not change at power outputs up to 50% of $\dot{V}O_{2max}$, but there was a doubling between 50 and 75% of $\dot{V}O_{2}max$ with a further 60% increase between 75 and 100% of $\dot{V}O_{2}max$ (Figure 2). In contrast, blood glutamine (Figure 2) and blood alanine concentrations (Figure 3) increased in a quasi-linear fashion with respect to aerobic power output during exercise and the extent of the change was much less than that for ammonia. Blood lactate concentrations showed a linear relationship with blood ammonia during exercise (Figure 4).

Discussion

One of the aims of the present study was to examine possible relationships between leucine oxidation and blood glutamine, alanine and ammonia concentrations in order to gain some insight into possible links between the pathways of their metabolism. In particular, we wanted to examine the extent of processes which may be important for the metabolism of the α-amino group from branched chain amino acids. The immediate fate of α-amino nitrogen from the branched chain amino acids is to be trans-aminated to produce glutamate. The fate of glutamate, however, is not clear. We have observed a fall in intramuscular glutamate concentrations

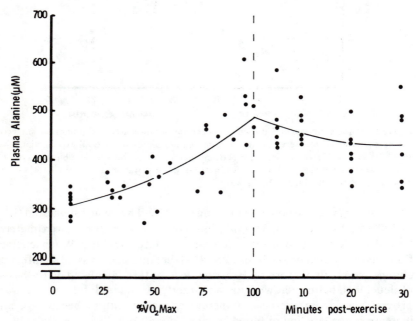

Figure 3—Blood alanine concentrations during exercise and recovery for seven normal subjects. Curve approximated to an exponential, $y = 287e^{0.005x}$ ($r^2 = 0.65$) where r^2 for individual subjects was 0.85.

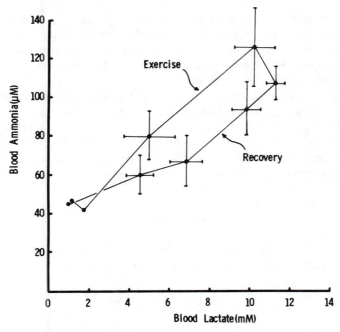

Figure 4—Relationship between blood ammonia and lactate during exercise and recovery for eight normal subjects, where $y = 35.6 + 8.37x (r^2 = 0.66)$ during exercise and r^2 for individual subjects was 0.87.

during exercise in man (Rennie et al., 1981) and there are a number of possible explanations for this. An obvious fate for the α-amino N of glutamate is transamination with pyruvate to form alanine, and the production of alanine by muscle is well documented. Another possibility is that glutamate is transaminated to produce aspartate (thus replenishing the α-ketoglutarate used to transaminate the branched chain amino acids) which then enters the purine nucleotide cycle, eventually to emerge as fumarate and ammonia. Since fumarate may enter the Krebs cycle, the steady state concentrations of Krebs cycle intermediates would thus be maintained during exercise (Lowenstein, 1972). Another possibility is that glutamate is amidated to produce glutamine which then leaves the muscle. The advantage of this for the organism would be in the associated net export of protons accumulated during exercise (Jones, 1980.) One problem here is to identify a source of ammonia for the glutamine synthetase reaction given the low activity of glutamate dehydrogenase in muscle (Lowenstein, 1972). The present results suggest that glutamine production occurs in direct proportion to metabolic rate during exercise whereas ammonia production by muscle only increases in intense exercise.

The pattern of metabolic change we have observed strongly suggests that there is little relationship between ammonia production and branched chain amino acid oxidation at any workload. The similarity in the pattern of change of lactate and ammonia during exercise suggests that the production of both of these metabolites is linked to mechanisms involved in rapid short-term energy provision.

There is, however, a strong relationship between the increases in blood alanine and glutamine and the extent of branched chain amino acid oxidation which all show a virtually linear relationship with metabolic rate during exercise. It therefore seems most likely that the fate of α-amino N derived from branched chain amino acid oxidation by muscle is linked to the production of alanine and glutamine by muscle.

Acknowledgments

This work was supported by The Wellcome Trust, Medical Research Council, Action Research and Abbott Laboratories Inc.

References

HALLIDAY, D. & W.W.C. Read. Mass spectrometric assay of stable isotopic enrichment for the estimation of protein turnover in man. The Proceedings of the Nutrition Society 40:321–334, 1981.

JONES, N.L., Hydrogen ion balance during exercise. Clinical Science 59:85–91, 1980.

LOWENSTEIN, J.M., Ammonia production in muscle and other tissues: the Purine Nucleotide Cycle. Physiological Reviews 52:2, 382–414, 1972.

MATTHEWS, D.E., E. Ben-Galim & D.M. Bier. Determination of stable isotope enrichment in individual amino acids by clinical ionisation mass spectrometry. Analytical Chemistry 51:80–88, 1979.

RENNIE, M.J., R.H.T. Edwards, S. Krywawych, C.T.M. Davies, D. Halliday, J.C. Waterlow & D.J. Millward. Effect of exercise on protein turnover in man. Clinical Science 61:627–639, 1981.

SCHWARZ, H.P., I.E. Karl & D.M. Bier. The α-keto acids of branched chain amino acids. Simplified derivatisation for physiological samples and complete separation as quinoxalinols by packed column gas chromatography. Analytical Biochemistry 108:360–366, 1980.

Potential Biochemical Basis of Muscle Atrophy during Prolonged Weightlessness

N.M. Cintrón-Treviño, C.S. Leach and P.C. Rambaut
National Aeronautics and Space Administration,
Johnson Space Center, Houston, Texas, U.S.A.

The biomedical results from the Skylab missions provided evidence that prolonged exposure to weightlessness induces muscle atrophy. The quantitative assessment of these alterations has been difficult due to the variability in the exercise and dietary regimens of each crew member from mission to mission and to the scarcity of subsequent space flight data in humans. Investigation into the underlying mechanisms and remedial treatments to the muscle imbalance has relied principally on human and animal research in experimental hypokinesia with the ground-based analog of disuse atrophy. The purpose of the present report is to compile and analyze relevant data accumulated to date from manned space flight and to correlate it with that from ground-based and corroborative biosatellite animal studies. Although abundant information exists regarding alterations in specific cellular components and processes, there is as yet no unifying hypothesis as to the events operative in the onset and progression of the atrophy process.

Effects of Weightlessness on Muscle

Insufficient loading of the muscle system in weightlessness, as in experimental hypokinesia, has been shown to induce a number of physiological and metabolic alterations which demonstrate the progressive development of muscle atrophy (Table 1). From a functional standpoint, this was first supported from the early observations in humans that significant decreases in muscle tone, muscle strength, tolerance, and physical work capacity occurred during the inflight and immediate postflight periods of a mission

Table 1

Summary of Supporting Evidence Indicative of Muscle Atrophy Development during Space Flight

	Morphological/physiological	Metabolic/biochemical
Humans	Decreases in: lean body mass leg volume & girth muscle strength & tone muscle endurance	Negative nitrogen balance Negative phosphorous balance Decrease in exchangeable body potassium Increase in excretion of: N-methylhistidine creatinine sarcosine
Animals	Decreases in: muscle mass muscle strength & elasticity muscle tolerance to fatigue muscle work capacity	Change in LDH isoenzyme spectrum: LDH-C→LDH-M Decreases in: myofiber area & cross-section sarcomer volume myofibrillar/sarcoplasmic proteins muscle synaptic vesicles and motorneuron function aerobic substrate oxidation

(Thornton & Rummel, 1976). The biomedical results from the Skylab program provided evidence suggesting that this muscular deconditioning was due in large part to an increased turnover and degradation of muscle tissue. In strong support of this proposal was the finding that the six crewmembers of the first two Skylab missions maintained a significant negative nitrogen balance inflight amounting to approximately -4.5 g/day while concomitantly experiencing a decrease in total body potassium (Whedon et al., 1976). That a large portion of this increased protein catabolism observed during space flight was specifically due to an increased net degradation of the contractile proteins of skeletal muscle was additionally supported by the elevated rates of creatinine, sarcosine, and 3-methylhistidine excretion during the period of weightlessness (Leach et al., 1978). The results although qualitatively similar to bedrest, exhibited a greater degree and a more continuous pattern of muscle loss during the inflight phase.

Corroborative data on experimental animals from the Cosmos Bio-satellite series have demonstrated further the development of muscle atrophy with prolonged exposure to weightlessness (Table 1). Moreover, they have provided for the first time insight into which cellular/biochemical processes potentially underlie the gross alterations in muscle function observed at the physiological level (Gazenko et al., 1980). As in humans, muscle mass loss, decreases in muscle strength and work capacity, and a decrement in muscle tolerance to fatigue have been observed in animals after space flight. These changes were manifested at the cell structural and metabolic levels by significant deterioration of myofibril, sarcomere and motorneuron structure and by decreases in the content of contractile proteins and in the activity of key energy-generating processes. Similar to reports from ground-based immobilization studies in both animals and humans, these changes occurred predominantly in muscles normally functional in antigravitational and postural activities (e.g., soleus).

Despite the available data accumulated to date, the elucidation of the mechanisms responsible for the space flight-induced deterioration of muscle remains unknown. This has principally been due to the limitations inherent in space flight investigations which have made it difficult to impose stringent controls over critical experimental variables such as spacecraft environmental conditions, nutrition, and efficiency of postflight sample collection.

Potential Mechanisms Operative in Atrophy

The regulation of the multiple cellular processes involved in the main-tenance of muscle integrity appear to be mediated by neural mechanisms involving nerve impulse activity and neurotrophic actions and by non-neural influences imposed by hormones and other humoral agents. The significant curtailment of the atrophy process experienced in the Skylab 4 crew after an intensive inflight exercise regiment indicated that, similar to the one-g situation, contractile activity represents a major factor determin-ing the dynamic properties of muscle in space (Thornton & Rummel, 1977). However, the fact that muscle deterioration was not completely eliminated in lieu of vigorous activity suggested that other control mechanisms were altered during weightlessness as well. Regardless of the external insult, the biochemical description of the cellular events active in the initial onset of atrophy remains central to the understanding of the etiology and progression of this metabolic imbalance. Using limb immobilization of rodents as a working model, ground-based research has generated evidence that early changes occur in such key processes as protein turnover, energy utilization, and contractile function and regulation. Decreases in protein synthesis rates have been recorded as early as 6 h after immobilization and as such represent the earliest alteration observed during these initial phases of disuse (Tucker et al., 1981). Despite these results, conflicting evidence still exists regarding the relative contribution and time sequence of change for the rates of synthesis and degradation in the overall increase of protein turnover.

The diminished load upon the muscular system during weightlessness or immobilization not only induces direct changes in protein metabolism which lead to muscle mass loss, but also imposes a reduced level of energy requirement to the system. This diminished energy need serves as a new homeostatic set point for which the body must adapt. It is not surprising then that major changes have also been observed in the normal energy-producing activities of muscle cells. A consistent finding under hypokinetic conditions has been a decrease in the rates of both glycolytic and oxidative enzyme activities with an apparent intensification of the relative contribution of glycolysis to the cellular energy supply (Edes et al., 1980). Similar results have also been obtained from biosatellite animal studies (Gazenko et al., 1980). This system response towards diminishing its bioenergetic capability no doubt contributes to the development of the protein metabolic imbalance and to the degeneration of muscle metabolism.

The principal hormonal controls which directly influence the rates of protein turnover in skeletal muscle and which are responsive to the overall homeostatic energy balance in a variety of diseased and altered body states are exerted by insulin, glucocorticoids, thyroid hormone, and growth hormone (Goldberg et al., 1980). During Skylab, significant elevations in plasma and urinary cortisol were observed inflight concomitant with decreases in circulating insulin concentrations (Leach & Rambaut, 1975). Interpretation of the isolated effects of these hormonal changes during space flight is difficult in light of the multiple physiological and hormonal alterations that occur in apparently indiscriminant directions and degrees. However, corroborative space flight data obtained suggests that cortisol, normally antagonized by insulin and proper nutrition, may have exerted metabolic effects on protein metabolism as evidenced by increases in plasma proteins, in blood and urinary amino acids and by the loss of muscle mass despite vigorous exercise. The low levels of insulin and the observed elevations of thyroid hormone upon recovery might also reflect additional controls of undefined nature operative in space flight.

What specific controls are altered during weightlessness and to what extent would they ultimately impact the integrity of muscle tissue if left unattended is at this time unclear. It appears, however, that information on the early biochemical changes, the regulation of cellular bioenergetics, and the combined effect of key hormones will provide a reliable framework from which to predict confidence limits of adaptation and reveal areas of further research.

References

EDES, I., I. Sohar, H. Mazarean, O. Takacs & F. Guba. Immobilization effects upon aerobic and anaerobic metabolism of the skeletal muscles. The Physiologist 23:S103–104, 1980.

GAZENKO, O.G., A.M. Genin, E.A. Ilyin, V.S. Oganov & L.V. Serova. Adaptation to weightlessness and its physiological mechanisms (Results of animal experiments aboard biosatellites). The Physiologist 23:S11–15, 1980.

GOLDBERG, A.L., M. Tischler, G. DeMartino & G. Griffin. Hormonal regulation of protein degradation and synthesis in skeletal muscle. Federation Proc. 39:31–36, 1980.

LEACH, C.S. & P.C. Rambaut. Biochemical observations of long duration manned orbital spaceflight. J. Am. Med. Women's Assoc. 30:153–166, 1975.

LEACH, C.S., P.C. Rambaut & N. Diferrante. Amino aciduria in weightlessness. Astronautica 6:1323–1333, 1979.

THORNTON, W.E. & J.A. Rummel. Muscular deconditioning and its prevention in space flight. In R.S. Johnston & L.F. Dietlein (Eds.), Biomedical Results from Skylab (NASA SP-377). Washington D.C., 1977.

TUCKER, K.R., M.J. Seider & F.W. Booth. Protein synthesis rates in atrophied gastrocnemius muscles after limb immobilization. J. Appl. Physiol. 51:73–77, 1981.

WHEDON, G.D., L. Lutwak, P. Rambaut, M.W. Whittle, M.C. Smith, J. Reid, C.S. Leach, C.R. Stadler & D.D. Sanford. Mineral and nitrogen metabolic studies. In R.S. Johnston & L.F. Dietlein (Eds.), Biomedical Results of Skylab (NASA SP-377). Washington D.C., 1977.

Acute Phase Proteins and Immune Complexes during Several Days of Severe Physical Exercise

B. Dufaux, K. Höffken, and W. Hollman

Institut fur Freislaufforschung und Sportmedizin, Köln,
Federal Republic of Germany, and Universitätsklinikum
Essen, Essen, Federal Republic of Germany

Intense and prolonged physical exercise is accompanied by a delayed elevation of acute phase plasma proteins (Liesen et al., 1977). This increase suggests an exercise-induced inflammatory reaction possibly as a consequence of a nonspecific mechanical tissue damage.

It has been demonstrated several times that low levels of immune complexes (IC) are detectable in the serum of healthy subjects (Hay et al., 1976; Morgan et al., 1979; Paganelli et al., 1981) and the possibility has been raised that IC might be formed secondary to tissue injury (Cooper et al., 1981; Farrell et al., 1977).

The purpose of the present study was to examine the serum concentrations of C-reactive protein (CRP) as a particularly sensitive indicator of an acute phase reaction and of circulating IC before, during and after 4 days of severe physical exercise.

Methods

Twenty-seven healthy male subjects (age 23 to 45, mean: 34 yr) ran 25 km a day (mean speed: 14.1 ± 1.2 km/h) for 4 days. During a preceding graded bicycle ergometer test the work load was determined at which the blood lactate reached a level of 4 mmol/l and was found to be 231 ± 35 (mean $\pm$ SD) W or 3.20 ± 0.51 (mean $\pm$ SD) W/kg. Blood samples were drawn before, after 3 days of running and 1 day after the end of the race. Serum was kept for 4 to 8 wk at $-60°$ C and was thawed once immediately before performing the IC assays. CRP, Immunoglobulin G (IgG) and the complement factors C4 and C3 were measured by an enzyme immunoassay according to the method of MacDonald et al. (1978). Monospecific antisera (Behring-Werke/Marburg) were used to coat microtiter plates and were labelled with horseradish peroxidase according to Nakane et al. (1974).

Circulating IC were assessed by two methods. A fluid phase C1q binding test was performed according to Zubler et al. (1976). In the C1q binding assay aggregated IgG (AHG) was used as a standard. The second method consisted of a polyethylene glycol (PEG, MW 6000) (Sigma) (35 g/l) precipitation of IC according to Digeon et al. (1977). The concentrations of IgG, C4 and C3 of the redissolved precipitates were determined by an enzyme immuno-assay. As a standard, freshly obtained normal human serum was incubated with various amounts of AHG for 30 min at 37° C and the mixtures were processed with PEG in the same way as the test samples. Statistical analysis of the results was performed by the Wilcoxon test considering $p < 0.05$ being significant.

Results

As shown in Figure 1 serum CRP was significantly elevated and serum total protein was significantly decreased on the third day of exercise as well as after the end of the race compared with the pre-race values. Evaluating the effective concentration changes of the serum components, it is important to consider the decrease of serum total protein during the 4 day race (Figures 1, 2, 3, 4). Serum C4 and C3 values were elevated on the third day and after the end of the race when compared with the values of serum total protein (Figure 3). Relative to serum total protein the results of the C1q binding assay did not demonstrate changes during or after the race (Figure 2). On the third day of the race IgG, C4 and C3 in the PEG precipitate were elevated compared with the pre-race values. At the end of the race C4 and C3 in the PEG precipitates were slightly below the pre-race levels (Figure 4).

Discussion

The present study confirms and extends our previous observation of an exercise induced elevation of CRP (Liesen et al., 1977). The highest level was found after 3 days of exercise and in spite of the maintenance of the same exercise intensity during the following days, CRP values decreased at the end of the race (Figure 1). This suggests a physiological adaptation that limits the inflammatory reaction after several days of severe exercise. The

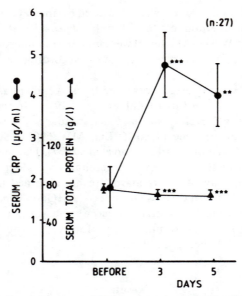

Figure 1—Serum C-reactive protein (CRP) and total protein (mean ± SD) in 27 subjects before, during (3 days) and after (5 days) a 4 day footrace. Significant differences between pre-race and 3 and 5 days value, respectively, are indicated (** = $p < 0.01$, *** = $p < 0.001$).

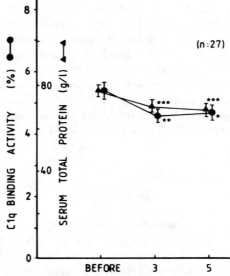

Figure 2—C1q binding assay of immune complexes (mean ± SE) as well as serum total protein (mean ± SE) in 27 subjects before, during (3 days) and after (5 days) a 4 day footrace. Significant differences between pre-race and 3 and 5 days values, respectively, are indicated (* = $p < 0.05$, ** = $p < 0.01$, *** = $p < 0.001$).

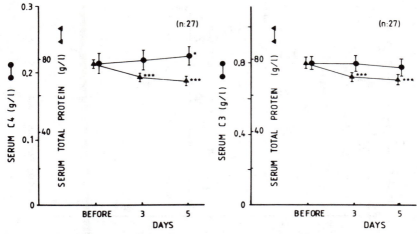

Figure 3—Serum concentrations of C4 and C3 (mean ± SE) measured by enzyme immunoassay and serum total protein (mean ± SE) in 27 subjects before, during (3 days) and after (5 days) a 4 day footrace. Significant differences between pre-race and 3 and 5 days values, respectively, are indicated (* = p < 0.05, *** = p < 0.001).

higher serum levels of C3 and C4 during and after the race, compared with serum total protein, provide further evidence for an exercise induced acute phase reaction (Figure 3). The size and molecular composition of ICs, which are influenced by, e.g., the concentration of antigen and antibody, the antibody class and the size and valence of the antigen, are important determinants of their detection by various methods. The C1q binding assay mainly quantifies complexes larger than 19 S in size and is particularly sensitive in measuring ICs formed at a near-equivalence antigen-antibody ratio. ICs of insufficient size may remain undetected by the test.

The PEG technique precipitates out of solution ICs of any immuno-globulin class and is capable of detecting antigen excess or antibody excess complexes. It has been shown repeatedly that the PEG precipitation method detects immune complexes as readily as the C1q binding assays (Digeon et al., 1977; Füst et al., 1980; Stevens et al., 1981). The sensitivity for heat-aggregated IgG (AHG), however, is low (Kilpatrick et al., 1981) and accounts for the high values of AHG equivalents found in our subjects. The observed changes of C4 and C3 in the PEG precipitates are not likely to be due to changes in serum C4 or C3 levels since the increase in those serum proteins on the third day of exercise was much less pronounced than that of C4 and C3 in the PEG precipitates.

Furthermore, the elevation of C3 and C4 serum concentration persisted beyond the end of the race whereas C4 and C3 in the PEG precipitates decreased (Figures 3 and 4). In addition, separation of redissolved PEG precipitates on a Sephacryl S-200 column resulted in the detection of IgG

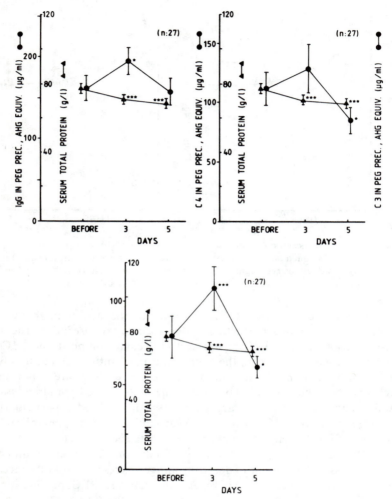

Figure 4—IgG, C4 and C3 (mean ± SE) measured by enzyme immunoassay in polyethylene glycol 6000 (35 g/l) precipitate and serum total protein (mean ± SE) in 27 subjects before, during (3 days) and after (5 days) a 4 day footrace. Significant differences between pre-race and 3 and 5 days values, respectively, are indicated (* = p < 0.05, *** = p < 0.001).

and C_4 almost exclusively in the height molecular weight region with little contaminating monomeric or free IgG and C_4 (Figure 5).

In summary, the increased content of IgG, C4 and C3 in the PEG precipitates reflects an elevation of IC levels during the first few days of severe exercise with a return to normal or subnormal levels afterwards. The absence of detectable IC changes in the C1q binding assay may have been caused by the size and composition of the IC involved.

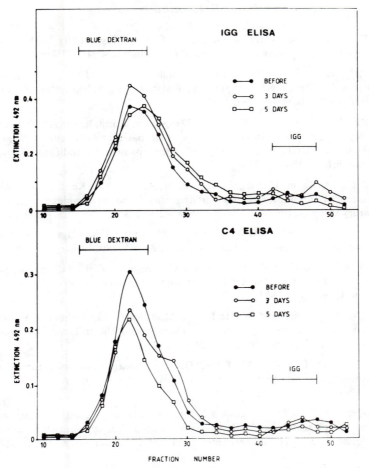

Figure 5—Redissolved PEG precipitates, obtained before, during (3 days) and after (5 days) a 4 day footrace, of one subject applied to a Sephacryl S-200 column (30 × 0.9 cm) and eluted with PBS (4 ml/h). IgG and C_4 were measured by enzyme immunoassay (ELISA) in the fractions (0.35 ml). The column was precalibrated with molecular wt markers consisting of blue dextran and IgG.

Acknowledgment

The study was supported by the Bundesinstitut fur Sportwissenschaft, Köln.

References

COOPER, K.M., M. Moore & A. Hilton. Clq binding activity in the sera of patients with chronic lung disease. Clin. Exp. Immunol. 45:18–28, 1981.

DIGEON, M., M. Laver, J. Riza & J.F. Bach. Detection of circulating immune complexes in human sera by simplified assays with polyethylene glycol. J. Immunol. Methods 16:165–183, 1977.

FARRELL, C., B. Bloth, H. Nielsen, H. Daugharty, T. Lundmann & S.E. Svehag. A survey for circulating immune complexes in patients with acute myocardial infarction: use of a Clq-binding assay with soluble protein A as indicator. Scand. J. Immunol. 6:1233–1240, 1977.

FÜST, G., M. Kavai, G. Szegedi, K. Merety, A. Falus, A. Kenkey & M. Misz. Evaluation of different methods for detecting immune complexes. An inter-laboratory study. J. Immunol. Meth. 38:281–289, 1980.

HAY, F.C., L.J. Nineham, G. Torrigiani & I.M. Roitt. 'Hidden' IgG antiglobulins in normal human serum. Clin. Exp. Immunol. 25:185–190, 1976.

KILPATRICK, D.C., J. Weston & W.J. Irvine. Some studies on the polyethylene glycol turbidity method for detecting immune complexes in serum. Ann. Clin. Biochem. 18:373–377, 1981.

LIESEN, H., B. Dufaux & W. Hollmann. Modifications of serum glycoproteins the days following prolonged physical exercise and the influence of physical training. Europ. J. Appl. Physiol. 37:243–25, 1977.

MacDONALD, D.J. & A.M. Kelly. The rapid quantitation of serum alpha-fetoprotein by two-site micro enzyme immunoassay. Clin. Chim. Acta 87:367–372, 1978.

MORGAN, A.C., R.D. Rossen & J.J. Twomey. Naturally occurring circulating immune complexes: normal human serum contains idiotype-antiidiotype complexes dissociable by certain IgG antiglobulins. J. of Immunol. 122(5):1672–1680, 1979.

NAKANE, P.K. & A. Kawaoi. Peroxidase-labeled and antibody a new method of conjugation. J. of Histochem. Cytochem. 22(12):1084–1091, 1974.

PAGANELLI, R., R.J. Levinsky & D.J. Atherton. Detection of specific antigen within circulating immune complexes: validation of the assay and its application to food antigen-antibody complexes formed in healthy and food-allergic subjects. Clin. Exp. Immunol. 46:44–53, 1981.

STEVENS, W.J. & C. Bridts. A method for rapid determination of IgG-containing circulating immune complexes using polyethylene glycol and radioactively labeled protein A. Immunol. Letters 3:1–4, 1981.

ZUBLER, R.H., U. Nydegger, L.H. Perrin, K. Fehr, J. McCormick, P.H. Lambert & P.A. Miescher. Circulating and intra-articular immune complexes in patients with rheumatoid arthritis. Correlation of 125 I Clq binding activity with clinical and biological features of the disease. J. of Clin. Invest. 57:1308–1319, 1976.

Alterations of the Content of Free Amino Acids in Skeletal Muscle during Prolonged Exercise

A.K. Eller and A.A. Viru

Tartu State University, Tartu, Estonian S.S.R., U.S.S.R.

An important role of the glucose-alanine cycle during exercise (Felig, 1973; Felig & Wahren, 1974) accentuates the attention to the amino acid metabolism in muscle tissue (Poortmans, 1975). During exercise the outflow of amino aids and particularly of alanine increases from muscles (Felig & Wahren, 1971). However, the dynamics of the amino acid content of working muscle during prolonged exercise is poorly studied (Christophe et al., 1971).

Methods

Wistar rats had to swim for 1.5 h (in water of 32°C) with a load of 6% of body w, or 12 h without additional load. In a working muscle (quadriceps femoris muscle) and also in the myocardium and liver the free amino acid content was determined with the aid of an amino acid detector (Hitachi KLA-3B).

Results

The total content of amino acids decreased in working muscle and also in liver during exercises (Figure 1). After 1.5 h of swimming both the content and percent of alanine decreased in skeletal muscle (Tables 1 and 2). However, after 12 h of swimming a further decrease of alanine content was not significant in comparison with the data obtained after 1.5 h of swimming. At the same time a relative increase in content of amino acids (glutamine, valine, isoleucine and leucine) connected with the alanine

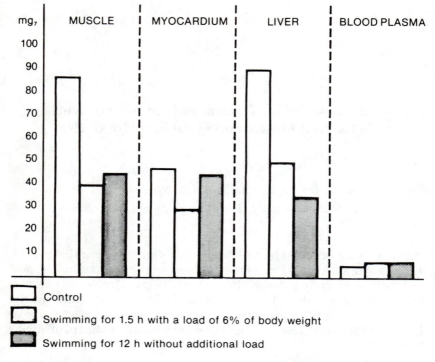

Control

Swimming for 1.5 h with a load of 6% of body weight

Swimming for 12 h without additional load

Figure 1—Percentage change of total amino acids in muscle, heart, liver, and plasma during control and after 1.5 h and 12 h of swimming.

synthesis (Goldberg & Cheng, 1978), was noted. The total content of amino acids that can be oxidized in muscle (alanine, aspargine, valine, glycine, isoleucine and leucine) decreased after 1.2 h but increased after 12 h of swimming.

Pronounced changes in the amino aid content of the myocardium were not observed.

It was suggested that after very prolonged muscular work arises an inhibition of the catabolism of amino aids in muscular tissue and also of the glucose-alanine cycle.

Table 1

Percentage of Various Free Amino Acids After Swimming for 1.5 h with a Load of 6% of Body W and for 12 h without Additional Load

	Skeletal muscle			Myocardium			Liver		
		Swimming			Swimming			Swimming	
	Control	1.5 h	12 h	Control	1.5 h	12 h	Control	1.5 h	12 h
Alanine	14.2	12.0	8.9	6.0	7.9	5.7	11.8	13.5	7.7
Aspargine	3.4	9.3	5.5	11.5	10.7	19.3	23.1	15.8	13.0
Valine	1.2	1.0	3.8	0.5	1.1	0.8	1.2	2.3	3.4
Histidine	8.8	9.0	4.6	4.0	2.0	1.2	1.7	2.3	2.7
Glycine	22.0	23.5	13.3	2.1	4.2	2.7	11.3	9.6	10.4
Gluthamine	7.1	6.0	15.2	39.0	38.1	33.6	12.1	12.8	19.7
Isoleucine	0.5	0.8	2.6	0.3	2.1	1.7	1.0	1.8	1.4
Leucine	1.2	1.9	3.5	1.6	2.7	2.2	1.8	3.3	3.0
Lysine	6.7	5.9	5.1	1.4	0.7	1.6	1.6	1.3	6.2
Methionine	0.6	—	2.8	—	—	1.4	1.2	0.4	0.9
Proline	1.7	3.2	1.8	1.1	—	—	0.6	2.6	2.5
Serine	10.0	7.5	5.9	4.3	4.8	4.7	9.9	14.1	9.2
Threonine	21.7	19.1	20.2	28.2	24.8	25.1	20.9	18.1	19.9
Tyrosine	0.8	0.8	3.1	—	0.8	—	1.1	1.4	—
Cystine	—	—	2.3	—	—	—	—	—	—
Phenylalanine	0.1	—	1.4	—	—	—	0.8	0.7	—

References

CHRISTOPHE, J., J. Winard, R. Kutzner & M. Hebbelnick. Amino acid levels in plasma liver muscle and kidney during and after exercise in fasted and fed rats. Am. J. Physiol. 221:453–457.

FELIG, P., The glucose alanine cycle. Metabolism 22:179–207.

FELIG, P. & J. Wahren. Amino acid metabolism in exercising man. J. Clin. Invest. 50:2703–2714, 1971.

Table 2

Changes of Free Amino Acid Contents (mg%) in Tissues during Swimming

Amino acid	Skeletal muscle		Myocardium		Liver	
	1.5 h	12 h	1.5 h	12 h	1.5 h	12 h
Alanine	− 7.6	− 8.2	−0.8	−0.4	− 3.8	− 8.1
Aspargine	+ 0.7	− 0.4	−2.7	+2.9	−12.7	−16.1
Histidine	− 4.2	− 5.5	−1.4	−1.4	− 0.3	− 0.5
Glycine	− 9.9	−13.0	−0.1	0	− 5.3	− 6.5
Gluthamine	− 3.8	+ 0.8	−8.4	−4.1	− 4.4	− 3.8
Lysine	− 3.6	− 3.5	0.5	+0.1	− 0.7	+ 0.9
Serine	− 5.5	− 6.0	−0.8	0	− 1.7	− 5.6
Threonine	−11.3	− 9.5	−6.9	−2.7	− 9.6	−11.7

Statistically significant changes ($P < 0.05$) are underlined.

FELIG, P. & J. Wahren. Protein turnover and amino acid metabolism in the regulation of gluconeogenesis. Fed. Proc. 33:1092–1037, 1974.

GOLDBERG, A.L. & T.W. Chang. Regulation and significance of amino acid metabolism in skeletal muscle. Fed. Proc. 37:2301–2307, 1978.

POORTMANS, J.R., Effects of long lasting physical exercise and training on protein metabolism. In H. Howald & J.R. Poortmans (Eds.), Metabolic Adaptation to Prolonged Physical Exercise. Basel: Birkhauser Verlag, 1975.

Effect of Intense Prolonged Running on Protein Catabolism

P.W.R. Lemon, D.G. Dolny and B.A. Sherman
Kent State University, Kent, Ohio, U.S.A.

Although it is generally believed that protein and/or amino aids do not contribute to the energy supply during exercise there is considerable recent evidence to the contrary (Dohm et al., 1977; Dohm et al., 1982; Lemon et al., 1980; Lemon & Nagle, 1981; Lemon et al., 1982; Rennie et al., 1981; Wolfe et al., 1982). When exercise is performed in the presence of low glycogen levels (Lemon & Mullin, 1980) or if the exercise is prolonged (Haralambie & Berg, 1976; Refsum et al., 1979) the protein degradation may become significant. Calculations indicate that protein may provide as much as 10 to 18% of the required energy under these conditions (Dohm et al., 1982; Lemon & Mullin, 1980). The importance of this increased exercise protein utilization to the endurance athlete who chronically engages in prolonged exercise while carbohydrate (CHO) loaded is presently unknown. The purpose of this investigation was to determine the magnitude of protein catabolism during a prolonged run.

Methods

Written informed consent was obtained from each subject as approved by the Kent State University Human Subjects Review Board. Ten healthy, well trained endurance runners completed a 26.6 ± 1.0 km run ($\bar{X} \pm SE$) in 128 ± 6 min at a pace equal to $75.5 \pm 2.3\%$ $\dot{V}O_{2max}$. The environmental conditions were $17.3 \pm 3.0°C$ and $63.8 \pm 2.5\%$ relative humidity. There were five males (age $- 29 \pm 2$ yr, $\dot{V}O_{2max} - 67 \pm 3$ ml/kg min$_{-1}$, best marathon time $- 176 \pm 13$ min) and five females (age $- 28 \pm 2$ yr, $\dot{V}O_{2max} - 62 \pm 4$ ml/kg min$_{-1}$, best marathon time $- 200 \pm 12$ min). Subjects were asked to avoid strenuous physical activity the day prior to and following the run and to maintain constant protein consumption throughout the experiment. Total urine volume was collected for 3 days (control, run, and post run 24 h samples) in plastic bottles containing toluene as a preservative. During the collection periods the urine bottles were kept refrigerated (5°C). Immediately following collection the daily samples were mixed, total volume recorded, and a sample frozen (-25°C). During the run all water intake was carefully measured and sweat samples were collected over 15 min time periods at ~ 30 min intervals by absorption into a pad of filter paper located inside water and air tight plastic capsules. These capsules (12.7 x 38.1 mm) were located over the infraspinous fossa of the scapula and held in place by an elastic net vest. The volume of sweat collected was determined by w gain of the capsules, total sweat rate estimated from body w changes with appropriate corrections, and both sweat and urine urea N concentrations measured by the diacetyl monoxime reaction (Barker, 1944). Total urea N excretion was calculated for the run (run day urine + sweat + post day urine) vs the control day (urine) and the differences converted to protein equivalents. These procedures and calculations have been described in detail previously (Lemon & Mullin, 1980). All data were analyzed by ANOVA procedures and where necessary with Tukey post hoc tests.

Results

There was no significant difference ($p > 0.05$) in water intake between males (452 ± 190 ml) and females (369 ± 123 ml); however, the males' total sweat loss ($2,112 \pm 202$ ml) was significantly greater ($p < 0.05$) than the females' ($1,424 \pm 222$ ml) despite the fact that the males completed the run in less time (117 ± 8 vs 139 ± 6 min).

Urine volumes are shown in Figure 1. On the control day these values were similar in both sexes. On the day of the run decreases in urine volume were observed in both sexes; however, only the decrease in the males was significant ($p < 0.05$). On the post run day while the females had returned to control values the volume was still reduced in the male group.

Total urea N excretion (urine + sweat) increased significantly ($p < 0.05$) over control as a result of the run (Figure 2). There were no apparent sex differences ($p > 0.05$) in sweat or urine urea N excretion. There were, however, considerable individual differences in the urea N excretion measures (four of five males and three of five females had increases). When converted to protein equivalents the total urea N excretion data from all 10 subjects indicated that protein contributed 6.2 ± 2.8 and $5.6 \pm 2.5\%$ of the total exercise kilojoules (kJ) in the males and females, respectively. For the seven subjects who demonstrated a protein breakdown this value increased to $8.4 \pm 2.1\%$.

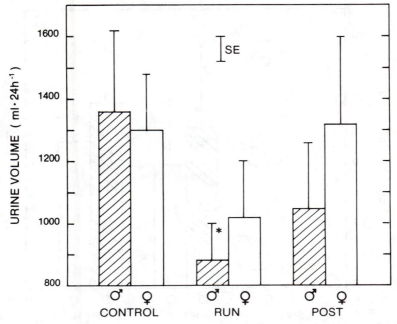

Figure 1—Comparison of control, run, and post run 24 h urine volumes in males and females. *indicates significant difference from control (p < 0.05).

Discussion

On the control day there was no significant sex difference in urine volumes and both were within the normal range of expected values (Muntwyler, 1968). The decreases in urine volume following the run (Figure 1) were not unusual and probably reflect several factors including increased anti-diuretic hormone levels and decreased blood flow to the kidneys. The greater decrease in urine volume for the males on both the run and following day suggests that they experienced a more significant water deficit than the females. This is not surprising because although fluid intake was similar in both sexes total sweat loss (2,112 vs 1.424 ml) and rate (1,103 vs 625 ml h^{-1}) were significant.

The increases in total urea N excretion (Figure 2) support our previous data (Lemon & Mullin, 1980) and together with accumulating recent evidence, strongly indicate that protein catabolism and/or amino acid oxidation increases during exercise (Dohm et al., 1977; Dohm et al., 1982; Lemon et al., 1980; Lemon et al., 1982; Rennie et al., 1981; Wolfe et al., 1982). Protein utilization in the present study was intermediate between that reported following 1 h bicycle exercise at 61% $\dot{V}_{O_2max}$ in subjects who

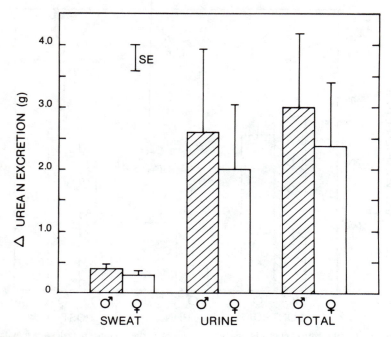

Figure 2—Comparison of run day sweat, urine, and total urea N excretion increases over control day values in males and females. Baseline represents control values.

were CHO loaded or depleted (Lemon & Mullin, 1980). Based on CHO intake it is reasonable to assume that the present subjects had high pre-run total CHO levels. Relative to the bicycle study, the present effort was longer and more intense. Therefore, one might expect a greater glycogen depletion and subsequently a greater protein utilization than in the CHO loaded cyclists but less than those who were CHO depleted.

Perhaps of greater importance than the mean contribution of protein to total kJ were the individual differences observed (< 1.0 to 15.6%). The reasons for these differences are presently unclear, but several possibilities exist. First, although the subjects consumed high amounts of CHO, their diets varied and significant individual differences in glycogen levels prior to and during the run might have occurred. If true, this further demonstrates the importance of CHO loading prior to endurance exercise. Secondly, protein degradation rates may differ in slow- vs fast-twitch muscle and despite the present subjects' homogeneous running ability subtle differences in fiber type may have been present. Thirdly, although all subjects studied were good runners the amount and type of training differed within the group. Training effects on protein utilization are largely unknown, but increased oxidation of the amino acid leucine has been reported in the rat

(Dohm et al., 1977). On the other hand, in another chronic situation where increased protein degradation occurs (starvation) there is a decrease in the protein catabolic rate with time. Therefore, it is possible that endurance training may actually decrease protein catabolism. Lastly, although unmeasured it is possible that individual catabolic hormone responses were different to this exercise stress.

The observed total urea N excretion (Figure 2) represents a protein degradation of 25 g. This is less than the 57 g recently reported by Dohm et al. (1982) in male subjects during a run ∼ 7 to 11 km and 44 min shorter than the present study. This discrepancy is difficult to explain. Despite similar mean running speeds (210 vs 209 m min$_{-1}$) it is possible that relative exercise intensities (% $\dot{V}O_{2max}$) differed. It would appear that sex differences are unimportant because these were absent in the present study. Perhaps the greater protein degradation in their study (Dohm et al., 1982) was due to CHO, fiber type, training, and/or hormonal differences as outlined above. In any event, the magnitude of protein utilization in the present study (25 g) is significant. For the runners studied this represents ∼ 1/2 the recommended daily requirement for protein (Food & Nutrition Board, 1980). Fortunately, this should not pose a serious nutritional problem, except possibly for some vegetarians, because most North Americans daily consume more protein than the requirement. However, increased protein intake may be necessary when endurance exercise is repeated more than once daily and especially in situations where protein needs are elevated for other reasons, i.e., during growth, recovery from injury, pregnancy, and/or lactation. Furthermore, in at least one study an increased oxidation of leucine was observed during exercise with no apparent increase in urea production (Wolfe et al., 1982). This suggests that urea excretion measures may underestimate the actual protein degradation. More study is needed to confirm these observations and to determine the fate of the NH_2 from the amino acids oxidized.

Acknowledgment

This study was supported in part by a grant from the Department of Research and Sponsored Programs, Kent State University.

References

BARKER, S.B., The direct colorimetric determination of urea in blood and urine. J. Biol. Chem. 152:453–463, 1944.

DOHM, G.L., A.L. Hecker, W.E. Brown et al. Adaptation of protein metabolism to endurance training. Biochem. J. 164:705–708, 1977.

DOHM, G.L., R.T. Williams, G.J. Kasperek & A.M. van Rig. Increased excretion of urea and N^T-methylhistidine by rats and humans after a bout of exercise. J. Appl. Physiol. 52:27–33, 1982.

FOOD & Nutrition Board. Recommended Dietary Allowances (9th ed.). Washington: Nat'l. Acad. Sci., 1980.

HARALAMBIE, G. & A. Berg. Serum urea and amino nitrogen changes with exercise duration. Eur. J. Appl. Physiol. 36:39–48, 1976.

LEMON, P.W.R. & J.P. Mullin. Effect of initial muscle glycogen levels on protein catabolism during exercise. J. Appl. Physiol. 48:624–629, 1980.

LEMON, P.W.R., J.P. Mullin, F.J. Nagle & N.J. Benevenga. Effect of daily exercise and food intake on leucine oxidation. Can. J. Appl. Sport Sciences 5(4):xi, 1980 (Abstract).

LEMON, P.W.R. & F.J. Nagle. Effects of exercise on protein and amino acid metabolism. Med. Sci. Sports Exercise 13:141–149, 1981.

LEMON, P.W.R., F.J. Nagle, J.P. Mullin & N.J. Benevenga. In vivo leucine oxidation at rest and during two intensities of exercise. J. Appl. Physiol. 53:947–954, 1982.

MUNTWYLER, E., Water and electrolyte metabolism and acid-base balance. St. Louis: C.V. Mosby Co., 1968.

REFSUM, H.E., R. Gjessing & S.B. Strömme. Changes in plasma amino acid distribution and urine amino acid excretion during prolonged heavy exercise. Scand. J. Clin. Lab. Invest. 39:407–413, 1979.

RENNIE, M.J., R.H.T. Edwards, S. Krywawych, et al. Effect of exercise on protein turnover in man. Clin. Sci. 61:627/09639, 1981.

WOLFE, R.R., R.D. Goodenough, M.H. Wolfe, et al. Isotopic analysis of leucine and urea metabolism in exercising humans. J. Appl. Physiol. 52:458–466, 1982.

Body Composition among College Wrestlers and Sedentary Students with Emphasis on 3-Methylhistidine Excretion and Calculation of Muscle Mass

J. Mendez, W. Vollrath and M. Druckemiller

The Pennsylvania State University, Pennsylvania, U.S.A.

Determination of total body muscle mass is of interest for studies of body composition, evaluation of nutritional status, growth and development and interpretation of disease. The assessment of muscle mass among sports competitors and body builders may be of special concern to these athletes and those who train them and care for their health needs. In the past, workers in the field have been satisfied with the determination of fat-free body weight (FFW) as representative of muscle weight (MW). Comparisons of MW among subjects, including athletes participating in different sports, have been made on this basis. The assumption was that MW constitutes the major component of FFW. FFW includes more than MW and FFW differs among people in the proportion of MW and non-muscle weight (NW) components. Bone weight, in particular, is a major variable. Thus, there is no true equivalence of MW to FFW. A more direct determination of MW is needed to better understand body composition.

Since the advent of total body neutron activation analysis (NAA), many elements in the body including total body nitrogen (TBN) can be determined with precision. When simultaneous determinations of total body potassium (TBK) and TBN are made, MW and NW can be calculated using the approach suggested by Burkinshaw et al. (1978). This model is based on differences in Potassium/Nitrogen (K/N) ratios between muscle and non-muscle components of the body, and the assumption that the K/N ratios are constant for different conditions and nutritional states.

The endogenous urinary excretion of the amino acid 3-methylhistidine (3MH), commonly used for the estimation of muscle protein degradation

(Young & Munro, 1978), has been reported to have significant relationship with MW and can be used for MW determination alone or in conjunction with TBN and TBK measurements (Lukaski et al., 1981). Thus, regression equations have been derived for the prediction of MW from 3 MH excretion and from TBN and TBK or all three variables combined. Determination of MW from 3 MH is possible in steady state conditions provided dietary restrictions are observed to limit exogenous 3 MH. Measurements of urinary 3 MH do not differentiate between endogenous and exogenous 3 MH. Urinary 3 MH excretion in humans appears to reach values attributable to endogenous sources only following 3 days on a meat-free diet (Lukaski et al., 1981).

In order to further explore the usefulness of urinary 3 MH excretion for the determination of MW in vivo, we studied 3 MH excretion in relation to other measurements of body composition among wrestlers and sedentary young men. A second objective was to determine if wrestlers, during training, changed their 3 MH excretion or experienced increased muscle protein catabolism.

Methods

Sixteen healthy young men, mean age 20 y, members of the Pennsylvania State University (PSU) varsity wrestling team participated in the study. Their data were compared to those of 14 sedentary (nonathletic) subjects (Lukaski et al., 1981). The wrestlers were selected from among the weight classes that ranged from 53.6 to 86.4 kg to provide a wide range of FFW. The study was done during the first 5 wk of training. Training activities consisted of running the football stadium steps 3 to 4 times per wk for 2 wk followed by practice in the wrestling room for another 2 to 3 wk. During the week following the 5 wk of training, the subjects were placed on a 4-day meat-free diet. They maintained a daily workout schedule that involved running 4 mi in the morning and wrestling for 2 h in the afternoon. They also lifted weights 3 times per wk. The subjects maintained their body weights during the week.

The meat-free diet consisted of three 12-oz portions of Sustacal (Mead-Johnson) per day supplemented with non-meat food products in quantities sufficient to maintain protein and calorie balance. The average amount of weight loss during the week was 0.5 kg; a value within the usual weight variation range for these men. During any given day, there was a weight shift of 2 to 3 kg before-to-after practice. Since it had been established previously that 3 MH excretion reached the endogenous value following 3 days on a meat-free diet, a 24-h urine sample was collected on the fourth day for 3 MH analysis.

Calculations of fat weight (FW), fat-free weight (FFW), and %FW and %FFW were made using body density (Akers & Buskirk, 1969) and an equation proposed by Brozek et al. (1963). Urine was analyzed for 3 MH by the method of Vielma et al. (1981). Creatinine (Cr), and urea nitrogen (UN) were measured by procedures described in the Technicon Manual. Muscle mass (MW) was calculated from 3 MH excretion using the regression equation proposed by Lukaski et al. (1981).

Results

The physical characteristics and body composition of the groups studied are presented in Table 1. Although height and weight were lower for the wrestlers than for the nonathletes, the wt/ht ratios were not significantly different, indicating that both groups had similar wt for ht. Fat weight was significantly lower for the wrestlers than for the nonathletes, but fat-free weight was not different. The wrestlers were only 1.2 kg heavier. The percent fat-free weight of the wrestlers averaged 4.4 higher than the nonathletes, a highly significant difference. Three-methylhistidine (3MH),

Table 1

Physical Characteristics, Body Composition, 3-Methylhistidine Excretion and Muscle Weight of Wrestlers (A) and Non-Athletes (NA)

	A (N = 16)	NA (N = 14)
Ht, cm	171.6 ± 7.4*	177.1 ± 7.9
Wt, kg	70.6 ± 9.1	73.0 ± 9.1
Wt/ht, kg cm^{-1}	0.41 ± 0.04	0.41 ± 0.04
FW, kg	7.3 ± 2.0*	10.9 ± 4.8
FFW, kg	63.3 ± 8.3	62.1 ± 7.0
%FFW	89.7 ± 2.4**	85.3 ± 5.2
3MH, μmol d^{-1}	286.2 ± 48.3**	221.3 ± 40.4
Creatinine (Cr), g d^{-1}	2.10 ± 0.32**	1.66 ± 0.27
3MH/Cr, μmol g^{-1}	136.8 ± 16.2	134.0 ± 20.0
Urea N (UN), g d^{-1}	12.4 ± 3.0	12.8 ± 1.1
MW, kg	30.3 ± 5.7**	22.7 ± 4.8
%MW	42.9 ± 3.8**	31.1 ± 4.2
%MFFW	47.9 ± 4.8**	36.5 ± 4.0

* and ** denote significant difference at 5 and 1% level of probability, respectively. Mean ± SD.

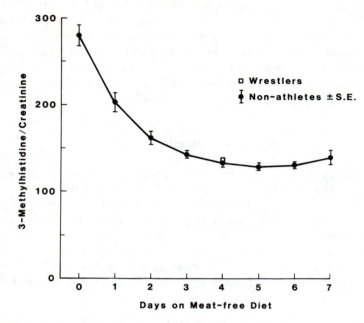

Figure 1—Urinary 3-methylhistidine/creatinine ratio (μmol g$_{-1}$) during a meat-free diet.

creatinine (Cr) and urea nitrogen (UN) urinary excretion, together with muscle mass and percent muscle mass in the body (%MW) and percent muscle mass in the fat-free mass (%MFFW) calculated from the 3 MH excretion are given in Table 1. Daily 3 MH and creatinine excretion were significantly greater among the wrestlers than the nonathletes, 286.2 vs 221.3 μmoles d^{-1} and 2.10 vs 1.66 g d$_{-1}$, respectively. The possibility that the greater excretion of these metabolites could be related to the exercise regimen of the wrestlers was ruled out by comparing the 3 MH/Cr ratios which differed by only 3 μmoles between groups. Also the urea nitrogen was similar, but became lower in the wrestlers not only in terms of total output, but when normalized with respect to creatinine excretion.

Muscle mass was 7.6 kg greater in the wrestlers and the difference increased to about 11 kg when expressed as a percentage of the body weight or the fat-free body weight.

Figure 1 presents the 3 MH/Cr ratios for the nonathletes over a period of 7 days on a meat-free diet (Lukaski et al., 1981). The mean 3 MH/Cr ratio for the wrestlers is plotted at day 4 to show that their mean value did not differ from the results obtained from the nonathletes.

Discussion

The results presented indicated that fat-free weight determined from densitometry cannot differentiate muscle mass development between wrestlers and nonathletes. In contrast, muscle mass calculated from 3 MH excretion appears to give a better appraisal of muscularity.

There are few values in the literature that provide information about skeletal muscle mass. The two most normal cadavers cited by Keys and Brozek (1953) provided values for muscle mass of 31.6 and 39.8% of body weight. When expressed in terms of fat-free weight, they became 32.7 and 42.6%. These values agree very well with those reported here, particularly for the nonathletes.

It is concluded that 3-methylhistidine can be used to evaluate muscle mass in man and that the information obtained is a better discriminator of muscle mass development than classical fat-free weight determinations.

References

AKERS, R., & E.R. Buskirk. An underwater weighing system utilizing "force cube" transducers. J. Appl. Physiol. 26:649–652, 1969.

BROZEK, J., F. Grande, J.T. Anderson & A. Keys. Densitometric analysis of body composition: Revision of some quantitative assumptions. Am. N.Y. Acas. Sci. 110:113–140, 1963.

BURKINSHAW, L., G.L. Hill & D.B. Morgan. Assessment of the distribution of protein in the human body by *in vivo* neutron activation analysis. In Nuclear Activation Techniques in the Life Science, pp. 787–798. Vienna: International Atomic Energy Agency, 1979.

KEYS, A. & J. Brozek. Body fat in adult man. Physiol. Rev. 33:245/09325, 1953.

LUKASKI, H., J. Mendez, E.R. Buskirk & S.H. Cohn. Relationship between endogenous 3-methylhistidine excretion and body composition. Am. J. Physiol. 240:E 302–E 307, 1981.

VIELMA, J., J. Mendez, M. Druckenmiller & H. Lukaski. A practical and reliable method for determination of urinary 3-methylhistidine. J. Biochem. and Biophys. Methods 5:75–82, 1981.

YOUNG, V.R. & H.N. Munro, $N^\tau\tau$-methylhistidine (3-methylhistidine) and muscle protein turnover. An overview. Fed. Proc. 37:2291–2300, 1978.

Cytochrome C Synthesis Rate is Decreased in the 6th Hour of Hindlimb Immobilization in the Rat

P.A. Watson, A. Srivastava and F.W. Booth
University of Texas Medical School at Houston,
Houston, Texas, U.S.A.

When hindlimb muscles are immobilized at a length which is less than their resting length, atrophy occurs (Booth, 1977; Summers & Hines, 1951). It has also been observed that the rate of protein synthesis is decreased significantly during the first 6 h of immobilization-induced atrophy in the gastrocnemius muscle of rats (Booth & Seider, 1979). The content of cytochrome c, a protein involved in electron transport, has been shown (in rat hindlimb muscles) to adapt to immobilization (Booth, 1977), by exponentially decaying to a new steady state. Other studies have offered some insight into the mechanism behind these observations. Studies concerning exercise-induced increases in cytochrome c content (and the subsequent decrease in content following cessation of exercise) reported that, as a result of exercise, no change in the rate of cytochrome c degradation occurred which implied an increased cytochrome c synthesis rate (Booth & Holloszy, 1977). Other studies have demonstrated an increase in δ-aminolevulinic acid synthetase activity following a bout of exercise which occurs prior to the resultant increase in skeletal muscle cytochrome c content (Holloszy & Winder, 1979). These studies would seem to indicate that a change in the rate of cytochrome c synthesis plays a major role in adapting content in various activity states. As a result of the implications of these studies the purpose of the present study was to determine if the synthesis rate of cytochrome c is decreased in red vastus lateralis muscles during the first 6 h of hindlimb immobilization.

Methods

Animal Care and Preparation

Adult female rats (Timco, Houston, TX) weighing 275 to 300 g, were provided Purina lab chow and water ad libitum. Animals were housed in quarters with controlled temperature and a 12 h light/12 h dark cycle. All animals had silastic catheters inserted surgically into the superior vena cava via the right jugular vein under sodium pentobarbital anesthesia (4 mg/100 g body wt, i.p.) one day prior to infusion.

Immobilization and Infusion Protocol

Hindlimb immobilization of animals in the immobilization group was accomplished as described previously (Booth, 1977). Briefly, animals were lightly anesthetized with ether, and then plaster of paris casts applied to hindlimbs so that quadriceps muscles were immobilized at resting length. Control animals were also placed under light ether anesthesia. While still anesthesized, animals (of both groups) were then connected to the infusion pump via tubing attached to a swivel. The tubing was then attached to the intravenous catheter, and secured with a shoulder harness. At this time no infusion was performed. Animals were placed in chambers which were ventilated to allow exiting air to pass through a dessicant trap. Three and one half hours following the beginning of immobilization, infusion of Krebs-Ringer lactate solution containing 5% glucose, 1.7 mM L-tyrosine, and 80 mM L-phenylalanine was begun at a rate of 2 ml/h (Booth & Seider, 1979). After 5 h of limb immobilization, 1.0 mCi of L-[ring-3,5-^{3}H]-tyrosine (New England Nuclear) was added to the infusate and infusion continued for 1 more h. During the last few min of infusion, animals were anesthetized with sodium pentobarbital, casts removed (when present), and red vastus lateralis muscles excised. Muscles were quickly cleaned of fat and connective tissue, and frozen with Wollenburger tongs cooled to the temperature of liquid nitrogen. Muscles were then stored at $-80°$ C until assayed.

Cytochrome C Isolation and Quantification

Cytochrome c was isolated as described by Booth and Holloszy (1977) with Biorex® 70 ion exchange resin used in place of Amberlite. Cytochrome c was then quantified by spectrophotometry, utilizing its absorbance and extinction coefficient (106.1×10^3 cn^2/mole) (Margoliash & Frohwirt, 1959). Purity of isolated cytochrome c samples was determined by SDS-urea polyacrylamide gel electrophoresis for small molecular weight proteins (Swank & Munkres, 1971).

Protein Synthesis

Incorporation of [^{3}H]-L-tyrosine into cytochrome c was determined by liquid scintillation spectrophotometry. The specific activity of the intracellular L-tyrosine pool was determined in the trichloroacetic acid supernatant of the muscle homogenate by L-tyrosine decarboxylase conversion of L-tyrosine to L-tyramine (Garlick & Marshal, 1972), fluorometric quantification of L-tyramine (Waalkes & Udenfriend, 1957), and subsequent liquid scintillation spectrophotometry.

The rate of protein synthesis was calculated by the following equation of Garlick (1973):

$$\frac{S_B}{S_i} = \frac{R}{(R-1)} \times \frac{(1-e^{-K_s t})}{(1-e^{-RK_s t})} - \frac{1}{(R-1)}$$

where S_B is the specific radioactivity of the cytochrome c, S_i is the specific activity of unincorporated intracellular L-tyrosine in the muscle TCA supernatant, R (calculated as 0.406) is the ratio of the amount of tyrosine in cytochrome c (nmoles cytochrome c $\times$ 4 nmoles tyrosine/nmole rat cytochrome c [Scarpulla et al., 1981]) to the amount of unincorporated free tyrosine (determined by L-tyrosine decarboxylase assay), K_s is fractional rate of cytochrome c synthesis, and t is infusion time of ^{3}H-L-tyrosine (0.0416 days). Results were analyzed statistically by one-tailed Student's t test. A probability level of $P < 0.05$ was selected as indicating significance.

Results

The fractional synthesis rate of cytochrome c in the red vastus lateralis muscle of immobilized rat hindlimbs is significantly decreased ($P < 0.05$) in the 6th h of limb immobilization as compared to muscles from non-immobilized control animals (Table 1).

Discussion

Skeletal muscle content of cytochrome c, a protein that is located on the inner membrane of mitochondria and is involved in electron transport, has been shown to be adaptable to various levels of muscular usage. Both increases (exercise) and decreases (limb-immobilization) in muscle usage have been shown to result in altered content of cytochrome c (Baldwin et

Table 1

**Cytochrome c Synthesis Rates in the Red Vastus Muscle
in the 6th H of Limb Immobilization**

	Control	Immobilized (6th h)
% of cytochrome c resynthesized daily	11.67 ± 0.3% (3)	*6.15 ± 0.9% (4)

*P < 0.05. Values given as mean ± SE. (   ) is the number of observations.

al., 1972; Booth, 1977). Cytochrome c concentration has been used as an index of the capacity of the muscle to produce ATP.

Two h of daily running has been shown to increase cytochrome c concentration in rat skeletal muscle from control levels to new elevated steady state levels. In fast-twitch muscle a percentage increase of 96% (Baldwin et al., 1972) was seen. In mixed fast-twitch red and white muscle, increases of 103% (Baldwin et al., 1972), 80% (Terjung, 1976), and 85% (Hickson & Rosenkoetter, 1981), while in mixed slow- and fast-twitch muscle and in predominantly slow-twitch muscles, increases of 72% and 90%, respectively, were observed in cytochrome c concentration (Hickson & Rosenkoetter, 1981). Such increases may result from either an increased cytochrome c synthesis rate, or a decreased cytochrome c degradation rate, or both in rats running 2 h daily. One procedure which has previously been employed to determine whether changes in synthesis or degradation rates played a role in exercise-induced increases in cytochrome c concentration was to estimate the $t_{1/2}$ of the change in cytochrome c content from one steady state to another. If a constant exercise stimulus was applied daily, then the $t_{1/2}$ of the resultant increase in cytochrome c concentration to a new "trained" steady state would provide data from which an estimation of cytochrome c $t_{1/2}$ during training could be made. Likewise, the $t_{1/2}$ of the decay of cytochrome c content from trained steady state to sedentary levels following the cessation of training would provide an estimate of the $t_{1/2}$ of cytochrome c in sedentary muscle. If upon comparison, these $t_{1/2}$ values were similar, one could conclude that training did not alter the $t_{1/2}$ of cytochrome c. This was shown to be the case, as Booth and Holloszy (1977) observed that the $t_{1/2}$ for the change in cytochrome c concentration during constant training with daily running, was similar to the $t_{1/2}$ of decay of cytochrome c content during detraining. It was therefore concluded that cytochrome c degradation rate in skeletal muscle was not altered during training. A similar observation has been reported (Terjung, 1979). However, no direct measurements of cytochrome c synthesis rates were made. It was concluded that since cytochrome c concentration increased while degradation rate remained unchanged, synthesis rate of cytochrome c must have increased. Support for this conclusion was later obtained when Holloszy and Winder (1979) observed that the activity of δ-aminolevulinate synthetase in vastus lateralis muscles was doubled 17 h after an acute bout of prolonged exercise. As δ-aminolevulinate synthetase is the rate limiting enzyme for heme synthesis, this observation supported the hypothesis that exercise increases the cytochrome c synthesis rate in skeletal muscle. However, no direct assessment of the effects of exercise on cytochrome c synthesis rate has been made.

Muscle atrophy produced by limb immobilization results in an exponential decline in cytochrome c content in fast-twitch muscle from a sedentary steady state to a new lower steady state (Booth, 1977). The decrease in cytochrome c content could be a result of a decreased cytochrome c synthesis rate or an increased degradation rate of cytochrome c, or both. Results of the present study indicate that the estimated rate of cytochrome c synthesis is decreased 47% in the 6th h of limb immobilization. Such an estimate is the first direct measurement showing a change in the rate of cytochrome c synthesis in response to an acute change in the amount of muscle usage. Thus, a decrease in the synthesis rate of cytochrome c plays an important role in causing the immobilization-induced decrease in cytochrome c synthesis.

An estimate of the $t_{1/2}$ for cytochrome c in the red vastus muscle of control rats in the present study was obtained as described next. Control animals were nongrowing and therefore muscle mass should remain relatively constant. Rates of synthesis and degradation must therefore be equal if the pool size of a given protein is unchanging. In control rats, the synthesis rate of cytochrome c was directly estimated and the degradation rate was presumed to be equal (Table 1). The estimate of 5.9 days for the $t_{1/2}$ of cytochrome c in the red vastus compares favorably with estimates made by the decay of cytochrome c content in the red vastus of rats undergoing detraining from a previous program of running training (Booth & Holloszy, 1977). The estimated $t_{1/2}$'s of cytochrome c of the red vastus in detrained red vastus muscles were 6.9 days (Booth & Holloszy, 1977), 7.5 days (Hickson & Rosenkoetter, 1981), and 9.4 days (Terjung, 1979) (Table 2).

It is difficult to make a definite statement concerning the effect of limb immobilization upon the degradation rate of cytochrome c in skeletal muscle. Using cytochrome c content decay data to estimate the $t_{1/2}$ of cytochrome c, values of 8.8 days in plantaris (Hickson & Rosenkoetter, 1981) and 8.1 days in gastrocnemius and plantaris muscles (Terjung, 1979) have been made. Estimations of cytochrome c $t_{1/2}$ obtained from limb-immobilization data using decay of cytochrome c content were 4.2 (gastrocnemius) and 5.7 (plantaris) days. It would therefore appear that the

Table 2

Estimated $t_{1/2}$ for Cytochrome C

	Detraining			Immobilized	Control
	Booth & Holloszy, 1977	Terjung, 1979	Hickson & Rosenkoetter, 1981	Booth, 1977	(This study)
Soleus	8.1 days	6.2 days	4.5 days		
Red vastus	6.9 days	9.4 days	7.5 days		5.9 days
Gastrocnemius or plantaris (mixed fibers)		8.1 days	8.8 days	3.8–4.2 days 5.7 days	

rate of degradation of cytochrome c was increased by 50% during limb immobilization. However, it seems unlikely that both synthesis and degradation rates changed by this magnitude during immobilization as this would result in a much more rapid decline in cytochrome c content with limb-immobilization than has actually been observed (Booth, 1977). It appears that one of these processes must therefore be overestimated. As our measurements of synthesis rate were made over a 1 h period, a possible explanation of this apparent discrepancy is that the determined synthesis rate is not representative of that which may be present during other time periods during immobilization. For example, cytochrome c synthesis rates may not be as depressed at 1 day of immobilization, as they are at 6 h.

In summary, present data is the first direct estimation of cytochrome c synthesis in skeletal muscle. Decreased muscle usage was associated with a decreased rate of cytochrome c synthesis. This effect was observed in the 6th h of limb immobilization. Rates of cytochrome c synthesis seem to be quickly adaptable to decreases in muscle usage. Since we previously noted a decreased state III respiratory rate in mitochondria isolated from muscle after 2, but not 1, days of limb immobilization (Krieger et al., 1980), the decrease in cytochrome c synthesis precedes a decrease in mitochondrial capacity to make ATP.

Acknowledgment

We thank Ms. Rafaelina Cintron for typing and the generous support from NIH 19393 and NASA contract NAS9-16478.

References

BALDWIN, K.M., G.H. Klinkerfuss, R.L. Terjung, P.A. Mole & J.O. Holloszy. Respiratory capacity of white, red and intermediate muscle: adaptive response to exercise. Am. J. Physiol. 222:373–378, 1972.

BOOTH, F.W., Time course of atrophy during disuse by immobilization. J. Appl. Physiol.: Respirat. Environ. Exercise Physiol. 43:656–661, 1977.

BOOTH, F.W. & J.O. Holloszy. Cytochrome c turnover in rat skeletal muscles. J. Biol. Chem. 252:416–419, 1977.

BOOTH, F.W. & M.J. Seider. Early changes in skeletal muscle protein synthesis after limb immobilization. J. Appl. Physiol.: Respirat. Environ. Exercise Physiol. 47:974–977, 1979.

GARLICK, P.J. & J. Marshal. A technique for measuring brain protein synthesis. J. Neurochem. 19:577–583, 1972.

GARLICK, P.J., D.J. Millward & W.P.T. James. The diurnal response of muscle and liver protein synthesis in vivo in fed rats. Biochem. J. 136:935–945, 1973.

HICKSON, R.C. & M.A. Rosenkoetter. Separate turnover of cytochrome c and myoglobin in the red types of skeletal muscle. Am. J. Physiol. 241:C-140–C-144, 1981.

HOLLOSZY, J.O. & W.W. Winder. Induction of δ-aminolevulinic acid synthetase in muscle by exercise or thyroxine. Am. J. Physiol. 236:R180–R183, 1979.

KRIEGER, D.A., C.A. Tate, J. McMillin-Wood & F.W. Booth. Populations of rat skeletal muscle mitochondria after exercise and immobilization. J. Appl. Physiol.: Respirat. Environ. Exercise Physiol. 48:23–28, 1980.

MARGOLIASH, E. & N. Frohwirt. Spectrum of horseheart cytochrome c. Biochem. J. 71:570–572, 1959.

SCARPULLA, R.C., K.M. Agne & R.W. Wu. Isolation and structure of a rat cytochrome c gene. J. Biol. Chem. 256:6480–6486, 1981.

SUMMERS, T.B. & H.M. Hines. Effects of immobilization in various positions upon the weight and strength of skeletal muscles. Arch. Phys. Med. 32:142–145, 1951.

SWANK, R.T. & K.D. Munkres. Molecular weight analysis of oligopeptides by electrophoresis in polyacrylamide gel with SDS. Anal. Biochem. 39:462–477, 1971.

TERJUNG, R.L., Cytochrome c turnover in skeletal muscle. Biochem. Biophys. Res. Comm. 66:173–178, 1975.

TERJUNG, R.L., The turnover of cytochrome c in different skeletal muscle fiber types of the rat. Biochem. J. 178:569–574, 1979.

Alanine Formation during Maximal Short-term Exercise

H. Weicker, H. Bert, A. Rettenmeier, U. Oettinger,
H. Hägele and U. Keilholz
Universität Heidelberg, Heidelberg,
Federal Republic of Germany

The importance of the alanine glucose cycle for the maintenance of glucose homeostatis during long-term exercise, especially in short periods of hunger, has been established by investigations of Ahlborg et al. (1974) and Felig et al. (1971). Little, however, is known about the mechanism of the enhanced muscular alanine production during maximal short-term work. Furthermore, the origin of pyruvate for the amination to alanine in the muscle is still contradictory. Felig et al. (1971) postulated the alanine formation from carbohydrate sources, e.g. muscular glycogen. Garber et al. (1976), however, showed in vitro that pyruvate is derived from muscular amino-acids by deamination. Both pathways are theoretically possible depending on the metabolic situation, whether or not the fuel supply demands a remetabolization or a de novo synthesis of glucose (Figure 1). The present knowledge about alanine as glucose precursor gave rise to a study of the production of this glucoblastic amino-acid during anaerobic work both in short-term running events and during two different ergometer-tests on a bicycle. Additional information was obtained by a tracer study in which ^{14}C glucose was applied as marker to guinea pigs exposed to short-term exercise on a rodent treadmill (Table 1).

Materials and Methods

1. Short term running events: 15 regional top sprinters (22.3 ± 2 yr) participated in the competition over 200 m distance.
2. Twenty sprinters and 15 middle distance runners (22.5 ± 3.4 yr) performed the 400 m distance.

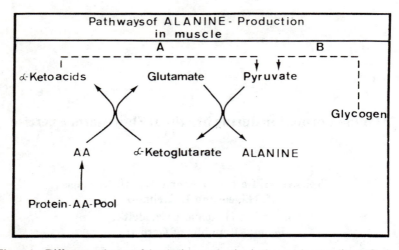

Figure 1—Different pathways of the alanine-production in the muscle according to Garber et al. (1976) (A) and Felig et al. (1971) (B).

Table 1

Aims

---

1. Alanine production during sprint and ergometer tests.
2. Comparison between lactate and alanine increase and their decline after short-term exercise.
3. Specific ^{14}C activity of alanine in serum and muscular free AA pool of guinea pigs after exhaustive run.

---

    3.   Fifteen all-round trained students of physical education (22.4 ± 3.2 yr) performed the 100 m distance, and one wk later the 1.500 m distance.

    4.   Nine all-round trained students of physical education (23.9 ± 2.5 yr) accomplished a vita maxima and an endurance test of 30 min at the anaerobic threshold. The ergometer tests were performed on a bicycle ergometer (Siregnost, Siemens) open system by the same students at an interval of 1 wk. Serum alanine was determined before and after the running events and the ergometer tests by gas liquid chromatography on a glass capillary column according to the enantiomer-labelling method of Frank et al. (1980). In addition, serum alanine was measured enzymatically according to Williamson (1974). Lactate was assayed fluorimetrically according to Olsen (1971).

In the tracer study with ^{14}C glucose, 10 male guinea pigs were exercised on a rodent treadmill with increasing speed up to the velocity of 30 m/min without inclination (Figure 2). After the last training session the animals received 10 $\mu C^{14}C$ glucose and 50 mg glucose intraperitoneal (i.p.). On the following day, the guinea pigs ran exhaustively at the maximal speed for 2 min. Immediately after this run, they were anesthetized and blood was drawn from the inferior vena cava and the abdominal aorta. The vastus lateralis, psoas major, and gastrocnemius muscles were removed. The white portions of these muscles were combined to the fast glycolytic (FG) muscle fiber pool and the red portion of the gastrocnemius to the fast oxidative glycolytic (FOG) muscle fiber pool.

Ten guinea pigs were exposed, as controls, to the same procedure, except for the training regimen. Serum alanine was separated from glucose, pyruvate, and lactate by ion-exchange column on an AA analyzer (Multichrome, Beckmann, Munich). The isolated alanine fraction was assayed as above. The ^{14}C activity was determined in a β-counter, Nuclear Chicago and the specific ^{14}C activity was calculated. The fraction containing glucose, pyruvate and lactate was separated in its single components by an additional ion-exchange passage on DOWEX, 4x1. The substrates were enzymatically determined. The ^{14}C activity was calculated. FG and FOG muscle fiber pools were homogenized in 0.6 N perchloric acid and centrifuged. In the supernatant perchloric acid was precipitated by $K_2 CO_3$. The separation and the determination of alanine, glucose, lactate, and pyruvate were carried out as mentioned above. Glycogen was enzymatically assayed in the supernatant according to Passoneau et al. (1974) using the amylo-glucosidase. Glycogen was precipitated in the supernatant by sodium chloride/methanol. The ^{14}C activity was determined in a β-counter, and the specific ^{14}C activity was calculated.

Results

A significant increase of alanine after each running event was found. The change of the alanine concentration after exercise, however, was not significantly correlated to the lactate increase or the running time. The decline of the augmented alanine level after exercise was slower than the decline of lactate. After the vita max ergometer test on bicycle, alanine was also significantly increased, but less than lactate. Neither alanine to lactate nor alanine to $\dot{V}O_2max$ were significantly correlated. The post-exercise decrease of alanine was also slower after the ergometer test than during the short-term running events (Figures 3,4). In the endurance test of 30 min at the anaerobic threshold on a cycle ergometer the alanine increase was relatively higher than that of lactate (Figure 4). In the ^{14}C glucose tracer

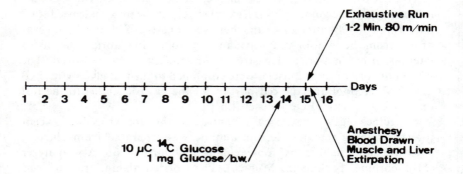

Material

Blood: Vena cava inf.
         Aorta abd.

Muscles: Fg muscle fibers
         Fog muscle fibers

Liver

Methods

Alanine: Separation on Ion Exchanger
          GC Enantiomer Method (Frank et al)
          14C Determination

Glycogen: Amyloglucosidase Method (Passoneau et al)

Figure 2—^{14}C Glucose Tracer Study in Trained (n = 10) and Untrained (n = 10) Guinea Pigs.

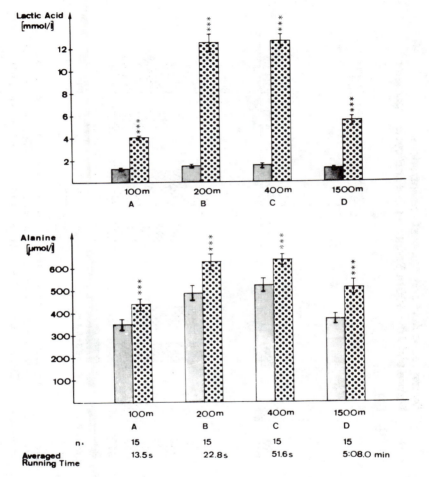

Serum Lactic Acid and Alanine Concentrations
Before and After 100m, 200m, 400m and 1500m Distances.

Figure 3—Lactic acid and alanine increase in serum after 100 m, 200 m, 400 m and 1500 m distances. A, D = students of physical education (n = 15, 22.4 ± 3,4 yr); B, C = top regional sprinters (n = 15, 22.3 ± 2 yr) + p < 0.05 ++p < 0.01 +++p < 0.001.

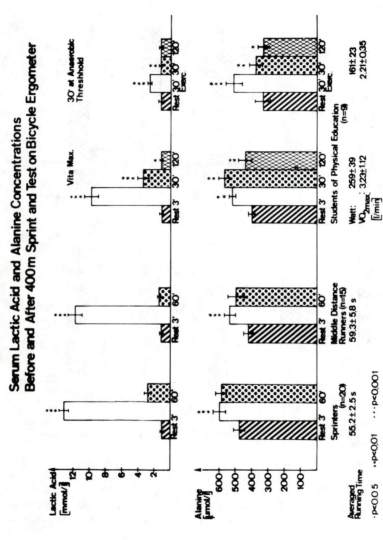

Figure 4— Lactic acid and alanine changes in serum at rest, 3' and 60' after 400 m sprint. A. short distance runners (n = 20, 22.4 ± 3,4 yr). C. vita max. ergometer test and endurance test for 30' on bicycle at the anaerobic threshold of 3–4 mmollactic acid. Students of physical education (n = 9, 23.9 ± 2,5 yr) +p < 0.05 ++p < 0.01 +++p < 0.001.

studies the guinea pigs were accustomed to running at a speed of 30 m/min on a rodent treadmill. This velocity can be considered as maximal short-term exercise for these animals. After the final exhaustive run, we found a higher specific ^{14}C activity of alanine in venous than in arterial blood which was higher in exercised than in sedentary animals (Figure 5). In the free amino acid pool of FOG muscle fibers the specific ^{14}C activity of alanine and glycogen was higher than in the amino acid pool of FG fibers (Figures 6 and 7). The specific ^{14}C activity of glycogen and alanine was higher in muscles than in the liver. In the exercised animals' muscles and livers we found a higher ^{14}C specific activity than in the sedentary ones, which had been exposed to the same trial conditions, except the training.

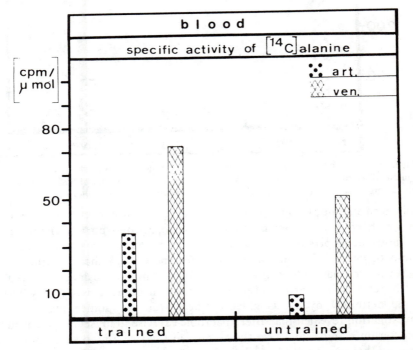

Figure 5—Specific activity of ^{14}C alanine in serum from venous and arterial blood of guinea pigs (n = 10) drawn after exhaustive short-term run on a rodent treadmill at a speed of 30 m/min. 10μC[14]C glucose and 50 mg glucose i.p. were applied 24 h before this run. Controls (n = 10) without training and exhaustive run.

Discussion

This finding demonstrated that during predominantly anaerobic work in which the muscular glycogen serves as fuel for the energy production, serum alanine rose significantly (Figures 3, 4). It is remarkable that this

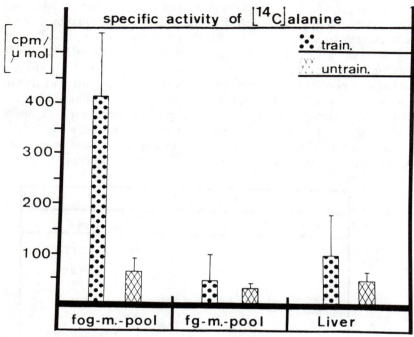

Figure 6—Specific activity of ^{14}C alanine derived from the free AA pool of FOG and FG muscle fibers and the liver after an exhaustive run and the application of 10 μC ^{14}C glucose and 50 mg glucose i.p. (Figure 5) exercised (n = 10) and sedentary animals (n = 10).

enhanced alanine production occurred during anaerobic energy formation in which neither glucose homeostatis is jeopardized nor an enhanced hepatic glucose production is necessary. This observation may support the thesis of Felig (1971) that alanine can be generated by the amination of pyruvate originating from muscular glycogen. This conclusion, evidenced by the results of the sprint and ergometer tests, encouraged us to clarify this topic definitely by a ^{14}C glucose tracer study in guinea pigs. This investigation revealed that after anaerobic exercise in muscular glycogen and serum alanine, and also the alanine in the free amino acid pool of muscles, became labeled by ^{14}C derived from ^{14}C glucose. The last finding is important and not yet reported, indicating that alanine of the free amino acid muscle pool can be recruited also during anaerobic exercise from muscular glycogen by the amination of pyruvate. These results of the ^{14}C glucose tracer study in animals confirmed the postulation of Felig et al. (1971) at least for short-term anaerobic exercise. Furthermore it raised some questions upon the finding of Garber et al. (1976) who postulated that alanine is produced by the amination of pyruvate originating from muscular amino acids by deamination.

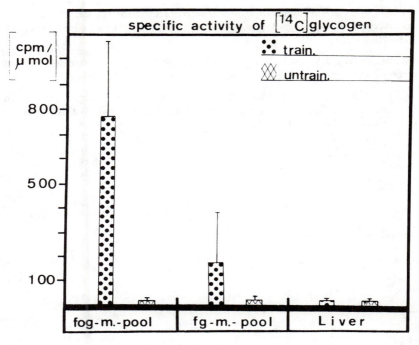

Figure 7—Specific activity of ^{14}C glycogen derived from the free AA pool of FOG and FG muscle fibers and liver (Figure 6).

References

AHLBORG, G., P. Felig, L. Hagenfeldt, R. Hendler & E.J. Wahren. Substrate turnover during prolonged exercise in man: Splanchnic and leg metabolism of glucose, free fatty acids and amino acids. J. Clin. Invest. 53:1080–1090, 1974.

FELIG, P. & E.J. Wahren. Amino acid metabolism in exercising man. J. Clin. Invest. 50:2703–2714, 1971.

FRANK, H., A. Rettenmeier, H. Weicker, G.J. Nicholson & E. Bayer. A new gas chromatographic method for determination of amino acid levels in human serum. Clin. Chim. Acta 105:201–211, 1980.

GARBER, A.J., I.E. Karl & D.M. Kipnis. Alanine and glutamine synthesis and release from skeletal muscle—II The precursor role of amino acids in alanine and glutamine synthesis. J. Biol. Chem. 251:836–843, 1976.

OLSEN, C., An enzymatic fluorimetric micromethod for the determination of acetoacetate, β-hydroxybutyrate, pyruvate and lactate. Clin. Chim. Acta 33: 293–300, 1971.

PASSONEAU, J.V. & V.R. Lauderdale. A comparison of three methods of glycogen measurement in tissues. Anal. Biochem. 60:405–412, 1974.

WILLIAMSON, D.H., L-Alanin-Bestimmung mit Alanin-Dehydrogenase. In H.U. Bergmeyer (Ed.), Methoden der enzymatischen Analyse. Weinheim/Bergstrabe: Verlag Chemie, 1974.

Lactate
Metabolism

Effects of Training on Oxidation of Injected [U-^{14}C]-Lactate in Rats during Exercise

G.A. Brooks and L. Divine-Spurgeon
University of California, Berkeley, California, U.S.A.

Lactate produced in skeletal muscle during exercise has traditionally been believed to be an end product of anaerobic glycogenolysis and glycolysis, with little potential as a substrate for oxidative catabolism (Hill & Lupton, 1923). However, studies utilizing radioactive tracers clearly indicate that lactate undergoes a significant turnover during rest, and rapid turnover during exercise and recovery from exercise. A major portion of the lactate flux is removed through oxidation (Depocas et al., 1969; Donovan and Brooks, 1982; Issekutz, 1976). There is at present insufficient data on how physical training might affect lactate oxidation during graded exercise. The purpose of this study was to assess the effects of endurance training on the oxidation of ^{14}C-lactate injected during rest and two levels of exercise in vivo.

Methods

Sixteen female Wistar rats (Simonsen Labs, Gilroy, CA) five wks old, were randomly divided into an untrained control group and an exercise trained group. Rats were individually caged, fed ad libitum on commercial diet (Feed-stuffs Processing, San Francisco, CA), and kept on a 7:00 am/7:00 pm, light/dark cycle.

Exercise training consisted of running on a Quinton Rodent Treadmill (Model 42-15). The training procedure involved five wks of incremental exercise and one wk of daily running for 1 h at a speed of 28.7 m min^{-1} on a 15% grade. After this level had been reached, training was maintained at 1 h per day, 5 days per wk throughout the 2 wk experimental period. Rats were habituated to treadmill running in a metabolism chamber (Brooks & White, 1978) by 5 to 8 min exposures, twice a wk. Untrained rats were habituated to running in the metabolism chamber by the same procedure. During this habituation period, the treadmill grade was held constant at 15% and the speed was gradually increased until 28.7 m min^{-1} was reached just prior to the testing period.

The open circuit, indirect calorimeter described previously by Brooks and White (1978) was used for $\dot{V}o_2$, $\dot{V}co_2$, and $^{14}Co_2$ determinations following an injection of ^{14}C-lactate. Rats were tested under the following three conditions for 1 h: 1) resting, 2) running at 14.3 m min^{-1} on a 1% grade (easy exercise), and 3) running at 28.7 m min^{-1} on a 15% grade (hard exercise). The order of testing for each rat was systematically varied, allowing animals 1 wk between tests. Animals were tested in the morning after an overnight fast. Immediately prior to testing the rat was given a 10 min rest period or warmup in the metabolism chamber at the assigned running speed and grade for that day. After the warm-up run or control rest period, the rat was administered an intraperitoneal (i.p.) pulse injection of 2 μCi of [U-^{14}C]-L-(+)-lactate (ICN, Irvine, CA) in 250 μl of phosphate-buffered 0.9% NaCl, pH 7.4.

Blood samples were taken for lactate and glucose concentrations in repeated tests under the same three metabolic conditions. During rest, or at selected times (15, 30, and 60 min) in exercise, rats were killed by a blow to the head and exsanguinated. Blood was collected in pre-weighed test tubes containing 6% perchloric acid. Deproteinized and neutralized blood extracts were assayed enzymatically for concentrations of lactate and glucose.

Results

Mean values for blood lactate and glucose levels during rest and exercise are given in Table 1. Blood lactate concentrations were similar for trained and untrained groups during rest. However, during both exercise conditions, blood lactate was significantly higher in untrained animals. No significant differences were found in blood glucose concentrations between groups or among testing conditions except in heavy exercise where untrained animals had lower blood glucose concentrations than the trained (p < .05).

Whereas trained animals could complete the one h run at 28.7 m min^{-1} on 15% grade, untrained animals could only endure ½ h at that speed and grade (Table 2). Steady-rate $\dot{V}o_2$ values were significantly different among

Table 1

**Blood, Glucose and Lactate Levels in Rats during Rest
and 2 Conditions of Exercise (See Table 2)†§**

	Glucose (mM)	Lactate (mM)
Rest		
Untrained, n = 8	3.68 ± 0.09	0.85 ± 0.13
Trained, n = 8	3.50 ± 0.11	0.97 ± 0.10
Easy exercise		
Untrained, n = 12	3.21 ± 0.21	$2.12 \pm 0.42*^G$
Trained, n = 12	3.42 ± 0.16	0.69 ± 0.06
Hard exercise		
Untrained, n = 8	$2.19 \pm 0.31*^G$	$5.44 \pm 0.32*^G$
Trained, n = 12	3.33 ± 0.20	2.13 ± 0.26
	$**C_2$	$**C_3$

$*p < .05$
$**p < .01$
† Mean ± SEM
G = Untrained and trained groups significantly different.
C_2 = Hard exercise significantly different from rest and easy exercise.
C_3 = All three metabolic conditions significantly different from each other.
§ Blood metabolite values for exercising animals represent a pooled mean of levels determined in four animals each at 15, 30, and 60 min.

the three metabolic conditions. Further, trained animals had a higher $\dot{V}O_2$ during easy exercise and untrained animals had a higher $\dot{V}O_2$ during hard exercise. The average values of R were significantly different among metabolic conditions, and during exercise R values for untrained rats exceeded those for the trained ($p < .05$) (Table 2).

Values for evolution of $^{14}CO_2$ following injection of ^{14}C-lactate into trained and untrained rats during the three metabolic conditions are given in Table 2. Cumulative recovery of ^{14}C as $^{14}CO_2$ differed among rest, easy, and heavy exercise (Table 2). Additionally, during easy exercise, a greater percentage of tracer was collected as $^{14}CO_2$ from untrained animals at both 30 and 60 min post-injection. During hard exercise, a greater percentage of

Table 2

Steady-rate§ Respiratory Determinations and Cumulative Percent of Tracer Injected as [U-^{14}C]-L-(+) Lactate Recovered as CO_2 in Untrained and Trained Rats during 1 H of Rest and Treadmill Running under 2 Different Conditions: 14.3 m min^{-1} on a 1% Grade (Easy Exercise) and 28.7 m min^{-1} on a 15% Grade (Hard Exercise)

	$\dot{V}_{O_2}$ (ml kg^{-1} min^{-1})	R $\dot{V}_{CO_2}/\dot{V}_{O_2}$	Run Time†† (min)	Cumulative Recovery of Tracer	
				% Recovery at 30 min	% Recovery at 60 min
Rest					
Untrained	22.6 ± 1.6*[G]	0.82 ± 0.03	— —	34.3 ± 3.2	56.1 ± 3.0
Trained	27.5 ± 1.2	0.82 ± 0.02	— —	39.5 ± 4.9	58.6 ± 4.3
Easy exercise					
Untrained	38.1 ± 1.0*[G]	0.99 ± 0.02*[G]	60	65.7 ± 5.2*[G]	84.4 ± 4.3*[G]
Trained	43.9 ± 0.5	0.86 ± 0.02	60	55.4 ± 6.4	74.8 ± 6.2

Hard exercise				
Untrained	64.4 ± 2.6*[G]	1.02 ± 0.02*[G]	27.5	75.8 ± 3.8*[G]	89.2 ± 1.8
Trained	58.7 ± 0.9	0.96 ± 0.02	60	67.0 ± 5.2	86.5 ± 5.6
	**C_3	**C_3	**C_2	**C_3	*C_3

n = 8 each group
*p < .05
**p < .01
†Mean ± SEM
††Mean Only
G = Untrained and trained groups significantly different from each other.
C_2 = Easy exercise significantly different from hard exercise.
C_3 = All three metabolic conditions significantly different from each other.
§Steady-rate defined as the steady level $\dot{V}O_2$ response to a constant work load.

injected tracer was collected from untrained animals within the first 30 min post-injection, the time when untrained animals ran.

Discussion

The present results suggest oxidation to be an active pathway of lactate removal during heavy exercise, and in untrained as well as trained rodents. Endurance training appears to depress the fraction of the lactate flux undergoing oxidation during sub-maximal exercise. In the present study no data were available on blood lactate specific activity, so lactate turnover rate could not be estimated. Therefore, in the present investigation it cannot be stated with certainty that training reduced the absolute lactate oxidation during exercise.

The results obtained with $^{14}CO_2$ production do, however, allow the conclusion that training reduces the fraction of the lactate turnover (flux) removed through oxidation during exercise. Together, the findings of a greater lactate pool size and a greater relative oxidation of injected lactate strongly implicate greater net lactate oxidation in untrained animals during exercise. Fortunately, the present results are corroborated by results obtained by Donovan and Brooks (1982) who studied lactate metabolism in trained and untrained animals by means of continuous infusion, dual-label technique. They observed a 6% reduction in lactate oxidation in trained rats during hard treadmill running.

The lesser lactate, and higher glucose levels in trained animals (Table 1), suggest glucose sparing or the enhancement of gluconeogenesis with training. This latter possibility is supported by data obtained by Donovan and Brooks (1982). Radiochromatograms developed by them from blood of trained rats exercising at 28.7 m min^{-1} on a 1% grade show significant incorporation of ^{14}C into glucose after injection with [U-^{14}C]–lactate. Krebs and Yoshida (1963) have previously observed an effect of physical training on gluconeogenesis from lactate, pyruvate, and fumarate in kidney cortex slices in vitro.

During hard exercise untrained animals had a higher $\dot{V}_{O_2}$ than did trained animals (Table 2). This difference may have been due to the lesser running skill of untrained animals. During easy exercise, however, trained animals had a higher $\dot{V}_{O_2}$ than untrained. This difference during easy exercise may reflect the energy cost of a metabolic process, such as gluconeogenesis in trained animals.

As the level of metabolic activity increases, the turnover of lactate also increases (Depocas et al., 1969; Donovan & Brooks, 1982; Issekutz et al., 1976). Consequently, blood lactate concentration does not represent the total amount of lactate formed, but rather reflects the balance of a dynamic steady-state.

Acknowledgments

Research supported by NIH grant AM19577.

References

BROOKS, G.A. & T.P. White. Determination of metabolic and cardiac frequency responses of laboratory rats to treadmill exercise. J. Appl. Physiol.: Respirat. Environ. Exercise Physiol. 45:1009–1015, 1978.

DEPOCAS, F., Y. Minaire & J. Chatonnet. Rates of formation and oxidation of lactic acid in dogs at rest and during moderate exercise. Can. J. Physiol. Pharmacol. 47:603–610, 1969.

DONOVAN, C.M. & G.A. Brooks. Endurance training effects lactate clearance, not lactate production. Med. Sci. Exercise Sports 14(2), 1982.

HILL, A.V. & H. Lupton. Muscular exercise, lactic acid and the supply and utilization of oxygen. Quart. J. Med. 16:135–171, 1923.

ISSEKUTZ, B., Jr., W.A.S. Shaw & A.C. Issekutz. Lactate metabolism in resting and exercising dogs. J. Appl. Physiol. 40:312–319, 1976.

KREBS, H.A. & T. Yoshida. Muscular exercise and gluconeogenesis. Biochem. Z. 338:241–244, 1963.

Blood Pressure Response in Relation to
Blood Lactate during Exercise

J. Karlsson, R. Dlin, F. Wahlberg, R. Sannerstedt
and C. Kaijser

Karolinska Hospital, Stockholm, Sweden, Wingate Institute,
Natanya, Israel, Occupational Health Center, Swedish
Commercial Bank, Stockholm, Sweden, and Occupational
Health Center, Medicar, Gothenberg, Sweden

It seems reasonable to assume that the high predictability of the onset of blood lactate accumulation (OBLA) concept for top athletic performance in endurance events not only depends on its integrating properties for circulatory and muscle metabolic potentials but that it also reflects regulatory mechanisms (Karlsson & Jacobs, 1982). Shepherd and colleagues have in a recent review introduced the term *ergoreceptors* for postulated sensory organs in the contracting muscle, which through afferents to the vasomotoric center activate the heart by sympathetic means (Shepherd et al., 1981). There are reasons to believe that these regulatory features are muscle fiber related. Thus, animal experiments have shown that an increased proportion of fast-twitch or type II muscle fibers is synonymous with a higher pressure response (Karlsson et al., 1983; Shepherd et al., 1981), while in humans it has been shown that individuals with a high percentage of slow-twitch or type I muscle fibers are more susceptible for adrenergic β-receptor blockade (Kaiser et al., 1981). Moreover it is clearly demonstrated that the work load corresponding to the definition of OBLA increases with the percentage of slow twitch muscle fibers and endurance training status, expressed as enzyme activities (Jacobs, 1981).

With this knowledge, it was thought of interest to study the blood pressure response at OBLA exercise intensities in subjects with different absolute work load capacities, especially since it has been found that patients with essential hypertension have a higher percentage of fast-twitch muscle fibers and a different central circulation regulation, and to include them in the protocol (Juhlin-Dannfelt et al., 1979).

Methods and Subjects

A total of 50 subjects was employed in the study. Thirty-four of them were patients with untreated "border line" or latent essential hypertension (WHO I), their mean age, height and weight being 43 yr, 179 cm and 79 kg, respectively. The controls were somewhat younger, with an average age of 37 yr. Their height and weight averaged 178 cm and 72 kg, respectively.

They reported to the laboratory during daytime and were advised not to eat, smoke or perform heavy exercise during the hours preceding the test. After 10 to 15 min of rest the ECG leads were attached and the cuff of the automatic blood pressure equipment (Cardionics AB, Stockholm) was positioned half way up the upper arm. The systolic blood pressure was then monitored during exercise with the arm hanging down in a relaxed fashion.

The exercise was performed on a mechanically-braked cycle ergometer supplied with a balance system, which allows precise power outputs with as low as 10 or 25 W increments (Cardionics AB, Stockholm). In healthy subjects the power output was increased by 50 W every 4th min. During the last 45 to 30 s heart rate, systolic blood pressure, blood lactate concentration and rated perceived exertion (RPE) were recorded at each load. For blood lactate determination 25 μl of blood was sampled from the finger tip and analyzed with a semi-automatic enzymatically based technique (Rydevik et al., 1982; Karlsson et al., 1983).

Results

The power output corresponding to a blood lactate concentration of 4 mmol $\times$ x^{-1} (W$_{OBLA}$) amounted to 154 $\pm$ 31 W in the patients with untreated essential hypertension and 168 $\pm$ 37 (SD) in the controls. The rated perceived exertion (RPE) amounted to 3.5 $\pm$ 1.3 and 4.1 $\pm$ 1.5, respectively.

Systolic blood pressure (SBP) at W$_{OBLA}$ amounted to 210 $\pm$ 19 in the patients as compared to 191 $\pm$ 24 mm Hg in the controls (p $<$ 0.01). The corresponding heart rates (HR) were 154 $\pm$ 31 beats/min (SD) and 168 $\pm$ 37 beats/min. Thus an elevated pressure was present in patients whereas HR and RPE were approximately the same. SBP during loadless pedaling

Increase in systolic blood
pressure between loadless
pedalling (W_0) and W_{OBLA}

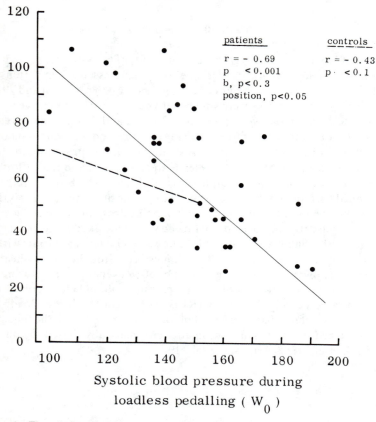

Figure 1—The relationship between the increase in systolic blood pressure (SBP) between loadless pedaling (W_O) and the work load corresponding to a blood lactate concentration of 4 mmol $\times$ l^{-1} (W_{OBLA}) and SQB at W_O. Two patient or subject materials are included: a) untreated patients with essential hypertension (WHO I) and healthy controls.

(W_O) averaged 147 $\pm$ 21 in the patients and 142 $\pm$ 15 mm Hg in the controls. The subsequent increases between W_O and W_{OBLA}, 63 $\pm$ 24 and 49 $\pm$ 12 mm Hg (p < 0.05), followed different functions in respect to SBP at W_O (Figure 1) (p < 0.05). If the increase in SBP for the patients was related to the corresponding increases in blood lactate concentration, it

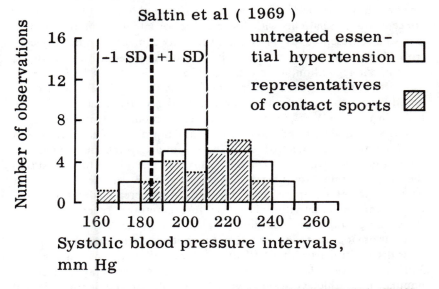

Figure 2—The number of SBPs at W_OBLA within the intervals 160 to 169, 170 to 179, 180 to 189, etc., for patients with untreated essential hypertension (Karlsson et al., 1983) and representatives for contact sports (Dlin & Karlsson, to be published). In addition the mean ± 1 SD of a population of sedentary, healthy middle aged men is presented for comparison (Saltin et al., 1969; Kilbom et al., 1969).

could be estimated that they ranged anywhere between approximately 5 and 35 mm Hg, whereas the corresponding values for the controls were around 20 mm Hg. It seems then reasonable to suggest that the hypertensive patients, who had a relatively low SBP at W_O, responded with an exaggerated SBP increase during exercise and that those who already had a high SBP at W_O were only able to increase it slightly in spite of a W_OBLA in the order of 150 W. There was no systematic relationship between final SBP or increase in SBP, respectively, with W_OBLA. When a histogram analysis was applied to the individual patient data it became obvious that only a few patients had SBPs at W_OBLA, which were within ± 1 SD of the mean of the present controls or a comparable population of healthy subjects (Saltin et al., 1969). It was also possible to conclude that the present data and distribution pattern were similar to that found in an ongoing study on athletes representing contact sport (Dlin & Karlsson, to be published) (Figure 2).

Discussion

The major finding in the present study was the exaggerated blood pressure response at a work load corresponding to 4 mmol $\times$ l^{-1} of blood lactate concentration in patients with "border line" or latent essential hypertension and representatives for contact sports as compared to healthy controls. This might be synonymous with a postulated desensitization of the sympathetic regulation in these groups (Karlsson et al., 1983) (Figure 3).

The continuous increase in plasma catecholamine levels for a given blood lactate concentration produced a relatively higher increase in SBP between W$_O$ and W$_{OBLA}$ in the patients with a moderately increased SBP at W$_O$ compared with endurance trained and untrained (Keul et al., 1982). On the other hand when SBP at W$_O$ corresponded to 160 mm Hg or more the SBP increase was subnormal. It seems then reasonable to suggest that the lack of ability to substantially increase SBP represents a further deterioration of the circulatory control due to an increased desensitization of the sympathetic activity and it is therefore tempting to speculate whether these changes represent a continuous alteration in the disease state.

It was obvious that representatives for contact sports seemed to have the same distribution pattern in SBP at W$_{OBLA}$ as the present patients (Figure 2). Dlin et al. (1981) have earlier reported that a high percentage of a similar population of athletes during a 5 yr period will develop clinical and treated hypertension. All these athletes have or have had normal blood pressure responses at rest but exaggerated during exercise in contrast to the present patients.

As pointed out above Shepherd et al. (1981) have suggested the presence of peripheral sensory organs—ergoreceptors—which through VMC activate central circulation and perhaps peripheral vasodilatation. These receptors would fire with, e.g., elevated extracellular concentrations of K$^+$, H$^+$ (i.e., decreased pH), lactate etc. It is possible that blood lactate as obtained in the OBLA test might reflect or mimic these conditions. An insufficient peripheral oxygen supply due to elevated peripheral resistance will result, for example, in elevated lactates, increased sympathetic activity and possibly elevated blood pressure. Lower peripheral blood flow and increased blood pressure has been experimentally coupled with muscles rich in fast-twitch muscle fibers. Patients with essential hypertension have also a higher percentage of fast-twitch fibers (Karlsson et al., 1983).

To summarize, the present data reveal that for the same metabolic conditions during exercise, patients with untreated, latent borderline hypertension have a higher systolic blood pressure than healthy controls. This is the result of either an exaggerated blood pressure response during exercise or an elevated blood pressure at onset of exercise (loadless

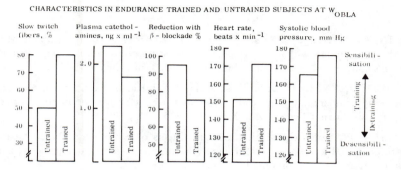

Figure 3a—A schematic presentation of some circulatory related features, which differentiates trained subjects from untrained (Keul et al., 1982; Karlsson et al., 1983).

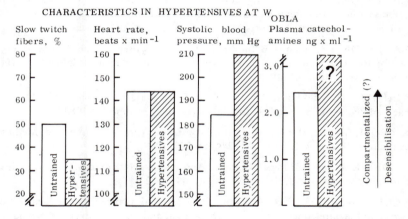

Figure 3b—The same as in Figure 3a but comparing patients with untreated essential hypertension with sedentaries.

pedaling), combined with a low pressure increase during exercise. It is suggested that the latter situation represents a further deterioration of the disease.

Acknowledgment

The study was supported by grants from ICI-Pharma AB, Gothenburg, Sweden.

References

DLIN, R., N. Hanne, D.S. Silverberg & O. Bar-Or. Follow up of apparently healthy individuals with exaggerated blood pressure response to exercise. The II world's congress of cardiac rehabilitation. p. 107, Jerusalem 1981.

JUHLIN-DANNFELT, A., M. Frisk-Holmberg, J. Karlsson & P. Tesch. Central and peripheral circulation in relation to muscle fibre composition in normo- and hypertensive man. Clin. Sc. Mol. Med. 56. 4:335–340, 1979.

KAISER, P., S. Rössner & J. Karlsson. Effects of B-adrenergic blockade on endurance and short-time performance in respect to individual muscle fiber composition. Int. J. Sports Med. 2:37–42, 1981.

KARLSSON, J., R. Dlin, P. Kaiser, P.A. Tesch & C. Kaijser. Muscle metabolism, regulation of circulation and B-blockade. J. Cardiovasc. Reh. Accepted for publication.

KARLSSON, J. & I. Jacobs. Onset of blood lactate accumulation (OBLA) during exercise and its significance in healthy untrained and trained subjects and some patient categories. Int. J. Sports Med. 1982. Accepted for publication.

KARLSSON, J., I. Jacobs, B. Sjödin, P. Tesch, P. Kaiser, O. Sahl & B. Karlberg. Semi-automatic blood lactate assay: experiences from an exercise laboratory. Int. J. Sports Med. 4. 45–48, 1983.

KEUL, J., H. Dickhuth, M. Lehmann & J. Staiger. The athlete's heart—haemodynamics and structure. Int. J. Sports Med. 3. Suppl. 1, 1983.

KILBOM, Å., L.H. Hartley, B. Saltin, J. Bjure, G. Grimby & I. Åstrand. Medical evaluation. Scand. J. Clin. Invest. 24. 315–323, 1969.

RYDEVIK, U., L. Nord & F. Ingman. Automatic lactate determination by flow injection analysis. Int. J. Sports Med. 3:47–49, 1982.

SALTIN, B., L.H. Hartley, Å. Kilbom & I. Åstrand. Oxygen uptake, heart rate, and blood lactate concentration at submaximal and maximal exercise. Scand. J. Clin. Invest. 24. 323–334, 1969.

SHEPHERD, J.T., C.G. Blomqvist, A.R. Lind, J.H. Mitchell & B. Saltin. Static (isometric) exercise. Retrospection and introspection. Circ. Res. 48(Suppl. 1): 179–188, 1981.

Maximal Blood Lactic Acid Concentration and its Recovery Course after Exhaustive Graded Treadmill Exercise in Young Men

J.-S. Lai and I-N. Lien
National Taiwan University Hospital, Taipei, Taiwan,
Republic of China

Åstrand (1960) found that the arterial lactic acid concentration increased during and after severe exercise of 2 min duration, followed by a slow decline back to the resting level. It was suggested by Åstrand and Rodahl (1977) that for a determination of peak lactic acid in the blood after severe exercise, samples must be taken at intervals during the first 5 to 10 min of the recovery period. The maximal blood lactate concentration is usually above 8 to 9 mM (Åstrand, 1960; Åstrand and Rodahl, 1977), when large muscle groups are involved in the exercise and the exercise duration exceeds 3 min.

Graded treadmill exercise has been widely used for detection of myocardial ischemia and evaluation of physical fitness. It is important and interesting to make clear whether the intensity and appearance time of maximal lactate concentration in venous blood after exhaustive graded treadmill exercise are similar to those after severe exercise of different duration, exercise mode, and sampling sites. In this study the authors also tried to observe the recovery course of maximal blood lactate (LA) and attempted to investigate the influence of endurance capacity and body size on the intensity of maximal blood LA and its recovery course after exhaustive graded treadmill exercise in young men.

Methods

Fifteen healthy young men with mean ($\pm$ SD) age of 21.6 ($\pm$ 2.6) yr, mean body weight of 62.7 ($\pm$ 7.2) kg, and mean body height of 170.0 ($\pm$ 6.4) cm were tested by continuous graded treadmill exercise using Bruce protocol (1973) until they were exhausted. During the recovery period they sat on a chair. A precordial lead (CM_5) of ECG was recorded at rest, during exercise and during recovery period. The heart rate (HR) was calculated by measuring four successive R-R intervals on the ECG record. The exercise duration was determined by a stop-watch from the beginning of exercise to the end point of complete exhaustion or severe muscular fatigue. Expired air was collected at the last min of each Bruce stage and during maximal treadmill running by the open circuit method. The gas volume was determined with a Tissot spirometer and the concentrations of expired CO_2 and O_2 were measured with a Scholander apparatus. The oxygen uptake was calculated by standard formula. Venous blood samples were taken from an indwelling catheter in the superficial vein of the upper extremity of each subject at fixed intervals, which were at rest, at the end of Bruce stage 1 to 3, immediately after exhaustive exercise and at the end of 5th, 10th, 15th, 20th, 30th, 45th, and 60th min of recovery period. The blood LA was measured by an enzymatic method using Calbiochem Rapid Lactate Reagents. The results were analyzed by Student's t-test and Pearson correlation study.

Results

The mean ($\pm$ SD) exercise duration of these subjects was 14.0 ($\pm$ 2.6) min. Their mean maximal HR was 196 ($\pm$ 8) beats/min, maximal oxygen uptake per kg body weight was 57 ($\pm$ 10) ml kg^{-1} min^{-1}, and maximal blood LA was 13.2 ($\pm$ 1.7) mM.

The resting LA was 1.3 ($\pm$ 0.4) mM and the mean LA at the end of Bruce stage 1, 2 and 3 were 1.3 ($\pm$ 0.2), 1.5 ($\pm$ 0.3), and 2.7 ($\pm$ 0.9) mM, respectively. The blood LA remarkably increased to 9.3 ($\pm$ 1.9) mM immediately after exhaustive exercise and reached its peak level at the end of the 5th or 10th min after exercise, then gradually returned toward resting level. At the end of the 1st h of recovery period the blood LA was 2.9 ($\pm$ 0.7) mM, which was two times higher than the resting level. The blood LA at the end of the 5th, 10th, 15th, 20th, 30th and 45th min of the recovery period were 12.7 ($\pm$ 1.4), 12.8 ($\pm$ 2.1), 11.3 ($\pm$ 2.4), 9.9 ($\pm$ 1.5), 7.2 ($\pm$ 2.0), and 4.4 ($\pm$ 1.4) mM, respectively.

The intensity of maximal blood LA was not significantly correlated with the subject's exercise duration, maximal oxygen uptake, maximal HR, body weight, or body height. The blood LA at the end of the 15th, 20th, 30th, 45th and 60th min after exercise as well as the removal rate of LA at the first 30 min of the recovery period were positively correlated with the intensity of maximal LA (r = 0.55 to 0.92, p < 0.05). However, there was no significant correlation between blood LA at the 1st h of the recovery period and exercise duration or maximal oxygen uptake. The recovery blood LA and removal rate of LA were not significantly correlated with endurance capacity, body size or maximal HR.

Discussion

Karlsson and Saltin (1970) found that the mean peak blood LA at exhaustion were 13.4 and 13.3 mM for the highest and medium loads, while at the lowest load the mean LA was definitely lower (9.1 mM). The durations of the heavy bicycle exercise bouts were 2 to 3 min (highest load), 5 to 7 min (medium load), and 15 to 20 min (lowest load). Although in this study blood samples were taken from the superficial vein instead of from the prewarmed and dry fingertip, the authors found that the intensity and appearance time of maximal LA after exhaustive graded treadmill exercise were similar to those after severe exercise of different duration, exercise mode and sampling sites (Åstrand, 1960; Åstrand & Rodahl, 1977; Karlsson & Saltin, 1970). In the authors' unpublished study using the same methods (except that blood samples were taken by repeated venipuncture) the maximal LA in 12 young female students was 12.3 (± 2.0) mM, which was very close to the mean value of our male subjects. This finding confirms the observation of Åstrand and Rodahl (1977) who found that the blood LA in men and women was, on the average, the same after maximal exercise and within the range of 11 to 14 mM for twenty- to forty-yr-old trained individuals.

The exercise duration achieved by graded treadmill exercise until exhaustion is dependent on the motivation of the subjects and their anaerobic and aerobic capacity (Cumming et al., 1978). In this study the mean exercise duration was 14 min and maximal oxygen uptake was 57 ml kg^{-1}, which suggested that most of these subjects belonged to the physically active group by our standard (Lien and Lai, 1980). The correlation studies shown in the previous reports (Bruce et al., 1973; Cumming et al., 1978; Lien and Lai, 1980) all indicate a very high positive correlation between exercise duration and maximal oxygen uptake per kg body weight. It was clearly shown in this study that the intensity and appearance time of maximal blood LA were not significantly correlated with exercise duration, maximal oxygen uptake, maximal HR, body weight or body height. The blood LA at any time of the 1st h of the recovery period and the removal rate of blood LA were not significantly correlated with these factors, either. So it is concluded that, in young men, the intensity and appearance time of maximal blood lactate and its recovery course after exhaustive graded treadmill exercise are not significantly influenced by one's endurance capacity and body size.

References

ÅSTRAND, I., Aerobic work capacity in men and women with special reference to age. Acta Physiol. Scand. 49(Suppl. 169), 1960.

ÅSTRAND, P-O. & K. Rodahl. Textbook of Work Physiology. New York: McGraw-Hill, 1977.

BRUCE, R.A., F. Kusumi & D. Hosmer. Maximal oxygen intake and nomographic assessment of functional aerobic impairment in cardiovascular disease. Am. Heart J. 85:546–562, 1973.

CUMMING, G.R., D. Everatt & L. Hastman. Bruce treadmill test in children: normal values in a clinic population. Am. J. Cardiol. 41:69–75, 1978.

KARLSSON, J. & B. Saltin. Lactate, ATP, and CP in working muscles during exhaustive exercise in man. J. Appl. Physiol. 29:598–602, 1970.

LIEN, I.-N. & J.-S. Lai. Endurance capacity and exercise response in young Chinese adults. J. Formosan Med. Assoc. 79:1109–1121, 1980.

Oxygen Uptake Kinetics and Lactate Accumulation in Heavy Submaximal Exercise with Normal and High Inspired Oxygen Fractions

P.K. Pedersen

Odense University, Odense, Denmark

A number of studies have shown that inhalation of gas mixtures with increased O_2 content (hyperoxia) gives rise to lower arterial blood lactate concentrations in heavy submaximal exercise (Ekblom et al., 1975; Welch et al., 1977). The reduced lactate concentration may be interpreted as evidence for an increased aerobic energy contribution with oxygen breathing, and hence less dependence on anaerobic energy sources. In a recent study (Welch & Pedersen, 1981) it could be demonstrated that at several levels of submaximal exercise the steady state O_2 consumptions were not significantly different in normoxia and hyperoxia. If the O_2 uptakes are similar, the observed lower lactate values in hyperoxia can hardly be ascribed to an increased aerobic energy yield unless the possible differences have appeared before steady state has been attained. Thus, from the onset of exercise, oxygen uptake ($\dot{V}O_2$) was measured min-by-min in order to compare the adjustment toward steady state while inspiring either atmospheric air or a hyperoxic gas mixture.

Procedures

Five male subjects (mean age, height and weight 27 yr, 180 cm and 70 kg, respectively) performed bicycle exercise at a work rate of 210 W (exact value 206 W), which corresponded to 70 to 80% of $\dot{V}O_2$max. The work was performed on a weight-braked ergometer, (Cardionics), which was preferred because the desired resistance (35 N for 60 revolutions per min [rpm]) was obtained from the very beginning of exercise with no need for subsequent adjusting. Pedal rate was recorded continuously over the 20 min work period.

415

The subjects reported to the laboratory in the morning in a fasting state and with no preceding physical effort. After 10 min supine rest, a 15 min acclimatization period was initiated with the subject seated on the ergometer breathing either atmosphere or 60% O_2 in N_2. The subjects were not aware of the composition of the actual inspired gas. Air was supplied to the subject from pressure cylinders via a 100 l reservoir bag. The administration of the gas mixtures was systematically varied in order to minimize any ordering effect. Each subject performed two experiments with each gas mixture, giving a total of 10 normoxic and 10 hyperoxic experiments.

$\dot{V}O_2$ and $\dot{V}CO_2$ were determined with open circuit spirometry utilizing a modified Douglas bag technique (Welch & Pedersen, 1981). Volumes were measured with a Tissot spirometer. Samples for analysis of air composition were drawn from a mixing chamber and analyzed with a Scholander apparatus by duplicate analysis. Analyses were accepted only if the samples differed by less than 0.06% (absolute values). Measurements were made at rest (last 5 min of acclimatization period), and at 1 min intervals over the first 10 min of exercise, and again at min 20 at the end of the work period. Preliminary experiments indicated that the $\dot{V}O_2$ values were similar at min 10 and min 20 so that the chosen observation period should allow for the attainment of steady state conditions. Heart rate was measured from ECG-recordings obtained at the same time as the ventilatory measurements. Blood samples for lactate determination were drawn from pre-warmed fingertips and analyzed with an enzymatic method (Boehringer). The samples were obtained at rest, at 2 min intervals during the first 10 minutes of exercise, and again at min 20.

The subjects' maximal working capacity while pedaling the cycle ergometer was tested on two separate occasions by means of a progressive exercise test regime. Standard criteria for the assessment of maximal effort were applied. The highest values obtained averaged for work rate 306 $\pm$ 25W, for $\dot{V}O_2$max 3.88 $\pm$ 0.35 l min^{-1} or 55 $\pm$ 5 ml min^{-1} per kg body wt, for blood lactate concentration 12.3 $\pm$ 2.0 mmols l^{-1}, for heart rate 183 $\pm$ 4 beats min^{-1}, for $\dot{V}E - \dot{V}O_2$ ratio 38 $\pm$ 6 l BTPS l STPD and for respiratory exchange ratio 1.17 $\pm$ 0.09 (means $\pm$ SD).

Conventional statistical methods were applied. Individual differences were evaluated by means of Student's t-test, utilizing p-values less than 0.05 to indicate statistical significance.

Results

Within 6 min after the onset of exercise $\dot{V}O_2$ had stabilized around a level typical for a work rate of 210 W (Figure 1). The attained steady state levels

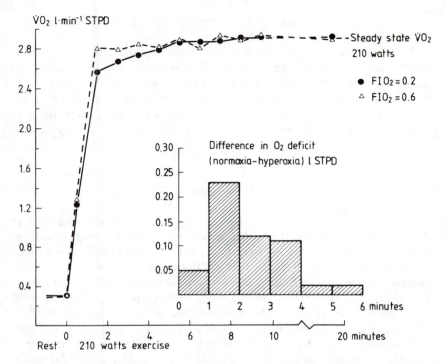

Figure 1—O_2 uptake as a function of time at 210 W cycle exercise while breathing either atmospheric air (●) or a 60% O_2 mixture (△). SEM (not indicated) averaged 0.02 to 0.05 l min^{-1}. The inserted figure elucidates the difference in O_2 deficit between the two gas mixtures during the first 6 min of exercise.

were similar in normoxia and hyperoxia (means ± 1 SEM 2.925 ± 0.03 l min^{-1} and 2.915 ± 0.04 l min^{-1}, respectively). The $\dot{V}_{O_2}$ increased to steady state at a faster rate when breathing 60% oxygen. Thus, from min 1 to 4 inclusive, the hyperoxic $\dot{V}_{O_2}$ values were significantly larger than the normoxic. The calculated O_2 deficit averaged 2.65 ± 0.13 l STPD in normoxia and 2.09 ± 0.18 l STPD in hyperoxia. The major part of the extra 0.56 l of oxygen taken up with hyperoxia could be located to min 1 to 2 (0.23 l STPD) and, to a lesser extent, min 2 to 4 (0.12 and 0.11 l STPD, respectively; see insertion, Figure 1). The O_2 deficit components from min 0 to 1 and 4 to 6 contributed only a little to the observed hyperoxic-normoxic difference.

The $\dot{V}_{CO_2}$ followed a pattern different from that of $\dot{V}_{O_2}$. From similar resting values in the two experimental conditions the hyperoxia $\dot{V}_{CO_2}$ increased at a slower rate and was significantly lower by 0.06 to 0.17 l min^{-1} between min 1 and 7. Although the difference was no longer significant, the average hyperoxic $\dot{V}_{CO_2}$ remained lower in the remaining exercise period so that the total CO_2 output was 1.49 l smaller with hyperoxia in spite of the fact that the O_2 uptake was 0.56 l larger.

The blood lactate concentrations at rest were similar in hyperoxia and normoxia (1.1 ± 0.1 vs 1.3 ± 0.1 mmols l^{-1}). With the onset of exercise the normoxia lactate concentration increased gradually to 4 to 4.5 mmols l^{-1} and remained at that level (Figure 2). In hyperoxia a peak value of approximately 3 mmols l^{-1} was attained after 4 to 6 min, after which a gradual decrease was seen to an average concentration of 2.1 ± 0.4 mmols l^{-1} at min 20.

Discussion

With the onset of exercise the working organism is confronted with a demand for ATP-resynthesis for which the aerobic energy yielding processes are initially inadequate. An O_2 deficit is created. The O_2 deficit accompanying heavy (more than 50% of individual $\dot{V}_{O_2}$max), steady level exercise has been described by a two-component model (Raynaud et al., 1974; Whipp & Wasserman, 1972). A fast component was assigned to the splitting of muscular creatine phosphate, a later occurring slow component to anaerobic glycogenolysis. In this study, the first min O_2 deficit was not significantly altered by oxygen breathing (1.64 ± 0.06 vs 1.69 ± 0.05 l STPD). A similar observation was made by Raynaud et al. (1974) when comparing sea level O_2 deficit with high altitude. Apparently, this O_2 deficit component is unaffected by oxygen availability. This may appear noticeable since anaerobic glycogenolysis is likely to have been accelerated in the last part of min 1 (Gollnick, & Hermansen, 1973), so that a possible difference between normoxia and hyperoxia might have been detectable. The explanation could be that the splitting of creatine phosphate dominates the energy release in this phase of exercise, so that possible differences related to glycogenolysis cannot be distinguished.

The fact that O_2 uptake rose more rapidly in min 1 to 4 with hyperoxia, reduced the O_2 deficit by about 0.5 l STPD. In comparison, for moderate exercise Linnarsson et al. (1974) have reported a 0.3 l decrease in O_2 deficit when breathing air at raised ambient pressure (1.4 ATA). The magnitude of the O_2 deficit is affected by several factors, i.e., work intensity (Knuttgen, 1970; Whipp & Wasserman, 1972), and state of training (Hickson et al., 1978). A significant positive correlation ($p < 0.025$) was observed in the present study between individual O_2 deficit values and relative work load (% $\dot{V}_{O_2}$ max) at the selected, fixed work rate. This is in accordance with earlier findings. The faster rise in O_2 uptake with hyperoxia may be based upon a faster adjustment of the circulatory transport of O_2 to exercise. Or it may reflect an inertia in the processes of the peripheral O_2 utilization

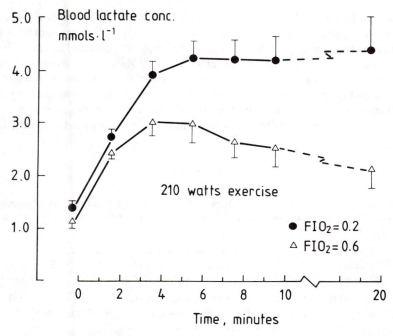

Figure 2—Blood lactate as a function of work time at 210 W while breathing either normoxic (●) or hyperoxic (△) air. Vertical lines indicate + or - 1 SEM.

(Linnarsson, 1974). Our present state of knowledge does not exclude either of these alternatives, although it is a common observation that oxygen saturation in venous effluent from exercising muscles rapidly becomes reduced, suggesting that the main delay occurs centrally.

The mean peak lactate concentrations obtained after 4 to 6 min of exercise differed by 1.3 mmols l^{-1} in hyperoxia and normoxia, a difference which presumably is related to a lower lactate release from the exercising muscles in hyperoxia (Welch et al., 1977). The question is then, can this difference be explained by the observed larger initial O_2 uptake with oxygen breathing? All estimates of total lactate formation based solely on lactate concentration values in blood are questionable. This is, in part, due to the lack of exact knowledge about the distribution space for lactate. Assuming an even distribution among all water compartments of the body, and taking into account the different water fractions for blood and for the body as a whole, Margaria et al. (1963) suggest a method based on multiplying the blood lactate concentration by 0.75 and body wt. Applied to the present material the observed difference would imply a difference in total amount of lactate of about 70 mmols. Depending on the substrate exploited, the extra 0.5 l, or 22 mmols, of oxygen consumed with hyperoxia may yield approximately 130 mmols of ATP. With 2 mmols of lactate obtained per 3 mmols of ATP, the formation of a similar amount of ATP via anaerobic glycogenolysis would bring about 90 mmols of lactate. Thus, the extra O_2 uptake at the onset of heavy submaximal exercise when hyperoxic air is inhaled is sufficient to explain the observed difference in peak lactate concentration. It remains to be explained, however, why the difference in blood lactate concentration increases over the later phases of exercise in spite of similar O_2 uptakes in the two experimental conditions.

References

EKBLOM, B., R. Huot, E.M. Stein & A.T. Thorstensson. Effect of changes in arterial oxygen content on circulation and physical performance. J. Appl. Physiol. 39:71–75, 1975.

GOLLNICK, P.E. & L. Hermansen. Biochemical adaptations to exercise: Anaerobic metabolism. In J.H. Wilmore (Ed.), Exercise and Sports Sciences Reviews. Vol. 1:1–43. New York and London: Academic Press, 1973.

HICKSON, R.C., H.A. Bomze & J.O. Holloszy. Faster adjustment of O_2 uptake to the energy requirement of exercise in the trained state. J. Appl. Physiol.: Respirat. Environ. Exercise Physiol. 44:877–881, 1978.

KNUTTGEN, H.G., Oxygen debt after submaximal physical exercise. J. Appl. Physiol. 29:651–657, 1970.

LINNARSSON, D., Dynamics of pulmonary gas exchange and heart rate changes at start and end of exercise. Acta Physiol. Scand. Suppl. 415:1–68, 1974.

LINNARSON, D., J. Karlsson, L. Fagraeus & B. Saltin. Muscle metabolites and oxygen deficit with exercise in hypoxia and hyperoxia. J. Appl. Physiol. 36:399–402, 1974.

MARGARIA, R., P. Cerretelli, P.E. DiPrampero, C. Massari & G. Torelli. Kinetics and mechanism of oxygen debt contraction in man. J. Appl. Physiol. 18:371–377, 1963.

RAYNAUD, J., J.P. Martineaud, J. Bordachar, M.C. Tillous & J. Durand. Oxygen deficit and debt in submaximal exercise at sea level and high altitude. J. Appl. Physiol. 37:43–48, 1974.

WELCH, H.G. & P.K. Pedersen. Measurement of metabolic rate in hyperoxia. J. Appl. Physiol.: Respirat. Environ. Exercise Physiol. 51:725–731, 1981.

WELCH, H.G., F.B. Petersen, T. Graham, K. Klausen & N. Secher. Effects of hyperoxia on leg blood flow and metabolism during exercise. J. Appl. Physiol.: Respirat. Environ. Exercise Physiol. 42:385–390, 1977.

WHIPP, B.J. & K. Wasserman. Oxygen uptake kinetics for various intensities of constant load work. J. Appl. Physiol. 33:351–356, 1972.

Oxygen Consumption and Lactate Production in Varanid and Iguanid Lizards: A Mammalian Relationship

H.J. Seeherman, R. Dmi'el and T.T. Gleeson
Harvard University, Cambridge, Massachusetts, U.S.A.

A general relationship between aerobic metabolism and anaerobic glycolysis during exercise has recently been described in a wide variety of mammals (Seeherman, 1981). The purpose of this study was to determine if a similar relationship occurs in two reptilian species.

Oxygen consumption ($\dot{V}_{O_2}$) in mammals was found to increase with running speed up to a maximal rate ($\dot{V}_{O_2}max$) that was not exceeded with further increases in speed. A continuous increase in the blood lactate concentration [lactate]$_b$ occurred only at running speeds where the rate of energy utilization ($\dot{E}$) exceeded that which could be supplied by $\dot{V}_{O_2}max$ ($\dot{E} > \dot{V}_{O_2}max$). When $\dot{V}_{O_2}$ was less than $\dot{V}_{O_2}max$ ($\dot{V}_{O_2} < \dot{V}_{O_2}max$), [lactate]$_b$ remained constant or declined after the initial minutes of exercise. Muscle lactate concentrations changed in a similar manner indicating that [lactate]$_b$ at least qualitatively reflects the changes in the lactate concentration of the body during these exercises. Constant or declining lactate concentrations indicate that whole animal anaerobic glycolysis (Lactate) is zero after the initial minutes of exercise where $\dot{E} < \dot{V}_{O_2}max$. $\dot{E}$ is therefore equal to $\dot{V}_{O_2}$ despite any local anaerobic glycolysis in active muscles. Additional $\dot{E}$ is supplied by Lactate at exercise intensities where $\dot{E} > \dot{V}_{O_2}max$. Carbon dioxide production ($\dot{V}_{CO_2}$) increased disproportionately to $\dot{V}_{O_2}$ at these exercise intensities, yielding respiratory quotients (R) greater than one reflecting respiratory compensation for acidosis.

To determine whether a similar relationship occurs in reptiles, $\dot{V}_{O_2}$, $\dot{V}_{CO_2}$, Lactate and arterial pH were measured during exercise in savannah moniters *(Varanus exanthematicus)* and green iguanas *(Iguana iguana)*.

Varanids are active predators with a large aerobic scope and cardiovascular adaptations that reflect their ability for sustained activity (Gleeson et al., 1980). Iguanas, in contrast, are sedentary herbivores with a low aerobic scope (Gleeson et al., 1980).

Materials and Methods

Animals

Four *Varanus exanthematicus* (mean body mass = 0.70 kg) and four *Iguana iguana* (mean body mass = 1.06 kg) were obtained from commercial dealers. Details concerning husbandry are in Gleeson et al. (1980).

Procedures

$\dot{V}_{O_2}$ and $\dot{V}_{CO_2}$ were measured during an initial training period while the lizards ran on a treadmill. They were considered "trained" after 2 to 4 wk when reproducible values were obtained at various speeds. $\dot{V}_{O_2}$ and $\dot{V}_{CO_2}$ were then measured as a function of running speed after equilibration for 2 h at 35° C. Arterial pH and $[lactate]_b$ were determined from blood samples drawn prior to, during and after each run.

Methods

$\dot{V}_{O_2}$ and $\dot{V}_{CO_2}$ were measured simultaneously using an open flow system. The lizards wore a light, loose fitting, plastic mask through which a metered amount of air was drawn. Details of this system are found in Seeherman et al. (1981) and Fedak et al. (1981). R values were calculated from $\dot{V}_{CO_2}$ and $\dot{V}_{O_2}$.

Arterial pH and $[lactate]_b$ were determined from samples taken through chronically implanted catheters in the external carotid artery. The sampling and analysis procedure is described in Gleeson et al. (1980). Lactate $(mmol\ kg^{-1}\ min^{-1})$ was estimated from the net increase in $[lactate]_b$ $(mmol\ l^{-1})$ averaged for the duration of the exercise (min) multiplied by a correction factor (0.86) equating the water content of blood $(0.831_{H_2O}$ $^1blood^{-1})$ to total body water $(0.71_{H_2O}\ kg^{-1})$. Hematocrits remained constant for iguana (35%) and varanus (30%).

Results

$\dot{V}_{O_2}$ increased with running speed up to a maximal rate $(\dot{V}_{O_2}max)$ and remained constant with further increases in speed in both the varanids and

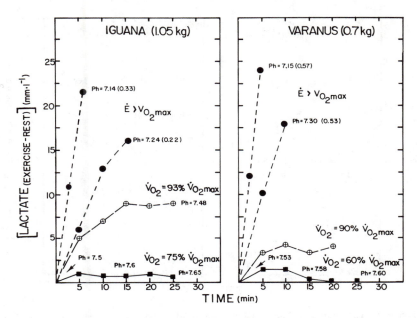

Figure 1—Mass specific oxygen consumption ($\dot{V}_{O_2}/M_b$,•) and lactate production ([Lactate$_{exercise-rest}$], ⊕) plotted as a function of running speed; and the mean respiratory quotient ($R = \dot{V}_{CO_2}/\dot{V}_{O_2}$, SEM, n.) for runs included within brackets for four iguanid and four varanid lizards. See methods for Lactate calculations. The least squares regressions of the functions relating $\dot{V}_{O_2}/M_b$ and speed (excluding $\dot{V}_{O_2}$max) are:

	slope (ml kg^{-1} min^{-1})	y-intercept (ml s^{-1} kg^{-1})	p	r^2	S$_{yx}$
Iguana	0.90	0.13	.001	96%	0.17
Varanus	0.69	0.08	.001	98%	0.27

For illustration $\dot{V}_{O_2}$max for one lizard was plotted. $\dot{V}_{O_2}$max for varanus (0.47 mlO$_2$ s^{-1} kg^{-1} ± .01, 4) was greater than $\dot{V}_{O_2}$max for iguana (0.28 mlO$_2$ s^{-1} kg^{-1} ± .01, 4, t = 9.5, d.f. = 6, p < .001). An order of magnitude comparison between $\dot{V}_{O_2}$ and Lactate was obtained by adjusting the axis such that 1 mlO$_2$ sec^{-1} kg^{-1} = 11 mmol lactate kg^{-1} min^{-1} (P/O ratio of 6, P/lactate ratio of 1.5).

the iguanids (Figure 1). The least squares regression of the function relating $\dot{V}_{O_2}$ (excluding $\dot{V}_{O_2}$max) and running speed are given in the legend for Figure 1. The mean value for $\dot{V}_{O_2}$max in varanids (0.47 mlO$_2$ s^{-1} kg^{-1}, 9.5 W kg^{-1}) was greater than $\dot{V}_{O_2}$max for iguanas (0.28 mlO$_2$ s^{-1} kg^{-1}, 5.7 W kg^{-1}, p < .001).

[Lactate]$_b$ changed in a similar manner in varanids and iguanids at equivalent exercise intensities relative to $\dot{V}_{O_2}$max (Figure 2). A continuous increase in [lactate]$_b$ occurred only at running speeds where E > $\dot{V}_{O_2}$max. Increases in [lactate]$_b$ were confined to the initial minutes of exercise where $\dot{V}_{O_2}$ < $\dot{V}_{O_2}$max. During the remainder of these exercises, [lactate]$_b$

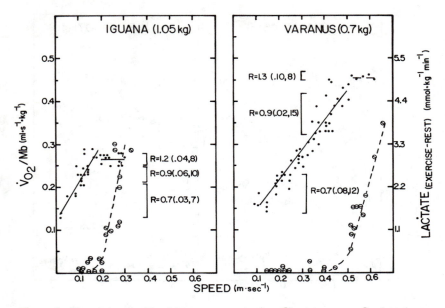

Figure 2—The change in blood lactate concentration ([Lactate_{exercise-rest}]) plotted as a function of time for iguana and varanus at three exercise intensities: 1) $\dot{V}_{O_2}$ well below $\dot{V}_{O_2}$max ($\bullet$); 2) $\dot{V}_{O_2}$ close to $\dot{V}_{O_2}$max ($\ominus$); and 3) two rates of energy utilization ($\dot{E}$) greater than $\dot{V}_{O_2}$max. Rest and exercise values for pH are also plotted (speed is given for $\dot{E} > \dot{V}_{O_2}$max in parentheses). Graphs are representative of multiple trials. The mean decrease in pH for iguana from rest ($7.5 \pm .04$, 5) to $\dot{E} > \dot{V}_{O_2}$max ($7.06 \pm .04$, 9) was significant (-0.45, t = 10.8, d.f. = 8, p < .01). Similar values for varanus from rest ($7.53 \pm .01$, 4) to $\dot{E} > \dot{V}_{O_2}$max ($7.14 \pm .08$, 4) were also significant (-0.39, t = 6.16, d.f. = 3, p < .02).

remained constant or declined. Lactate increased dramatically at running speeds where $\dot{E} > \dot{V}_{O_2}$max (Figure 1).

The mean difference in arterial pH decreased in iguana and varanus at exercise intensities where $\dot{E} > \dot{V}_{O_2}$max (-0.45, p < .01; -0.39, p < .02) (Figure 2). Minimal changes in arterial pH occurred at slower speeds (Figure 2).

R values for iguana and varanus were greater than one at running speeds where $\dot{E} > \dot{V}_{O_2}$max ($1.2 \pm .04$; $1.3 \pm .10$ (Figure 1). R was less than 1.9 when $\dot{V}_{O_2} < \dot{V}_{O_2}$max.

Time to exhaustion (inability to maintain a constant speed) decreased with increasing speed where $\dot{E} > \dot{V}_{O_2}$max (5 min at the highest speeds). Experiments were terminated after 15 to 30 minutes at running speeds where $\dot{V}_{O_2} < \dot{V}_{O_2}$max.

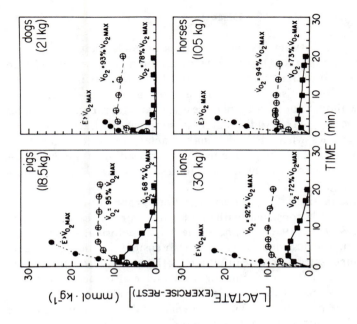

Figure 4—Blood lactate concentrations [Lactate(exercise-rest)] plotted as a function of time for four species of mammals at three exercise intensities (from Seeherman et al., 1981).

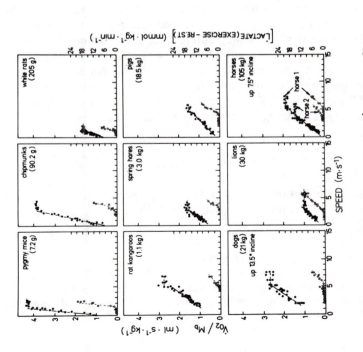

Figure 3—Mass specific oxygen consumption $\dot{V}_{O_2}/M_b$ and [Lactate(exercise-rest)] plotted as a function of speed for nine species of mammals (from Seeherman et al., 1981).

Discussion

The relationship between oxygen consumption and lactate production during exercise appears to be the same in varanus and iguana as has been previously described in mammals (Figures 3 & 4). In particular, these two reptilian species are also capable of sustained aerobic exercise, as are mammals, without a continuous increase in $[lactate]_b$ at exercise intensities where $\dot{V}_{O_2} < \dot{V}_{O_2}max$. Additional $\dot{E}$ is supplied by Lactate only at exercise intensities where $\dot{E} > \dot{V}_{O_2}max$. Lactate values calculated for speeds where $\dot{V}_{O_2} < \dot{V}_{O_2}max$ (Figure 1) are misleading in the sense that Lactate becomes zero after the initial minutes of exercise when $[lactate]_b$ stops increasing.

The decline in arterial pH observed in both groups of lizards during maximal exercise is consistent with similar results reported for humans (Hermansen & Osnes, 1972). It appears that lactic acidosis also plays a role in causing fatigue in lizards at these exercise intensities. Lactic acidosis also caused a compensatory increase in $\dot{V}_{CO_2}$ in both groups of lizards resulting in R values greater than one. Similar results have also been reported in mammals (Seeherman et al., 1981). In this regard, R values can be useful in determining the proximity of $\dot{V}_{O_2}$ to $\dot{V}_{O_2}max$ and offers a non-invasive estimate of the contribution of lactate at a given intensity.

In summary, the relationship between aerobic metabolism and anaerobic glycolysis during exercise appears to be identical in mammals and these two species of reptiles. The major difference is that $\dot{V}_{O_2}max$ of lizards is only 1/5 that of mammals of similar body mass (Taylor et al., 1981). Higher values of $\dot{V}_{O_2}max$ allow mammals to maintain aerobic conditions at much higher running speeds [up to 5.5 m s^{-1} for 1 kg rat kangaroos vs 0.5 m s^{-1} for varanids (Figures 1 & 3)]. Resting rates of oxygen consumption of lizards are also 1/5 those of mammals of the same body mass.

Acknowledgments

We thank Gayle Kaufman for her help in preparing this manuscript. Research support was provided by NSF PCM75-22684.

References

FEDAK, M.A., L. Rome & H.J. Seeherman. One-step N_2 dilution technique for calibrating open-circuit $\dot{V}_{O_2}$ measuring systems. J. Appl. Physiol.: Respirat. Environ. Exercise Physiol. 51(3):772–776, 1981.

GLEESON, T.T., G.S. Mitchell & A.F. Bennett. Cardiovascular responses to graded activity in the lizards *Varanus* and *Iguana*. Am. J. Physiol.: Regulatory, Integrative Comp. Physiol. 8:R174–179, 1980.

HERMANSEN, L. & J.-B. Osnes. Blood and muscle pH after maximal exercise in man. J. Appl. Physiol. 32:304–308, 1972.

SEEHERMAN, H.J., C.R. Taylor, G.M.O. Maloiy & R.B. Armstrong. Design of the mammalian respiratory system. II. Measuring maximal aerobic capacity. Resp. Physiol. 44:11–23, 1981.

TAYLOR, C.R., G.M.O. Maloiy, E.R. Weibel, V.A. Langman, J.M.Z. Kamau, H.J. Seeherman & N.C. Heglund. Design of the mammalian respiratory system. III. Scaling maximal aerobic capacity to body mass: wild and domestic mammals. Resp. Physiol. 44:25–37, 1981.

Mathematical Approach to Lactate Kinetics in Short Strenuous Exercise

X. Sturbois and P. Jacqmin

Unit EDPH—UCL—1, Louvain-la-Neuve, Belgium,
and Laboratoire de Pharmacologie,
Bruxelles, Belgium

In sport, anaerobic metabolism is a major source of muscular energy for many events. Blood lactate concentration is used by Cerretell and Cantone (1976) in exercise physiology as a measure of the anaerobic capacity of athletes. This measure is very interesting for the physiologist who studies energetic production and acid base balance in the muscle. This is proposed by Hermansen and Osnes (1972) as a limiting factor of effort. The aim of this study is to show the kinetic evolution of blood lactate concentration (rest value corrected) and to estimate the initial production of lactate in muscle.

Methods

The first experiments were performed on 21 male volunteer athletes (age 19 to 21 yr, wt 66–74 kg). The strenuous short exercise consisted of cycling at 400 W and 128 rpm to exhaustion ($\pm$ 45 s). Blood samples were taken from the antecubital vein for lactate determination 3 times before the exercise, immediately after it concluded, and at 2, 4, 6, 8, 10, 12, 15, 20, 25,...55, and 60 min.

The individual approach of the lactate kinetics is done with slow perfusion of lactate solution (3.5 M) through the antecubital vein. The catheter and the perfusion were connected to an automatic pump (Secan). The blood samples were taken at short intervals in a time long enough to lead for a full exploitation of the experiment. Lactate concentration was

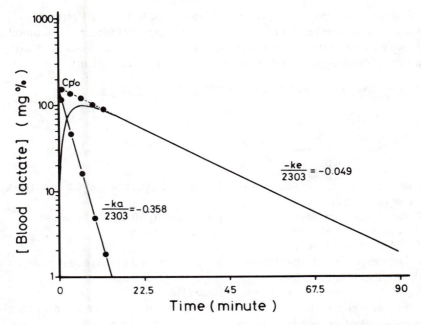

Figure 1—Kinetic evolution of blood lactate concentration (rest value corrected) on 21 male volunteer athletes after a strenuous short exercise (400 W, ± 45 s.) Graphical determination of Cp'_o, ka and ke.

determined by an enzymatic method (Boehringer) at the same time of the investigations.

Results

Evolution of Plasma Lactate Concentration in 21 Male Volunteer Athletes after One Strenous Short Exercise

The time course of the plasma lactate concentration (Figure 1) shows two steps:

1. the first step represents the diffusion of lactate from muscle in vascular compartment;
2. the second step represents the blood lactate elimination by metabolism.

This can accurately be represented by a kinetic mathematical model (open one compartment) similar to one intramuscular injection of drug. The form of equation proposed by Wagner (1975) or Gilbaldi and Perrier (1975) shows one exponential function associated with a lactate increase in blood compartment (diffusion from muscle space) and one other associated with a lactate decrease in blood compartment (lactate metabolism).

$$Cp_t = Cp_o \; [ka/(ka-ke)] \; (e^{-ke \, t} - e^{-ka \, t})$$

$$\underbrace{}_{} \quad\quad\quad (1)$$

$$\frac{F \, D}{Vd}$$

$$\underbrace{}_{Cp'_o}$$

Cp_t: plasma lactate concentration (in mg%) for the time "t" (in min) (rest value corrected).
Cp'_o: plasma lactate concentration (mg%) extrapolated to the initial time.
Ka and Ke: the absorption and elimination constants.
D: initial dose of lactate in muscle (g)
$F = 1$: hypothesis is that lactate in muscle at the end of exercise represents the total amount produced.
V_d: lactate space or distribution volume (1)

Figure 1 allows the graphical determination of Cp'_o, ka and ke and the equation for this case is:

$$Cp_t = 131 * [.358 - .049)] * (e^{-.049 \, * \, t} - e^{-.358 \, * \, t}) \quad\quad (2)$$
$$Cp_t = 151{,}77 * (e^{-.049 \, * \, t} - e^{-.358 \, * \, t})$$

Individual Parameters Evaluation

The individual kinetic curves are not exactly the same for all the subjects. We find two forms of responses which correspond to sprinter athletes (high ke) and to mild distance runners (low ke). It seems thus that an individual approach is indicated in order to know the part that anaerobic energy plays in effort. This is possible if we can determine the lactate space (Vd) for each athlete. The method used is the long time constant perfusion (venus perfusion) after a short strenuous exercise (0.50 to 0.75 time of maximal strenuous exercise).

For the illustrated athlete, we obtain a constant level of plasma lactate concentration (Figure 2). In this time, we can use the equation (in accordance with Wagner, 1975; and Gilbaldo & Perrier, 1975).

$$Cp_{steady \; state} \; K/(Vd * ke); \quad\quad (3)$$
$$K \text{ is the constant of perfusion}$$

and we can so obtain Vd. In this case, the value of Vd is 30.9 l. The

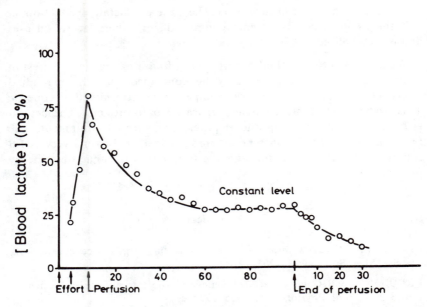

Figure 2—Time course of the plasma lactate concentration after strenuous short exercise (400 W, ± 30 s.) and constant perfusion (0.545 g min⁻¹).

produced dose in lactate is now easily calculated since we know Cp_0 and Vd following dose = $Cp_0 * Vd$.

For this subject: dose = 1310 mg/l * 30.91 = 40.48g.

Discussion and Conclusion

The proposed method can give a good estimation of the lactate quantity produced by the muscle after a short strenuous exercise. We have repeated the measure three times for this illustrated athlete and the variation of the results is compatible with other biological measures.

We have not verified the constance of parameters with changes in training. This point is very important for the anaerobic energy part determination. Indeed, if ka, ke and Vd stay constant, then the maximal plasma concentration is shown in the same time (t max) and is in good relation with the dose. If not, the determination of the kinetic parameters is needed at short intervals.

Many authors, like Freund and Gendry (1978), propose an open two-compartment model. The open two-compartment model is in best accordance with the two lactate space:

1. the first one is the muscle space where there is lactate production;
2. the second one is the "central compartment" where lactate diffuses and is metabolized.

We have used this method but we do not find it of practical interest in estimating the parameter. It may be concluded that the individual approach of the anaerobic energy source can be determined in athletes. This can be of major interest in the preparation of top athletes. The method is unfortunately very long and requires sophisticated equipment. The results are not strongly constant but are sufficiently close for a very useful follow up of training and for a good estimation of the anaerobic energy production.

References

CERRETELL, P. & A. Cantone. The role of different energy sources in exhaustive exercise. Med. Sport 29:199–205, 1976.

FREUND, W. & P. Gendry. Lactate kinetics after short strenuous exercise in man. Europ. J. of Appl. Physiol. 39:123–135, 1978.

GILBALDI, M. & D. Perrier. Pharmacokinetics. Ed. Marcel Dekker, New York, p. 329, 1975.

HERMANSEN, L. & J.B. Osnes. Blood and muscle pH after maximal exercise in man. J. Appl. Physiol. 32:304–308, 1972.

WAGNER, J.G., Fundamentals of Clinical Pharmacokinetics. Ed. Drug Intelligence Publications (Hamilton), p. 461, 1975.

Physical Training in Humans:
A Central or Peripheral Effect

K. Young and R.J. Maughan

University Medical School, Queen's Medical
Centre, Nottingham, England, and University
Medical School, Aberdeen, Scotland

The improvement in exercise capacity following endurance training is accompanied by cardiovascular changes, manifested as a decreased heart rate at any given exercise intensity, and by metabolic changes within the skeletal muscles which result in a decreased blood lactate concentration (BLa) at any given exercise intensity (Gollnick & Saltin, 1982). In bicycle exercise an increase in BLa corresponds to the anaerobic threshold and is first observed when the exercise intensity exceeds approximately 50 to 60% of an individual's maximum oxygen uptake ($\dot{V}o_2$max) (Hermansen & Stensvold, 1972). This lactate production has been interpreted as either a hypoxic state within the muscle (Wasserman, 1967) or, alternatively, as a result of pyruvate production by glycolysis exceeding pyruvate mobilization by the TCA cycle (Jobsis & Stainsby, 1968).

BLa has been shown to be lower in endurance trained individuals compared to untrained, not only at the same absolute exercise intensity, but also at the same exercise intensity when expressed as a percentage of $\dot{V}o_2$max.

The purpose of the present investigation was to investigate whether the changes in BLa response to exercise, which occur with training, result from changes in the capacity of the cardiovascular system or from local changes in the trained muscle. This has been achieved by studying individuals who had undergone extensive training of specific muscle groups.

Methods

A total of 16 subjects took part in the investigation, consisting of six racing cyclists (leg-trained group), four slalom canoeists (arm-trained group) and six active physical education students. All subjects were male except for one female canoeist. Physical characteristics of the subjects are presented in Table 1.

$\dot{V}O_2$max was established for all subjects using an intermittent exercise protocol. For leg exercise, subjects were seated on a cycle ergometer; for arm exercise, the same ergometer was mounted on a table and used for hand-cranking.

Arterialized finger prick blood samples (20 μl) for lactic acid determination (Maughan, 1982) were taken at rest and 4 min after the finish of each of four 5-min exercise periods corresponding to approximately 25, 50, 75 and 95% of $\dot{V}O_2$max for each subject. Five min of rest separated each exercise bout. Expired air samples for $\dot{V}O_2$max determination were collected during the last min of each work period and heart rate was measured over the last 30 s. All measurements were made during arm and leg exercise for each subject, at least 1 being allowed between tests. Differences between groups were assessed using Student's t-test.

Results

The $\dot{V}O_2$max recorded for the cyclists during leg exercise was higher than that of the canoeists ($p < 0.001$) and of the control subjects ($p < 0.05$). The $\dot{V}O_2$max achieved by the physical education students was considerably

Table 1

Physical Characteristics and $\dot{V}O_{2max}$ (Arms and Legs) for the Control Subjects, Cyclists and Canoeists (male only)

Subjects	Age (yrs)	Height (cm)	Weight (kg)	$\dot{V}O_{2max}$ – Legs		$\dot{V}O_{2max}$ – Arms	
				l/min	ml/kg/min	l/min	ml/kg/min
Control	20.3	177	71.3	4.50	62.8	2.65	37.1
n = 6	± .7	± 5.9	± 3.0	± .59	± 6.3	± .73	± 8.8
Cyclists	23.7	177	75.7	5.42	72.1	2.76	36.5
n = 6	± 5.3	± 5.4	± 8.2	± .46	± 6.7	± .61	± 8.2
Canoeists	19.3	179	69.5	3.86	55.8	2.91	43.6
n = 3	± 1.5	± 4.5	± 7.0	± .21	± 5.2	± .11	± 3.7

(Mean ± SD)

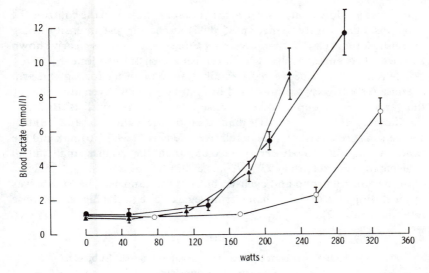

Figure 1—Relationship of blood lactate concentration to absolute exercise intensity in leg exercise.

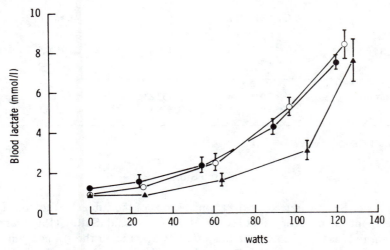

Figure 2—Relationship of blood lactate concentration to absolute exercise intensity in arm exercise.

higher than would be expected for an untrained population and reflects their high level of general physical activity (Åstrand & Rodahl, 1977). No significant differences in $\dot{V}O_{2max}$ were found between the groups in arm exercise, although the canoeists recorded the highest mean value.

Figures 1 and 2 present the relationships between BLa and absolute intensity in leg and arm exercise, respectively. For all subjects, BLa began to increase at a lower intensity in arm exercise, although in the controls the BLa was significantly higher ($p < 0.05$) at the highest intensity in leg exercise. A reduced BLa at equivalent exercise intensities is clearly shown by the cyclists and canoeists when exercising their trained limbs.

The relationships between BLa and relative intensity for leg and arm exercise respectively, are presented in Figures 3 and 4. Even allowing for the variation in group relative exercise intensities, the cyclists' BLa was significantly lower ($p < .01$) than that of the canoeists and control subjects in leg exercise at a mean relative intensity of 69% $\dot{V}O_2$max and the canoeists significantly lower ($p < 0.05$) than the cyclists and control subjects in arm exercise at 73% $\dot{V}O_2$max.

In the control group and cyclists, but not the canoeists, heart rate was lower during arm exercise than during leg exercise at the highest exercise intensity (Table 2).

Table 2

Maximal Heart Rates (Arms and Legs) in Control Subjects, Cyclists and Canoeists

	Control	Cyclists	Canoeists
Legs	177.8 ± 5.5	169.8 ± 6.3	180.0 ± 6.3
Arms	167.0** ± 5.9	152.8** ± 8.1	180.2 ± 6.8

(Mean ± SD)
**$P < .01$

Discussion

The lower BLa produced by the trained limbs of the subjects studied is in agreement with earlier reports by Hermansen and Saltin (1967) and Ekblom (1969). The fact that this effect is specific to the trained limbs suggests that the improvement in physical exercise capacity associated with endurance training is brought about primarily by changes occurring within the trained muscles. The changes in cardiovascular dimensions which take place during the training period do not appear to cause an increase in $\dot{V}O_2$max or in exercise performance if the exercise involves muscle groups other than those involved in the training program.

The results also support the concept that lactate production by working muscles is a result of an imbalance between the rates of pyruvate formation and oxidative decarboxylation. Endurance training has been shown to

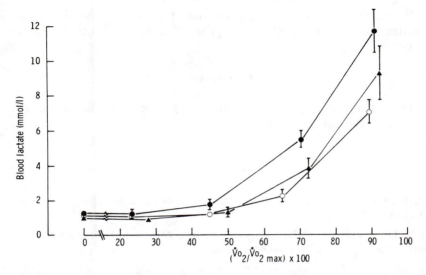

Figure 3—Relationship of blood lactate concentration to relative exercise intensity in leg exercise.

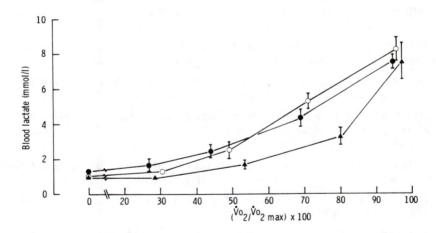

Figure 4—Relationship of blood lactate concentration to relative exercise intensity in arm exercise.

result in an increase in the activity of enzymes of the TCA cycle in the trained muscle (Holloszy, 1967 & 1975). Gollnick et al. (1972) showed that the adaptive response of the TCA cycle enzyme succinate dehydrogenase was restricted to the trained muscle groups. In contrast to these effects, endurance training has little effect on the enzymes of the glycolytic pathway (Costill et al, 1976). These observations may explain the reduced rate of carbohydrate oxidation and enhanced fat oxidation which occur in submaximal exercise in the trained state (Henrikssen, 1977) and can account for the decreased lactate production by trained limbs during exercise.

Conclusions

The reduction in BLa associated with long term endurance training is specific to the trained limbs.

References

ÅSTRAND, P.-O. & K. Rodahl. Textbook of Work Physiology, 2nd ed. London: McGraw-Hill, 1977.

COSTILL, D.L., W.J. Fink & M.L. Pollock. Muscle fibre composition and enzyme activities of elite distance runners. Med. Sci. Sports 8:96–100, 1976.

EKBLOM, B., Effect of physical training on O_2 transport system in man. Acta. Physiol. Scand. Suppl. 328, 1969.

GOLLNICK, P.D., R.B. Armstrong, C.W. Saubert, K. Piehl & B. Saltin. Enzyme activity and fibre composition in skeletal muscle of untrained and trained men. J. Appl. Physiol. 33:312–319, 1972.

GOLLNICK, P.D. & B. Saltin. Significance of skeletal muscle oxidative enzyme enhancement with endurance training. Clin. Physiol. 2:1–12, 1982.

HENRIKSSEN, J., Training induced adaptations of skeletal muscle and metabolism during submaximal exercise. J. Physiol. 270:661–675, 1977.

HERMANSEN, L. & B. Saltin. Blood lactate concentration during exercise at acute exposure to altitude. In Exercise at Altitude. Amsterdam: Excerpta Medica Foundation, pp. 48–53, 1967.

HERMANSEN, L. & I. Stensvold. Production and removal of lactate during exercise in man. Acta. Physiol. Scand. 86:191–201, 1972.

HOLLOSZY, J.O., Biochemical adaptations in muscle. J. Biol. Chem. 242:2278–2282, 1967.

HOLLOSZY, J.O., Adaptations of skeletal muscle to endurance exercise. Med. Sci. Sports 7:155–164, 1975.

JOBSIS, F.F. & W. Stainsby. Oxidation of NADH during contractions of circulated mammalian skeletal muscle. Respir. Physiol. 4:292–300, 1968.

MAUGHAN, R.J., A rapid, simple method for the determination of glucose, lactate, pyruvate, alanine, 3-hydroxybutyrate and acetoacetate on a single 20 μl blood sample. Clin. Chim. Acta (in press), 1982.

WASSERMAN, K., A.L. Van Kessel & G.G. Burton. Interaction of physiological mechanisms during exercise. J. Appl. Physiol. 2:71–85, 1967.

Enzymes

Exercise-Induced Loss of Muscle Enzymes

F.J. Cerny and G. Haralambie
Freiburg University, Freiburg,
Federal Republic of Germany

The exercise-induced release of muscle enzymes into the blood has been related to work intensity (Bratton, 1962; Hunter 1971; Keif, 1972; Schwartz, 1971; Shapiro, 1973; Thomson, 1975) and duration (Fowler, 1968; Griffiths, 1966). Much of the confusion could be attributed to problems associated with obtaining data under field conditions (Siest, 1974). This study examined, in the laboratory, the questions of whether the enzyme loss was related more to work intensity or duration.

Methods

Subjects were eight active but untrained healthy male university students ($\overline{X} \pm SD$, height 185 ± 6.0 cm, weight 78 ± 4.1 kg, $\dot{V}o_2max$ 3.3 ± 0.43 l/min). $\dot{V}o_2max$ was determined during a progressive cycle ergometer test. One wk later and at successive one-wk intervals, four tests were performed in random order: a) 50% $\dot{V}o_2max$, 60' and b) 100'; c) 80% $\dot{V}o_2max$, 60' and d) 100'. This protocol allowed comparisons between exercise intensity and duration. In addition, total energy cost was equal for b and c which enabled examination of the possibility that the *total* energy expended may determine the loss. In order for 4 subjects to complete d, the workload was decreased. For c and d, 82% and 75.6% $\dot{V}o_2max$, respectively, were maintained. Subjects drank water ad libitum.

Venous blood samples were obtained without stasis before, 10', 60' and 24 h after each test. Samples were allowed at least 20' but no more than 45' to clot, centrifuged at 3000 rpm for 10', serum transferred, again centrifuged and pipetted for analysis. All samples were free of hemolysis. Analysis of pyruvate kinase (PK) was done within 2 h, all others within 8 h. Analyses were as follows: creatine phosphokinase (CPK, activated) (Szasz,

1970); phosphohexose isomerase (PHI) (Schwartz, 1971); lactate dehydrogenase (LDH) (Elliott, 1963); α-hydroxybutyrate (HBDH) (Elliott, 1963); pyruvate kinase (PK) (Beisenher, 1953); NAD-malate dehydrogenase (MD) (Bergmeyer, 1965), all at 25° C, and aldolase (ALD) (Beisenher, 1953) at 37° C. These enzymes were chosen to allow comparison of the free cytoplasmic (CPK, PHI, LDH, HBDH, PK) with the bound enzymes ALD and MDH. They represent a wide range of molecular weights. Specific activities were known, allowing calculation of mg protein present. Data were analyzed using ANOVA and post hoc tests for linear contrasts in test or time. The $p < 0.05$ level of significance was used.

Results

Figure 1: Intracondition

Compared to resting values, post-exercise activities of the cytoplasmic enzymes exhibited three patterns of change. In all conditions PHI significantly increased immediately after exercise and with the exception of condition d had returned to resting levels 24 h later. Elevations in CPK and PK were observed later, with the most significant increases taking place 24 h after exercise. LDH and PK were significantly elevated immediately and for 24 h after condition d. MDH was significantly elevated 60' after c and ALD 24 h after d.

Intercondition

Duration effects (a vs b and c vs d) were noted only for MDH (a vs b) and PK (c vs d). Intensity effects (a vs c and b vs d) were much greater with c increasing CPK, LDH, HBDH and PK more than a, and d increasing PHI and PK more than b. Comparing conditions where total energy expenditure was equal, (b vs c) showed that condition c resulted in significantly higher blood CPK, PK, LDH, HBDH, ALD and MDH levels than b.

Discussion

These results indicate that exercise intensity was the predominant factor underlying exercise-induced muscle enzyme redistribution. Interpretation has been based on the assumption that: 1) the specific activity of the enzyme was not significantly altered by the exercise, 2) the increase was not due to a simple activation of protein already present in the serum and 3) the rate of elimination or inactivation and synthesis remained constant over the period studied.

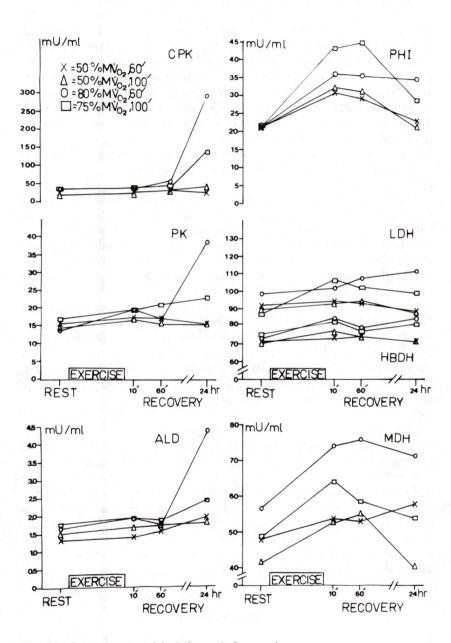

Figure 1—Serum enzyme activity before and after exercise.

Factors which could increase the specific activity or activate previously inactive serum enzymes included temperature and catecholamines. Although catecholamine-induced increases in activity could not be ignored, Loegering et al. (1975) suggested that the increase in plasma CPK with epinephrine-injected rats was due to a release from muscle tissue. Where possible, "activated," optimal or near-optimal analytical methods were employed in the present study. Finally, disappearance rates reported by Rapaport (1975) for CPK fractions indicated that higher concentrations of plasma enzymes will increase the rate of disappearance. This would result in an underestimate of the amount of protein being lost. Dale (1972) found that only 50% of the injection-induced increases in plasma LDH were eliminated in 25 h, suggesting that a relatively constant rate of elimination could be assumed.

The loss of enzymes from the working muscles could occur as a result of tissue destruction or of changes in the membrane permeability (Hansen, 1982; Haralambie, 1981). If the release were related to a simple process of membrane disruption, the amount appearing in serum should be related to the molecular size as estimated from molecular wt (Young, 1974). Calculations indicate that the **amount** of enzyme lost, in mg, is not related to molecular wt ($r = 0.07$).

Electron microscopic examination of the muscle membrane after various types of exercise (Gale, 1981) have been unable to find evidence of irreversible tissue destruction. The different time course of appearance in the blood also would suggest that extreme membrane disruption does not occur. Based on measurements of intramuscular concentrations (Lujf, 1973; personal data), the redistribution seemed to be related to the intra- to extra-cellular concentration gradient, so active transport does not play a role.

Permeability of the membrane is not fixed, but is subject to metabolic control. Among those factors which may lead to an increase in cellular permeability are hypoglycemia, anoxia, changes in pH and altered ionic concentrations. By decreasing the amount of ATP available for the maintenance of membrane integrity, these factors could eventually lead to a loss of the cellular content (Kröner, 1971). The present study, showing that exercise intensity was the primary influence on enzyme loss, supports the idea that competition between ATP needed for working muscles and for maintenance of cell integrity may result in the loss of cellular enzymes during intense exercise lasting from 60 to 100 min.

Acknowledgment

This research was supported by the Bundesinstitut für Sportwissenschaften, Köln, and Dr. J. Keul, Freiburg.

References

BEISENHERZ, G.H., H.J. Boltze, T. Bücher, R. Czock, K. Garbade, E. Meyerarndt & G. Pflerderer. De-hydrogenase, Milchsäuer -Dehydrogenase, Glycerophosphat-Dehydrogenase und Pyruvate-Kinase aus Kaninchen-muskulatur in einem Arbeitsgang. Z. Naturforsch 8b:55, 1953.

BERGMEYER, H.U., Method of enzymatic analysis. Verlag Chemie Weinheim, 2nd edition, p. 757, 1965.

BRATTON, R.D., S.R. Chowdhury, W.M. Fowler, G.W. Gardner & C.M. Pearson. Effect of exercise on serum enzyme levels in untrained males. Res. Quart. 33:182–193, 1962.

DALE, J., E. Myhre & K. Rasmussen. Elimination of hemoglobin and lactate dehydrogenase from plasma in normals and patients with intravascular hemolysis. Scand. J. Clin. Lab. Invest. 29:389–395, 1972.

ELLIOTT, B.A. & J.H. Wilkinson. The serum "α-hydroxybutyrate dehydrogenase." Clin. Sci. 24:343, 1963.

FOWLER, W.M., Jr., G. Gardner, H. Kazerunian & W. Lauvstad. The effect of exercise on serum enzymes. Arch. Phys. Med. Rehabil. 49:554–565, 1968.

GALE, J.B., The use of electron microscopy in the investigation of exercise-related muscle failure. In F. Nagle & H. Montoya (Eds.), Exercise in Health and Disease. Springfield: C.C. Thomas, 1981.

GRIFFITHS, P.D., Serum levels of ATP creatine phosphotransferase (Creatine Kinase). The normal range and effect of muscular activity. Clin. Chem. Acta 13:413–420, 1966.

HANSEN, K., J. Bjerre-Knudsen, U. Brodthagen, R. Jordal & P.E. Paulev. Muscle cell leakage due to long distance training. Eur. J. Appl. Physiol. 48:177–188, 1982.

HARALAMBIE, G., Serum aldolase isoenzymes in athletes at rest and after long-lasting exercise. Int. J. Sports Med. 2:31–36, 1981.

HUNTER, J.B. & J.B. Critz. Effect of training on plasma enzyme levels in man. J. Appl. Physiol. 31:20–23, 1971.

KIEF, W., B. Klein & E. Möller. Enzymbewegungen unter körperlicher Belastung bei Trainierten und Untrainierten Probanden. Med. Klin. 67:195–199, 1972.

KRÖNER, H., Cell damage and serum enzymes. Disch. med. Wschr. 96:551, 1971.

LOEGERING, D.J., M.L. Bonin & J.J. Smith. Effect of exercise, hypoxia, and epinephrine on Lysosomes and plasma enzymes. Exp. Mol. Path. 22:242–251, 1975.

LUJF, A., Veränderungen von-Muskel und Serumenzymaktivitäten bei Erkrankungen menschlicher Skelettmuskulstur. Wein. Klin. Wschr. 85:3, 1973.

RAPAPORT, E., The fractional disappearance rate of the separate isoenzymes of creatine phosphokinase in the dog. Card. Res. 9:473–477, 1975.

SCHWARTZ, M.K., V.A. Bethune, D.L. Bach & J.E. Woodbridge. New assay for measuring Phosphohexose Isomerase (PHI) activity. Clin. Chem. 17:656, 1971.

SCHWARTZ, P.L., H.W. Carroll & J.S. Douglas. Exercise-induced changes in serum enzyme activities and their relationship to Max $\dot{V}O_2$. Int. Z. angew. Physiol. 30:20–33, 1971.

SHAPIRO, Y., A. Magazonik, E. Sohar & C.B. Reich. Serum enzyme changes in untrained subjects following a prolonged march. Can. J. Physiol. Pharmacol. 51:271–276, 1973.

SIEST, A. & M.M. Galteau. Variations of plasmatic enzymes during exercise. Enzyme 17:179–195, 1974.

SZASZ, G., E.W. Busch & H.B. Farohs. Serum-Kreatinkinase. I. Methodische Erfahrungen. Und. Normalwerte mit einem neuen handlesüblichen Test. Dtsch. med Wschr. 95:829–835, 1970.

THOMSON, W.H.S., J.C. Sweetin & I.J.D. Hamilton. ATP and muscle enzyme effects after physical exertion. Clin. Chem. Acta 59:241–245, 1975.

YOUNG, D., The origin of serum enzymes and the basis for their variation. Blume & Frier (Eds.), The Practice of Laboratory Medicine. New York: Academic Press, 1974.

Enzyme Profiles in Type I, IIA, and IIB
Fiber Populations of Human Skeletal Muscle

B. Essén-Gustavsson and J. Henriksson
Swedish University of Agricultural Sciences,
Uppsala, Sweden, and Karolinska Institutet,
Stockholm, Sweden

Skeletal muscle is composed of fibers of different types with different contractile and metabolic properties. Based on stainings for myofibrillar ATPase, fibers can be differentiated into Type I (slow-twitch) and Type IIB (fast-twitch) fibers (Brooke & Kaiser, 1970). Most quantitative information regarding the characteristics of the different fiber types has been obtained from studies on animals in which certain muscles or parts of muscles are composed exclusively of one fiber type (Peter et al., 1972).

In human muscle, the different fiber types are arranged in a mosaic pattern which makes quantitative analysis of fiber type properties difficult. Most of this information has therefore come from histochemical studies, which in human muscle have shown that Type I fibers have the highest oxidative and Type IIA and IIB fibers the highest glycolytic capacity. Furthermore, the stains indicate a somewhat higher oxidative capacity in Type IIA fibers than Type IIB fibers. However, in endurance trained subjects there seems to be a smaller difference in oxidative capacity between Type I, IIA and IIB fibers than in untrained subjects (Jansson & Kaiser, 1977). Histochemical stains are, however, only semiquantitative. The purpose of the present study was therefore to analyze enzyme profiles in pools of histochemically identified Type I, IIA and IIB fibers of the human thigh muscle using a recently developed method (Essén et al., 1975). Furthermore, in order to study the effects of training, muscle was obtained from individuals with different physical activity levels.

Subjects and Methods

Eleven healthy young men participated in the study. The control group (C-group) consisted of six subjects. Four of these had not performed any regular physical training during the preceding 2 to 3 yr. The other two were sprint runners (100 to 200 m). The endurance trained group (E-group) consisted of five subjects; three long distance runners (75 to 160 km/wk) and two cyclists (500 to 600 km/wk). Mean value and range of age, ht and wt were 27 (23 to 32) yr, 182 (178 to 194) cm and 78 (65 to 95) kg for group C and 23 (18 to 27) yr, 185 (179 to 190) cm and 71 (66 to 79) kg for group E. Muscle tissue was obtained from the thigh muscle (vastus lateralis) with the needle biopsy technique. Two muscle biopsies were taken. The sample for enzyme analysis was immediately frozen in liquid nitrogen and the sample for histo-chemical analysis was mounted in embedding medium and frozen in isopentane cooled with liquid nitrogen. Both pieces of muscle were stored at -80° C prior to analysis.

Enzyme Analysis

The sample was freeze-dried and dissected free from connective tissue, fat and blood before fragments of single muscle fibers were dissected out. After staining of a small part of the fiber for myofibrillar ATPase the fiber fragments were classified and pooled into groups of Type I, IIA and IIB fibers (Essén et al., 1975). These pools (23 to 417 μg) were then weighed on a Cahn electrobalance and homogenized in ice-chilled 0.1 M potassium phosphate buffer; pH 7.0 (dilution 1:400) with an ultrasound disintegrator. A larger piece of freeze-dried muscle was also homogenized in the same way (muscle homogenate). The activities of citrate synthase (CS) as a measure of citric acid cycle capacity, 3-OH-acyl-CoA dehydrogenase (HAD) as a measure of the capacity for lipid oxidation and glyceraldehyde phosphate dehydrogenase (GAPDH) as a measure of glycolytic capacity were determined by previously described fluorimetric methods (Essén et al., 1980). As a measure of the capacity for lactate production from pyruvate, lactate dehydrogenase (LDH) was determined. (TEA-buffer 50 mM ph 7.6, EDTA 6 mM, NADH 0.05 mM, pyruvate 1 mM.) The analyses were performed at 25° C and enzyme activities are expressed as mkatal/kg d.w. The coefficients of variation for a single value (C.V.) calculated from analyses on duplicate pools of fibers from the same sample (same fiber type, n − 19) were (CS) 12%, (HAD) 18%, (GAPDH) 12% and (LDH) 13%.

Histochemistry

Serial transverse sections (10 μ) were cut in a cryostat at -20° C and stained for myofibrillar ATPase to allow fiber identification as Type I, IIA, IIB or

IIC (Brooke & Kaiser, 1970). A minimum of 200 fibers were counted on each sample.

Results

Fiber Type Proportion

The relative occurrence of fiber types in the lateral part of the thigh muscle differed in the two groups. In the E-group the means and ranges were for Type I 68 (54 to 97)%, Type IIA 23 (3 to 32)%, Type IIB 8 (0 to 12)% and Type IIC 1 (0 to 3)%. In one of the cyclists no Type IIB fibers were noted. In the C-group the corresponding values were for Type I 35 (21 to 50)%, Type IIA 40 (25 to 51)%, Type IIB 20 (7 to 31)% and Type IIC 7 (0 to 17)%. Three of the control subjects had an unusually high percentage of Type IIC fibers (7.8 and 17%).

Analyses on Muscle Homogenates

The enzymatic analyses on muscle homogenates revealed significantly higher (CS and HAD) activity levels in the E-group as compared to the C-group (CS 2x, HAD 1.4 x)(Figure 1). For GAPDH and LDH enzymes

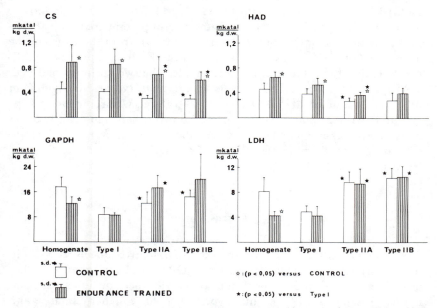

Figure 1—Enzyme activity levels in muscle homogenates and in pools of type I, IIA and IIB fibers from thigh muscle biopsy samples of endurance trained (n = 5) and control subjects (n = 6).

the opposite pattern was noted with higher activity levels in the C-group (GAPDH 1.4x, LDH 1.9x). There was a strong correlation between both CS and HAD activities in the homogenates and the Type I fiber percentage of muscle (r = -0.91 and r = -0.84, respectively). In contrast GAPDH and LDH were negatively correlated with the type I fiber percentage (r = -0.83 and r = -0.78, respectively).

Analyses on Fiber Pools; CS and HAD

Similar to the pattern found for muscle homogenates the E-group had higher mean activity levels of CS and HAD in both Type I (CS 2.1x, HAD 1.4x), Type IIA (CS 2.3x, HAD 1.4x) and Type IIB fibers (CS 2.0x, HAD 1.5x)(Figure 1). CS and HAD activities were significantly higher in Type I as compared to Type II fibers. These differences (I/IIA) were of a similar magnitude in both the E-group (CS 1.2x, HAD 1.5x) and C-group (CS 1.4x, HAD 1.4x). A comparison of CS and HAD activities in the Type II subgroups revealed no significant differences.

Analyses on Fiber Pools; GAPDH and LDH

The GAPDH and LDH activities in Type I fibers were similar in the E- and C-groups (Figure 1). This was also true for the LDH activity in Type IIA and IIB fibers. However, in both Type IIA and IIB fibers GAPDH activity levels tended to be higher (1.5x) in the E-group compared with the C-group. If the two sprint runners were withdrawn from the C-group the Type IIA GAPDH (1.5x) activity level in the E-group would be significantly higher than in untrained individuals. The two sprinters had high GAPDH and LDH activity levels in all fiber types. No significant difference was detected in GAPDH and LDH levels when comparing Type IIA and IIB fiber pools. Type I fibers, however, had significantly lower activity levels as compared with Type II fibers. This difference (I/IIA) was similar for the E-group (GAPDH 0.5x, LDH 0.5x) and the C-group (GAPDH 0.7x, LDH0.5x).

Discussion

The present results on the enzyme markers of oxidative capacity (CS), lipid oxidation (HAD) and glycolytic capacity (GAPDH, LDH) confirm earlier histochemical and biochemical data on single fibers of human skeletal muscle (Essén et al., 1975). Thus Type I fibers have higher oxidative and lower glycolytic capacity, whereas the reverse pattern is seen in Type II fibers. Furthermore, this study supports earlier findings indicating that a pronounced increase in oxidative capacity may occur in all fiber types in response to increased muscle usage (Jansson & Kaiser, 1977; Sjoogaard et al., 1978).

Most earlier studies on muscle homogenates imply unchanged or even slightly decreased levels of glycolytic enzymes with endurance training (for ref. see Salmons & Henriksson, 1981). Thus an interesting observation is the increased GAPDH level in Type II fibers of the E-group as compared to the four untrained subjects. As the increased GAPDH levels in the E-group were accompanied by similar LDH levels as in the C-group, this suggests a facilitation of carbohydrate oxidation rather than of lactate production. According to the present results, the low glycolytic capacity usually seen in homogenates of muscle obtained from endurance trained individuals is to a great extent explained by the predominance of Type I fibers in their muscles. This is, however, not the major explanation for the high oxidative capacity in endurance trained muscle as all fiber types increase in oxidative enzyme levels.

Most histochemical data indicate a difference in enzyme profiles between Type IIA and IIB fibers. In contrast, no differences in enzyme activities between pools of Type IIA and IIB fibers were seen in the present study. For three of the subjects who had a substantial proportion of Type IIC fibers, these results should be viewed with some caution. In these subjects some Type IIC fibers might erroneously have been typed as IIB fibers as our typing procedure of single fibers did not allow identification of Type IIC fibers. However, in subjects with a low percentage of Type IIC fibers, Type IIA and IIB fiber pools had similar enzyme activities. Furthermore, small differences between the Type II subgroups might be impossible to detect as the enzymes varied as much as 12 to 15% (C.V.) in two pools of fibers from the same individual.

As shown by Lowry et al. (1978) for the human biceps muscle there is a great variation in enzyme pattern within fibers of the same type. While this has also been shown for malate dehydrogenase activity (MDH) of the rat EDL muscle, fibers that belonged to the same motorunit had almost identical MDH levels (Nemeth & Pette, 1980). Since the fibers in two pools most likely do not originate from the same motorunits some variation in enzyme levels between pools of fibers from the same individual is to be expected.

As indicated above, all fiber types have a high capacity for adaptation with regard to oxidative capacity. In the present study the oxidative capacity in Type II fibers of the endurance trained subjects was almost twice that in Type I fibers of the control subjects. This shows that it may be misleading to refer to Type I fibers as high oxidative and Type II fibers as low oxidative.

Acknowledgment

This study was supported by grants from the Swedish Medical Research Council, the Forestry and Agricultural Research Council and the Research Council of the Swedish Sports Federation.

References

BROOKE, M.H. & K.K. Kaiser. Muscle fibre types: How many and what kind? Arch. of Neurol. 23:369/09379, 1970.

ESSÉN, B., E. Jansson, J. Henriksson, A.W. Taylor & B. Saltin. Metabolic characteristics of fibre types in human skeletal muscle. Acta Physiol. Scand. 95:153–165, 1975.

ESSÉN, B., A. Lindholm & J. Thornton. Histochemical properties of muscle fibre types and enzyme activities in skeletal muscles of Standardbred trotters of different ages. Equine Vet. J. 12:175–180, 1980.

JANSSON, E. & L. Kaiser. Muscle adaptation to extreme endurance training in man. Acta Physiol. Scand. 100:315–324, 1977.

LOWRY, C.V., J.S. Kimmey, S. Felder, M.M.-Y Chi, K.K. Kaiser, P.N. Passonneau, K.K. Kirk & O.H. Lowry. Enzyme patterns in single human muscle fibres. J. Biol. Chem. 253:8269–8277, 1978.

NEMETH, P.M., D. Pette & G. Vrbova. MDH homogeneity of single fibers of the motor unit. In Plasticity of Muscle. Berlin: New York: de Gruyter, 1980.

PETER, J.B., R.J. Bernard, V.R. Edgerton, G.A. Gillespie & K.E. Stempel. Metabolic profiles of three fibre types of skeletal muscle in guinea pigs and rabbits. Biochemistry II:2627–2633, 1972.

SALMONS, S. & J. Henriksson. The adaptive response of skeletal muscle to increased use. Muscle and Nerve 4:94–105, 1981.

SJOOGAARD, G., M.E. Houston, E. Nygaard & B. Saltin. Subgrouping of fast twitch fibres in skeletal muscles of man. Histochemistry 58:79–87, 1978.

Cyclic Adenosine Monophosphate in Skeletal Muscle in Response to Exercise

A.H. Goldfarb and J.F. Bruno

University of Maryland, College Park, Maryland, U.S.A.

Skeletal muscle glycogen can be degraded as a result of exercise depending on the intensity and duration of the work. The exact mechanisms responsible for skeletal muscle glycogenolysis during exercise are not well defined. The control of skeletal muscle glycogenolysis has been studied in vitro, in perfused hind limbs and in vivo. From these studies there appears to be more than one mechanism controlling skeletal muscle glycogenolysis. β-adrenergic activation of adenylate cyclase by catecholamines which increases intracellular adenosine 3',5'-cyclic monophosphate (cAMP) has been shown to stimulate skeletal muscle glycogenolysis. The increase in cAMP in turn activates a protein kinase which catalyzes the phosphorylation of inactive phosphorylase kinase. Active phosphorylase kinase converts the less active phosphorylase β to the more active phosphorylase α. Phosphorylase catalyzes the phosphorylation of glycogen resulting in glycogenolysis.

During exercise cAMP-stimulating hormones such as epinephrine are elevated. This suggests that cAMP should be elevated in skeletal muscle during exercise resulting in glycogenolysis. Recently, Palmer and co-workers (1980) have reported an increase in cAMP in fast-twitch white muscle after a 2 h swim. No differences were reported for either fast-twitch red or soleus muscle cAMP levels. This could mean that different mechanisms are responsible for glycogenolysis in the red and white fiber types.

It is not clear whether cAMP is responsible for glycogenolysis in both red and white fiber types during exercise. Therefore, the purpose of this study was to determine whether or not cAMP does, in fact, increase during the course of an exercise bout in both red and white skeletal muscles.

Methods

Male Sprague-Dawley rats (Flo-Labs Inc., Dublin, VA) weighing 200 to 280 g were housed two to a cage in a temperature ($22 \pm 1°C$) and light (light-dark cycle 12:12) controlled room. Food (Purina Laboratory Rat Chow) and water were provided ad libitum.

Animals were randomly assigned to nonexercised control or exercised groups. Rats were familiarized with the motor driven rodent treadmill (Quinton, Seattle, WA) by exposing them to two 10 min/day runs at 10 m/min at least 5 days prior to the experiment. All animals were handled in a similar manner. Rats were run for 0, 5, 10, 15 or 30 min (at 15 m/min) and at the end of the run were killed by decapitation. The quadriceps muscles were dissected out and separated into a superficial white portion and a deep red portion. The middle portion containing mainly mixed fibers was discarded as described by Baldwin et al. (1972). Immediately the piece of muscle was frozen between aluminum block tongs cooled in liquid nitrogen. The procedure from treadmill to freeze clamping the tissue required less than 30 s. Frozen skeletal muscles were kept in a -70° freezer until analyzed.

Glycogen was determined by the method described by Lo et al. (1970). cAMP was extracted by the method described by Weller et al.(1972), as modified by Goldfarb and Kendrick (1979). cAMP was measured by the method of Gilman (1970) as modified by Brown et al. (1971) using charcoal to bind the unbound free nucleotide.

Results are expressed as means ± SE. A one-way analysis of variance was utilized to determine homogeneity of variance, and localization of pairwise differences was achieved by using the Tukey A post hoc method where appropriate. The 0.05 level of significance was used for all tests.

Results

White skeletal muscle glycogen was significantly depleted after 30 min of the treadmill run (Figure 1). White muscle glycogen seemed to be utilized at a relatively constant rate during the exercise. Resting glycogen content in white muscle was 4.47 ± 0.31 mg/g wet wt and was depleted 54% after 30 min of running.

White muscle cAMP was significantly increased after 5 min of the exercise and continued to remain elevated at each of the time points (Figure 1). Control white muscle cAMP was 0.26 ± 0.02 pmol/mg wet wt and reached 0.51 ± 0.03 pmol/mg wet wt at 5 min of the treadmill run. After 30 min white cAMP was 0.50 ± 0.02 pmol/mg wet wt.

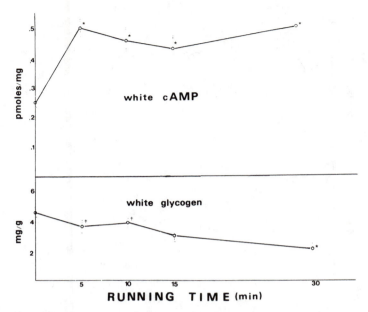

Figure 1—White skeletal muscle glycogen and cAMP during submaximal treadmill run (15 m/min). Asterisks indicate significant differences from preexercise control value and † indicates significant difference from 30 min value (P < 0.05). Values are means ± SE for 7 rats for all points.

Red skeletal muscle glycogen also demonstrated a significant depletion after 30 min of the exercise (Figure 2). Red muscle glycogen did not elicit a similar constant decrease as was observed in white muscle. Resting glycogen in red muscle was 4.72 ± 0.62 mg/g wet wt and was depleted 61% after the 30 min of exercise.

cAMP in red skeletal muscle was significantly elevated after 5 min of exercise (Figure 2). Red muscle cAMP remained at an increased level at each of the time points. Resting cAMP in red muscle was 0.22 ± 0.02 pmol/mg wet wt and was 0.48 ± 0.03 pmol/mg wet wt at 5 min of the run. After 30 min cAMP in red muscle was still elevated at 0.41 ± .04 pmol/mg wet wt.

Discussion

The present study has shown that a moderately intense run of 30 min can result in a degradation of glycogen in red and white skeletal muscle in untrained rats. During the 30 min run cAMP levels were significantly elevated in both fiber types compared to nonexercised rats. These findings

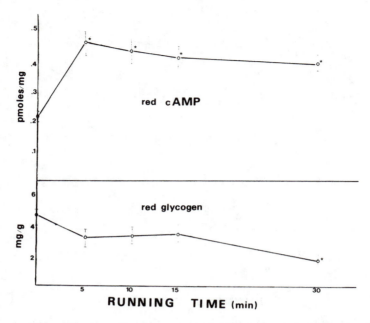

Figure 2—Red skeletal muscle glycogen and cAMP during submaximal treadmill run (15 m/min). Asterisks indicate significant differences from preexercise control value (P < 0.05). Values are means ± SE for 7 rats for all points.

seem to suggest that glycogenolysis in red and white skeletal muscle is controlled by cAMP resulting in activation of phosphorylase during the exercise run. However, on examination of the time course data comparing glycogen content with cAMP levels, there appear to be differences in regulation of glycogenolysis in the two fiber types.

White skeletal muscle demonstrated a linear decrease (r = − .97) in muscle glycogen content which was accompanied by elevations in cAMP throughout the 30 min treadmill run. This implies that there is a relationship between glycogenolysis and cAMP during an exercise run in white muscle. There was a 97% increase in cAMP after 5 min which decreased to 68% after 15 min. After 30 min cAMP levels were again almost double control values. Our values are similar to what has been observed by Mayer and Stull (1971) for fast-twitch rabbit muscles which were injected with isoproteronol in situ. They reported increases in cAMP from 0.17 ± 0.04 μMol/kg to .53 ± 0.14 μMol/kg after 60 s. This suggests that epinephrine was probably elevated during the treadmill run to stimulate the increases observed in cAMP. Recently, Rennie and co-workers (1981) have shown that epinephrine administration could reverse phosphorylase inactivation in soleus and plantaris muscle following tetanic stimulation by elevating cAMP levels which resulted in a resumption of glycogenolysis.

Palmer et al. (1980) have reported a 33% increase in cAMP in fast-twitch white muscle following a 2 h swim. Ivy (1977), however, showed no differences in white muscle cAMP immediately after a swim to exhaustion. It is possible that these differences are a result of the type of exercise, the method of sacrifice or the length of time of the exercise. Ivy (1977) utilized ether to extract cAMP which Kimura et al. (1974) has shown to elevate cAMP levels in muscle. There is a possibility that this might have masked an increase in cAMP in this tissue. Another possible explanation for some of these differences is that there might be a decrease in the elevation of cAMP in white muscle as the duration of the exercise is extended. The rate of glycogenolysis would probably decrease as the duration of the exercise was continued. This could partially be explained by hormonal adjustments at the beginning of the exercise which would result in a decrease in cAMP content. There is a possibility that cAMP-independent mechanisms could be at least partially responsible for controlling glycogenolysis; we therefore cannot conclude without question that cAMP was involved in exercise-induced glycogenolysis.

Red muscle did not demonstrate a similar glycogenolytic response to the exercise as did white muscle. There was not a consistent reduction in glycogen content at all time points measured although cAMP was consistently elevated. This indicates that there may be a cAMP-independent mechanism responsible for controlling glycogenolysis in red muscle to modulate the increased cAMP which was observed. Red muscle showed an initial decline in glycogen content in the first 5 min of the run which coincided with a significant elevation in cAMP. These initial changes in both muscle glycogen and cAMP were similar to those observed in white muscle. It is possible that glycogenolysis was stimulated at the start of the exercise in both fiber types to compensate for the increased energy demands. Following this adjustment period, increased blood flow to the red muscle could have decreased the requirement for glycogen as a major energy source.

There was a linear decrease (r = .90) in cAMP levels from 5 to 30 min in red skeletal muscle. This suggests that if this trend continued, cAMP levels in red muscle would slowly return toward control levels if the duration of the exercise was extended. This might partially explain why Palmer et al. (1980) did not observe any differences in cAMP content in fast-twitch red muscle and soleus muscle following a 2 h swim. There is a possibility that cAMP was elevated at the onset of the exercise and declined to control levels by the 2 h. When we project the regression equation for cAMP content to 90 min or 120 min, cAMP content would not be significantly different from control values. There is a possibility that the differences in the exercise protocol could also have influenced cAMP content.

There is a possibility that cAMP-independent mechanisms could be at least partially responsible for regulating glycogenolysis. Exton et al. (1978)

have reported an α-adrenergic receptor-mediated increase in catecholamine-initiated liver glycogenolysis which does not affect cAMP levels. It has been shown that calcium mediates the α-effects of catecholamines, probably allosterically activating phosphorylase kinase to activate phosphorylase. Whether this mechanism occurs to regulate glycogenolysis in muscle is not clear. It is possible that both calcium mediated and cAMP mediated processes are required to regulate glycogenolysis in working muscle.

In summary, we have demonstrated that during a moderate run of 30 min cAMP is elevated in red and white skeletal muscle which resulted in a significant degradation of glycogen in both fiber types after 30 min.

Acknowledgment

This work was supported in part by Biomedical Research Grant No. RR-07042 to the University of Maryland from the Division of Research Resources, NIH, PHS.

References

BALDWIN, K.M., G.H. Klinterfuss, R.L. Terjung, P.A. Mole & J.O. Holloszy. Respiratory capacity of white, red, and intermediate muscle adaptive response to exercise. Am. J. Physiol. 222:373–378, 1972.

BROWN, B.L., J.D.M. Albano, R.P. Ekins & A.M. Agherzi. A simple and sensitive saturation assay method for the measurement of adenosine 3',5'-cyclic monophosphate. Biochemical Journal 121:561–562, 1971.

GILMAN, A.G., A protein binding assay for adenosine 3',5'-monophosphate. Proc. Nat. Acad. Sci., U.S.A. 67:305–312, 1970.

GOLDFARB, A.H. & Z.V. Kendrick. Effect of an exercise run to exhaustion on cAMP in the rat heart. J. Appl. Physiol. 51:1539–1542, 1981.

IVY, J.L., Role of insulin during exercise-induced glycogenolysis in muscle: effect on cyclic AMP. Am. J. Physiol. 2:E509–513, 1977.

KIMURA, H., E. Thomas & F. Murad. Effects of decapitation, ether and pentobarbital on guanosine 3;,5;-phosphate and adenosine 3',5'-phosphate levels in rat tissues. Biochem. at Biophysica Acta 343:519–528, 1974.

LO, S., T.C. Russell & A.W. Taylor. Determination of glycogen in small tissue samples. J. Appl. Physiol. 28:234–236, 1970.

MAYER, S.E. & J.T. Stull. Cyclic AMP in skeletal muscle. In G.A. Robinson, G.G. Nahas & L. Triner (Eds.), Cyclic AMP and Cell Function. 1971.

PALMER, W.K., T.A. Studney & S. Pikramenos. Tissue cyclic AMP following an acute bout of exercise. Med. & Sci. in Sports and Exercise 12:2, 110, 1980.

RENNIE, M.J., R.D. Fell, J.L. Ivy & T.O. Holloszy. Reversal of phosphorylase activation during contractile activity; reactivation by adrenaline. Fed. Proc. 40:3, 614, 1981.

WELLER, M., R. Rodnight & D. Carrera. Determination of adenosine 3',5'-cyclic monophosphate in cerebral tissue by saturation analysis. Bioch. J. 129:113–121, 1972.

Kinetic Potential of Succinate Dehydrogenase during Localized Muscular Exercise and Training

S. Govindappa, P. Reddanna and
C.V.N. Murthy

SRI Venkateswara University, Tirupati, India

Electrical stimulation of muscle results in the general changes in the organism, similar to the involuntary movements such as changes in pulse rate, blood pressure and O_2 consumption (Gutmann, 1962) and thereby this has been treated as analogous to that of exercise (Hirche et al., 1975; Narasimha Murthy et al., 1981). In our recent findings induction of training effects into the muscle through in vivo electrical stimulations has been demonstrated (Narasimha Murthy et al., 1981). The purpose of the present study was to understand changes in the muscular protein ionizable groups in exercise and training and their impact on the kinetic parameters of succinate dehydrogenase.

Materials and Methods

Experimental Materials

The right gastrocnemius of intact and conscious frogs was stimulated with an electronic stimulator (INCO/CS10 Research Stimulator, Ambala, India) in a special chamber. Two platinum electrodes were placed 1 cm apart from each other directly on the skin of the gastrocnemius muscles stimulated with a series of impulses (biphasic-rectangular) of 5 V at a frequency of 2 C/s for 30 min per day for 1 day in one batch of experimental animals (exercise) and for 10 successive days (trained) as described by Narasimha-murthy et al. (1981). Immediately after electrical stimulations, the frogs were sacrificed by double pithing, the muscles were isolated rapidly and chilled in freezing mixture. The muscles from normal animals were taken as controls.

Titration Curves of Muscle Homogenates

The dialyzed sucrose homogenates of control and experimental muscles and sucrose medium were titrated against 1×10^{-2} N HCl and NaOH separately and the changes in pH were recorded. The titration curves of both sucrose and tissue homogenates were plotted. The area between the two curves was traced on a uniformly thick paper and different ionizable groups were marked on it as per their respective pKa values. The paper was trimmed according to the respective ionizable groups and they were represented in terms of mg of paper wt.

Determination of SDH activity (EC : 1.3.99.1)

The activity of succinate dehydrogenase was estimated by the method of Nachlas et al. (1960) as modified by Reddanna & Govindappa (1978a). Substrate and pH dependency activities of the enzyme were determined at different substrate concentrations and pH media respectively.

Results

Exercise led to drastic decreases in the muscular acidic protein ionizable groups with an elevation in basic protein ionizable groups over the controls (Table 1). But in trained muscle there was considerable elevation in acidic ionizable groups with depletion in protein ionizable groups over the controls. In exercised muscles V_{max} of the SDH was decreased with an elevation in K_m over the controls. But in trained muscles the opposite trend was witnessed, where V_{max} was elevated with depletion in K_m (Table 2). The double reciprocal curves revealed higher K_m with lesser V_{max} in

Table 1

Pattern of Protein Ionizable Groups in Control and Experimental Muscles

S. No.	Protein ionizable groups	Control	Exercised	Trained
1.	Total acidic groups	29.55 ± 0.6	13.7 ± 1.17 – 53.84 $p < 0.001$	34.8 ± 0.8 + 7.61 $p < 0.001$
2.	Total basic groups	26.0 ± 1.9	39.0 ± 2.5 + 50.0 $p < 0.001$	19.8 ± 2.1 – 23.85 $p < 0.001$

Values are represented in terms of mg paper weight (Mean of 6 observations).

Table 2

Kinetic Parameters of SDH in Control and Experimental Muscles

S. No.	Kinetic parameters	Control	Exercised	Trained
1.	Slope	1.16 ± 0.09	1.7 ± 0.08 $+ 46.55$ $p < 0.001$	0.912 ± 0.005 $- 21.38$ $p < 0.001$
2.	Intercept	8.47 ± 0.9	11.49 ± 0.5 $+ 35.65$ $p < 0.001$	8.8 ± 0.4 $+ 39$ NS
3.	Vmax	0.119 ± 0.005	0.084 ± 0.012 $- 29.41$ $p < 0.001$	0.128 ± 0.002 $+ 7.56$ $p < 0.001$
4.	Km	0.134 ± 0.008	0.151 ± 0.005 $+ 12.69$ $p < 0.001$	0.113 ± 0.007 $- 15.67$ $p < 0.001$

Values are mean of 6 observations.

exercised muscles (Figure 1a) and lesser K_m and higher V_{max} in trained muscles (Figure 1b) in comparison to the controls. The pH dependency curves (Figures 2a & b) showed a shift in the optimum pH of SDH towards the basic side in exercised muscles and towards the acidic side in trained muscles in comparison to controls.

Discussion

Exercised muscles had elevated basic protein ionizable groups with simultaneous depletion in acidic ionizable groups indicating the prevalence of basic protein medium in this muscle. The exercised muscles actively accumulate bicarbonate ions from the blood (Hirche et al. 1973; Mainwood & Brown, 1971) and hence such a situation not only leads to the deionization of the acid protein ionizable groups but also leads to the ionization of basic protein ionizable groups. The observed decrease in the lactic and pyruvic acid levels (Reddana, 1979) and increase in intramuscular pH in the acid levels (Reddanna, 1979) and increase in intramuscular pH in the exercised muscles (pH 7.8) also suggest a basic protein medium in these muscles. Decreased SDH activity in exercised muscles was indicative of decreased succinate oxidation in particular and oxidative metabolism in general. Since basic protein ionizable groups exert inhibitory influence on SDH Activity (Govindappa & Swami, 1969), the observed inhibition of

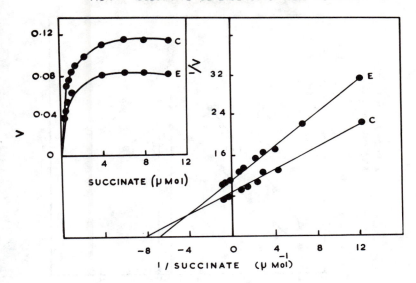

FIG.1 A. SUBSTRATE DEPENDENCY OF SDH ACTIVITY

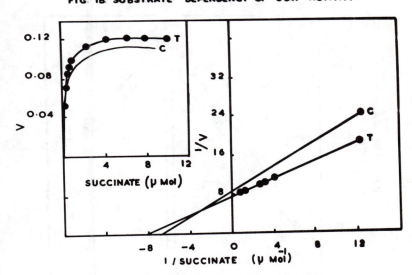

FIG. 1B. SUBSTRATE DEPENDENCY OF SDH ACTIVITY

Figure 1—Lineweaver—Burkeplots of SDH activity in control, exercised and trained muscles.
C — Control muscle
E — Exercised muscle
T — Trained muscle

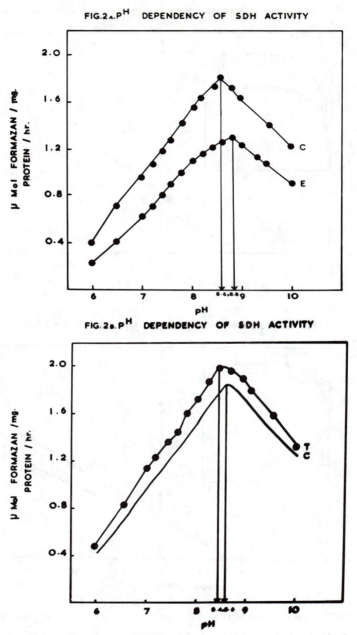

Figure 2—pH dependency curves of SDH activity in control, exercised and trained muscles.
C — Control muscle
E — Exercised muscle
T — Trained muscle

SDH in exercised muscles might be due to accumulated basic protein ionizable groups. Since K_m of the enzyme was increased considerably over the control, decreased affinity between substrate and enzyme and thereby diminished enzyme kinetic efficiency can be expected. The shift in the optimum pH of the enzyme towards the alkaline side must be responsible for the inhibition of enzyme activity in the exercised muscles.

In contrast to the above situation, further days of exercise leading to the trained conditions resulted in the development of a quite opposite trend in the protein ionizable groups and kinetic parameters of the enzyme. The observed activation of SDH in trained muscles might be due to the influence of acidic protein groups, since these groups activate SDH (Govindappa & Swami, 1969). K_m of the enzyme was considerably depleted with elevation in V_{max} over the controls and hence increased kinetic potential of the enzyme can also be expected. The shift in the optimum pH of the enzyme towards the acidic side denotes the acidic-side-oriented activation pattern of the enzyme. Hence, from these observations it can be concluded that the training program was responsible for the gradual setting in of a congenial protein environment in the muscle leading to improved catalytic efficiency of the enzyme.

References

GOVINDAPPA, S. & K.S. Swami. Succinate dehydrogenase activity and protein charges in cell free systems. Ind. J. Expl. Biol. 7:221–224, 1969.

GUTMANN, E., The Denervated Muscle. Prague: Czechoslovak Academy of Sciences, 1960.

HIRCHE, H.J., K.U. Boven, J. Bassu, V. Homobach & J. Manthey. Factors influencing lactic acid production in working skeletal muscle. Scand. J. Clin. Lab. Invest. 31:11–15, 1973.

HIRCHE, H.J., V. Homobach, H.D. Longowr, U. Walker & J. Bassu. Lactic acid permeation rate in working gastrocnemius of dog during metabolic alkalosis and acidosis. Pflugers Arch. 356:209–222, 1975.

MAINWOOD, G.W. & P.W. Brown. Fatigue and recovery in isolated skeletal muscle. In Proc. XXV Int. Cong. Physiol. Sciences, Munich Abstract No. 1070, pp. 361.

NACHLAS, M.M., S.P. Morgulis & A.M. Seligman. A colorimetric method for estimation of succinate dehydrogenase activity. J. Biol. Chem. 235:2739–2743, 1960.

NARASIMHA MURTHY, C.V., P. Reddanna & S. Govindappa. Muscular training through localized in vivo electrical stimulations. Ind. J. Physiol. & Pharmacol. 25:229–236, 1981.

REDDANNA, P., Some aspects of carbohydrate metabolism in amphibian muscle under the influence of electrical stimulation. Ph.D. Thesis submitted to Sri Venkateswara University, Tirupati, India, 1979.

REDDANNA, P. & S. Govindappa. Effect of in vivo muscular stimulations. III. Some aspects of carbohydrate metabolism of cardiac tissue. Curr. Sci. 47:531–533, 1978a.

The Role of Superoxide Dismutase and Catalase in Muscle Fatigue

R.R. Jenkins

Ithaca College, Ithaca, New York, U.S.A.

Our lab has been involved for some time now in research related to the effect of both use and disuse on the proteins involved in protecting the body from the reaction products associated with the univalent reduction of molecular oxygen. It is now known that the superoxide radical is a common product of the reduction of O_2 in biological systems (Hassan & Fridovich, 1978). The proteins which have evolved to protect against the potentially toxic reaction products of the univalent reduction of oxygen include Superoxide dismutase (SOD = superoxide oxidoreductase EC 1.15.1.1) which scavenges two O_2^- and combines them with two hydrogen ions to form hydrogen peroxide (H_2O_2) and catalase (H_2O_2) oxido-reductase, EC1.2.1.6) which combines two H_2O_2 to yield two H_2O and O_2. These enzymes are ubiquitous to aerobic cells and are of interest to those of us involved in the study of the body's adaptation to exercise. First, theoretically, it may be possible to recycle some O_2 back to the tissue from the catalase reaction. Secondly, Chance et al. (1979) in their review of hydroperoxide metabolism have pointed out that a transition from two electron to one electron transfer occurs at the succinate dehydrogenase-cytochrome b segment of the cytochrome chain and that this is a major physiological source of superoxide anion. Since an efficient cytochrome shuttle is essential to endurance exercise, we might expect that it would be important to protect the organism against a build up of the superoxide anion which might occur with endurance training. We (Jenkins, R.R.; H. Howald) have found that in humans, there is a significant correlation between $\dot{V}_{O_2max}$ and the activity of both SOD and catalase of the EDL muscle. For these reasons we set out to further investigate the relationship between the level of fitness of the organism and the hydroperoxide enzyme activity. Our studies have sought to answer the question whether 1) the state of aerobic fitness was related to hydroperoxide enzyme activity and 2) whether a chemical inhibition of the hydroperoxide enzymes would alter the resistance of the organism to fatigue.

Methods

Herein are reported a series of studies designed to determine the relationship of the hydroperoxide enzymes to endurance and fatigue resistance in skeletal muscle.

Experiment 1

In our first study 18 male Sprague Dawley-derived rats were assigned to either a sedentary control group, an 8 wk endurance trained group or a 12 wk endurance group. The training consisted of treadmill running according to the protocol of Holloszy (1967). At the termination of the training period rats from the sedentary control group and experimental group were anesthetized. One soleus muscle was removed immediately for enzyme analysis and the muscle of the opposite leg was set up for the measurement of the contractile properties according to the method of Fitts and Holloszy (1977). The enzyme analysis included catalase (Goldstin, 1968; Leighton et al., 1968), superoxide dismutase (Sun & Zigman, 1978), succinate dehydrogenase and cytochrome oxidase (oxygen cathode adaptation of Potter, 1964). Samples of each homogenate were examined for erthrocyte contamination by the method of Sinet (1980).

Experiment 2

Frogs or rats were injected intraperitoneally with either 3-amino-1,2,4-triazole, (ATA), a catalase inhibitor, or diethyldithiocarbamate (DDC), an inhibitor of superoxide dismutase. The dose of ATA or DDC was at the level of 1 gm/kg body weight and was carried in isotonic saline. Two h after the injection, the peak tetanic tension of the muscle was recorded as in the first experiment.

Results

The effects of 8 and 12 wk of endurance training are shown in Table 1. The activities of both cytochrome oxidase and succinate dehydrogenase were significantly increased above control levels by the 8th wk and the level of activity remained elevated at the 12th wk but was not significantly different

Table 1

The Effects of 8 and 12 Wk of Endurance Training on the Enzyme Activities of Rat Soleus Muscle

Group	Cytochrome Oxidase μl O_2/min/g	Succinate Dehydrogenase μl O_2/min/g	Superoxide Dismutase μ/mg Protein	Catalase μ/mg Protein
Sedentary (6)	408 ± 18	89 ± 8	4.74 ± 2	5.68 ± 0.44
Exercise-8 wk (6)	$600 \pm 28^*$	$105 \pm 8^*$	5.82 ± 2	6.72 ± 0.37
Exercise-12 wk (16)	$680 \pm 31^*$	148 ± 6	$7.20 \pm 3^*$	$8.24 \pm 0.30^*$

Values are mean $\pm$ SE. Number of animals is given in parenthesis. *Control versus trained, $p < 0.05$.

from the 8th wk. Although both catalase and superoxide dismutase were slightly increased by the 8th wk, there was no significant difference from control values until the 12th wk. The data in Table 2 demonstrate that the trained rats were able to maintain a peak tetanic tension after 30 min of electrical stimulation while the value for the sedentary rats decreased by 32%. The training imposed in this study did not alter the animals' ability to maintain twitch tension. These findings are comparable to those of Fitts and Holloszy (1977).

Table 2

Effect of Endurance Training and Fatigue on the Contractile Charactistics of Rat Soleus Muscle

Condition	Group	Peak tetanic tension	Twitch tension
		g/cm^2	g/cm^2
Base line, before fatigue	Control (6)	2108 ± 52	190 ± 24
	Trained (6)	1999 ± 47	184 ± 25
Post 30 min stimulation (110 stimuli/min)	Control (6)	$1433 \pm 46\dagger$	$70 \pm 28\dagger$
	Trained (6)	$1840 \pm 34*$	$57 \pm 31\dagger$

Values are means $\pm$ SE. Number of animals is given in parenthesis. *Control versus trained, $p < 0.05$. †Baseline versus post 30 min stimulation, $p < 0.05$.

We have found that ATA will depress catalase and DDC will depress SOD in skeletal muscle (Table 3). These compounds reduced the enzyme activity in both frog and rat muscle. The depression of catalase and SOD in the frog sartorius muscle was without effect on peak tetanic tension. However, ATA reduced peak tetanic tension in rat muscle by 28% and DDC produced a 52% reduction. The differences in the effects on the two muscles was at first thought to be due to the fact that the frog sartorius consists primarily of twitch fibers while the rat soleus consists predominantly of Type I fibers. However, preliminary findings on a study of rat muscles consisting predominantly of Type I or Type II fibers have not supported that conclusion.

In summary, we have found that endurance training of sufficient intensity and duration does result in an increase in both muscle catalase and superoxide dismutase. We have also found that, at least in rat muscle, depressed levels of these enzymes reduced the muscle's ability to withstand fatigue.

Table 3

Effect of 3-amino-1,2,4-triazole (ATA) and Diethyldithiocarbamate (DDC) on the Maintenance of the Peak Tetanic Tension of Frog and Rat muscle

Condition	Frog sartorius	Rat soleus
ATA	102%	27%*
DDC	97%	48%*

Values are percent of baseline data. *Baseline versus post 30 min stimulation, p < 0.05.

References

CHANCE, B., H. Sies & A. Boveris. Hydroperoxide metabolism in mammalian organs. Physiol. Rev. 59:527–605, 1979.

FITTS, R.H. & J.O. Holloszy. Contractile properties of rat soleus muscle effects of training and fatigue. Am. J. Physiol. 233:C86–91, 1977.

GOLDSTEIN, D.B., A method for assay of catalase with the oxygen cathode. Anal. Biochem. 24:431–437, 1968.

HASSAN, H.M. & I. Fridovich. Regulation and role of superoxide dismutase. Biochem. Soc. Trans. 6:356–361, 1978.

HOLLOSZY, J.O., Biochemical adaptations in muscle. J. Biol Chem. 242:2278–2282, 1967.

JENKINS, R.R. & H. Howald. The relationship of maximum oxygen uptake to superoxide dismutase and catalase activity in human skeletal muscle. Accepted by International Journal of Sports Medicine.

LEIGHTON, F., B. Poole, H. Beaufay, P. Baudhuin, J.W. Coffay, S. Fowler & C. DeDuve. The large-scale separation of peroxisomes, mitochondria, and lysosomes from the livers of rats injected with Triton WR-1339. J. Cell Biol. 37:482–513, 1968.

POTTER, V.R., Manometric techniques. In W.W. Umbreit, R.H. Burris, & J.F. Staufers (Eds.). Burgess Publishing Company, 1964, p. 162.

SINET, P.M., R.E. Heikkila & G. Cohen. Hydrogen peroxide production by rat brain in vitro. J. Neurochem. 34:1421–1428, 1980.

SUN, M. & S. Zigman. An improved spectrophotometric assay for superoxide dismutase based on epinephrine autoxidation. Anal. Biochem. 90:81–89, 1978.

Alterations in Substrate Supply
and Circulatory Responses to Exercise
in Myophosphorylase Deficiency

S.F. Lewis, R.G. Haller, J.D. Cook and C.G. Blomqvist
University of Texas Health Science Center,
Dallas, Texas, U.S.A.

In normal man cardiac output ($\dot{Q}$) and oxygen uptake ($\dot{V}_{O_2}$) are tightly coupled during dynamic exercise. For each liter of increase in $\dot{V}_{O_2}$, $\dot{Q}$ increases approximately 5 liters (i.e., $\Delta\dot{Q}/\Delta\dot{V}_{O_2} \simeq 5$) (Faulkner et al., 1977). The underlying control mechanisms are unknown.

We recently identified three patients with muscle phosphorylase deficiency (McArdle's disease) and a hyperkinetic circulation specific to exercise, i.e., $\dot{Q}$ was normal at rest but $\Delta\dot{Q}/\Delta\dot{V}_{O_2}$ excessive during exercise. In order to test the hypothesis that the hyperkinetic circulatory response to exercise in McArdle's disease is linked to the abnormal metabolic state and can be normalized by increasing free fatty acid (FFA) or glucose supply to muscle or worsened by FFA deprivation, the patients were studied during interventions known to specifically modify the supply of these substrates to exercising muscle.

Methods

Three patients (2 males, ages 17 and 22 and 1 female, age 28) with absent muscle phosphorylase were studied. Nine healthy physically inactive individuals (8 males, 1 female, ages 22 to 39 yrs) served as control subjects.

Six types of experiments were performed: 1) control—following a 2 h fast after a normal mixed diet, 2) fasting—following a 12 to 14 h fast, 3) fasting plus prolonged exercise—during 45 min of exercise following the 12 to 14 h fast, 4) nicotinic acid—approximately 650 mg nicotinic acid (an antilipolytic agent) were taken orally over the duration of the experiment,

5) glucose—a total of 500 ml of a 10% glucose solution was infused intravenously over the duration of the experiment (1½ to 2 h), 6) nicotinic acid plus glucose—experiment 5 was performed following experiment 4.

In each experiment exercise was performed on a cycle ergometer and all measurements were obtained in the sitting position. In experiments 1, 2 and 4 to 6 exercise was performed for 5 to 6 min at workloads requiring approximately 50, 75 and 100% of maximal oxygen uptake. The workloads were performed in ascending order of intensity and were separated by 15 min rest periods. In experiment 3 moderately heavy (50 to 80% $\dot{V}_{O_2max}$) exercise was performed continuously for 45 min. $\dot{Q}$ was measured by our modification of the acetylene rebreathing technique (Triebwasser et al., 1977). Heart rate (HR) was obtained from electrocardiographic tracings. Oxygen uptake was measured with the Douglas bag technique and a Tissot spirometer, and corrected to STPD. Plasma FFA concentration was measured according to Lauwerys (1969).

Results

Rest

Under all conditions resting HR and $\dot{Q}$ were similar in the patients and the healthy subjects. No intervention altered resting HR or $\dot{Q}$.

Control Experiments

The mean (± SE) maximal oxygen uptake in the myophosphorylase deficient patients was approximately one-third that of the healthy subjects (13.5 ± 3.4 vs 39.0 ± 1.6 ml/kg/min). The slope of the relation between $\dot{Q}$ and $\dot{V}_{O_2}$ in the patients was more than twice that of the healthy subjects (11.08 vs 4.46) (Figure 1).

Fasting

Resting plasma FFA concentrations were greater than in the control experiments in both the patients and in the healthy subjects. In the McArdle's patients the $\Delta\dot{Q}/\Delta\dot{V}_{O_2}$ ratio obtained in min 5 to 6 of exercise under fasting conditions (9.6 ± 2.0) was lower than in control experiments at the same workloads (12.2 ± 1.1). In the healthy subjects fasting had no effect on $\dot{Q}$ or HR during exercise.

Fasting Plus Prolonged Exercise

This intervention most effectively normalized the $\Delta\dot{Q}/\Delta\dot{V}_{O_2}$ ratio in the myophosphorylase deficient patients. $\dot{Q}$ dropped by an average of

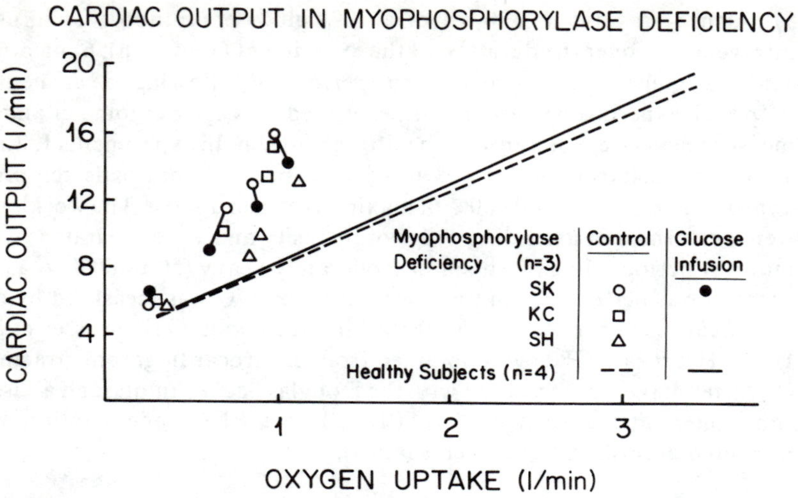

CARDIAC OUTPUT IN MYOPHOSPHORYLASE DEFICIENCY

OXYGEN UPTAKE (l/min)

Figure 1—Cardiac output in relation to oxygen uptake at rest and during submaximal and maximal exercise in the myophosphorylase deficient patients (individual points) and healthy subjects (regression lines) under control conditions and during infusion of 10% glucose. The regression equation for the patients under control conditions was $y = 2.82 + 11.08x$; $r^2 = 0.87$. The regression equation for the healthy subjects 1) under control conditions was $y = 3.88 + 4.46x$; $r^2 = 0.96$ and 2) during glucose infusion was $y = 3.85 + 4.64x$; $r^2 = 0.94$.

approximately 2 l/min and heart rate fell by more than 20 beats/min between mins 5 and 15, each remaining at the lower levels through min 45 (Figure 2, Table 1). In the healthy subjects $\dot{Q}$ remained essentially constant over the duration of prolonged exercise.

Glucose

Plasma glucose levels rose by approximately 70% in the patients and the healthy subjects as a result of the glucose infusion. Hyperglycemia was associated with partial normalization of the $\dot{Q}$-$\dot{V}O_2$ relationship as shown in patient S.K. in Figure 1. There were no significant effects in the healthy subjects.

Nicotinic Acid

In both the patients and the healthy subjects plasma FFA concentrations were lower at rest and during exercise after nicotinic acid administration. Nicotinic acid did not affect $\dot{Q}$ or HR during exercise in the healthy subjects. In patients S.H. and K.C., $\dot{Q}$ was an average of 2 l/min higher and HR 22 beats/min higher at 50 W after nicotinic acid. In patient S.H. nicotinic acid also was administered following fasting plus prolonged

$\Delta\dot{Q}/\Delta\dot{V}O_2$ RATIO IN MYOPHOSPHORYLASE DEFICIENCY DURING PROLONGED EXERCISE

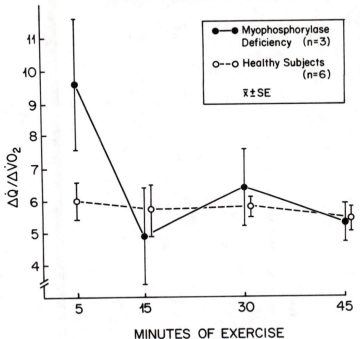

MINUTES OF EXERCISE

Figure 2—The ratio of the increase in cardiac output ($\dot{Q}$; l/min) from rest to exercise in relation to that of oxygen uptake ($\dot{V}O_2$; l/min) from rest to exercise ($\Delta\dot{Q}/\Delta\dot{V}O_2$). The data shown were obtained during prolonged exercise after a 12 to 14 h fast.

exercise. Remarkably, the $\Delta\dot{Q}/\Delta\dot{V}O_2$ ratio increased from 4.6 to 13.7 and $\dot{Q}$ rose by 5 l/min (Table 2).

Nicotinic Acid Plus Glucose

In patient K.C. dramatic effects were obtained with a glucose infusion begun following the administration of nicotinic acid. At a workload of 50 W, $\dot{Q}$ and HR were reduced by 4.5 l/min (from 17.1 to 12.6 l/min) and 61 beats/min (from 160 to 99 beats/min).

Discussion

An excessive exercise tachycardia has been noted in several cases of myophosphorylase deficiency (Lange Andersen et al., 1969) but cardio-

Table 1

Responses to Prolonged Exercise in Patients with Myophosphorylase Deficiency

Group[A]	Rest			Min of Exercise												
				5			15			30			45		
	$\dot{Q}^{B}_{E}$	HR[C]	FFA[D]	$\dot{Q}$	HR	FFA	$\dot{Q}$	HR	FFA	$\dot{Q}$	HR	FFA	$\dot{Q}$	HR	FFA
MD	5.5	81	0.61	10.9	135	0.48	8.8	114	0.58	9.1	118	0.67	8.8	114	0.79
(N = 3)	± 0.2	± 4	± 0.10	± 1.5	± 17	± 0.06	± 1.1	± 10	± 0.05	± 0.6	± 10	± 0.04	± 0.9	± 13	± 0.04
HS	5.2	78	0.66	12.9	123	0.53	13.0	134	0.53	12.6	139	0.68	12.5	144	0.91
(N = 6)	± 0.3	± 4	± 0.05	± 1.4	± 5	± 0.05	± 1.6	± 4	± 0.08	± 1.1	± 6	± 0.15	± 1.3	± 5	± 0.19

[A]MD = myophosphorylase deficient patients, HS = healthy subjects; [B]$\dot{Q}$ = cardiac output (l/min); [C]HR = heart rate (beats/min); [D]FFA = plasma free fatty acid concentration (mEq/l); [E]Data are expressed as $\bar{X} \pm$ SE.

Table 2

Effects of Nicotinic Acid Administration after Prolonged Exercise in Patient S.G.[a]

	Prolonged exercise				Nicotinic acid		
$\dot{Q}^b$ (l/min)	$\Delta\dot{Q}/\Delta\dot{V}_{O_2}$	HR (beats/min)	FFA (mEq/l)	$\dot{Q}$ (l/min)	$\Delta\dot{Q}/\Delta\dot{V}_{O_2}$	HR (beats/min)	FFA (mEq/l)
9.3	4.6	135	0.77	14.4	13.7	176	0.41

[a]Values are from min 45 of prolonged exercise at a workload of 50 W and 2 h later during min 5 of exercise at an identical workload after oral administration of a total of 400 mg nicotinic acid.
[b]Abbreviations as in Table 1 and text.

vascular regulation has not been studied in detail. The present results clearly demonstrate a hyperkinetic circulatory response to exercise in myophosphorylase deficiency. Although McArdle's patients sometimes complain of muscle pain associated with exercise, the observation of a normal $\Delta\dot{Q}/\Delta\dot{V}_{O_2}$ during exercise in patients with muscle aches and pains but normal muscle metabolism and in patients with muscular dystrophy (Lewis et al., 1981) indicates that muscle pain, weakness or atrophy are not causes of the hyperkinetic circulation. The present results show unequivocally that the hyperkinetic circulatory response to exercise in McArdle's disease can be totally or partially normalized by increasing the availability of FFA or glucose to active muscle or worsened by inhibition of lipolysis. These data firmly link the excessive $\Delta\dot{Q}/\Delta\dot{V}_{O_2}$ to an abnormal local muscle metabolic state and are consistent with the evidence for a critical role for metabolically sensitive muscle receptors in cardiovascular regulation during exercise (Longhurst & Mitchell, 1979).

Acknowledgment

Supported by grants from the Muscular Dystrophy Association and the American Heart Association, Texas Affiliate. Dr. Lewis was the recipient of a New Investigator Research Award (HL 26958).

References

FAULKNER, J.A., G.F. Heigenhauser & M.A. Schork. The cardiac output-oxygen uptake relationship of men during graded bicycle ergometry. Med. Sci. Sports 9:148-154, 1977.

LANGE ANDERSEN, K., P. Lund-Johansen & G. Clausen. Metabolic and circulatory responses to exercise in a subject with glycogen storage disease (McArdle's disease). Scan. J. Clin. Lab. Invest. 24:105-113, 1969.

LAUWERYS, R.R., Colormetric determination of free fatty acids. Anl. Biochem. 32:331-333, 1969.

LEWIS, S.F., R.G. Haller, J.D. Cook, R.M. Graham, W.A. Pettinger & C.G. Blomqvist. Hyperkinetic circulation during dynamic exercise in skeletal muscle disease. Circulation 64(Suppl IV):200, 1981.

LONGHURST, J.C. & J.H. Mitchell. Reflex control of the circulation by afferents from skeletal muscle. Internat. Rev. Physiol. 18:125-148, 1979.

TRIEBWASSER, J.H., R.L. Johnson, Jr., R.P. Burpo, J.C. Campbell, W.C. Reardon & C.G. Blomqvist. Noninvasive determination of cardiac output by a modified acetylene rebreathing procedure utilizing mass spectrometer measurements. Aviat. Space Environ. Med. 48:203-209, 1977.

Effect of Physical Exercise on Erythrocyte Zinc and Carbonic Anhydrase Isozyme in Men

**H. Ohno, F. Hirata, K. Terayama, T. Kawarabayashi,
R. Doi, T. Kondo and N. Taniguchi**
Asahikawa Medical College, Asahikawa, Japan, and
Hokkaido University School of Medicine, Sapporo, Japan

Carbonic anhydrase (CA, EC 4.2.1.1.) plays an important role in the acid/base equilibrium in the tissues and the transport of CO_2. In erythrocytes, two major types of isozymes designated as the B (CA-B) and C (CA-C) types occur. These two types of isozymes have been distinguished genetically and immunologically (Funakoshi & Deutsch, 1970). On the other hand, CA is a major zinc metalloenzyme in erythrocytes and most of the erythrocyte zinc ions are known to be derived from CA (Funakoshi & Deutsch, 1970). Our recent studies (Ohno et al., 1981; Ohno et al., 1982) showed that although the levels of CA-B as well as CA-B-dependent activity in human erythrocytes are decreased after acute physical exercise, no significant changes occur in the levels of CA-C or CA-C-dependent activity. In the present study, experiments on the influences of physical exercise on CA-B, CA-C and zinc concentrations in human erythrocytes were performed. A possible significance of CA isozyme under heavy physical exercise of short duration is discussed in relation to the zinc level.

Methods

Subjects: Five untrained healthy male volunteers, aged 21 to 23 yr, participated in this study. After fasting for about 12 h, the subjects worked on a bicycle ergometer (Monark, Sweden) with a load of 200 W for 30 min. They were made to take a rest in supine position for 30 min before and after the exercise. Heparinized blood samples were withdrawn from the antecubital vein at the beginning and end of exercise session and then 30 min later. To remove white cells and platelets from whole blood, dry α-cellulose and dry microcrystalline cellulose (Sigmacell type 50, U.S.A.) were used. The hemoglobin (Hb) content of each hemolysate was measured by the cyanmethohemoglobin method and then adjusted to approximately 5% for use in enzymic and immunological assays.

Measurement of the Levels of CA-B and CA-C

CA-B and CA-C isozymes were assayed according to a single radio-immunodiffusion technique with a slight modification of the method described by Funakoshi and Deutsch (1970).

Assay of Enzyme Activity

The esterase activity of CA was measured by the method described previously (Kondo et al., 1978), using a Hitachi model 624 spectrophotom-eter (Japan). The specific immunoadsorbent for CA-B was prepared by coupling Sepharose 4-B (Pharmacia Fine Chemicals, Sweden) with the antihuman CA-B horse IgG fraction as described previously (Kondo et al., 1978). CA-B dependent esterase activity was calculated by subtracting CA-C-dependent esterase activity (the activity after immunoadsorbent) from total CA esterase activity (the activity of untreated hemolysate). The specific activities of CA-B and CA-C were expressed as units per mg of isozyme.

Experiment of Zn^{2+} Addition

Effects of Zn^{2+} addition on total CA and CA-C-dependent activities were investigated by the method described previously (Kondo et al., 1978). CA-B-dependent activity as well as the specific activities of the two isozymes after Zn^{2+} addition were calculated according to the above-mentioned method.

Determination of Zinc Concentration

The concentrations of zinc in hemolysates were determined using a Shimadzu model AA 640-13 atomic absorption spectrophotometer (Japan) after an ashing process.

Results

The Levels and Specific Activities of CA-B and CA-C

In agreement with the findings of our previous studies (Ohno et al., 1981; Ohno et al., 1982), there were significant decreases in the levels of CA-B, total CA esterase activity and CA-B-dependent esterase activity immediately after exercise, as shown in Table 1 (p $<$ 0.001, p $<$0.01 and p $<$0.05, respectively). After 30 min of rest, they returned to their pre-exercise levels. The specific activity of CA-B tended to decrease immediately after exercise (p $<$ 0.1), and returned to its pre-exercise value after 30 min of rest. On the other hand, no significant change in CA-C level, CA-C-dependent esterase activity or the specific activity of CA-C was found after exercise.

Effect of Zn^{2+} Addition

Immediately after exercise, total CA activity and Ca-B-dependent activity following the addition of Zn^{2+} showed significant increases (p $<$ 0.05) and the specific activity of CA-B after Zn^{2+} addition seemed to increase (p $<$ 0.1), compared with their respective activities without Zn^{2+} addition. However, no such effects were observed just before exercise or after 30 min of rest. On the other hand, the addition of Zn^{2+} had no effect on CA-C-dependent activity or the specific activity of CA-C at any time.

Zinc Concentration

There was a significant decrease in erythrocyte zinc content immediately after exercise (p $<$ 0.05) and a return to its pre-exercise level after 30 min of rest.

Discussion

The present study describes the decrease in zinc level and the effect of Zn^{2+} addition on CA-B-dependent activity in human erythrocytes under heavy physical exercise of short duration.

It is known that the levels of human erythrocyte CA-B vary under certain physiologic or pathologic conditions, and that CA-C levels appear to be essentially unchanged in those conditions (Kondo et al., 1978). Likewise, it is apparent in peripheral blood that acute physical exercise has some effects on alternations in the levels of CA-B enzyme protein; however, the mechanisms of their actions is not fully understood (Ohno et al., 1981). On the other hand, the findings of the present study indicated that the active CA-B enzymes are converted in part to inactive enzymes during acute physical exercise, possibly by decreased zinc binding. CA has one atom zinc per one molecule enzyme in its active center (Funakoshi & Deutsch, 1970). In this study, a significant decrease in venous pH level was observed immediately after exercise (p $<$ 0.05; data not shown). One may speculate that adaptive decreases in the levels of CA-B and CA-B-dependent activity in erythrocytes occur with increased acidification in blood during exercise.

Table 1

Effect of Zn^{2+} Addition on the Activities of Carbonic Anhydrase Isozymes in Human Erythrocytes before and after a 30 Min Session of Physical Exercise

Subject	Hematocrit	Total Zn	Immunological level		Added Zn^{2+}	Esterase activity					
			CA-B	CA-C		Total activity	Dependent activity		Specific activity		
							CA-B	CA-C	CA-B	CA-C
	%	$\mu g/gHb$	mg/gHb				units/gHb		units/mg isozyme	
Just before exercise	45.4 ± 1.08	50.2 ± 1.91	14.1 ± 0.48	1.74 ± 0.095	−	17.5 ± 1.96	10.7 ± 1.24	6.76 ± 0.731	0.754 ± 0.0686	3.88 ± 0.389
					+	17.5 ± 1.88	10.7 ± 1.19	6.75 ± 0.707	0.753 ± 0.0658	3.88 ± 0.377
Immediately after exercise	48.5 ± 1.21[a]	43.4 ± 2.65[a]	11.9 ± 0.36[c]	1.76 ± 0.096	−	13.8 ± 1.43[b]	6.98 ± 1.036[a]	6.83 ± 0.685	0.582 ± 0.0697	3.91 ± 0.388
					+	17.2 ± 1.78*	10.4 ± 1.18*	6.89 ± 0.705	0.864 ± 0.0801	3.94 ± 0.399
After 30 min of rest	45.4 ± 1.07	50.3 ± 3.52	14.9 ± 0.56	1.75 ± 0.091	−	17.3 ± 1.41	10.6 ± 0.84	6.66 ± 0.575	0.711 ± 0.0562	3.83 ± 0.378
					+	18.5 ± 1.72	11.8 ± 1.30	6.68 ± 0.588	0.781 ± 0.0624	3.84 ± 0.382

The incubation mixture was composed of hemolysate containing 50 μM CA-B, 0.45% bovine serum albumin, pH 7.4, 4.5 mM Tris-HCl buffer, 100 μM $ZnCl_2$ in a final volume of 1 ml. The incubation was carried out for 6 h at 37°C before assay.

Number of cases = 5.

Values are expressed as mean ± SE.

Significantly different from "before" value: [a] $p < 0.05$, [b] $p < 0.01$, [c] $p < 0.001$.

*Significantly different from the respective activity without Zn^{2+} added: $p < 0.05$.

Differences between pre- and post-exercise values were tested for significance using a paired t-test.

Oelshlegel et al. (1974) reported that zinc binding to erythrocyte 2,3-diphosphoglycerate (2,3-DPG) can cause an increase in erythrocyte oxygen affinity since 2,3-DPG plays a major role in modulating erythrocyte oxygen affinity through binding to Hb. On the other hand, Hořejši and Komárková (1960) indicated that CA causes a decrease in erythrocyte oxygen affinity. In our previous reports (Ohno et al., 1981), we found that there are negative correlations between the changes in the CA-B and 2,3-DPG levels as well as in the total CA activity and 2,3-DPG levels after acute physical exercise. Under physical exercise, however, whether or not direct co-operation is present among CA, zinc, 2,3-DPG and Hb in erythrocytes remains to be elucidated. As demonstrated in the study of prolonged physical exercise by Hetland et al. (1975), the differences in exercise intensity may have some effects on the changes in these factors.

References

FUNAKOSHI, S. & H.F. Deutsch. Human carbonic anhydrase. III. Immunological studies. J. Biol. Chem. 245:2852-2856, 1970.

HETLAND, Ø., E.A. Brubak, H.E. Refsum & B. Strømme. Serum and erythrocyte zinc concentrations after prolonged heavy exercise. In H. Howald & J. Poortmans (Eds.), Metabolic Adaptation to Prolonged Physical Exercise. Basel: Birkhäuser, 1975.

HOŘEJŠI, J. & A. Komárková. The influence of some factors of the red blood cells on the oxygen-binding capacity of hemoglobin. Clin. Chim. Acta 5:392-395, 1960.

KONDO, T., N. Taniguchi, K. Taniguchi, I. Matsuda & M. Murao. Inactive form of erythrocyte carbonic anhydrase B in patients with primary renal tubular acidosis. J. Clin. Invest. 62:610-617, 1978.

OELSHLEGEL, F.J., Jr., G.J. Brewer, C. Knutsen, A.S. Prasad & E.B. Schoomaker. Studies on the interaction of zinc with human hemoglobin. Arch. Biochem. Biophys. 163:742-748, 1974.

OHNO, H., N. Taniguchi, T. Kondo, E. Takakuwa, K. Terayama & T. Kawarabayashi. Effect of physical exercise on the specific activity of carbonic anhydrase isozyme in human erythrocytes. Experientia (in press) 1982.

OHNO, H., N. Taniguchi, T. Kondo, K. Terayama, F. Hirata & T. Kawarabayashi. Effect of physical exercise on erythrocyte carbonic anhydrase isozymes and 2,3-diphosphoglycerate in men. Int. J. Sports Med. 2:231-235, 1981.

Nutrition

Some Effects of the Quality of Protein, Food Restriction, and Physical Exercise on Liver Development: Hepatic Total Lipids

M.C.R. Belda and S.M. Zucas

Faculty of Pharmaceutical Sciences of Araraquara, São Paulo, Brazil, and University of São Paulo, São Paulo, Brazil

All the functions of the systems of the organism depend on the action of outside factors such as nutrients. In the normal process, during the embryonic development and post birth periods, the nutrients from foods are incorporated into the cells of the organism to promote its development, differentiation, energy production and nervous activity. Nutritional deficiencies occurring in these stages of life can determine health problems, such as: body weight decrease, growth interruption, and nervous system retardation, which can cause a reduction in the capacity of educational retention, besides other problems.

A great number of scientific papers have demonstrated the effects of malnutrition on development. So, the authors have found much information which indicates the index of animal development. Among the most studied organs are the brain and the liver, due to the great importance that they have in the animal's development and metabolism, respectively, and to their sensitivity to deficiency in diets. Several authors have also tried to verify the effects of food recovery in development after an undernourishment period as well as to observe the effects of protein deficiency in the ribosomal and cellular contents distributions.

All those publications emphasize the importance of the quantity and quality of protein in the diet in the regulation of the synthesis and degradation of the hepatic RNA as well as the importance of the food restriction in the animal's development, in general.

With the purpose of associating physical activity, diet and development, much research has been done with humans and laboratory animals (Askew & Hecker, 1976; Askew et al., 1975; Borer et al., 1979; Carlson, 1967; Deb

& Martin, 1975; Hanley, 1979; Holm et al., 1978; Wijn et al., 1979; Yakoviev, 1975).

The association of the nutrition-development and nutrition-physical activity parameters has a great importance in several countries where physical exercise is an obligation in grade and high school.

Despite all those publications, nothing is found in consulted literature that relates the effects of protein quality on physical activities.

In the function of populational groups eating principally foods with a low biological value, protein and dietary quantity under normal values, it seemed of great importance to study the effects of these protein sources parallel to the quantitative observation and an association with physical activity.

Soybean was the low biological protein value choice, because of its deficiency of methionine, classically known, and because its use has been very recognized in the last years.

Our objective was to determine the effects of the quality of dietary protein (casein or soybean) with restriction (at the level of 60%) and of physical exercise on development of the liver of 120 male rats during the weaned period up to sexual maturation (1st period), some groups being followed by a food recovery period (2nd period). Physical exercise was performed during the 2nd period, 30 min daily. Eight groups of animals were not exercised (Table 1).

After the 1st period and the end of the experiment (adult phase), the animals were killed and the livers were analyzed for lipids.

Procedures

Composition of the Rations

The compositions of the rations are in Table 2.

Biochemistry Determinations

After the animal's death (with ether) the liver was removed and total lipid content determined by the method of Folch et al. (1957).

Physical Exercise Realization

An activity box with a counter (photo) was used for physical exercise realization.

Statistical Analysis

The statistical analysis was done according to Morrison (1967).

Table 1

Distribution of the 16 Animal Groups according to the Assayed Diet and Type of Administration in the Different Experimental Periods

Group	1st period Initial		2nd period (Recovery for some groups)		Specification used in text
	Protein dietary source	Type of adm	Protein dietary source	Type of adm	
1 9*	Casein	Ad libitum	Casein	Ad libitum	(CAL)
2 10*	Casein	Restriction	Casein	Restriction	(CR + CAL)
3 11*	Casein	Restriction	Casein	Restriction	(CR)
4 12*	Soybean	Ad libitum	Casein	Ad libitum	(SAL + CAL)
5 13*	Soybean	Ad libitum	Soybean	Ad libitum	(SAL)
6 14*	Soybean	Restriction	Casein	Ad libitum	(SR + CAL)
7 15*	Soybean	Restriction	Soybean	Ad libitum	(SR)
8 16*	Soybean	Restriction	Soybean	Restriction	(SR)

*Animals that were submitted to the same dietary conditions but were exercised in the 2nd experimental period.

Table 2

Composition of the Rations (g/100 g)

Rations	Rations according to protein source	
	Casein	Soybean
Casein[a]	12.50	—
Isolated soy protein[b]	—	11.11
Starch	62.40	63.79
Saccharose	10.00	10.00
Soy oil	8.00	8.00
Fiber (corncob)	3.00	3.00
Salt mixture	3.00	3.00
Vitamin mixture	1.00	1.00
Sodium benzoate	0.10	0.10

[a]with 80% of protein
[b]with 90% of protein

Results

The average results observed in the analysis of total lipids for the different groups studied are expressed in Table 3.

Discussion

The lipids are structural components of all cellular membranes, plasmatic, nuclear wrappers and membranes of the endoplasmatic reticulum, Golgi's apparatus, mitochondria and lysosomes. Many of the membranes occur from the chemical and physical characteristics of the structural lipids.

The liver is a very important organ in metabolism; it is sensitive to nutritional alterations such as the deficiency of essential amino acids in diets and has an important position in lipid metabolism. The liver removes approximately 30% of free plasmatic fatty acids and 30% of chilomicron glycerides which are quickly transformed into fatty acids through hydrolysis. These fatty acids are oxidized with the purpose of providing the necessary energy to the liver itself or they can be esterified, remaining as triglycerides. A fraction of these esters can be part of an accumulation of a relatively low turnover where, under normal conditions, they are in part incorporated into lipoprotein structures of several organic cells and, under

Table 3

Average ($\overline{X}$), Standard Deviation(s) and Variation Ratio (CV) of the Liver Weights and Total Lipids of the Various Experimental Groups in the Different Periods

| | | First period | | Second period (recovery for some groups) | | | |
| | | | | Without exercise | | With exercise | |
Groups	Estima-tion	*	**	*	**	*	**
G₁	$\overline{X}$	5.505	58.56	5.718	81.12	6.845	52.68
	s	0.9300	4.629	1.4928	63.284	1.1661	15.085
CAL	CV	16.89	7.91	26.11	78.01	17.03	28.63
G₂	$\overline{X}$	3.478	73.92	5.238	50.64	5.946	45.96
	s	0.0852	4.794	0.5798	9.624	0.9094	4.494
CR + CAL	CV	2.45	6.49	11.07	19.00	15.29	9.77
G₃	$\overline{X}$	3.478	73.92	4.857	65.64	4.878	48.96
	s	0.0852	4.794	0.5608	27.531	0.5102	3.943
CR	CV	2.45	6.49	11.55	41.94	10.46	8.05
G₄	$\overline{X}$	2.822	95.76	5.822	62.76	4.771	57.36
	s	0.2729	15.086	1.2240	13.942	16.753	14.633
SAL+CAL	CV	9.67	15.76	20.96	22.21	35.12	25.51
G₅	$\overline{X}$	2.822	95.76	5.552	160.70	4.275	70.41
	s	0.2729	15.086	1.3149	64.424	1.5886	41.450
SAL	CV	2.45	15.76	23.68	40.08	37.16	58.86
G₆	$\overline{X}$	2.351	92.88	6.060	88.68	6.620	56.04
	s	0.2195	12.615	0.7451	66.216	0.7557	24.392
SR + CAL	CV	9.34	13.58	12.30	74.66	11.42	43.52
G₇	$\overline{X}$	2.351	92.88	4.921	115.92	4.268	113.40
	s	0.2195	12.615	0.8673	53.575	1.0782	47.824
SR + SAL	CV	9.34	13.58	17.62	46.21	25.27	42.17
G₈	$\overline{X}$	2.351	92.88	3.343	85.32	3.185	69.69
	s	0.2195	12.615	0.173	33.088	0.2078	17.778
SR	CV	9.34	13.58	5.17	38.78	6.53	25.50

*liver weight (g)
**liver total lipids (mg/g)

abnormal conditions, they are deposited and stored, constituting a hepatic steatosis.

On the other hand, a second fraction combines itself with specific proteins and consequently constitutes the low density lipoproteins that are liberated in the circulation.

From the observation of the results of the averages of total lipids (mg/g) in the different experimental groups, in Table 3 and Figure 1, we verify that in the 1st experimental period there was a significant increase in the groups fed with soybeans either in ad libitum (SAL) or with restriction diets when compared with the control group (CAL).

In the animals from the 2nd experimental period, not submitted to exercise, we can verify the same observation: the group fed with soybean ad libitum presented higher results than the other groups with significant differences when compared to the control groups (CAL) and to the soybean restriction groups, in the initial period, followed by a recovery with the same diet ad libitum (SR + SAL) and soybean restriction group during the whole experimental period (SR). It is important to emphasize that the soybean ad libitum group, which presented higher results than the other ones, showed a large variation among the individual results, expressed by a high standard deviation, that was observed in almost all the groups.

We cannot correlate all the levels obtained for the different groups with any kind of diet effects, qualitatively or quantitatively.

In the animals of the 2nd experimental period, submitted to exercise, we can verify by the results that there was no influence in these parameters in the restriction groups. The group that presented the highest result that received soybean was the one with restriction in the initial period, with a recovery with soybean ad libitum (SR + SAL). We also verify that the animals which received the soybean diet presented higher results than those obtained in casein groups.

In general, the results were lower than those obtained with the animals in the experimental groups with no exercise.

Since the quantity of a determined substance in an organ is dependent on its own size and the size also is a function until a determined moment of the somatic development of the animal, the expression of the results in the concentration of the substance in the organ, many times, is an important factor to interpret the biological phenomenon that occurred and can be obtained by associating the quantity of the substance determined with the wt of the organ.

Many authors have studied the factors which cause hepatesteatosis. Among these factors we find choline and total sulfur amino acids in the diet. As we know, soybean protein is deficient in total sulfur amino acids that can interfere in the formation of the low density lipoproteins at hepatic

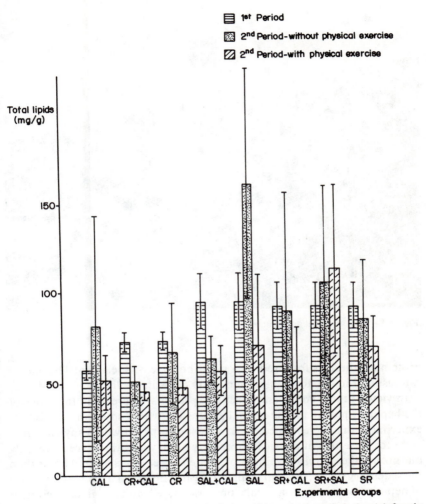

Figure 1—Average values and standard deviation of hepatic total lipds (mg/g) for the experimental groups.

lysosome level. This deficiency was demonstrated by Dianzani (1976) who studied liver and toxic injuries caused by inhibitors of protein synthesis and demonstrated that the administration of ethionine, an antagonist of methionine, can be a primary agent in the inhibition of amino acid synthesis, actuating in the inhibition of adenosine monophosphate formation, a precursor of proteic synthesis. The decrease of ATP is rapidly followed by an accentuated inhibition of RNA synthesis. The absence of

Figure 2—Activity wheel and counter.

protein synthesis is followed by a polyribosome dissociation. This factor can resound in the accumulation of fatty acids in the liver.

In view of these considerations we expected an increase in the levels of the hepatic lipids in all the animal groups fed with a soybean diet. In our experiment this increase occurred in most groups fed with a soybean diet.

As for the effect of physical exercise, we can verify that except for only one group, the exercised animals presented lower levels of total lipids than the sedentary ones. Narayan et al. (1975) also observed a decrease in levels of hepatic lipids in racer animals, pre scratch, fed with a high fatty content when compared to the controls.

After studying the effect of exercise on lipid metabolism in rats fed with a high carbohydrate diet, Herbert et al. (1975) verified that the diet did not influence the levels of the serum triglycerides, but the exercise decreased significantly the levels of serum triglycerides in relation to the levels of the sedentary groups.

Studying effects of prolonged physical exercise on human serum lipids, after running, Thompson et al. (1980) verified that the triglycerides and cholesterol levels had decreased.

Walimaki et al. (1980) observed in school children the influence of physical activities on serum lipids and concluded that in sedentary ones, the levels of cholesterol were reduced.

The results obtained by us in different determinations allowed the following considerations:

1. The ingestion of an inferior quality protein diet in both alimentation forms caused an increase in the content of the total lipids (mg/g) in all animals of the first experimental period.

2. The ingestion of inferior quality protein diet in the 1st experimental period, followed by recovery or during all experimental observation independent of ad libitum or restricted intake, determined an increase of total lipid content (mg/g).

3. Physical exercise caused decrease of the contents of the hepatic total lipid contents (mg/g).

References

ASKEW, E.W., G.L. Dohm & R.L. Huston. Fatty acid and ketone body metabolism in the rat: response to diet and exercise. J. Nutr. 105:1422–32, 1975.

ASKEW, E.W. & A.L. Hecker. Adipose tissue cell size and lipolysis in the rat: response to exercise intensity and food restriction. J. Nutr. 106:1351–60, 1976.

BORER, K.T., J. Hallfrisca, A.C. Tsai & C. Hallfrisch. The effect of exercise and dietary protein levels on somatic growth, body composition, and serum lipid levels in adult hamsters. J. Nutri. 109:222–28, 1979.

CARLSON, L.A., Lipid metabolism and muscular work. Fed. Proc. 26:1755–59, 1967.

DEB, S. & R.J. Martin. Effects of exercise and of food restriction on the development of spontaneous obesity in rats. Nutrition 105:543–49, 1975.

DIANZANI, M.U., Toxic liver injury by protein synthesis inhibitors. Prog. Liver. Dis. 5:232–45, 1976.

FOLCH, J., M. Less & G.H.S. Stanley. A simple method for isolation and purification of total lipids from animal tissues. J. Biol. Chem. 226:497–509, 1957.

HANLEY, D.L. Basic diet guidance for athletes. Nutr. Today 14(6):22–23, 1979.

HERBERT, J.A., L. Kerkohoff, L. Bell & S. Lopes. Effect of exercise on lipid metabolism of rats fed high carbohydrate diets. J. Nutr. 105:716–25, 1975.

HOLM, G., P. Bjorntorp & R. Jagenburg. Carbohydrate, lipid and amino acid metabolism following physical exercise in man. J. Appl. Physiol. 45:128–31, 1978.

MORRISON, D.F., Multivariate statistical methods. New York: Macgraw Hill, 1967.

NARAYAN, K.A., J.J. McMullen, D.P. Butler, T. Wakefield & W.K. Calhoun. Effect of exercise on tissue lipids and serum lipoproteins of rats fed two levels of fat. J. Nutr. 105:581–87, 1975.

THOMPSON, D.P., E. Cullinane, L.O. Henderson & P.N. Herbert. Acute effects of prolonged exercise on serum lipids. Metabolism 29:662–65, 1980.

VALIMAKI, I., M.L. Hursti, L. Pihlakoski & J. Viikari. Exercise performance and serum lipids in relation to physical activity in school children. Int. J. Sport. Med. 5:542–46, 1964.

WIJN, J.F. de, J. Leusink & G.B. Post. Diet body composition and physical condition of champion rowers during periods of training and out of training. Bibl. Nutr. Dieta. 27:143–48, 1979.

YAKOVIEV, N.N., Biochemistry of sport in the Soviet Union: beginning development and present status. Med. Sci. Sports 7:237–47, 1975.

Changes in Skeletal Muscle Metabolism Induced by a Eucaloric Ketogenic Diet

E.C. Fisher, W.J. Evans, S.D. Phinney, G.L. Blackburn,
B.R. Bistrian and V.R. Young

Boston University, Boston, Massachusetts, U.S.A.,
and Massachusetts Institute of Technology,
Cambridge, Massachusetts, U.S.A.

A previous study by Phinney et al. (1980) investigated the effects of a six-week adaptation to a hypocaloric ketogenic diet in obese untrained female subjects. The subjects demonstrated a shift in substrate utilization from carbohydrate to free fatty acids during submaximal exercise without compromise in their endurance performance. These results cannot be attributed to the state of ketosis alone because all subjects experienced weight loss. Therefore, the present study was designed to determine whether well-trained male subjects adapted to a eucaloric ketogenic diet (EKD) would demonstrate similar changes during submaximal exercise. Tolerance to submaximal exercise training and performance was observed following 4 wks of EKD. Exercise and its relationship to substrate utilization and skeletal muscle enzyme changes were quantified following adaptation to the carbohydrate-restricted diet.

Methods

Five well-trained male subjects ($\dot{V}O_{2max} = 5.10 \pm 181$ min^{-1}) were chosen for participation on the basis of their current state of training. The subjects were required to maintain their usual training program throughout the duration of the study. All subjects were informed of the risks and hazards of the experiment and signed a consent form previously approved by the M.I.T. and Boston University Human Experimentation Committees.

The subjects were fed a eucaloric ketogenic diet over a 4-wk period. The diet provided 35 to 50 kcal kg^{-1} day^{-1} with 1.75 g protein kg^{-1} day^{-1}, less than 20 g carbohydrate day^{-1} and the remainder of kcal as fat. The diet was supplemented to meet RDA for vitamins and minerals.

Maximal oxygen uptake ($\dot{V}O_{2max}$) was determined for each subject prior to and after adaptation to the ketogenic diet. Expired gas samples were collected using a semi-automated system while subjects pedaled on an electrically-braked cycle ergometer. Resistance was increased to a self-determined end-point of fatigue. Before and after adaptation to the diet, the subjects were endurance exercised at 60% of this pre-determined $\dot{V}O_{2max}$ until exhaustion.

A primed intravenous infusion of U–^{13}C-glucose was used to determine glucose oxidation rates immediately before and during the endurance exercise periods. Blood samples and expired gas samples were collected throughout the infusion period for later analysis of label enrichment (Wolfe et al., 1979). Expired gas samples were also analyzed using the semi-automated system in order to determine $\dot{V}O_2$, $\dot{V}CO_2$, and respiratory exchange ratio (RER).

Muscle biopsy samples were collected from the vastus lateralis before and after each endurance exercise period. The samples were frozen in liquid nitrogen for later determination of muscle glycogen and lactic dehydrogenase (LDH), malate dehydrogenase (MDH), glycogen phosphorylase (GP), hexokinase (HK), and carnitine palmityl transferase (CPT) activities.

Differences between the means of the pre- and post-diet enzyme activities were determined using the paired t-test. Significance was set at the 95% level of confidence.

Results

The ability to maintain endurance exercise was not compromised in duration or intensity by the diet. There was no significant difference in exercise time to exhaustion resulting from adaptation to EKD (pre-EKD = 147 min, post-EKD = 150 min). Following adaptation to the diet, blood-borne glucose supplied 9% of the energy expenditure during exercise as compared to 28% before adaptation. During exercise and after the dietary adaptation, average muscle glycogen utilization was reduced to 21% of the pre-diet value. The mean RER during exercise before dietary adaptation was .83. After adaptation, the corresponding RER was 172 (Table 1). The specific activities of CPT increased 35% while HK decreased 46% in skeletal muscle tissue following a four-week adaptation to the eucaloric ketogenic diet. In contrast, a significant change was not observed in the

Table 1

Exercise Results

	$\dot{V}O_{2\,max}$ (1 min^{-1})	Duration (min)	Endurance Test		
			Glucose oxidation rate (mg kg^{-1} min^{-1})	RER	$\dot{V}O_2$ (1 min^{-1})
pre-EKG	5.10	147	15.1	.83	3.18
post-EKD	5.00	151	5.1	.72	3.21

Muscle glycogen (μmol g^{-1}) diet and endurance exercise effects

	pre-ex	post-ex
pre-EKD	143	53
post-EKD	76	56

Table 2

Mean Enzyme Activities (μmoles g^{-1} min^{-1} $\pm$ SEM)

	MOH	LDH	GP	HK	CPT
pre-EKD	85.12 $\pm$ 4.72	116.7 $\pm$ 19.5	3.95 $\pm$ 1.09	2.12 $\pm$ .39	.43 $\pm$ .10
post-EKD	89.70 $\pm$ 29.83	117.4 $\pm$ 19.8	3.49 $\pm$ .61	1.15* $\pm$ .14	.62* $\pm$ .05

*Significantly different from pre-EKD; $p < .05$

specific activities of GP, MDH and LDH from these skeletal muscle tissues (Table 2).

Discussion

The fact that highly trained endurance athletes were able to maintain training and endurance performance during ketosis is not in agreement with previous investigations examining exercise performance and ketosis (Bergstrom et al., 1967; Christensen et al., 1939; Galbo et al., 1979; Pruett, 1970). In these four previous studies, 14 days or less were allowed for dietary adaptation, indicating that exercise performance is impaired during the early stages of adaptation to a carbohydrate restricted diet. A 4 wk EKD regime, as was characteristic of the present investigation, appears to be a more adequate period of adaptation. The mechanism of adaptation appears to involve a simultaneous maximization of glucose storage during periods of rest and limitation of its mobilization during exercise. The degree of net carbohydrate sparing after 4 wks of EKD is exemplified by the more than four-fold reduction in muscle glycogen mobilization and the three-fold decrease in blood glucose oxidation during the endurance exercise. The shift in the carbon source for the citric acid cycle, from carbohydrate to free fatty acids, is observed in the decreased RER value post-EKD. An evaluation in β-oxidation allowed for an increase in glycerol release from triglycerides and this, coupled with gluconeogenic carbon salvaged from lactate, pyruvate and gluconeogenic amino acids, provides substrate for the glucose consumed during exercise.

The dramatic rise in the specific activities of CPT with adaptation to the EKD is consistant with an increased use of free fatty acids as a metabolic substrate during ketosis. The reduced resting and exercise RER values following the post-dietary adaptation confirm an increased dependence on fatty acid oxidation. McGarry et al. (1975) have suggested that the regulation of fatty acid oxidation is controlled primarily by the CPT system. These authors further propose that this regulation is achieved by CPT in catalyzing the facilitated transport of acylcarnitine to the site of β-oxidation within the mitochondria.

Previous studies (Mitchell, 1978; Stearns, 1980) have shown that CPT activities can be altered by adaptation to nutrient deficiency as well as by the diabetic condition. CPT activities were stimulated in rats treated with alloxan (Stearns, 1980) or streptozotocin (McGarry et al., 1975) to induce diabetes. In both of these cases, ketone production was elevated due to an augmented fatty acid oxidation.

Muscle hexokinase plays an important role in the catabolism of glucose. Hexokinase is responsible for the phosphorylation to glucose of glucose-6-phosphate. Thus it can serve to control the rate of net glucose retention in the muscle cell and its subsequent oxidation. The activity of this enzyme has also been shown to be altered by dietary adaptation and diabetes. Kazan et al. (1970) demonstrated that in stretozotocin-induced diabetic rats, there was a decrease in HK activity from heart, diaphragm, and gastrocnemius muscle homogenates when compared to normals. Blumenthal et al. (1964) displayed a similar decrease in hepatic glucokinase (GK) activities from alloxan diabetic and fasted rats. Sols et al. (1966) propose a regulatory model for the hepatic GK enzyme. In this model, elevated insulin levels will induce GK synthesis and high glucose concentrations will inhibit GK degradation. The data suggest that, according to the model of Sols et al., the reduced muscle HK activities may be induced by the lowered insulin levels observed during ketosis.

Results obtained from previous animal investigations exhibit the importance of the metabolic state on the specific activities of enzymes. Under conditions of severe dietary carbohydrate restriction, there is a reduced glucose and glycogen oxidation, increased fatty acid oxidation, and elevated ketone production in humans. These shifts in substrate use can result in an alteration in HK and CPT activities. These results demonstrate that the ability for athletes to maintain both maximal oxygen uptake and aerobic endurance is not compromised after 4 wks of EKD. This is accomplished by a conservation of carbohydrate stores during exercise, accompanied by an increased lipid oxidation.

References

BERGSTROM, J., L. Hermansen, E. Hultman et al. Diet, muscle glycogen and physical performance. Acta Physiol. Scan. 71:140–150, 1967.

BLUMENTHAL, M.P., S. Abraham & I.L. Charkoff. Adaptive behavior of hepatic glucokinase in the alloxan-diabetic rat. Archives of Biochemistry and Biophysics 104:225–230, 1964.

CHRISTENSEN, E.H. & O. Hansen. Zur methodik der respiratorischen quotient bestimmung in Ruhe und bei Arbeit. III. Arbeits Faehigkeit und Ernaehrung. Skand. Arch. Physiol. 81:160–171, 1939.

GALBO, H., J.J. Holst & N.T. Christensen. The effect of different diets and of insulin on the hormonal response to prolonged exercise. Acta Physiol. Scand. 107:19–32, 1979.

KATZEN, H.M., D.D. Soderman & C.E. Wiley. Multiple forms of hexokinase. The Journal of Biological Chemistry 245:4081–4096, 1970.

McGARRY, J.D., Robles-Valdes & D.W. Foster. Role of carnitine in hepatic ketogenesis. Proc. Nat. Acad. Sci. USA. 72:4385–4388, 1975.

MITCHELL, M.E., Carnitine metabolism in human subjects: Metabolism in disease. Am. J. Clin. Nutrition 31:645–659, 1978.

NEWSHOLME, E.A., Use of enzyme activity measurements in studies on the biochemistry of exercise. International Journal of Sports Medicine 1:100–102, 1980.

PRUETT, E.D.R., Glucose and insulin during prolonged work stress in men living on different diets. J. Appl. Physiol. 28:199–208, 1970.

PHINNEY, S.D., E.S. Horton, J. Hanson et al. The capacity for moderate exercise in obese subjects after a hypocaloric ketogenic diet. J. Clin. Invest. 66:1152–1161, 1980.

SOLS, A., A. Sillers & J. Salas. Insulin-dependent synthesis of glucokinase. J. Cell and Comp. Physiol. 66:23–38, 1966.

STEARNS, S.B., Carnitine content of skeletal muscle from diabetic and insulin-treated diabetic rats. Biochem Medicine 24:33–38, 1980.

WOLFE, R.R., J.R. Allsop & J.F. Burke. Glucose metabolism in man: Response to intravenous glucose infusion. Metabolism 28:210–220, 1979.

Effect of Fat and Carbohydrate Supply on Myocardial Substrate Utilization during Prolonged Exercise

L. Kaijser, M.L. Wahlqvist, S. Rössner and L.A. Carlson

Karolinska Hospital, Stockholm, Sweden

While skeletal muscle utilizes the possibility of storing substrates during rest and covers a major fraction of its substrate demand from these stores during exercise, the heart muscle, which works continually throughout life, must rely mainly upon a continuous substrate provision from the bloodstream. Thus, it has been shown that in healthy humans under a variety of conditions, myocardial substrate extraction, from a quantitative point of view, equals simultaneous myocardial substrate oxidation (Lassers et al., 1972). However, some experiments with prolonged exercise in rats have resulted in the suggestion that this may lead to myocardial glycogen depletion (Blount & Meyer, 1959), and in humans signs of utilization of intramyocardial triglycerides have been found when exercise exceeds 1½ to 2 h (Kaijser et al., 1972).

Most studies of myocardial substrate metabolism in humans have been performed in the postabsorptive state while competitive exercise is most often performed with optimal nutrition. Since it has been shown that one of the major factors determining myocardial extraction of a substrate is its blood concentration (Wahlqvist et al., 1972) it was considered of interest to study to what extent increased availability of both carbohydrates and fat in blood affected myocardial substrate metabolism during prolonged exercise. To maintain constant substrate supply, studies were done during an artificial steady-fed state produced by constant rate infusion of glucose and a fat emulsion.

Methods and Procedures

Myocardial substrate utilization was estimated by simultaneous blood sampling from percutaneously introduced catheters in an artery (a) and the coronary sinus (cs) with analyses of oxygen, glucose, lactate, pyruvate, free fatty acid (FFA) and triglyceride (TG) concentrations. To facilitate estimation of myocardial FFA extraction ^{3}H labelled palmitate was infused i.v. and fractional extraction of labelled fatty acid determined. Five healthy young male volunteers performed cycle exercise in the supine position for 2 h at 40% of Vo_2max during constant rate i.v. infusion of glucose (0.32 g/min) and a fat emulsion, a modified form of Intralipid^R (0.17 g/min) in which TG appears in chylomicron-size particles, and which contained only 0.8 mmol/l glycerol. Results were compared with those of 15 men performing the same work after an over-night fast (some of these data have been presented previously [Kaijser et al., 1972]).

Details regarding analytical methods have been presented elsewhere (Carlson et al., 1973; Kaijser et al., 1972). In the following the term *extraction* of a substrate will be used to denote its removal from the blood stream in μmol/l blood, i.e., in case it is based on chemical analysis of a and cs blood it equals the a-cs difference. When calculated from the isotope data it is the fractional extraction of labelled fatty acid multiplied by the arterial concentration. *Uptake* denotes the removal in μmol/unit time (i.e., extraction multiplied by blood flow).

Results

At Rest

The infusion doubled the blood glucose concentration and increased the TG concentration four-fold, but did not significantly alter the FFA concentration. This led up to markedly increased myocardial TG extraction: if fully oxidized, the removal of TG would have covered 40 to 45% of myocardial oxidative metabolism. Also myocardial glucose extraction was significantly increased, while FFA extraction was significantly decreased (Figure 1).

A significant release of glycerol from the heart, (which was not found in the overnight fasted subjects) suggested that triglycerides indeed underwent lipolysis on passage across the heart, which is a prerequisite for their myocardial uptake.

During Exercise

Shortly after the onset of exercise, blood glucose concentration was reduced to the normal resting preinfusion level, in spite of the continuous infusion. In the fasting subjects blood glucose tended to decrease and after 2 h of exercise they had slightly lower blood glucose concentration than i.v. fed subjects. TG concentration remained unaltered throughout the 2 h of

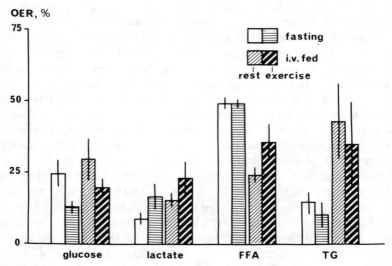

Figure 1—Myocardial extraction of glucose, lactate, free fatty acids (FFA) and triglycerides (TG) in fasting and i.v. fed subjects at rest and after 2 h exercise. Mean ± SE.

exercise. Measurements during the last 10 min of the exercise period showed a myocardial TG extraction, which was greater than during exercise in the postabsorptive state, although in relation to the oxygen extraction smaller (about half) than at rest with i.v. feeding. Arterial FFA concentrations increased towards the end of exercise similarly as in the fasting subjects. Myocardial extraction of FFA remained lower than in the fasting subjects, while extraction of glucose was insignificantly higher. Myocardial glycerol release (μmol/l blood) decreased to half the resting value, like the TG extraction.

Discussion

In the present study myocardial metabolism was measured in a condition where glucose and a chylomicron-like fat emulsion was infused i.v. to create a steady-fed state. Comparing exercise with resting conditions it may be noted that arterial concentrations of glucose rapidly decreased to preinfusion levels, suggesting that the infused glucose was taken care of, most probably by exercising skeletal muscle, since the a-cs difference of glucose was decreased by 40% at the same time as the coronary blood flow had a 2 to 3 fold increase (own unpublished data), indicating moderately increased myocardial glucose uptake which together with the small heart muscle mass makes myocardial removal insufficient to account for more than a small fraction of the infused glucose. Blood TG concentration, on the other hand, remained unaltered throughout work, suggesting that skeletal muscle did not increase its TG removal during exercise. This is in accordance with previous findings of significant removal of TG in skeletal muscle of a magnitude corresponding at rest, to about 20% of the oxidative substrate utilization, with no increase in uptake during exercise of moderate duration (Kaijser & Rössner, 1974). There are, on the other hand, indications that with exercise of very long duration (5 to 10 hours) the prerequisite for increased TG removal in the form of increased activity of lipoprotein lipase is at hand (Lithell et al., 1979).

In the heart muscle the decrease in blood glucose concentration led to a decrease in glucose extraction of similar magnitude, i.e., the fractional extraction remained unaltered. Myocardial TG extraction, on the other hand, decreased to ¾ the resting value in spite of unaltered blood concentration. The decrease was numerically slightly less than the calculated simultaneous increase in coronary blood flow, but the number of observations is too small to permit a conclusion as to whether myocardial TG uptake increased or remained unaltered during exercise. The significant myocardial glycerol release during exercise, with a negative a-cs difference which had decreased to the same extent as the positive a-cs TG difference, indicated that TG removed by the heart underwent lipolysis, which is a prerequisite for the TG uptake and utilization by the heart muscle, during exercise as at rest.

In comparing fed and fasting subjects it may be noted that myocardial FFA extraction was smaller in the fed subjects, during exercise as well as at rest. This is most probably a reflection of competition between FFA and fatty acids released from TG during passage in the heart vasculature. That the smaller a-cs difference of FFA in the fed state represents a true decrease in myocardial extraction of FFA, and not only reflects lipolysis of plasma TG without myocardial uptake of the released fatty acids is apparent from the finding of smaller myocardial FFA extractions in the fed state as calculated from radio-isotope data.

However, there was a difference between chemically and radio-isotopically estimated myocardial extraction of FFA at rest which tended to disappear during exercise suggesting more complete myocardial uptake of fatty acids released from TG during exercise compared to at rest (Kaijser et al., in press).

While at rest myocardial glucose extraction was greater in fed than in fasting subjects, this difference became insignificant during exercise, in accordance with the disappearance of the difference in arterial glucose concentration, i.e., the fractional extraction of glucose was the same in the fed as in the fasting state.

With regard to the total myocardial fraction of blood-borne substrates it may be noted that while in the fasting state the appearance after 2 h of work of significant myocardial glycerol release indicated that the heart had started to utilize intra-myocardially stored TG, no indication of net utilization of intra-myocardial substrates was found in the fed state. On the contrary, extractions of glucose, lactate, FFA and TG together exceeded 100% of substrate oxidation suggesting that after 2 h of work the heart muscle still took up more substrate than it utilized.

Conclusions

Prolonged exercise in the fasting state may lead to partial depletion of myocardial substrate stores. This may be prevented by the supply of substrate, either i.v. or orally. Not only glucose or FFA, but also triglycerides may then serve as additional substrate. With regard to the ability to utilize additional amounts of triglycerides, the capacity of the heart muscle seems to exceed that of skeletal muscle.

Acknowledgments

This study was supported by grants from the Swedish Medical Research Council (204 and 4494).

References

BLOUNT, D.H. & D.K. Meyer. Effects of cardiac work on glycogen fractions of the heart. Amer. J. Physiol. 197:1013–1018, 1959.

CARLSON, L.A., L. Kaijser, S. Rössner & M.L. Wahlqvist. Myocardial metabolism of exogenous plasma triglycerides in resting man. Acta Med. Scand. 193:233–245, 1973.

KAIJSER, L., B.W. Lassers, M.L. Wahlqvist & L.A. Carlson. Myocardial lipid and carbohydrate metabolism in fasting men during prolonged exercise. J. Appl. Physiol. 32:847–858, 1972.

KAIJSER, L. & S. Rössner. Removal of exogenous triglycerides in human forearm muscle and subcutaneous tissue. Acta Med. Scand. 197:289–294, 1975.

KAIJSER, L., M.L. Wahlqvist, S. Rössner & L.A. Carlson. Myocardial substrate metabolism in healthy men during infusion of glucose and a fat emulsion at rest. Clin. Physiol. (in press)

LASSERS, B.W., L.A. Carlson, L. Kaijser & M. Wahlqvist. The nature and control of myocardial substrate metabolism in healthy man. In M.F. Oliver, D.G.

Julian & K.W. Donald (Eds.), Effect of Acute Ischemia on Myocardial Function. Edinburgh & London: Churchill Livingstone, 1972.

LITHELL, H., J. Örlander, R. Schéle, B. Sjödin & J. Karlsson. Changes in lipoprotein lipase activity and lipid stores in human skeletal muscle with prolonged heavy exercise. Acta Physiol. Scand. 107:257–261, 1979.

WAHLQVIST, M.L., L. Kaijser, B.W. Lassers, H. Löw & L.A. Carlson. The role of fatty acid and of hormones in the regulation of carbohydrate metabolism by the human heart. Europ. J. Clin. Invest. 2:311, 1972.

Lipoprotein Lipase Activity and Intramuscular Triglyceride Stores in Conditioned Men: Effects of High Fat and Low Fat Diets

B. Kiens, B. Essén, P. Gad and H. Lithell

University of Copenhagen, Copenhagen, Denmark, and
University of Uppsala, Uppsala, Sweden

The metabolic responses in human skeletal muscle have mostly been evaluated after induced exercise and/or physical training. An alternative way of changing the metabolic response is to change the substrate availability by dietary restrictions. Most dietary studies, however, have dealt with extreme diets of short durations. Previously we demonstrated (Kiens et al., 1981) that long-term changes of the fat/carbohydrate composition of the diet did not influence the low serum triacylglycerol (TG) level at rest in trained men. Those findings indicated an excess capacity of hydrolysis of circulating TG by an increased lipoprotein lipase (LPL) activity and/or a variation of the capacity as a response to the change of diet. The purpose of the present study was to evaluate the long-term effects of a fat-rich and a carbohydrate-rich (CHO) diet on the metabolic response at rest in men, who trained regularly.

Material and Methods

Nineteen healthy subjects, with regular training habits for several yrs, age 30 to 44 yr, weight 63 to 90 kg, and $\dot{V}_{O_2}$max 3.2 to 5.0 $1 \times min^{-1}$, volunteered for the study. Initially, the composition and caloric intake of the ordinary diet was established both through a dietary recall interview, conducted by dietitians, and self-registration for a week. After an overnight fast of 12 h and no training the last 48 h, venous blood samples were drawn after rest for 10 to 15 min in the supine position and muscle biopsies were taken from the vastus lateralis muscle. A questionnaire was used to assess the habitual physical activity levels. The subjects were divided into an *experimental group* (n = 10) and a *control group* (n = 9).

During the next 4 wk the subjects in the experimental group consumed a fat-rich diet. This period was followed by a CHO-diet for another 4 wk. Calculations of the dietary intake revealed that the energy derived from fat increased from 43% in the ordinary diet to 54% in the fat diet (p < 0.05) and decreased to 29% during the CHO diet (p < 0.05). The energy-% of carbohydrates decreased from 42% in the ordinary diet to 29% during the fat-rich period (p < 0.05) and increased to 51% (p < 0.05) during the CHO diet. For all the diets the P/S ratio and the protein content were constant. A more detailed description has been given (Kiens et al., 1981). No changes were seen in training habits and the body weight remained constant.

Analysis

Muscle tissue obtained for biochemical evaluation was immediately frozen at −80° C until analyzed. Glycogen and the activities of citrate synthase (CS) and β-OH-acyl-CoA-dehydrogenase (HAD) were analyzed on freeze-dried material using a fluorometric technique (Éssen et al., 1980). The TG content was analyzed according to Chernick (1969) and Éssen (1978). The LPL activity was measured as described by Lithell and Broberg (1978). Mounted muscle tissue was cut in serial cross sections for histochemical stainings to identify fiber types (for the method see Andersen & Henrikson, 1977) and the amylase PAS method was used in order to visualize capillaries (Andersen, 1975).

The TG concentration in the serum was determined in isopropanol extracts fluorometrically in a Technicon Auto Analyzer II in accordance with the Lipid Research Clinics (LRC) manual (1974). Serum insulin was measured using the Phadebas insulin method (Pharmacia, Uppsala, Sweden).

Values are given as mean ± SE. Differences between means were tested for significance using the appropriate Student's t-test. Significance was set at the 0.05 level of confidence.

Results

The initial mean value for LPL activity in the vastus lateralis muscle was 67 ± 8 mmol FA/g w wt/min for the whole group studied (n = 17). The LPLA was related to the capillarization of the muscle (r = 0.87, n = 5, p < 0.05). The mean number of capillaries per mm^2 was 309 (range 200 to

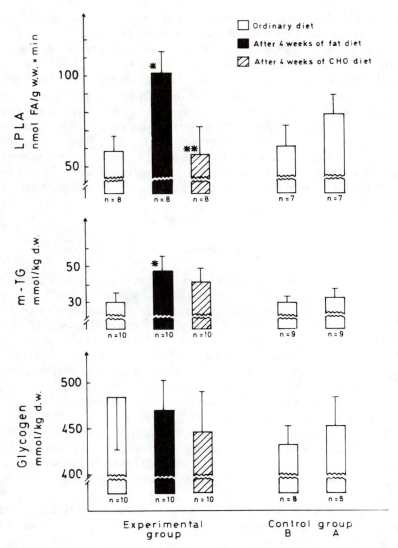

Figure 1—Muscle concentrations of triacylglycerols (TG), glycogen and the activity of lipoprotein lipase (LPLA) at rest after the fat and the carbohydrate diet and in the control group before (B) and after (A) 8 wk.

444). The proportion of slow-twitch (ST) fibers averaged 61% (range 48 to 85) and no relationship was found between percentage of ST-fibers and LPLA. After the fat-rich diet the LPLA was increased in the experimental group from 59 ± 8 to 106 ± 12 mmol FA/g w wt/min (p < 0.05) and decreased again (p < 0.05) following the CHO diet to the level of the

Table 1

The Enzyme Activity at Rest of β-OH-acyl-CoA-dehydrogenase (HAD) and Citrate Synthase (CS) after Different Dietary Periods and in the Control Group before and after 8 wk

nmol/kg d.w.	Experimental Group			Control Group	
	ordinary diet	fat diet 4 wk	CHO diet 4 wk	ordinary diet 0 wk	8 wk
HAD	30 ± 2	30 ± 3	30 ± 3	30 ± 1	29 ± 2
CS	31 ± 2	31 ± 2	30 ± 2	27 ± 2	28 ± 2

ordinary diet (Figure 1). None of the mitochondrial enzymes studied changed significantly with the different dietary regimens (Table 1). The muscle-TG level amounted initially to 30 ± 3 mmol/d.kg for the whole group (n = 19). An increase in muscle TG was observed after the fat-rich diet from 30 ± 4 to 47 ± 8 mmol/d.kg (p < 0.05) (Figure 1). After the CHO diet the muscle TG level was 41 ± 7 mmol/d.kg. The glycogen concentration in the skeletal muscle averaged 485 ± 68 mmol/d.kg in the experimental group before the dietary treatments. Small and insignificant decreases were observed both after the fat and CHO dietary periods (Figure 1). The serum TG and insulin remained unchanged during either of the diets (Table 2). In the *control group* no changes were observed in any of the variables studied before and after the 8 wk period.

Discussion

Studies have demonstrated that extreme variation in diet for short periods affects the muscle metabolic response. Such diet induced changes were not seen in the present study after longer lasting dietary manipulations. Following the CHO diet, the activity of m-LPL was similar to the initial values. This is in contrast to findings of a 3-day CHO diet, where m-LPLA was depressed by 50%. This was associated with increased serum TG and insulin concentrations discussed as being possible consequences of, and causes of, the depressed LPL activity, respectively (Jacobs et al., 1981). Such marked elevations are diminishing with time and in the present study neither serum TG nor insulin differed significantly from the pre-trial values. The fat diet, however, was followed by an increase in m-LPLA of

Table 2

Serum Triacylglycerol (TG) and Insulin Concentrations at Rest after Different Dietary Periods

	ordinary diet	fat diet 4 wk	CHO diet 4 wk
S-TG (nM)	1.02 ± 0.12	1.07 ± 0.13	1.29 ± 0.21
S-insulin (nU/l)	3.5 ± 0.7	3.5 ± 0.8	3.9 ± 0.4

78% but the serum insulin level was not decreased. This means that mechanisms other than insulin variations, which have been discussed as a possible regulator of LPLA in muscle, must be of importance in the regulation of m-LPLA. The content of TG in the muscle tissue after the fat diet was increased. Also high intramuscular TG-values were obtained after a short-term fat-rich diet (Jansson, 1980). In both cases insignificant decreases were found following the CHO-diet. Also, 5 days on a fat diet resulted in an increase in the activity of oxidative enzymes (HAD and SDH) of 15% compared to a CHO-diet (Jansson, 1981). However, the enzymatic response in HAD and CS to the long-term dietary manipulations, in our study, did not change during either of the diets. Thus, the present study reveals that various dietary regimens affected the metabolic response differently in the human muscle. It seems as if one month's adaptation to a fat-rich diet is associated with a higher capacity for uptake of circulating serum TG and with larger stores of TG in the muscle. This may indicate a larger turnover of fat in this tissue under these conditions.

Acknowledgments

The financial support from the Research Councils of the Danish and Swedish Sports Federations, the Swedish Diabetes Association and the Swedish Medical Research Council, (No. 5640) is gratefully acknowledged.

References

ANDERSEN, P., Capillary density in skeletal muscle of man. Acta Physiol. Scand. 95:203–205, 1975.

ANDERSEN, P. & J. Henriksson. Capillary supply of the quadriceps femoris muscle of man: Adaptive response to exercise. J. Physiol. 270:677–691, 1977.

CHERNICK, S.S., Determination of glycerol and acylglycerol. Methods in Enzymology 14:627–630, 1969.

ÉSSEN, B., Studies on the regulation of metabolism in human skeletal muscle using intermittent exercise as an experimental model. Acta Physiol. Scand. 102(Suppl):454, 1978.

ÉSSEN, B., A. Lindholm & J. Thornton. Histochemical properties of muscle fibre types and enzyme activities in skeletal muscle of Standard bred trotters of different ages. Equine Vet. J. 12:175–180, 1980.

LITHELL, H. & J. Broberg. Determination of lipoprotein-lipase activity in human skeletal muscle tissue. Biochem. Biophys. Acta. 528:58–68, 1978.

JACOBS, I., Lactate, muscle glycogen and exercise performance in man. Acta Physiol. Scand. (Suppl):495, 1981.

JANSSON, E., Diet and muscle metabolism in man with reference to fat and carbohydrate utilization and its regulation. Acta Physiol. Scand. (Suppl):487, 1980.

JANSSON, E., Muscle enzyme adaptation to diet in man. In J. Poortmans & G. Niset (Eds.), Biochemistry of Exercise IV-B. Baltimore: University Park Press, 1981.

KIENS, B., P. Gad, H. Lithell & B. Vessby. Minor dietary effects on HDL in physically active men. Eur. J. Clin. Invest. 11:265–271, 1981.

MANUAL of Laboratory Operations, Lipid Research Clinics Program. Vol 1. Lipid and lipoprotein analysis. Bethesda, Maryland. DHEW publication NO (NIH) 75-628, 1974.

Influence of Caffeine on Serum Substrate Changes during Running in Trained and Untrained Individuals

J.J. Knapik, B.H. Jones, M.M. Toner,
W.L. Daniels and W.J. Evans
US Army Research Institute of Environmental Medicine,
Natick, Massachusetts, U.S.A.

Previous studies have indicated that caffeine (CAF) increases the mobilization and metabolism of free fatty acids (FFA) during prolonged exercise while sparing muscle glycogen (Costill et al., 1978; Essig et al., 1980; Ivy et al., 1979). It has also been shown that differences exist between trained and untrained individuals in terms of substrate utilization and enzyme activity (Evans et al., 1979; Holloszy et al., 1975). Therefore, examining trained and untrained individuals may be a way to further elucidate the mechanism of action of CAF on lipid and carbohydrate metabolism during exercise and this was the purpose of the present study.

Methods

Ten male subjects (five trained runners [T] and five untrained individuals [UT]) volunteered to participate in this investigation after being informed of the nature and risks of the study. The T ran an average of about 32 miles a wk while the UT did not exercise habitually nor had they performed any regular exercise of a long continuous nature in over 3 mo. The average height and weight ($\pm$ SD) of the T was 177.3 ± 4.5 cm and 70.1 ± 7.8 kg and that of the UT was 175.5 ± 4.6 cm and 81.5 ± 9.8 kg. $\dot{V}o_2$max of the T was 62.4 ± 3.0 ml/kg/min and that of the UT was 46.5 ± 5.1 ml/kg/min.

Preliminary testing consisted of a maximal oxygen uptake determination ($\dot{V}o_2$max) and a submaximal discontinuous test to determine the onset of blood lactate accumulation (OBLA). $\dot{V}o_2$max was determined using an interrupted uphill running treadmill protocol. The criterion for $\dot{V}o_2$max was a change of less than 1.5 ml/kg/min in $\dot{V}o_2$ despite a 2.5% increase in grade. One wk later the OBLA was determined by having subjects run on a treadmill at 0% grade for 8 to 12, 4 min intervals. Near the end of each interval, $\dot{V}_{O_2}$ values were collected and the subject walked for 2 min while blood samples were collected and analyzed for lactates. Treadmill speeds were increased 0.8 km/h on each successive interval and $\dot{V}_{O_2}$ and lactate values were collected as above. These studies indicated that no subject began accumulating a significant amount of lactate (2 mmoles/l) until greater than 60% $\dot{V}_{O_2}$max. Therefore, a speed corresponding to 60% of each subject's $\dot{V}_{O_2}$max was extrapolated to within 0.4 km/h from the $\dot{V}_{O_2}$ data and this speed was used in the subsequent 1 h runs.

The experimental sessions consisted of three 1 h runs separated by about 2 wks each. Shortly after reporting to the laboratory, subjects ingested a premixed beverage containing either 0, 5 or 9 mg/kg body weight of anhydrous caffeine dissolved in a lemon-lime flavored drink sweetened with saccharin. Administration of the beverage was conducted in a double-blind fashion. One h after ingesting the drink, subjects began running for 1 h at 60% $\dot{V}_{O_2}$max. During the run an 11 ml blood sample was obtained from an indwelling catheter at 0 min, 5 min and 10 min, and every subsequent 10 min. Respiratory exchange ratios (R values) were obtained at each time period just prior to blood collection.

Data were analyzed using a three way fixed model analyses of variance with repeated measures. Post hoc analysis was performed using the Tukey test. The level of statistical significance was set at $p < 0.05$.

Results

Figure 1 shows the R values during the 1 h run. The R values for the T were lower than those of the UT and they declined significantly over time; however, there were no differences among the three sessions within the two groups.

The serum and blood substrate values during the 1 h runs are shown in Figure 2. Serum glycerol values increased progressively over time. The T had higher glycerol levels than the UT except during the 9 mg/kg session. Although within each group no significant dosage effect was seen, a tendency toward higher levels was noted in the CAF sessions ($F[2,16] = 2.86$, $p < .09$). Serum FFA declined at 5 min of exercise then rose significantly in both groups in all sessions. No significant differences were found among sessions.

Serum glucose rose significantly over time for the T but was elevated to a greater extent in the CAF sessions. In the UT a significant decline occurred over time during the placebo session. After 20 min the serum glucose of the UT during the CAF sessions was significantly elevated over that of the

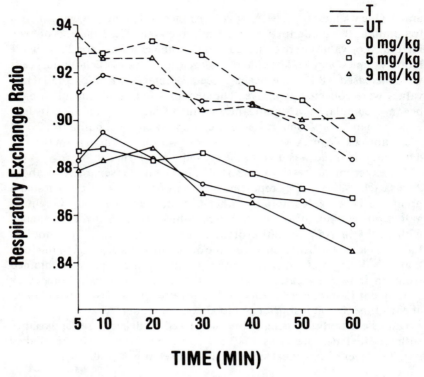

Figure 1—Respiratory exchange ratios during the 1 h run.

placebo session and this difference was maintained throughout the remainder of the exercise. Blood lactates declined significantly over time at all dosages for the T and during the placebo session for the UT. However, lactates rose in the UT during both CAF sessions.

Discussion

Previous studies have noted that CAF stimulates mobilization and oxidation of FFA. In the present study there was a trend towards increased mobilization as suggested by the glycerol data. However, the R values did not change, indicating that the effect of CAF on fat oxidation was not substantial. Previous studies have used a cycle ergometer (Costill et al., 1978; Essig et al., 1980) or a Fitron (Ivy et al., 1979) which required a smaller muscle mass than the treadmill utilized in the present study. Any increase in circulating FFA would be distributed to this additional muscle

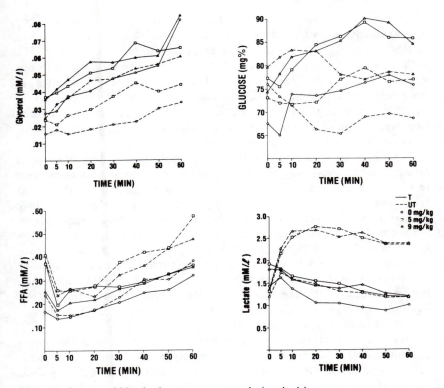

Figure 2—Serum and blood substrate parameters during the 1 h run.

mass and thus a significant increase in fat oxidation may not be seen. A variable that confounds this comparison of involved muscle mass is the exercise intensity used in two of the previous studies. Costill et al. (1978) had subjects exercise at 69% $\dot{V}_{O_2}$max while Essig et al. (1980) used 80% $\dot{V}_{O_2}$max. These were both higher than the 60% $\dot{V}_{O_2}$max of our study. However, the higher exercise intensities themselves may be an alternate explanation for the differences in fat mobilization between previous studies and the present one. Higher exercise intensities are known to cause a greater epinephrine release (Terjung, 1980). While CAF stimulates epinephrine release it also potentiates the effects of existent epinephrine on FFA mobilization (Vaughan & Steinberg, 1963). Thus, the lower exercise intensity of the present study may have resulted in lower amounts of total circulating epinephrine on which CAF could exert a potentiation effect.

The dramatic increase in blood lactates seen in the UT with CAF and the higher serum glucose levels may suggest a glycogenolytic effect of CAF. CAF is known to stimulate glycogenolysis in both liver and muscle

(Sutherland et al., 1968). Artificially elevated blood glucose levels induced by oral glucose ingestion increases the active muscle uptake of glucose and increases glucose oxidation during exercise (Ahlborg & Felig, 1977). Although no evidence for increased glucose oxidation was found in our study, the possibility that elevated blood glucose levels resulted in increased uptake by the muscle cannot be discounted. Regardless of whether or not increased glucose uptake occurred, the lactate accumulation seen in the UT indicates increased glycolytic flux during the CAF sessions compared to the placebo session.

The failure to show a CAF effect on lactates in the T despite elevated serum glucose levels may reflect enzymatic adaptations that occur as a result of training. Training results in less lactate accumulation at the same relative exercise intensity when rates of glycogenolysis are similar (Saltin & Karlsson, 1971). Levels of aspartate transaminase and malate dehydrogenase are higher in T (Holloszy et al., 1975). These enzymes are used in the malate-aspartate shuttle to transfer reducing compounds across the mitochondrial membrane so less need be produced in the lactate dehydrogenase (LDH) reaction. There is a reduction in total LDH in muscle with training but an increase in the heart isozyme of the enzyme (Sjodin et al., 1976). Training also results in an increase in the ability to transaminate pyruvates to alanine (Holloszy et al., 1975; Mole et al., 1973).

To summarize, there was no evidence for increased fat oxidation in the present study. However, there was some suggestion of increased glycogenolysis. The elevated serum glucose levels during the CAF sessions indicated increased liver glycogenolysis. The higher blood lactate and serum glucose levels in the UT during the CAF sessions suggested an increased glycolytic flux in the active muscles. While the elevated serum glucose levels may also indicate increased glycolytic flux in the T, the failure to find lactate accumulation in this group may reflect enzymatic adaptations induced by training.

Acknowledgments

The technical assistance of Doris Jackson, Linda Suek and Nancy Seaver is gratefully acknowledged. Thanks also to Julie Cyphers for word processing of the manuscript.

References

AHLBORG, G. & P. Felig. Substrate utilization during prolonged exercise preceded by ingestion of glucose. Am. J. Physiol. 233:E188–E194, 1977.

COSTILL, D.L., G.P. Dalsky & W.J. Fink. Effects of caffeine ingestion on metabolism and exercise performance. Med. Sci. Sports 10:155–158, 1978.

ESSIG, D., D.L. Costill & P.J. VanHandel. Effects of caffeine ingestion on utilization of muscle glycogen and lipid during leg ergometer cycling. Int. J. Sports Med. 1:86–90, 1980.

EVANS, W.J., A.S. Bennett, D.L. Costill & W.J. Fink. Leg muscle metabolism in trained and untrained men. Res. Quart. 50:350–359, 1979.

HOLLOSZY, J.O., F.W. Booth, W.W. Winder & R.H. Fitts. Biochemical adaptation of skeletal muscle to prolonged physical exercise. In H. Howald & J.R. Poortmans (Eds.), Metabolic Adaptation to Prolonged Physical Exercise. Switzerland: Birkhauser, 1975.

IVY, J.L., D.L. Costill, W.J. Fink & R.W. Lower. Influence of caffeine and carbohydrate feedings on endurance performance. Med. Sci. Sports 11:6–11, 1979.

MOLE, P.A., K.M. Baldwin, R.L. Terjung & J.O. Holloszy. Enzymatic pathways of pyruvate metabolism in skeletal muscle: adaptations to exercise. Am. J. Physiol. 224:50–54, 1973.

SALTIN, B. & J. Karlsson. Muscle ATP, CP and lactate during exercise after physical conditioning. In B. Pernow & B. Saltin (Eds.), Muscle Metabolism During Exercise. New York: Plenum, 1971.

SJODIN, B., A. Thorstensson, K. Frith & J. Karlsson. Effect of physical training on LDH activity and LDH isozyme pattern in human skeletal muscle. Acta Physiol. Scand. 97:150–157, 1976.

TERJUNG, R., Endocrine response to exercise. In R.S. Hutton & D.I. Miller, Exercise and Sports Science Reviews. Vol 7, Franklin Institute Press, 1980.

VAUGHAN, M. & D. Steinberg. Effect of hormones on lipolysis and esterification of free fatty acids during incubation of adipose tissues *in vitro*. J. Lipid Res. 4:193–199, 1963.

Glycogen Overloading in Rats: Effects of Various Sugar Intakes

M. Ledoux, L. Voghel, L. Brassard,
G. Brisson and F. Peronnet
Université de Montréal, Montréal, Quebec, Canada, and
Universite du Quebec a Trois-Rivières, Trois-Rivières,
Quebec, Canada

The value of glycogen overloading for endurance type exercise is well known. However, few studies have examined the influence of the type of dietary carbohydrate consumed on glycogen resynthesis while using the original technique described by Bergström and Hultman (1967) and modified by Saltin and Hermansen (1967). Bergström and Hultman suggested that glucose and fructose infusion resulted in different resynthesis only when glycogen depletion was performed. Costill (1979), studying glycogen resynthesis for 2 days after muscle depletion, observed similar effects of simple (glucose and fructose) and complex carbohydrate intakes.

In rats, muscle depletion has been obtained using swimming or running exercises. Increased resynthesis after depletion was observed after fasting (Fell et al., 1980) or after a high carbohydrate diet (Conlee et al., 1978). Types of carbohydrates used were not identified and the different efficiencies of specific sugar intakes on glycogen overloading remains unclear. Therefore, the purpose of this study was to compare the efficiencies of d-glucose, d-fructose, d-galactose and starch on glycogen synthesis when given as the unique source of carbohydrates in the 3 last days of glycogen overloading.

Methods and Procedures

Seventy (70) male Sprague-Dawley rats were divided into seven groups (10/group) and fed ad libitum the various test diets. The experiment was divided into three phases. Phase 1 (or adaptation) where all the rats were fed a mixed rat chow (Charles River). Phase II where groups I to V were fed for 3 days a high-protein, high-fat, no carbohydrate diet coupled with daily training (30 min—1,6 km/h) was completed by one session of exercise prolonged to exhaustion on a treadmill (2 km—2.4 km/h). Group VI and VII were not exercised and were fed the Charles River rat chow. Rats from groups V and VII were sacrificed at the end of this phase. In Phase III the rats were fed for 3 days with one of the test sugars: d-glucose (group I), starch (group II), d-galactose (group III), d-fructose (group IV) or with the Charles River rat chow (group VI). Rats of all groups were not exercised during this phase and were sacrificed at the end of the 3 day period. On the last day of the experiment, the rats were anesthetized by intraperitoneal injection of sodium pentobarbital. After removal of connective tissue, 35 to 70 mg of liver, diaphragm, plantaris and rectus femoris muscles were collected and frozen in liquid nitrogen before analysis according to the method of Lo et al. (1970), which was modified in our laboratory.

Mean data were analyzed for differences between trials by use of variance and covariance analyses followed by an *a posteriori* Scheffé test and a paired-T test was used for *plantaris* and *rectus femoris* data analyses. Statistical significance was determined to be at the 0.05 level of confidence.

Results

A positive correlation in glycogen content of the plantaris and the rectus femoris muscles is found for groups I, II, III, IV and VI (Table 1). Results of a paired-T test show a significant difference between both muscles, the increase in glycogen content being statistically higher (p < 0.05) in the *plantaris* muscle. It seems therefore that the *plantaris* muscle responds better to glycogen overloading than the *rectus femoris* even if both muscles have a better overall response than the control muscle, the diaphragm.

A statistically significant difference was shown between the effect of glucose and galactose on glycogen loading in the plantaris muscle, glucose being more efficient than galactose for that purpose. No significant differences were found between glucose and fructose feeding even if glucose ingestion resulted in a higher glycogen resynthesis in both muscles. The effect of exercise alone was shown to be significant when comparing results from group VI (exercise-mixed diet) and VII (no exercise-mixed diet). Liver glycogen was statistically different between groups I (glucose) and III (galactose). No significant difference was observed between the glucose and fructose groups.

Table 1

Mean Glycogen Content of Muscles and Liver for Each Group of Rats (g/100 g Wet Tissue)

Groups Tissue	I Glucose	II Starch	III Galactose	IV Fructose	V Protein-fat	VI Mixed	VII Control
Plantaris	1.04	0.90	0.82	0.88	0.89	0.86	0.78
Rectus femoris	0.84	0.72	0.66	0.73	0.86	0.82	0.68
Diaphragm	0.75	0.60	0.46	0.58	0.64	0.76	0.62
Liver	6.72	4.16	1.99	3.28	2.48	6.00	5.42

Discussion

As previously reported by Fell et al. (1980), gluconeogenesis may have occurred in the protein-fat group since a delay of 6 h was observed between the exercise test and sample collection. This seems sufficient for gluconeogenesis to occur and to hide more significant differences between glycogen content of the various samples. This may explain why no significant difference was observed between this group and the carbohydrates groups. However, all sugars tested are more efficient than a mixed diet for glycogen overloading. This efficiency is improved by a concurrent exercise during the first phase of the diet. Overloading of muscle and liver glycogen stores is minimal in rats. Pawan (1973) previously demonstrated glucose superiority over fructose for muscle glycogen resynthesis after exhaustion. This was confirmed by Bergström and Hultman (1967) who used an infusion rather than an ingestion technique. However, in our study, glucose ingestion results in the highest increase in glycogen content in both muscles, liver and diaphragm as compared to the other sugars. This was not related to any difference in total caloric intake in our study.

References

BERGSTRÖM, J., L. Hermansen, E. Hultman & B. Saltin. Diet, muscle glycogen and physical performance. Acta Physiol. Scand. 71:140–150, 1967.

BERGSTRÖM, J. & E. Hultman. Muscle glycogen synthesis after exercise: an enhancing-factor localized to the muscle cells in man. Nature (Lond.) 210:309, 1966.

CONLEE, R.K., R.C. Hickson, W.W. Winder, J.M. Hagberg & J.O. Holloszy. Regulation of glycogen resynthesis in muscles of rats following exercise. Am. J. Physiol. 235(3):R145–R150, 1978.

COSTILL, D.L., A scientific approach to distance running. Track and Field News, 1979.

FELL, R.D., J.A. McLane, W.W. Winder & J.O. Holloszy. Preferential resynthesis of muscle glycogen in fasting rats after exhausting exercise. Am. J. Physiol. R238:R238–R332, 1980.

LO, S., J.C. Russell & A.W. Taylor. Determination of glycogen in small tissue samples. J. Appl. Physiol. 28:234–236, 1970.

PAWAN, G.L.S., Fructose. In Molecular Structure and Function of Food Carbohydrate. Toronto: John Wiley and Sons, 1973.

SALTIN, B. & L. Hermansen. Glycogen stores and prolonged severe exercise. In Nutrition and Physical Activity. Almqvist and Uppsala, 1967.

Effects of Exercise Training and Sucrose Intake on Cellular Proliferation in Rat Parametrial Adipose Tissue

J. Lupien, G. Côté, E. Cardinal,
A.L. Vallerand and L. Bukowiecki
Laval University Medical School, Québec, Canada

We have recently reported that exercise training inhibited cellular proliferation and decreased adipocyte size in gonadal adipose tissue of young (2-month-old) female rats (Bukowiecki et al., 1980). Such an effect was not observed in epididymal adipose tissue of male rats where hyperplasia stops at an earlier age than in females. In male rats, exercise training reduced adipocyte size without affecting adipocyte number. It therefore appears that exercise training may inhibit adipocyte proliferation in growing adipose tissue, but is ineffective in reducing adipocyte number once tissue hyperplasia ceased.

On the other hand, it has also been demonstrated that, in contrast to males, female rats lose little or no weight after intensive exercise training because they compensate the calories lost during exercise by increasing their food intake. Thus, exercise training increases appetite in females but not in males. Considering that sucrose consumption has been shown to increase the total calorie intake in sedentary rats (Castonguay et al., 1981), we analyzed the interactions between sucrose consumption and exercise training on total calorie consumption, body weight gain efficiency, parametrial white adipose tissue (PWAT) cellularity and adipocyte size in young growing female rats.

Procedures

Female Wistar rats weighing approximately 140 to 150 g were obtained from Charles River. They were housed in individual cages and were maintained at 25° C with 12 h intervals of light and dark. The animals were divided into four groups of 10 animals each: sedentary controls receiving Purina chow and water ad libitum (C-P), sedentary controls receiving a 32% sucrose solution in addition to the Purina diet (C-S), and two corresponding groups that were submitted to a training program consisting of 2 h of daily swimming at 36° C during 10 wk (E-P and E-S). Adipocytes were isolated from parametrial white adipose tissue exactly as previously described (Bukowiecki et al., 1980). Triplicate samples of the final cellular suspension were counted in Neubauer's hemacytometer. The cellularity of PWAT, expressed in millions of adipocytes present in both left and right PWAT depots was obtained by dividing the total PWAT triglyceride content by the triglyceride content of one million isolated adipocytes (Bukowiecki et al., 1980).

Results

Figure 1 shows that sucrose consumption and exercise training synergistically increased calorie consumption in female rats, the total energy intake of exercising animals offered sucrose being approximately double that of sedentary controls. The animals receiving 32% sucrose ad libitum partially compensated the extra calories taken in sucrose by consuming less Purina chow, but nevertheless remained hyperphagic as far as total calorie intake is concerned. Thus, exercise training increases appetite in female rats, particularly in the presence of palatable food.

Remarkably, there was no significant difference in body weight gain between all four groups of animals during the entire experimental period (10 wk). The increased energy consumption accompanied by a normal growth in exercising and/or sucrose consuming animals resulted in significant decreases in body weight gain efficiency (the quantity of energy an animal has to consume to increase its body weight by 1 g) (Table 1). This phenomenon was particularly striking in E-S rats where body weight gain efficiency decreased by as much as 44% compared to controls. This value represents more than the sum of the effects of exercise training (−13%) and sucrose consumption in sedentary rats (−20%).

Exercise training significantly decreased PWAT weight, cellularity and adipose size in normal as well as in hyperphagic animals, whereas sucrose consumption markedly increased the same parameters (Table 1). Although exercise training inhibited the stimulatory effects of sucrose consumption on PWAT weight and cellularity, it did not significantly reduce adipocyte size.

Table 1

Effects of Exercise Training and Sucrose Consumption on Body WT Gain and Adipose Tissue Composition in Female Rats

	Controls/ Purina Chow (C-P)	Exercise-trained/ Purina Chow (E-P)	Controls/ sucrose (C-S)	Exercise-trained/ sucrose (E-S)
Final body weight (g)	232.94 ± 3.77	238.26 ± 3.54 (102%)	240.33 ± 4.23 (103%)	240.8 ± 2.46 (103%)
Body weight gain (g gained in 10 wk)	91.16 ± 3.06	92.39 ± 3.66 (101%)	99.94 ± 4.27 (110%)	95.42 ± 2.68 (105%)
Total energy intake (MJ/10 wk)	15.23 ± 0.21	17.85 ± 0.29** (117%)	20.81 ± 0.55## (136%)	28.45 ± 0.53**## (187%)
Body weight gain efficiency (g gained/MJ eaten)	5.99 ± 0.21	5.20 ± 0.23** (87%)	4.82 ± 0.19## (80%)	3.36 ± 0.08**## (56%)

Parametrial fat weight (g)	3.86 ± 0.25	$2.17 \pm 0.16^{**}$ (56%)	$8.66 \pm 0.37^{\#\#}$ (224%)	$6.02 \pm 0.38^{***\#\#}$ (156%)
Total triglyceride content (g)	1.90 ± 0.12	$0.99 \pm 0.06^{**}$ (52%)	$4.77 \pm 0.23^{\#\#}$ (251%)	$3.42 \pm 0.34^{**\#\#}$ (180%)
Total number of adipocytes ($\times 10^{6}$)	17.14 ± 1.20	$13.28 \pm 1.23^{*}$ (77%)	$21.10 \pm 1.23^{\#}$ (123%)	$15.77 \pm 0.59^{*}$ (92%)
Triglyceride content per adipocyte (ng)	121 ± 8	$70 \pm 5^{**}$ (58%)	$232 \pm 16^{\#\#}$ (192%)	$219 \pm 21^{**\#\#}$ (181%)

Note: The values represent the means ± SEM of 8 to 10 individual experiments.

%: Denotes % of controls/Purina Chow values.

* or **: Significant effect of exercise training (*: $p < 0.05$; **: $p < 0.01$)

or ##: Significant effect of sucrose consumption (#: $p < 0.05$; ##: $p < 0.01$)

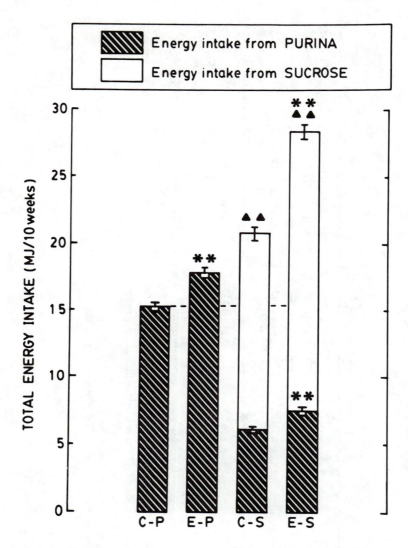

Figure 1—Effect of sucrose consumption and exercise training on the total energy intake from Purina chow and 32% sucrose during the entire 10 wk experimental period. The bars represent the means ± SEM of data collected on 10 animals for each experimental group. Data were analyzed by variance analysis. The double asterisks and triangles indicate highly significant effects (P < 0.01) of exercise training or sucrose consumption, respectively.

Discussion

The present results demonstrate that exercise training increases calorie consumption in female rats, especially when palatable sucrose solutions are offered to the animals. Indeed, total energy intake was increased by 17% in the E-P group and by 36% in C-S animals whereas it nearly doubled in exercising animals offered sucrose and Purina chow (E-S group) (Figure 1). Paradoxically, there was little or no difference in body weight gain among all four groups during the entire experimental period (Table 1). The decreased body weight gain efficiency of exercising animals on Purina chow resulted most probably from the increased calorie expenditure during exercise training. The explanation for the decreased body weight efficiency in animals consuming sucrose is not as evident as for exercising animals. The most likely mechanism is an enhanced calorie expenditure by diet-induced thermogenesis induced by excessive sucrose intake. Indeed, sucrose is very effective in stimulating the activity of the sympathetic nervous system (Landsberg & Young, 1978). Moreover, we recently showed that hyperphagia induced by palatable nutrients increases the thermic response of rats to catecholamines, improves their resistance to cold and stimulates adipocyte proliferation in brown adipose tissue, thereby increasing the capacity of the animals for calorie expenditure by diet-induced thermogenesis (Bukowiecki et al., 1982).

Exercise training reduced the normal adipocyte proliferation in sedentary rats, confirming our previous results (Bukowiecki et al., 1980). It also inhibited the stimulatory effects of hyperphagia induced by sucrose consumption (Table 1). These data demonstrate that adipocyte number in PWAT can be modulated by at least two factors with antagonistic effects, exercise training inhibiting adipocyte proliferation and hyperphagia stimulating the same process.

References

BUKOWIECKI, L., A.J. Collet, N. Folléa, G. Guay & L. Jahjah. Brown adipose tissue hyperplasia: a fundamental mechanism of adaptation to cold and hyperphagia. Am. J. Physiol. 242 (Endocrinol. Metab. 5) (in press), 1982.

BUKOWIECKI, L., J. Lupien, N. Folléa, A. Paradis, D. Richard & J. LeBlanc. Mechanism of enhanced lipolysis in adipose tissue of exercise-trained rats. Am. J. Physiol. (Endocrinol. Metab. 2):E422-429, 1980.

CASTONGUAY, T.W., E. Hirsch & G. Collier. Palatability of sugar solutions and dietary selection. Physiol. Behav. 27(1):7–12, 1981.

LANDSBERG, L. & J.B. Young. Fasting, feeding and regulation of the sympathetic nervous system. N. Engl. J. Med. 298:1295–1301, 1978.

Effects of Physical Training and High Energy Diet on Glucose Homeostasis in Male Rats

D. Richard, A. Labrie, A. Tremblay and J. Leblanc

Laval University, Québec, Canada

It has been demonstrated during recent years that exercise-training could reduce insulin requirements in both man and rat (LeBlanc et al., 1979; Richard & LeBlanc, 1980). Similarly a close relationship between obesity and hyperinsulinemia has been well-established (Bray & York, 1971). The present study was aimed at investigating the role of exercise-training on glucose tolerance and insulin secretion in rats fed an obesity-inducing diet (Scalfani & Springer, 1976).

Methods

Male rats (200 g) were divided into exercise-trained or sedentary groups which received either a palatable high energy diet (Richard et al., in press) or merely standard stock diet. The mean energy content of the high energy diet was 18.8 kJ/g and its content in fat, carbohydrate and protein was respectively 30, 35 and 11%. The corresponding values for the stock diet were 14.4 kJ/g for energy content and 4.5, 49 and 23% for its respective constituents. Periodically during the study, the quantity of each food item consumed by the rats was measured and, according to food tables, the amount of protein, fat, carbohydrate and energy intake were determined. After 10 wk of training which consisted of 2 h per day of swimming sessions, the animals were submitted to an intravenous glucose tolerance test (Richard & LeBlanc, 1980). Plasma insulin was determined according to Hales and Randle (1963). Total areas under the glucose and insulin response curves were calculated by trapezoidal rules. The insulinogenic index was expressed as the total area of insulin to total area of glucose. The fractional clearance rate of glucose (K-value) was calculated according to

Wahlberg (1966). Adipocyte size and number were measured as previously described (Richard et al., in press). Finally, an analysis of variance combined with a Duncan multiple-range test was used as statistics.

Results

The results reported in Figure 1 show that high energy feeding caused a greater body weight gain as well as greater fat cell enlargement in both sedentary and active animals. On the other hand, exercise-training significantly reduced body weight and adipose cell size of rats, whatever the diet used. No differences in the number of fat cells were found among the various groups studied. Table 1 illustrates that total energy intake was increased in rats fed on supermarket foods. This table also shows that the ingestion of fat was much greater in animals fed on the high energy diet.

Both total area of glucose (Figure 2) and K-value (Figure 4) show that glucose tolerance was impaired in sedentary but not in trained rats fed the high energy diet. Similarly, basal values for plasma insulin, insulin

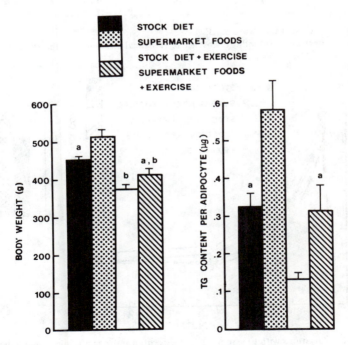

Figure 1—Body wt and size of adipocytes at the end of experiment. Common superscript indicates absence of significant differences at the .05 level.

Table 1

Proteins, Fat, Carbohydrate and Energy Ingested Each Day by the Different Groups*,†

	Protein (g)	Fat (g)	Carbohydrate (g)	Energy intake (kJ)
Stock diet	5.40[a]	0.516[a]	11.69[a]	303.67[a]
	0.201	0.022	0.436	11.339
Supermarket foods	5.49[a]	7.43	10.09[a]	540.95
	0.212	0.258	0.567	16.121
Stock diet and exercise	5.15[ab]	0.44[a]	11.15[a]	289.78[a]
	0.182	0.015	0.394	10.242
Supermarket foods and exercise	4.64[b]	6.12	8.08[a]	452.37
	0.120	0.171	0.788	20.510

*The values represent the means ± SEM.
†Common superscripts (a) indicate absence of statistical differences at the .05 level.

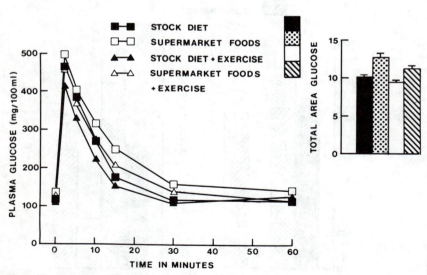

Figure 2—Variations with time in plasma glucose after intravenous injection of glucose and total areas under the glucose response curves (mg/dl × min × 10₃). The total areas only were considered by statistical analysis. Absence of a superscript indicates that all groups differ from each other at the .05 level (Richard et al., J. Nutr. [in press]).

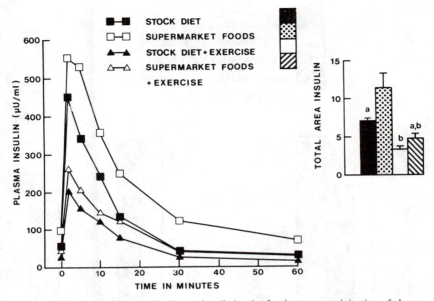

Figure 3—Variations with time in plasma insulin levels after intravenous injection of glucose with total areas under the insulin response curves ($\mu U / ml \times min \times 10_3$). The total areas only were considered by statistical analysis. Common superscripts indicate absence of significant differences at the .05 level (Richard et al., J. Nutr. [in press]).

response to a glucose load and insulinogenic index were also influenced by both diet and level of activity (Figures 3 & 4). The high energy diet increased the values of these variables in sedentary but not in trained rats, while exercise-training lowered them no matter what the diet used.

Discussion

The present study shows that it is possible to prevent impaired glucose tolerance and hyperinsulinemia which ordinarily result from a high energy diet (Bray & York, 1971). Similarly, recent findings (Zavaroni et al., 1981) have demonstrated that exercise training could prevent hyperinsulinemia caused by a high sucrose diet. However, the mechanisms by which exercise-training modulate insulin levels are not entirely known yet. Some studies have suggested a role for reduced body weight or body fat which occurs during training (LeBlanc et al., 1979; Richard & LeBlanc, 1980). In addition, in a recent study we have measured insulin response to glucose in female rats fed supermarket foods. Interestingly, contrary to males, female rats fed a high energy diet while training become heavier, fatter and more

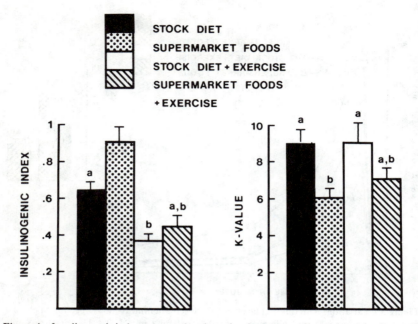

Figure 4—Insulinogenic index expressed as the ratio of total area of insulin over total area of glucose and K values. Common superscripts indicate absence of significant differences at the .05 level (Richard et al., J. Nutr. [in press]).

insulinemic than sedentary rats fed a stock diet. Considering these findings, it seems that the ability to resist hyperinsulinemia is related to the capacity of trained male rats to resist obesity when fed supermarket foods.

References

BRAY, G.A. & D.A. York. Genetically transmitted obesity in rodents. Physiol. Rev. 51:598–646, 1971.

HALES, C.N. & P.J. Randle. Immunoassay of insulin with insulin antibody precipitate. Biochem. J. 88:137–146, 1963.

LEBLANC, J., A. Nadeau, M. Boulay & S. Rousseau-Migneron. Effects of physical training and adiposity on glucose metabolism and ^{125}I-insulin binding. J. Appl. Physiol. 46:235–239, 1979.

RICHARD, D., A. Labrie, D. Lupien, A. Tremblay & J. LeBlanc. Interactions between dietary obesity and exercise-training on carbohydrate metabolism. Int. J. Obes. (in press)

RICHARD, D., A. Labrie, D. Lupien, A. Tremblay & J. LeBlanc. Role of exercise-training in the prevention of hyperinsulinemia caused by high energy diet. J. Nutr. (in press)

RICHARD, D. & J. LeBlanc. Effects of physical training and food restriction on insulin secretion and glucose tolerance in male and female rats. Am. J. Clin. Nutr. 33:2588–2594, 1980.

SCLAFANI, A. & D. Springer. Dietary obesity in adult rats: Similarities to hypothalamic and human obesity syndromes. Physiol. Behav. 17:461–471, 1976.

WAHLBERG, F., Intravenous glucose tolerance in myocardial infarction, angina pectoris and intermittent claudication. Acta Med. Scand. 180(suppl.):453, 1966.

ZAVARONI, I., Y.I' Chen, C.E. Mondon & G.M. Reaven. Ability of exercise to inhibit carbohydrate-induced hypertriglyceridemia in rats. Met. Clin. Exptl. 30:476–480, 1981.

Substrate Utilization during Normal and Loading Diet Treadmill Marathons

S.E.H. Hall, J.T. Braaten, T. Bolton, M. Vranic and J. Thoden
Ottawa Civic Hospital, Ottawa, Ontario, Canada,
The University of Ottawa, Ottawa, Ontario, Canada, and
The University of Toronto, Toronto, Ontario, Canada

The purpose of these experiments was, using radiosotopes, to measure the contribution of two major blood borne fuels, glucose and free fatty acids (FFA) in good (2½ to 3 h) marathoners during a marathon run: a) on their normal diet (ND) and b) after an isocaloric high carbohydrate diet (HCD), immediately following a period of glycogen depletion.

Methods

Subjects

Informed written consent was obtained from one female and six male marathoners who had run at least one competitive marathon in under 3 h within the last yr. All were in training (wt 71.0 ± 3.4 kg; age 36.9 ± 2.3 yr; $\dot{V}_{O_2max}$, 63.4 ± 1.9 ml/min·kg, range 56.8 to 71.4). They recorded their dietary intake for 3 days and an 80% CHO isocaloric diet was designed for each subject.

Experimental Protocol

Each subject was investigated twice in the postabsorptive state: a) on ND, (45.5% CHO, 2140 ± 289 cal) b) after 3 days of HCD, (77.2% CHO, 2688 ± 220 cal), immediately following 3 days of strenuous training and a high protein and fat diet. The order of the runs was varied and the pace of the second run was matched to that of the first for the first 2 h. Each subject received a primed infusion of ^{14}C-1-glucose and ^{3}H-9-10 palmitic acid, via a superficial arm vein. Two h were allowed for tracer equilibration. Twelve blood samples were taken from a contralateral antecubital vein, at rest and at regular intervals during a treadmill run at a 3 h marathon pace. Open circuit measurements of O_2 consumption and CO_2 production were made at similar intervals.

Chemical Methods

Labeled and unlabeled glucose and FFA concentrations were measured as described by Hall (1979), except that unlabeled FFA was measured spectrophotometrically (Mossinger, 1965). Hormones measured were: norepinephrine (NE) and epinephrine (E) (Sole, 1977), free insulin (Kuzuya, 1977) and triiodothyronine (T3), using a Tri-Tab RIA kit purchased from General Diagnostics.

Calculations

The rates of appearance (Ra) and disappearance (Rd) for glucose (G) and FFA (F) were calculated during running using a form of the Steele equation (Steele, 1956).

Results

Most runners stopped exhausted after: 21.4 ± 1.3 miles in 149.7 ± 9.5 min on ND and 24.0 ± 0.91 miles in 163.6 ± 4.6 min on HCD.

Glucose Kinetics

Figure 1 shows the changes in rest and exercise in glucose concentration, Ra, Rd and MCR, for both runs. After HCD the fasting glucose concentration at rest was lower than on ND (4.6 ± 0.13 vs 5.1 mmol/L) and in five runners the peak running value was lower as well. There were no significant differences in resting glucose turnover (RT) (11.08 ± 2.0 and $13.2 \pm 1.06 \mu$mol/min·kg) or MCR (2.15 ± 0.34 and 2.85 ± 0.2 ml/min·kg) despite resting MCR_G being increased after HCD in four subjects. Initially the Ra_G began to rise to exceed Rd_G so that blood glucose concentration rose, reaching a peak on ND after approximately 50 min. At this time Rd_G rose to exceed Ra_G, returning the concentration to normal. MCR_G rose progressively to three times the resting level with a greater volume cleared during the HCD than ND (978.4 ± 72.3 vs 713.0 ± 109.7 ml/kg), p < 0.02.

FFA Kinetics

Resting FFA concentration and turnover, Figure 1, were lower after HCD than ND (0.54 ± 0.035 vs 0.79 ± 0.08 mmol/L), p < 0.02, and 3.4 ± 0.46 vs 6.6 ± 1.07 μmmol/min kg, p < 0.05. In contrast to glucose both Ra_F and Rd_F fell initially with Rd_F exceeding Ra_F. After approximately 40 min RT_F rose progressively with Ra_F being marginally higher than Rd_F, leading to a steadily increasing concentration in the blood while MCR_F remained constant. Ra_F increased markedly 10 min after the end of the run, while Rd_F remained similar to the previous running value, leading to a sharp rise in FFA concentration. However, 10 min later Rd_F rose and the concentration declined.

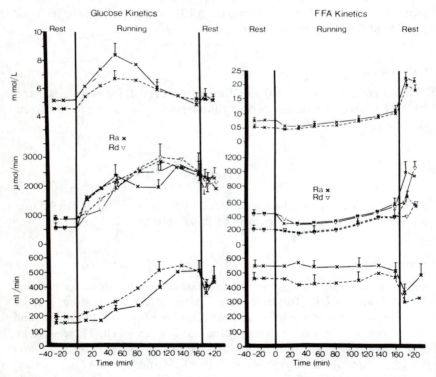

Figure 1—Average changes at rest and running in plasma glucose concentration (mmol/L), rates of appearance x, disappearance Δ (μmol/min) and metabolic clearance (ml/min) after normal diet (solid lines) and high carbohydrate diet (dashed lines). Lines and bars denote standard errors of the mean.

Energy Balance

Resting $\dot{V}_{O_2}$ was unchanged by HCD (3.93 ± 0.21 and 4.1 ± 0.19 ml/min·kg). A similar volume of O_2 (Q_{O_2}) was consumed on both runs (7.26 ± 0.5 and 7.23 ± 0.4 L/kg), and 42.7 ± 9.1% (ND) and 46.6 ± 3.3% (HCD) of resting Q_{O_2} was required to oxidize blood glucose. Had all FFA removed from the blood been oxidized that would have accounted for 8.6 ± 12.6% resting Q_{O_2} after ND and, unlike glucose, this was reduced to 42.6 ± 5.6% by HCD. Thus, at rest all Q_{O_2} (129.3 ± 15.6% ND, 89.1 ± 75% HCD) could be accounted for by oxidation of blood glucose and FFA. In contrast, while running, irrespective of the previous diet, only 15.8 ± 1.1% (ND) or 15.96 ± 1.1% (HCD) of Q_{O_2} could be attributed to oxidation of blood borne FFA or glucose, with the proportion attributable to FFA being reduced by HCD from 6.2 ± 0.74% to 4.5 ± 0.88% (p < 0.3).

Hormone Concentrations

Figure 2 shows that NE and E exhibited a 7 to 8-fold rise during the run (p < 0.001 and 0.01). Free insulin concentration was lower at rest and running after HCD and the changes were opposite to those in plasma glucose. Thus insulin fell at the start of exercise (p < 0.02), rose to a peak (ND p < 0.02, HCD p < .05) and fell again towards the end of the run (p < .05) and T_3 was higher at the end of the run (ND p < .02, HCD p < .05).

Glucose Kinetics

Glucose was the preferred blood borne substrate, as running caused a three fold increase in the amount of glucose used, i.e., to 4.8 ± 0.63 (ND) and 5.7 ± 0.32 mmol/kg (HCD), compared with the 1.75 ± 0.35 and 2.2 ± 0.19 mmol/kg, which would have been used for an equal rest time.

The most notable change in exercise was the considerable fluctuation in plasma glucose. The kinetic measurements show that the initial rise occurred as a result of the rapid increase in Ra_G, presumably as the result of hepatic glycogenolysis, induced by the rising NE, E and T_3 in the face of falling insulin concentration. The latter in turn was likely influential in raising MCR_G so that the concentration was returned to normal. However, MCR_G was at its highest when insulin levels were depressed indicating, according to Goldstein (1961), the involvement of other factors. Despite the popular belief that hypoglycemia is responsible for exhaustion in marathoners, no hypoglycemia was observed in any of the runners, presumably because counter-regulatory mechanisms were stimulated in response to the compensatory fall in plasma insulin. Issekutz (1970)

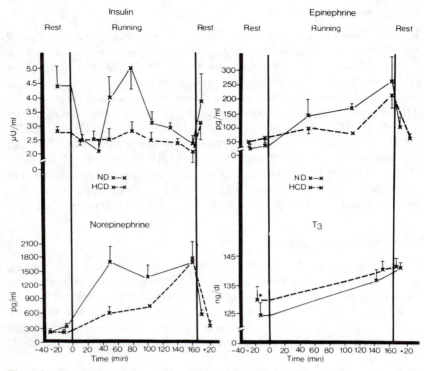

Figure 2—Average changes in norepinephrine, epinephrine, insulin and T_3 at rest and running after normal diet (solid lines) and high carbohydrate diet (dashed lines). Lines and bars denote standard errors of the mean. *ND vs HCD t = 1.77 p < .02.

suggested that when MCR_G increases to the extent seen here, working muscles compete with the CNS for glucose, leading to exhaustion. Certainly the rapid rise in MCR coincided with feelings of exhaustion and signalled the end of the run.

FFA Kinetics

Blood FFA concentration and turnover have been observed to decline in humans (Pruett, 1970) and dogs (Issekutz, 1963) at the start of high intensity and rise as exercise continues. In humans this fall is frequently of short duration but in our runners FFA concentration did not return to normal for 90 to 100 min. A fall of similar magnitude has been reported previously by Issekutz (1963) in exercising dogs and by Havel (1967) in humans.

At first glance the prolonged fall in FFA appears paradoxical because the hormonal milieu would be expected to promote lipolysis. It is well accepted that blood flow is diverted from non-working tissues; the low Ra_F is thus more likely a function of reduced adipose tissue blood flow than the rate of lypolysis. This is supported by the marked rise in Ra_F and concentration which occurred immediately after exercise, when normal blood distribution was restored.

Remarkably, as a result of the initial fall in RT_F, FFA utilization was unchanged overall, indicating that it is of lesser importance as an energy source than glucose. Furthermore, the reliance on blood borne FFA was even less after HCD.

Energy Balance

As described by Owen (1971) oxidation of plasma FFA and glucose each accounted for approximately half of the resting oxygen consumption. In contrast blood borne fuels supplied only 16% of the energy needed for the run with a maximum of 12.2% as the contribution of blood glucose, agreeing well with estimates by Paul (1967). We concluded that 84% of the energy for the run must have been met by other metabolites or stored fuels. If, as has been described by Bergstrom (1967), the remainder of the energy requirements were supplied by glycogen, the QO_2 of our runners suggests that they would have needed 540 g glycogen, which is within observed limits.

Acknowledgments

Supported by the C.D.A., M.R.C. (Canada) and the M. McNulty Fund.

References

BERGSTROM, J., L. Hermansen, E. Hultman & B. Saltin. Diet, muscle glycogen and physical performance. Acta Physiol. Scand. 71:140–150, 1967.

GOLDSTEIN, M.S., Humoral nature of hypoglycemia in exercise. Am. J. Physiol. 200(1):67–70, 1961.

HALL, S.E.H., J. Saunders & P.H. Sonksen. Glucose and free fatty acid turnover in normal subjects and in diabetic patients before and after insulin treatment. Diabetologia 16:297–306, 1979.

HAVEL, R.J., B. Pernow & N.L. Jones. Uptake and release of free fatty acids and other metabolites in the legs of exercising men. J. Appl. Physiol. 23(1):90–99, 1967.

ISSEKUTZ, B., Jr., A.C. Issekutz & D. Nash. Mobilization of energy sources in exercising dogs. J. Appl. Physiol. 29(5):691–697, 1970.

ISSEKUTZ, B., Jr., H.I. Millier & K. Rodahl. Effect of exercise on FFA metabolism and pancreatectomized dogs. Am. J. Physiol. 205(4):645–650, 1963.

KUZUYA, H., P.N. Blix, D.L. Horowitz et al. Determination of free and total insulin and C peptide in insulin treated diabetes. Diabetes 26:22–29, 1977.

MOSINGER, F., Photometric adaptation of Doles Microdetermination of free fatty acids. J. Lipid Res. 6:159, 1965.

OWEN, O.E. & G.A. Reichard, Jr. Fuels consumed by man; the interplay between carbohydrates and fatty acids. Progr. Biochem. Pharmacol. 6:177–213, 1971.

PAUL, P. & B. Issekutz. Role of extramuscular sources in the metabolism of the exercising dog. J. Appl. Physiol. 22:615–622, 1967.

PRUETT. FFA Mobilization during and after prolonged severe muscular work in men. J. Appl. Physiol. 29(6):809–815, 1970.

SOLE, M.J. & M.N. Hussain. A simple specific radioenzymatic assay for the simultaneous measurement of picogram quantities of norepinephrine, epinephrine and dopamine in plasma and tissues. Biochem. Med. 18:301–307, 1977.

STEELE, R., J.S. Wall, R.C. DeBodo & N. Altzuley. Measurements of size and turnover rate of body glucose pool by the isotope dilution method. Am. J. Physiol. 187:15–24, 1956.

Electrolytes

Calcium Regulation of Myofibril ATPase Activity at Exhaustion and Recovery

A.N. Belcastro, M.M. Sopper and M.P. Low
Dalhousie University, Halifax, Nova Scotia, Canada, and
University of Alberta, Edmonton, Alberta, Canada

It appears that associated with fatigue, there is a decrease in the force-generating capabilities of skeletal muscle. Witzmann et al. (1979), employing swim-exercised rats (fatigued by 7 h) have reported decreases in isometric contraction and half relaxation time to 77% and 66% of control values. A similar pattern was noted for twitch tension (P_t), tetanic tension (P_o), and the rate of tension development and decline.

Since a correlation coefficient of 0.82 has been reported for tension production and ATPase activity, it has been suggested that the decrement in force production with fatigue may be due to an interruption in the function of the ATPase enzyme (Fitts & Holloszy, 1977). Furthermore, Ca^{++} control of the myofibril ATPase activity may be influenced, since the Ca^{++} transport ability of fast-twitch muscle from exhausted animals is altered (Belcastro et al., 1981).

Thus, the purpose of this study was to observe the response of the myofibril ATPase activity to calcium with exhaustion and ascertain whether the rate of cross-bridge cycling is affected in muscles from fatigued animals.

Method

Forty-eight male wistar rats (200 g) were randomly assigned to either a control (C) group (N = 24) or an exhausted (E) group (N = 24). The experimental animals were familiarized with the testing procedure for 2 days prior to the actual test day. Animals in the E group were then run on a motor driven treadmill at a speed of 25 m/min with a 10% grade.

Termination of exercise was voluntary and the inability for the animals to right themselves was used as a criterion for exhaustion. Following the experimental protocol, animals were killed by decapitation. Preliminary studies revealed that this exercise protocol results in 90% glycogen depletion of fast twitch muscles and depressed Ca^{++} transport (Belcastro et al., 1981).

Plantaris muscles were homogenized as previously described (Belcastro et al., 1980). All protein concentrations were determined by the Lowry method and are expressed relative to wet muscle weight.

Myofibril ATPase activity was determined with 50 mM KCL, 2 mM Tris HCl (pH 7.4) and 2 mM $MgCl_2$, at 30° C for 5 min, with 0.5 mg/ml of myofibril protein. After this pre-incubation, 5 mM Mg ATP was added and the sample mixed thoroughly and allowed to incubate for another 5 min at 30° C. The myofibril ATPase activities were run with 0.1, 0.5, 1.0, 2.5, 5.0, and 10.0 mM free Ca^{++} and 3, 5, 10 and 15 uM monomeric vanadate (Vi). The Ca^{++} binding was terminated by rapid filtration through millipore filters and an aliquot of the filtrate added to Bray's solution for liquid scintillation counting.

Results

The myofibril protein yield was not affected by the exhaustive exercise.

Myofibril protein isolated from muscles of exhausted animals (E) did not respond in a similar manner as the C group to Ca^{++} activation. Although the full range of Ca^{++} concentrations did enhance the myofibril ATPases activity, the ATPase activities at lower Ca^{++} concentrations were considerably greater than the values observed for the C group ($p < 0.05$) (Figure 1). A similar observation was noted during the recovery period following exhaustion (30 min). With respect to Ca^{++} binding, exhaustion did not alter the normal pattern of greater myofibril Ca^{++} binding with increasing Ca^{++} concentrations in the incubation medium and minimal differences were observed between the C and E groups (Table 1). The myofibril ATPase activities for C and E animals with varying monomeric vanadate levels were similar (Table 2).

Discussion

The myofibril ATPase response of control and exhausted muscle to vanadate incubation was similar. Since monomeric vanadate reflects the rate of cross-bridge cycling without influence on Ca^{++} binding (Herzig et al., 1981), it appears that exhaustive exercise does not alter the cross-bridge

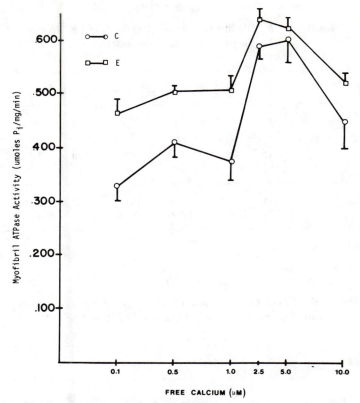

Figure 1—Calcium regulation of myofibril ATPase activity of skeletal muscle from control and exhausted animals. Results are means ± S.E.M.

cycling potential of skeletal muscle. However, the force generating potential of skeletal muscle, at the moment of fatigue, may still be involved, particularly in light of the observations that accumulation of inorganic phosphate may reduce cross-bridge cycling of skeletal muscle (Jacobus et al., 1977).

The response of the myofibril ATPase activity of the E group to increased Ca^{++} concentrations was different from the results obtained for the C group. At the 0.1, 0.5 and 1.0 μM Ca^{++} concentrations, the ATPase activities were greater for the E compared to the C group. It appears that the E group has become more sensitive to Ca^{++} activation, since a greater ATPase activity was observed at the lower Ca^{++} concentrations. However, the changes in ATPase activity are not related to the Ca^{++} binding capacity of myofibril protein. The enhanced activation of myofibril ATPase activity in exhausted muscle to Ca^{++} may be due to alterations in the ionic strength

Table 1

Myofibril Ca^{++} Binding (umol Ca^{++} 1 g) for C and E Groups with Varying Free Ca^{++} Concentrations

	Ca^{++} (μM)					
	0.1	0.5	1.0	2.5	5.0	10.0
Control	0.12 ± 0.03	0.25 ± 0.06	0.43 ± 0.10	0.75 ± 0.11	1.20 ± 0.17	2.97 ± 0.30
Exhaust	0.13 ± 0.04	0.27 ± 0.05	0.51 ± 0.09	0.78 ± 0.14	1.50 ± 0.19	3.41 ± 0.42

($\overline{X}$ ± S.E.M.)

Table 2

Myofibril ATPase Activity (μmol Pi mg^{-1} min^{-1}) for C and E Groups with Varying Monomeric Vanadate Concentration (Vi)

	Vi (uM)				
	0	3	5	10	15
Control	0.652 ± 0.032	0.627 ± 0.037	0.609 ± 0.041	0.613 ± 0.029	0.628 ± 0.034
Exhaust	0.649 ± 0.041	0.585 ± 0.034	0.575 ± 0.039	0.603 ± 0.033	0.617 ± 0.028

($\overline{X}$ ± S.E.M.)

(I) sensitivity of the contractile proteins. It has been suggested that the I of the cellular environment may regulate muscular contraction. In vitro, at lower ionic strengths the myofibril ATPase activity is activated to a greater extent compared to higher I conditions at similar Ca^{++} concentrations (Portzehl et al., 1969). Thus, Ca^{++} sensitivity of the ATPase activity may be modulated by ionic imbalance during exhaustive exercise (Sembrowich et al., 1979). In addition, the greater sarcoplasmic Ca^{++} concentration which has been suggested to occur as a result of exhaustion may activate Ca^{+2} dependent myosin light chains kinases. These specific kinases may enhance myosin light chain phosphorylation which has been reported to increase ATPase activity, without necessarily altering the Ca^{++} binding capacity of myofibril protein.

In summary, exhaustive exercise results in an elevated myofibril ATPase activity which is related to an enhanced Ca^{++} sensitivity and no alteration in the potential for rate of cross-bridge cycling. The mechanism involved with the change requires further investigation.

References

BELCASTRO, A.N., M. Rossiter, M.P. Low & M.M. Sopper. Calcium activation of sarcoplasmic reticulum ATPase following strenuous activity. Can. J. Physiol. Pharmacol. 59:1214–1218, 1981.

BELCASTRO, A.N., H. Wenger, T. Nihei, D. Secord & A. Bonen. Functional overload of rat fast-twitch skeletal muscle during development. J. Appl. Physiol.: Resp. Environ. Exercise. Physiol. 49:583–588, 1980.

FITTS, R.H. & J.O. Holloszy. Contractile properties of rat soleus muscle: effects of training and fatigue. Am. J. Physiol. 233(3):C86–C91, 1977.

HERZIG, J.W., J.W. Peterson, J.C. Ruegg & R.J. Solaro. Vanadate and phosphate ions reduce tension and increase cross-bridge kinetics in chemically skinned heart muscle. Biochem. Biophys. Acta 672:191–196, 1981.

JACOBUS, W.E., G.J. Taylor, D.P. Hollis & R.L. Nunnally. Phosphorus nuclear magnetic resonance of perfused working rat hearts. Nature 265:756–758, 1977.

PORTZEHL, H., P. Zaoralek & J. Gaudin. The activation by Ca^{++} of the ATPase of extracted muscle fibrils with variations of ionic strength, pH and concentrations of MgATP. Biochem. Biophys. Acta 189:671–687, 1969.

SEMBROWICH, W.L., E. Wang, T.E. Hutchinson & D. Johnson. Electron microprobe analysis of myofilaments, mitochondria and sarcoplasmic reticulum of soleus muscle from exhausted rats. Med. Sci. Sport Exercise 12(2):131, 1979.

WITZMANN, F.A., D.H. Kim & R.H. Fitts. The effect of prolonged activity on the contractile properties of fast and slow skeletal muscle. The Physiologist 22(4):135, 1979.

The Changes in the Concentration and Total Amount of the Electrolytes in the Blood Serum at Various Muscular Exercises

O. Imelik

Tartu State University, Tartu, Estonian S.S.R., U.S.S.R.

The investigations of the changes in the blood serum electrolyte concentrations at muscular exercise have given various results. After prolonged exercise either an increase in the concentration of sodium and potassium (Åstrand & Saltin, 1964) or no changes (Refsum et al., 1973) or a tardy increase of potassium (Afar et al., 1981) have been found. [After an exercise of short duration an increase of the concentration of calcium has been shown and alternating and not typical changes in the concentration of sodium and potassium (Thörner, 1960).] About the dynamics of the changes during exercise and about the changes in the total amount of serum electrolytes we have found no evidence.

In the present investigation the concentration of sodium, potassium and in some series of calcium was determined by flame emission spectrophotometry before and after exercises of various character and duration (Table 1). At prolonged exercise on the cycle ergometer the estimations were carried out before the exercise, in the 10th, 30th and 60th min of exercise and 30 min after the exercise. In the last 4 series (Table 1) simultaneously with the determination of the serum electrolytes, the blood volume was estimated: in the 5th, 7th and 8th series by means of ^{131}I albumine (Malov, 1970), and in the 6th series by means of ^{51}Cr labelled erythrocytes (Moore et al., 1963). Using the plasma volume and serum electrolyte concentration the total amount of electrolytes was calculated. In the 5th, 7th and 8th series, in addition, the blood pH changes, and in 6th series the concentration of serum lactate and of the aldosterone (by means of RIA) were estimated.

The arithmetic means of the data, the standard deviation of the mean and significance of the difference from the rest value (Dixon & Mood, 1946) are presented in Tables 1 and 2.

At strenuous exercise of short duration the concentration and total amount of the electrolytes changed in opposite directions. The decrease in the total amount of sodium and potassium can be explained by diminution of plasma volume due to increased filtration from the capillaries into the interstitial space, called forth by the increased capillary blood pressure. The renal sodium retention, induced by the renin-angio-tensin-aldosterone mechanism that is started by the blood volume diminution could be regarded as one reason for the increase in the serum sodium concentration. Though an increase of the aldosterone concentration in the blood serum could be stated already in the initial stage of strenuous prolonged exercise (Table 2), the retained sodium can scarcely become apparent as sodium concentration increase in the blood serum during a 3 min exercise. In the increase of plasma potassium concentration, the diminished pH that induces the flux of potassium from the cells could be taken into consideration, although for a noticeable rise in serum potassium concentration also a longer time interval would be expected. Apparently an increase in the plasma electrolyte concentration simultaneously with a decrease in its total amount merely indicates to the part of osmotic pressure at the plasma volume changes: the increased osmotic pressure in the tissues in the initial stage of exercise leads to a filtration of water from the capillaries and thus to an increased electrolyte concentration in the blood serum. Hence the changes in the serum sodium and potassium content at a short term exercise are largely secondary, depending on the interchange of fluid between the intravasal and extravasal space and can scarcely give an idea about the changes in the electrolytes metabolism. Since the changes in the plasma volume are individually different in various stages of exercise (Imelik, 1978), the "variable and not typical" changes in the serum sodium and potassium concentration at a short term exercise (Thörner, 1960) are self evident.

After prolonged exercises on the bicycle ergometer the changes in the concentration and total amount of sodium were small. Though the increase of the aldosterone concentration during the strenuous prolonged exercise is individually largely variable (Imelik, 1982), a considerable and, during the exercise, progressing increase could be found in all subjects. After an hour's exercise the retention of sodium could be fixable as a change in the serum concentration, but in many cases apparently, the increase is compensated by the loss of sodium through perspiration (the subjects lost 1.2 kg of body weight on the average) and by the dilution of plasma by water from the processes of oxidation and glycogenolysis (Olsson & Saltin,

Table 1

The Changes in the Electrolyte Concentration and Total Amount at Various Exercises

Series	Exercise	Duration in min	Subjects	Number of subjects	Sodium mmol/l				Potassium mmol/l				Calcium mmol/l			
					Before exercise	After exercise	*g	D s.%	Before exercise	After exercise	*g	D s.%	Before exercise	After exercise	*g	D s.%
1.	Skiing competition 10 km	up to 50	Male students of physical education	11	118.7 ± 3.70	122.2 ± 5.13		< 95	5.2 ± 0.28	5.0 ± 0.20		< 95	3.0 ± 0.09	2.9 ± 0.10		< 95
2.	Skiing competition 5 km	up to 20	Female students of physical education	6	115.7 ± 1.26	109.1 ± 4.61		< 95	4.8 ± 0.16	4.9 ± 0.18		< 95	2.9 ± 0.13	2.9 ± 0.10		< 95
3.	Swimming session	45	Pupils of sports-school, boys	12	137.7 ± 313.	133.8 ± 1.87		> 99.5	5.1 ± 0.18	5.4 ± 0.17		< 95	3.0 ± 0.16	2.9 ± 0.05		> 97.5

No.	Description	n									
4.	Swimming session	45	138.6 ± 1.01	135.3 ± 1.24	< 95	4.8 ± 0.14	5.0 ± 0.15	< 95	2.9 ± 0.08	2.9 ± 0.04	> 99.5
5.	Exercise on bicycle ergometer 60 rev per min, 75 W	60	139.6 ± 1.28	141.9 ± 1.18	> 97.5	4.1 ± 0.15	4.1 ± 0.09	< 95	—	—	—
			*11.8 ± 0.78	*12.7 ± 0.87	> 97.5	*0.6 ± 0.05	*0.6 ± 0.05	< 95	—	—	—
6.	Exercise on bicycle ergometer 70 rev per min, 225W	60	137.0 ± 0.60	136.5 ± 1.65	> 97.5	4.3 ± 0.06	5.0 ± 0.08	> 99.5	2.7 ± 0.05	2.7 ± 0.04	< 95
			*10.0 ± 0.39	*9.8 ± 0.44	< 95	*0.5 ± 0.02	*0.6 ± 0.02	> 99.5	*0.3 ± 0.02	*0.3 ± 0.02	< 95
7.	Male athletes and students	3	140.0 ± 0.58	143.1 ± 0.72	> 99.5	3.7 ± 0.06	4.1 ± 0.06	> 99.5	—	—	—
			*10.0 ± 0.34	*8.7 ± 0.28	< 99.5	*0.5 ± 0.02	*0.4 ± 0.14	> 99.0	—	—	—
8.	Female athletes and students	3	138.9 ± 0.79	141.7 ± 1.00	< 95	3.8 ± 0.08	4.2 ± 0.09	< 95	—	—	—
			*6.8 ± 0.24	*6.3 ± 0.18	< 95	*0.3 ± 0.02	*0.3 ± 0.01	< 95	—	—	—

D.s.% = difference significance (%) from the rest value after Dixon & Mood (1946).

Table 2

The Changes in the Serum Electrolyte Concentration and Total Amount during Prolonged Exercises on the Bicycle Ergometer

Exercise	Substance	Unit	Before exercise	10th min	D.s.%	During the exercise 30th min	D.s.%	60th min	D.s.%	30 min after the exercise	D.s.%
Bicycle ergometer 60 rev per min, 75 W. 1 h	Sodium	mmol/l	139.6 ± 1.28	140.6 ± 1.85	< 95	142.0 ± 1.52	> 97.5	141.9 ± 1.18	> 97.5	143.7 ± 0.93	> 97.5
		g	11.8 ± 0.78	12.1 ± 0.10	< 95	12.3 ± 0.93	< 95	12.7 ± 0.87	> 97.5	12.1 ± 0.16	> 99.5
	Potassium	mmol/l	4.1 ± 0.15	4.1 ± 0.09	< 95	4.2 ± 0.10	< 95	4.1 ± 0.09	< 95	4.1 ± 0.04	< 95
		g	0.59 ± 0.053	0.60 ± 0.056	< 95	0.62 ± 0.061	< 95	0.62 ± 0.051	< 95	0.59 ± 0.052	< 95
	pH		7.35 ± 0.010	7.34 ± 0.013	< 95	7.34 ± 0.008	< 95	7.36 ± 0.009	< 95	7.38 ± 0.008	< 95
Bicycle ergometer 70 rev per min, 225 W. 1 h	Sodium	mmol/l	137.0 ± 0.60	137.3 ± 1.46	> 97.5	137.8 ± 0.87	> 97.5	136.5 ± 1.60	> 97.5	135.2 ± 1.30	< 95
		g	10.0 ± 0.39	9.6 ± 0.34	> 97.5	9.5 ± 0.28	> 99	9.8 ± 0.44	< 95	10.4 ± 0.34	< 95
	Potassium	mmol/l	4.3 ± 0.06	5.0 ± 0.07	> 99.5	5.1 ± 0.06	> 99.5	5.0 ± 0.08	> 99.5	4.6 ± 0.06	> 97.5
		g	0.53 ± 0.019	0.60 ± 0.021	> 99.5	0.60 ± 0.017	> 99.5	0.60 ± 0.022	> 99.5	0.58 ± 0.020	> 97.5
	Calcium	mmol/l	2.6 ± 0.09	2.6 ± 0.10	< 95	2.6 ± 0.09	< 95	2.7 ± 0.07	< 95	2.6 ± 0.08	< 95
		g	0.34 ± 0.015	0.32 ± 0.013	> 99	0.32 ± 0.013	< 95	0.33 ± 0.015	< 95	0.35 ± 0.011	> 99
	Aldosterone	g/ml	104 ± 33.8	203 ± 29.3	> 99.5	422 ± 61.1	> 99.5	576 ± 71.0	> 99.5	611 ± 123.5	> 99.5
		mg	3.5 ± 0.97	5.9 ± 0.86	> 99.5	13.5 ± 1.80	> 99.5	18.5 ± 2.01	> 99.5	22.8 ± 6.60	> 99.5
	Lactate	mg/100 ml	16.8 ± 3.47	41.8 ± 8.19	> 99	47.8 ± 5.46	> 99.5	34.3 ± 5.79	> 99.5	19.6 ± 2.96	> 97.5

D.s.% = Difference significance (%) from the rest value after Dixon & Mood (1946).

1969). The decrease of sodium concentration after the swimming session accentuates the role of the last mentioned factor.

The changes in the serum potassium content at prolonged exercise were dependent upon the intensity of exercise. At a less strenuous exercise where the pH remained unchanged, the changes in the potassium content were small and statistically insignificant. After the strenuous prolonged exercise, the concentration and total amount of potassium were increased. These increases can be explained by the acidosis that induced a flux of potassium from the cells.

A statistically significant shift in the serum calcium concentration could be found only after the swimming session.

The results obtained at skiing competitions are influenced to a certain extent by the relatively long interval between the effort and taking of the blood samples (10 to 15 min), and by the transport of the samples from the training camp to the laboratory before separation of serum.

The mean data of serum electrolyte content obtained in different stages of the exercise (Table 2) could give an erroneous idea of their constancy. That is caused by the individually non-coinciding fluctuations of the serum electrolyte content during the exercise.

References

AFAR, J., T. Djarova, D. Stefanova, B. Zaharieva, A. Ilkov & N. Popova, Hormonal, Enzyme, electrolyte and microelement changes in bicycle training. In J. Poortmans & G. Niset (Eds.), Biochemistry of Exercise IV-B. Baltimore: University Park Press, 1981.

ÅSTRAND, P.O. & B. Saltin. Plasma and red cell volume after prolonged severe exercise. J. Appl. Physiol. 19(5):829–832, 1964.

DIXON, W.J. & A.M. Mood. The statistical sign test. J. Amer. Statist. Assoc. 41:557–566, 1946.

IMELIK, O., Veränderungen des roten Blutblides und Blutvolumes bei Muskelarbeit. Transactions of Estonian Academy of Sciences. Biology 27/3:212–222, 1978.

IMELIK, O., The interrelation of the changes in the content of the hormones of hypophysis, thyroid gland and suprarenal cortex in the blood serum at prolonged strenuous exercise. Acta et commentationes Universitatis Tartuensis, (in press) 1982.

MALOV, G.A., An express method of determining the circulating blood volume with the aid of radioiodine-labelled albumin. Medical Radiology (Moscow) 11:31–34, 1970.

MOORE, F.D., K.H. Oleson & J.D. McMurrey. Body Cell Mass and its Supporting Environment. Dublin, Philadelphia, London: W.B. Saunders Co., 1963.

OLSSON, K.E. & B. Saltin. Variation in total body water with muscle glycogen changes in man. Medicine and Sport Basel-New York 3:159–162, 1969.

REFSUM, H.E., B. Tveit, H.D. Meen & S.B. Strømme. Serum electrolyte, fluid and acid-base balance after prolonged heavy exercise at low environmental temperature. Scand. J. Clin. Lab. Invest. 32:117–122, 1973.

THÖRNER, W., Blut und Blutbildungsstätten beim Sport. In A. Arnold (Ed.), Lehrbuch der Sportmedizin. Leipzig: Johan Ambrosius Barth, 1960.

Muscle Electrolyte Changes in Young Exercised Rats

M.M. Jaweed, R.C. DeGroof, G.J. Herbison, J.F. Ditunno, Jr.
and C.P. Bianchi
Thomas Jefferson University, Philadelphia, Pennsylvania,
U.S.A.

Adaptation of rat skeletal muscle to long term endurance exercises, such as low intensity running, manifests an increased capillarization (Faulkner et al., 1980; Holloszy, 1976; Salmons & Henricksson, 1981) and enhanced oxidative metabolism of the muscle. There is an elevation of the sarcoplasmic proteins (Gordon et al., 1967; Jaweed et al., 1974) and the capacity of the muscle to generate ATP. There is no increase in the muscle mass, glycolytic capacity, muscle fiber type pattern or in the isometric contractile properties. The role of low-intensity endurance exercises in improving the muscle resistance to fatigue has not been established (Faulkner et al., 1980; Holloszy, 1967). All the above parameters are directly or indirectly dependent on the maintenance of the appropriate concentration of the intra- and extracellular ions.

The objective of this study was to measure the total content of Na, K, Mg and Ca in the extensor digitorum longus (EDL) muscles of young exercised rats, and to test whether the endurance exercise-induced changes are general or specific to certain electrolytes.

Methods

Two groups (n = 5 to 7) of young female Wistar rats (3 to 4 weeks of age and 50 to 60 g in body weight) were acclimated under standard laboratory conditions (65° F and 75% humidity, 12 h dark/light cycle) on ad libitum Purina chow and water. After 3 days, one group was transferred to cages attached with non-motorized running wheels (Gordon et al., 1967, Jaweed et al., 1974). Daily records of the running animals showed that on average,

Table 1

Body Weights (BW) and Wet Muscle Weights (MW) of the Rat EDL

Group	n	Pre BW (g)	Post BW (g)	Δ BW (g)	MW (mg)	MW/BW Ratio
Normal	5	61.0 ± 2.1	243.7 ± 5.3	182.7 ± 5.2	129.5 ± 11.1	5.31 ± 0.55
Exercised	7	69.0 ± 0.6	195.7 ± 2.2	126.7 ± 2.5	125.8 ± 7.9	6.43 ± 0.36
% Difference††		13.1*	-19.7**	-30.7**	-2.9	21.1

*$P < 0.05$; **$P < 0.01$; † % Difference = $\dfrac{\text{Exercised} - \text{Control}}{\text{Control}} \times 100$

Δ BW = Post BW - Pre BW; MW/BW = $\dfrac{MW(mg)}{BW(g)} \times 10^4$

$(\overline{X} \pm S.E.)$

the exercising rats ran 3 to 4 km per day. After 12 weeks, the two animal groups were sacrificed under the influence of anesthesia (chloral hydrate: 400 mg/kg BW, intra-peritoneal [i.p.]) and the extensor digitorum longus (EDL) muscles were carefully removed, cleaned of fascia, and immediately weighed in electrolyte-free quartz crucibles and dried overnight at 110° C. The muscles were then ashed in a muffle furnace at 450° C for 20 h (Bianchi & Shanes, 1959) with 0.5 ml of electrograde concentrated HNO_3 to aid in ashing. The ashed muscles were dissolved in 0% $SrCl_2$ (0.01 M) solution and were used for analyzing the total content of Na, K, Mg, and Ca as per the methods detailed elsewhere (DeGroof et al., 1980, 1981). The data were expressed as micromoles (μM) of electrolytes per g of wet muscle tissue.

The body weights (BW), wet muscle weight (MW) and muscle weight/ body weight (MW/BW) ratios, as well as the total electrolyte content of the control and the experimental animals were compared by student 't' tests. Values having a p < 0.05 were considered significant.

Results

The pre-experimental BWs of the exercised animals were 13% larger than control animals. After 12 weeks of ad libitum feeding, the control animals gained in BW more (30.6%, p < 0.01) than the exercise-trained rats. In contrast to the post-experimental BW, the MW of the EDL in the two groups were almost identical. The MW/BW ratio of the exercised group was not significantly different from that of the control group (Table 1).

In EDL muscles of the two groups, the total content of K was the highest, followed by Na, Mg and Ca, in that order. Compared to the control group, the exercised muscles showed significant (p < 0.01) elevations in the content of Na (63.5%), K (41.7%), Mg (46.1%) and Ca (66.7%). The highest increases were found in Na and Ca (Table 2). However, the Na/K and Ca/Mg ratios of the two groups were similar (Figure 1).

Discussion

It has been known that cations, such as Na, K, Mg and Ca are transported across membranes. During steady state, these cationic gradients are maintained by a passive influx of Na^+ ions and passive efflux of K^+ ions. The extrusion of Ca^{++} as well as Na^+ from inside the cell is dependent on an active transport system involving metabolic energy and the Na^+/K^+ - ATPase (Hasselbach, 1981; Schurmans-Stekoven, 1981). Presently, it is not clear whether a certain type of exercise may influence a specific ionic gradient. Several investigators have shown that during exercise, K level

Table 2

Total Electrolyte Content (Micromoles/g Muscle) in the rat EDL Muscle
After Exercise

Electrolyte	Normal (n = 5)	Runners (n = 7)	Absolute difference+	% Difference
Na	11.5 ± 0.8	18.8 ± 0.4	7.3**	63.5**
K	57.8 ± 5.6	81.9 ± 2.9	24.1**	41.7**
Mg	6.5 ± 0.5	9.5 ± 0.2	3.0**	46.1**
Ca	0.9 ± 0.1	1.5 ± 0.1	0.6**	66.7**

**p < 0.01
+Absolute difference = Runners – Normals
For % Difference see foot note under Table 1
($\overline{\times}$ ± S.E.)

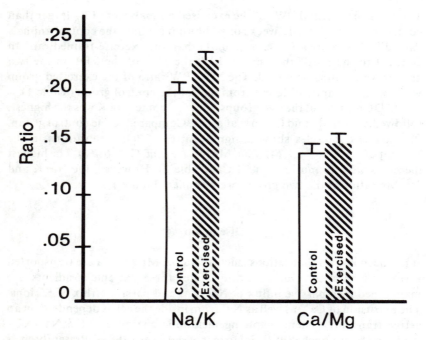

Figure 1—Changes in Na/K and Ca/Mg ratios of the control and exercised EDL muscles. Note the high Na/K ratio compared to Ca/Mg ratio. The differences between the control and the exercised muscles are non-significant.

decreases in the muscle, and increases in the venous blood flow (Sréter & Friedman, 1963; Korge & Viru, 1971). Similarly, it has been reported that Na content increases in the exercised muscles (Ahlborg et al., 1967). In in vitro studies, Bianchi and associates (Bianchi & Shanes, 1959; Bianchi & Narayan, 1982) observed a marked influx of Ca during twitch and tetanic tension development. However, Kim et al. (1981) were unable to demonstrate an increased or a decreased uptake of Ca^{++} and peak amount of Ca^{++} sequestered in the predominantly type IIB, type IIA or type I fibers of exercised muscles. In the present study, it is noteworthy that the prolonged exercise of EDL muscle increased the concentration of Na, K, Mg and Ca but neither the Na/K nor the Mg/Ca ratio is significantly altered (Figure 1). It seems that exercise may foster or enhance the growth of the muscle but it may not significantly alter the membrane gradients.

Recently we have reported that during growth of albino rats (between 16 and 24 weeks of age) the concentration of Na, K and Mg are also increased; but, unlike the findings of the present study, Ca content is not significantly elevated (Jaweed et al., 1982a). In the present study, total Ca content of the exercised EDL muscles increased more than 66%, compared to the control muscles (Table 2). It appears that the growing and exercising muscles differ with each other in regards to calcium accumulation. An increased level of muscle Ca may be a specific physiological response of the chronically exercised rat fast-twitch muscle as found after low frequency (10 Hz) electrical stimulation of the normal and the denervated rat soleus muscles (Jaweed et al., 1982b). Other investigators (Bianchi & Shanes, 1959; Bianchi & Narayan, 1982) also have observed, in vitro, an increased influx (and efflux) of Ca across the membrane due to electrical stimulation of muscle. This suggests that during electrical stimulation or muscular exercise there is an increased amount of intracellular calcium available for mobilization from the intracellular compartments, such as the sarcoplasmic reticulum and the sub-sarcolemmal Ca vesicles (Bianchi & Narayan, 1982; Somlyo et al., 1980). Thus the exercised muscle may be able to sequester Ca more efficiently than the control because of a high oxidative capacity or availability of ATP. The removal of intracellular Ca^{++} may require an increased oxidative capacity by the muscle. Wrogemann and Pena (1976) have proposed that an increased level of intracellular Ca^{++}, which exceeds the capacity of the sarcoplasmic reticulum, may cause muscle necrosis.

In summary, the results of this study have indicated that if low-intensity chronic exercise is initiated at an early stage of life, it may produce a specific adaptation in the fast-twitch muscle, EDL, of the rat: to efficiently sequester and recycle Ca^{++} into and from the intracellular compartments so as to meet the demands of overactivity. This is achieved because of a high oxidative capacity and more availability of the ATP to the exercised muscles.

Acknowledgment

This research was supported, in part, by a PHS grant BR 56 and RR 5414.

References

AHLBORG, B., J. Bergstrom, L.G. Ekelund & E. Hultman. Muscle glycogen and muscle electrolytes during prolonged physical exercise. Acta Physiol. Scan. 70:129–142, 1967.

BIANCHI, C.P. & S. Narayan. Muscle fatigue and the role of transverse tubules. Science 215:295–296, 1982.

BIANCHI, C.P. & A.M. Shanes. Calcium influx in skeletal muscle at rest, during activity and during potassium contracture. J. Gen. Physiol. 42:803–813, 1959.

DeGROOF, R.C., C.P. Bianchi & S. Narayan. The effect of diazepam on tension and electrolyte distribution in frog muscle. Europ. J. Pharmacol. 66:193–199, 1980.

DeGROOF, R.C., M.M. Jaweed, G.J. Herbison & J.F. Ditunno. Alteration of electrolytes in rat soleus following spinal nerve (L_4 & L_5) denervation. Fed. Proc. 40:672, 1981.

FAULKNER, J.A., L.C. Maxwell & T.P. White. Adaptations in skeletal muscle. In J.A. McNamara Jr. & D.S. Carlson (Eds.), Muscle Adaptations in the Craniofacial Region. Ann Arbor: Center for Human Growth & Development, 1978.

GORDON, E.E., K. Kowalski & M. Fritts. Protein changes in quadriceps muscle of rat with repetitive exercises. Arch. Phys. Med. Rehabil. 48:577–582, 1967.

HASSELBACH, W., Calcium-activated APTase of the sarcoplasmic reticulum membranes. In Bonting & DePond (Eds.), Membrane Transport. Elsevier Biomedical Press, pp. 183–205, 1981.

HOLLOSZY, J.D., Adaptations of muscular tissue to training. Prog. Cardiovasc. Dis. 18:445–458, 1976.

JAWEED, M.M., E.E. Gordon, G.J. Herbison & K. Kowalski. Endurance and strengthening exercise adaptations: protein changes in skeletal muscle. Arch. Phys. Med. Rehabil. 513–517, 1974.

JAWEED, M.M., C.P. Bianchi, R.C. DeGroof, G.J. Herbison & J.F. Ditunno. Muscle electrolyte changes in aging Lewis-Wistar rats. Fed. Proc. 41:1321, 1982a.

JAWEED, M.M., G.J. Herbison, J.F. Ditunno & R.C. DeGroof. Direct electrical stimulation of muscle: changes in Ca and Mg content. Arch. Phys. Med. Rehabil. 63:Dec. 1982b.

KIM, D.H., G.S. Wible, F.A. Witzmann & R.H. Fitts. The effect of exercise training on sarcoplasmic reticulum function in fast and slow skeletal muscle. Life Sci. 28:2671–2677, 1981.

KORGE, P. & A. Viru. Water and electrolyte metabolism in skeletal muscle of exercising rats. J. Appl. Physiol. 31:1–4, 1971.

SALMONS, S. & J. Henricksson. The adaptive response of skeletal muscle to increased use. Muscle & Nerve 4:94–105, 1981.

SCHURMANS-STEKHOVEN, F.M. & S.L. Bonting. Sodium-potassium activated adenosinetriphosphatase in membrane transport. In Bonting & DePond (Eds.), Membrane Transport. Elsevier Biomedical Press, pp. 159–179, 1981.

SOMLYO, A.P., A.V. Somlyo, H. Shuman, B. Sloan & A. Scarpa. Electron probe analysis of calcium compartments in cryosections of striated and smooth muscles. Ann. NY Acad. Sci. 30(7):523–544, 1978.

SRÉTER, F.A. & S.M. Friedman. Distribution of water, sodium and potassium in resting and stimulated mammalian muscle. Can. J. Biochem. 41:1035–1045, 1963.

WROGEMANN, K. & S.D. Pena. Mitochondrial calcium overload: a general mechanism for cell necrosis in muscle diseases. Lancet 1:672–674, 1976.

Sweat Electrolyte Losses during Prolonged Exercise in the Horse

M.G. Kerr and D.H. Snow

University of Glasgow Veterinary School, Glasgow, Scotland

It has been reported by several authors that horse sweat, in marked contrast to human sweat, is hypertonic relative to plasma (Carlson & Ocen, 1979). The volume of sweat produced by the exercising horse is similar to that produced by the human (on a body weight basis) but the consequences of this in terms of electrolyte loss have not been investigated. It is not known whether these high electrolyte concentrations are maintained throughout sweating nor whether the composition of exercise-induced sweat differs significantly from heat or adrenaline-induced sweat. Sweating in the horse is mediated via β_2 adrenoceptors (Snow, 1977), and it has been shown that there are fundamental physiological differences between heat and exercise induced sweating (Robertshaw & Taylor, 1969).

This study investigates the sweat composition of a group of thoroughbred horses during prolonged exercise, heat exposure and adrenaline infusion, and estimates total sweat electrolyte losses occurring during exercise.

Methods

A group of six thoroughbred horses was used in this study, but only four animals were involved in each section. Sweating was induced by exercise, 80 km at 18 km/h, with each horse exercised twice (i.e. n = 8), adrenaline infusion, 0.13 to 0.31 μg/kg/h for 3 h (n = 4), and heat exposure, 41°C (33°C wet bulb) for 5 h (n = 4). Sweat was collected during exercise in absorbent pads placed under the saddles and changed every 16 to 24 km, and during heat exposure and adrenaline infusion by scraping directly from the coat. Blood was also collected at various times before, during and after exercise. Full details of these procedures and techniques are discussed elsewhere (Kerr & Snow, 1982; Snow et al., 1982).

Results

The Cl^-, Na^+ and K^+ concentrations of the sweat collected during each of the four stages of the endurance ride are shown in Figure 1. Cl^- and Na^+ fluctuated somewhat but neither fell below 155 mmol/1 at any time. K^+ decreased steadily with time but again did not fall below 20 mmol/1. The

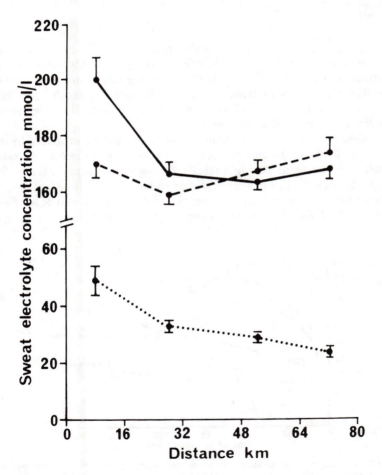

Figure 1—Chloride (continuous line), sodium (broken line) and potassium (dotted line) concentrations of sweat collected during long distance exercise.

Table 1

Sweat Electrolyte Concentrations in Exercise, Heat and Adrenaline Induced Sweat

	Cl^-(mmol/l)			Na^+(mmol/l)			K^+(mmol/l)		
Exercise	167 ± 3	**		166 ± 2	***		30 ± 2	**	
Heat	194 ± 8		**	149 ± 4	***	*	50 ± 8	**	*
Adrenaline	207 ± 9			194 ± 6			24 ± 2		

Significant differences between groups: $*p < 0.05$, $**p < 0.01$, $***p < 0.001$.
(mean $\pm$ S.E.)

sweat produced by the other two stimuli showed some significant differences in electrolyte concentration but again levels were always hypertonic compared to plasma. Table 1 gives the mean concentrations of electrolytes in the sweat from all three stimuli, using results obtained during the central period of sweating when concentrations remained relatively constant in all experiments.

The most striking difference in the composition of sweat produced by the three stimuli was in the changes occurring with time in the Na^+/K^+ ratio (Figure 2). The ratio during exercise followed a course intermediate between those seen during heat exposure and adrenaline infusion.

Water loss during exercise was estimated from the weight loss of the horses as 37 ± 2.6 liter. From this it can be calculated that sweat electrolyte

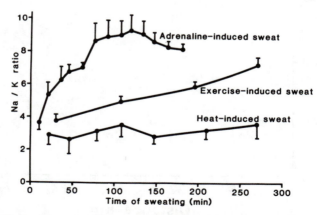

Figure 2—Change in sweat sodium: potassium ratio with time of sweating—comparison between heat, exercise and adrenaline-induced sweat. For comparative purposes 16 km distance is considered approximately equivalent to 60 min.

losses were 5.3 ± 0.4 moles Cl^-, 5.1 ± 0.4 moles Na^+ and 1.0 ± 0.07 moles K^+. Expressed as a percentage of total body content these are 12% water, 41% Cl^-, 28% Na^+ and 4% K^+. The changes in total plasma protein and electrolyte concentrations which occurred during and after the exercise period are shown in Figure 3.

Certain other electrolytes were also measured in the sweat. Ca^{2+} was only determined in the adrenaline experiment, when it was found to decrease with time from 4.4 ± 0.5 mmol/l to 0.9 ± 0.2 mmol/l. Mg^{2+} decreased with time in all experiments; in the exercise-induced sweat the decrease was from 4.1 ± 0.36 mmol/l to 0.7 ± 0.07 mmol/l. There was a close correlation in all cases between sweat Mg^{2+} and sweat protein concentration, which showed an identical pattern of decrease with time. Initial sweat protein concentrations (7.7 ± 0.9 g/l in exercise-induced sweat) were much higher than those reported for human sweat.

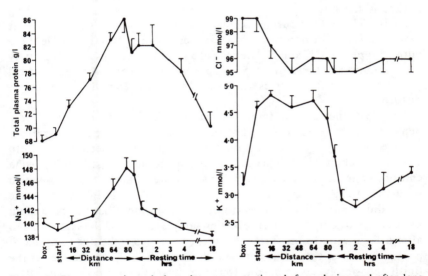

Figure 3—Plasma protein and electrolyte concentrations before, during and after long distance exercise. The "Box" sample was collected from the horses when at rest in the stable and the "Start" sample after saddling up and walking 1 km to the starting line.

Discussion

This study confirms previous reports (Carlson & Ocen, 1979; Rose et al., 1980) that horse sweat is hypertonic compared to plasma for Cl^-, Na^+ and K^+ and that this is true for sweat induced by exercise, heat and adrenaline infusion. The results of this study indicate that these high concentrations are maintained during prolonged periods of sweating.

The differences in electrolyte concentrations between the three stimuli suggest that neither heat exposure nor adrenaline infusion are ideal models for the study of sweat electrolyte losses in the exercising animal. Robertshaw and Taylor (1969) showed that while sweating can be induced by circulating adrenaline alone, heat-induced sweating is controlled solely by the sympathetic nervous system and exercise-induced sweating involves both circulating adrenaline and sympathetic nervous activity. This may provide an explanation for the finding that Na^+ and K^+ concentrations and Na^+/K^+ ratio in exercise-induced sweat were intermediate between those of heat and adrenaline-induced sweat.

These observations demonstrate that the physiology and biochemistry of sweating in the horse are very different from those of humans. The most important consequence from the point of view of the equine endurance athlete is the much greater degree of electrolyte loss sustained during exercise. Water loss during exercise in this study was similar to that described for marathon runners (Costill & Miller, 1980), but while these authors reported Cl^- and Na^+ deficits of 5 to 70%, deficits seen in this study were around six times greater. In spite of this, there were no medical problems among these horses which could be attributed to fluid/electrolyte disturbances.

The 25% increase in total plasma protein is indicative of severe dehydration, but despite this, plasma sodium and chloride concentrations changed by only 5 to 6%. Although the sweat was hypertonic for both sodium and chloride, plasma sodium concentration increased while plasma chloride concentration decreased. This indicates that the total loss which occurred, sweat plus respiratory water loss, was in effect slightly hypotonic for sodium and slightly hypertonic for chloride. Plasma potassium increased markedly between the resting sample (box) and the sample taken after a 1 km walk to the start. This was considered to be due to a flux of potassium out of the active muscle cells, and the sharp decrease after the finish may simply have been due to the reversal of that process. Any effect of sweat potassium loss was not readily appreciable against the background of these large alterations.

It appears that the horse's methods of dealing with the fluid/electrolyte losses consequent on sweating differ from those of man. In man electrolyte losses are kept to a minimum by the secretion of hypotonic sweat, which implies that if no water is drunk plasma electrolyte concentrations will tend to rise and if pure water is drunk they will tend to fall. In the horse the hypertonic sweat and the respiratory water loss combine to give as near to an isotonic loss of sodium and chloride is as possible, considering that these ions are equimolar in sweat but not in plasma. Thus a large volume loss may cause severe dehydration with comparatively little change in plasma electrolyte concentrations. It might be expected that the horse would be more susceptible than the human to hemodilution on drinking pure water, but this appears not to be so. In this study the horses did not drink during the ride and were allowed tap water ad libitum after the finish. No hemodilution was observed, with sodium and potassium returning to pre-ride levels while chloride remained constant. Total plasma protein concentration decreased only gradually, which suggests that the large volume of water drunk was not all absorbed immediately. In addition, the horse possesses a large cecum which normally contains semi-digested fermenting vegetable matter. It is possible that the cecum may contain some sodium and chloride which acts as a reservoir to prevent any serious fall in plasma concentrations during the recovery period. This hypothesis, however, requires further investigation.

It is difficult to make direct comparisons between the human and the horse in terms of fatigue, particularly in endurance exercise. This is mainly due to the rules of equine endurance events which require competitors to pass a rigorous veterinary inspection after finishing in order to be eligible for a prize. It does, however, appear that in British weather conditions, horses may be less liable to develop fluid/electrolyte problems than human endurance athletes.

Acknowledgment

This study was supported by a grant from the Horserace Betting Levy Board.

References

CARLSON, G.P. & P.O. Ocen. Composition of equine sweat following exercise in high environmental temperatures and in response to intravenous epinephrine administration. Journal of Equine Medicine and Surgery 3:27–31, 1979.

COSTILL, D.L. & J.M. Miller. Nutrition for endurance sport: carbohydrate and fluid balance. International Journal of Sports Medicine 1:2–14, 1980.

KERR, M.G. & D.H. Snow. The composition of equine sweat during prolonged adrenaline infusion, heat exposure and exercise. American Journal of Veterinary Research. (in press)

ROBERTSHAW, D. & C.R. Taylor. Sweat gland function of the donkey (Equus asinus). Journal of Physiology 205:79–89, 1969.

ROSE, R.J., K.S. Arnold, S. Church & R. Paris. Plasma and sweat electrolyte concentrations in the horse during long distance exercise. Equine Veterinary Journal 12:19–22, 1980.

SNOW, D.H., Identification of the receptor involved in adrenaline mediated sweating in the horse. Research in Veterinary Science 23:246–247, 1977.

SNOW, D.H., M.G. Kerr, M.A. Nimmo & E.M. Abbott. Alterations in blood, sweat, urine and muscle composition during prolonged exercise in the horse. Veterinary Record 110:357–354, 1982.

Electron Microprobe Analysis of Fatigued
Fast- and Slow-twitch Muscle

W.L. Sembrowich, D. Johnson, E. Wang and T.E. Hutchinson
University of Washington, Seattle, Washington, U.S.A., and
Cardiac Pacemakers, Inc., St. Paul, Minnesota, U.S.A.

Despite intense study over the last century, the process of muscular fatigue has never been elucidated. Studies on muscle mechanics, metabolism, biochemistry and electron microscopy have all failed to identify a cause of muscular fatigue. There has, however, been progress in many fields. Germane to the present investigations are studies which have demonstrated ultrastructural alterations in tissues of animals run to exhaustion (Gollnick & King, 1968) and elemental and ultrastructural changes in tetanized frog muscle (Gonzalez-Serratos et al., 1978; Somlyo et al., 1978). Such experiments suggest a relationship with muscular fatigue. In addition, some biochemical evidence demonstrates that functional changes occur in subcellular organelles. Moreover, fatigue resistant and easily fatigable muscle fibers co-exist in skeletal muscle which makes interpretation of results with whole muscles difficult. In the present experiments, we have studied the effects of electrical stimulation to fatigue on the intracellular elemental composition of fatigue resistant and rapidly fatiguable muscle.

Methods

Male Sprague-Dawley rats weighing approximately 350 g were used in these experiments. Following induction of anesthesia (pentobarbital, 6 mg/kg body weight, i.p.), the skin of the hindlimb was removed and the muscle exposed. Either the gastrocnemius or soleus muscle was isolated with its nerve and blood supply intact. Nerves to all other muscles were cut and a length of sciatic nerve cleared. The distal attachment was removed along with a chip of the calcaneus and secured by a silk ligature to a force transducer. The knee joint and lower leg were secured in a specially designed clamp (Dr. T.R. Nichols) and the entire lower leg and isolated muscle were lowered into a bath which contained circulating Krebs-Ringer phosphate buffer. Temperature was maintained at 37° C. Muscle length was adjusted with a micrometer until maximum twitch tension was obtained. The muscle was then stimulated at either 40 Hz (gastroc) or 20 Hz (soleus) for three 10 min bursts. A 30 s rest period was spaced between each stimulus train. Muscles were unable to generate tension by the end of the stimulation period. At the conclusions of stimulation, the level of the water bath was lowered and a small strip of muscle was cut in a direction parallel to the muscle fibers using fine dissecting scissors and a magnifying headset. The sample was taken as the muscle was contracting. Contralateral muscles served as controls and were handled in the same manner as experimental muscles except for the stimulation.

The muscle strips ($\approx$ 1 $\times$ 3 mm) were laid over stainless steel chucks and rapidly plunged into stirred Freon 22 kept at –156° C in a liquid nitrogen bath. Samples were stored under liquid nitrogen until sectioned on a cryoultramicrotome (Sorvall MT2-B) using a glass knife. Dry cut sections were transferred to sandwich grids precoated with formvar and carbon and freeze dried in an Edwards high vacuum evaporator. Freeze dried sections were examined in the electron microscope (JEOL-100C) using both conventional transmission and STEM modes. Energy dispersive X-ray analysis of various tissue compartments was performed with a 50 nm beam size using a Kevex 7000 quantitative analysis system.

Results

Electrical stimulation caused a decline in tension to nearly zero in both types of muscle used. Shifts in elemental concentrations occurred in all three tissue compartments examined (Table I). There was a loss of potassium (K) in each compartment as a result of muscular stimulation. This K loss ranged from 8% to 27%. Sodium (Na) generally tended to increase in fatigued muscles, while no clear patterns emerged for magnesium (Mg), phosphorous (P) and sulfur (S). Calcium (Ca) was elevated in the cytoplasm of fatigued soleus muscle by over 400%, while it declined by some 20% in gastrocnemius muscle. Similarly, Ca was elevated in soleus mitochondria by over 3 fold, while decreasing by 50% in gastrocnemius muscle. The most prominent change was in the Ca content of the SR. Tremendous increases were observed in the Ca content of SR of both soleus and gastrocnemius muscles at fatigue. Tissues examined were well preserved and allowed identification of major landmarks. Membrane components of SR are not always easy to identify in freeze dried cryosections and thus our estimates are based on the presumed location of SR because of its location relative to the T-system. In tissue sections from fatigued muscles, apparent swelling of T-tubules was observed. However, this seemed to be an exaggeration of the normal state of the muscle as we could observe similar T-tubule swelling in rested muscle tissue sections. The swelling was not as pronounced nor as widespread in rested tissue but could be observed in focalized areas.

Discussion

The elemental composition of the muscles examined is in reasonable agreement with published literature values on the rat (Campion, 1974; Drahota, 1961; Gonzalez-Serratos et al., 1978). However, direct comparisons are difficult since we have compartmentalized our analysis. Some trends were noted in control samples of soleus and gastrocnemius muscle. In the cytoplasm, Na, Mg and Ca concentrations tended to be lower in soleus muscle (slow twitch) than in gastrocnemius muscle (fast-twitch), while K concentrations tended to be higher in the slow-twitch muscle (Table I). These trends were not followed in other cellular compartments. In mitochondria, Na was considerably higher in slow-twitch muscle compared with fast-twitch muscle, but Ca levels were lower. The relatively high Ca level observed in the cytoplasm of gastrocnemius muscle (11.09 ± 2.16 μmoles/g dry weight) is difficult to explain. This may have been brought about by our choice of background model for the computer analysis of X-ray spectra. The detection of Ca is compounded by its overlap with the potassium K-beta peak. However, the presence of the potassium K-beta probably did not exert a major effect since potassium levels did not vary much while Ca levels tended to fluctuate a considerable amount. The physiological meaning of these differences in element distribution is not clear although they may be related to differences in resting membrane potential which depends on an ion distribution across the muscle membranes (Campion, 1974).

Electrical stimulation to fatigue resulted in large shifts of elements within muscle types and within all compartments. There was a loss of K in all compartments with fatigue, which completely agrees with earlier work (Gonzalez-Serratos et al.; Hnik, 1976; Sréter, 1963). This K loss is generally associated with exercise and fatigue. A low K level within muscle may be related to reduced contractility of the muscle (Sréter, 1963). The mechanism of this action is not known. Concomitant with the loss of K is a rise in Na. This phenomenon has also been noted previously (Sréter, 1963) although not in all cell compartments. The combination of a loss in K which is partly compensated by a gain in Na may have drastic effects on the

Table 1
Cytoplasm, SR and Mitochondrial Elemental Concentrations of Soleus and Gastrocnemius Muscles Following Electrical Stimulation to Fatigue

(μmoles/g dry weight)

| | | Soleus | | | | | | | Gastroc | | | | | |
	Na	Mg	P	S	Cl	K	Ca	Na	Mg	P	S	Cl	K	Ca
Cytoplasm (N = 6) Control	— —	50.71 ±31.14	253.20 ±20.07	197.81 ±24.38	110.90 ±5.49	500.60 ±31.42	1.23 ±1.23	83.08 ±43.43	61.58 ±4.07	224.70 ±22.87	129.37 ±4.99	99.54 ±27.12	427.10 ±7.35	11.09 ±2.16
Stim	118.1 ±12.95	40.75 ±3.17	78.62 ±6.66	115.83 ±7.31	164.20 ±5.86	355.10 ±10.87	5.86 ±0.75	113.83 ±28.75	51.71 ±4.10	194.54 ±13.01	124.76 ±13.13	120.78 ±8.41	394.98 ±15.61	26.78 ±3.79
SR (N = 6) Control	67.04 ±1.52	54.34 ±2.26	251.86 ±11.77	174.99 ±3.74	112.49 ±3.70	455.62 ±11.87	0.62	76.18	37.90 ±10.50	342.58 ±19.62	145.04 ±5.44	109.20 ±24.49	518.04 ±16.97	7.64 ±1.43
Stim	82.28 ±8.38	38.94 ±4.65	123.49 ±7.55	24.94 ±1.27	145.78 ±2.32	380.91 ±3.52	9.86 ±1.23	139.31 ±7.62	69.21 ±4.53	382.32 ±48.47	103.30 ±7.26	104.35 ±5.97	383.19 ±14.93	135.61 ±32.55
MITO (N = 10) Control	108.20 ±14.04	63.39 ±4.49	295.60 ±24.21	99.08 ±11.04	124.60 ±11.56	477.20 ±47.31	2.26 ±1.24	29.71 ±15.24	80.59 ±9.05	304.70 ±8.43	123.31 ±2.85	88.29 ±21.26	476.75 ±5.21	39.75 ±2.47
Stim	95.71 ±18.18	41.14 ±4.86	288.45 ±16.39	32.08 ±2.59	123.47 ±8.62	388.87 ±39.67	10.21 ±2.41	133.33 ±19.05	99.62 ±39.39	305.83 ±34.43	114.05 ±26.37	110.64 ±19.26	369.43 ±40.87	18.34 ±4.01

Values are the mean ± S.E.M.

membrane potential and may be related to the excitability of the sarcolemma. This intriguing problem requires further research work.

Perhaps one of the most striking observations we have made is the tremendous increase in the Ca content of the presumed SR at fatigue in fast-twitch muscle (over an order of magnitude). This elevated Ca can be visualized as calcium phosphate granules in an area of presumed SR. Such granules have been observed in mitochondria but have not been reported previously in presumed SR locations. The Ca content of SR in slow-twitch muscle was elevated by the same percentage, but the concentration levels were much lower. Ca was also elevated in the cytoplasm of both muscle types and in the mitochondria of soleus muscle. Elevated Ca levels in SR have been reported (Somlyo et al., 1978). The elevated Ca concentrations we report here for cytoplasm and mitochondria are new observations and suggest that at fatigue there is a general elevated Ca level throughout the various cell compartments. These high concentrations of Ca may be related to the fatigue state, although no mechanism is suggested at this time. Further research is indicated.

In summary, we have demonstrated the efficacy of using electron microprobe analysis to study muscular fatigue in at least two skeletal muscle types. The results reported are consistent with the concept that muscular fatigue is associated with variations and shifts in the intracellular elemental composition. The precise physiological consequences of these element shifts remains to be established.

Acknowledgment

This work was supported by N.I.H. Grant HL-25439 and N.S.F. Grant PCM 7921657.

References

CAMPION, D.S., Resting membrane potential and ionic distribution in fast and slow twitch mammalian muscle. J. Clin. Invest. 54:514–518, 1974.

DRAHOTA, Z., Ionic composition of various types of muscle in relation to their functional activity. In A. Kleinzeller & A. Kotyk (Eds.), Membrane Transport and Metabolism. London & New York: Academic Press, 1961.

GONZALEZ-SERRATOS, H., A.V. Somlyo, L.M. Borrero, G. McClellan, H. Shuman & A.P. Somlyo. The composition of vacuoles and sarcoplasmic reticulum in fatigued muscle: electron probe analysis. Proc. Nat. Acad. Sci. 75:1329–1333, 1978.

GOLLNICK, D.D. and D.W. King. Effect of exercise and training on mito-chondria of rat skeletal muscle. Am. J. Physiol. 216:1502–1509, 1969.

HNÍK, P., M. Holas, I. Krekule, N. Kríz, S. Mejsnar, V.S. Miesko, E. Ujec & F. Vyskocil. Work induced potassium changes in skeletal muscle and effluent venous blood assessed by liquid ion exchanger micro-electrodes. Pfluegers Arch. 362: 85–94, 1976.

SOMLYO, A.V., H. Gonzalez-Serratos, G. McClellan, H. Shuman, L.M. Borrero & A.P. Somlyo. Electron microprobe analysis of the sarcoplasmic reticulum and vacuolated t-tubule system of fatigued frog muscles. Ann. N.Y. Acad. Sci. 307:232–234, 1978.

SRÉTER, F.A., Cell water, sodium, and potassium in stimulated red and white mammalian muscles. Am. J. Physiol. 205:1295–1298, 1963.

Acid-Base Balance

Leg Muscle pH following Sprint Running

A. Katz, A. Barnett, D.L. Costill, W.J. Fink and R.L. Sharp

Ball State University, Muncie, Indiana, U.S.A.

The limiting factors responsible for fatigue in highly anaerobic exercise are not fully understood. One factor considered responsible for fatigue during such exercise is the accumulation of metabolites, primarily hydrogen ions and lactic acid, within the exercising muscle. It has been demonstrated that the rise in intramuscular hydrogen ion concentration (pH_m) inhibits both energy production, via glycolysis, and the normal contractile process (Sahlin, 1978). Most previous investigations concerning muscle metabolism during intense short term exercise have been limited to cycling exercise. Hence, the purpose of this study was to examine the effects of sprint type running (125% $\dot{V}O_{2max}$) on skeletal muscle pH. Specifically, the pH changes in two leg muscles, the gastrocnemius muscle (G), and the vastus lateralis muscle (V), have been compared in order to determine which muscle undergoes the greatest disturbance in pH.

Subjects

Six males whose average age, height, weight, and maximal oxygen uptake were 26.3 yrs, 181.3 cm, 71.1 kg, and 54.1 ml/kg/min, respectively, gave written consent to participate in the study, after being informed of the risks involved. Three of the subjects ran, on the average, more than 45 km/wk and were, therefore, classified as endurance trained (ET). The remaining subjects were classified as non endurance trained (NET), and had competed in sprint running (100 and 200 m events) within the past 5 years.

Methods

On the day of the trial, the subjects reported to the laboratory in the post absorptive state. Resting blood samples were obtained from an antecubital vein. Resting biopsy samples were obtained from the lateral aspects of the G and V muscles. The subjects then performed a treadmill run at a speed equivalent to 125% of their aerobic capacity. Exercise was terminated when the subjects expressed an inability to maintain the predetermined speed. Immediately following the run, the subjects were placed in the supine position, in which they remained throughout the 30 min recovery period. Biopsy samples were again obtained from the G and V muscles. The time delay between the end of exercise and freezing of the biopsy samples averaged 27 s (range = 16 to 50 s). Previous studies from this laboratory have demonstrated that pH_m and muscle lactate (LA_m) remain unchanged for 45 to 60 s following such an exercise bout (Costill et al., 1982). Respiratory exchange values were collected throughout 30 min of recovery. Blood samples were taken 1, 3, 5, 7, 9, 11, 13, 15, 20 and 30 min post exercise. Several days later four of the subjects performed a 400 m run on an outdoor track. Immediately following the run, a biopsy sample was obtained from the lateral aspect of the G muscle. A blood sample was taken 5 min after the run.

All blood samples were analyzed for lactate (LA_b) and pH (pH_b), as were the muscle samples. pH_m was determined by the homogenate technique, as described elsewhere (Costill et al., 1982). pH was measured at $37°$ C using a Radiometer BMS MK.2 Blood Micro System and PH M73 pH/Blood Gas Monitor. Student's paired t-test was used to determine statistically significant differences between values for the G and V muscles.

Results

Subgroup mean values and overall mean ($\pm$ SE) values for pH_m and LA_m for both G and V muscles measured prior to and following the treadmill run are shown in Table 1. Mean ($\pm$ SE) values for resting muscle pH and LA were 7.03 ($\pm$ 0.02) and 1.5 ($\pm$ 0.1) mmol/kg ww (wet weight), and 7.04 ($\pm$ 0.01) and 1.3 ($\pm$ 0.01) mmol/kg ww for the G and V muscles, respectively. After the treadmill run, pH_m decreased to 6.88 ($\pm$ 0.05) and 6.86 ($\pm$ 0.03), while LA_m increased to 12.5 ($\pm$ 0.4) and 10.8 ($\pm$ 0.4) mmol/kg ww for the G and V muscles, respectively. There were no significant differences between the muscles in either pH or LA before or after the treadmill run ($p > 0.05$). However, the ET group had lower LA levels (7.6 vs 14.0 mmol/kg ww) and higher pH values (6.91 vs 6.80) in the V muscle following the treadmill run than the NET group.

Table 1

Pre and Post Treadmill Run Values for pH_m and LA_m

	Pre pH_m		Post pH_m		Pre LA_m (mmol/kg ww)		Post LA_m (mmol/kg ww)	
	G	V	G	V	G	V	G	V
ET (n = 3)	7.01	7.04	6.84	6.91	1.2	1.0	12.5	7.6
NET (n = 3)	7.05	7.04	6.92	6.80	1.7	1.5	12.5	14.0
$\overline{X}$ (n = 6)	7.03	7.04	6.88	6.86	1.5	1.3	12.5	10.8
± SE	0.02	0.01	0.05	0.03	0.0	0.0	0.4	0.4

The average (± SE) pH_b at rest was 7.35 (± 0.01), reaching a low of 7.10 (± 0.03) 5 min into recovery. Three subjects demonstrated lowest pH_b values at 1 min post exercise, whereas the other three subjects reached lowest values 7 to 9 min after exercise. Generally, the ET group reached their lowest pH_b values earlier than the NET group, while, however, demonstrating higher mean values throughout recovery.

LA_b (1.6 mM at rest), 5 min after the treadmill sprint run, averaged 11.3 mM. Throughout the 30 min recovery period, the ET group showed lower LA_b levels than the NET group. This is consistent with the differences observed between the groups' LA levels in the V muscle after the treadmill run. In addition, peak LA_b levels occurred at 3 min post exercise for the ET group, while occurring at 11 min post exercise for the NET group.

Since no statistically significant difference between the G and V muscles was found regarding pH changes following the treadmill sprint run, the gastrocnemius was chosen as the biopsy site for the 400 m run. Pre-exercise values for pH_m, LA_m, pH_b and LA_b were assumed to be similar to the values obtained prior to the treadmill run. Following the 400 m run, values (mean ± SE) for pH_m, LA_m, pH_b and LA_b were 6.63 ± 0.03, 19.7 ± 0.1 mmol/kg ww, 7.10 ± 0.03 and 12.3 ± 0.8 mmol/liter, respectively. There appeared to be no consistent relationship between sprinting ability and pH_m values.

Recovery $\dot{V}O_2$, following the treadmill run, averaged 234 and 237 ml/kg/30 min for the ET and NET groups, respectively. Thus, under the circumstances, there appears to be little relationship between oxygen uptake during recovery and LA or pH in either muscle or blood.

Discussion

There are conflicting data regarding the pH of resting muscle (Roos & Boron, 1981). This is probably due to variations in the technique of measuring pH_m. Resting pH_m values have been reported to be in the range of 6.8 to 7.1 (Sahlin, 1978). Thus, mean values for resting pH_m in the present report (7.03 to 7.04) are consistent with previously published findings.

The present study attempted to examine the pH_m changes incurred by sprint running. Termination of the treadmill sprint run occurred when the subjects demonstrated an inability to continue running at the predetermined speed. Although the subjects were fatigued at this point, they were by no means exhausted. Even in the 400 m trial, the state of exhaustion was not reached. In sprint running (e.g., a competitive 400 m run) the participant stops running, because the finish line has been reached or due to relative fatigue. Regardless of the reason, the state of exhaustion is usually not reached. In sprint running (e.g., a competitive 400 m run), the participant stops running because the finish line has been reached or due to supported, the exercise can be continued to exhaustion without fear of bodily injury. This may explain the discrepancy between the average pH_m values reported in the present study, 6.88 (G) and 6.86 (V) following the treadmill run, and 6.63 (G) following the 400 m run and those reported by Hermansen and Osnes (1972), who exercised cyclists to exhaustion (mean ± SE value for pH of V muscle was 6.41 ± 0.04).

The changes in pH_m and LA_m in the G muscle following the treadmill run were not significantly different ($P > 0.05$) from those observed in the V muscle. However, there seemed to be a difference between the ET and NET subjects, regarding the acidification of the G and V muscles. It was our impression that the ET runners tended towards greater decreases in the pH of the G muscle, whereas the NET group showed larger pH changes in the V muscle. Generally, distance runners have a relatively short stride, which tends to stress the lower leg muscles more than the thigh muscles (Costill et al., 1974). In addition, it has been shown that distance runners accumulate less LA_b than sprinters and middle distance runners during maximal treadmill tests (unpublished findings). The data from the present study are in agreement with the idea that distance runners accumulate less LA in muscles that are employed in running. Hence, the differences in pH_b and LA_b values following the treadmill sprint run are likely due to the higher pH and lower LA values in the V muscle of the ET group. Thus, the findings of this study seem to indicate differences in the metabolic and biomechanical involvement of the G and V muscles between the two groups.

The differences in the time at which peak LA_b levels were observed between the two groups during the recovery period (3 min for the ET group vs 11 min for the NET group) may have been due to lesser LA accumulation in the thigh muscles of the ET runners, assuming that the V muscle is indicative of the metabolic state of other active muscles in the thigh. An additional factor that should be considered is the fiber type distribution of the subjects. It has been suggested that muscles rich in high oxidative fibers (ST and/or FTa) are capable of greater release, probably due to a more developed capillary network (Andersen, 1975). Moreover, the capillary supply of skeletal muscle fibers in ET men may be significantly greater than in untrained men (Brodal et al., 1977). In the present study, mean ($\pm$ SE) values for % ST fibers in the G and V muscles of the ET group were 66 $\pm$ 11% and 59 $\pm$ 3%, compared to 39 $\pm$ 5% and 33 $\pm$ 4% in the G and V muscles of the NET group, thus, lending support to the possibility of a fiber type influence.

The results of this study indicate that either the G or V muscle can be used as an appropriate biopsy site for studying pH_m in sprint running. However, although this may seem to be the case in general, we feel that the training regimen (endurance vs sprint) need also be considered.

References

ANDERSEN, P., Capillary density in skeletal muscle of man. Acta Physiol. Scand. 95:203–205, 1975.

BRODAL, P., F. Inger & L. Hermansen. Capillary supply of skeletal muscle fibers in untrained and endurance-trained men. Am. J. Physiol. 232(6):705–712, 1977.

COSTILL, D.L., E. Jansson, P.D. Gollnick & B. Saltin. Glycogen utilization in leg muscles of man during level and uphill running. Acta Physiol. Scand. 91:475–481, 1974.

COSTILL, D.L., R.L. Sharp, W.J. Fink & A. Katz. Determination of human muscle pH in needle biopsy specimens. J. Appl. Physiol.: Respirat. Environ. Exercise Physiol. (in press)

HERMANSEN, L. & J.B. Osnes. Blood and muscle pH after maximal exercise in man. J. Appl. Physiol. 32:304–308, 1972.

ROOS, A. & W.F. Boron. Intracellular pH. Physiol. Rev. 61(2):296–434, 1981.

SAHLIN, K., Intracellular pH and energy metabolism in skeletal muscle of man. Acta Physiol. Scand. Suppl. 455, 1978.

Skeletal Muscle Buffering Capacity in Elite Athletes

D.C. McKenzie, W.S. Parkhouse, E.C. Rhodes,
P.W. Hochochka, W.K. Ovalle, T.P. Mommsen
and S.L. Shinn
The University of British Columbia, Vancouver, B.C., Canada

During intense physical activity, lactate accumulates within the muscle; hydrogen ions are formed in an equivalent amount to lactate and the intracellular pH decreases. The alteration in skeletal muscle pH in response to an anaerobic task is profound. Hermansen and Osnes (1972) and Sahlin et al., (1976) have reported values between 6.4 and 6.6 at exhaustion. This acidosis in the muscle has been thought to be an important factor in the development of fatigue; lowering cellular pH can inhibit the glycolytic regulatory enzymes as well as induce changes in the muscle's contractile mechanism (Fuchs et al., 1970; Sutton et al., 1981).

Anaerobically trained athletes demonstrate an enhanced capability to perform high intensity work with resultant large increases in lactate. The ability of the skeletal muscle to resist changes in pH associated with intense activity may be a significant factor in the limitation of anaerobic performance. Therefore, this study was designed to investigate the skeletal muscle buffering capacity (β) of the elite aerobic and anaerobically trained athletes.

Methods

Three groups of five subjects volunteered for the study: marathon runners, 800m runners, and untrained controls. All subjects were informed of the procedures and risks involved and informed consent was obtained. Physical characteristics were recorded. A progressive treadmill run to fatigue was used to determine $\dot{V}_{O_2max}$; the initial treadmill velocity was 2.22 m s^{-1} increasing by 0.22 m s^{-1} each min. Expired gases were sampled and analyzed by a Beckman Metabolic Measurement Cart interfaced into a Hewlett Packard Data Acquisition System which determined respiratory exchange variables every 15 s. $\dot{V}O_{2max}$ was determined by the mean of the four highest consecutive 15 s values. Anaerobic performance was evaluated by the Anaerobic Speed Test (Cunningham & Faulkner, 1969). The subjects ran on a treadmill at 3.0 m s^{-1}, 20° incline, with time to fatigue as the dependent variable.

Needle biopsies were obtained at rest from the vastus lateralis muscle (Bergstrom, 1962). Samples for histochemical analysis were immediately placed in isopentane immersed in liquid nitrogen. Another sample for biochemical determination was frozen in liquid nitrogen. The myosin ATPase method (Dubowitz & Brooke, 1973) was used for muscle fiber classification. To determine β, the sample was homogenized at 25° C in 10 volumes of a salt solution containing (in mM) 145 KCl, 10 NaCl and 5 iodoacetic acid. This was deproteinized with the addition of 3% solid sulfosalicylic acid. β was determined by a modification of the method of Davey (1960). Supernatant extracts were adjusted by pH 7.00 ± 0.05 with 0.1 N NaOH. 100 μl aliquots were titrated to pH 6.00 ± 0.05 with 0.01 N HCl. Measurements were made at 38° C with microelectrode equipment (MI 410, Microelectrodes Inc.). β was expressed in μ moles HCl g^{-1} pH^{-1} ww (wet weight). Analysis of variance and post hoc comparison analysis were used to determine significant differences (p < 0.05).

Results

The means and standard deviation for each physiological and biochemical parameter investigated are presented in Table 1. The marathon runners were significantly (p < 0.05) older than the 800 m runners or the control group. The control group was heavier with a greater percent body fat and reduced $\dot{V}O_{2max}$. Time to fatigue, anaerobic speed test (AST) was significantly different between all groups (p < 0.05). The marathon runners exhibited a significant difference in percent fast-twitch fibers. There were no differences in β of the controls and marathon runners. However, β of the 800 m runners was significantly elevated (p < 0.01).

Discussion

The marathon and 800 m runners were highly trained competitive athletes with physiological profiles comparable to similar trained subjects. The major finding of this study is the significant increase in β in the 800 m runners with no significant difference between the untrained subjects and

Table 1

Physical Characteristics, Aerobic and Anaerobic Parameters, Percent Fast-twitch Fibers

Group	Age (yrs)	Height (cm)	Weight (kg)	Body Fat %	$\dot{V}_{O_{2}max}$ (ml mg^{-1} min^{-1})	AST (s)	Fast-twitch Fibers %	β (μmol g^{-1} pH^{-1})
800 meter runners	20.6 ± 2.3	180.6 ± 4.9	68.6 ± 3.4	6.5 ± 3.4	63.2 ± 3.1	115[c] ± 18	56.6 ± 7.0	30.0[b] ± 5.6
Marathon runners	35.2[a] ± 7.3	175.2 ± 2.9	71.4 ± 3.8	10.1 ± 3.3	59.1 ± 5.5	54 ± 15	26.6[b] ± 7.1	20.4 ± 4.1
Untrained	22.6 ± 0.9	182.4 ± 4.3	81.7[a] ± 5.1	21.1[b] ± 4.9	46.9[a] ± 3.3	38 ± 9	50.6 ± 9.9	21.3 ± 5.0

(mean ± S.D.)

[a] p < 0.05 vs other groups
[b] p < 0.01 vs other groups
[c] p < 0.05 all groups different

marathon runners. This is descriptive data and although it is not possible to state that the changes in β are a function of anaerobic training it is tempting to speculate that, as β is largely a biochemical parameter, there may be room for adaptation on the basis of repetitive anaerobic work. Castellini and Somero (1981) have examined β in several species and report a significant relationship between β and the muscles glycolytic potential. The present investigation also reflects a significant positive correlation between anaerobic performance, as measured by the AST, and β (Figure 1). As the fast-twitch fiber is considered to have a higher capacity for anaerobic work it is not surprising to find the positive relationship between % FT and β (Figure 2).

The ability of the skeletal muscle to resist the changes in pH is dependent upon several complex buffering processes. Siesjo and Messeter (1971) suggest that three important factors determine the pH in a living tissue during acid-base changes: 1. physico-chemical buffering, 2. consumption or production of non-volatile acids, and 3. transmembrane fluxes of H^+ or HCO_3^- ions. The contribution of the dipeptide carnosine to the physico-chemical buffering mechanism forms the basis of another paper presented at this symposium.

The values of β in this study are comparable to those reported by Davey (1966) on deproteinized rabbit psoas muscle. The contribution of proteins to cell buffering is important and deproteinizing the homogenate results in a significant reduction in β. Hultman and Sahlin (1980) have calculated that proteins are responsible for approximately 50% of the physico-chemical buffering during exhaustive bicycle exercise.

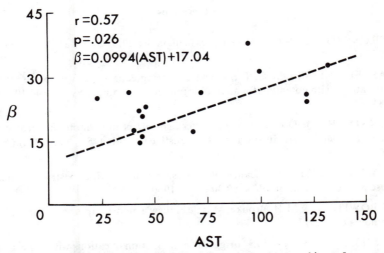

Figure 1—The relationship between buffering capacity and anaerobic performance as determined by the anaerobic speed test (AST).

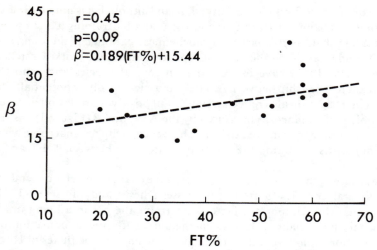

Figure 2—The relationship between buffering capacity and percentage fast-twitch fibers (% FT).

In summary, β was significantly elevated in highly trained anaerobic athletes in comparison to elite aerobic athletes and controls. There was a significant relationship between β and an anaerobic performance test (AST) and the % FT fibers. Further research is necessary to determine if the increase in β is a result of repetitive anaerobic work.

References

BERGSTROM, J., Muscle electrolytes in man. Scand. J. Clin. Lab. Invest. Suppl. 68, 1962.

CASTELLINI, M.A. & G.N. Somero. Buffering capacity of vertebrate muscle: correlations with potentials for anaerobic function. J. Comp. Physiol. 143:191–198, 1981.

CUNNINGHAM, D.A. & J.A. Faulkner. The effect of training on aerobic and anaerobic metabolism during a short exhaustive run. Med. Sci. Sport 1(2):65–69, 1969.

DAVEY, C.L., The significance of carnosine and anserine in striated skeletal muscle. Arch. Bioch. Biophys. 89:303–308, 1960.

DUBOWITZ, V. & M.H. Brooke. Muscle biopsy: a modern approach. Vol. 2. W.B. Saunders Co., 1973.

FUCHS, F., Y. Reddy & F.N. Briggs. The interaction of cations with the calcium-binding site of troponin. Biochem. Biophys. Aeta. 221:407–409, 1970.

HERMANSEN, L. & J.B. Osnes. Blood and muscle pH after maximal exercise in man. J. Appl. Physiol. 32:304–308, 1972.

HULTMAN, E. & K. Sahlin. Acid-base balance during exercise. Exer. Sp. Sci. Rev. 7:41–128, 1980.

SAHLIN, K., R.C. Harris, B. Nylind & E. Hultman. Lactate content and pH in muscle samples obtained after dynamic exercise. Pflugers Archiv. 367:143–149, 1976.

SIESJO, B.K. & K. Messeter. Factors determining intracellular pH. In B.K. Siesjo and S.C. Sorenson (Eds.), Ion Homeostasis of the Brain. New York: Academic Press, pp. 244–262, 1971.

SUTTON, J.R., N.L. Jones & C.J. Toews. Effect of pH on muscle glycolysis during exercise. Clin. Sci. 61:331–338, 1981.

The Relationship between Carnosine Levels, Buffering Capacity, Fiber Type and Anaerobic Capacity in Elite Athletes

W.S. Parkhouse, D.C. McKenzie, P.W. Hochochka,
T.P. Mommsen, W.K. Ovalle, S.L. Shinn and E.C. Rhodes
The University of British Columbia, Vancouver, B.C., Canada

Sprint trained athletes, as a group, demonstrate a remarkable ability to perform high intensity, short duration exercise. The energy requirements are met principally by anaerobic glycolysis resulting in a rapid production of ATP, an accumulation of protons, and an intracellular pH decrement. A reduction in pH, if of sufficient magnitude, has been associated with reduced rates of glycolysis (Roos & Boron, 1981), correlations with fatigue (Fitts & Holloszy, 1976; Stevens, 1980), inverse relationships between force generation and proton accumulation (Dawson et al., 1978) and inhibition of the contractile mechanism (Katz, 1970). During a short exhaustive exercise bout, intramuscular pH was reported to fall to approximately 6.60 at fatigue resulting in a proton production of approximately 3.5 umoles 1^{-1} ICH_2O (Hultman & Sahlin, 1980). This would decrease pH to approximately 1.5 in an unbuffered solution, thus emphasizing the physiological importance of the skeletal muscle buffer capacity (β). The buffering ability of a tissue in relation to its glycolytic capacity has been investigated, demonstrating a high correlation (Castellini & Somero, 1981), but until recently no investigations have examined the ability of human skeletal muscle to buffer in relation to training specificity and fiber composition. The most important buffers within skeletal muscle were identified by Somero (1981) and Burton (1978) as: the protein bound histidyl residues, histidine-containing dipeptides and free histidine. Davey (1960) suggested that the histidine-containing depiptide carnosine, found principally within fast-twitch fibers (Tamaki et al., 1976) and in much greater concentrations than histidine (Bergstrom et al., 1978) could contribute up to 40% of total buffering. Relatively little information on carnosine levels within human skeletal muscle exists and no attempts to equate carnosine levels with training specificity have been made. Although many physiological roles for carnosine have been proposed, the only universally accepted function is that of a physiological buffer. Therefore, the purpose of the present investigation was to examine the relationships between fast-twitch fiber percentage, anaerobic performance, buffer capacity and carnosine levels within human skeletal muscle subjected to different training programs.

Methods

Eighteen men comprising four groups (S-sprinters; R-rowers; M-marathoners; UT-untrained) volunteered for the investigation and were made aware of the risks involved. Anthropometric and physical characteristics and pulmonary function were determined using standard techniques. Maximal oxygen uptake ($\dot{V}O_{2max}$) was assessed by a continuous treadmill test initiated at 2.22 m s^{-1} with increments in speed of 0.22 m s^{-1} every min until fatigue. Expired gases were continuously sampled and analyzed (Beckman Metabolic Measurement Chart) with $\dot{V}O_{2max}$ determined by averaging the four highest consecutive 15 s oxygen uptake values. Anaerobic performance was assessed by the Anaerobic Speed Test (AST) of Cunningham and Faulkner (1969), employing time to fatigue as the performance index. Needle biopsies were obtained from the vastus lateralis muscle at rest by the technique of Bergstrom et al. (1962). Fast-twitch percentage (FT %) was determined by staining intensity for myosin ATPase (pH 4.6). Buffer capacity (β) was determined at 25° C by a modification of the method of Davey (1960) involving homogenization of the sample in 10 volumes salt solution (145 mM KCl, 10 mM Nacl, 5 mM IAA) and deproteinization with a 3% solid sulfosalicylic acid. pH adjusted supernatant extracts were repeatedly titrated over the pH range of 7.0 to 6.0 with 0.01N HCl. One hundred μl aliquots of the supernatant were utilized for determination of carnosine levels, using an amino acid autoanalyzer (Beckman 118C) with a single column lithium hydroxide buffer system and manual integration of the area under the curve. Univariate comparisons were performed on all the variables. Post-hoc Scheffe Tests were performed on the variables demonstrating significant omnibus F values. Regression equations were calculated for the relationships between AST, FT %, β, and carnosine.

Table 1

Physical Characteristic, Physiological, Histochemical and Biochemical Analyses

Group	N	Age (yrs)	Height (cm)	Weight (kg)	Percent Body Fat	Maximal Oxygen Uptake (ml kg^{-1} min^{-1})	Anaerobic Speed Test (s)	Fast Twitch Percent	Buffer Capacity (μmol g^{-1} pH^{-1})	Carnosine (μmol g^{-1})
S	5	20.6 ± 2.3	180.6 ± 4.8	68.6 ± 3.4	6.5 ± 3.4	63.2 ± 3.1	115[b] ± 18	56.6 ± 7.0	30.0[c] ± 5.6	4.9[c] ± 0.8
R	4	21.0 ± 1.8	177.1 ± 3.9	70.4 ± 3.1	8.2 ± 3.3	61.8 ± 1.2	73[c] ± 5	45.8 ± 7.5	28.6[c] ± 2.1	5.0[c] ± 0.8
M	5	37.8[a] ± 9.3	176.5 ± 4.2	70.0 ± 4.4	10.6 ± 4.0	60.1 ± 4.2	53 ± 15	33.0[a] ± 12.2	20.8 ± 4.4	2.8 ± 0.7
UT	4	22.3 ± 0.5	183.9 ± 3.1	83.7[a] ± 3.1	22.3[a] ± 4.8	46.2[a] ± 3.3	37 ± 10	50.5 ± 11.3	22.3 ± 5.1	3.8 ± 1.0

(mean ± S.D.)

[a] $p < 0.05$ significantly different from other 3 groups
[b] $p < 0.01$ significantly greater than all other groups
[c] $p < 0.05$ significantly greater than M and UT groups

Results and Discussion

The major findings of the present investigation were that: first, carnosine levels were significantly related to β (r = 0.69) and anaerobic performance (r = 0.60); second, carnosine levels were correlated with FT %(r = 0.44) and third, increased carnosine levels, β, and anaerobic performances were exhibited by the anaerobically trained groups (Table 1). Thus, it appears that the biochemical parameters, β and carnosine levels, may either be augmented by training specificity or be an inherent factor influencing the athlete's choice of activity. In agreement with the findings of Castellini and Somero (1981), elevated β values were found in the tissues of those subjects demonstrating enhanced anaerobic capacities which suggests that the ability of skeletal muscle to sequestor protons may enhance anaerobic performance. The suggestion by Somero (1981) and Burton (1978) that the imidazole containing compounds were important contributors to β was substantiated for human skeletal muscle; carnosine accounted for 48% of the variance in β. The relationship between carnosine levels and FT % agrees with the finding of Tamaki et al. (1976) who reported enhanced carnosine levels within fast-twitch fibers of animal tissues. Thus, within human skeletal muscle, carnosine levels may be related to the glycolytic capacity of muscle contributing substantially to β. It appears, therefore, that the ability of skeletal muscle to sequester the protons which accumulate during anaerobic glycolysis may augment anaerobic performance. In conclusion, within human skeletal muscle, carnosine appears to be related to β, anaerobic performance and FT %, suggesting a relationship to glycolytic capacity and training specificity. Thus, the ability of carnosine to act as a physiological buffer may influence anaerobic performance.

References

BERGSTROM, J., Muscle electrolytes in man. Scand. J. Clin. Lab. Invest. Suppl. 68, 1962.

BERGSTROM, J., P. Furst, L-O. Noree & E. Vinnars. Intracellular free amino acids in muscle tissue of patients with chronic uraemia: effect of peritoneal dialysis and infusion of essential amino acids. Clin. Sci. Mol. Med. 54:51–60, 1978.

BURTON, R.F., Intracellular buffering. Resp. Physiol. 33:51–58, 1978.

CASTELLINI, M.A. & G.N. Somero. Buffering capacity of vertebrate muscle: correlations with potentials for anaerobic function. J. Comp. Physiol. 143: 191–198, 1981.

CUNNINGHAM, D.A. & J.A. Faulkner. The effect of training on aerobic and anaerobic metabolism during a short exhaustive run. Med. Sci. Sp. 1(2):65–69, 1969.

DAVEY, C.L., The significance of carnosine and anserine in striated skeletal muscle. Arch. Bioch. Biophys. 89:303–308, 1960.

DAWSON, M.J., D.C. Grandian & D.R. Wilkie. Muscular fatigue investigated by phosphorus nuclear magnetic resonance. Nature 274:861, 1978.

FITTS, R.H. & J.O. Hollaszy. Lactate and contractile force in frog muscle during development of fatigue and recovery. Am. J. Physiol. 231:430–433, 1976.

HULTMAN, E. & K. Sahlin. Acid-base balance during exercise. Exer. Sp. Sci. Rev. 7:41–128, 1980.

KATZ, A.M., Contractile proteins of the heart. Physiol. Rev. 50:63–158, 1970.

ROOS, A. & W.F. Boron. Intracellular pH. Physiol. Rev. 61(2):296–434, 1981.

STEVENS, E.D., Effect of pH on muscle fatigue in isolated frog sartorious muscle. Can. J. Physiol. Pharm. 58:568–570, 1980.

TAMAKI, N., M. Nakamura, M. Harada, K. Kimura, H. Kawano & T. Hama. Anserine and carnosine contents in muscular tissue of rat and rabbit. J. Nutr. Sci. Vitaminol. 23:213–219, 1977.

Buffer Capacity of Blood in Trained and Untrained Males

R.L. Sharp, L.E. Armstrong, D.S. King and
D.L. Costill

Ball State University, Muncie, Indiana, U.S.A.

It is known that endurance training results in an improved ability to delay lactic acid (LA) accumulation in blood during progressive exercise, so that for a given submaximal workload the untrained individual will show a greater blood LA concentration than the endurance trained individual (Costill, 1973). It is not known, however, if the ability of blood to buffer LA is related to training. Early studies found no difference in standard bicarbonate ($[HCO^3]$ at actual pH and $PCO^2 = 40$ mmHg) of blood between trained and untrained subjects (Robinson, 1941; Steinhaus, 1933) and took this as evidence that buffer capacity (BC) of blood was not improved by training. Since bicarbonate is but one of several buffers of the blood and since it is regulated within rather narrow limits by the kidneys and respiratory system, standard bicarbonate may not be a valid measure of the functional ability of blood to buffer metabolic acids.

The purpose of the present study was to determine if a difference in BC exists between trained and untrained subjects, using as an index of BC, an in vivo exercise titration of blood.

Subjects

Six endurance trained male cyclists (ET) and 6 untrained (UT) males volunteered for this study. The subjects were informed of the purpose, benefits and risks of the study prior to giving their written consent to participate. The grouping of subjects was determined by placing the subjects in the group classification that corresponded to their daily physical training habits. The fact that the mean $\dot{V}o_2max$ (ml/kg) was significantly different between the groups, indicates that, in fact, the two groups were sampled from different populations (ET: 65.74 ± 3.25 ml/kg; UT: 47.96 ± 1.70 ml/kg).

Methods

The exercise test consisted of cycling on a Collins electrically braked ergometer at 90 rpm starting at 25 W and increasing power by 25 W each min until voluntary exhaustion. Blood samples were taken from a forearm vein at rest, during the last 5 s of each load and each min of static recovery for 5 min. The resting blood sample was analyzed for hemoglobin, hematocrit, total protein (Biuret), LA and acid-base status. Blood samples taken during exercise and recovery were analyzed for LA and acid-base status. The acid-base variables were measured at 37° C using a Radiometer BMS 3 MK.2 Blood Micro System and PH M73 pH/Blood Gas Monitor.

The BC of blood in these subjects was expressed in two ways: with respiratory compensation (BC1) and without respiratory compensation (BC2). BC1 was determined by using a linear regression equation with LA as the independent variable and pH as the dependent variable. This regression equation was constructed for each subject using the values obtained from samples taken at each load and during recovery. From each subject's equation, the amount of LA required to cause a change in pH of 1.0 units was calculated. The units of BC were slykes (Woodbury, 1965) and were expressed as mmols/liter/pH unit. BC2 was determined in the same manner but PCO_2 was included in the regression equation as another independent variable (Davenport, 1974). This allowed calculation of BC while statistically holding PCO_2 constant at 40 mmHg. By thus holding PCO_2 at a constant value, the effect of LA on blood pH without intervention of respiratory compensation (i.e., lowered PCO_2 in response to metabolic acidosis) was calculated.

Differences between groups were tested for significance using t-tests for hemoglobin, hematocrit, VO_2max, total protein and resting bicarbonate. Variables collected during exercise and recovery were analyzed for differences using a two-way analysis of variance (ANOVA) with group and load as the two factors. The level of probability considered sufficient to reject the null hypothesis was set at $P < 0.05$. Means are reported ± SE.

Results

There were no significant differences between groups in blood protein concentration (ET: 7.56 ± 0.25 g%; UT: 8.13 ± 0.41 g%), hematocrit (ET:

44.47 ± 1.01 %; UT: 45.75 ± 1.05 %) or resting bicarbonate (ET: 25.46 ± 0.65 mEq/L; UT: 26.08 ± 0.71 mEq/L). The UT group did, however, have a significantly higher hemoglobin concentration than the ET group (ET: 15.02 ± 0.22 g/100 ml; UT: 16.05 ± 0.15 g/100 ml).

No significant differences were observed in resting LA, resting blood pH, or any of the other acid-base variables between the groups. During the exercise there was no significant difference in LA values between groups until 80% Vo_2max (ET: 2.82 ± 0.29 mM; UT: 4.27 ± 0.40 mM). This was followed by an abrupt increase in LA concentration for the ET group such that the difference in absolute concentration was not significant at 90% (ET: 4.15 ± 0.36 mM; UT: 5.10 ± 0.40 mM) but was significantly higher in ET from 100% of Vo_2max through 4 min post exercise. The first significant deviation from resting LA occurred at 80% $\dot{V}o_{2max}$ in the ET group, whereas the break from resting LA level in the UT group occurred at 70%. It is interesting to note that the rate of increase in LA following this breakpoint was significantly greater in the ET than in the UT group when expressed as ΔLA/10% increase in $\dot{V}o_2$ (ET: 2.88 ± 0.46 mM/10%; UT: 1.05 ± 0.13 mM/10%).

The blood pH values followed the same pattern of decrease during exercise as did the increase in LA. During post exercise, however, the blood pH showed a different response between the groups. In the UT group, pH continued a steady drop in proportion to the amount of LA increase, while in the ET group, pH began to increase toward normal resting values from 1 min to 5 min post exercise.

The PCO_2 values of blood during exercise were not significantly different between the groups (Figure 1). All values obtained during post exercise, however, were significantly lower in the ET group. In both groups PCO_2 dropped significantly from 100% $\dot{V}o_{2max}$ to 5 min post exercise. This decrease in PCO_2 was significantly greater in the ET than in the UT group.

BC during exercise with respiratory compensation (BC1) was calculated for each subject based on the relationship of LA and pH observed at each load. This relationship was linear with correlation coefficients ranging from -0.82 to -0.98 in UT and -0.85 to -0.99 in ET. From each subject's regression equation, the amount of change in LA required to cause a pH drop of 1.0 units was calculated (slykes) and group means were determined. No significant difference between groups was found (ET: 47.17 ± 6.77 slykes; UT: 55.40 ± 6.28 slykes).

BC during exercise without respiratory compensation (BC2) was calculated by including PCO_2 as an independent variable in the regression equation. The multiple correlation coefficients were improved by this procedure and ranged from -0.94 to -0.98 in UT and -0.96 to -1.00 in ET. Slykes were thus determined based on the amount of LA required to cause a pH drop of 1.0 units while PCO_2 was held constant at 40 mmHg. Again,

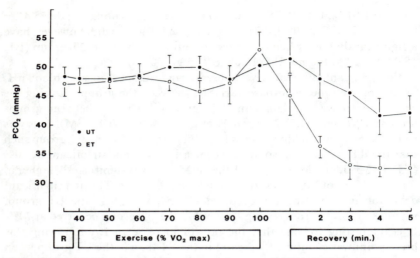

Figure 1—Blood PCO_2 at rest, during exercise and recovery in ET and UT groups. Means are reported ± SE.

no significant difference was found between groups in BC2 (ET: 56.61 ± 5.97 slykes; UT; 57.80 ± 9.83 slykes).

BC1 and BC2 were also determined for the post exercise period. BC1 was significantly different between groups (ET: -9.71 ± 16.85 slykes; UT: 641.29 ± 538.70 slykes). The negative value of slykes for the ET group represents a greater BC than the high positive value for the UT group since, in the ET group, the pH increased in spite of an increasing LA concentration. This resulted in a negative value for slykes ($\Delta LA/\Delta pH$), which actually represents a greater BC. This effect, which is opposite from what one would expect from increased LA, could be due either to effective respiratory compensation on the part of the ET group or to metabolic buffering of the LA. The former seems to be the case since BC2, which factors out the effect of respiratory compensation, was found to be not significantly different between the groups (ET: 46.36 ± 5.60 slykes; UT: 55.52 ± 13.21 slykes).

Discussion

The results of this study indicate that there is no difference in the blood's ability to buffer LA during exercise in endurance trained and untrained subjects. This finding is supported by the observation that there was no difference in concentration of actual bicarbonate or protein, both of which are primary buffers of blood. The third major buffer of blood, hemoglobin, was higher in the untrained subjects but was not, apparently, high enough to allow any difference in buffer capacity. This difference is likely due to a plasma volume expansion in the ET group, thereby causing a slight hemodilution effect.

The higher BC1 of the ET group during the post exercise period can be accounted for by a significant respiratory compensation. This is indicated by the fact that when PCO_2 was statistically held constant, the difference in BC between the groups was negated. Respiratory compensation is stimulated to a large extent by the pH of the blood, and it is very likely that the lower pH found in the ET group at 100% $\dot{V}O_{2max}$ persists on the arterial side of the circulation and thereby accounts for the increased respiratory compensation in the ET group. The PCO_2 values reported in Figure 1 also indicate an elevated respiratory compensation in the ET group.

Based on these findings, it seems the trained individual possesses no advantage over the untrained person in the ability of blood to withstand large increases in LA content. This serves as an indication that the BC of blood, both during exercise and during recovery, may not be altered by prolonged physical training. The possibility still exists, however, that some adaptation may occur as a result of other forms of training that would continually overload the system with large quantities of LA. The training done by the ET group of the present study was primarily endurance training with only a small amount of sprint type training. As such, these subjects have likely trained their systems to delay LA accumulation as long as possible rather than to buffer it. Perhaps it is possible that the more anaerobically trained athlete adapts differently by improving his ability to nullify the pH disturbance created by LA. Further studies in this area should examine whether BC of blood is indeed improved by such highly intense training.

References

COSTILL, D.L., H. Thomason & E. Roberts. Fractional utilization of the aerobic capacity during distance running. Med. Sci. Sports and Exercise 5:248–252, 1973.

DAVENPORT, H.W., The ABC of Acid-Base Chemistry. 6th edition. Chicago: The University of Chicago Press, 1974.

ROBINSON, S. & P.M. Harmon. The lactic acid mechanism and certain properties of the blood in relation to training. Am. J. Physiol. 132:757, 1941.

STEINHAUS, A.H., Chronic effects of exercise. Physiol. Rev. 13:103, 1933.

WOODBURY, J.W., Regulation of pH. In T.C. Ruch & H.D. Patton (Eds.), Physiology and Biophysics. Philadelphia: Saunders, 1965.

Metabolic and Acid-base Exchange during Repetitive Twitch Contractions of in Situ Dog Skeletal Muscle

W.N. Stainsby and R.W. Barbee
University of Florida, Gainesville, Florida, U.S.A.

These experiments were designed to measure oxygen, carbon dioxide, lactate and acid-base exchange by mammalian skeletal muscle during repetitive twitch contractions. While some of these variables have been measured previously during contractions, others have not been measured before and they have not been measured together in the same experiment so the interactions among them can be seen.

Maximal twitch contractions at a frequency of 4/s were chosen because they induce nearly maximal metabolic rate and allow relatively unhindered blood flow through the muscles. The tension developed during the contractions was measured to assess the magnitude of fatigue.

The results reveal an output of acid which cannot be accounted for by CO_2 or lactate. This has not been reported previously and may be associated with the development of fatigue.

Methods

Eight mongrel dogs of both sexes weighing between 12 and 18 kg were anesthetized with sodium pentobarbital, 30 mg/kg, given intravenously. They were connected to a respirator which was adjusted to maintain the CO_2 in expired gas at $5 \pm 0.5\%$. Additional anesthetic was given when needed. Blood coagulation was prevented with heparin, 2000 to 3000 units/kg. Esophageal temperature was measured and kept at $37°C \pm 0.2$ by heating pad and radiant heat. The right femoral artery was cannulated to monitor arterial blood pressure and to obtain samples of arterial blood.

The muscle studied was the gastrocnemius-plantaris group of the left hind leg. It was exposed via a medial incision from mid-thigh to ankle. The overlying muscles were doubly ligated, and cut between the ties. The venous outflow from the muscle was isolated by ligating all vascular connections to the popliteal vein except those from the muscle group and by ligating the few veins from the muscle which did not drain directly into the popliteal vein. The popliteal vein was cannulated and the venous blood conducted via silastic tubing through a 3 mm cannulating type electromagnetic flowmeter and back to the animal via the left jugular vein. The flowmeter was calibrated by timed collections of the venous effluent at intervals during each experiment. A PE 90 polyethylene tube was passed through the wall of the venous blood tubing and threaded to the tip of the cannula in the popliteal vein. The other end had a y connection to two sampling ports, so two venous blood samples could be obtained simultaneously.

The sciatic nerve was isolated and cut between ties. The distal stump was put into a close fitting tubular electrode holder. The stimulating pulses were square, 4 volts and 0.2 ms. duration. The insertion tendon was freed, clamped and connected to an isometric lever system. The system was calibrated with weights after each experiment. The muscle was removed and weighed after every experiment.

Samples of arterial and venous blood were drawn at regular intervals. Two arterial blood samples were drawn, one immediately after the other, at rest before the contractions and at 2, 5 and 10 min of contractions. The two venous blood samples were drawn simultaneously at rest and at 15, 30, 45, 60, 120, 300 and 600 s of contraction. All the samples were drawn just past the 1 ml mark of 1 ml glass tuberculin syringes. The first sample of each doublet was adjusted exactly to the 1 ml mark and squirted into iced perchloric acid. The tube was agitated and stored in ice until analyzed for lactate. The other sample of the doublet was fitted with a mercury-containing cap and stored in ice. Part of this sample was analyzed for PO_2, PCO_2 and pH at 37° C using a Radiometer or an IL blood gas analyzer. The remainder, about 0.8 to 0.9 ml, was analyzed for O_2 and CO_2 concentrations by the Van Slyke manometric method. Blood acid-base values were obtained as base excess using a calculator program (Ruiz et al., 1975). Oxygen uptake ($\dot{V}O_2$), CO_2 output ($\dot{V}CO_2$), lactate output ($\dot{L}$), and non-CO_2 acid output ($\dot{H}A$), were calculated using the Fick equation.

Results

The tension developed during the contractions increased about 30% during the first min of contractions. It then decreased linearly during the remaining 9 min to an average of 80% of the 1 min value. In other words, fatigue resulted in a 20% reduction of developed tension.

The average venous outflow increased throughout the period of contractions. The increase was rapid during the first min and was progressively slower thereafter as shown in the top panel of Figure 1.

The average arterial concentrations of all the variables were nearly constant. The average venous concentrations changed rapidly during the first 1 to 2 min and were fairly stable during the remainder of the contraction period. To illustrate this, the concentrations of oxygen, lactate and (negative) base excess are shown in the lower 3 panels of Figure 1.

As shown in Figure 2, $\dot{V}O_2$ and $\dot{V}CO_2$ rose rapidly, reaching a plateau in 2 to 3 min. However, they did not change at precisely the same rate. The differences are illustrated by the changes in R, the respiratory exchange ratio, which shows a striking dip early in the contraction period. Lactate output rose during the first five min of contractions and then declined. The output of non-CO_2 acid ($\dot{H}A$) showed a sharp dip early in the contraction period which correspond to the dip in R. Thereafter $\dot{H}A$ increased during

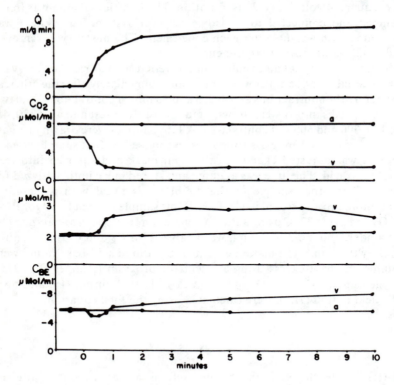

Figure 1—Blood flow, arterial and venous blood O_2 concentration, arterial and venous blood lactate concentration and arterial and venous blood (negative) base excess at rest and during ten min of contractions at 4 Tw/s. The contractions begin at time 0.

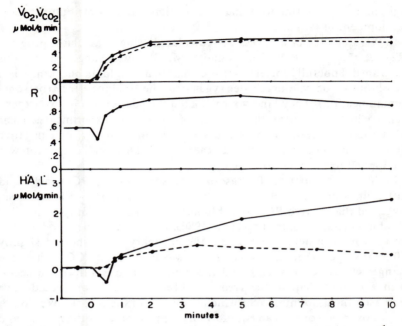

Figure 2—The $\dot{V}_{O_2}$ •——•, $\dot{V}_{CO_2}$ •---•, respiratory exchange ratio, R, $\dot{V}_{CO_2}/\dot{V}_{O_2}$, $\dot{L}$ •---•
and $\dot{H}A$ •——• during rest and ten min of twitch contractions.

the remainder of the contraction period. It diverges clearly from $\dot{L}$ after
about 5 min of contractions.

Discussion

The observation which is most pertinent to the topic of this symposium is
the exchange of non-CO_2 acid ($\dot{H}A$) during the contraction period. The
average net exchange of $\dot{H}A$ shows a brief period of uptake during the first
min of contractions. After that there is output of $\dot{H}A$ which increased
during the remainder of the 10 min contraction period. Lactate output by
the muscles would be expected to be the major contributor to acid output
during contractions, but $\dot{L}$ is rarely equal or parallel to $\dot{H}A$ during the
contraction period. The identity $\dot{H}A$ is unknown.

We assume that the progressive output of $\dot{H}A$ from these muscles during
contractions reflects a progressive increase in the intracellular H^+ concentration. Since it has already been shown that increasing intracellular acid
concentrations causes a decrease in the contractile capacity of muscle
(Roos & Boron, 1981), progressive intracellular acidosis during continuing

high intensity contractions may be a significant contributor to the development of fatigue.

In this study the base excess was calculated using a calculator program (Ruiz et al., 1975). R.B. Reeves and A.M. Olszowka suggested that this program did not fully correct for the change in hemoglobin buffering when the venous blood was largely deoxygenated (the Haldane effect) (personal communication). They kindly recalculated the base excess in two experiments which were near the average of all of the experiments. Their calculations changed the base excess values only slightly, in a negative direction, the venous more so than the arterial. This small correction was not applied to the data presented.

Both of the calculations for base excess assume that the PCO_2-pH blood buffer lines are parallel as in vitro buffer lines must be. It has been suggested that in vivo PCO_2-pH blood buffer lines are not quite parallel (Bouhuys et al., 1966; Francis et al., 1980). In particular the blood lines of blood from a systemic artery and the coronary sinus showed slightly different slopes when the whole animal was titrated with CO_2. These are changes which occur when blood-tissue equilibrium is allowed to occur. Such a change, had it occurred and been corrected for, would have decreased the amount of acid which came out of the muscle. We do not know how much of such an equilibrium is appropriate for the muscle, and we did not attempt to make the correction. Further evaluation of these findings requires evaluation of the presence or absence of blood-muscle acid-base equilibrium and other factors which might alter the buffer capacity of blood passing through muscle.

Acknowledgment

This study was supported in part by NIH Grant HL 22258.

References

BOUHUYS, A., J. Pool, R.A. Binkhorst & P. Van Leeuwen. Metabolic acidosis of exercise in healthy males. J. Appl. Physiol. 21:1040–1046, 1966.

FRANCIS, C.M., P. Foëx & W.A. Ryder. A comparison of carbon dioxide titration curves of arterial and mixed venous and coronary sinus blood. Resp. Physiol. 40:149–164, 1980.

ROOS, A. & W.F. Boron. Intracellular pH. Physiol. Rev. 61:296–434, 1981.

RUIZ, B.C., W.K. Tucker & R. Kirby. A program for calculation of intrapulmonary shunts, blood-gas and acid-base values with a programmable calculator. Anesthesiology 42:88–95, 1975.

Temperature

Economy of Isometric Exercise vs Temperature in the Cat Soleus: An Isolated, Perfused Muscle Preparation

A.A. Biewener, R. Karas, G. Goldspink and C.R. Taylor
Harvard University, Cambridge, Massachusetts, U.S.A., and
University of Hull, England

The cross-bridge cycling rate of striated muscle depends on fiber type as well as temperature. For a variety of animals, myosin ATPase activity has been shown to correlate closely with maximum rates of shortening in different muscles (Barany, 1967; Close, 1972). With an increase in muscle temperature, an increase in the speed of shortening and myosin ATPase activity was observed. Recent work on the sartorius muscle of frogs (Rome, 1982) shows that the energy expended by a muscle to maintain isometric tension also increases with increasing temperature. However, to date there has been no direct measure of the energy expenditure of a mammalian muscle as a function of temperature. It seems likely that muscle economy (Ns/J) should increase with a decrease in the cross-bridge cycling rate. To test this, we have developed an isolated, perfused preparation of the cat soleus to measure oxygen consumption ($\dot{V}o_2$) during exercise. The cross-bridge cycling rate was estimated by measuring the time to peak twitch tension (TPT) and was varied by changing temperature. The relationships between $\dot{V}o_2$ and TPT versus temperature were then compared. Because of the homogeneous population of slow-oxidative, or fatigue resistant (type I) fibers in the soleus, interpretations of the data can be directly related to the slow-oxidative fibers of this muscle.

Methods

Ten cats (1.5 to 2.5 kg) were anesthetized with sodium pentobarbitol (30 mg/kg) injected intravenously. The arterial supply and venous drainage of the soleus were isolated by ligation of all peripheral vessels branching from the tibial artery and vein. The proximal ends of these vessels were cannulated and the muscle perfused with heparinized saline. The muscle and its nerve were then removed from the animal's limb and placed in a waterjacketed chamber containing paraffin oil. The chamber temperature was regulated to $\pm 0.5°$ C. The distal tendon was attached to an isometric force transducer (Grass, FT03B) and its proximal attachment to the fibula clamped to a rigid bar. The arterial and venous cannulae were then connected to a flow through, perfusion system.

The perfusate consisted of rejuvenated red blood cells (Valeri, 1974) suspended in a phosphate buffered Krebs Ringers (pH 7.1). This was pumped through gas permeable tubing (Silastic, Dow Corning) and equilibrated with 95% O_2/5% CO_2 in a constant temperature bath before entering the muscle. Pyruvate (p.15mM), papaverin (30 mg/l), and bovine serum albumin (1% by vol) were also added to increase colloidal oncoctic pressure and lower smooth muscle tension. Arterial blood pressure was monitored continuously with a Statham (P23CB) pressure transducer. Venous blood was collected in 1 cc syringes over known periods of time. This provided a measure of blood flow and a time-integrated determination of venous O_2 content for each collection period. These were compared with the O_2 content of arterial samples taken at corresponding intervals. O_2 content was determined using the method previously described by Tucker (1967). This method is accurate to 0.2 vol %. Samples were analyzed within 5 min of the collection period. This was well within the 25 min time period for which no appreciable change in O_2 content (due to contamination by room air) was observed. Steady-state oxygen consumption was then determined using the Fick principle.

Stainless steel electrodes were attached to the nerve. The minimum stimulation voltage required to elicit maximum twitch tension was used (0.3 to 1.0 V). Following two preconditioning tetani of 1 s duration, the muscle was adjusted to its optimal length (maximum twitch tension) and the time to peak twitch tension determined. The muscle was allowed to rest for 30 min before exercise. Baseline activity (considered to be resting) for the muscle was one twitch every 16 s. This was necessary to prevent venous pooling within the muscle.

Once resting venous and arterial samples were collected, the muscle was given a series of nine tetani (40 Hz, 1 ms pulse duration at 0.3 to 1.0 v). Each tetanus lasted from 6 to 20 s (mean: 10 s), with an interval of 150 s between successive tetani. Tetanus duration was chosen to minimize the effects of activation and development of tension on the net cost measured. Force was integrated to determine the tension-time integral of each tetanus. Time-integrated force was monitored by a microprocessor via an A/D converter. The value obtained for the first tetanus of a run established the target level for succeeding tetani. Stimulation was triggered externally by the microprocessor. The stimulator (Grass S44) was turned on each time the integrator was reset. Once the time-integrated force reached the target level established by the first tetanus, the stimulator was turned off. Each tetanus then achieved the same integral of tension over the course of the run. Runs in which the stimulation period increased more than two-fold or the level of tetanus force fell to below 50% of the initial tetanic force were discarded.

Three separate collections of venous blood were made during exercise. Each collection represented the time-averaged O_2 content of venous blood over the period of two tetani. To allow for the lag before steady state oxygen consumption was attained, collection of the first venous sample was not begun until the third tetanus. Oxygen consumption ($\dot{V}O_2$) at rest was subtracted from the mean $\dot{V}O_2$ for the three samples taken during the run to calculate net $\dot{V}O_2$. Economy of maintaining isometric tension (Ns/ J), then, was calculated as $\int Fdt/$ net $\dot{V}O_2 \times 20.08$ (J/ ml O_2). The muscle was allowed to rest 30 min between runs at a given temperature or for 30 min following equilibration at a new temperature. Generally, three to four runs were obtained for each muscle; lasting over a period of 5 to 6 h. Lactate concentration was determined for all venous samples (Boehringer-Mannheim) to establish whether there was a significant anaerobic component to energy expenditure. Values are given as the mean ± SD. Least squares linear regression of the data was used to determine the slope ± 95% confidence interval and the correlation coefficient.

Results

Varying blood flow from 1.0 to 3.0 μl/ s and tetanus duration from 6 to 20 s had no effect on measured rates of oxygen consumption. Arterial pressure ranged from 50 to 80 mmHg at these flows. Generally, there was a slight rise in arterial pressure over the course of an experiment. Arterial pressure did not rise above 120 mmHg, however, without a corresponding decline in muscle performance. Rates of lactate production during exercise (0.11 ± 0.04 μM/ min) were not significantly greater than those at rest (0.09 ± 0.03 μM/ min). Net lactate production accounted for less than 4% of the total energy expended by the muscles. Thus, only aerobic metabolism was considered in the calculation of economy.

A significant decrease in economy was observed with an increase in temperature from 15 to 35° C (Figure 1). A slope of –0.144 ± 0.073 (R = 0.65) was obtained for these data. This represents Q_{10} of 0.72. Correspondingly, the time to peak twitch tension also decreased at higher temperatures (Figure 2), with a slope of –15.9 ± 3.4 (R = 0.92). This represents a Q_{10} of 0.54.

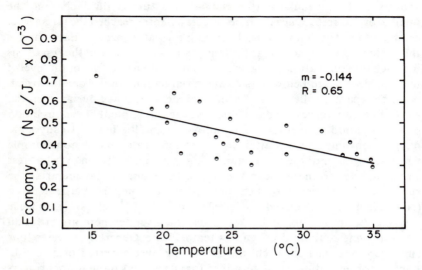

Figure 1—Isometric economy vs temperature for the cat soleus. Twenty-four runs obtained from the soleus muscles of 10 cats are included in the graph. The slope and its correlation coefficient are shown. The 95% confidence interval is –0.22 to –0.07. The Q_{10} is 0.72.

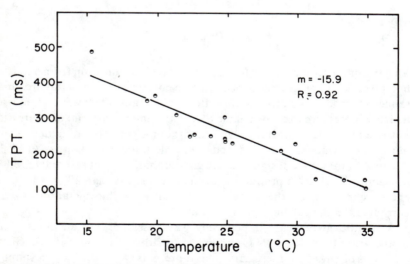

Figure 2—Time to peak twitch tension (TPT) vs temperature for the cat soleus. The slope and its correlation coefficient are shown. The 95% confidence interval is –19.3 to –12.5. N = 19. The Q_{10} is 0.54.

Discussion

The results of this study support the hypothesis that a decrease in cross-bridge cycling rate leads to an increase in isometric economy. We used time to peak twitch tension (TPT) to estimate cross-bridge cycling rate because TPT is a good indicator of intrinsic velocity, which correlates with myosin ATPase activity (Barany, 1967); the presumed rate determining step in the cycling of cross-bridge. The Q_{10} for the decrease in TPT is less than that for economy (i.e., the rate of decrease is greater for TPT than economy with increasing temperature). It seems unlikely that this difference is due to a temperature independent cost, such as Ca^{++} pumping, since the duration of the tetani (6 to 20 s) was great enough to make the activation cost insignificant compared to the maintenance cost. These data are consistent with measurements of the energy expended during isometric contractions in fast and slow muscles of the hamster (Goldspink et al., 1970).

The isolated, perfused preparation described here for the soleus can be usefully applied to study the effect of temperature on economy and TPT in a fast muscle, such as cat EDL. Finally, this technique provides a basis for the development of a method to selectively recruit different motor unit populations within an heterogeneous muscle (e.g., medial gastrocnemius). This would allow a direct measurement of the influence of recruitment order on the energetics and the development of tension.

Acknowledgment

This work was supported by NIH Training Grant T32GM07117 and NIH Grant AM18140 to C.R. Taylor.

References

BARANY, M., ATPase activity of myosin correlated with speed of muscle shortening. J. Gen. Physiol. 50:197–216, 1967.

CLOSE, R.E., Dynamic properties of mammalian skeletal muscles. Phys. Rev. 52:129–197, 1972.

GOLDSPINK, G., R.F. Larson & R.E. Davies. The immediately energy supply and cost of maintenance of tension for different muscles of the hamster. Z. Vergleich. Physiol. 66:389–397, 1970.

ROME, L., Energetic cost of generating force as a function of temperature in frog sartorius muscle. Am. J. Physiol. (in press)

TUCKER, V.A., Method for oxygen content and dissociation curves on microliter blood samples. J. Appl. Physiol. 23(3):410–414, 1967.

VALERI, C., in T.J. Greenwalt & G.A. Jamieson (Eds.), Human Red Blood Cell in Vitro. New York: Grune & Stratton. pp. 281–321, 1974.

Velocity Dependent Effect of Muscle Temperature on Short Term Power Output in Humans

A.J. Sargeant

Polytechnic of North London, London, England

The effect of muscle temperature on the capacity of human muscle to develop power during short-term dynamic exercise lasting for a few seconds has received surprisingly little systematic attention. Asmussen and Bøje (1945) examined the effect of warming the leg muscles on power output during a sprint effort lasting 14 s, performed on a cycle ergometer. More recently Asmussen, Bonde-Petersen and Jørgensen (1976) reported on the effect of changes in muscle temperature on vertical jump performance. Binkhorst, Hoofd and Vissers (1977) have studied the effect on hand grip muscles. Bergh and Ekblom (1979) presented data on the effect of changes in muscle temperature on maximal dynamic strength and power, and jumping and sprinting performance. In this last study, measurement of leg muscle force and power was made under different constant velocity conditions using a Cybex dynamometer. However, the authors were unable to demonstrate any velocity dependent effect of changes in muscle temperature on muscle force or power, at least within the velocity limitations of the Cybex equipment used. However, the recent development of an isokinetic cycle ergometer instrumented with strain gauges as described by Sargeant, Hoinville and Young (1981) overcomes the velocity limitations inherent in studying knee extension movements using the Cybex equipment. We have thus applied this new technique in order to study the effect of changes in muscle temperature on power output over a wider range of constant velocity conditions than has previously been possible.

Methods

Two young male subjects who were active but untrained were studied. The subjects exercised on an isokinetic cycle ergometer. During maximal 20 s efforts performed pedaling at a preset constant velocity the 'effective' force and power generated on the cranks of the ergometer were continuously monitored. The technique and definition of the terms used are fully described in our previous papers to which readers are referred for a complete account (Sargeant, Hoinville & Young, 1981). The present paper will describe data in terms of: Peak (effective) Force, that is, the greatest force exerted in each revolution tangentially to the crank arc of rotation; and Peak Power, that is, the power as calculated at the instant of Peak Force.

Experiments were performed at three pre-set pedalling rates of 54, 95, and 140 revs.min^{-1}. At each of these speeds measurements of maximal short-term power output were made under three temperature conditions: After resting for at least 1 h at room temperature (21 to 22° C); and after 45 min immersion to the level of the gluteal fold in a stirred water bath at 18° C and 44° C. Muscle temperature (T_m) was measured with a needle thermo-couple inserted to a depth of 3 cm midway along the anterior surface of the thigh.

Results

Under all conditions and pedal rates Peak Force increased during the first few revolutions before reaching maximal levels and thereafter declined over the 20 s period of measurement. The maximal level attained (PF_{max})at any given pedal rate was systematically associated with water bath and hence muscle temperature (Figure 1).

Maximal Peak Force was inversely and linearly related to crank velocity. The effect of manipulating muscle temperature by cooling in a water bath at 18° C was to shift this relationship to the left: the effect of warming by immersion in a water bath at 44° C was to shift the relationship to the right (Figure 2).

The relationship between Maximal Peak Power and crank velocity was parabolic under each of the three temperature conditions. The effect of cooling the muscle was to decrease the amplitude of the parabola and shift the zero force intercept at the theoretical maximal velocity to the left; the effect of warming was to increase the amplitude and shift the intercept to the right (Figure 3).

Thus the magnitude of the effect of the changes in muscle temperature on power output was shown to be velocity dependent. At a crank velocity of 54 revs.min^{-1} Maximal Peak Power increased by 2% per ° C rise in T_m; at 140 revs.min^{-1} the power output increased by 10% per ° C rise in T_m.

Associated with the higher maximum power outputs with increased T_m at any given velocity was a faster rate of fatigue. Thus at 95 revs/min^{-1} Peak Power declined by 4.6 W s^{-1} l$_{ulv}$$^{-1}$ (ulv = upper leg volume) after 45

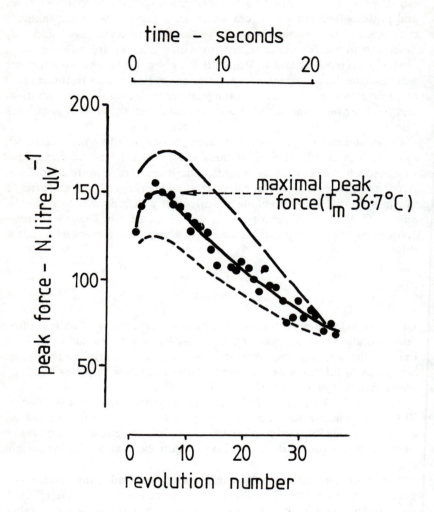

Figure 1—Changes in Peak Force during 20 s maximum efforts. Subject A, crank velocity 95 revs/min^{-1}. Data is expressed in Newtons and standardized for upper leg muscle volume (Sargeant, Hoinville & Young, 1981).

•——• Normal conditions (T$_m$ 36.7° C);
△---△ Water bath at 18° C (T$_m$ 31.3° C);
○—·—○ Water bath at 44° C (T$_m$ 39.3° C):

Data points have been omitted from the immersion experiments for clarity.

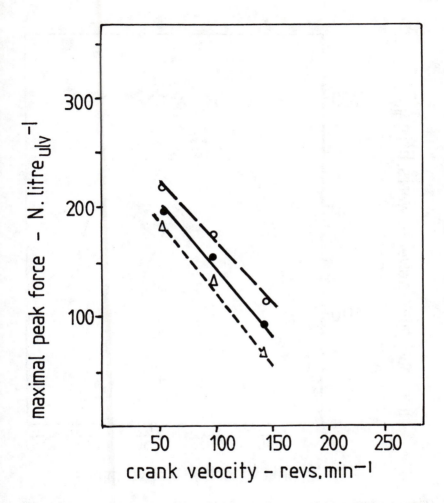

Figure 2—Relationship of Maximal Peak Force to crank velocity under the three temperature conditions studied. The calculated regression lines are shown with mean data for each condition and speed. Symbols and units as in Figure 1.

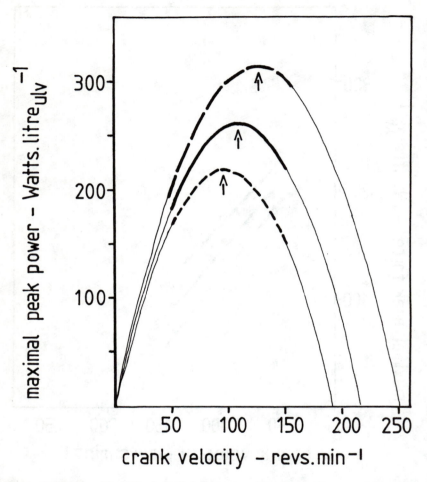

Figure 3—Relationship of Maximal Peak Power to crank velocity. The arrow indicates the optimal velocity for achieving Maximal Peak Power under each condition. Symbols as for Figure 1. Power expressed in watts and standardized for upper leg muscle volume.

min in the 18° C water bath when muscle temperature had fallen to 31.3° C; this compares with 6.8 watts $sec^{-1} l_{ulv}^{-1}$ after immersion in the water bath at 44° C when muscle temperature was 39.3° C.

Discussion

This paper demonstrates the significant effect that muscle temperature has on force and power output during short-term dynamic exercise lasting 20 s or less and thus confirms the work of previous investigators (Asmussen & Bøje, 1945; Asmussen, Bonde-Petersen & Jorgensen, 1976; Bergh & Ekblom, 1979; Binkhorst, Hoofd & Vissers, 1975). In addition, by studying the muscle power output under isokinetic conditions at relatively high velocities (i.e., beyond the optimal velocity for maximal power output—see Sargeant, Hoinville & Young, 1981) we have been able to demonstrate the velocity dependent effect of muscle temperature. This is in agreement with the observations of Binkhorst, Hoofd and Vissers (1975) who studied hand grip muscles, but in contrast to the observations of Bergh and Ekblom (1979) who used the Cybex apparatus to study knee extension. However, in the light of our present observations the findings of this latter study are not perhaps surprising since at the relatively low velocities attained with the Cybex, muscle temperature will have a rather small effect (see Figure 3). Indeed it is not until optimal velocity for maximal power output is achieved, or exceeded, that the most dramatic effects can be seen and clearly demonstrated.

The observation of a faster rate of fatigue with increased muscle temperature at any given crank velocity is consistent with increased substrate utilization as previously described by Edwards et al. (1972).

References

ASMUSSEN, E. & O. Bøje. Body temperature and capacity for work. Acta Physiol. Scand. 10:1–22, 1945.

ASMUSSEN, E., F. Bonde-Petersen & K. Jørgensen. Mechanoelastic properties of human muscles at different temperatures. Acta Physiol. Scand. 96:83–93, 1976.

BINKHORST, R.A., L. Hoofd & A.C.A. Vissers. Temperature and force velocity relationship of human muscles. J. Appl. Physiol. 42:471–475, 1975.

EDWARDS, R.H.T., R.C. Harris, E. Hultman, L. Kaiser & L-D. Nordesjö. Effect of temperature on muscle energy metabolism and endurance during successive isometric contractions. J. Physiol. 220:335–352, 1972.

SARGEANT, A.J., E. Hoinville & A. Young. Maximum leg force and power output during short-term dynamic exercise. J. Appl. Physiol. 51:1175–1182, 1981.

Exercise and Heat-induced Sweat

T. Verde, R.J. Shephard, P. Corey and R. Moore
University of Toronto, and Sunnybrook Hospital,
Toronto, Ontario, Canada

The present experiment was designed 1) to develop a simple technique of regional sweat collection, 2) to study regional differences in sweat composition, 3) to examine the impact of sweat flow rates upon mineral ion concentrations, and 4) to compare the composition of sweat secreted during exercise and sauna exposure.

Method

Subjects

The subjects were male volunteers, seven aged 20 to 26 years, and one aged 52 years. Physical characteristics were unremarkable (height 176.1 ± 5.2 cm; body mass 78.6 ± 8.6 kg; skinfold prediction of body fat $17.7 \pm 4.5\%$; predicted maximum oxygen intake 47.7 ± 6.7 ml $kg^{-1}min^{-1}$ STPD). One member of the group practiced sauna bathing twice per wk but none of the others were heat acclimated.

Sweat Collection and Analysis

Sweat samples were collected on pads of 12-ply gauze sponge (5 cm x 5 cm), taped to skin cleaned with 70% isopropyl alcohol, and covered with alcohol-cleansed plastic. On the arms, lifting of the tape was further prevented by a length of elastic strapping.

Sponges were weighed in air-tight conical tubes before and after sampling. Preliminary trials established that a 20 min collection yielded an adequate sweat volume for biochemical analysis, without saturating the pads. The sweat was separated by centrifuging pads at 2,000 rpm for 10 min. Sodium and potassium ion concentrations were determined by flame emission photometer, calcium and magnesium ion concentrations by atomic absorption spectrophotometry, and chloride ion concentrations by Buchler-Cutlove titration chloridometer. Long-term quality control was assured by monthly analysis of standard solutions.

Blank tests were carried out by applying 1 ml of distilled water to the gauze pads. Ten samples showed no contamination by Na^+ or Mg^{++}; three of 10 samples contained 0.1 mEq l^{-1} of K^+, two of 10 samples contained 0.1 mEQ l^{-1} of Cl^-, and all samples contained 0.08 or 0.09 mEq l^{-1} of Ca^{++} (mean contamination 0.084 mEq l^{-1}).

Recovery tests were carried out by applying 1 ml of "artificial sweat" in similar fashion:

	Na^+	K^+	Ca^{++}	Mg^{++}	Cl^-
Initial composition (mEq l^{-1})	39	17.7	6.00	3.00	25.20
Recovery Mean (mEq l^{-1})	38.6	17.55	5.77	2.93	24.88
± SD	± 1.2	± 0.11	± 0.16	± 0.16	± 0.12
Percent	99	99	96	98	99

Environmental Conditions

Outdoor exercise was performed at a temperature ranging from 14 to 23° C, and a wind-speed of 50 to 1,000 m min^{-1}. Indoor temperatures were 20 to 23° C, with a low relative humidity (23 to 47%). Subjects wore standard sweat suits for all exercise experiments. The sauna was controlled thermostatically at 93° C, with a low relative humidity. Subjects were nude while in the sauna.

Results and Discussion

Validity of Gauze-pad Technique

In contrast to the experience of Lobeck and McSherry (1967) with filter paper discs, our preliminary tests established minimal contamination of sweat specimens. Mineral ion recovery was also satisfactory and reproducible.

Were sweat volume and composition modified by the impermeable covering? Local forearm sweat rates amounted to 1.3 to 2.0 mg $min^{-1}cm^{-2}$, as compared with values of 0 to 0.4 mg $min^{-1}cm^{-2}$ reported by Nadel et al. (1971) for the seventh to the tenth min of exercise. The total decrease of body mass for the 20 min period (0.6 kg, equivalent to a sweat output of 1.65 mg min⁻¹cm⁻²) also seems in good agreement with the pad data. Moreover, our forearm Na^+ and Cl^- concentrations (Table 1) agree well with data obtained by the standard washdown technique. Possibly, the gauze pads behave as though permeable until they have become saturated with sweat.

Regional Differences of Sweat Secretion

A two-way analysis of variance of environment (outdoor exercise, indoor exercise or sauna) with respect to collection site (proximal or distal forearm) confirmed a symmetrical sweating response (Kuno, 1956), but proximal pads contained more sweat than those from the distal forearm ($P < 0.001$).

Samples collected from the right armpit and the abdomen yielded as great a sweat mass as the forearm, but volumes from the thigh were generally smaller. Samples from the hands yielded only limited amounts of sweat, with a high concentration of all ions (Mickelson & Keys, 1943). Possibly, the thick stratum corneum imbibes appreciable amounts of water. Alternatively, the outer layers of the epidermis may have contaminated the sample. Active mechanisms for conserving mineral ions may also be less well developed in the glands of the hands.

Impact of Sweat Flow Rates

There were substantial inter-individual differences in sweat composition. Adjusting for this factor, only magnesium and chloride ion concentrations varied significantly with sweat mass (Table 2). Adjusting also for the effects of environment, only the magnesium ion concentration showed any relationship to sweat production ($P < 0.06$). Apparently, sweat volume had a larger effect on Mg^{++} concentrations in the sauna than when exercising.

Vellar and Askevold (1968) previously found no relationship of magnesium concentration to sweat output over a 1 h period, but Strømme et al., (1973) observed a similar relationship to that we have seen. Possibly, magnesium output stabilizes over 30 min, as hormonal controls become effective.

Influence of Environment

The volume of sweat secreted was significantly larger for indoor exercise than for the other two experimental conditions ($P < 0.05$). Calcium ion values ($P < 0.002$) and magnesium ion ($P < 0.02$) concentrations were

Table 1

A Comparison of Sweat Composition from the Three Environments Tested

Environment	Sweat mass (mg/20 min)	Ion concentrations (mEq l^{-1})				
		[1]Na^+	[1]K^+	[1]Ca^{++}	[1,2]Mg^{++}	[1]Cl^-
Outdoor exercise	706	59	9.4	2.7	0.8	39
		± 20	± 2.9	± 2.3	± 0.2	± 19
Indoor exercise	931	65	8.8	2.0	0.6	58
	(P < 0.05)	± 12	± 2.2	± 1.0	± 0.2	± 19
Sauna	775	61	7.9	4.7	1.8	46
		± 12	± 1.1	± 2.3	± 1.0	± 12
				(P < 0.002)	(P < 0.02)	

(1) Omitting one subject with incomplete data.
(2) Omitting two subjects with incomplete data.

Table 2

Influence of Sweat Output (mg 20 min^{-1}) upon the Ionic Composition of Sweat (m Eq l^{-1}).

Variable	Na^+	K^+	Ca^{++}	Mg^{++}	Cl^-
Significance of departure from zero slope	P < 0.006	n.s.	P < 0.009	P < 0.0001	P < 0.0001
Significance after adjusting for interindividual differences of sweat output	n.s.	n.s.	n.s.	P < 0.0001	P < 0.03
Significance after adjusting for interindividual differences and effect of differences in environment	n.s.	n.s.	n.s.	P < 0.06	n.s.

All data analyzed by linear regression techniques, with sweat output as the independent variable.

significantly greater in the sauna. Since calcium and magnesium ions play key roles in sustained exercise, it may be that specific mechanisms conserve these ions during physical activity. Certainly, there is some evidence that sweating responses differ for a peripheral, skin stimulus such as sauna bathing, and a relatively pure central stimulus such as running in a cool environment (Nadel et al., 1971).

References

KUNO, Y., Human Perspiration. Springfield, Ill.: C.C. Thomas, 1956.

LOBECK, C.C. & N.R. McSherry. The ionic composition of pilocarpine induced sweat in relation to gland output during aging and in cystic fibrosis. In E. Rossi and E.S. Stoll (Eds.), Proceedings of the 4th International Conference on Cystic Fibrosis of Pancreas (Mucoviscidosis). Berne/Grindelwald, Part I. Mod. Probl. Pediat. 10:41–57. New York: Karger, Basel, 1967.

MICKELSON, O. & A. Keys. The composition of sweat with specific reference to the vitamins. J. Biol. Chem. 149:470–490, 1943.

NADEL, E.R., J.W. Mitchell, B. Saltin & J.A. Stolwijk. Peripheral modifications to the central drive for sweating. J. Appl. Physiol. 31:828–833, 1971.

STRØMME, S.B., I.C. Stensvold, H.O. Meen & H.E. Refsum. Magnesium metabolism during prolonged heavy exercise. In H. Howald & J.R. Poortmans (Eds.), Metabolic Adaptation to Prolonged Physical Exercise. p. 361. Basel: Birkhauser Verlag, 1973.

VELLAR, O.D. & R. Askevold. Studies on sweat losses of nutrients III. Calcium, magnesium and chloride content of whole-body cell-free sweat in healthy unacclimatized men under controlled environmental conditions. Scand. J. Clin. Lab. Invest. 22:65–71, 1968.

Hormones—
General

Effect of Exercise and Testosterone on the Active Form of Glycogen Synthase in Human Skeletal Muscle

K. Allenberg, N. Holmquist, S.G. Johnsen, P. Bennett,
J. Nielsen, H. Galbo, and N.H. Secher
August Krogh Institute, University of Copenhagen,
Copenhagen, Denmark, Hormone Department, Statens
Seruminstitut, Copenhagen, Denmark, Institute of Medical
Physiology B, University of Copenhagen, Denmark,
and Department K, Frederiksberg Hospital,
Frederiksberg, Denmark

Repletion of muscle glycogen after exercise is dependent on the availability of substrate and on the activity of the rate limiting enzyme glycogen synthase (GS), which exists in a dephosphorylated active I-form and a phosphorylated inactive D-form interconvertible through intermediates. The enzyme activity is supposed to be dependent on the total GS-level, on the relative amount of the I-form, and of some of the intermediates (Roach & Larner, 1977). It has been suggested that a low content of intramuscular glycogen stimulates the conversion of the D-form to the I-form, and the GS-activity increases in the presence of insulin (Larner, 1976) and testosterone (Adolfsson, 1973) and also increases during recovery from exercise (Kochan et al., 1979). The purpose of this work was to study the interaction between muscle glycogen content, levels of serum testosterone, cortisol, insulin, plasma glucose, and the effect of exercise on GS-activity in male subjects on a controlled diet.

Procedure and Methods

Four healthy male subjects, age 24 to 30 years, weight 76 to 89 kg, height 178 to 192 cm, and $\dot{V}_{O_2max}$ 3.1 to 4.5 1 $\times$ min^{-1}, were studied after an informed consent was obtained.

Testosterone or placebo was administered in a double blind cross-over design involving two 5-day periods, D1 to D5, with an intermission of 10 days. The code revealed that three subjects ingested the drug during the first period. Three 100 mg microminized testosterone tablets or placebo were given four times a day on D2 to D5.

Each day a controlled mixed diet was administered at 6 am, at noon, and at 6 pm, the final meal given at 6 am on D5. No breakfast was allowed on the exercise day (D3), whereas water intake was unrestricted. Exercise was continued until exhaustion on a Krogh bicycle ergometer using an intensity corresponding to 68% of $\dot{V}_{O_2max}$, decreasing to 64% after 75 min, and to 55% after 105 min of exercise.

Expiratory gas was collected in Douglas bags and was analyzed using a wet Tissot spirometer for volume determination, an infrared Beckman LB-2, and a paramagnetic Servomex OA 184 System for CO_2 and O_2 analysis, respectively. Cubital vein blood samples and muscle biopsies from the vastus lateralis muscle were taken at least 3 h after food ingestion with the subjects resting in the supine position. On D1 blood samples were drawn at 9 am, 3 pm and 9 pm for determination of diurnal variations. On D3 blood samples and muscle biopsies were taken fasting at 9 am (O h = before exercise), 45 min after beginning of exercise (0.75 h), at cessation of exercise (2 h), and after 8, 14, 26 and 50 h following the onset of exercise. Blood samples were analyzed for testosterone (Coyotupa et al., 1972), cortisol (Nielsen, 1979), insulin (Albano et al., 1972), and glucose (Lowry & Passønneau, 1972). Muscle biopsies were obtained, immediately frozen in liquid nitrogen, and stored at –80° C until analyzed for glycogen (Lowry & Passønneau, 1972), and for GS-activity by a modification of the filter paper method (Thomas et al., 1968). The frozen biopsy was homogenized in Tris-buffer 1:20 wxv^{-1}. The reagent contained either 15 mM G6P or 15 mM Na_2SO_4. A sulfate concentration of 2 mM has been described to cause a 20% increase of GS-I % (Schlender & Larner, 1973). We have found this effect unchanged when varying the sulfate concentration from 2 to 20 mM. Apparently a sulfate concentration of 15 mM activated more of the enzyme than is often reported (Kochan et al., 1979). Diurnal urine was analyzed for fractionated 17-ketosteroids (Kampmann et al., 1976). Values are given as mean ± SEM. Observations were compared using Student's paired t-test. Significance was set at the 0.05 level of confidence.

Results

Resting plasma glucose vlaues increased after 45 min of exercise from 3.7 ± 0.23 mM to 4.6 ± 0.29 mM, but returned to 3.7 ± 0.44 mM at cessation of

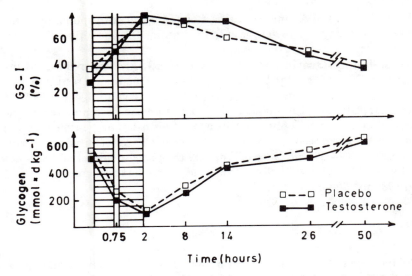

Figure 1—The active form of glycogen synthase, GS-I % (upper panel) and glycogen concentration in the vastus lateralis muscle (lower panel) followed before, during and after 2 h of exhaustive dynamic exercise and for the following 48 h of restitution.

exercise. During 48 h of recovery there was an increase to 4.40 ± 0.36 mM after 12 h and then a gradual decrease to 4.0 ± 0.31 mM.

Muscle glycogen decreased from 540 ± 57 mmol $\times$ dwkg^{-1} (dry weight in kg^{-1}) to 110 ± 27 mmol $\times$ dwkg^{-1} at cessation of exercise (Figure 1). During the 48 h of restitution the concentration increased to 640 ± 50 mmol $\times$ dwkg^{-1}. The pre-exercise level was reached after 24 h.

Total GS-activity amounted to 3.3 ± 0.10 mmol $\times$ wwkg^{-1} (wet weight in kg^{-1}) $\times$ min^{-1} and was neither influenced by exercise nor by testosterone. GS-I increased from $31 \pm 4.5\%$ to $53 \pm 8.0\%$ after 45 min of exercise and to $76 \pm 4.5\%$ at cessation of work. After 48 h of restitution the GS-I % deceased gradually to $39 \pm 1.9\%$.

During the placebo period the serum testosterone showed a diurnal variation from 25 ± 2.8 nM at 9 am decreasing to 23 ± 1.3 at 3 pm and to 19 ± 1.2 at 9 pm. The concentration increased during the workout from 28 ± 3.1 nM to 37 ± 4.4 nM, but the initial level was found again at the cessation of exercise. During the 48 h of recovery a diurnal variation was also seen. Testosterone treatment caused a similar pattern in serum testosterone, but the values were on an average at an 80 (43 to 112)% higher level.

Serum cortisol also showed a diurnal variation from 470 ± 26 nM at 9 am, to 270 ± 29 nM at 3 pm and to 200 ± 100 nM at 9 pm. Exercise induced a progressive increase in serum cortisol from 440 ± 76 nM to 890 ± 63 nM.

After exercise the resting values were found again corresponding to the time of the day. Testosterone administration did not influence serum cortisol levels.

Non-fasting serum insulin on D1 was 180 ± 53 pM compared to a fasting value on D3 of 60 ± 3.1 pM. Exercise induced an increase in resting insulin to 170 ± 88 pM. During the 48 h of recovery a peak value of 440 ± 199 pM was reached after 12 h, but after 48 h the value had decreased to the one seen after exercise (Figure 2). Throughout testosterone treatment serum insulin was reduced 17 to 63%. 11-desoxy-17-ketosteroids in urine amounted 32 – 37 μmol $\times$ day^{-1} during the placebo period and was not influenced by exercise. In the testosterone period the ketosteroids increased to 1980 to 2700 μmol $\times$ day^{-1}.

Discussion

An increase in serum testosterone by 80% coincided for unknown reasons with a decrease of the serum insulin level by 40%. Testosterone influences GS-I formation by GS-phosphotase activation and by increasing the muscle cell glucose-6-phosphate level (Adolfsson, 1973). Insulin increases the GS-I % by an inhibition of the deactivating enzyme GS-kinase (Roach & Larner, 1977). Thus the decrease in serum insulin during testosterone administration may explain the unchanged GS-I%. However, when taking the glycogen concentration into account (Figure 2) it seems as if testosterone has potentiated the relationship between GS-I % and insulin. Cortisol may compete with testosterone for the receptor sites (Viru & Korge, 1979). Cortisol concentration increased during exercise only, but the GS-I % relative to the glycogen concentration during and after exercise appeared to be identical (Figure 2) and, therefore, a separate influence of cortisol on GS-I % could not be detected. Throughout the 2 days of recovery studied, blood glucose levels were larger than before exercise. Thus the present findings conform to a dominating influence of muscle glycogen on GS-I %, but it appears also to be rigorously regulated by insulin and testosterone, resulting in a higher muscle glycogen level after recovery from exhaustive dynamic exercise when ample glucose is available because of an increase in serum insulin for at least 2 days of recovery.

Acknowledgments

The authors wish to thank S. Molbech, the Danish-Anti-Polio Society, DK—2900 Hellerup, Denmark, and O. Grønnerød, Institute of Muscle

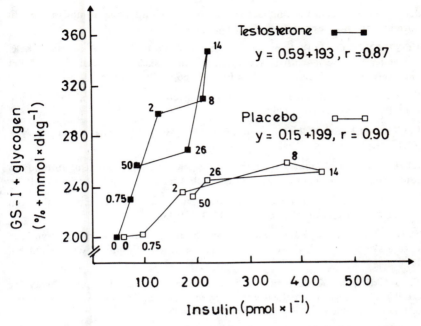

Figure 2—The sum of GS-I % and muscle glycogen concentration plotted against the serum insulin concentration. Time of samples indicated, h. To express changes the values are given in relative units to the resting values obtained before exercise ("O" h). The sum is increasing with increasing serum insulin levels, and decreasing with declining concentrations, in both testosterone and placebo periods indicating a minor decrease in GS-I % than would be expected at a given glycogen level. During testosterone administration the slope is increased by a factor 3.9.

Physiology, Oslo, Norway for help during the preparation of the manuscript.

References

ADOLFSSON, S., Effects of insulin and testosterone on glycogen synthase activity in rat levator ani muscle. Acta Physiol. Scand. 88:234–247, 1973.

ALBANO, J.D.M., R.P. Ekins, G. Maritz & R.C. Turner. A sensitive precise radioimmunoassay of serum insulin relaying on a charcoal separation of bound and free hormone moieties. Acta Endocrinol. 70:487–509, 1972.

COYOTUPA, J., A.F. Parlow & G.E. Abraham. Simultaneous radioimmunoassay of plasma testosterone and dihydrotestosterone. Anal. Lett. 5(6):329–340, 1972.

KAMPMANN, J.P., F. Schønau-Jensen, E.P. Bennett & S.G. Johnsen. Rise in dehydroepiandrosterone and oestrogens during clomiphene administration in normal men. Acta Endocrinol. (Kbh.) 83:166–172, 1976.

KOCHAN, R.G., D.R. Lamb, S.A. Lutz, C.V. Perrill, E.M. Reimann & K.K. Schlender. Glycogen synthase activation in human skeletal muscle: effects of diet and exercise. Am. J. Physiol. 236(6):E660–E666, 1979.

LARNER, J., Mechanisms of regulation of glycogen synthesis and degradation. Circ. Res. 38(5):I-2–I-7, 1976.

LOWRY, O.H. & J.V. Passønneau. A Flexible System of Enzymatic Analysis. New York: Academic Press, 1972.

NIELSEN, J., Poly-maleic acid anhydride (PMA) coupled cortisol antibody in routine determination of urinary free cortisol. Acta Endocrinol. 91:680–691, 1979.

ROACH, P.J. & J. Larner. Covalent phosphorylation in the regulation of glycogen synthase activity. Mol. Cell. Biochem. 15(3):179–200, 1977.

SCHLENDER, K.K. & J. Larner. Purification and properties of glycogen synthase I from skeletal muscle: Two kinetic forms. Biochem. Biophys. Acta 293:73–83, 1973.

THOMAS, J.A., K.K. Schlender & J. Larner. A rapid filter paper assay for UDPG-glucose-glycogen glucosyltransferase, including an improved biosynthesis of UDP-^{14}C-glucose. Anal. Biochem. 25:486–499, 1968.

VILLAR-PALASI, C. & J. Larner. Feedback control of glycogen metabolism in muscle. Fed. Proc. 25:583, 1966.

VIRU, A. & P. Korge. The role of androgenic anabolic steroids in regulation of the skeletal muscle adaption. J. Ster. Biol. 11:931–932, 1979.

Dysadrenarche as a Possible Explanation for Delayed Onset of Menarche in Gymnasts

G.R. Brisson, M. Ledoux, S. Dulac and F. Peronnet
Université du Québec à Trois-Rivières et Université
de Montréal, Québec, Canada

Several studies have shown that physical exercise could bring about a delay in the onset of menarche (Druss et al., 1979; Frisch et al., 1980; Malina et al., 1973, 1978, 1979; Ross et al., 1976; Vincent, 1979; Warren, 1980). But in pubertal progression, menarche is a late event that may be significantly distant from anterior modifications associated with the initiation of puberty (Brisson et al., 1982a). Since adrenarche appears to be an early phenomenon in the maturational process (Brisson et al., 1982b; Ducharme, 1981), we want here to put in evidence of the existence of a possible dysadrenarche in certain athletes in order to explain, at least partially, a late menarche.

Methods

Two groups of pre-menarched girls volunteered for the study. Informed consents were obtained. A physical examination confirmed the absence of pathology in every subject. The trained group ($N = 17$) was recruited from a gymnastics club necessitating 12 to 20 weekly h of training. The untrained control group ($N = 23$) was issued from healthy sedentary girls. Pairing was based on chronological age, since matching for pubarche and/or thelarche, both resulting from endocrine modifications, would have reduced any existing difference.

Every subject was submitted to a 20-min submaximal bicycle exercise consisting of a PWC_{170} test (Howell et al., 1968) in which the last working level was extended for an extra 8 min. Venous blood was sampled immediately before and after the submaximal test from an antecubital vessel. Body fat (Siri, 1956) and body density (Durnin et al., 1974) were

Table 1
Values (x̄ ± 1 SD) for Selected Anthropometric Measurements

Class (months)	Group	N		Age (months)	Weight (kg)	Height (m)	Body fat %	Body fat kg	Lean kg	Body density	PWC /kg
110 to 126	trained	7	x̄	117.6	26.4	1.29	17.7	21.8	4.7	1.0584	13.2
			± sd	6.8	3.3	.79	1.6	2.9	.6	.0037	2.0
	untrained	6	x̄	117.4	37.8	1.42	21.9	28.8	8.7	1.0489	12.8
			± sd	8.5	8.4	.50	5.9	4.2	4.6	.0129	2.7
127 to 150	trained	5	x̄	135.6	30.2	1.37	16.5	25.1	5.0	1.0612	14.7
			± sd	8.3	6.6	.10	1.9	5.1	1.6	.0042	4.0
	untrained	7	x̄	138.9	39.8	1.47	22.6	30.6	9.2	1.0476	12.7
			± sd	4.2	5.6	.51	5.2	2.6	3.4	.0114	2.7
151 to 190	trained	5	x̄	169.2	40.4	1.51	16.1	33.8	6.6	1.0622	16.4
			± sd	12.0	4.0	.53	3.8	2.1	2.1	.0088	2.6
	untrained	10	x̄	162.7	40.7	1.49	21.5	32.8	8.7	1.0499	13.7
			± sd	11.0	4.1	.67	2.3	5.4	1.4	.0051	3.2

estimated from anthropometric measurements (Table 1) in accordance with the Programme Biologique International.

Hemoconcentration was evaluated using an automated measurement of serum total proteins (K.D.A.; Amer. Monitor Corp., Indianapolis). Hormonal standards were purchased from Sigma Chem. Co. (St. Louis). Estrone (E_1), 17β-estradiol (E_2) and dehydroepiandrosterone (DHEA) were extracted from sera with a hexane: ethyl acetate (9:1) solvent. Labeled tritiated hormones were obtained from NEN (Boston). Androgen anti-serum was purchased from RSL (Carson, Calif.) and estrogen anti-sera from Accurate Chem. (New York). Cross-reactives were non-significant (0.1% of E_2 in E_1; 0.7% of E_1 in E_2). Sensitivities were sufficient in all four systems; precision and reproducibility of methodologies were found acceptable. Statistical treatments included analysis of variance for repeated measures (BMDP2V; Biomed. Comp. Pgm, Univ. Calif. Press, 1975).

Results

Every hormonal value reported in the study has been corrected for exercise hemoconcentration using modifications in circulating total proteins. In neither case the imposed submaximal exercise was able, when measured by analysis of variance, to significantly modify the evolution of both circulating estrogens and androgens (P= .493 E_1, .520 E_2, .484 E_1/E_2, .688 DHEA and .264 DHEA-SO_4). On the other hand, as illustrated in Figure 1, every measured dependent variable was significantly modified by the chronological age factor (P= .003 E_1, .015 E_2, .050 E_1/E_2, .014 DHEA and .005 DHEA-SO_4). The influence of physical training manifested itself only in the evolution of circulating E_1/E_2 ratio (P= .016) and DHEA-SO_4 (P= .047).

Discussion

One cannot but realize when examining Figure 1 that puberty is a transition period between two states, one of physiological re-adjustments which brings about sexual maturity. Therefore, the observation that chronological age exerts a very significant influence upon each of the studied variables comes as no surprise. Indeed, E_1 as well as E_2, or DHEA and its conjugate, are significantly different when one compares the various age groups under study. However, the submaximal physical exercise turns out to be an apparently insignificant factor: no marked differences were observed between pre- and post-effort blood levels for each studied dependent variable.

634 Brisson, Ledoux, Dulac and Peronnet

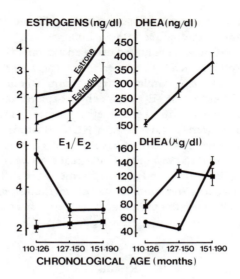

Figure 1—Values ($\bar{x}$) for selected circulating estrogenic and androgenic hormones (▲———▲: all subjects; ●———●: trained; ■———■: untrained).

Nevertheless, the influence of physical training appears in the evolution of the E_1/E_2 ratio and of DHEA-SO$_4$ in the blood. If one keeps in mind that E_1 is, during pre-pubescence, mainly adrenal in origin and that blood E_1 during puberty increases more slowly than blood E_2 (the latter reflecting incipient ovarian activity), it is normal that a gradual decrease occurs in the E_1/E_2 ratio with pubertal progression (Angsusingha et al., 1974; Gupta et al., 1975). The E_1/E_2 ratio has also been shown to increase with menopause (Hammond & Maxson, 1982). Thus, a higher (P= .016) estrogenic index (E_1/E_2) in trained subjects than in sedentary subjects could represent a cortico-adrenal physiological delay, particularly in young subjects (Figure 1).

However, adrenarche is much better characterized by its androgenic rather than its estrogenic changes (Ducharme, 1981). Indeed, we observe with pubescence a marked increase in the adrenal release of androgens. Our observations (Figure 1) suggest, on the other hand, that adrenal androgenesis, as measured by the variations in circulating DHEA-SO$_4$ (rather than in DHEA) is slower (P= .047) in trained than in control subjects: the circulating DHEA-SO$_4$ levels eventually reached by the untrained subjects within the 125 to 150 month age class are attained only much later (approximately 1.5 yr) by their trained counterparts.

Adrenarche being an integral part of the evolution towards menarche, the specific behavior of these steroids in young athletes could indicate a dysadrenarche which might explain an eventual delay in menarche.

Acknowledgment

The authors are indebted to D. Lacroix-Sylvestre and J.P. De Chezet for their respective scientific and professional contributions to the preparation of this manuscript.

References

ANGSUSINGHA, K., F.M. Kenny, H.R. Nankin & F.H. Taylor. Unconjugated estrone, estradiol and FSH and LH in prepubertal and pubertal males and females. J. Clin. Endocrinol. Metab. 39:63–68, 1974.

BRISSON, G.R., F. Péronnet, M. Ledoux & S. Dulac. The onset of menarche: a late event in pubertal progression to be affected by physical training. Can. J. Appl. Sp. Sci. 7:61–67, 1982a.

BRISSON, G.R., F. Péronnet, M. Ledoux & S. Dulac. L'âge à la ménarche en médecine sportive. Méd Québec 17:137–141, 1982b.

DRUSS, R.G. & J.A. Silverman. Body image and perfectionism of ballerinas. Comparison and contrast with anorexia nervosa. Gen. Hosp. Psychiatr. 1:115–121, 1979.

DUCHARME, J.R., Normal puberty: clinical manifestations and their endocrine control. In R. Collu, J.R. Ducharme & H. Guyda (Eds.), Pediatric Endocrinology, pp. 293–326. New York: Raven Press, 1981.

DURNIN, J.V.G.A. & J. Womersley. Body fat assessed from total body density and its estimation from skinfold thickness. Brit. J. Nutr. 32:77–97, 1974.

FRISCH, R.E., G. Wyshak & L. Vincent. Delayed menarche and amenorrhea in ballet dancers. N. Engl. J. Med. 303:17–19, 1980.

GUPTA, D., A. Attanasio & S. Raaf. Plasma estrogen and androgen concentrations in children during adolescence. J. Clin. Endocrinol. Metab. 40:636–643, 1975.

HAMMOND, C.B. & W.S. Maxson. Current status of estrogene therapy for the menopause. Fert. Steril. 37:5–25, 1982.

HOWELL, M.L. & R.B.J. Mcnab. The physical work capacity of Canadian children aged 7 to 17 years. Toronto: Canadian Association for Health, Physical education and Recreation, 1968.

MALINA, R.M., C. Bouchard, R.F. Shoup, A. Demirjian & G. Larivière. Age at menarche, family size, and birth order in athletes at the Montreal Olympic Games, 1976. Med. Sci. Sports 11:354–358, 1979.

MALINA, R.M., A.B. Harper, H.H. Avent & D.E. Campbell. Age at menarche in athletes and non-athletes. Med. Sci. Sports 5:11–13, 1973.

MALINA, R.M., W.W. Spirduso, C. Tate & A.M. Baylor. Age at menarche and selected menstrual characteristics in athletes at different competitive levels and in different sports. Med. Sci. Sports 10:218–222, 1978.

ROSS, W.D., S.R. Brown, R.A. Faulker & M.V. Savage. Age at menarche of elite Canadian skaters and skiers. Can. J. Appl. Sp. Sci. 1:191–193, 1976.

SIRI, W.E., Body composition from fluid spaces and density. Donner Lab. Med. Physics, Univ. of Calif. Rept. of 19 March, 1956.

VINCENT, L.M., Competing with the Sylph. New York: Andrews & Mc Meel, 1979.

WARREN, M.P., The effect of exercise on pubertal progression and reproductive function in girls. J. Clin. Endocr. Metab. 51:1150–1157, 1980.

Plasma Leucine Enkephalin-like Radioreceptor Activity and Tension-Anxiety before and after Competitive Running

P.A. Farrell, W.K. Gates, W.P. Morgan and C.B. Pert
University of Wisconsin-Milwaukee, Wisconsin, U.S.A.,
University of Wisconsin-Madison, Madison,
Wisconsin, U.S.A., and National Institute of Mental Health,
Bethesda, Maryland, U.S.A.

Several recent studies have shown that Methionine and Leucine-Enkephalin are present in pg/ml concentrations in human plasma (Ryder & Eng, 1981). The origin of these pentapeptides is not known, however, they do not seem to be breakdown products of Beta-endorphin (B_h-E_p) (Smith et al., 1981) or Beta-Lipotropin (B_h-LPH) (Shanks et al., 1981). These peptides have been found in the adrenal medulla (Lewis et al., 1980; Linnoila et al., 1980; Smith et al., 1981). Since it is well-established that the adrenal medulla is hormonally active during exercise (Hartley et al., 1972) we felt it possible that plasma levels of leucine enkephalin may be elevated during exercise.

It has previously been demonstrated that acute physical activity of a vigorous nature is associated with decreases in anxiety. The characteristic tension reduction following such exercise has been observed in both athletic and non-athletic populations (Morgan, 1979). It is not known, however, whether exercise actually causes anxiety reduction or merely covaries with the improved affect. The endogenous opiates are a possible candidate to effect mood due to the established effects of exogenous opiates. Several recent studies have shown increased plasma levels of endogenous opiates after exercise (Carr et al., 1981; Colt et al., 1981; Farrell et al., 1982) and it was our intention to continue this line of research using an assay which reflects radio-receptor activity (RRA) as opposed to immunological activity since immunological activity may not coincide with biological activity (Li et al., 1977).

Methods

Nine experienced male runners and five experienced female runners volunteered to participate in the study. All aspects of the study were verbally explained to each subject prior to signing a written consent. A local 10 mile road race was chosen as the experimental setting.

The subjects arrived at the race site at 0600 h and reported to a secluded indoor area where they immediately completed the Profile of Mood States (POMS) developed by McNair, Lorr and Droppleman (1971). The subjects were instructed to respond in terms of "How you have been feeling during the past week including today," when completing this pre-race questionnaire. Following the completion of the questionnaire, approximately 11 ml of venous forearm blood was obtained by venipuncture. Prior to this blood draw, the subjects were required to refrain from any type of warm up.

Forearm venous blood (9 ml) was immediately placed into an ice cold tube containing 0.5 ml Bacitracin (2 mg/ml) and EDTA (1.4 mg/ml). The tube was shaken, centrifuged (20° C, 2000 g, 8 min) and the plasma frozen (-20° C). The mean time between the blood draw and freezing was 12 min. When not being centrifuged the samples were kept on ice. Plasma remained frozen at -20° C until being placed in the laboratory freezer at -70° C. Samples were thawed only once at the time of assay. Two ml of whole blood were used for the determination of Hct (triplicates), Hb and lactic acid by methods previously described (Farrell et al., 1979).

After the 10 mile race, the subjects reported as quickly as possible to the secluded area where identical procedures were followed with the exception that the POMS instrument was administered in 5 min after the post race blood draw. In accordance with the theoretical basis of the POMS, however, the subjects were instructed to respond in terms of how "you are feeling right now at this time" during the post-race testing. The mean time between the end of the race and the blood draw was 4.5 min with a range of 2.5 to 8.0 min.

The radioreceptor assay used was based on the assay described by Naber et al. (1980). The extraction procedure was modified to include the use of Cep PaK C_{18} chromatography in place of the acetic acid, alumnina procedure. Three ml of plasma were added to 8 ml of Trifluoroacetic acid buffer (TFA), shaken, vortexed and centrifuged at 2,000 g, 4° C, 10 min. The supernatant was decanted and saved. The pellet was resuspended with 7 ml TFA buffer and shaken, vortexed and centrifuged again. The supernatants were combined and passed through Cep PaK C_{18} three times after the cartridges had been prepared with 5 ml MeOH and 5 ml H_2O. This modification increased the recovery of L-Enk to a mean of 65% as compared to the 15% reported by Naber. $[H^3]$ - $[D-Ala^2]$ enkephalin (L-leuamide)[5] (Leu-ERA) was used as the exogenous competitive ligand.

Receptor tissue was prepared from rat brains according to Naber et al. (1980). This RRA has been structured to utilize the GTP-resistant "Type 2" opiate receptor. This results in a greater sensitivity of these receptors to opiate peptides (Pert & Taylor, 1980). The characteristics of this assay, using duplicate samples, were an intra-assay coefficient of variation of 6.9% and an inter-assay coefficient of variation of 13.2%. The lower limit of detection was 0.5 pmol/tube. All samples for this study were run in the same assay.

Results

The physical characteristics of the subjects are provided in Table 1. Plasma volume was calculated using pre- and post-race Hct and Hb according to Dill and Costill (1974). Mean ($\pm$ SD) plasma volume decreased by -6.17% ($\pm$ 4.06) due to an increase in Hct from 43.07 ($\pm$ 1.93) to 43.85 ($\pm$ 2.30) and an increase in Hb from 14.37 to 14.94 ($\pm$ 1.14). Correlation analysis failed to reveal significant relationships between PV and Leu-ERA. Plasma lactic acid increased significantly $p < 0.05$ from a mean pre-race value of 1.02 mM ($\pm$ PR 1.42) to 5.00 mM ($\pm$ 1.74) after the race with both faster and slower runners accumulating similar amounts of lactic acid.

The mean plasma Leu-ERA was 22.2 ($\pm$ 13.73) pmol/ml prior to the race and increased to 26.05 ($\pm$ 21.5) pmol/ml after the race ($p > 0.05$). The standard deviations indicate a large individual variation in the responses as shown in Figure 1. This figure is arranged so that subject A was the fastest and subject N the slowest runners. When only the post-race Leu-ERA values are plotted, Figure 2, a trend ($r = -0.52$, $p < 0.05$) of higher values for faster runners seems to exist.

Table 1

Mean Characteristics of the Subjects and Tension
Before and After the Race. N = 14

	Age (yr)	Ht (cm)	Wt (kg)	Running Experience	Tension (pre-race)	(post-race)
Mean	35.0	170.6	61.8	7.6	10.8	5.6*
SD	10.8	8.4	11.1	5.2	7.2	3.9
Range	(21–53)	(157–183)	(40.8–82.1)	(1–18)		

*Post-race value significantly lower than pre-race value, $p < 0.05$.

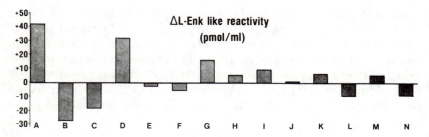

Figure 1—Post-pre race (10 mile) leucine-enkephalin radioreceptor activity, pmol/ml. Subjects are arranged so that subject A was the fastest and subject N, the slowest runner.

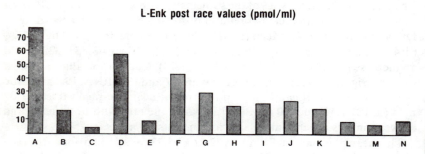

Figure 2—Post 10 mile race values for leucine-enkephalin radioreceptor activity, pmol/ml. Subjects are arranged as in Figure 1. Post-race values were significantly (p < 0.05) correlated to race time, r = -0.52.

The POMS yields six mood state scores, but the present investigation was only concerned with the tension-anxiety factor. Table 1 provides a summary of these data, and it will be noted that tension-anxiety decreased from a mean value of 10.8 prior to the race, to a value of 5.6 following the race. This reduction in tension was statistically significant as measured by a paired-t test (P < .05). This finding is in agreement with a series of studies which indicate that vigorous physical activity is associated with a decrease in tension (Markoff, Ryan & Young, 1982; Morgan, 1979; Morgan, Horstman, Cymerman & Stokes, 1980). Eleven of the 13 subjects experienced decreases in tension, one did not change, and one had a slight increase. The mean tension reduction for those subjects (N = 7) characterized by elevated post-exercise Leu-ERA was 6.3 in comparison to 2.8 for those subjects (N = 7) who did not experience an elevation in Leu-ERA. This difference in tension, however, was not statistically significant (P > .05).

Discussion

The mean value for Leu-ERA is an order of magnitude higher than values previously reported by Ryder and Eng (1981). This discrepancy is partially attributed to the fact that an RRA will detect all biologically active ligands whereas an RIA will bind only those ligands containing antigenic amino acid sequences which may or may not be biologically meaningful (Gros et al., 1978; Naber et al., 1980). The source of the detected Leu-ERA is not known but may be the adrenal medulla since Lundberg et al. (1979) have demonstrated the existence of enkephalin-like immunoreactivity in this gland and Schultzberg et al. (1978) further suggested the possibility of enkephalins being released from the adrenal gland into the blood.

It was not possible to judge the relative stress of the 10 mile run for all subjects since we did not have the opportunity to test all subjects on a treadmill. However, we had oxygen consumption ($\dot{V}_{O_2}$) versus treadmill velocity data as well as maximal oxygen consumption ($\dot{V}_{O_2max}$) values for six subjects who had participated in a previous study. The mean %$\dot{V}_{O_2max}$ for these six subjects was 82% ($\pm$ 2.9) with a range of 78 to 86%, for the average pace over the 10 miles. This mean value agrees well with previously reported values for this distance race (Farrell, 1979). The small range for %$\dot{V}_{O_2max}$ suggests that the relative intensity (%$\dot{V}_{O_2max}$) of the race was similar between slow and fast runners. (As an example, subject M ran the race in 85 min which required 82% of $\dot{V}_{O_2max}$, while subject B ran the race in 56 min and used 83% of $\dot{V}_{O_2max}$.)

A large individual variation in the Leu-ERA response to competitive running is noteworthy. As shown in Figure 1, the response, both increases and decreases, is much more pronounced in the faster runners. A large individual variation in the Beta-endorphin (B_h-E_p) response to running was reported by Colt et al. (1981) in that during moderate exercise, which was not further defined, only 45% of the runners demonstrated increases in plasma levels of B_h-E_p. In this same report after a more intense run, again not further defined, only 80% of the runners had increased plasma levels of B_h-E_p. We have also reported a large individual variation in the B_h-e_p in response to treadmill running in human plasma (Farrell et al., 1982).

In light of the growing evidence that peripheral levels of endorphins may not reflect central levels (Pardridge et al., 1981) and that the B-E_p and enkephalin systems are probably distinct, it is difficult to address the physiological significance of either the increased or decreased plasma levels of Leu-ERA demonstrated in this study. It is also important to note that these peptides may have peripheral effects since enkephalin receptors have been identified in numerous peripheral tissues. It should also be noted that many peripherally administered peptides, including enkephaln analogues, have potent CNS effects in humans (Graffenried et al., 1978). In the only animal study (to our knowledge) that has investigated rat brain opiate receptor occupancy before and after exercise, Pert and Bowie (1979) found decreased exogenous binding after short duration running and swimming, which was interpreted as an increased turnover of the endogenous ligand.

While not statistically significant, it is interesting that the subjects who experienced the greatest tension reduction also had higher post race Leu-ERA when compared to the pre race values. We agree with Colt et al. (1981) that the increases in plasma B_h-E_p after exercise are not of a large enough magnitude to justify the conclusion that endogenous opiates are responsible for significant mood elevation sometimes reported by distance runners. Further investigations are necessary, however, to exclude the possibility that these small increases in plasma levels reflect large changes of these opiates in the brain. On the other hand, the recent report by Markoff et al. (1982) suggests that this may not be the case. These investigators observed decreases in tension-anxiety following distance runs where the narcotic antagonist, naloxone, was given in a double-blind fashion, and naloxone did not inhibit the development of positive mood following the runs.

Summary

It is important to note that this study differs from previous studies which have used radioimmunoassays in that we measured only biologically active ligands as opposed to immunologically active ligands. As an example, our assay does not crossreact with beta-lipotropin whereas all the previous studies of endorphins and exercise used antibodies which did not totally differentiate between beta-endorphin and beta-lipotropin. Since beta-lipotropin has no opiate activity, the chances of our measuring non-opiate-like ligands is greatly reduced when compared to radioimmunoassays.

Plasma levels of Leu-ERA (radioreceptor assay) and tension-anxiety (POMS) were measured in competitive runners before and after a 10 mile road race. Tension was observed to decrease significantly as reported in earlier research, and was not related to Leu-ERA. Mean plasma levels of Leu-ERA did not change; however, a large individual variation was noted with some subjects markedly increasing while others demonstrated variable decreases. The change in Leu-ERA was not related to changes in plasma volume nor the time to complete the race; however, the post race Leu-ERA values were significantly related to the time to complete the race. The physiological significance of these alterations requires further investigation.

References

CARR, D.B., B.A. Bullen, G.S. Skriner, M.A. Arnold, M. Rosenblatt, I.Z. Beitins, J.B. Martin & J.W. McArthur. Physical conditioning facilitates the exercise-induced secretion of Beta-Endorphin and Beta-Lipotropin in women. New Eng. J. Med. 305:560–563, 1981.

COLT, E.W.D., S.L. Wardlow & A.G. Frantz. The effect of running on plasma B-endorphin. Life Sci. 28:1637–1640, 1981.

DILL, D.B. & D.L. Costill. Calculation of percentage charges in volumes of blood, plasma, and red cells in dehydration. J. Appl. Physiol. 37:247–248, 1974.

FARRELL, P.A., W.K. Gates, M.G. Maksud & W.P. Morgan. Increases in plasma Beta-Endorphin/Beta-Lipotropin immunoreactivity after treadmill running in man. J. Appl. Physiol.: Respir. Environ. and Ex. Physiol. 52:1245–1249, 1982.

FARRELL, P.A., J.H. Wilmore, E.F. Coyle, J.E. Billing & D.L. Costill. Plasma lactate accumulation and distance running performance. Med. Sci. Sports 11(4):338–344, 1979.

GRAFFENRIED, B., E.D. Pozo, J. Roubicek, E. Krebs, W. Poldinger, P. Burmeister & L. Kerp. Effects of the synthetic enkephalin analogue, FK 33-824 in Man. Nature 270:729–730, 1978.

GROS, C., P. Pradelles, C. Rouget, O. Bepoldin, F. Dray, M.C. Fournie-Zaluski, B.P. Rogues, H. Pollard, C. Llorens-Cortes & J.C. Schwartz. Radio immunoassay of Methionine and Leucine-Enkephaline in regions of rat brain and comparison with endorphins estimated by a radioreceptor assay. J. Neurochemistry 31:29–39, 1978.

HARTLEY, L.H., J.W. Mason, R.P. Hogan, L.G. Jones, T.A. Kotchen, E.H. Mougey, F.E. Wherry, L.L. Pennington & P.T. Ricketts. Multiple hormonal responses to prolonged exercise in relation to physical training. J. Appl. Physiol. 33:607–610, 1972.

LEWIS, R.V., A.S. Stern, S. Kimura, J. Rossier, S. Stein & S. Udenfriend. An about 50,000 dalton-protein in adrenal medulla: A common precursor of [Met] - and [Leu] Enkephalin. Science 208:1459–1461, 1980.

LI, C.H., A.J. Rao, B.A Doneen & D. Yamashiro. B-endorphin: lack of correlation between opiate activity and immunoreactivity by radioimmunoassay. Biochem. Biophys. Res. Commun. 75:576–580, 1977.

LINNOILA, R.I., R.P. Diaugustine, A. Hervonen & R.J. Miller. Distribution of [Met[5]] - and [Leu[5]] - Enkephalin-, Vasoactive intestinal polypeptide - and Substance P-like immunoreactivities in human adrenal glands. Neuroscience 5:2247–2259, 1980.

LUNDBERG, J.M., B. Hamberger, M. Schultzberg, T. Hokfelt, P.O. Grandberg,

S. Efendic, L. Terenius, M. Goldstein & R. Luft. Enkephalin and somatostatin-like immunoreactivities in human adrenal medulla and pheochromocytoma. Proc. Natl. Acad. Sci. 76:4079–4083, 1979.

MARKOFF, R.A., P. Ryan & T. Young. Endorphins and mood changes in long-distance running. Med. Sci. Sports & Exercise 14:11–15, 1982.

MCNAIR, D.M., M. Lorr & L.F. Droppleman. Profile of Mood States Manual. Educational and Industrial Testing Service, San Diego, CA, 1971.

MORGAN, W.P., Anxiety reduction following acute physical activity. Psychiatric Annals 9:36–45, 1979.

MORGAN, W.P., D.H. Horstman, A. Cymerman & J. Stokes. Exercise as a relaxation technique. Primary Cardiology 6:48–57, 1980.

NABER, D., D.P. Pickar, R.A. Dionne, D.L. Bowie, B.A. Ewols, T.W. Moody, M.G. Soble & C.B. Pert. Assay of endogenous opiate receptor ligands in human CSF and plasma. Sub. Alcohol Act./Misuse 1:113–118, 1980.

PARDRIDGE, W.M. & L.J. Mietus. Enkephalin and blood-brain barrier: Studies of binding degradation in isolated brain microvessels. Endocrinology 109:1138–1143, 1981.

PERT, C.B. & D.L. Bowie. Behavioral manipulation of rate causes alterations in opiate receptor occupancy. In E. Usdin, W.E. Bunney & N.S. Kline (Eds.), Endorphins in Mental Health Research, pp. 93-104. New York: Oxford University Press, 1979.

PERT, C.B. & D. Taylor. Type 1 and Type 2 opiate receptors: a sub-classification scheme based upon GTP's differential effects on binding. In E.C. Way (Ed.), Endogenous and Exogenous Opiate Agonists and Antagonists, pp. 87–90. Oxford: Pergamon Press, 1980.

RYDER, S.W. & J. Eng. Radioimmunoassay of Leucine-Enkephalin-like substance in human and canine plasma. J. Clin. Endocrin. Metab. 52:367–369, 1981.

SCHULTZBERG, M., J.M. Lundberg, T. Hokfelt, L. Terenius, J. Brandt, R.P. Elde & M. Goldstein. Enkephalin-like immunoreactivity in gland cells and nerve terminals of the adrenal medulla. Neuroscience 3:1169–1186, 1978.

SHANKS, M.F., V.C. Jones, C.J. Linsell, P.E. Mullen, L.H. Rees & G.M. Besser. A study of 24-hour profiles of plasma Met-Enkephalin in man. Brain Res. 212:403–419, 1981.

SMITH, R., A. Grossman, R. Gaillard, V.C. Jones, S. Ratter, J. Mallinson, P.J. Lowry, G.M. Besser & L.H. Rees. Studies on circulating Met-Enkephalin and B-Endorphin: Normal subjects and patients with renal and adrenal disease. Clin. Endocrin. 15:291–300, 1981.

Influence of the Training Level on the Dynamics of Plasma Androgens at Rest and during Exercise in Male Dog

J. Gagnon, D. DeCarufel, G.R. Brisson and R.R. Tremblay
Laval University and University of Quebec at
Trois-Rivières, Québec, Canada

It is recognized that the endocrine system reacts to physical exercise. However, several factors have led to various interpretations in the behavior of circulating androgens during exercise; the type of exercise (running, cycling, etc), its intensity (light to exhausting loads), its duration (10 to 180 min, 300 m to 42.2 km) and differences in age, sex and degree of physical fitness between the subjects have contributed to the diversity of the reported results (Brisson et al., 1981; Kuoppasalmi et al., 1981; Sutton et al., 1978). Oppenheimer and Gurpide (1975) have also noted that the clinical effects of a hormone were related to certain kinetic factors not directly reflected by the amounts of the hormone present in the circulatory flow. In view of these difficulties, a convenient model based upon the constant infusion of labelled steroids (Gurpide, 1975) has been developed to study the dynamics of testosterone and androstenedione during rest or standardized exercise in male dogs, because our knowledge of the effects of acute and chronic exercise on androgen metabolism remains scarce.

Methods

Animals

Fifteen adult male mongrel dogs weighing 12.3 ± 1.0 kg were used in this study. The animals were selected on the basis for their ability to run on a motor driven treadmill (Quinton). They were housed individually with respect of a diurnal rhythm lighting schedule.

Treadmill Running Test

All the animals were subjected to a submaximal exercise test (SMT). Tipton et al., (1974) has developed a seven-stage, 21-min progressive exercise test with each stage of 3 min duration. The resting and exercise heart rates were recorded at each minute for a duration of 20 s. Analysis of the SMT was performed in order to estimate the surface area under the curve corresponding to the heart rate (bpm) vs time (min) for each animal.

Training Program

Eight of the animals were subjected to a progressive training program. Animals were considered trained when a statistically significant reduction in area under the curve (bpm min) had been achieved during the SMT-Tipton performed before and after the training program. The exercise program was a modification of the regimen described by Tipton et al. (1974). The progressive treadmill running program lasted 6 weeks and consisted of five sessions per week; 2 days were devoted to exercise consisting of high speed and low grade (10 km/h; 5% incline; 15 min) and 3 days were allotted for low speed and high grade (6.4 km/h; 12% incline; 30 to 60 min).

Experimental Design

Immediately after the dogs were anesthetized, a modified radiopaque double-lumen catheter (7 fr) was inserted into the left external jugular vein and its distal lumen was carefully localized under fluoroscopy in the caudal vena cava for blood sampling at regular time intervals; at 10 cm ahead from the distal part of the catheter, a 1 mm hole in the second probe allowed the tracer infusion in the right atrium. After catheterization, a minimum of 2 days elapsed prior to the beginning of the infusion of either labelled hormones.

Constant Infusion of Labelled Steroids

$[1\beta\text{-}2\beta\text{-}^3H]$ testosterone (40 Ci/mmol; NEN) or $[1,2\text{-}^3H]$ androstenedione (40 Ci/mmol, NEN) were infused separately, at a 3-day interval, at a constant rate (300,000 dpm/min). The delivery of the pump was adjusted to 2 ml/min during 165 min. Blood samples of 15 ml were collected into heparinized tubes every 15 min from the caudal vena cava to reach hormonal equilibrium at rest (90 to 105 min) or following 1 h treadmill running (150 to 165 min) at 6.4 km/h with 10% grade. Two different

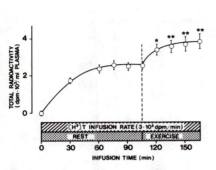
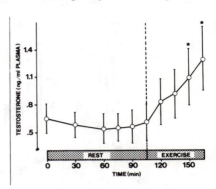

Figure 1—Experimental procedure used in the rest and exercise studies to measure the dynamics of androgens. Testosterone is infused at a constant rate to reach hormonal steady-states at rest (90 to 105 min) or exercise (150 to 165 min) (left panel). Evolution of hormonal plasma levels are represented (right panel). The open circles (mean ± SEM) correspond to the blood sampling at indicated time intervals. Similar procedure is used to measure the dynamics of androstenedione. Significant difference between mean and rest value (105 min) (*p < .05; **p < .01) is indicated.

steady-states were observed with T and A and the behavior of T is illustrated, as example, in Figure 1. Hormonal concentrations at steady-state were used to calculate the metabolic clearance rate (MCR) of both steroids.

Hormonal Determinations

Total plasma testosterone (T) and androstenedione (A) were determined by RIA as described by Brisson et al. (1981). Plasma samples containing labelled steroids were extracted with ethyl ether (3 vol. χ for total radioactivity measurements. The ether extract was applied on thin layer chromatography (Merck, Germany) for separation of T and A (Benzene: methanol 9:1 (I) and benzene: ethyl acetate 3:2 (II). A and T recoveries obtained by addition of [^{14}C] A and [^{14}C] T averaged respectively 73.3 ± 11.3% (I) and 49.1 ± 12% (I + II). Radioactivity was measured by double isotope counting (Beckman LS9000) and each hormonal value was corrected by reference to plasma protein concentration to take into account modifications of plasma volumes. Metabolic clearance rates (MCR^T and MCR^A), conversion ratio of precursor to product CR^{A-T} and CR^{T-A}), blood production rates (PB^T and PB^A) of testosterone and androstenedione were calculated as described by Gurpide (1975). Statistical analyses were accomplished using a paired-t-test and correlation from linear regression.

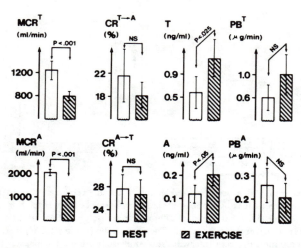

Figure 2—Effects of 1 h treadmill running (6.4 km/h; 10% incline) on the kinetics variables of testosterone (T) and androstenedione (A) in ten dogs. Mean values ± SEM for metabolic clearance rates (MCR) of T & A, conversion rates of T → A and A → T, plasma levels of T & A and blood production rates of T & A were calculated, at rest and exercise, by a non-compartmental model.

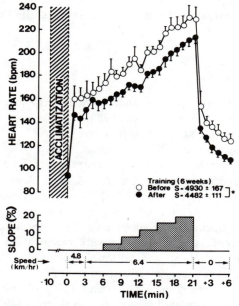

Figure 3—Resting and exercise heart rates during submaximal treadmill test (SMT) of Tipton before "o" and after a progressive training program for 6 weeks (5 days/week) "●" in the same group of dogs (n = 8). The area under the curve from the submaximal test of Tipton (bpm min) is calculated before and after the training program.
*Significant difference between mean values (p < .05).

Results

Kinetics of Androgens in Untrained Dogs

As can be seen from Figure 2, acute exercise was accompanied by a significant decrease ($p < 0.001$) of MCR of both steroids, while conversion ratio between T and A remained unchanged. As expected, peripheral plasma concentration increased from 61% (A) to 105% (T) following exercise but no statistical difference could be demonstrated in the production rates of both hormones at their respective steady-states.

Kinetics of Androgens in Trained Dogs

Trained vs untrained animals were selected according to their performance when submitted to SMT of Tipton (Figure 3). Surfaces ranging between 4 to 5×10^3 were estimated as representative of trained (4×10^3) to untrained (5×10^3) animals. Under these circumstances, none of the kinetics parameters of androgens in dog at rest and exercise were significantly affected by their level of training (Figure 4).

Discussion

It has been postulated by several authors that the increase in plasma testosterone (T) and androstenedione (A) following short-term exercise in humans was probably attributable to a decrease in hepatic metabolic clearance rate (MCR) of both steroids (Kuoppasalmi et al., 1981). This observation found support in the data of the present study which was undertaken in dogs for a precise assessment of the role of the hepatic clearance of T and A. The constant infusion technique of labelled steroids which has been used in highly controlled conditions allowed the demonstration that sustained exercise led to a new steady-state of blood steroids, significantly higher than the one measured in resting conditions, and then the MCR decrease appeared responsible for the peripheral changes in circulating T; the parallel plasma elevation of A, which is produced by the adrenal cortex, and secreted along with cortisol in the course of physical exercise, might very well reflect decreased MCR^A in association with a slight elevation of glandular secretion. In our experimental conditions, PB^T and PB^A were, nevertheless, unaffected by exercise. Recent reports in humans indicated that the kinetics of estradiol (Keiser et al., 1981) and testosterone (Jurkowski et al., 1978; Sutton et al., 1978) followed a pattern similar to the one observed in dogs submitted to physical exercise.

In an attempt to determine whether long-term training could cause a more favorable androgenic balance in trained vs sedentary dogs, plasma

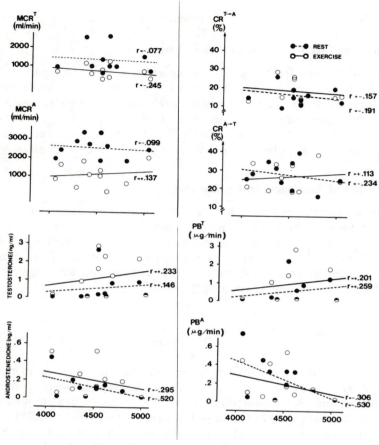

SMT – TIPTON (area under the curve : bpm·min)

Figure 4—Correlation between kinetics variables for testosterone (T) and androstenedione (A) and the level of training expressed as the area under the curve from the submaximal test of Tipton (Figure 3) for each dog at rest and exercise. The correlation coefficient (r) and the equation of the curve were calculated by linear regression analysis.

kinetics of T and A was measured after a 6-week program of controlled exercise whose intrinsic value was estimated by the SMT-Tipton (1974) test. None of the kinetics parameters of androgens, at rest or exercise, were significantly affected by the level of training of our animals. These data strongly suggest that the changes in circulating plasma testosterone (decreased following prolonged exercise; increased during short-term exercise) occurring in different mammals subjected to training are transitory in nature. Thus, in term of androgenic-catabolic interactions in the whole body, intense training does not accelerate the hepatic clearance of androgens although non-measurable extrahepatic clearance might be slightly altered as part of an adaptive process which contributes to a more active utilization of testosterone by muscles.

References

BRISSON, G.R., D. DeCarufel, J. Brault, M.A. Volle, A. Audet, M. Desharnais & C. Lefrancois. Circulating Δ^4-androgen levels and bicycle exercise in trained young men. In J. Poortmans & G. Niset (Eds.), Biochemistry of Exercise IV-B. Baltimore: University Park Press, 1981.

GURPIDE, E., Tracer methods in hormone research. (Monograph in Endocrinology; V. 8). New York: Springer-Verlag, 1975.

JURKOWSKI, J.E., N.L. Jones, W.C. Walker, E.V. Younglai, & J.R. Sutton. Ovarian hormonal responses to exercise. J. Appl. Physiol. 44:109–114, 1978.

KEISER, H.A., J. Poortmans & G.S.J. Bunnik. Influence of physical exercise on sex hormone metabolism. In J. Poortmans & G. Niset (Eds.), Biochemistry of Exercise IV-B. Baltimore: University Park Press, 1981.

KUOPPASALMI, K., H. Näveri, K. Kosunen, M. Härkönen & H. Adlercreutz. Plasma steroid levels in muscular exercise. In J. Poortmans & G. Niset (Eds.), Biochemistry of Exercise IV-B. Baltimore: University Park Press, 1981.

OPPENHEIMER, J.H. & E. Gurpide. Quantitation of the production, distribution, and interconversion of hormones. In L.J. DeGroot et al. (Eds.), Endocrinology, Vol. 3. New York: Grune & Stratton, 1979.

SUTTON, J.R., M.J. Coleman & J.H. Casey. Testosterone production rate during exercise. In F. Landry & W.A.R. Orban (Eds.), Third International Symposium of Biochemistry of Exercise, Vol. 3. Quebec: Editeur officiel, 1978.

TIPTON, C.M., R.A. Carey, W.E. Eastin & H.H. Erickson. A submaximal test for dogs: evaluation of effects of training, detraining, and cage confinement. J. Appl. Physiol. 37:271–275, 1974.

Changes in Plasma Gut Hormones during Exercise in the Horse

G.M. Hall, J.N. Lucke, T.E. Adrian and S.R. Bloom
Royal Postgraduate Medical School, London, and
University of Bristol, Langford, Bristol, England

Prolonged exercise in humans has been used to study changes in circulating gut hormones in an attempt to establish their physiological roles and the stimuli responsible for the release of these peptides (Galbo et al., 1975; Hilsted et al., 1980). In a recent study, Hilsted et al. (1980), found that prolonged exercise was associated with a significant increase in plasma concentrations of vasoactive intestinal polypeptide (VIP), pancreatic polypeptide (PP), somatostatin (SRIF) and secretin. Additionally, these authors observed a significant correlation between the hypoglycemia of exercise and plasma VIP, PP and pancreatic glucagon (PG) values.

The purpose of the present study was twofold: First, to determine whether similar changes in circulating gut hormones occurred during exercise in the horse; secondly, by studying two metabolically different types of exercise, (hypo- and hyperglycemia) to determine if changes in these were related to circulating metabolites.

Methods

The two types of exercise investigated were an 80 km long-distance ride, (Golden Horseshoe Ride) and a 42 km race (Arab Horse Society's Marathon Race).

80 km Ride

Venous blood was collected from seven horses on the morning before the 80 km ride, 3 to 4 h after the last feed. A second sample was obtained immediately after the ride was completed. The mean speed of the horses studied was 12.1 km/h.

42 km Race

Venous blood was collected from five horses during the late evening before the race, 1 to 2 h after feeding, and from the same horses immediately after completion of the race. The mean speed of the animals studied was 25.2 km/h.

Analysis

Plasma concentrations of gut peptides were measured by specific radio-immunoassays. Blood glucose and lactate concentrations and plasma non-esterified fatty acids (NEFA) were measured as described previously (Lucke & Hall, 1980).

80 km Ride

The 80 km ride produced a profound hypoglycemia with a decrease in blood glucose from 3.47 to 2.11 mmol/l ($p < 0.01$), together with an eightfold rise in NEFA to 2.28 mmol/l($p < 0.001$). There was only a small increase in blood lactate from 0.75 to 1.92 mmol/l($p < 0.05$) showing that aerobic muscle metabolism predominated during the ride. Plasma SRIF, PP, VIP and gastrin increased significantly ($p < 0.05$) during the ride, whereas there was an insignificant decrease in gastric inhibitory peptide (GIP) from 75.0 to 50.3 pmol/l. The hypoglycemia and increased lipolysis were associated with a significant increase in PG from 3.0 to 142.6 pmol/l ($p < 0.01$) and decrease in insulin from 78.7 to 13.1 pmol/l ($p < 0.01$).

42 km Race

There was a significant rise in blood glucose from 2.39 to 8.29 mmol/l ($p < 0.01$) and in plasma NEFA from 0.10 to 1.55 mmol/l ($p < 0.01$). Blood lactate increased to 7.60 mmol/l ($p < 0.01$) which demonstrated the importance of anaerobic muscle metabolism during the race. Plasma SRIF and VIP values increased during the race ($p < 0.05$ and < 0.01 respectively) but the rise in PP from 15.8 to 27.8 pmol/l was not significant ($p < 0.1$). Circulating gastrin and GIP values did not change significantly. PG increased to 65.4 pmol/l ($p < 0.01$) but there was no change in plasma insulin concentration. The recent ingestion of food before the control samples were collected was evident from the higher pre-race gastrin, GIP and insulin, and lower NEFA values compared with the 80 km ride.

Table 1

Mean (and) SEM Plasma Hormone and Metabolite Values before and after an 80 km Ride and a 42 km Race

		80 km ride		42 km race	
		Pre-ride	Post-ride	Pre-ride	Post-ride
Insulin	(pmol/l)	78.7 ± 18.5	13.1 ± 4.8**	174.6 ± 46.7	148.8 ± 45.0
Pancreatic glucagon	(pmol/l)	3.0 ± 1.2	142.6 ± 37.0***	3.5 ± 1.3	65.4 ± 11.1**
Somatostatin	(pmol/l)	10.0 ± 2.9	16.4 ± 2.5*	15.0 ± 1.9	19.7 ± 1.7*
Panreatic polypeptide	(pmol/l)	24.7 ± 4.7	101.6 ± 26.7*	15.8 ± 5.0	27.8 ± 6.8
Vasoactive intestinal polypeptide	(pmol/l)	11.1 ± 2.3	24.0 ± 2.6*	7.1 ± 2.6	16.4 ± 3.1**
Gastrin	(pmol/l)	1.5 ± 0.3	5.1 ± 1.0*	7.0 ± 1.1	6.0 ± 2.4
Gastric inhibitory peptide	(pmol/l)	75.0 ± 24.0	50.3 ± 5.0	142.0 ± 38.8	139.4 ± 15.9
Glucose	(mmol/l)	3.47 ± 0.20	2.11 ± 0.22**	2.39 ± 0.45	8.29 ± 0.81**
Non-esterified fatty acids	(mmol/l)	0.29 ± 0.02	2.28 ± 0.13***	0.10 ± 0.04	1.55 ± 0.27**
Lactate	(mmol/l)	0.75 ± 0.07	1.92 ± 0.34	0.56 ± 0.14	7.60 ± 1.18**

Significance of difference of means from pre-ride values shown by *p < 0.05, **p < 0.01, and ***p < 0.001.

Discussion

The results of this study show that there are significant increases in plasma VIP, SRIF and PG values during exercise which are not dependent, either on the relative contributions of aerobic and anaerobic muscle metabolism, or on associated changes in blood glucose. The hyperglucagonemia response to exercise has been well documented (Galbo et al., 1975), and reflects increased adrenergic activity. It is likely that the severe hypoglycemia observed during the 80 km ride further exacerbated the increase in PG.

The results of the 80 km ride confirmed the changes in plasma PP, VIP and SRIF found by Hilsted et al. (1980), during prolonged exercise in humans. Hypoglycemia was probably the main stimulus to PP secretion during the ride, since there was only a small increase during the 42 km race. The main physiological effects of PP are the inhibition of pancreatic exocrine secretion and bile secretion (Adrian & Bloom, 1982), changes which are clearly appropriate during prolonged exercise in which no food is taken.

VIP exerts a general vasodilatory effect and has been shown to evoke atropine-resistant dilation of blood vessels in skeletal muscle (Jarhult et al., 1980). It is tempting to speculate that the increase in plasma VIP during exercise has a role in regulating the increase in muscle blood flow.

There is controversy about whether SRIF acts as a circulating hormone (Adrian & Bloom, 1982) but there is general agreement that it has a role as a local hormone or paracrine substance. Induced hypoglycemia elicits a twofold increase in plasma SRIF, but surgical stress with hyperglycemia had no effect on SRIF secretion (Wass et al., 1980). It is difficult to provide an explanation for the significant increases in SRIF found in both the 80 km ride and 42 km race. SRIF may restrain the rate of absorption of nutrients from gut to circulation; a suitable adaptive response to the decrease in splanchnic blood flow found during exercise.

In conclusion, we have shown significant changes in circulating gut peptides in two different forms of exercise in the horse. The precise function of this hormonal response is not apparent, but we suggest that it may be part of the general physiological adaptation to exercise.

References

ADRIAN,T.E. & S.R. Bloom. Gut hormones and related neuropeptides: physiology and clinical implications. In J.L.H. O'Riordan (Ed.), Recent Advances in Endocrinology and Metabolism-2. London: Churchill Livingstone, 1982.

GALBO, H., J.J. Holst & N.J. Christensen. Glucagon and plasma catecholamine responses to graded and prolonged exercise in man. J. Appl. Physiol. 38:70–76, 1975.

HILSTED, J., H. Galbo, B. Sonne, T. Schwartz, J. Fahrenkrug, O.B. Schaffalitzky-de Muckadell, K.B. Lauritsen & B. Tronier. Gastroenteropancreatic hormonal change during exercise. Am. J. Physiol. 239:G136–140, 1980.

JARHULT, J., P. Hellstrand & F. Sundler. Immunohistochemical localization and vascular effects of vasoactive intestinal polypeptide in skeletal muscle of cat. Cell Tissue Research 207:55–64, 1980.

LUCKE, J.N. & G.M. Hall. Further studies on the metabolic effects of long riding: Golden Horseshoe Ride 1979. Equine Vet. J. 12:189–193, 1980.

WASS, J.A.H., E. Perman, S. Medbak, A.M. Dawson, D. Tsiolakis, V. Marks, G.M. Besser & L.H. Rees. Immunoreactive somastostatin changes during insulin-induced hypoglycaemia and operative stress. Clin. Endocrinology 12:269–275, 1980.

Effects of Exhaustive Exercise on Prostaglandin Metabolism in SHR Rat Kidney

I. Hashimoto, M. Higuchi and K. Yamakawa

National Institute of Nutrition, Tokyo, Japan

Vigorous physical exercise increases plasma and urinary prostaglandin concentrations in man (Greaves et al., 1972; Hashimoto & Lamb, 1980; Norwak & Wennmalm, 1978) and laboratory animals (Zambraski & Dunn, 1980). Because prostaglandins (PG) can have important effects on blood pressure, this study was designed to examine the effects of exhaustive exercise on PGE concentrations in the kidneys of sedentary and voluntarily trained spontaneously hypertensive rats (SHR).

Methods

Male SHR rats (90 to 130 g body weight) were randomly assigned, seven per group to one of four groups: 1) Sedentary-Rest, 2) Sedentary-Exhausted, 3) Trained-Rest, 4) Trained-Exhausted. The animals were trained by voluntary exercise in running wheels for 6 wk. Animals were housed in individual cages with or without running wheels and fed rat chow and water ad libitum. Food was withdrawn 12 h prior to an exhaustive run, which was done on a motor-driven treadmill at a speed of 21.5 m min on 5% slope. SHR rats were killed by decapitation immediately after the exercise or control treatments, and both kidneys from each animal were removed, frozen in 95% ethanol cooled with dry ice, and stored at $-85°$ C for 1 day until extraction of PGE. Samples of heart, liver, soleus and gastrocnemius muscles from both hind limbs were frozen in liquid nitrogen, and stored in a freezer at $-85°$ C until later enzyme analysis. Other samples of liver, heart and skeletal muscles were placed in tubes containing 0.5 ml 2N HCl for glycogen assay. Each frozen kidney was minced and homogenized in 10 ml Tris buffered saline solution with glass Potter-Elvehjem homogenizers

Table 1

Effect of Voluntary Exercise Training on Succinate Dehydrogenase Activity and Tissue Glycogen Concentrations of Spontaneously Hypertensive Rats

	Acute exercise	No. of rats	Succinate dehydrogenase activity (μmol/g/min)				Glycogen concentration (μmol/g)				
			Heart	Soleus	R.G.[a]	W.G.[b]	Heart	Liver	Soleus	R.G.	W.G.
Sedentary control	Rest	7	20.97 ± 0.63	7.32 ± 0.35	7.65 ± 0.37	2.29 ± 0.19	27.48 ± 0.71	40.07 ± 5.38	19.95 ± 1.26	19.44 ± 0.85	25.70 ± 1.18
	Exhausted	7	20.02 ± 1.01	7.48 ± 0.25	7.51 ± 0.41	2.15 ± 0.17	4.24 ± 0.86	6.54 ± 0.65	2.05 ± 0.15	1.34 ± 0.13	3.12 ± 0.34
Voluntary trained	Rest	7	20.06 ± 0.35	7.55 ± 0.25	10.59** ± 0.30	2.16 ± 0.19	26.08 ± 1.38	26.45 ± 4.67	17.64 ± 0.75	23.27 ± 1.86	23.87 ± 0.71
	Exhausted	7	21.44 ± 0.75	7.78 ± 0.21	10.07** ± 0.56	2.30 ± 0.18	7.72 ± 1.61	7.65 ± 0.51	4.51** ± 0.71	3.84 ± 1.67	8.02* ± 2.13

Values are means ± SEM. Differences between sedentary control vs trained group by voluntary exercise are significant. * p < 0.05, ** p < 0.01.
[a]Red gastrocnemius
[b]White gastrocnemius

that were immersed in ice water during homogenization. Recoveries of tritium-labelled PGE_1 (3,000 cpm, New England Nuclear, Boston, Massachusetts) added to an aliquot of the kidney homogenate averaged 72%. PGE was extracted on silicic acid columns (Dray et al., 1975), and dried in a nitrogen gas stream converted to prostaglandin B (PGB)(Zusman, 1972) and subjected to radioimmunoassay (Jaffe et al., 1972) with antisera to PGB provided by Clinical Assays, Cambridge, Massachusetts, U.S.A.

Glycogen concentrations and succinate dehydrogenase (SDH) activities of rat tissues were assayed by following the fluorometric method of Sembrowich et al. (1977). Blood pressure was measured on the tail by using a plethysmographic device produced by Natsume Factory Co., Tokyo, Japan.

Results

There was no measurable change in blood pressure of trained SHR rats as a result of voluntary training. SDH activity of the red gastrocnemius muscle was 36% greater ($p < 0.01$) for the trained animals than the sedentary controls, whereas SDH activities of heart, soleus and white gastrocnemius muscles were unchanged, as shown in Table 1. Sedentary and voluntarily trained animals ran for 2.39 ± 0.22 and 10.08 ± 1.17 h (mean $\pm$ SEM), respectively. Colonic temperature of the exhausted rats was 3.8° C higher than that of rats which did not run on the treadmill. Exercise reduced the glycogen reserves of the liver, myocardium and skeletal muscles by 70 to 95% for all groups (Table 1). The kidney PGE concentration of trained SHR rats at rest was 12% ($p < 0.05$) greater than that of sedentary SHR rats. After exhaustion, the kidney PGE concentration of sedentary SHR rats was 39% ($p < 0.01$) higher than that of rats sacrificed at rest, but that of trained SHR rats was unaffected by the exhaustive exercise (Figure 1).

Discussion

Renal PG synthesis is influenced by many factors such as renal nerve stimulation, angiotensin II, norepinephrine, and genetic background (Dunn & Hood, 1977; Hashimoto et al., 1981; McGiff et al., 1972). Some of these changes also occur during strenuous exercise. The mechanism of exercise-induced increase in renal PGE concentration remains unknown. We reported that a trained SHR rat had higher concentrations of kidney PGE than those of a sedentary animal, whereas there was measured no change in Wistar Kyoto Rat (WKY) as a result of training (Hashimoto et al., 1981). This may be related with the findings of Pace-Asciak (1976), who

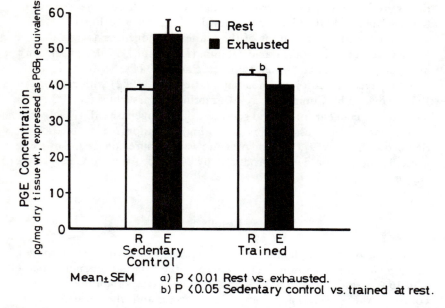

Figure 1—Kidney prostaglandin E concentrations following exhaustive exercise in male sedentary and trained SHR rats.

reported that young pre-hypertensive SHR have greater synthetic capacity for PGE_2 and $PGF_{2\alpha}$ in the kidney than that of normotensive WKY rats. The effect of exercise on renal PGE was expected in light of the finding of an exercise-induced increase in urinary PGE excretion, presumably derived from renal PGE (Zambraski & Dunn, 1980). The kidney PGE concentration of sedentary SHR rats increased, but that of trained SHR rats was not influenced by the exhaustive running. The results of these experiments demonstrate that the exercise-induced PGE production in the kidney may be associated with the relative intensity of the exercise and/or other factors related to the state of training. It may be concluded that both acute and chronic exercise can modify PG metabolism in SHR rats. This may be a clue to the regulation of blood pressure.

Acknowledgment

This work was supported by a grant from the Agency of Science and Technology.

References

DRAY, F., B. Charbonnel & J. Maclouf. Radioimmunoassay of prostaglandins F, E_1 and E_2 in human plasma. Europ. J. Clin. Invest. 5:311–318, 1975.

DUNN, M.J. & V.L. Hood. Prostaglandins and the kidney. Am. J. Physiol. 1977.

GREAVES, M.W., W.J. MacDonald-Gibson & R.G. MacDonald-Gibson. The effect of venous occlusion, starvation, and exercise on prostaglandin activity in whole human blood. Life Sci. 1:919–924, 1972.

HASHIMOTO,I., M. Higuchi, K. Yamakawa & S. Suzuki. Effect of exercise on spontaneously hypertensive rats. Med. Sci. Sports 13:138, 1981.

HASHIMOTO, I. & P.R. Lamb. Immunoreactive plasma prostaglandins in men. Jpn. J. Physical Fit. Sports Med. 29:1–4, 1980.

JAFFE, B.M. & H.R Behrman. Prostaglandins E, A, F. In B.M. Jaffe & H.R. Behrman (Eds.), Methods of Hormone Radioimmunoassay. New York: Academic Press, 1974.

MCGIFF, J.C., K. Crowshaw, N.A. Terragno, K.U. Malik & A.J. Lonigro. Differential effect of noradrenaline and renal nerve stimulation on vascular resistance in the dog kidney and the release of a prostaglandin E-like substance. Sci. 42:223–233, 1972.

NOWAK, J. & Å. Wennmalm. Effect of exercise on human arterial and regional venous plasma concentrations of prostaglandin E. Prostaglandins and Medicine: 1: 489–497, 1978.

PACE-ASCIAK, C.R., Decreased renal prostaglandin catabolism precedes onset of hypertension in the developing spontaneously hypertensive rat. Nature 263:510–512, 1976.

SEMBROWICH, W.L., M.B. Knudson & P.D. Gollnick. Muscle metabolism and cardiac function of the myopathic hamster following training. J. Appl. Physiol. 43:936–941, 1977.

ZAMBRASKI, E.J. & M.J. Dunn. Renal Prostaglandin E_2PR and F_2 secretion and in exercising conscious dogs. Prostaglandins and Medicine 4:311–324, 1980.

ZUSMAN, R.M., Quantitative conversion of PGA or PGE to PGB. Prostaglandins 1:167–168, 1972.

Exercise-induced Changes in Plasma Prostaglandin E and Fα, Renin and Catecholamines in Hypertensive Patients

P. Lijnen, R. Fagard, J. Staessen and A. Amery

Hypertension and Cardiovascular Rehabilitation Unit
University of Leuven, K.U. Leuven, Belgium

In various experimental conditions a positive correlation between plasma levels of prostaglandin E and plasma renin activity (PRA) has been described (Terragno et al., 1977). On the contrary in borderline and essential hypertensive patients, Hornych et al. (1980) reported a negative correlation between PRA and venous $PGF_{2\alpha}$ (r = -0.93; p $<$ 0.01) or arterial PGF_2 (r = -0.65; p $<$ 0.05).

In the present study we evaluate the behavior of arterial plasma prostaglandin E and Fα, renin activity and catecholamines during exercise in hypertensive patients.

Methods

Eleven hypertensive patients (nine men, two women) with an average age of 30.7 $\pm$ 4.4 (SE) years and weighing 71.7 $\pm$ 4.3 kg were studied, their informed consent having been obtained. Seven had essential hypertension, one had hypertension with renal artery stenosis and three borderline hypertension.

The patients were studied in the morning after a light breakfast in the laboratory where room temperature was 18 to 22° C and humidity 40 to 60%. A small catheter (Vygon 115.09) was introduced into the brachial artery for sampling of blood and for recording the intra-arterial pressure using an Elema Schonander EMT pressure transducer; pressure was recorded continuously on a Mingograph 81 recorder and mean arterial pressure (MAP) was obtained by electrical damping. After the insertion of the catheters the patients were allowed to rest in the recumbent position for 30 min and then assumed the sitting position on the bicycle ergometer for 15 min. The exercise test was started at a work load of 20 W for 4 min and the load was increased by 30 W every 4 min until exhaustion. Arterial blood (30 ml) was drawn after 30 min rest recumbent (RR), after 15 min rest sitting (RS) and at the final work load of the uninterrupted graded exercise test for assay of plasma prostaglandins, renin activity and catecholamines.

The plasma concentrations of prostaglandin E (PGE) and of prostaglandin F_α(PGF) were measured radioimmunologically after extraction of the plasma samples with petroleum ether and with ethyl acetate and after chromatography of the extracts on silicic acid (Lijnen et al., 1981). The plasma renin activity (PRA) was measured by radioimmunoassay according to the method of Fyhrquist and Puutula (1978). Plasma noradrenaline (PNA) and adrenaline (PA), concentration were measured by a radio-enzymatic assay (Peuler & Johnson, 1977).

Student's two-tailed t-test for paired data was used for the statistical analysis. The dispersion of the data is given by standard error of the arithmetic mean (SEM). Except for the plasma PGE and PGFα concentration the biochemical parameters were transformed to logarithms since the log distribution was closer to Gaussian: the mean was derived from the logarithmic values (geometric mean).

Results

The patients were characterized as follows. The severity of the hypertension was assessed by criteria of the World Health Organization: 8 were at stage I and 3 at stage II. ECG was normal in 10 and one had left ventricular hypertrophy. Their creatinine clearance averaged 73.8 ± 7.8 ml/min^{-1}. The 24-h urinary sodium and potassium excretion averaged respectively 124.3 $\pm$ 21.4 meq and 44.7 $\pm$ 5.0 meq.

The highest external work load averaged 143.0 ± 16.5 W. As shown in Table 1, significant increases during exercise were observed in arterial blood pressure (systolic, diastolic and mean) and heart rate.

Table 2 shows the plasma arterial levels of PGE and of PGFα at rest recumbent, rest sitting and at the final work load of the exercise test in the brachial artery. Compared to the rest recumbent values no changes (p > 0.10) in arterial plasma PGE or PGFα concentrations were observed during upright posture nor during exercise.

On the contrary a significant increase in arterial PRA was already noted at RS and at the final work load (Table 1). Also a significant rise in arterial PNA and PA was found during upright posture and during exercise (Table 2).

Table 1

Blood Pressure and Heart Rate at Rest and during Exercise in 11 Hypertensive Patients

	Rest recumbent	Rest sitting	Final work load
SBAP (mm Hg)	163.2 ± 9.6	171.0 ± 11.3	226.1 ± 12.4[xxx]
DBAP (mm Hg)	83.6 ± 4.4	94.0 ± 4.4[x]	96.8 ± 4.7[xxx]
MBAP (mm Hg)	112.9 ± 5.7	120.2 ± 6.5	134.1 ± 6.5[xxx]
HR (beats/min)	84 ± 4	95 ± 6[xxx]	179 ± 4[xxx]

[xxx]$p < 0.001$
[x]$p < 0.05$ compared to rest recumbent values

Table 2

Plasma Prostaglandin E and F$_\alpha$; Plasma Renin Activity, Noradrenaline and Adrenaline Concentration at Rest and during Exercise in 12 Hypertensive Patients

	Rest recumbent	Rest sitting	Final work load
PGE (pg/ml)	73.8 ± 9.8	89.7 ± 15.7	82.3 ± 9.0
PGF$_\alpha$ (pg/ml)	28.8 ± 3.2	28.6 ± 2.9	36.9 ± 4.3
PRA (ng/ml/h)			
range	0.23 to 8.04	0.34 to 12.49	0.68 to 15.9
Geometric mean	1.38	2.03[xx]	3.94[xx]
PNA (pg/ml)			
range	191 to 1235	355 to 1806	804 to 5699
Geometric mean	460	797[xxx]	2698[xxx]
PA (pg/ml)			
range	62 to 395	113 to 415	93 to 988
Geometric mean	160	230[x]	311[xxx]

[xxx]$p < 0.001$
[xx]$p < 0.01$ compared to rest recumbent values
[x]$p < 0.05$

Discussion

In the present study during the graded uninterrupted exercise test on a bicycle ergometer several well-known hemodynamic adaptations occur (Table 1). This exercise also provoked an increase in plasma renin activity and plasma concentration of catecholamines while no effect on plasma concentration of PGE or PGF$_\alpha$ was observed. This increase in plasma renin activity and the plasma concentration of catecholamnes is in accordance with several previous reports (Bozovic et al., 1967; Fasola et al., 1966; Kotchen et al., 1971). The magnitude of the renin and catecholamine response to exercise is related to the intensity of exercise, occurs rapidly and is short-lived after the completion of exercise.

The data, however, on the influence of exercise on the plasma prostaglandin concentrations are scarce. During vigorous exercise like a marathon race Demers et al. (1981) found an increase in venous plasma PGE (of 46%), PGF$_{2\alpha}$ (of 54%) and 6-keto-PGF$_{1\alpha}$ (of 106%). Nowak and Wennmalm (1978) found that leg exercise in the supine position at 130 W did not change the arterial level of PGE nor the hepatic venous concentration of PGE. On the contrary during exercise a significant increase in the renal and femoral venous concentration of PGE was noted, suggesting that vascular prostaglandins play a role in the "local" adaptation of the circulatory system to exercise. Indeed increased concentrations of PG-like substances have also been found in the venous blood from canine working skeletal muscle (Herbaczynska-Cedro et al., 1974). In the human forearm the PG synthesis inhibitor indomethacin has been found to decrease functional hyperemia (Kilbom & Wennmalm, 1976).

Because of this possibility that regional differenes may occur in plasma prostaglandin levels at exercise we measured the prostaglandin levels in the brachial and pulmonary artery since the former is likely to be the best indicator of the action in different tissues and the latter can be considered as the best approximation of the average prostaglandins released in the different tissues. No exercise-induced increase was observed although the exercise intensity was rather high, reaching an average maximum load of 143 W and a maximum heart rate of 179 beats per min, which is close to the maximum which can be reached in these subjects. Since the prostaglandin E and F$_\alpha$ did not change it is unlikely that they are important determinants of PRA release during exercise. It is possible, however, that the renin release can be regulated by metabolites of cyclic endoperoxide other than PGE or PGF$_\alpha$ such as prostacyclin PGI$_2$ or 6-keto-PGE$_1$.

References

BOZOVIC, L., J. Castenfors & M. Piscator. Effect of prolonged, heavy exercise on urinary protein excretion and plasma renin activity. Acta Physiol. Scand. 70:143–146, 1967.

DEMERS, L.M., T.S. Harrison, D.R. Halbert & R.J. Santen. Effect of prolonged exercise on plasma prostaglandin levels. Prostaglandins and Medicine 6:413–418, 1981.

FASOLA, A.F., B.L. Martz & O.M. Helmer. Renin activity during supine exercise in normotensives and hypertensives. J. Appl. Physiol. 21:1709–1712, 1966.

FYHRQUIST, F. & L. Puutula. Faster radioimmunoassay of angiotensin I at 37°C. Clin. Chem. 24:115–118, 1978.

HERBACZYNSKA-CEDRO, K., J. Staszenska-Barczak & H. Jinczenska. The release of prostaglandin-like substances during reactive and functional hyperemia in the hind leg of the dog. Pol. J. Pharmacol. 26:167, 1974.

HORNYCH, A., G. London, M. Safar, Y. Weiss, A. Simon, T.T. Guyene, J. Bariety & P. Milliez. Prostaglandins and plasma renin activity in hypertensive patients. In B. Samuelsson, P.W. Ramwell & R. Paoletti (Eds.), Advances in Prostaglandin and Thromboxane Research, pp. 1119–1122. New York: Raven Press, 1980.

KILBOM, Å. & Å. Wennmalm. Endogenous prostaglandins as local regulators of blood flow in man: effect of indomethacin on reactive and functional hyperaemia. J. Physiol. 257:109, 1976.

KOTCHEN, T.A., L.H. Hartley, T.W. Rice, E.H. Mougey, L.G. Jones & J.W. Mason. Renin, norepinephrine and epinephrine responses to graded exercise. J. Appl. Physiol. 31:178–184, 1971.

LIJNEN, P., L.J. Verschueren & A. Amery. Radioimmunoassay of 11α, 15-dihydroxy-9-keto-prosta-5-13-dienoic acid and 9α, 11α, 15-trihydroxyprosta-5-13-dienoic acid in human plasma. Bull. Soc. Chim. Belg. 90:229–239, 1981.

NOWAK, J. & Å. Wennmalm. Effect of exercise on human arterial and regional venous plasma concentrations of prostaglandin E. Prostaglandins and Medicine 1:489–497, 1978.

PEULER, J. & G.A. Johnson. Simultaneous single isotope radioenzymatic assay of plasma norepinephrine, epinephrine and dopamine. Life Sci. 21:625–636, 1977.

TERRAGNO, N.A., D.A. Terragno & J.C. McGiff. Contribution of prostaglandins to the renal circulation in conscious, anesthetized and laparotomized dogs. Circ. Res. 40:590–595, 1977.

The Effects of an Acute Exercise Bout on the Serum Level of Testosterone and Luteinizing Hormone in Male Subjects above Forty Years of Age

G. Métivier

University of Ottawa, Ottawa, Ontario, Canada

The effects of acute physical exercise on endocrine secretory mechanism in young men has been studied. We know much about the role played by the thyroid, the adrenal glands, the pancreas and the hypophysis cerebri (Métivier, 1981). Furthermore, the androgens' response to physical work has been reported in young athletes (Métivier, 1980; Morville, 1979; Sutton, 1973). Submitting elderly men to acute physical exercise sometimes constitutes a certain risk factor—for this reason, therefore, few studies have been reported concerning the role of androgens during physical exercise in that population. Also, the rate of secretion of testosterone at the age of 50 in men is about half way down the curve for secretion between 25 and 80 (Guyton, 1966). Consequently, the present study purported to determine whether acute exercise could stimulate the production of testosterone in men over 40 years of age. It was also the purpose of the study to yield information on the role luteinizing hormone (LH) plays in the regulation of testosterone secretion in that age group.

Methods and Subjects

The 17 male subjects with mean age of 50.3 yr who were chosen for the experiment, reported at a scheduled time in our laboratory between eight thirty and nine o'clock in the morning. Each subject, while in a post-prandial state was made to rest in bed in a supine position, for 15 min after which time blood was collected from the anticubital vein. Hematocrit values were determined at the time of sampling and the remainder of the serum was frozen rapidly at -10 C° and hormonal determinations made within a 30 day period. A second sample was collected immediately after the exhaustive physical effort and again 60 min later.

The exercise consisted of walking on a motor-driven treadmill according to the protocol developed by Bruce and described by Wilson and associates (1981). The subjects' ECG tracing was monitored throughout the exercise until a minimal value of 90% of predicted maximal heart rate for age was reached. Respiratory gases were collected at each third min of the exercise and analyzed by means of an automatic open circuit system. Subcutaneous fat layers were measured by skinfold techniques and percentage body fat was calculated by the densiometry technique.

The hormone testosterone was analyzed according to the method described by Diagnostic Biochem Canada Inc. using antitestosterone serum T-169 as a specific reagent. The luteinizing hormone (LH) was determined according to the method of Bio-Ria (1975) radioimmunoassay technique based on the double antibody method of Midgley (1966). Analysis of variance was applied to the data.

Results

The physiological and physical descriptions of the subjects appear in Table 1. Accordingly, the oxygen intake capacity reached was 38.3 ± 10.1 ml/kg/min $\dot{V}o_2$ and a terminal pulse rate of 174.5 ± 17.8 beats/min. The subjects were in average 67.9 ± 3.0 inches tall and weighed an average of 164.4 ± 20.3 lbs. The mean percentage body fat was calculated to be 17.0 ± 3.5%.

In Table 2 appear the results of the hormones testosterone, LH and hematocrit values of blood before, immediately after and 60 min after the exercise.

Testosterone values rose significantly (P < .05) from 290.6 ± 97.9 mg/100 ml to 360.0 ± 118 mg/100 ml after the exercise, to fall significantly (P < .05) to 276.6 ± 91.17 mg/100 ml 60 min thereafter. Luteinizing hormone (LH) increased from 10.7 ± 4.3 miu/ml to 11.4 ± 4.4 miu/ml after the work bout. It fell thereafter to 11.7 ± 4.2 miu/ml 60 min after the work. These changes in LH were, however, not significant.

Significant rises (P < .05) were also noted in hematocrit values. Values of 44.2 ± 2.7% to 48.4 ± 3.1% were recorded after the exercise and 45.3 ± 2.4% 60 min later.

Discussion

The significant increase in testosterone of 69.4 mg/100 ml during the exercise represents a ninefold increase of normal daily output, or a 24%

Table 1

Physical and Physiological Characteristics of Subjects

	Age (yrs)	Weight (lbs)	Height (ins)	$\dot{V}_{O_2}$(ml/kg/min)	MHR (beats/min)	% Fat
X	50.3	164.4	67.9	38.3	174.5	17.0
S	7.6	20.3	3.0	10.1	17.8	3.5
N	17	17	17	17	17	17

MRH = Maximal heart rate reached

Table 2

Blood Testerone, Luteinizing Hormone (LH) Level and Hematocrit Value before, after, and 60 Min Following Exercise

	Testosterone (mg/100 ml)			LH miu/ml			Hematocrit (%)		
	Pr	Post	Post (60)	Pr	Post	Post (60)	Pr	Post	Post (60)
X	290.6	360.0*	276.6*	10.7	11.4	11.7	44.2	48.4*	45.3*
S	97.9	118.6	91.17	4.3	4.4	4.2	2.7	3.1	2.4
N	17	17	17	17	17	17	17	17	17

*Significant $p < .05$
Pr = pre-exercise
Post = immediately after exercise
Post (60) = 60 min post exercise

change which could not be explained entirely by a plasma volume change, as the hematocrit value varied only 8.8%. Similar results were recorded in young men (Métivier, 1980). These findings demonstrate well that testosterone can still be produced in large quantities in male subjects 50 years old when the aforesaid are submitted to physical work. It is unlikely that the hormone comes from other tissues, as pointed out by Sutton (1978).

We are of the opinion that the physical stress experienced by the work was sufficient to initiate the production of a secretory stimulator thereby increasing the production of testosterone. The sudden drop in the hormone level in blood following the exercise could have been the result of increased metabolic degradation or rapidly reversible binding to carrier protein (Plager, 1965; Sandberg, 1957).

The role this hormone plays during physical work was not answered by the present experiment but it could be associated with energy metabolism. Testosterone appears to facilitate the capture of glucose (Aloia, 1976) by maintaining the enzymatic concentration necessary for the oxidizing metabolism of carbohydrates and lipids (Aloia, 1976; Janda, 1976).

Testosterone is also known to increase the muscle glycogen concentrations by a saving effect resulting from the inhibition of its breakdown (Bergamini, 1975; Gillespie et al., 1970). In doing so testosterone could enhance physical performance. Our work bout was relatively short (less than 30 min) and testosterone production was very high. If the work performed had been for a longer duration on our subjects, we could have observed drops in the hormone (Morville, 1979).

Luteinizing hormone, as in the case of younger males (Métivier, 1980), did not change, thus adding to the evidence that this hormone is of no significance in the control of testosterone during physical exercise.

Based on the data, we concluded that the secretory tissues for testosterone were still active in the older male subjects (50.3 years of age) and, as in younger age groups, LH has no significant regulatory mechanism for testosterone secretion.

References

ALOIA, J.F. & R.A. Field. The effect of androgens and related substances on carbohydrate metabolism lipid metabolism and diabetes mellitus. Pharmac. Ther. C. 1:241–257, 1976.

DIAGNOSTIC Biochem. Canada Inc. Anti-testosterone serum NO T-169. Technical data. London & Ontario, 249 Wortley Road, N6C 3P9.

BIO-RIA (a division of Bio-Endo) HLH Radioimmunoassay kit double antibody method using 125 inclinated HLH as tracer. Procedures for the radioimmunoassay of HLH levels in urine and serum, 1975.

GILLESPIE, C.A. & V.R. Edgerton. The role of testosterone in exercise-induced glycogen supercompensation. Horm. Metab. Res. 2:364–366, 1970.

GUYTON, A.C., Textbook of Medical Physiology. Philadelphia & London: W.B. Saunders Company, 1966.

JANDA, J., O. Mrhova, D. Urbanova, & J. Linhart. The effect of anabolic hormone 19-Nortestosterone propionate on the metabolism of striated muscle during experimental ischaemia. Pfügers Arch. 361:159–163, 1976.

MÉTIVIER, G., R. Gauthier, J. de La chevrotière & D. Grymala. The effect of acute exercise on the serum levels of testosterone and luteinizing (LH) hormone in human male athletes. J. Sports Med. Phys. Fit. 20:(3) 235–238, 1980.

MÉTIVIER, G. Hormonal behavior and regulation during physical exercise. In P.E. DiPrampero and J.R. Poortmans (Eds.), Physiological Chemistry of Exercise and Training. Basel: S. Karger, 1981.

MORVILLE, R., P.C. Pesquies, C.Y. Guezennec, D.D. Serrurier & M. Guignard. Plasma variations in testicular and adrenal androgens during prolonged physical exercise in man. Ann. D'Endocrinolo. (Paris) 40:501–510, 1979.

PLAGER, J.E., The binding of androsterone sulfate, etiocholanolone sulfate, and dehydroisoandrosterone sulfate by human plasma protein. J. Clin. Invest. 44:1234, 1965.

SANDBERG, A., W.R. Slaunwhite Jr. & H.N. Antoniades. The binding of steroids and steroid conjugates to human plasma proteins. Rec. Progr. Hormone Res. 13:209–1957.

SUTTON, J.R., M.J. Coleman & J.H. Casey. Testosterone production rate during exercise. In F. Landry & W.A.R. Orban (Eds.), 3rd International Symposium on Biochemistry of Exercise, 1978.

SUTTON, J.R., M.J. Coleman, J. Casey & L. Lazarus. Androgen responses during physical exercise. Brit. Med. Jour. 1:520–522, 1973.

Hormones—
Insulin

The Influence of Physical Training on Glucose Turnover and Hormonal Responses in Insulin-induced Hypoglycemia

H. Galbo, M. Kjaer, K.J. Mikines, N.J. Christensen,
B. Tronier, B. Sonne, J. Hilsted and E.A. Richter
University of Copenhagen, Copenhagen, Denmark,
Herlev Hospital, Herlev, Denmark, and
NOVO Research Institute, Copenhagen, Denmark

Physical training influences a number of hormonal and metabolic responses. For example, during exercise the concentrations in plasma of the counterregulatory hormones, whose common effect is to increase the plasma glucose level, are lower in trained than in untrained subjects. Nevertheless, compared to untrained subjects, trained subjects have a more pronounced increase in plasma glucose during short term exercise and less tendency to hypoglycemia during prolonged exercise. This may at least partly be due to training-induced enzymatic changes increasing carbohydrate stores in liver and muscle, and enhancing mobilization and combustion of fat during exercise (Galbo, 1982). Physical training has also been shown to reduce the increase in plasma insulin concentration seen during glucose loading. This effect probably reflects that training may increase insulin sensitivity in muscle and fat tissue (Galbo, 1981).

In the clinic hypoglycemia is induced by administration of insulin to evaluate the function of the endocrine glands producing counterregulatory hormones as well as the capability of the organism to recover from hypoglycemia by increasing glucose delivery to and decreasing glucose uptake from plasma. In light of the above mentioned differences between trained and untrained subjects as regards endocrine responses to exercise, size of glycogen depots, and sensitivity of target tissues to insulin and glucoregulatory hormones, it might be expected that these groups respond differently to an insulin challenge. The decrease as well as the subsequent increase of plasma glucose concentrations, the underlying changes in rates of glucose production and uptake, and also the accompanying hormonal and metabolic changes might be influenced by training. Accordingly we found it of theoretical as well as clinical interest to study the influence of training on glucose turnover and hormonal changes during insulin-induced hypoglycemia.

Methods

Eight healthy, sedentary male students (25 [24 to 26] years [mean and range]) and eight athletes competing in elite class endurance sports (24 [22 to 25] years [mean and range]) gave their informed consent to participate in the study. The $\dot{V}_{O_2}$max, weight and height were 49 (41 to 53) ml kg^{-1} min^{-1}, 72 (69 to 74) kg and 180 (176 to 184) cm, respectively, in the untrained, and 65 (60 to 72), 74 (71 to 76) and 184 (181 to 186), respectively, in the trained subjects. For 3 and 2 days, respectively, prior to the experiment the subjects ate more than 250 g carbohydrate per day, and avoided training, tobacco and alcohol. The subjects arrived in the laboratory after an overnight fast (10 h). During the experiment they rested recumbent with the upper part of the body at an angle of 45° to the bed. They had two cannulas (for infusions) inserted intravenously in one forearm and one (for blood sampling) in the other. The electrocardiogram was registered with precordial electrodes and arterial blood pressure was measured by the indirect ausculatory method. Turnover of the glucose pool was measured with the primed, constant rate tracer infusion technique which allows determination in steady state as well as in non-steady state of rates of appearance (R_a) and disappearance (R_d) of glucose (Hetenyi & Norwich, 1974). A modified single compartment analysis was applied assuming that the pool fraction in which rapid changes in concentration and specific activity of glucose take place is 0.65. After a priming dose (0.2 μCi kg^{-1}) 3-^{3}H-glucose (specific activity 2 Ci mol^{-1}, dissolved in isotonic saline) was infused at a rate of 0.002 μCi kg^{-1} min^{-1}. The glucose pool was calculated from concentrations and specific activities of glucose obtained 60 to 180 min after start of infusion. After 180 min insulin (Actrapid^R, NOVO, dissolved in isotonic saline containing in addition 5% human serum albumin) was infused at 0.15 IU kg^{-1} h^{-1}. The insulin infusion was stopped when the blood glucose concentration (measured every 5 min with a reflectance meter [Ames] as well as with a rapid spectrophotometric procedure yielding results within 3 min from blood drawing) was below 2.5 mM. Before, during and after insulin infusion blood was sampled at intervals for analysis of hormones and substrates. Catecholamine concentrations were determined by a single-isotope derivative method. Other hormones and substrates were determined by radioimmunoassays and conventional procedures, respectively. Statistical evaluation was made by Mann-Whitney's rank sum test, the confidence level used being 5%.

Results

The basal C-peptide (0.45 ± 0.04 [SE] vs 0.69 ± 0.06 pmol ml^{-1}) and glucagon concentrations were lower, and the basal epinephrine level as well as the glucose production significantly higher in trained (T) than in untrained (C) subjects. During infusion insulin increased about 24 fold to an average of 1300 pmol l^{-1} in both groups. The rate of decrease of plasma glucose concentration, and accordingly the duration of the insulin infusion were identical in the groups (Figure 1). Also the increases in R_d and metabolic clearance of glucose, and the decrease in R_a were similar in time as well as in magnitude (Figure 1). Recovery of plasma glucose concentrations were identical in T- and C-subjects, reflecting that increases in R_a as well as decreases in R_d were similar (Figure 1). However, in response to hypoglycemia, epinephrine increased significantly more and glucagon less

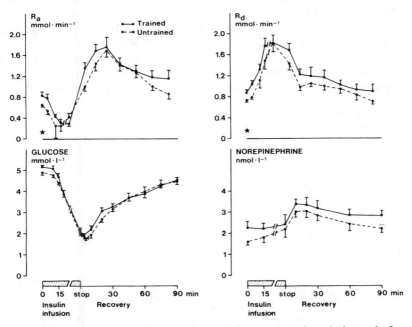

Figure 1—Glucose kinetics and plasma norepinephrine concentrations during and after infusion of insulin (0.15 IU kg^{-1} h^{-1}) in eight trained and eight untrained male students. Values are means ± SE. Asterisks denote significant differences (p < 0.05) between groups.

in T- than in C-subjects (Figure 2). In the former group GH and pancreatic polypeptide increments were augmented (p < 0.05) while glycerol increased and heart rate declined significantly faster to basal values during glucose recovery. Responses of C-peptide, norepinephrine, cortisol, TSH, FFA, β-hydroxybutyrate, lactate, alanine and arterial blood pressure were similar in the two groups.

Discussion

A main conclusion of the present study is that the state of training does not influence glucose turnover during insulin-induced hypoglycemia. This is surprising for the reason among others that a training-induced increase in insulin sensitivity has been demonstrated during glucose loading as well as in insulin clamp experiments in intact man, and, furthermore, in muscle and fat cells in vitro (Galbo, 1981). In the present study a higher insulin sensitivity in trained compared to untrained subjects was apparent neither during nor after insulin infusion. Yet, in the recovery phase the effect of a higher insulin sensitivity in the trained subjects might be counteracted by the higher concentrations of the insulin-antagonistic hormones epinephrine and GH, and also by the higher fat combustion, which as indicated by the higher glycerol levels existed in these subjects. Still, the mentioned differences occurred too late to explain that peak glucose clearance values obtained at identical plasma insulin concentrations at the end of infusion did not differ between the groups. Recent evidence, however, has suggested

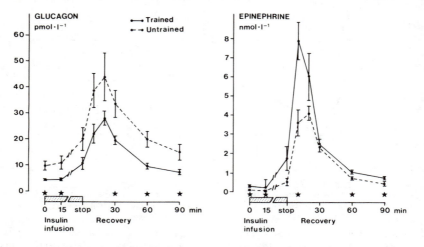

Figure 2—Concentrations of hormones in peripheral venous plasma during and after infusion of insulin (0.15 IU kg^{-1} h^{-1}) in eight trained and eight untrained male students. Values are means ± SE. Asterisks denote significant differences (p < 0.05) between groups.

that enhanced insulin sensitivity in trained subjects, being quickly acquired and also vanishing rapidly, to a great extent reflects effects of the last exercise bout rather than a proper adaptation to repeated exercise (LeBlanc et al., 1981; Ivy & Holloszy, 1981). Accordingly, in the present study an effect of the level of physical activity on insulin sensitivity may have disappeared during the two days which elapsed between the last training session and experiment. Another possibility which might explain the present findings is that training may influence the K_M rather than the V_{max} of insulin-mediated glucose assimilation. In accordance with this view we found a higher basal glucose clearance in trained than in untrained subjects, although the former had lower basal insulin secretion as judged from C-peptide levels. Furthermore, peak clearance values obtained at an insulin concentration possibly being maximally effective, were identical in the two groups. Finally, in previous studies of isolated muscle and intact man enhancement of insulin sensitivity by training was found at insulin concentrations which were reported to be submaximal or were lower than in the present study, respectively (Galbo, 1981).

Another main finding of the present study is that in trained subjects glucose recovery depends more on epinephrine and less upon glucagon than in untrained subjects. In normal man glucagon secretion plays a primary role for glucose counterregulation whereas epinephrine secretion becomes important only when glucagon secretion is inhibited (Cryer, 1981). The two hormones readily substitute for each other in experimentally induced states of hormone deficiency as well as in endocrine disease. Accordingly, the higher epinephrine levels in trained subjects compared to untrained subjects might reflect an increased discharge in sympathetic nerves to the adrenal glands secondary to the lower glucagon levels in the former subjects. However, the higher epinephrine levels (which approached values similar to those found in pheochromocytoma and which possibly accounted for the increased secretion of pancreatic polypeptide) may also result from an increased capacity of the adrenal glands to synthesize epinephrine in trained subjects, a view supported by analysis of adrenal glands (Galbo, 1982). Finally, it may be recalled that during prolonged exercise the increase in epinephrine secretion is in part elicited by a decrease in plasma glucose concentration. It appears from the present study that the lower epinephrine response to prolonged exercise in trained compared to untrained subjects is probably not due to a lower sensitivity to hypoglycemia in the former group.

Acknowledgment

The study was supported by The Danish Medical Research Council (12-1851).

References

CRYER, P.E., Glucose counterregulation in man. Diabetes 30:261–264, 1981.

GALBO, H., Endocrinology and metabolism in exercise. Int. J. Sports Med. 2:203–211, 1981.

GALBO, H., Hormonal and Metabolic Adaptation to Exercise. Stuttgart & New York: Georg Thieme Verlag, 1982.

HETENYI, G. & K.H. Norwich. Validity of the rates of production and utilization of metabolites as determined by tracer methods in intact animals. Federation Proc. 33:1841–1848, 1974.

IVY, J.L. & J.O. Holloszy. Effect of training on glucose uptake by striated muscle. Med. Sci. Sports and Exercise 13:90, 1981.

LEBLANC, J., A. Nadeau, D. Richard & A. Tremblay. Studies on the sparing effect of exercise on insulin requirements in human subjects. Metabolism 30: 1119–1124, 1981.

Enhanced Insulin Sensitivity of Skeletal Muscle following Exercise

L.P. Garetto, E.A. Richter, M.N. Goodman
and N.B. Ruderman
Boston University Medical Center, Boston, Massachusetts,
U.S.A., and University of Copenhagen, Copenhagen,
Denmark

Although it is well known that following exercise, glucose utilization by muscle is enhanced and its glycogen stores are rapidly repleted, the mechanism for this is unclear. This report explores the notion that these events are due to an increase in the sensitivity or responsiveness of the muscle cell to insulin. For the purposes of this discussion an increase in sensitivity is defined as a shift in the dose-response curve for insulin to the left and an increase in responsiveness as an increase in the maximum effect of the hormone (Kahn, 1978).

Methods

Male Sprague-Dawley rats weighing between 200 and 250 g fed ad libitum were used in all experiments. After a treadmill familiarization period of 3 to 5 days, the rats were randomly divided into exercised and control (non-exercised) groups on the morning of the experiment. The exercise group ran for 45 min on the treadmill at 18 m/min, except for the last 2 min during which the speed was increased to 31 m/min. At various times following the completion of the run the rats were anesthetized and surgically prepared for either hindquarter perfusion or soleus muscle incubation.

Hindquarters were perfused by the method of Ruderman et al. (1980). The perfusion medium was composed of Krebs-Henseleit solution, rejuvenated aged human red cells (30% hematocrit), 4% albumin, 6 mM glucose, glucose-U-^{14}C and physiological concentrations of lactate and pyruvate.

Insulin, when present, was added to achieve concentrations ranging from 10 to 40,000 μU/ml. Glucose uptake was determined by the disappearance of glucose from the media and glycogen synthesis from the incorporation of U-^{14}C-glucose into glycogen. Strips of soleus muscle were isolated by a modification of the method of Crettaz (1980). They were incubated in Krebs-Hanseleit solution containing 1.5% albumin, 5 mM glucose and 3-^{3}H glucose (Maizels et al., 1977). Insulin binding to the soleus was assessed according to the technique of LeMarchand et al. (1978).

Results

Glycogen Depletion and Repletion, in Vivo

The treadmill running caused a 50% depletion of muscle glycogen in the soleus, a 30% decrease in the red portion of the gastrocnemius and no change in the white portion of the gastrocnemius (Table 1). As shown by a number of investigators, this pattern of glycogen depletion is indicative of exercise of moderate intensity (Armstrong, 1974). Pre-exercise glycogen levels were restored in the soleus and red fibers of the gastrocnemius within 1 h of the termination of exercise and by 2 h glycogen levels in these muscles were even higher than their initial values. The white gastrocnemius did not demonstrate these changes.

Glucose Metabolism in the Perfused Hindquarter following Exercise

To determine whether prior exercise alters the ability of insulin to stimulate glucose utilization, studies were carried out with the isolated perfused

Table 1

Muscle Glycogen following Treadmill Exercise

Muscle	Control	Hours following exercise			
		0	1	2	4
Soleus	26.7	13.3*	23.0	31.9*	28.9*
Red gastrocnemius	30.5	20.7*	33.8	38.3*	34.0
White gastrocnemius	35.7	35.5	31.3	33.3	31.6

Results are expressed as μmol/g wet wt and are the means of 2 to 5 observations per group. Adapted from Richter et al. (1982). *p $<$ 0.05 compared with controls.

hindquarter. As shown in Table 2, in the absence of added insulin, glucose utilization was not enhanced in hindquarters of previously exercised rats. On the other hand, when insulin was present at concentrations as low as 30 μU/ml, glucose uptake was significantly higher following exercise. Although there was a modest increase in the maximum response to insulin, the enhanced effect of the hormone appeared to be due principally to an increase in sensitivity. Thus the concentration of insulin necessary to produce a half maximal stimulation of glucose uptake was 480 μU/ml in control animals and 150 μU/ml following exercise. In experiments not shown here, it was found that the increase in glucose utilization is the result of an increase in glucose transport into the muscle cell. In addition, we have found that the increase in insulin sensitivity persists for at least 4 h, whereas the modest increase in maximum responsiveness is transient (Richter et al., 1982).

The incorporation of U-^{14}C-glucose into glycogen wa also greater following exercise. The increase was most apparent when insulin was present in the perfusate, although a trend to increased glycogen synthesis was present even in its absence (Table 3). Of particular note, the greatest enhancement of insulin-stimulated glucose incorporation into glycogen following exercise was observed in the red gastrocnemius, which was partially depleted of glycogen by the run, and the smallest increase in the white gastrocnemius, which was not glycogen depleted. An intermediate response was observed in the soleus.

Glucose Metabolism and Insulin Binding to the Soleus Muscle

To assess whether the increase in insulin sensitivity following exercise is due to enhanced binding of the hormone to its receptor, studies were carried out using the incubated soleus muscle. Since rats weighing 200 to 250 g were used, it was necessary to utilize small strips of the soleus to allow

Table 2

Glucose Utilization by the Perfused Rat Hindquarter following Treadmill Exercise

| | Insulin μU/ml | | | | | | |
	0	10	30	75	500	20,000	40,000
Control	1.8	2.2	1.9	3.2	6.3	10.2	10.6
Exercise	1.9	2.8	2.7*	6.1*	9.4*	12.6*	12.1*

Glucose utilization was measured during a 45 min perfusion and results are expressed as μmol/g/h. Adapted from Richter et al. (1982). *p $<$ 0.05 compared to control perfused with same insulin concentration.

for adequate diffusion of O_2 and other substances (Crettaz, 1980). Also it was necessary to determine whether exercise enhanced the stimulation of glucose utilization by insulin as it did in the hindquarter. As shown in Table 4, glucose utilization (defined as the sum of ^3H-glucose incorporation into glycogen and ^3H_2O production) was significantly more enhanced by insulin (100 $\mu U/ml$) following exercise than in control rats, although the effect was not as marked as in the perfused hindquarter. The increase in glucose utilization was principally due to an increase in glucose incorporation into glycogen; however, at an insulin concentration of 100 $\mu U/ml$, glycolysis as estimated from ^3H_2O production, also tended to be increased (10.0 vs 11.5 $\mu moles/g/hr$). At times an increase in glucose incorporation into glycogen was observed following exercise in the absence of insulin. The magnitude of the increase was similar to that observed in the hindquarter (see Table 3); however, in the incubated soleus it was statistically significant, due to the greater number of experiments performed.

Despite the apparent increase in its sensitivity to insulin following exercise, we found no difference in the binding of tracer quantities of insulin by solei from previously exercised and control rats (Exercise— 60,489 ± 4304 dpm/g (n = 14); Control—60,274 ± 3740 dpm/g (n = 9). These values are equivalent to 0.02% of total insulin bound per mg of muscle. The reason for this low specific binding remains to be determined.

Discussion

The results suggest that the enhancement of glucose utilization in skeletal muscle and the rapid repletion of its glycogen stores following exercise are due, at least in part, to an increased sensitivity of the muscle cell to insulin.

In the present study glucose utilization was not increased in either the hindquarters or soleus muscles of exercised rats when insulin was not added to the medium. This contrasts with the recently reported findings of Ivy and Holloszy (1981) who, in rats following a swim, found a large increase in both glucose uptake and glycogen resynthesis in the perfused hindquarter even when insulin was not added. They also observed no increase in insulin sensitivity. The rats in their study were fasted overnight rather than fed and they were almost certainly more glycogen depleted at the beginning of the perfusion than those used by us. Whether this accounts for the differing results remains to be determined.

Increases in insulin binding to muscle have been observed in 48 h fasted rats (LeMarchand-Brustel, 1979) in association with an increase in the sensitivity of muscle to insulin (Brady et al., 1981). Likewise, increased insulin binding to monocytes and red blood cells has been shown to occur

Table 3

Effect of Prior Exercise and Insulin on Glucose Incorporation into Glycogen by the Isolated Perfused Rat Hindquarter

	Soleus	Red gastrocnemius	White gastrocnemius
	μmol/g/h		
No insulin			
Control	0.17	0.10	0.06
Exercise	0.43	0.27	0.10
Insulin (75 μU/ml)			
Control	0.79	0.39	0.18
Exercise	2.06*	3.08*	0.37*

Results are means of 4 to 5 observations without insulin present and 9 to 10 when insulin was added. Adapted from Richter et al. (1982). *p $<$ 0.05 compared to control perfused with same insulin concentration.

Table 4

Effect of Prior Exercise on Glucose Incorporation into Glycogen and Glucose Utilization in Rat Soleus Muscle

Insulin μU/ml	Glucose incorporation into glycogen		Glucose utilization	
	Control	Exercise	Control	Exercise
0	1.41 ± 0.15	2.02 ± 0.18*	9.38 ± 0.92	9.74 ± 0.52
100	1.91 ± 0.24	3.09 ± 0.20*	11.92 ± 0.96	14.58 ± 0.73*
1000	2.13 ± 0.25	3.34 ± 0.18*	12.89 ± 0.87	14.07 ± 0.97

Results are expressed as μmol/g/h and are means ± SE of 10 to 17 observations. Glucose utilization is calculated from the sum of ^{3}H-glucose incorporation into glycogen and ^{3}H$_2$O production during a 2 h incubation at 37° C. Muscles were placed into incubation flasks $\simeq$ 15 min after the end of exercise. *p $<$ 0.05 compared to controls.

in humans following exercise, and it has been suggested that this reflects a generalized increase in insulin binding (Kovisto et al., 1980; Petersen et al., 1980). Data obtained here do not support this notion as the ability of insulin to stimulate glucose appeared to be confined predominantly to those muscles that had been deglycogenated. In addition we found no increase in insulin binding to soleus muscle strips from exercised rats. Thus the available evidence suggests that the enhancement of insulin sensitivity in skeletal muscle following exercise is the result of an as yet unidentified post receptor event.

Acknowledgment

Supported in part by U.S.P.H.S. grants AM 19514 and AM 19469. The authors would like to thank Ms. Eva Belur and Susan Dluz for their expert technical assistance.

References

ARMSTRONG, R.B., C.W. Saubert IV, W.L. Sembrowich, R.E. Shepheard & P.D. Gollnick, Glycogen depletion in rat skeletal muscle fibers at different intensities and duration of exercise. Pfluegers Arch. 352:243–256, 1974.

BRADY, L.B., M.N. Goodman, F.R. Kalish & N.B. Ruderman. Insulin binding and sensitivity in rat skeletal muscle: Effect of starvation. Am. J. Physiol. 240:E184–E190, 1981.

CRETTAZ, M., M. Prentki, D. Zaninetti & B. Jeanrenaud. Insulin resistance in soleus muscle from obese zucker rats. Biochem. J. 186:525–534, 1980.

IVY, J. & J. Holloszy. Persistant increase in glucose uptake by rat skeletal muscle following exercise. Am. J. Physiol. 241:C200–C203, 1981.

KAHN, C.R., Insulin resistance, insulin insensitivity, and insulin unresponsiveness: A necessary distinction. Metab. Clin. Exp. 27:1893–1902, 1978.

KOIVISTO, V.A., V.R. Soman & P. Felig. Effects of acute exercise on insulin binding to monocytes in obesity. Metab. Clin. Exp. 29:168–172, 1980.

LeMARCHAND-BRUSTEL, Y. & P. Freychet. Effect of fasting and streptozotocin diabetes on insulin binding and action in the isolated mouse soleus muscle. J. Clin. Invest. 64:1505–1515, 1979.

MAIZELS, E.Z., N.B. Ruderman, M.N. Goodman & D. Lau. Effect of acetoacetate on glucose metabolism in the soleus and exensor digitorum longus muscles of the rat. Biochem. J. 162:557–568, 1977.

PEDERSEN, O., H. Beck-Nielsen & L. Heding. Increased insulin receptors after exercise in patients with insulin-dependent diabetes mellitus. N. Engl. J. Med. 302:886–892, 1980.

RICHTER, E., L. Garetto, M. Goodman & N. Ruderman. Muscle glucose metabolism following exercise in the rat. Increased sensitivity to insulin. J. Clin. Invest. In press, 1982.

RUDERMAN, N.B., F.W. Kemmer, M.N. Goodman & M. Berger. Oxygen consumption in perfused skeletal muscle. Effect of perfusion with aged—fresh and aged—rejuvenated erythrocytes on oxygen consumption, tissue metabolites, and inhibition of glucose utilization by acetoacetate. Biochem. J. 190:57–64, 1980.

Endocrine Function of the Pancreas during Exercise

V.V. Menshikov, E.P. Gitel, T.D. Bolshakova,
V.G. Cukes and O.B. Dobrovolsky
State Central Institute of Physical Education, Moscow, USSR

Physical exercise which substantially increases the energy turnover of the human organism requires a clear-cut reconstruction of metabolic processes to establish optimal correlation between the energy substrate utilization of an active muscle and its absorption into the blood from a depot. The endocrine system plays an important role in the realization of that reconstruction. Insulin and glucagon (the hormones of the pancreas) are of great importance. They regulate the rate of utilization of the main energy substrates (glucose and non-esterized fatty acids) by an active muscle and the rate of their mobilization from the liver and adipose tissue.

It is well known that, during intense long-term exercise, there is an insulin concentration reduction (Felig & Wahren, 1975; Pmett, 1970; Wirth et al., 1981) and glucagon concentration increase in blood plasma. It also been demonstrated that well-trained persons have a reduced basal level of blood-plasma compared to untrained persons (Koivisto et al., 1979; Lohman et al., 1978). Glucose injections given to trained persons result in insulin secretion (Lohman et al., 1978). However, the data are contradictory on the question of the dependence of insulin concentration alterations exercise of the physically fit person (Galbo et al., 1977; Sutton, 1978; Wirth et al., 1981). It is necessary to note that the question on the correlation between insulin secretion alterations and its metabolism during exercise which results in its peripheral concentration, has not been ascertained. Radioimmunological estimations of insulin and C-peptide in blood give the opportunity to appreciate the metabolism of these hormones indirectly and separately.

The purpose of this study was to investigate the function of alpha and beta cells of the pancreas in elite athletes, in the process of maximal aerobic testing on a veloergometer and their dependence on individual capacities exercise.

Methods

Twenty-one elite athletes (male academic rowers) were tested while in a post-absorptive state. The subjects were from 16 to 22 years of age. Samples of venous blood were taken by a catheter from a forearm vein immediately after the cessation of progressively increasing maximal exercise with the initial effort of 1 watt/kg and with the initial value increasing every 3 min. Non-esterized fatty acids were determined by Dole's method, glucose by the orthotoluidine method, and triglycerides by an enzymatic method. The hormones of the pancreas, insulin and glucagon, as well as C-peptide (which is the index of secretion activity of beta cells) were determined radioimmunologically. Molar concentration correlations of insulin and glucagon were estimated in every sample.

Discussion and Results

The average duration of exercise was 13.2 ± 0.8 min. The athletes' exercise capacity averaged 219.2 ± 5.41 kgm/kg which corresponded to 16.86 ± 0.199 kgm/kg min. A marked glucose blood concentration increase which amounted to 20.9 ± 4.1 mg/100 ml in response to exercise was found as well as statistically insignificant increases in non-esterized fatty acids amounting to 72.41 ± 33.13 meg/l in triglycerides of 21.9 ± 7.39 mg/100 ml. After the cessation of exercise, a marked insulin blood-plasma concentration reduction averaged 3.53 ± 1.35 ($p < 0.05$) and a C-peptide reduction averaged 1.48 ± 0.45 ng/ml ($p < 0.05$). The level of glucagon in plasma ($p < 0.05$) increased from 64.8 ± 5.03 to 82.58 ± 7.69 kg/ml, whereas the molar concentration correlation of insulin and glucagon dropped from 7.92 to 4.31 at the point of departure after exercise ($p < 0.01$). The individual analyses of the response to exercise revealed that changes in C-peptide blood concentration were characterized by the lowest variability. In 91% of the cases, the dynamics in response to exercise coincided with the average value of the group and in the same direction. A reduction in blood insulin level was found in 71% of the athletes, with the remaining athletes characterized by either unchanged insulin levels or insignificantly higher levels than at the point of departure. In the same number of cases (71%), the increase of glucagon level was found together with its invariability or moderate degree of reduction. All changes in molar insulin and glucagon concentration correlations were directed towards its lowering.

Analyzing the changes in the indices under question as to their dependence on the individual exercise capacity (Figure 1), it has been demonstrated that higher exercise capacity athletes (left columns) in comparison with the lower exercise capacity athletes (right columns) have

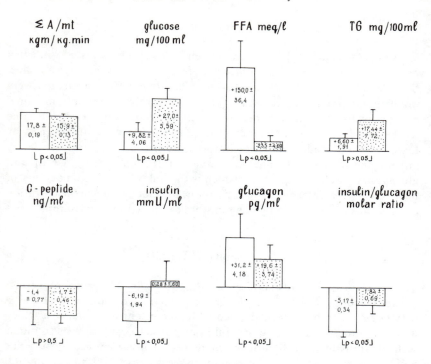

Figure 1—Dynamics of the concentration of certain energy yielding substances and pancreatic hormones in response to maximal physical loading of athletes with different capacities for exercise. The left columns (white) represent the higher exercise capacity athletes while the right columns (stippled) represent the lower exercise capacity athletes.

the same dynamics of C-peptide blood concentrations caused by exercise and have meaningful differences in the concentrations of insulin. Higher exercise capacity athletes were characterized by marked blood insulin concentration reduction ($p < 0.05$) compared to initial values while lower exercise capacity athletes had a greater increase in plasma glucagon level and a more marked decrease in insulin/glucagon relationship. Differences were also found in glucose and non-esterized fatty acid concentration changes. Athletes with higher exercise capacity are noted for less pronounced non-esterized fatty acid concentration increases.

The results obtained indicate that athletes are noted for the inhibition of insulin secretion during a veloergometer maximal exercise test. That was confirmed by the reduction of C-peptide blood-plasma level in the overwhelming majority of subjects. Taking into consideration the above mentioned factor, it is possible to presume that differences in direction and degree of pronouncement of insulin plasma concentration changes are explained by differences in its peripheral metabolism. Higher exercise

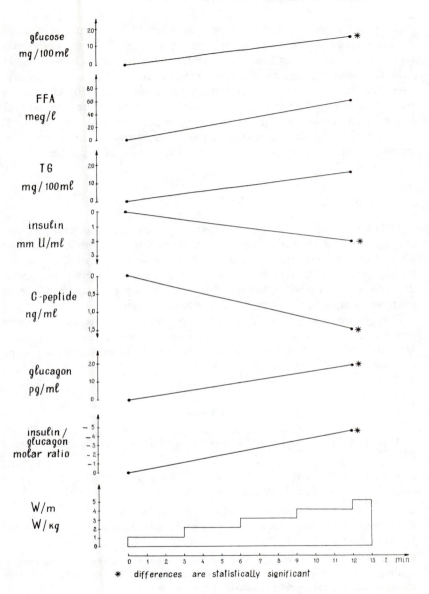

Figure 2—Dynamics of the concentration of some energetic substances and pancreatic hormones on a group of athletes in response to maximal physical loading.

capacity athletes are noted for more pronounced insulin binding in tissue during exercise compared to lower capacity athletes. That presumption is confirmed by glucose utilization prevalence over its production in higher exercise capacity athletes compared to the lower exercise capacity group. The data on the importance of insulin/glucagon correlation changes in plasma in response to maximal exercise are of great interest to us. Our observations show, that, although there is a possibility for an individual insulin concentration increase or a glucagon concentration decrease in plasma, both molar concentrations always lessen during exercise.

Conclusions

1) The reaction of the human organism to maximal exercise includes changes in insulin and glucagon concentrations. In addition to the alteration in the correlation of these hormone concentrations in plasma, athletes with higher exercise capacity are noted for a more pronounced decrease in the insulin/glucagon relationship in plasma in response to exercise, compared to lower exercise capacity athletes.

2) Higher exercise capacity athletes are noted for more active insulin binding in tissues during exercise compared to lower exercise capacity athletes, while the inhibition of insulin secretion is the same in both groups of athletes.

References

FELIG, P & J. Wahren. Fuel homeostasis during exercise. N. Eng. J. Med. 293:1078–1084, 1975.

FELIG, P., J. Wahren, R. Hendler & G. Alberg. Plasma glucagon levels in exercising man. N. Eng. J. Med. 287:184–185, 1972.

GALBO, H., E.A. Richter, J. Hilsted, J.J. Holst, N.J. Christensen & J. Henrickssen. Hormonal regulation during prolonged exercise. Ann. N. Y. Acad. Sci. 301:72–80, 1977.

KOIVISTO, V.A., V. Soman, P. Conrad, R. Hendler, E. Nadel & P. Felig. Insulin binding to monocytes in trained athletes: changing in the resting state and after exercise. J. Clin. Invest. 64:1011–1015, 1979.

LOHMAN, D., F. Liebold, W. Heilman, H. Senger & A. Pohl. Diminished insulin response in highly trained athletes. Metabolism 27:521–524, 1978.

PRUETT, E.D.R., Plasma insulin concentration during prolonged work at near maximal oxygen uptake. J. Appl. Physiol. 29:155–158, 1970.

SUTTON, J.R., Hormonal and metabolic responses to exercise in subjects of high and low work capacities. Med. Sci. Sports 10:1–6, 1978.

WIRTH, A., C. Diehm, H. Mayer, H. Mörl, I. Vogel, P. Björntorp & G. Schlierf. Plasma insulin and C-peptide in trained and untrained subjects. J. Appl. Physiol. 50(1):71–77, 1981.

Regulation of Insulin Receptor Affinity during Exercise

G. Michel, T. Vocke and W. Bieger

University of Heidelberg, Heidelberg,
Federal Republic of Germany

Physical exercise induces significant alterations in the composition of human peripheral blood. Serum levels of insulin fall rapidly after the onset of exercise (Pruett, 1970), whereas the counterregulatory hormones glucagon, catecholamines, growth hormone, and cortisol are increased (Galbo et al., 1975). The rise in glucotropic hormones is understood as a mechanism to restore plasma glucose concentration by stimulation of hepatic glucose production (Rowell et al., 1965). Insulin, despite its decreasing plasma level, is believed to have a permissive effect on muscular glucose uptake (Vranic & Berger, 1979). Insulin binding affinity has been found to be increased in monocytes and erythrocytes after exercise (Bieger et al., 1981; Pederson et al., 1980; Soman et al., 1978). From recent findings in our laboratory (Michel et al., 1981) we have evidence that exercise-induced serum factors can alter cellular insulin binding affinity.

Methods

Exercise Program

Fourteen nonobese volunteers, aged 25 ± 3 years, participated in the study. None of them was engaged in regular physical training. After an overnight fast the bicycle ergometer test was started with a 30 min rest period at 8 a.m.: 5 min exercise at 1 W/kg, 5 min at 2 W/kg, thereafter stepwise acceleration until exhaustion with 30 W increments every min followed by an active recovery period. Exercise performance was continuously monitored on a Siregnost FD-85s (Siemens, FRG). Blood samples were drawn after 30 min rest, 5 min exercise, exhaustion, and the recovery period. Except the catecholamines, which were determined radioenzymatically

(Da Prada & Zürcher, 1976), hormones and metabolites were determined by routine methods. All results were corrected for changes in hemoconcentration according to Beaumont (1973).

Cell Preparations

After sedimentation by dextrane 500, the mononuclear cells were separated over Ficoll-Hypaque. Monocytes, identified by unspecific esterase (Tucker et al., 1977) and peroxidase staining (Kaplow, 1965), yielded on the average 21.8%. Erythrocytes from the dextrane sediments were purified two times over Ficoll-Hypaque.

Insulin Binding

Purified mononuclear cells were suspended in binding buffer pH 7.6 at a final concentraiton of 4.5×10^7 (3.1 to 5.2) cells per ml. Using A 14-^{125}I-Insulin (Novo Institute, Denmark; 285 μCi/μg) as tracer, cells were incubated with 0.2 ng/ml ^{125}I-Insulin in the absence or presence of increasing concentrations of unlabeled insulin for 90 min at 15° C. Binding was terminated by centrifuging the cell suspensions through chilled newborn calf serum. All binding data were corrected for unspecific binding determined in the presence of 5×10^4 ng/ml unlabeled insulin. The erythrocyte binding assay (3.5 ± 0.9 RBC/ml, 200 min at 15° C) was carried out according to Gambhir (1978).

In Vitro Studies

Sera were prepared from blood samples obtained before (A) and after (B) exhaustive exercise. In a series of experiments resting cells were incubated (15 min, 37° C) in 80% serum A or B, respectively. After washing in cold binding buffer insulin binding was performed as described above. Control experiments revealed that incubation of resting cells in serum A had no effect on insulin binding characteristics. Autologous serum A was further used as preincubation medium to test the influence of glucotropic hormones and ketone bodies on insulin binding (Table 2, Figure 3). Cells were suspended in serum and the hormones or metabolites added at the concentrations indicated (Table 2). In additional experiments the described compounds were tested in the insulin binding assay without preincubation.

Data Analysis

Insulin binding data were analyzed according to Scatchard (1949) and De Meyts & Roth (1975). The student's t-test was used for statistical analyses.

Results

During the ergometer test the mean heart rate rose from 86 to 192 beats/min, oxygen consumption increased to 3440 ml O_2/min, the aerobic power (VO_2max) reached 46.7 ml/kg/min. As shown in Table 1, insulin fell till the end of the exhaustion phase, followed by a rapid increase in the recovery period exceeding basal values. Counterregulatory hormones, somatostatin and glucose increased and remained elevated during the active recovery time (not shown). A slight rise could also be seen in the serum concentrations of free fatty acids and, more accentuated, in circulating ketone bodies. Lactate and pyruvate showed the expected increments.

Insulin binding to monocytes was significantly reduced after exhaustive exercise (Figure 1). Maximum binding measured at insulin tracer concentration was decreased by 26.4% (p ≤ 0.01). Although less pronounced, the same result could be obtained by incubating resting cells in post-exercise serum, prepared from blood drawn immediately after the exhaustion phase of the exercise test. After incubation for 15 min at 37° C insulin binding was impaired by 12.5% (p ≤ 0.05). Similar results were observed using erythrocytes. Here the decrement in insulin binding after exercise was 10.4% (p ≤ 0.05), but no statistically significant difference could be seen after preincubation in post-exercise serum.

Scatchard analysis revealed curvilinear competition curves indicating differences in insulin binding affinity without affecting the total number per cell (Figures 1,2).

No effects of catecholamines, growth hormone, glucagon, and cortisol on insulin binding to monocytes and erythrocytes could be detected after preincubation or direct incubation (data not shown). The result of incubation-experiments with ketone bodies are summarized in Table 2, and Figure 3. Statistically significant changes were entirely due to altered binding affinity (Scatchard plots not shown).

Discussion

Changes in cellular insulin receptors have been observed in a variety of disease states and under various conditions (Olefsky, 1981; Felig & Soman, 1979; Olefsky, 1981). Insulin binding was also studied under conditions of acute exercise. Soman and coworkers (1978) have observed about a 36% increase in insulin binding affinity of monocytes after 3 hr bicycle exercise at about 40% VO_2max. Using similar exercise models insulin binding was reported to decrease by 31% in endurance trained athletes and to increase slightly in obese subjects (Koivisto et al., 1980) and in insulin dependent

Table 1

Effects of Exercise on Serum Factors

Serum factor	Exercise			
	0 min	5 min	15 min	15 + 5 min
Glucose (mg%)	87.6 ± 3.0	89.1 ± 3.2	106.4 ± 2.1	107.8 ± 9.9
Insulin (μU/ml)	14.3 ± 2.4	11.1 ± 2.1	8.1 ± 3.2	17.3 ± 3.5
Free fatty acids (μmol/l)	333 ± 37	327 ± 39	341 ± 45	345 ± 57
β-Hydroxybutyrate (μmol/l)	307 ± 144	284 ± 193	763 ± 628	854 ± 674
Acetoacetate (μmol/l)	64 ± 11	63 ± 17	68 ± 14	73 ± 15
Lactate (mmol/l)	1.82 ± 0.15	3.62 ± 0.19	8.0 ± 0.35	7.9 ± 0.45
Pyruvate (mmol/l)	0.076 ± 0.010	0.101 ± 0.025	0.167 ± 0.020	0.159 ± 0.022
pH	7.375 ± 0.036	—	7.196 ± 0.060	—
Hematocrit	46.0 ± 2.3	—	49.5 ± 2.1	—

(mean ± SEM)

diabetics (Pedersen et al., 1980). Employing shorter (30 min) though heavier exercise increased binding could be observed in untrained persons (Bieger et al., 1981). Now we have demonstrated the opposite effect after 15 min exhaustive exercise. Both effects, the increase after 30 min (Michel et al., 1981) and the decrease after 15 min could be reproduced by incubating resting cells in 'post-exercise' serum indicating involvement of serum components.

At the first sight insulin antagonistic hormones might be considered as candidates for modulation of insulin receptor binding. But in vitro experiments using various concentrations of these hormones showed no effect on insulin binding to monocytes and erythrocytes (Michel et al., 1982). Numerous metabolite-concentrations are changed by heavy muscular work. In vitro experiments on IM-9 lymphocytes (Merimee et al., 1976; Misbin et al., 1978) and fibroblasts (Hidaka et al., 1981) have revealed that

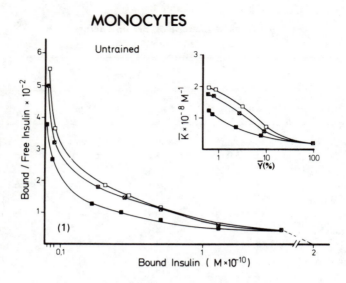

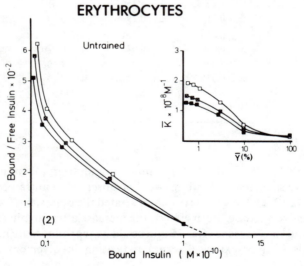

Figures 1,2—Scatchard analysis of insulin receptor binding in untrained subjects. The inset at the upper right shows the average affinity profile plot of the respective data. Average affinity $\overline{K}$ is plotted against the average receptor occupancy $\overline{y}$.

☐ Before exercise

☒ After incubation in post-exercise serum

■ After exhaustive exercise

Table 2

Effect of Ketone Bodies on Insulin Binding to Monocytes at Insulin Tracer Concentration in Percent of Control

Ketone bodies[1] (mmol/l)	Preincubation pH 7.38	Direct incubation pH 7.60	pH 7.10
Control	100	100	100
1,8	98	104	102
9,0	96	103	110
18,0	85*	97	142*

[1]80% β-Hydroxybutyrate = 20% acetoacetate
*p $\leq$ 0.05 (n = 7)

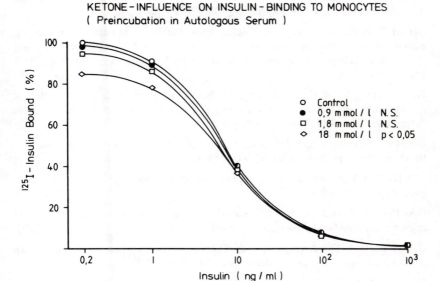

Figure 3—Specific insulin binding to monocytes after preincubation in autologous serum supplemented with ketone bodies.

ketone bodies are capable to restore insulin binding to normal after reduction by lowered pH. Our findings using human monocytes confirm these results. The molecular basis of short term receptor regulation is not clear. Receptor mediated degradation of insulin is well documented for monocytes (Goldstein & Livingston, 1981), but not existing in IM-9 lymphocytes (Sonne & Gliemann, 1980). Therefore it is unlikely that this mechanism is involved in the phenomena observed in the presence of ketones. Using ketone bodies in binding buffer at pH 7.6, insulin binding was not affected, whereas preincubation with serum induced significant changes in binding affinity which could be explained by interaction of serum components with the ketones. We conclude that pH and ketone bodies are among the factors modulating insulin binding to circulating cells. However, they cannot alone explain the biphasic changes in insulin receptor binding observed during short-term exercise. The physiological role of the exercise effects at the receptor level remains to be established.

References

BEAUMONT, W. van, J.C. Strand, J.S. Petrofsky, S.G. Hipskind & J.E. Greenleaf. Changes in total plasma content of electrolytes and proteins with maximal exercise. J. Appl. Physiol. 34:102–106, 1973.

BIEGER, W.P., M. Weiss, & H. Weicker. Insulin affinity and hormone-dependent activity of human circulating monocytes after exercise. In J. Poortsmans & G. Niset (Eds.), Biochemistry of Exercise IV-A, pp. 163–171. Baltimore: University Park Press, 1981.

DA PRADA, M., & G. Zürcher. Simultaneous radioenzymatic determination of plasma and tissue adrenaline, noradrenaline, and dopamine within the femtomole range. Life. Sci. 19:1161–1174, 1976.

DE MEYTS, P. & J. Roth. Cooperativity in ligand binding: a new graphic analysis. Biochem. Biophys. Res. Commun. 66:1118–1126, 1976.

FELIG, P. & V. Soman. Insulin receptors in diabetes and other conditions. Am. J. Med. 67:913–915, 1979.

GALBO, H., J. Holst & N.J. Christensen. Glucagon and plasma catecholamine response to graded and prolonged exercise in man. J. Appl. Physiol. 38:70–76, 1975.

GAMBHIR, K.K., J.A. Archer & C.J. Bradley. Characteristics of human erythrocyte insulin receptors. Diabetes 27:701–708, 1978.

GOLDSTEIN, B.J. & N. Livingston. Insulin degradation by insulin target cells. Metabolism 30:825–835, 1981.

HIDAKA, H., B.V. Howard, F. Shibahi, F.C. Kosmakos, J.W. Craig, P.H. Bennett & J. Larnes. Effect of pH and β-hydroxybutyrate on insulin binding and action in cultured human fibroblasts. Diabetes 30:402–406, 1981.

KAPLOW, L.S., Simplified myeloperoxidase stain using bencidine dihydrochloride. Blood 26:215–219, 1965.

KOIVISTO, V., V. Soman, E. Nadel, W.V. Tamborlane & P. Felig. Exercise and insulin: insulin binding, insulin mobilisation, and counterregulatory hormone secretion. Federation Proc. 39:1481–1486, 1980.

MERIMEE, T.J., A.J. Pulkkinen & S. Lofton. Increased insulin binding by lymphocyte receptors induced by β-OH-butyrate. J. Clin. Endocrinol. Metab. 43:1190–1192, 1976.

MICHEL, G., K. Biehl, H. Weicker & W. Bieger. Increased insulin binding to human monocytes after exposure to post-exercise serum. In D. Andreani, R. De Pirro, R. Lauro, J. Olefsky & J. Roth (Eds.), Current Views on Insulin Receptors, pp. 429–437. London: Academic Press, 1981.

MICHEL, G., T. Vocke, W. Fiehn, H. Weicker & W. Bieger. Regulation of insulin receptor affinity in circulating monocytes and erythrocytes. Decreased affinity after short-term exhaustive exercise. 1982. (to be published)

MISBIN, R.I., A.J. Pulkinen, S.A. Lofton & T.J. Merimee. Ketoacids and the insulin receptor. Diabetes 27:539–542, 1978.

OLEFSKY, J.M., Insulin resistance and insulin action. Diabetes 30:148–162, 1981.

PEDERSEN, O., H. Beck-Nielsen & L. Heding. Increased insulin receptors after exercise in patients with insulin-dependent diabetes mellitus. N. Engl. J. Med. 302:886–892, 1980.

PRUETT, E.D.R., Plasma insulin concentrations during prolonged work at near maximal oxygen uptake. J. Appl. Physiol. 29:155–158, 1970.

ROWELL, L.B., E.J. Masoro & M.J. Spencer. Splanchnic metabolism in exercising man. J. Appl. Physiol. 20:1032–1040, 1965.

SCATCHARD, G., The attractions of proteins for small molecules and ions. Ann. N.Y. Acad. Sci. 51:660–672, 1949.

SOMAN, V.R., V.A. Koivisto, P. Grantham & P. Felig. Increased insulin binding to monocytes after acute exercise in normal man. J. Clin. Endocrinol. Metab. 47:216–218, 1978.

SONNE, O. & J. Gliemann. Insulin receptors of cultured human lymphocytes (IM-9). Lack of receptor mediated degradation. J. Biol. Chem. 255:7449–7454, 1980.

TUCKER, S.R., R.V. Pierre & R.E. Jordon. Rapid identification of monocytes in a mixed mononuclear cell preparation J. Immunol. Meth. 14:267–269, 1977.

VRANIC, M. & M. Berger. Exercise and diabetes mellitus. Diabetes 28:147–163, 1979.

The Influence of Physical Training and High Energy Intake on Glucose Tolerance and Insulin in Human Subjects

A. Tremblay, A. Nadeau, D. Richard and J. Leblanc

Laval University, Québec, Canada

Highly trained subjects demonstrate a normal tolerance to injected glucose in spite of a lower insulin response than that observed in non-trained subjects (LeBlanc et al., 1979). However, we have shown recently that this insulin sparing effect may depend substantially on the level of food intake and physical activity (LeBlanc et al., 1981). Indeed, trained subjects inactive during three days maintained their low insulin levels when they ate 2076 kcal/day while an elevation, making their insulin values comparable to those of non-trained subjects, was noted when they consumed 3291 kcal/day. Energy intake, energy balance, and physical activity were then suspected to exert an influence on insulin variations. In order to obtain more information about the influence of these variables on plasma insulin, the present study was designed and consisted in measuring the effects of high energy intake with or without exercise on insulin.

Methods

Three groups of eight subjects gave their written consent to participate in this study. Two groups were trained subjects and one group non-trained subjects. In a first experiment, one group of trained subjects received a 5122 kcal/day diet during 3 days while running 16 km/day. The second experiment was done on the other group of trained subjects and the non-trained subjects. They consumed 4262 kcal/day and were inactive during 3 days. The carbohydrate content of both these diets was 641 and 537 g/day, respectively. The energy content of each diet was calculated to impose a comparable excessive intake of 1500 kcal/day. All these 3-day protocols were followed by an intravenous glucose tolerance test (IVGTT) which was performed during the morning of the fourth day in the fasting state. The insulin and glucose values obtained after high energy intake diet were compared to previous observations made when subjects maintained usual habits of food intake and physical activity (LeBlanc et al., 1981). In order to obtain an overall indication of insulin and glucose concentration during IVGTT, total area under the curve was calculated for both of these parameters.

Results

Fasting plasma glucose and insulin were comparable among the three groups. The glucose tolerance was comparable after the three high energy diets (Figure 1). However, all these treatments, independent of the level of fitness or physical activity, induced an improvement of glucose tolerance as compared to values obtained when subjects maintained their usual life habits. This was reflected by lower total area under the glucose curve (Figure 2).

The insulin response to glucose injection following high energy diet was not similar in the three groups of subjects. In active trained and non-trained subjects, insulin level during IVGTT was not affected by high energy diet (Figure 2). On the other hand, it was observed that inactive trained subjects exhibited a substantial rise of insulin which made their scores comparable to those of non-trained subjects (Figure 2).

Discussion

The high energy diet during 3 days produced a considerable improvement of glucose tolerance in active trained and inactive trained and non-trained subjects. It would then seem that the level of fitness and daily physical activity could not explain this effect. Recently, it has been reported that feeding a high carbohydrate diet for 10 to 20 days accelerated glucose transport in adipocyte and activated lipogenesis and glucose oxydation (Salans et al., 1981). In the present study, the active trained subjects consumed 641 g carbohydrate/day while the ingestion of inactive trained and non-trained subjects amounted to 537 g/day, which may be considered in both cases as a high carbohydrate diet. On the other hand, an impairment of glucose tolerance was noted in animals after 10 weeks of the high-energy diet. This was associated with an increase of the adipocyte size (Richard et al., in press). Thus, it appears that the high-energy, high-carbohydrate diet induced a short-term improvement of glucose tolerance

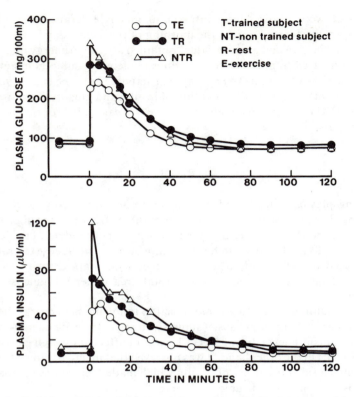

Figure 1—Variations with time in plasma glucose and insulin after intravenous injection of glucose. One group of trained subjects were tested after 3 days of exercise with the high energy diet while the other groups rested during the high energy diet.

and this might be due to the activation of some metabolic pathways concerned with glucose disposal. The long term effect of such a diet is an enlargement of the adipocytes associated with insulin resistance and glucose intolerance (Salans et al., 1978). In conclusion, our results, in agreement with those of other investigators, suggest that a high energy intake has a beneficial transitory effect on glucose tolerance. This effect would disappear when substantial alteration in the storage of triglycerides occurs in the adipocytes.

The insulin response to glucose injection was not modified by excessive food intake in active trained and non-trained subjects while inactive trained subjects demonstrated substantial increase of insulin level. In a previous experiment, similar increases of plasma insulin were noted in trained subjects submitted to a 3-day rest and consuming 3291 kcal/day (LeBlanc et al., 1981). Energy balance was then considered to be a variable which may explain insulin variations. In the present study, this possibility cannot be evoked since the three groups of subjects consumed a comparable

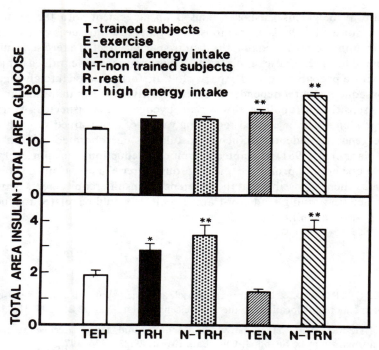

Figure 2—Total area under the curve of glucose and insulin after intravenous injection of glucose. Data obtained after high-energy diets are compared to those obtained when subjects maintain their usual food intake and exercise habits. Total area glucose units are mg/ 100 ml X min X 10^3. ** Glucose area after TEN treatments was higher (P < 0.01) than after TEH treatment. Glucose area after NTRN was higher (P < 0.01) than after all other treatments. Total area insulin units are μU/ 100 ml X min X 10^3. ** Insulin area after NTRH and NTRN treatments were higher (P < 0.01) than after TEN and TEH treatments. * Insulin area after TRH treatment was higher (P < 0.05) than after TEN and TEH treatments.

estimated excess of 1500 kcal/day. A factor which may be related to insulin variations observed in the present study concerns the difference in glycogen formation and utilization between sedentary and active subjects. Bergström et al. (1972) have shown that the exercise-induced glycogen synthetase activity increase disappears 2 days after exercise, a time corresponding to almost complete replenishment of glycogen stores (Bergström & Hultman, 1966). In our inactive trained subjects, it is probable that 3 days of rest with 4262 kcal/day consumption were sufficient to allow saturation of glycogen resynthesis. On the other hand, even if active trained subjects have also excessive food intake, we can suspect that their daily 16 km running provided some glucose disposal availability to form glycogen. Recently, Ivy and Holloszy (1981) showed that, in absence or with standard levels of insulin, muscle moderately depleted of its glycogen content by exercise demonstrates greater glucose uptake. Glycogenesis was then the main pathway of glucose disposal. This is an agreement with the results of Mondon et al. (1980) who demonstrated that muscle tissue is very sensitive to insulin and proposed that the insulin sparing effect of exercise would be related to a greater use of this tissue. Thus, in our inactive trained subjects, having a probable reduced capacity to transform glucose into glycogen, a consequent greater dependence of less sensitive tissue than muscle and/or less sensitive metabolic pathway than glycogenesis is suspected. A similar dependence can be evoked regarding values of non-trained subjects who are generally sedentary. Such a postulate is corroborated by results of LeBlanc et al. (1981) who noted a significant reduction of insulin level after only one bout of prolonged and vigorous exercise in non-trained subjects. It may then be possible that the difference in insulin requirements between trained (in training) and non-trained subjects would be related to glucose disposal availability for glycogen synthesis.

References

BERGSTRÖM, J. & E. Hultman. Muscle glycogen synthesis after exercise. An enhancing factor localized to the muscle cells in man. Nature 210:309–310, 1966.

BERGSTRÖM, J., E. Hultman & A.E. Roch-Norlund. Muscle glycogen synthetase in normal subjects. Scand. J. Clin. Lab. Invest. 29:231–236, 1972.

IVY, J.L. & J. O. Holloszy. Persistent increase in glucose uptake by rat skeletal muscle following exercise. Am. J. Physiol. 241:C200–C203, 1981.

LEBLANC, J., A. Nadeau, M. Boulay & S. Rousseau-Migneron. Effects of physical training and adiposity on glucose metabolism and ^{125}I-insulin binding. J. Appl. Physiol. 46:235–239, 1979.

LEBLANC, J., A. Nadeau, D. Richard & A. Tremblay. Studies on the sparing effect of exercise on insulin requirements in human subjects. Metabolism 30:1119–1124, 1981.

MONDON, C.E., C.B. Dolkas & G.M. Reaven. Site of enhanced insulin sensitivity in exercise-trained rats at rest. Am. J. Physiol. 239:E169–E177, 1980.

RICHARD, D., A. Labrie, D. Lupien, A. Tremblay & J. LeBlanc. Interactions between dietary and exercise training on carbohydrate metabolism. Int. J. Obes. (in press)

SALANS, L.B., J.E. Foley & S.W. Cushman. The adipose cell size and insulin resistance. In G. Bray (Ed.), Recent Advances in Obesity Research: II. Westport: Food and Nutrition Press, 1978.

SALANS, L.B., J.E. Foley, L.J. Wardzala & S.W. Cushman. Effect of dietary composition on glucose metabolism in rat adipose cells. Am. J. Physiol. 240: E175–E183, 1981.

Effect of Exercise on Glucose Homeostasis in Humans with Insulin and Glucagon Clamped

R.R. Wolfe, E.R. Nadel and J.H. Shaw

Harvard Medical School, Boston, Massachusetts, U.S.A., and
John B. Pierce Foundation, New Haven, Connecticut, U.S.A.

In the first 60 min of light (30% $\dot{V}_{O_2}$max) exercise, plasma glucose homeostasis is maintained by an increase in hepatic glucose production that balances the increase in glucose uptake by working muscles (Ahlborg et al., 1974). Increased sensitivity to insulin has been cited as the explanation for the increased peripheral clearance of glucose, perhaps because of increased binding of insulin to receptors (Koivisto et al., 1980). The mechanism responsible for the increase in hepatic glucose production in light exercise is not clear, although it has been proposed recently that there is a central neural stimulation of glucose production in exercise that is sensitive to small changes in arterial blood glucose (Chisholm et al., 1982).

The role of insulin and glucagon as mediators of the changes in glucose kinetics over the first 60 min of light exercise is generally considered to be minimal, since there are no measureable changes in those hormone levels (Ahlborg et al., 1974). However, the absence of a statistically significant change in peripheral hormone levels does not rule out the possibility of a small but physiologically important change in hormone release occurring. For example, altered hepatic clearance may minimize a change in peripheral insulin concentration despite different levels existing at the liver. We have therefore approached the problem of determining the role of insulin and glucagon in maintaining glucose homeostasis during mild exercise and recovery by preventing changes in these hormones. We have clamped insulin and glucagon by inhibiting their secretion with somato-statin, and infusing insulin and glucagon at constant rates throughout rest, exercise and recovery. The results of subjects exercising with insulin clamped have been compared to control studies in which hormones were not clamped.

Our results indicate that blood glucose homeostasis in light exercise and recovery is lost when changes in insulin and glucagon are prevented. Furthermore, our experiment provided unexpected but convincing data that hypoglycemia can both increase perceived exertion and induce exhaustion.

Methods

Five normal male volunteers were studied at rest and during light (30 to 40% $\dot{V}_{O_2}$max) exercise. All were highly motivated and experienced subjects. In one study, hormones were not manipulated. In the second study, somatostatin (.1 ug/kg min), glucagon (1.4 mg/kg min), and insulin (.2 mU/kg min) were infused throughout rest, exercise and recovery in order to clamp those hormones at constant, physiological levels. In all studies, the stable isotopes $U^{13}C$-glucose and/or [6,6-d_2]-glucose were given as primed-constant infusions throughout in order to determine rates of glucose production and oxidation. When hormones were not clamped, the exercise (300 to 400 kpm/min) lasted 105 min. The planned exercise period for the clamped experiment was 60 min, but two of the five subjects could not complete the full 60 min due to exhaustion.

In two subjects, the rest period of the hormonal clamp protocol was repeated, but during exercise and recovery enough glucose was infused to maintain euglycemia. In the two subjects who were not able to complete the 60 min of exercise when hormonally clamped, the clamp study was repeated with an additional infusion of lipid emulsion (Liposyn) and heparin maintained throughout in order to elevate FFA levels.

Results

Without hormonal clamp, all subjects completed 105 min of exercise easily, with the average heart rate being 110. Mean insulin levels decreased slightly, but not significantly, by 60 min of exercise (15.3 ± 2.0 to 11.0 ± 3.1 uU/ml), and glucagon concentrations were constant. There was a significant increase in the rate of glucose production during exercise which basically matched the increase in glucose clearance, such that plasma glucose concentration did not fall significantly. Plasma FFA levels approximately doubled. Plasma levels of norepinephrine and epinephrine did not change.

When insulin and glucagon were clamped, plasma glucose fell precipitously at the onset of exercise (Figure 1), due to a higher rate of glucose clearance than in the unclamped study as well as a failure of hepatic glucose

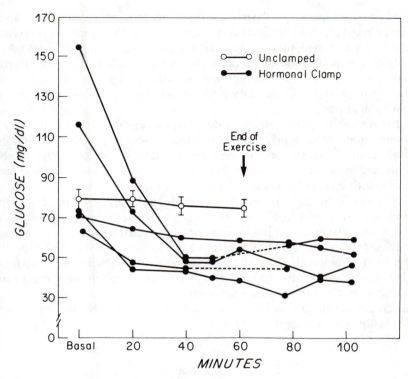

Figure 1—Plasma glucose levels in light exercise. Individual values are presented for the hormonal control studies, whereas mean values are presented for the studies without hormonal control. Two of the five subjects in the clamped study could not complete 60 min of exercise.

Table 1

Mean Characteristics of the Subjects and Tension before and after the Race

	Age (yr)	Ht (cm)	Wt (kg)	Running Experience (yr)	Tension (pre-race)	(post-race)
Mean	35.0	170.6	61.8	7.6	10.8	5.6*
SD	10.8	8.4	11.1	5.2	7.2	3.9
Range	(21 to 53)	(157 to 183)	(40.8 to 82.1)	(1 to 18)		

*Post-race value significantly lower than pre-race value, $p < 0.05$.
N = 14.

production to increase. Plasma glucose concentration failed to rebound during recovery. Two subjects could not complete the 60 min of exercise, and two additional subjects became symptomatic during recovery. All subjects complained of total exhaustion, despite the fact that mean heart rate was only 111 ± 4 during exercise. Plasma FFA levels were low throughout rest and exercise despite a 10-fold increase in norepinephrine and epinephrine.

In the two subjects infused with dextrose during exercise with hormonal clamp, more than 4 mg/kg min of glucose was required to maintain euglycemia. When euglycemia was maintained in this manner, FFA levels remained very low, but the subjects were able to complete the exercise easily, with minimal stress or fatigue.

The two subjects who could not finish the exercise in the required clamped study were able to complete the 60 min of exercise when FFA levels were elevated by the Liposyn and heparin. In these studies, plasma glucose fell in exercise, but never to the hypoglycemic levels seen without the lipid infusion. In one of these subjects infused with lipid, plasma glucose reached a nadir of 53 mg/dl at 20 min, at which time he complained of exhaustion and felt he could not complete the exercise period. However, his glucose level rebounded to 75 mg/dl by the end of exercise, and he had no trouble completing the full 60 min.

Discussion

There is little doubt from our experiment that hypoglycemia develops in light exercise when changes in insulin and glucagon are prevented. We know it was not a direct effect of the somatostatin, because it has recently been reported (Chalmers et al., 1980) that somatostatin infusion alone during exercise only causes a modest drop in blood glucose. Further, since the glucagon levels were higher in our study than when somatostatin was infused without hormone replacement (Chalmers et al., 1980), it can be concluded that the hypoglycemia in our study could not be attributed to glucagon lack. By deduction, it would seem that the failure of insulin to fall in the hormonal clamp was the crucial factor in causing the hypoglycemia. There appeared to be not only an increased sensitivity to the insulin in terms of peripheral clearance, but also in terms of the suppressive effect of insulin or hepatic glucose output. Despite the enormous rise in catecholamine activity in the hormonal clamp study and the profound hypoglycemia, glucose production did not increase.

The failure of FFA levels to increase during the clamp study was surprising in light of the pronounced sympathetic stimulation. It would seem there was an enhanced sensitivity to the antilipolytic effect of insulin in exercise in addition to the enhancement of insulin's effects on glucose kinetics. The failure of FFA levels to increase presumably played a role in the extent to which hypoglycemia developed, since when FFA levels were increased by infusion of Liposyn and heparin, plasma glucose was spared. Our study was not intended to provide an assessment of factors contributing to exhaustion during exercise. The exercise load was clearly below a level that would normally be expected to cause fatigue, as evidenced by the maximal heart rate of 110 and 111 in the two studies. It was therefore quite striking that when hypoglycemia developed, subjects could either not complete the exercise or could complete it only with great effort. We know it was the hypoglycemia and not the low FFA levels that caused the exhaustion, since when euglycemia was maintained with glucose infusion, the exercise was completed quite easily despite low FFA levels. The lipid infusion presumably enabled the subjects to complete the exercise by sparing the plasma glucose sufficiently so that hypoglycemia was avoided.

The situation of the "unclamped" runner exercising to exhaustion is not strictly analogous to this experiment, since we prevented hormonal changes that would allow FFA levels to increase and minimize the fall in glucose. However, our results enable us to conclude that hypoglycemia can induce exhaustion and increase perceived exertion, and that an elevation of FFA levels can minimize the effect by sparing glucose and thereby preventing the development of marked hypoglycemia.

References

AHLBORG G., P. Felig, L. Hagenfeldt, R. Handler & J. Wahren. Substrate turnover during prolonged exercise in man: splanchnic and leg metabolism of glucose, free fatty acids and amino acids. J. Clin. Invest. 53:1080–1090, 1974.

CHALMERS, R.J., S.R. Bloom, G. Duncan, R.H. Johnson & W.R. Sulaiman. The effect of somatostatin on metabolic and hormonal changes during and after exercise. Clin. Endo 10:451–458, 1979.

CHISOHOLM, D.J., A.B. Jenkins, E.W. Kraeger & D.E. James. Hepatic glucose output during exercise is precisely regulated by systemic glucose supply. Diabetes 31(Suppl. 2):33, 1982.

KOIVISTO, V., V. Soman, E. Nadel, W.V. Tamborlane & P. Felig. Exercise and insulin: studies on insulin binding, insulin mobilization and counterregulatory hormone secretion. Fed. Proc. 39:1481–1486, 1980.

Hormones—Catecholamines

Effect of Physical Activity on β-Receptor Activity

W.P. Bieger and R. Zittel

University of Heidelberg, Heidelberg,
Federal Republic of Germany

Catecholamines are essential in the metabolic and cardiovascular adaptation to physical work. During acute exercise norepinephrine and epinephrine both increase their plasma levels depending on the duration and intensity of work (Hartley et al., 1972). Although there are major local disparities in concentration and activity the plasma levels of the catecholamines may reflect their regulatory role. After physical training (Galbo, 1981) basal catecholamine concentrations in plasma are lower than in untrained subjects and the response to a given work load is diminished. The basis of this adaptational process is under investigation. The decrease in norepinephrine response is paralleled by decreased cardiovascular sensitivity (Peronnet et al., 1981; Wiegman et al., 1981; Williams et al., 1981). The decrease in epinephrine response is accompanied by enhanced metabolic effectiveness (Bukowiecki et al., 1980). As has been demonstrated before for insulin (Koivisto et al., 1980) the increase in tissue sensitivity may be due to increased β-receptor activity. We have therefore investigated the effect of acute exercise and training on β-receptor binding in intact granulocytes using the β-adrenergic antagonist dihydroalprenolol.

Methods

Cell Preparation

Human polymorphonuclear leukocytes were prepared from heparinized blood. After dextrane sedimentation the buffy coat was centrifuged over lymphoprep and the bottom layer, containing the granulocytes and erythrocytes, was resuspended in distilled water (3 × 20 s each). Purified granulocytes were identified by peroxidase stain and the viability tested by acridine orange stain ($\geq$ 98%).

Binding Assay

The methodology described in detail by Dulis and Wilson (1980) was used with minor modifications: cells were suspended in binding buffer with deoxyribonuclease (50 μg/ml) added to prevent cell clumping (10^7 cells/ml). Specific binding of [^{3}H]-dihydroalprenolol (^{3}HDHA; 35 Ci/mmol; Amersham Buchler, Braunschweig) was determined in the presence of 50 μM chloroquine to reduce excessive nonreceptor uptake. After 20 min at 37° C cells were placed on ice and filtered through Gelman glass fiber Type A-E (Gelman Sci., Ann Arbor, Mich) presoaked in phosphate buffered saline containing 0.1 mM DL-propranolol (PBSP). During storage at 0° C specific binding remained unchanged up to 60 min. The filters were immediately washed with excessive cold PBSP and placed directly in scintillation fluid (Rotiszint 21; Roth, Karlsruhe, FRG). Nonspecific binding was determined in the presence of 0.1 μM DL-propranolol and was a linear function of the added ^{3}HDHA.

Hormone Measurements

Plasmaimmunoreactive insulin was determined by use of a specific antibody (Behring Inst. Marburg; FRG) and polyethylene glycol precipitation, human growth hormone (HGH) by a double antibody method and glucagon with the C-terminal specific antiserum 30 k after polyethylene glycol precipitation. Samples for epinephrine and norepinephrine assays were collected in tubes containing EGTA (9 mg/ml) and glutathione (6 mg/dl). The hormones were assayed radioenzymatically according to DaPrada and Zürcher (1977). Venous blood concentrations of lactate and glucose were determined using standard procedures (Boehringer, Mannheim, FRG). All serumvalues were corrected for hemoconcentration on the basis of hematocrit changes (van Beaumont et al., 1973).

Exercise Program

The tests were performed on a bicycle ergometer with continuous monitoring of physical performance on a Siergnot FD 85-S (Siemens, Erlangen; FRG). After 30 min rest in the morning, blood was taken from an antecubital vein, bicycle exercise was started at 100 W with stepwise increasing work load every 3 min until exhaustion. The second blood sample was taken immediately at the end of exercise. Oxygen uptake and heart rate were recorded.

Table 1

Anthropometric Data

Subjects	n		age (years)	Broca (%)	$\dot{V}O_2$max (ml min^{-1} kg^{-1})	W/kg	HR beats/min
Untrained	10	x̄	29.7	95.4	42.3	3.4	184
		SD	10.7	6.5	6.3	0.4	9
Trained	12	x̄	23.4	88.1	56.8	4.2	187
		SD	4.8	7.1	9.4	0.7	12

Results

Dihydroalprenolol binding was studied in 10 normal male volunteers (N) without regular physical activity and in 12 endurance trained athletes (T), mainly long distance runners. Anthropometric values from both groups are listed in Table 1. Both had almost normal weight, and the trained group had a 34% higher maximum oxygen uptake. During exercise the heart rate rose to an average maximum of 184 beats/min in the untrained and to 187/min in the trained group. The hematocrit changed from 46.5 to 48.4 (N) and from 45.7 to 47.6 in the athletes. The total number of leukocytes increased by about 70% in both groups with a 45% increase in the number of PMN leukocytes.

Basal insulin and catecholamine concentrations were slightly lower in the trained compared to the untrained group (Table 2). Significant changes in circulating hormone concentrations were obtained in both groups as a result of exhaustive exercise. Insulin decreased by about 34% in the untrained as well as in the trained subjects. Epinephrine and norepinephrine increased 6 to 20 times until the end of exercise with a smaller increase throughout in the trained group.

Acute Exercise

Binding of ^{3}H-dihydroalprenolol to PMN leukocytes approached a rapid equilibrium within 15 to 20 min incubation at 37° C. With increasing concentrations of DHA specific binding reached saturation at 1.6 to 1.8 nM (Figure 1). Scatchard analysis of the binding data revealed a linear relationship between B and B/F, indicating a single class of receptors

Table 2

Exercise Effect on Plasma Concentrations of Hormones and Metabolites

	Untrained 0 min	(n = 10) 20 min	Trained 0 min	(n = 12) 20 min
Insulin (μE/ml)	11.4	7.8[++]	9.0	4.7[+++]
Glucagon (ng/ml)	0.248	0.271[+]	0.256	0.280[+]
HGH (ng/ml)	0.8	9.6[+++]	1.9	8.7[+++]
Cortisol (ng/ml)	221	271[++]	211	290[++]
Epinephrine (pg/ml)	56	345[+++]	29	208[+++]
Norepinephrine (ng/ml)	0.18	3.14[+++]	0.11	2.96[+++]
Glucose (ng/dl)	81	108[++]	90	111[++]
Lactate (mmol/l)	1.1	11.7[+++]	1.4	9.2[+++]

[+] $p \leq 0.05$ [++] $p \leq 0.01$ [+++] $p \leq 0.001$

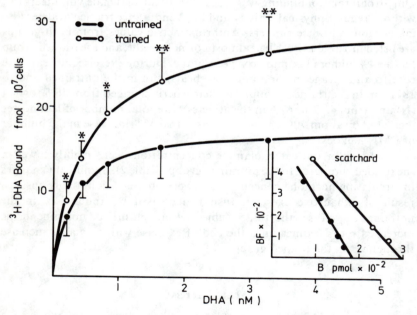

Figure 1—Specific binding of ^{3}H-dihydroalprenolol (DHA) to intact neutrophils as a function of DHA concentration. Neutrophils (10^7/ml) were incubated before and after 20 min exercise for 20 min at 37° C with 0.2 to 3.2 nM DHA. Unspecific binding was determined in the presence of excess propranolol and increased from 28% of total binding at 0.2 nM to 65% at 3.2 nM DHA. The asterisk (*) indicates the only statistically significant difference obtained. Inset: Scatchard plot analysis of the mean values.

EXERCISE EFFECT ON SPECIFIC β-RECEPTOR BINDING

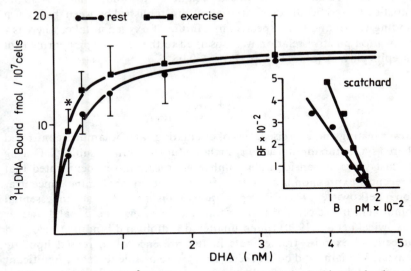

Figure 2—Specific binding of ³HDHA to neutrophils from untrained subjects and endurance trained athletes as a function of DHA concentration. +: p ≤ 0.05; ++: p ≤ 0.01. Inset: Scatchard plot analysis of the mean values.

without interaction. In neutrophils from untrained subjects the total number of binding sites was on the average 1023/cell. Acute exercise increased the binding of ³H-DHA from 3.3 to 4.7% at 0.2 nM concentration with a probability of 95% (p = 0.047). With increasing ligand concentration the binding values before and after exercise moved together, the differences were without statistical significance. The total number of binding sites was unchanged after exercise (1053/cell). These values are in agreement with previously published data of Galant and Allred (1981) who found 864 sites/cell in intact neutrophiles, but are considerably lower than those observed by Dulis and Wilson (1980), 1800 sites/cell.

Training

In contrast to acute exercise, training obviously has a marked effect on β-receptor activity. Specific binding reached saturation in neutrophils from the highly endurance trained group at 2.5 to 3 nM (Figure 2), the maximum attained was about 70% above the level in the untrained subjects. Scatchard analysis revealed a total number of 1746 binding sites cell in the trained group. The specific binding at 0.2 nM ligand concentration was 4.35 compared to 3.30% (p ≤ 0.05). Both values were not significantly related to the Broca weight of the individuals. Also no statistically significant correlation could be established between the highest binding values and the respective maximum of oxygen uptake, whereas a fairly good positive relation was observed to the resting epinephrine (not norepinephrine) values (r = 0.57; p ≤ 0.05).

Discussion

Lower resting and exercise levels of circulating catecholamines confirmed altered sympathoadrenal activity in the endurance trained subjects. During the last years, a considerable number of studies have documented that β-receptor expression is an inverse function of catecholamine concentration (Lefkowitz, 1979). Receptor activity is regarded as an essential component in the cellular transmittance of the hormone signal, however Bukowiecki et al. (1980) found unaltered radioligand binding to isolated fat cells of exercise trained rats in the presence of increased lipolytic activity. Williams and co-workers (1981) were unable to detect significant changes in lymphocyte β-adrenergic receptors when they performed a cross-sectional and a longitudinal 6 week training study.

In contrast to these findings our own results suggest that the model of catecholamine dependent receptor expression has implications for the exercise situation as well. The discrepancy could at least partly result from considerable differences in the experimental approach. We used intact circulating cells instead of membrane preparations (Williams et al., 1981) or enzymatically treated cells (Bukowiecki et al., 1980) for the binding experiments, thus ensuring the presence of all regulatory components required for normal receptor function. In addition, β-receptors on fat cells may be regulated differently from neutrophil receptors. Williams and co-workers, in their longitudinal training study, observed an increase in specific [3]HDHA binding to lymphocyte membranes in 60% of their test persons; however, because of considerable variation among the individuals the difference was not significant. It is conceivable that β-receptor expression is not exclusively controlled by the concentration of catecholamines. Like polypeptide hormone receptors they may be influenced by various circulating agents or by change in pH and osmolality. Further studies will be necessary to clarify this conflicting situation.

The effect of acute exercise on β-adrenoreceptor expression has not been studied before. Despite increasing concentrations of circulating catecholamines we obtained an increase in cellular β-receptor binding after exhaustive exercise. Only few observations on the short-term regulation of β-receptors have been published so far. Tohmeh and Cryer (1980) saw a similar increase in β-receptor binding during 30 to 60 min infusion of isoproterenol. Although using particulate fractions of lymphocytes they were able to relate the increase in receptor activity to an increase in cellular sensitivity. It is uncertain however to what extent the acute alterations in receptor expression are due to changes in the circulating leukocyte pool itself. Considerable work will be needed to prove their mechanism and their physiological significance.

Acknowledgment

This study was supported by a grant from the DFG (Bi 236/3-2).

References

BEAUMONT, W. van, J.C. Strand, J.S. Petrofsky, S.G. Hipskind & J.E. Greenleaf. Changes in total plasma content of electrolytes and proteins with maximal exercise. J. Appl. Physiol. 34:102–106, 1973.

BUKOWIECKI, L., J. Lupien, N. Follea, A. Paradies, D. Richard & J. Le Blanc. Mechanism of enhanced lipolysis in adipose tissue of exercise-trained rats. Am. J. Physiol. 239:E422–E429, 1980.

DA PRADA, M. & G. Zürcher. Simultaneous radioenzymatic determination of plasma and tissue adrenaline, noradrenaline and dopamine within the femtomole range. Life Sci. 19:1161–1174, 1976.

DULIS, B.H. & J.B. Wilson. The β-adrenergic receptor of live human polymorpho-nuclear leukocytes. J. Biol. Chem. 255:1043–1048, 1980.

GALANT, S.P. & S. Allred. Binding and functional characteristics of beta adrenergic receptors in the intact neutrophil. J. Lab. Clin. Med. 98:227–237, 1981.

GALBO, H., Endocrinology and metabolism in exercise. Int. J. Sports Medicine 2:203–211, 1981.

HARTLEY, L.H., J.W. Mason, R.P. Hogan, L.G. Jones, T.A. Kotchen, E.H. Mongey, F.E. Wherry, L.L. Pennington & P.T. Ricketts. Multiple hormonal responses to prolonged exercise in relation to physical training. J. Appl. Physiol. 33:607–610, 1972.

KOIVISTO, V., V. Soman, E. Nadel, W.V. Tamborlane & P. Felig. Exercise and insulin: insulin binding, insulin mobilization, and counterregulatory hormone secretion. Federation Proc. 39:1481–1486, 1980.

LEFKOWITZ, R.J., Direct binding studies of adrenergic receptors: biochemical, physiologic, and clinical implications. Ann. Int. Med. 9:450–458, 1979.

PERONNET, F., J. Clerouc, M. Perrault, D. Cousineau, J. de Champlain & R. Nadeau. Plasma norepinephrine response to exercise before and after training in humans. J. Appl. Physiol. 51:812–815, 1981.

TOHMEH, J.F. & P.E. Cryer. Biphasic adrenergic modulation of β-adrenergic receptors in man. Agonist induced early increment and late decrement in β-adrenergic receptor numer. J. Clin. Invest. 65:836–840, 1980.

WIEGMAN, D.L., P.D. Harris, J.G. Joshna & F.N. Miller. Decreased vascular sensitivity to norepinephrine following exercise training. J. Appl. Physiol. 51:282–287, 1981.

Regulation of Glycogenolysis in Human Muscle during Epinephrine Infusion and during Exercise

D. Chasiotis, K. Sahlin and E. Hultman
Huddinge University Hospital, Huddinge, Sweden

Glycogen is the main energy source in skeletal muscle during intense muscular activity. Two key enzymes are involved in the breakdown and resynthesis of glycogen: glycogen phosphorylase (EC 2.4.1.1) and glycogen synthetase (EC 2.4.1.11). These enzymes exist in two interconvertable forms: *a* and *b*, and I and D, respectively. At least three main mechanisms regulate the phosphorylase-synthetase system:

1. Transformation of phosphorylase *b* to *a* and/or synthetase I to D is mediated by a hormonal mechanism involving cAMP or by Ca^{++} increase in the sarcoplasm during contraction.
2. Activation of the enzymes by changes in the concentration of their substrates.
3. Allosteric activation by other metabolic changes.

It is generally believed that increased glycogenolysis in skeletal muscle is initiated by a transformation of phosphorylase *b* to *a*. The purpose of the present investigation was to study the relationship between phosphorylase *a* activity and glycogen breakdown during epinephrine infusion and exercise.

Materials and Methods

Thirty-three healthy volunteers, 21 males and 12 females, participated in these experiments. To 14 subjects, epinephrine was given by continuous intravenous infusion at a rate of 0.15 μg epinephrine kg^{-1} body weight min^{-1} and the infusion was maintained for 6 to 20 min. Six subjects performed an isometric contraction at 66% of their maximal voluntary contraction force (MVC) for 25 s or to fatigue. Ten subjects worked on an ergometer with a work load corresponding to about 100% of their maximum oxygen uptake capacity for 30 s or to exhaustion.

Muscle samples were taken before and after epinephrine infusion, and before and after exercise. All muscle samples were taken from quadriceps femoris muscle using the needle biopsy technique and were frozen in liquid freon maintained at its melting point with liquid N_2. Samples were analyzed for cAMP, glycogen phosphorylase and synthetase, hexosemonophosphate and lactate. Values given are means $\pm$ SD.

Results

Phosphorylase Activity and P_i Concentration

Based upon the analysis of samples from 4 subjects, half maximal velocity of phosphorylase a in the absence of AMP was observed at a P_i concentration (K_m) of 26.2 ± 1.2 mmol l^{-1}. In the presence of AMP the K_m of phosphorylase $a + b$ for P_i was 6.8 ± 2.0 mmol l^{-1}. The K_m values of phosphorylase a and $a + b$ for P_i were not affected by epinephrine infusion.

Effects of Epinephrine Infusion upon Enzyme Activities and Metabolite Concentrations

The content of cAMP in resting muscle was 2.7 ± 0.7 μmol (kg d m)$^{-1}$; it increased approximately 3-fold within 5 min of infusion and remained elevated at a mean content of 5.4 ± 0.8 μmol (kg d m)$^{-1}$ until the end of infusion. In estimating in vivo phosphorylase activity, a P_i concentration of 11 mmol l^{-1} was used in the assays. This concentration is similar to that found in perchloric acid extracts of resting human skeletal muscle. During infusion phosphorylase a increased from a basal value of 11.5 ± 2.6 to 47.9 ± 10.9 mmol glycosyl units (kg d m)$^{-1}$ min^{-1} and synthetase I decreased from 4.0 ± 1.5 to 1.2 ± 0.8. Total enzyme activities were unchanged during infusion. Muscle glycogen content decreased slightly in all subjects during infusion of epinephrine. The average rate of glycogen decrease was 11.3 ± 1.8 mmol glycosyl units (kg d m)$^{-1}$ min^{-1} during the first 2 min of infusion, but decreased to 2.0 ± 2.3 during prolonged infusion despite continuation of high phosphorylase a activity. There was no significant change in the muscle content of ATP, PCr and AMP during 6 min infusion. There was a trend towards lower P_i values during infusion and this was of the same order as the increase in hexosemonophosphate. This trend however, was not statistically significant.

Effects of Exercise upon Enzyme and Metabolite Contents

The concentration of cAMP in muscle was unchanged during isometric contraction but increased approximately 100% during dynamic exercise. No significant changes in total enzyme activities were observed after isometric contraction or bicycle exercise. Phosphorylase a increased during isometric contraction from 7.9 at rest to 19.1 mmol glycosyl units $(kg\ d\ m)^{-1}min^{-1}$ but the increase was shortlasting and values returned to the basal level when the contraction was continued to fatigue (after 60 s). A similar pattern of change was observed during bicycle exercise when phosphorylase a increased initially from 9.8 at rest to 15.8 mmol glycosyl units $(kg\ d\ m)^{-1}min^{-1}$. At exhaustion phosphorylase a activity was only 4.0 mmol glycosyl units $(kg\ d\ m)^{-1}min^{-1}$. All subjects showed a decrease in synthetase I during exercise. Synthetase I was maintained at a low value when isometric contraction was sustained to fatigue or when dynamic exercise was performed to exhaustion. During both isometric and dynamic exercise lactate and hexosemonophosphates accumulated in the muscle. An average glycogenolytic rate of 58.4 ± 26.6 mmol glycosyl units $(kg\ d\ m)^{-1}min^{-1}$ was calculated over the first 25 s of isometric exercise.

Discussion

Phosphorylase at Rest

Based upon the results of all 21 subjects, phosphorylase a and $a+b$ activity at rest assayed at 11 mmol l^{-1} P_i, was 9.5 ± 3.8 and 89.5 ± 19.9 mmol glycosyl units $(kg\ d\ m)^{-1}min^{-1}$. Using the K_m values given previously, $V_{max}\ a$ and $V_{max}\ a+b$ were in these samples 32.0 ± 12.8 and 144.2 ± 32.0 mmol glycosyl units $(kg\ d\ m)^{-1}min^{-1}$, and the ratio of $V_{max}\ a$: $V_{max}\ a+b$ 0.211 ± 0.059. This ratio may be used as a lower estimate of the proportion of phosphorylase in the a form, provided that the specific activity per mole of phosphorylase a is the same as that for the b form and $V_{max}\ a$ in the presence of AMP is *not less* than $V_{max}\ a$ in the absence of AMP (Chasiotis et al., accepted for publication). The present results indicate, therefore, that at least 20% of phosphorylase is in the a form in resting muscle. The $V_{max}\ a$: $V_{max}\ a+b$ ratio must not be confused with the ratio of a to total activity measured at different P_i concentration. For instance at 11 mmol l^{-1} P_i, despite at least 20% of the molecules being in the a form, the activity ratio is only 10%. It is evident that samples assayed in the direction of glycogen breakdown, the ratio of a to total activity as a measure of the state of phosphorylase in vivo has little or no physiological meaning and is critically dependent upon the conditions of assay used. The most useful ratio is that of $V_{max}\ a$: $V_{max}\ a+b$ expressing the proportion of phosphorylase molecules in the a form.

The State of Phosphorylase during Epinephrine Infusion and during Exercise

Again using the previous estimates of K_m to calculate V_{max} for phosphorylase 2a and $a + b$, the mole fraction of phosphorylase in the a form—estimated from V_{max}: V_{max} $a + b$ (Chasiotis et al., submitted for publication)—was 0.807 ± 0.210, 0.904 ± 0.263 and 0.791 after 2, 6 and 15 min infusion, respectively. The results suggests that after 6 min infusion 90% or more of the enzyme was in the a form. In the isometric exercise study the observed ratio was 0.519 ± 0.112 after 25 s and 18.8 at fatigue. Thus, at the end of 25 s contraction more than 50% of the phosphorylase molecules were in the a form. During dynamic exercise the V_{max} ratio was 0.339 ± 0.051 after 30 s and 0.090 ± 0.032 at exhaustion.

Phosphorylase Activity and Rate of Glycogenolysis

As shown in Figure 1 the glycogenolytic rate during epinephrine infusion was extremely low despite phosphorylase being almost entirely in the a form. This rate was 11.3 during the first 2 min and decreased to 2.0 mmol glycosyl units $(kg\ d\ m)^{-1} min^{-1}$ during prolonged infusion. In contrast, during isometric exercise the glycogenolytic rate was twice that predicted by phosphorylase a activity (Figure 1). The activities of phosphorylase a in the above studies, assayed in vitro, do not, however, take into account changes in P_i and other metabolites within the muscle. In the isometric exercise study re-adjustment of phosphorylase a to the in vivo P_i concentration leads to a far better agreement between phosphorylase a activity and glycogenolytic rates (Figure 1—filled squares). In this case a activity accounted for 68% $\pm$ 21 of the average glycogenolytic rate. In practice a activity in vivo is probably even higher due to the presence of small amounts of AMP which will increase its affinity for P_i.

On the other hand the low glycogenolytic rate during epinephrine infusion, despite most of the phosphorylase being in the a form, can probably be accounted for by a low concentration of P_i at the active site of the enzyme. Recent NMR studies have indicated that this may be of the order of 1 mmol l^{-1} (Dawson et al., 1977). We have estimated from the present data that P_i concentrations of 1.9 and 0.3 mmol l^{-1} would be sufficient, given the amounts of phosphorylase a observed, to sustain rates of glycogen breakdown equal to those found during 0 to 2 and 2 to 6 min infusion. It is concluded that transformation of phosphorylase to the a form may occur in muscle without major change in activity; that it is the change in P_i concentration which regulates activity. In this model P_i release from PCr during exercise is paramount, providing a link between phosphagen utilization and glycogen breakdown.

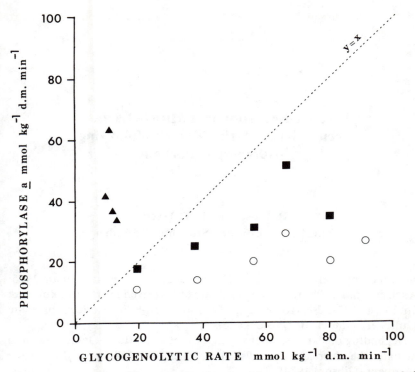

Figure 1—The relation between phosphorylase *a* activity and the observed glycogenolytic rate. Filled triangles (▲) denote phosphorylase *a* activity during epinephrine infusion measured at 11 mmol l^{-1}. Open circles (O) denote phosphorylase *a* activity measured at 11 mmol l^{-1} P$_i$, and filled squares (■) activity after adjustment for the increase in muscle P$_i$ concentration during exercise. Muscle P$_i$ concentration for each biopsy was calculated from the content per kg dry muscle with an assumed intracellular water content of 3.0 l per 4.3 kg wet muscle.

References

CHASIOTIS, D., K. Sahlin & E. Hultman. Regulation of glycogenolysis in human muscle at rest and during exercise. Accepted for publication in J. Appl. Physiol.

CHASIOTIS, D., K. Sahlin & E. Hultman. Regulation of glycogenolysis in human muscle in response to epinephrine infusion. Submitted for publication in J. Appl. Physiol.

DAWSON, M.J., D.G. Gadian & D.R. Wilkie. Contraction and recovery of living muscles studies by [22]P nuclear magnetic resonance. J. Physiol. London. 267:703–735, 1977.

Perceived Exertion and Muscle Lactate Accumulation during Exercise following β-Adrenergic Blockade

P. Kaiser and P.A. Tesch

Karolinska Institutet, Stockholm, Sweden

Impaired physical performance (Ekblom et al., 1972; Epstein et al., 1965; Tesch & Kaiser, 1981) and increased rate of perceived exertion (Ekblom & Goldbarg, 1971; Pearson et al., 1981) are well-known effects of acute β-adrenergic blockade (β-blockade).

Depending on exercise intensity and duration, different factors seem to mediate the decrease in physical performance following β-blockade treatment (Gibson, 1974).

In patients with essential hypertension treated with acute β-blockade and performing submaximal exercise, blood lactate concentration decreased and muscle lactate concentration increased in most cases (Frisk-Holmberg et al., 1979). Since an impaired efflux of lactate from skeletal muscle during exercise could be implied, which might explain the increased experience of physical fatigue, the present investigation was undertaken to study lactate accumulation in muscle and blood during submaximal exercise on β-blockade.

Methods and Procedures

Fifteen healthy male volunteers, used to physical exercise, served as subjects. Their mean ($\pm$ SD) age, height, weight and $\dot{V}o_{2max}$ (cycling) were 26 ± 5 yrs, 181 ± 5 cm, 75 ± 5 kg and 3.81 ± 0.33 l min^{-1} (51 ± 4 ml kg^{-1}min^{-1}), respectively.

In β-blockade experiments, 80 mg of propranolol, (Inderal®) was administered orally 120 min before exercise in order to induce β-blockade. Venous blood samples were taken for drug concentration determinations immediately before exercise for subsequent analyses (McAinsch, 1978).

In the first set of experiments the following protocol was applied: After a warm up period, continuous cycle exercise was performed at increasing 4 min exercise intensities at a pedaling frequency of 60 rpm until near exhaustion. Blood samples from the finger tip were collected at the end of each exercise intensity and analyzed for lactate concentrations using an automatic fluorometric method (Karlsson et al., 1982; Rydevik et al., 1982). The exercise intensity which corresponded to a lactate concentration of 4 mmol l^{-1} blood was calculated (W_{OBLA}) (Tesch et al., 1982). The same protocol was repeated on a separate day after β-blockade.

In a second set of experiments (n = 12) exercise was performed stepwise up to a final exercise intensity corresponding W_{OBLA} (the same exercise was performed in control and β-blockade experiments). Muscle biopsies were taken from the vastus lateralis muscle after 4 min performance at the final exercise intensity and analyzed for lactate according to Karlsson (1971). At each exercise intensity, the rate of local and central perceived exertion (RPE), according to Borg (1970) was obtained.

Results

During β-blockade heart rate was reduced at all work loads ($p < 0.001$). For instance, at 180 W, heart rate averaged 149 ± 12 and 112 ± 7 beats min^{-1}, which corresponded to a 25% decrease. The blood lactate concentrations at 180 and 210 W were significantly higher following β-blockade (3.6 and 5.6 mmol l^{-1}) than in the control experiment (2.9 and 4.6 mmol l^{-1}) ($p < 0.05$ and $p < 0.01$). At lower exercise intensities, no difference was observed.

In later experiments performed before and after β-blockade, where exercise was terminated at the same absolute exercise intensity level, muscle lactate concentration averaged 8.3 ± 3.0 and 8.2 ± 4.0 mmol kg^{-1} wet weight, respectively (Figure 1). The corresponding blood lactate values were 3.7 ± 0.5 and 4.3 ± 0.9 mmol l^{-1} ($p < 0.05$). The ratio between muscle and blood lactate concentrations did not change significantly with blockade.

In the same experiments, local RPE averaged 13.9 ± 2.5 and 15.6 ± 2.5, respectively ($p < 0.05$, Figure 2). The corresponding values for central RPE were 12.7 ± 2.9 and 13.8 ± 3.1, respectively (n.s., Figure 3).

Discussion

The major finding was that no drug effect on the translocation of lactate from muscle cell to blood was present. In contrast, Frisk-Holmberg et al. (1979), studying patients with hypertension, demonstrated that alprenolol

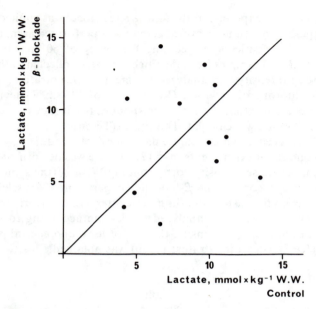

Figure 1—Muscle lactate concentration, at a certain submaximal intensity before and after β-blockade. Line of identity is plotted.

(an unselective blocker) caused lower blood lactate levels and in some of their subjects higher muscle lactate values following β-blockade. Based on this finding they speculated on the occurrence of drug-induced translocation hindrances between muscle and blood in addition to that existing under normal exercising conditions (Jorfeldt et al., 1978).

At high intensity of exercise, perceived exertion is higher following β-blockade. This effect was evidently specifically localized in the exercising muscles, since a difference was observed between the drug-induced response on local and central RPE. It is therefore tempting to suggest that events taking place within the muscle, such as accumulation or depletion of certain metabolites, are the cause for the increased sensation of muscular fatigue associated with β-blockade. In light of the present results, lactate accumulation has not been proved to be the "fatigue" substance.

To summarize, with acute β-adrenergic blockade the exercise intensity corresponding to a blood lactate concentration of 4 mmol l^{-1} is decreased and the rate of perceived exertion increased. However, the muscle lactate level at a certain exercise intensity was not increased, indicating that the increased sensation of fatigue, previously reported to occur during β-blockade and in the present study most markedly localized to the exercising muscles, can probably not be attributed to accumulation of lactate.

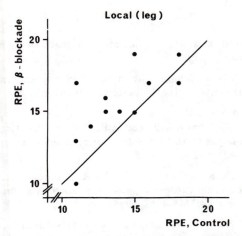

Figure 2—Local (leg) rate of perceived exertion (RPE), at a certain submaximal exercise intensity before and after β-blockade. Line of identity is plotted.

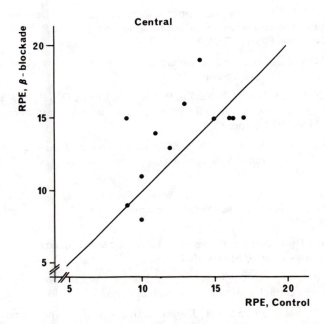

Figure 3—Central rate of perceived exertion (RPE), at a certain submaximal exercise intensity before and after β-blockade. Line of identity is plotted.

References

BORG, G., Perceived exertion as an indicator of somatic stress. Scand. J. Rehab. Med. 2:92–98, 1970.

EKBLOM, B. & A.N. Goldbarg. The influence of physical training and other factors on the subjective rating of perceived exertion. Acta Physiol. Scand. 83:399–406, 1971.

EKBLOM, B., A.N. Goldbarg, A. Kilbom & P.-O. Åstrand. Effects of atropine and propranolol on the oxygen transport system during exercise in man. Scand. J. Clin. Lab. Invest. 30:35–42, 1972.

EPSTEIN, S.E., B.F. Robinson, R.L. Kahler & E. Braunwald. Effects of beta-adrenergic blockade on the cardiac response to maximal and submaximal exercise in man. J. Clin. Invest. 44:1745–1753, 1965.

FRISK-HOLMBERG, M., L. Jorfeldt, A. Juhlin-Dannfelt & J. Karlsson. Metabolic changes in muscle on long-term alprenolol therapy. Clin. Pharmacol. Ther. 26:566–571, 1979.

GIBSON, D.G., Pharmacodynamic properties of β-adrenergic receptor blocking drugs in man. Drugs 7:8–30, 1974.

JORFELDT, L., A. Juhlin-Dannfelt & J. Karlsson. Lactate release in relation to tissue lactate in human skeletal muscle during exercise. J. Appl. Physiol.: Respirat. Environ. Exercise Physiol. 44:350–352, 1978.

KARLSSON, J., Lactate and phosphagen concentrations in working muscle of man. Acta Physiol. Scand. Suppl. 358, 1971.

KARLSSON, J., I. Jacobs, B. Sjödin, P. Tesch, P. Kaiser, O. Sahl & B. Karlberg. Semi-automatic blood lactate assay: Experiences from an exercise laboratory. Int. J. Sports Med. 1982. In press.

MCAINSH, J., N.S. Baber, R. Smith & J. Young. Pharmacokinetic and pharmacodynamic studies with long-acting propranolol. Br. J. Clin. Pharmacol. 6:115–121, 1978.

PEARSON, S.B., D.C. Banks & J.M. Patrick. The effect of adrenoceptor blockade on factors affecting exercise tolerance in normal man. Br. J. Clin. Pharmacol. 8:143–148, 1979.

RYDEVIK, U., L. Nord & F. Ingman. Automatic lactate determination by flow injection analysis. Int. J. Sports Med. 1982.

TESCH, P.A. & P. Kaiser. Effect of β-adrenergic blockade on maximal oxygen uptake in trained males. Acta Physiol. Scand. 112:351–352, 1981.

TESCH, P.A., D.S. Sharp & W.L. Daniels. Influence of fiber type composition and capillary density on onset of blood lactate accumulation. Int. J. Sports Med. 2:252–255, 1981.

The Role of Catecholamines and Triiodothyronine on the Calorigenic Response to Norepinephrine in Cold-Adapted and Exercise-Trained Rats

J. Leblanc, A. Labrie
D. Lupien and D. Richard
Laval University, Québec City, Canada

When the secretion of catecholamines is increased by cold exposure (Hsieh & Carlson, 1957) or if animals are injected repeatedly with catecholamines (LeBlanc et al., 1972), an enhanced thermogenic response to norepinephrine of cold-adapted and of exercise-trained rats has been reported. Preliminary studies were made to obtain comparable levels of energy expenditure and sympathetic stimulation in the two groups studied. To assess the role of triiodothyronine and corticosterone, these hormones were also determined in the adapted animals.

Methods and Procedures

Rats were adapted to cold by placing them at $-15°$ C for 2 h per day for a period of 4 weeks. Another group was adapted to exercise by swimming 2 h per day in water (36° C) also for 4 weeks. At the end of the adaptation period, oxygen consumption, plasma catecholamines, and triiodothyronine were determined before and during the swimming or cold exposure periods. The colonic temperature response to injected norepinephrine (300 μg/kg s.c.) was measured in both groups. When the animals were sacrificed, the interscapular brown adipose tissue was dissected and weighted. Various methods for catecholamines and triiodothyronine (T_3) have been described elsewhere (LeBlanc et al., in press).

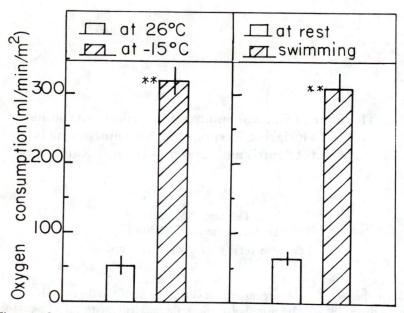

Figure 1—Oxygen consumption in rats at 26°C inactive, in rats at –15°C for 1 h, and in rats swimming in water at 35°C for 1 h. *** = P < 0.01.

Results

As reported in Figure 1, rats swimming in an agitated tank or exposed to cold at –15°C show a six-fold increase in O_2 consumed. Figure 2 shows a three to four-fold increase in catecholamines in rats exposed to cold or swimming for the first time. In the cold adapted rats a larger increase in plasma norepinephrine was observed whereas a fall in epinephrine was found in exercise-trained rats. As seen in Figure 3, a marked increase in norepinephrine-induced thermogenesis and in the size of the brown adipose tissue was found in the cold-adapted animals; no such changes were observed in the exercise-trained rats. The results obtained on plasma T_3 determinations are illustrated in Figure 4. Cold exposure caused a significant increase in plasma T_3 and exercise produced a diminution of this hormone in the blood.

Discussion

We have confirmed the enhanced thermogenic response to norepinephrine in cold-adapted rats (Hsieh & Carlson, 1957; LeBlanc et al., 1972). This effect has been shown to be related to the hypertrophy of the brown adipose

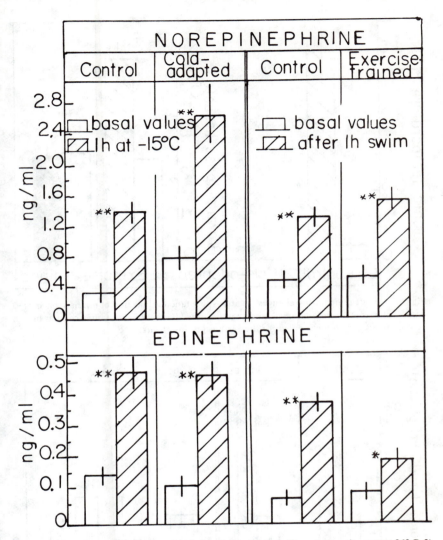

Figure 2—Norepinephrine and epinephrine plasma levels in 1) control resting rats at 26° C, 2) in rats acutely exposed for 1 at –15° C, 3) in rats swimming for 1 h for the first time, 4) in rats cold-adapted rats (–15° C for 2 h per day for 4 weeks), and exposed to –15° C for 1 h, 5) in rats exercise-trained (swimming for 2 h per day for 4 weeks), and swimming for 1 h.** = P < 0.01; * = P < 0.05.

tissue and to the increased norepinephrine secretion (Foster & Frydman, 1977). The exercise-trained animals do not become thermogenic in spite of an enhanced catecholamine secretion. It is suggested that the lack of brown adipose tissue hypertrophy in the exercise-trained animals and the marked hypertrophy of this tissue in the cold-adapted rats is related to the

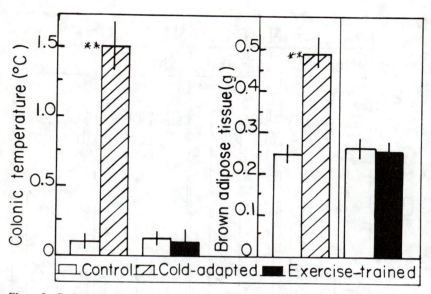

Figure 3—Body temperature changes 1 h after subcutaneous injection of norepinephrine (30 μg/ 100 g) in control, cold-adapted, and exercise-trained rats. Weight of interscapular brown adipose tissue is also represented for the various groups of rats.

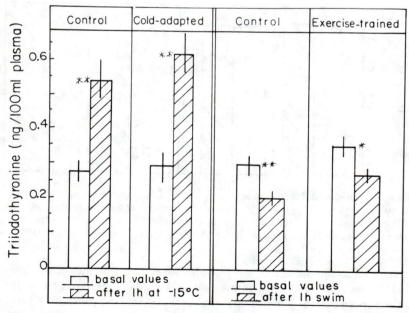

Figure 4—Plasma triiodothyronine variations in various groups of rats as described in Figure 2. ** = P < 0.01; * = P < 0.05.

difference in the thermogenic response to norepinephrine observed in these two situations. The fact that T_3 is increased in cold adaptation and decreased in exercise training also points to a possible relationship between thermogenesis and thyroid hormone secretion. It should also be mentioned that a synergism between catecholamine action and thyroid activity has been described (Williams, 1963; Wurtman et al., 1963).

The increased thermogenic effect of norepinephrine observed in cold adapted rats is associated with increased T_3 secretion and with a marked hypertrophy of the brown adipose tissue. In the exercise-trained rats norepinephrine fails to induce thermogenesis while the secretion of T_3 is diminished and the brown adipose tissue remains unchanged. A causal relationship between these various factors is suggested but is not completely understood.

References

FOSTER, D. & L. Frydman. Nonshivering thermogenesis in the rat. Can. J. Physiol. and Pharmacol. 56:110–122, 1977.

HAGGENDAL, J., L.H. Hartly & B. Saltin. Arterial norepinephrine concentration during exercise in relation to the relative work levels. Scand. J. Clin. Lab. Invest. 26:337–342, 1970.

HSIEH, A.C.L & L.D. Carlson. Role of adrenaline in chemical regulation of heat production. Am. J. Physiol. 190: 242–246, 1957.

LEBLANC, J., A. Labrie, D. Lupien & D. Richard. Catecholamines and the calorigenic response to norepinephrine in cold-adapted and exercise-trained rats. Can. J. Physiol. and Pharmacol. (in press)

LEBLANC, J., J. Vallières & C. Vachon. Beta receptor sensitization by repeated injections of isoproterenol. Am. J. Physiol. 222:1043–1046, 1972.

WILLIAMS, R.H., The Thyroid, 3rd ed. R.H. Williams & W.B. Saunders (Eds.). Philadelphia, 1963.

WURTMAN, R.J., I.J. Koplin & J. Axelrod. Thyroid function and the cardiac disposition of catecholamines. Endocrinology 73:63–74, 1963.

Time and Intensity Dependent Catecholamine Responses during Graduated Exercise as an Indicator of Fatigue and Exhaustion

M. Lehmann, R. Kapp, M. Himmelsbach, J. Keul
University of Freiburg, Freiburg,
Federal Republic of Germany

The free catecholamines, adrenaline and noradrenaline, in blood and urine are indicative of sympathetic activity and may serve in the evaluation of ergotropic demands of the organism (Euler, 1974; Hertting, 1979; Lehmann et al., 1981b; Lehmann et al., 1981f; Lehmann et al., 1981a; Lehmann et al., 1981d). The present investigation should contribute to the following:

1. behavior of dopamine, in addition to noradrenaline and adrenaline during exercise;

2. differences in catecholamine response during "cardiac exhaustion" or following "muscular exhaustion";

3. time and intensity dependent catecholamine responses during cumulative, graduated exercise;

4. conditions for steady state behavior of the plasma catecholamines during cumulative, graduated exercise;

5. initial half-life of the endogenous catecholamines following strenuous exercise.

Procedure

To produce a predominate cardiac exhaustion, nine healthy male subjects performed a graduated bicycle spiroergometric test with a 50 W increase every 3 min (short-term experiment). The catecholamine levels were estimated every min (Table 1) and assayed radioenzymatically (Da Prada & Zürcher, 1976). A more peripheral muscular exhaustion was expected during a graduated bicycle ergometric test with increases of 50 W every 15 min (prolonged experiment). During the prolonged experiment, the catecholamines were measured every third min, in the same nine subjects. The following additional measurements were analyzed as follows: Oxygen intake using the open system, ERGOPNEUMOTEST; lactate, using an enzymatic method (Hohorst, 1962); glucose, using an enzymatic method (Slein, 1962); and heart rate, using ECG.

Table 1

Behavior of Performance Data during Maximal Short-term and Prolonged Graduated Exercise

	Graduated exercise	
	50 W/3rd min	50 W/15th min
n	9	9
Vo$_2$max (ml/kg × min) $\bar{x}$	48.4	
Heart rate (/min) $\bar{x}$	185	*	174
Noradrenaline (nmol/l) $\bar{x}$	30.7	**	11.0
Adrenaline (nmol/l) $\bar{x}$	6.9	**	2.0
Lactate (mmol/l) $\bar{x}$	9.2	**	6.6
Glucose (mmol/l) $\bar{x}$	5.7	**	4.78
Performance (W) $\bar{x}$	305	**	233
Time (min) $\bar{x}$	17.5	***	51.0

* $p < 0.05$; ** $p < 0.01$; *** $p < 0.001$.

Results and Discussion

During the "short-term experiment" (Figure 1), adrenaline and noradrenaline showed parabolic increases as described before (Euler, 1974; Haggendal et al., 1970; Lehmann et al., 1981a-f). The mean dopamine concentrations were nearly constant (Van Loon & Sole, 1980), with only slight linear increases. During aerobic exercise, below a 4 mmol lactate level and below 70% of $\dot{V}_{O_2max}$, there occurred predominantly, a small time dependent increase of adrenaline and noradrenaline; however, in the anaerobic range a strong time and intensity dependent increase was observed.

During the short-term experiment, the individual lactate and catecholamine levels (adrenaline and noradrenaline) changed in a parallel manner (Figure 2). Therefore, direct linear descriptive correlations between individual catecholamine and lactate concentrations were observed ($r = 0.84$ and $r = 0.91$), at the same submaximal exercise levels (Figure 3). A direct correlation was also observed between heart rate and the catecholamine concentrations. The relationship between the catecholamine and lactate concentrations or the heart rate responses may be an indicator for the sympathetic influence on glycolysis (Gollnick, 1973) and heart rate regulation (Ekblom et al., 1968).

During the "prolonged experiment" no time or intensity dependent change of dopamine was observed (Figure 4). Noradrenaline and adrenaline showed such a dependency, only at higher exercise levels. A rapid degradation was found following exhaustion, as reported for noradrenaline following strenuous exercise by Hagberg et al. (1979). The mean increase in lactate concentration was approximately three-fold above the resting value, the mean decrease of blood glucose was approximately 20%, and the heart rate elevation approximately 100 beats/min (Figure 5).

During the prolonged experiment such strong correlations were not observed between the catecholamine and lactate concentrations (Figure 6). Possible causes include, a decreased pyruvate excess and an increased hepatic glucose resynthesis from alanine (Felig, 1973) and lactate during prolonged exercise. However as in the short-term experiment, a direct relationship between heart rate response and catecholamine concentrations ($r = 0.91$) occurred, and in addition a direct relationship between adrenaline increase and the depression of blood glucose was observed ($r = 0.75$).

During the prolonged experiment, noradrenaline, adrenaline, lactate and the heart rate showed steady state behavior only under aerobic exercise conditions (Figure 7).

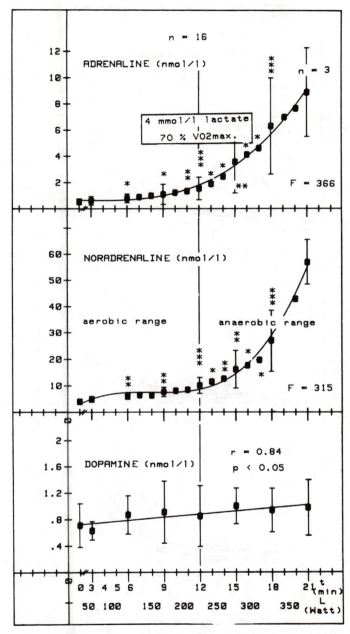

Figure 1—Catecholamine responses during the short-term experiment, in relation to the 4 mmol lactate threshold.

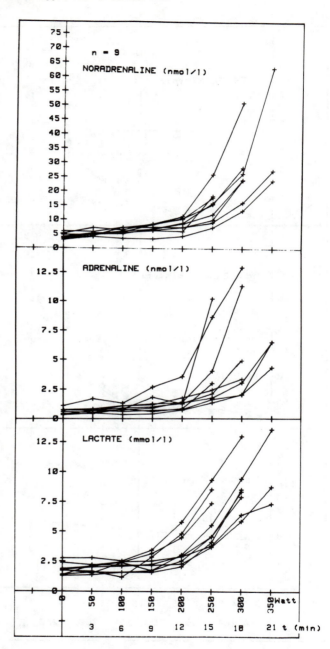

Figure 2—Individual noradrenaline, adrenaline, and lactate responses during the short-term experiment.

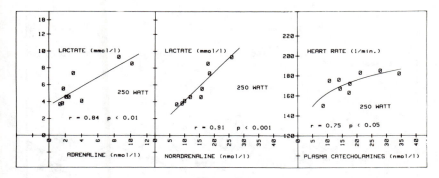

Figure 3—Descriptive corrrelations between the submaximal lactate response, heart rate response, and the catecholamine concentration during the short-term experiment.

In summary the following conclusions may be stated:

1. The dopamine levels remained nearly constant during short-term or prolonged graduated exercise.

2. Noradrenaline and adrenaline levels were three times higher during cardiac exhaustion than during peripheral muscle exhaustion.

3. Below 70% $\dot{V}O_{2max}$, noradrenaline and adrenaline levels showed slight time dependency, and a steady state behavior between the 3rd and 15th min.

4. Above 70% $\dot{V}O_{2max}$, noradrenaline and adrenaline levels showed a strong time and intensity dependency and no steady state behavior.

5. The mean initial half-life of noradrenaline and adrenaline following strenuous exercise was approximately 3 min ($e^{-0.048 \text{ min}}$).

Acknowledgment

This research was supported by Bundesinstitute für Sportwissenschaft. Kölm-Lövenich.

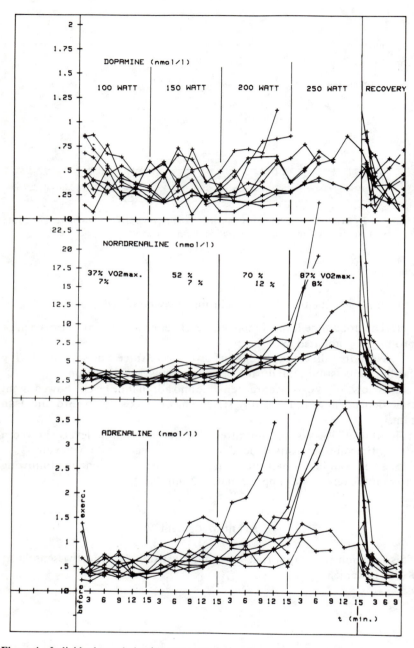

Figure 4—Individual catecholamine responses during the prolonged experiment.

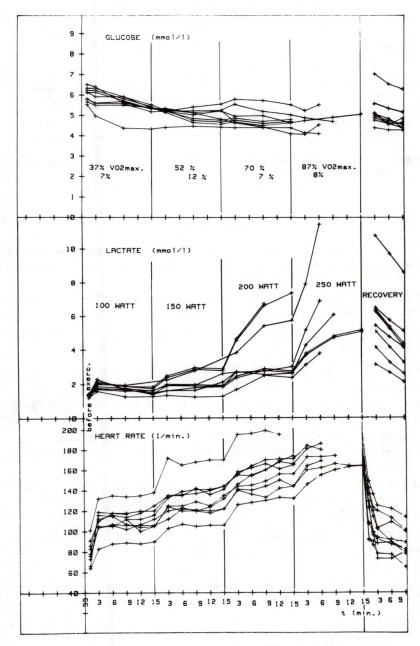

Figure 5—Individual glucose, lactate, and heart rate responses during the prolonged experiment.

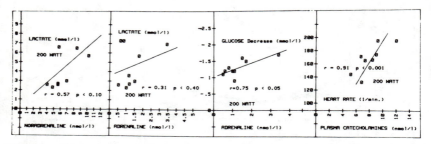

Figure 6—Descriptive correlations between lactate, glucose, heart rate responses, and the catecholamine concentrations.

References

DA PRADA, M. & G. Zürcher. Simultaneous radioenzymatic determination of plasma and tissue adrenaline, noradrenaline and dopamine within the fentomole range. Life Sci. 19:1161–1174, 1976.

EKBLOM, B., P.O. Astrand, B. Saltin, J. Steinberg & B.M. Wallström. Effect of training on circulatory response to exercise. J. Appl. Physiol. 24:518–528, 1968.

EULER, U.S. von, Sympathoadrenal activity in physical exercise. Med. Sci. Sports 3:165–173, 1974.

FELIG, P., The glucose-alanine cycle. Metabolism 22:179–207, 1973.

GOLLNICK, P.O., Factors controlling glycogenolysis and lipolysis during exercise. In J. Keul (Ed.), Limiting Factors of Physical Performance, pp. 81–93. Stuttgart: Georg. Thieme. Verlag, 1973.

HAGBERG, J.M., R.C. Hickson, J.A. McLane, A.A. Ehsani & W.W. Winder. Disappearance of norepinephrine from the circulation following strenuous exercise. J. Appl. Physiol. 47:1311–1314, 1979.

HAGGENDAL, J., H.L. Hartley & B. Saltin. Arterial noradrenaline concentration during exercise in relation to the relative work levels. Scand. J. Clin. Lab. Invest. 26:337–342, 1970.

HERTTING, G., Das sympathische Neuron: Regelwege für Synthese, Freisetzung und Wirkung des Transmitters. Klin. Wochenschr. 57:593–598, 1979.

HOHORST, H.J., L-(+)-Lactat. Bestimmung mit Lactatdehydrogenase und DPN. In H.U. Bergmeyer (Ed.), Methoden der enzymatischen Analyse, pp. 266–277. Verlag. Chemie. Weinheim., 1962.

LEHMANN, M., J. Keul, A. Berg & S. Stippig. Plasmacatecholamine und metabolische Veränderungen bei Frauen während Laufbandergometrie. Eur. J. Appl. Physiol. 46:305–315, 1981a.

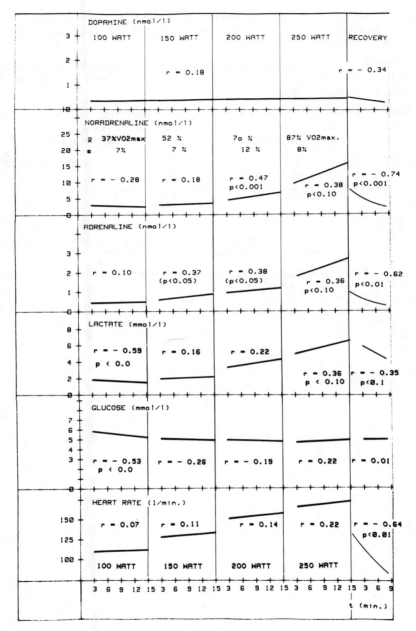

Figure 7—Calculated regressions for each exercise level with recovery, based on the individual values during the prolonged experiment for evaluation of steady state or non-steady state behavior.

LEHMANN, M., J. Keul, G. Huber, N. Bachl & G. Simon. Alters- und belastungsabhängiges Verhalten der Plasmacatecholamine. Klin. Wochenschr. 59:19–24, 1981b.

LEHMANN, M., J. Keul, G. Huber & M. Da Prada. Plasma catecholamines in trained and untrained volunteers during graduated exercise. Int. J. Sports Medicine 2:143–147, 1981c.

LEHMANN, M., J. Keul & U. Korsten-Reck. Einfluß einer stufenweisen Laufbandergometrie bei Kindern und Erwachsenen auf die Plasmacatecholamine, die aerobe und anaerobe Kapazität. Eur. J. Appl. Physiol. 47:301–311, 1981d.

LEHMANN, M., J. Keul, U. Korsten-Reck, & H. Fischer. Einfluß der Ergometerarbeit im Liegen und Sitzen auf Plasmacatecholamine, metabolische Substrate sowie Sauerstoffaufnahme und Herzfrequenz. Klin. Wochenschr. 59:1237–1242, 1981e.

LEHMANN, M., J. Keul, & K. Wybitul. Einfluß einer stufenweisen Laufband- und Fahrradergometrie auf die Plasmacatecholamine, energiereichen Substrate. Aerobe und anaerobe Kapazität. Klin. Wochenschr. 59:553–559, 1981f.

SLEIN, M.W., D-Glucose, Bestimmung mit Hexokinase und Glucose-6-Phosphat-Dehydrogenase. In Bergmeyer, H.N. (Ed.) Methoden der enzymatischen Analyse, pp. 117–123. Verlag. Chemie. Weinheim: 1962.

VAN LOON, G.R. & M.J. Sole. Plasma dopamine: Source, regulation and significance. Metabolism 29:1119–1123 (Suppl I), 1980.

Catecholamine Response to Maximal Anaerobic Exercise

I.A. Macdonald, S.A. Wootton, B. Muñoz,
P.H. Fentem and C. Williams
Queen's Medical Centre, Nottingham, England,
and University of Technology,
Loughborough, England

Exercise leads to an activation of the sympathoadrenal system and an increase in plasma catecholamine levels. Approximately fivefold increases in norepinephrine and epinephrine concentrations have been observed in prolonged submaximal exercise and in short duration progressive exercise up to maximal aerobic capacity (Galbo et al., 1975). Although one would expect an increase in sympathoadrenal activity to occur during anaerobic exercise, the effect of high intensity exercise of less than 1 min duration on plasma catecholamine levels has not been determined. Knowledge of the extent of the change in plasma catecholamines during sudden bursts of heavy exercise may be important for assessing the possible risks associated with and explaining the intensity of the symptoms which sometimes accompany such exercise.

The aim of this study was to measure plasma norepinephrine and epinephrine concentrations immediately after 30 s of maximal anaerobic exercise and during a 30 min recovery period.

Methods

Protocol

Four healthy male subjects (aged 25 to 39), who had fasted for 12 to 18 h, were well acquainted with the exercise protocol and had given their informed consent, took part in this study. A polythene cannula with three-way tap was inserted into a forearm vein and maintained patent by the infusion of sodium chloride solution (150 mmol/l, 0.4 ml/min). Following cannulation, each subject rested supine on a couch for 30 min before a baseline blood sample was taken. The subject was then seated on the bicycle ergometer and further blood samples were taken after the 'warm-up' period and immediately after the 30 s sprint. The subject then returned to the couch and remained supine for 30 min, during which time further blood samples were taken.

Exercise

A Monark cycle ergometer, with a modification to the braking mechanism to allow accurate selection of load and rapid loading, was used. Each subject performed a 'warm-up', two 30 s rides at 120 and 150 W (against a 14.7 N load), which was designed to associate the subject with the leg speeds and provide data for calibration while causing minimal metabolic disturbance. The subject then performed the Anaerobic Work Test (AWT) by pedaling maximally, from an initial pedal frequency of 75 rpm, against a pre-determined loading for 30 s. The load was employed within 2 s and was pre-determined to ensure that each subject would achieve the maximal power output attainable in this experimental model, while pedaling within specific ranges of pedal frequency.

In order to discriminate the rapid changes in pedal frequency, a photo-optic counting device was utilized (Lakomy & Wootton, 1981). The voltage output, proportional to pedal frequency, was recorded on a pen recorder as a fatigue profile. The trace was digitized in order to calculate peak, mean 0 to 30 s and instantaneous power outputs for each individual.

Lean upper leg volumes (LULV) were determined (Jones & Pearson, 1969) and the power outputs were corrected for LULV to provide a qualitative description of the work performed (i.e., $W \, l \, LULV^{-1}$), thereby reducing the intersubject variability in size.

Analysis

From each venous sample, 10 ml whole blood was placed in tubes containing lithium-heparin and 100 EGTA/glutathione to minimize oxidation of the catecholamines. Samples were immediately centrifuged at 4° C, the plasma aspirated and stored at −80° C prior to analysis. Plasma norepinephrine and epinephrine concentrations were determined by high performance liquid chromatography with electrochemical detection following alumina extraction of duplicate 1 ml plasma samples (Green & Macdonald, 1981). Following deproteinization of whole blood with perchloric acid, blood glucose was determined by the glucose oxidase method (Boehringer GOD-Perid) and blood lactate by a modification of the method of Olsen (1971).

Table 1

Body Weight, Leg Volume and Power Output

	Body weight (kg)	LULV (l)	Load (N)	Peak power (W)	Peak power (W lLULV^{-1})	Mean power (W)	Mean power (W lLULV^{-1})
Mean	76.2	10.16	55.2	914	90.3	645	63.8
SEM	3.8	0.56	2.6	45	4.5	29	3.1

Table 2

Blood Glucose and Lactate and Plasma Norepinephrine and Epinephrine

	Glucose (mmol/l)		Lactate (mmol/l)		Norepinephrine (nmol/l)		Epinephrine (nmol/l)	
	Mean	SEM	Mean	SEM	Mean	SEM	Mean	SEM
Rest	3.92	0.15	0.60	0.07	1.97	0.13	0.83	0.19
Warm-up	3.96	0.13	1.13	0.11	3.73	0.59	1.13	0.20
AWT Recovery (min)								
0	4.35	0.8	5.35	0.45	17.00	3.60	7.40	2.75
3	4.86	0.41	10.48	0.59	16.70	5.69	8.24	3.66
5	5.20	0.57	12.00	0.80	10.0	2.47	2.92	0.77
10	5.16	0.52	11.80	0.99	7.15	2.58	1.93	0.59
15	5.02	*	10.82	*	4.92	1.77	1.52	0.75
20	4.97	*	9.74	*	6.25	*	2.23	*
30	4.88	0.51	7.55	1.07	5.62	*	1.25	*

(* n = 3, otherwise n = 4)

Results

The power outputs achieved during the AWT are presented in Table 1. Power output rapidly rose to a peak of 914 ± 45 (SEM) W within the first 5 s and then diminished such that the mean value for 30 s was 645 ± 29 W.

The influence of the AWT on lactate, glucose, norepinephrine and epinephrine levels is presented in Table 2 and Figures 1 and 2. Blood glucose was unaffected by the 'warm-up' period, rose after the AWT,

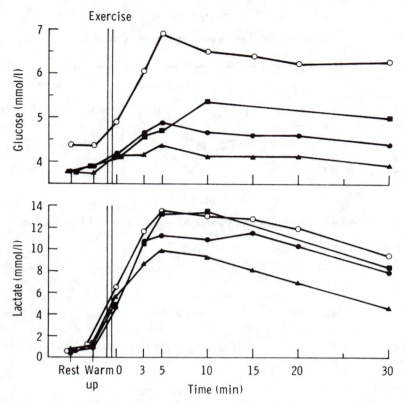

Figure 1—Blood glucose and lactate levels for individual subjects (● = Subject 1, ■ = 2, ▲ = 3, O = 4) at rest, following warming-up and then after 30 s maximal anaerobic exercise.

reaching a peak level between 5 and 10 min, and was higher than the resting level at 30 min. Following the warm-up, blood lactate and plasma norepinephrine doubled and plasma epinephrine increased slightly. Following AWT, the pattern of change for blood lactate was similar to that for glucose, peak levels of the metabolites being recorded at the same time for each subject. After 30 min recovery, lactate levels were still markedly elevated. Plasma catecholamine levels rose rapidly with the AWT, maximal values being recorded immediately after the exercise or following 3 min recovery (Figure 2). Norepinephrine levels increased four- to five-fold in all subjects, whereas the range of increase in epinephrine was two- to ten-fold. During the remainder of the recovery period epinephrine levels decreased and approached resting values in all subjects. A similar pattern was observed for norepinephrine in two subjects, but with the other two the levels were still markedly elevated after 30 min recovery.

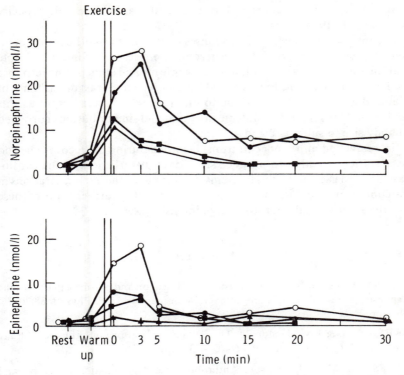

Figure 2—Plasma norepinephrine and epinephrine levels for individual subjects (key as for figure 1).

Discussion

This study was designed to establish the effect of maximal anaerobic exercise on the plasma concentrations of norepinephrine and epinephrine. The power outputs achieved by the subjects were 2 to 3.5 times greater than the power output that would elicit a maximal oxygen consumption; the large and sustained increases in blood lactate concentration indicate the degree of metabolic stress imposed by this form of exercise. This maximal anaerobic exercise was associated with increases in both norepinephrine and epinephrine immediately after the 30 s exercise period. The first post-exercise blood sample was obtained between 30 and 90 s after the end of the AWT. It thus seems likely that the maximal peripheral venous concentrations of the two catecholamines would have occurred sometime between the end of the AWT and after 3 min of recovery. All subjects showed similar increases in plasma norepinephrine, responses which are also similar to those reported during tests of maximal aerobic capacity (Galbo et al., 1975).

In contrast, the epinephrine responses were more variable but in general were higher than observed during aerobic exercise. It is of interest to note that the subject with the smallest norepinephrine and epinephrine response (Subject 3) also had the lowest blood lactate levels and least disturbance of blood glucose. This was not due to differences in power output as this subject had the greatest overall mean output and similar outputs corrected to body size to subjects 2 and 4.

The catecholamine concentrations recorded during the recovery period were sufficiently high to produce significant disturbances of the cardiovascular and gastrointestinal systems. In particular, the likely alterations in the control of blood pressure and gut motility justify further study of these variables during and after brief, high intensity exercise.

References

GALBO, H., J.J. Holst & N.J. Christensen. Glucagon and plasma catecholamine responses to graded and prolonged exercise in man. J. Appl. Physiol. 38:70–76, 1975.

GREEN, J.H. & I.A. Macdonald. The influence of intravenous glucose on body temperature. Q.J. Exp. Physiol. 66:465–473, 1981.

JONES, P.R.M. & J. Pearson. Anthropometric determination of leg fat and muscle plus bone volumes in young male and female adults. J. Physiol. 204:63–660, 1969.

LAKOMY, H.K. & S.A. Wootton. Discrimination of rapid changes in pedal frequency. J. Physiol. 316:1P, 1981.

OLSEN, C., An enzymatic fluorometric micromethod for the determination of acetoacetate, -hydroxybutyrate, pyruvate and lactate. Clin. Chim. Acta 33:293–300, 1971.

The Influence of β-Adrenoceptor Blockade on the Hemodynamic Response and Physical Performance of Middle and Long Distance Runners

D.D.J. Malan, J.T. Fritz,
A.C. Dreyer, and G.L. Strydom
University of Potchefstroom for C.H.E.,
Potchefstroom, South Africa

The efficacy of β-adrenoceptor blocking drugs in the treatment of various cardiovascular disorders is well documented. The beneficial effects of these drugs in angina pectoris have been ascribed to a decrease in cardiac work, thereby reducing oxygen demand, and the longer filling time associated with slower heart rate which allows for better coronary perfusion. It has also been proposed that these drugs are useful in the prevention and therapy of acute myocardial infarction (Marshall et al., 1981).

Williams (1979) observed that propranolol adversely affected the performance of marathon runners. This was confirmed by Kayser et al. (1980) who warned that hypertensive individuals, previously accustomed to physical activity, could expect a certain degree of deterioration of their performance capacity when treated with β-adrenoceptor blocking drugs. Regular exercise is thought to reduce blood pressure (Boyer & Kasch, 1970) and many hypertensives are advised by their doctors to take up physical training. It would therefore be important to ascertain to what extent concurrent β-adrenolytic therapy affects the physical capacity of such patients.

Clinically there is little to choose between the various β-adrenoceptor blocking drugs when used for hypertension, except possibly for the preference of cardiac selective agents in asthmatic patients. The β-adrenoceptor blocking drug, pindolol, is claimed to possess a certain degree of intrinsic sympathomimetic activity which may constitute a safer therapy in patients at risk from β-blockade (Choquet et al., 1972; Frishman et al., 1979). It was therefore decided to assess the effect of this drug on the performance of well trained athletes, compared with that obtained with

755

propranolol, a β-adrenoceptor blocking drug devoid of intrinsic sympatho-mimetic activity (Connelly et al., 1976). Side-effects of the β-adrenoceptor blocking agents include hypoglycemia in diabetics (Kotler et al., 1966) and bronchoconstriction in patients with obstrutive airway disease (McNiell, 1964). Although resting plasma glucose levels and airway resistance seem to be unaffected in normal individuals (Frishman, 1980), it is uncertain how these parameters would be influenced in trained athletes during exercise. As these side-effects may affect physical performance, blood glucose levels at various work intensities as well as certain lung function parameters were determined.

Method

Ten well trained male middle and long distance runners (mean age 18.5 years) were selected for this study.

The experimental design for the evaluation of both drugs consisted of an initial study, acting as a control procedure for the double-blind cross-over administration of the active component or placebo. The doses for propranolol (80 mg) and pindolol (10 mg) were chosen according to the average therapeutic doses employed in hypertension therapy (Aellig, 1976).

The evaluation protocol in both experimental designs included determination of the following parameters: heart rate, blood pressure, vital capacity (VC), forced expiratory volume (FEV), forced expiratory ratio (FER), maximal voluntary volume (MVV) and peak flow (Vitalograph and Wright peak flow) venous blood glucose (Unitest model 250: Oxidase-peroxidase method). This was followed by a physical work capacity test (Collins bicycle ergometer) consisting of four workloads (200 W, 250 W, 300 W, 350 W) of 5 min each, at a pedaling rate of 60 rpm. During the last minute of each workload the hemodynamic parameters (heart rate and blood pressure) were registered and a blood sample for blood glucose determination was taken. After the termination of the PWC-procedure the recovery heart rate was determined.

A seven day wash-period was allowed between tests. To compare the effects of propranolol and pindolol on the above mentioned parameters, and to evaluate their influence on the physical work capacity, a statistical analysis of variance was employed.

Results

Heart Rate

The effects of the two drugs and placebo on the heart rate response are outlined in Figure 1, which demonstrates the extent of the negative chronotropic effect of the β-adrenoceptor blocking drugs.

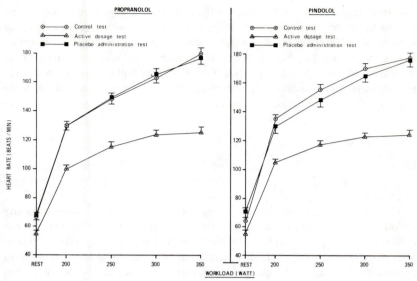

Figure 1—The effect of propranolol and pindolol on the heart rate response during rest and a PWC test.

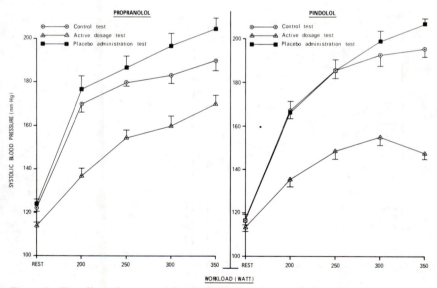

Figure 2—The effect of propranolol and pindolol on the systolic blood pressure response during rest and PWC test.

Comparison between the effect of the active drugs and control values, as well as those between the active drugs and placebo values, revealed highly significant differences (p < 0.0001 in each case). An average decrease of 16.4% in resting heart rate occurred during drug administration. This increased to 30% in exercise heart rate at the highest work intensity. Statistically insignificant differences (p > 0.45) were obtained between control and placebo heart rate values in each case.

Blood Pressure

Figures 2 and 3 illustrate systolic and diastolic blood pressures during a control, drug- and placebo-administered (double-blind cross-over) exercise test.

These results indicate that both drugs caused a slight reduction in resting blood pressure. A statistically significant difference (p < 0.0001) occurred between control and drug administered values in the systolic blood pressure during exercise. It also appeared that pindolol induced a more pronounced reduction in blood pressure at the higher levels (300 W & 350 W) of exercise than propranolol. These changes were significant (p < 0.0001; systolic and p < 0.05 diastolic) when compared with the control values.

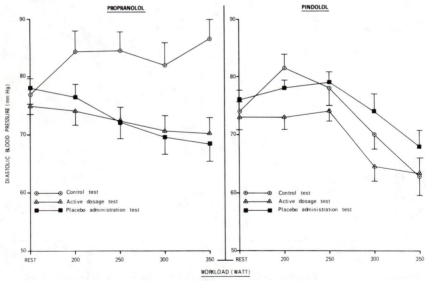

Figure 3—The effect of propranolol and pindolol on the diastolic blood pressure response during rest and a PWC test.

Recovery Heart Rate

As can be seen from Figure 4, a highly significant difference (p < 0.0001) occurred in the recovery heart rate values of the control versus drug administered and placebo versus drug administered tests for each drug.

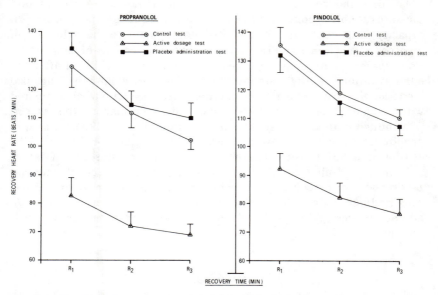

Figure 4—The effect of propranolol and pindolol on a 3 minute recovery heart rate response after a PWC test.

No statistical difference in recovery heart rate was observed between the two β-adrenoceptor blocking agents nor between control and placebo values.

Blood Glucose and Resting Lung Functions

No evidence of a statistically significant effect could be found with either β-adrenoceptor blocking agent on the venous blood glucose concentrations and resting lung functions.

Discussion

The marked reduction in resting heart rate and exercise induced tachycardia found with administration of propranol and pindolol is in accordance with findings of Epstein et al. (1965) and Sjöberg et al. (1979).

These results also demonstrate that not only exercise induced tachycardia, but also the exercise induced increase in systolic blood pressure showed a pronounced reduction with each β-adrenoceptor blocking drug. No indication of any significant effect on resting or exercise induced diastolic blood pressure values were evident with either of the two drugs. The drug-induced reduction in both the heart rate and systolic blood pressure (double product) led to a pronounced reduction in physical work capacity. This could also be an attributable factor to the reduction in myocardial function, as well as a drop in physical effectivity of the subjects. Four subjects were unable to complete the highest workload after the administration of both β-adrenoceptor blocking drugs. Both the drugs also increased the observed perceived exertion during exercise when compared with the control and placebo tests, and were distinguishable in this respect.

Although statistically insignificant, propranolol seemed to have a more pronounced effect on recovery heart rate than pindolol. This may be attributed to the intrinsic sympathomimetic activity possessed by the latter.

From the results presented in this study, it may be concluded that β-adrenoceptor blocking agents reduce physical performance in athletes. This finding may also be of clinical importance to hypertensives on β-adrenolytic therapy, advised by doctors, to take up exercise as part of their therapy.

References

AELLIG, W.H., β-adrenoceptor blocking activity and duration of action of Pindolol and Propranolol in healthy volunteers. Br. J. Clin. Pharmacol. 3:251–257, 1976.

BOYER, J.L. & F.W. Kasch. Exercise therapy in hypertensive men. J.A.M.A. 211:1668, 1970.

CHOQUET, Y., R.J. Capone, D.T. Mason, E.A. Amsterdam & R. Zelis. Comparison of the beta-adrenergic blocking properties and negative inotropic effects of oxprenolol and propranolol in patients. American Journal of Cardiology 29:257, 1972.

CONNOLLY, M.E., F. Keusting, C.T. Dollery. The clinical pharmacology of beta-adrenoceptor blocking drugs. Prog. Cardiovasc. Dis. 19:203–234. 1976.

EPSTEIN, S.E., B.F. Robinson, R.L. Kahler & Braunwald. Effects of beta-adrenergic blockade on the cardiac response to maximal and submaximal exercise in man. J. Clin. Invest. 44:1745–1753, 1965.

FRISHMAN, W., R. Davis, & J. Strom. Clinical pharmacology of the beta-adrenergic blocking drugs. American Heart Journal 98:393, 1979.

FRISHMAN, W. In Clinical Pharmacology of the Beta-adrenergic Blocking Drugs. New York: Appleton-Century-Crofts, p. 26, 1980.

KAYSER, L., P. Kayser, J. Karlson & S. Rossner. Beta blockers and running. American Heart Journal 100(6):943, 1980.

KOTLER, M.N., L. Berinda & A.H. Rubenstein.: Hypoglycemia precipitated by Propanolol. Lancet 2:1389–1390, 1966.

MARSHALL, R.C., G. Wisenberg, H.R. Schelbert & E. Henze. Effect of oral propranolol on rest, exercise and post-exercise left ventricular performance in normal subjects and patients with coronary artery disease. Circulation 63(3):572–583, 1981.

MCNEILL, R.S., Effect of a beta-adrenergic blocking agent, propranolol, in asthmatics. Lancet 2:1101, 1964.

SJÖBERG, H., Frankenhauser & H. Bjurstedt: Interactions between heart rate, psychomotor performance and perceived effort during physical work as influenced by beta-adrenergic blockade. Biol. Psyc. 8:31–43, 1979.

WILLIAMS, J.W., Propranolol and marathon running. American Heart Journal 98(4):542, 1979.

Endurance and Metabolic Adjustments to Exercise in Sympathectomized (6-OHDA) Rats

F. Peronnet and A. Imbach
Université de Montréal,
Montreal, Quebec, Canada

Chemical sympathectomy with 6-hydroxydopamine (6-OHDA) impairs metabolic adjustments to exercise in dogs (Nadeau et al., 1981). In order to explain the mechanisms by which degeneration of the sympathetic nerve endings produces such modifications of the metabolic response to exercise, liver, skeletal muscle and cardiac glycogen contents as well as plasma substrate concentrations were measured at rest and following exercise prolonged to exhaustion, in 6-OHDA treated rats.

Methods

The study was conducted in male Sprague-Dawley rats (320 ± 20 g; $\overline{X}$ ± SEM). Glycogen content of liver, heart and quadriceps femoris (Carrol et al., 1956) as well as plasma glucose, lactate (Calbiochem reagents) and FFA concentrations (Duncombe, 1964) were measured at rest in nine control and eight sympathectomized rats (100 mg kg^{-1}, 6-OHDA from Labkemi, administered in the vein of the tail, 24 h before experiment). These measurements were repeated on nine control and eight sympathectomized rats at the end of an exercise (Quinton Rodent Treadmill; 28 m min^{-1}, 8% slope) conducted to exhaustion. The effectiveness of the 6-OHDA treatment was evidenced by depletion of the heart (apex) norepinephrine stores from 442 ± 55 to 9 ± 1 ng g^{-1} in 6-OHDA treated rats (de Champlain et al., 1967).

Results

Chemical sympathectomy increased plasma glucose and FFA concentrations (Figure 1) as well as heart and skeletal muscle glycogen contents at rest. Resting plasma lactate concentration and liver glycogen content were not modified. Chemical sympathectomy markedly reduced running time to exhaustion from 121 ± 13 min to 42 ± 6 min. This decreased endurance was associated with an increase in the rate of glycogen utilization (Figure 1), and in the plasma lactate concentration, and with a fall in the plasma FFA concentration.

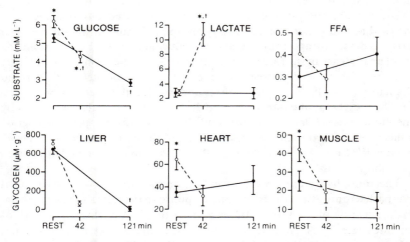

Figure 1—Plasma substrate concentrations and glycogen contents at rest and at the time of exhaustion (42 ± 6 min and 121 ± 13 min) control (filled symbols, n = 9) and sympathectomized rats (open symbols, n = 8) (* = different from from control value, † = different from resting value; P < 0.05).

Discussion

Treatment with 6-OHDA, which selectively destroys the sympathetic nerve endings but does not affect the adrenal medulla (Kostrzewa & Jocobowitz, 1974), impairs the resting metabolic function in both rats (Imbach, 1971) and dogs (Nadeau et al., 1981; Porlier et al., 1977) and the metabolic adjustments to exercise in dogs (Nadeau et al., 1981). Modifications of the resting metabolic function include an increase in the plasma substrate concentrations which may presumably be accounted for by the compensatory increase in plasma catecholamine concentration (Nadeau et al., 1981; Porlier et al., 1977) and the denervation hypersensitivity of the target tissues (Kostrzewa & Jacobowitz, 1974; Nadeau et al., 1971). Modifications of the metabolic adjustments to prolonged exercise include an increase in the plasma glucose and lactate responses and a decrease in the plasma FFA response (Nadeau et al., 1981). This suggests that the sympathetic nerve endings play a predominant role in the control of substrate mobilization and utilization during prolonged exercise.

Results from the present investigation confirm that the degeneration of the sympathetic nerve endings following 6-OHDA treatment, alters metabolic functions. Elevated resting plasma glucose and FFA concentrations presumably explained by the high plasma catecholamine values (Nadeau et al., 1981; Porlier et al., 1977) are associated with an increase in the cardiac and skeletal muscle glycogen stores (Figure 1). This increase in carbohydrate storage could be due to a glycogen-sparing effect resulting from an increase in the rate of lipolysis and FFA utilization at rest. During exercise, the marked decrease in endurance is associated with a rapid depletion of the liver, skeletal muscle and cardiac glycogen stores (Figure 1). The increase in the rate of carbohydrate utilization is also evidenced by the high plasma lactate values and the drop in the plasma glucose concentration at the time of exhaustion. On the contrary, the decrease in plasma FFA concentration suggests that lipid mobilization and utilization are inhibited.

The most likely explanation for the mechanisms by which chemical sympathectomy alters the metabolic response to exercise, is the following: 1) denervation of the adipose tissue impairs the FFA release; 2) the lack of an adequate FFA supply, as well as the circulating catecholamines (Nadeau et al., 1981) promote a compensatory utilization of the glycogen stores; 3) the rapid depletion of the glycogen stores reduces endurance. Moreover the increase in muscle glycogen utilization leads to an increase in the plasma lactate concentration, which could further decrease the FFA release from the adipose tissue. Finally insulin secretion could be high in 6-OHDA treated rats due to the denervation of the endocrine pancreas and to the high plasma glucose concentration. This could further inhibit the FFA release from the adipose tissue.

Previous studies conducted in 6-OHDA treated rats (Richter et al., 1980) have reported little if any modification in the mobilization and utilization of substrates during exercise. However, the evaluations were made in a late stage following 6-OHDA administration, at a time when a significant regrowth of the sympathetic nerve endings may have occurred (de Champlain & Nadeau, 1971). On the contrary, the present study was conducted in the acute phase of chemical sympathectomy and indicates that norepinephrine locally released in the adipose tissue is of prime importance for promoting an adequate FFA supply during prolonged exercise. On the contrary, sympathetic denervation of the heart, liver and skeletal muscles does not impair mobilization of their glycogen stores, which would rather appear to be under the control of local factors (Richter et al., 1982) and circulating epinephrine (Richter et al., 1980).

References

CARROL, N.V., R.W. Longley & J.H. Roe. The determination of glycogen in liver and muscle by use of enthrone reagent. J. Biol. Chem. 220:583–593, 1956.

DE CHAMPLAIN, J., L.R. Krakoff & J. Axelrod. The metabolism of norepinephrine in experimental hypertension in rats. Circ. Res. 20:136–144, 1967.

DE CHAMPLAIN, J. & R.A. Nadeau. 6-hydroxydopamine, 6 hydroxdopa and degeneration of adrenergic nerves. Fed. Proc. 39:877–885, 1971.

DUNCOMBE, W.G., The colorimetric micro-determination of nonesterified fatty acids in plasma. Clin. Chim. Acta 9:122–125, 1964.

IMBACH, A., Metabolic alterations following chemical sympathectomy with 6-hydroxydopamine in the rat. In P.E. Roy & V.S. Dhalla (Eds.), Recent Advances in Studies on Cardiac Structure and Metabolism, Vol. 9, The Sarcolemma, pp. 109–112. Baltimore: University Park Press, 1971.

KOSTRZEWA, R.M. & D.M. Jacobowitz. Pharmacological actions of 6-OHDA. Pharmacol. Rev. 26:199–288, 1974.

NADEAU, R., F. Péronnet, D. Cousineau, A. Imbach & J. de Champlain. Plasma catecholamine and metabolite responses to exercise in chemically sympathectomized dogs (Abstract). In J. Poortmans & G. Niset (Eds.), Biochemistry of Exercise IV-B, p. 123. International Series on Sports Sciences. Volume 11-B. Baltimore: University Park Press, 1981.

PORLIER, G.A., R.A. Nadeau, J. de Champlain & D.G. Bichet. Increased circulating plasma catecholamines and plasma renin activity in dogs after chemical sympathectomy with 6-hydroxydopamine. Can. J. Physiol. Pharmacol. 58:724–733, 1977.

RICHTER, E.A., H. Galbo, B. Sonne, J.J. Holst & N.J. Christensen. Adrenal medullary control of muscular and hepatic glycogenolysis and of pancreatic hormonal secretion in exercising rats. Acta Physiol. Scand. 108:235–242, 1980.

RICHTER, E.A., N.B. Ruderman, H. Gavras, E.R. Belur & H. Galbo. Muscle glycogenolysis during exercise: dual control by epinephrine and contractions. Am. J. Physiol. 242:E25–E32, 1982.

Alpha and Beta Adrenergic Effects on Muscle Metabolism in Contracting, Perfused Muscle

E.A. Richter, N.B. Ruderman and H. Galbo
Boston University Medical Center,
Boston, Massachusetts, U.S.A., and
University of Copenhagen, Copenhagen, Denmark

It has recently been shown that glycogenolysis in contracting, perfused skeletal muscle is under the dual control of a contraction-induced mechanism and of epinephrine: Contractions by themselves increase the activity of glycogen phosphorylase and glycogenolysis for a brief period only, whereas epinephrine causes a sustained increase in phosphorylase activity and glycogenoloysis (Richter et al., 1982a). The present study was carried out to determine whether in contracting muscle, effects of epinephrine (an α- as well as β-adrenergic agonist) on metabolism can be attributed to α- and/or β-adrenergic stimulation. Furthermore, we studied the role of cAMP and of alterations in the activities of glycogen phosphorylase and synthase for epinephrine-induced enhancement of glycogenolysis during contractions.

Methods

The isolated, perfused rat hindquarter (Richter et al., 1982a) was studied at rest and during 10 min of contractions (60 sub-tetanic contractions/min). After 15 min of perfusion with perfusate containing 75 μU insulin/ml cell free perfusate, 6mM glucose and erythrocytes at a hematocrit of 30%, muscle samples were taken from the right leg, and the left leg was subsequently prepared for muscle stimulation. Then the left hindleg was stimulated through the sciatic nerve or left resting. During contractions, tension developed by the gastrocnemius-soleus-plantaris muscle group was continuously recorded. At the end of the experimental period (rest or contractions) muscle samples were taken from the left leg. Flow of perfusate was at the onset of stimulation increased to secure adequate oxygenation of the working muscles.

Glucose uptake and lactate release were calculated from the changes in concentrations in the recirculating medium, and oxygen uptake was calculated from the arteriovenous oxygen concentration difference and the flow. Adrenergic stimulation was obtained by epinephrine in a high, physiological concentration (2.4×10^{-8} M) and α- and β-adrenergic blockade by 10^{-5} M phentolamine and propranolol, respectively.

Results

Effects of Electrical Stimulation

Electrical stimulation increased glucose uptake, lactate release and oxygen uptake by the hindquarter (Figure 1). Glycogen breakdown was in slow-twitch red fibers not significantly different from zero (Figure 2). In fast-twitch red and fast-twitch white fibers glycogen breakdown was 13.9 ± 1.3 and 24.1 ± 1.5 mmol/kg (mean $\pm$ SE, n = 9), respectively, whereas glycogen breakdown in resting hindquarters was not different from zero in all fiber types. After 10 min of contractions, phosphorylase a activity and cAMP concentration in slow-twitch red fibers were not significantly different from resting values (Figure 2), whereas synthase activity ratio was increased (Figure 2).

Effect of Epinephrine during Contractions

When epinephrine was present during contractions, glucose uptake tended to increase (p > 0.05), and lactate release and oxygen uptake were increased (Figure 1). Glycogen breakdown was significantly increased in slow-twitch red fibers (Figure 2), and in fast-twitch red and fast-twitch white fibers by 50 and 20%, respectively. In slow-twitch fibers epinephrine increased phosphorylase a activity, cAMP concentration and decreased synthase activity ratio (Figure 2). Epinephrine also had a positive inotropic effect on the contracting muscle.

Effect of Epinephrine + β-blockade

When the β-adrenergic effects of epinephrine were blocked by propranolol glucose uptake was increased, lactate release was decreased, and oxygen uptake was unchanged (Figure 1) compared to experiments with epinephrine alone. The stimulatory effect of epinephrine on glycogen

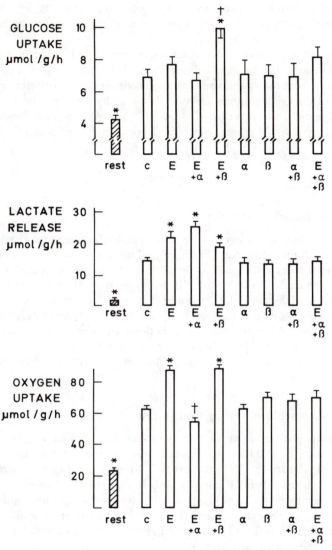

Figure 1—Effects of electrical stimulation, and α- and/or β-adrenergic blockade on glucose uptake, lactate release and oxygen uptake. Hindquarters were perfused at rest for 20 min at a flow of 9 ml/min or during electrical stimulation (60 twin-pulses/min) for 10 min at a flow of 22 ml/min. Additions to perfusate: C: none, E: epinephrine (2.4×10^{-8} M), α: phentolamine (10^{-5} M), β: propranolol (10^{-5} M). Results are means for 13 to 14 observations at rest and 8 to 10 during electrical stimulation. Bars represent SE.

* p < 0.05 when compared to C-values.

† p < 0.05 when E + α or E + β are compared to E-values.

SOLEUS MUSCLE

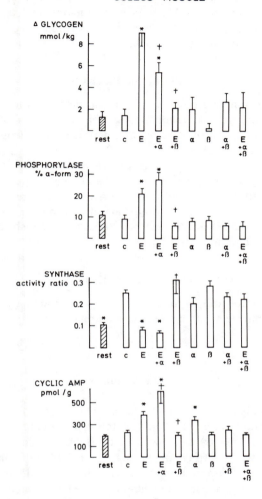

Figure 2—Effects of electrical stimulation, epinephrine and α and/or β-adrenergic blockade on glycogen breakdown, phosphorylase- and synthase activity, and on cyclic AMP concentration in soleus muscle (slow-twitch red fibers). Hindquarters were perfused at rest for 20 min at a flow of 9 ml/min or during electrical stimulation (60 twin-pulses/min) for 10 min at a flow of 22 ml/min. The concentration of glycogen was measured in the right leg before and in the left leg after the experimental period, and the difference (Δ) is shown in the figure. Initial mean glycogen level was 28 mmol/kg. Phosphorylase and synthase activity and cyclic AMP concentration were measured at the end of the experimental period only. Phosphorylase activity is expressed as active enzyme in percent of total activity. Synthase activity is expressed as activity ratio (−Glucose-6-phosphate/+Glucose-6-phosphate). Results are means and bars represent SE. At rest, n = 14 for Δ glycogen, n = 6 for phosphorylase and synthase, and n = 3 for cyclic AMP. Electrical stimulation values are means of 8 to 10 observations. For explanation of symbols, see legend to Figure 1.

breakdown was abolished in all fiber types as were the effects on phosphorylase a activity, cAMP concentration and synthase activity ratio in slow-twitch fibers (Figure 2). The positive inotropic effect of epinephrine was abolished by β-adrenergic blockade.

Effect of Epinephrine + α-Adrenergic Blockade

When the α-adrenergic effects of epinephrine were blocked by phentolamine, lactate release tended to be higher ($p > 0.05$) than with epinephrine alone, and the effects of epinephrine on glucose and oxygen uptake were completely blocked (Figure 1). α-adrenergic blockade lessened the glycogenolytic effect of epinephrine in slow-twitch fibers (Figure 2), but not significantly so in fast-twitch fibers. In slow-twitch fibers phentolamine increased the effect of epinephrine on cAMP concentration (Figure 2), whereas phosphorylase a and synthase activity ratio were not significantly changed (Figure 2). α-adrenergic blockade also abolished the positive inotropic effect of epinephrine.

Effect of Adrenergic Blockers by Themselves

When propranolol and/or phentolamine were added to the perfusate without epinephrine no effects were seen compared to control experiments without any additions.

Discussion

The present study suggests that effects of a physiological concentration of epinephrine (Galbo et al., 1977) on metabolism in contracting, perfused muscle are mediated by α- as well as by β-adrenergic mechanisms. In contracting fast-twitch fibers the glycogenolytic effect of epinephrine was due to β-adrenergic stimulation, whereas in the more epinephrine sensitive slow-twitch fibers the epinephrine-induced enhancement of glycogen breakdown was lessened by α-adrenergic blockade. These findings suggest that the glycogenolytic effect of epinephrine is triggered off by stimulation of β-adrenergic receptors and that simultaneous stimulation of α-adrenergic receptors potentiates the effect of β-adrenergic stimulation. This is to our knowledge the first study suggesting that glycogenolysis in contracting skeletal muscle may be under α-adrenergic influence.

The β-adrenergic mechanism by which epinephrine enhances glycogen depletion in contracting slow- and fast-twitch muscle may be ascribed to an increase in the concentration of cAMP (Figure 2), which in turn increases phosphorylase a activity and decreases synthase activity ratio (Figure 2). In heart (Keely et al., 1977) and in liver (Hems & Whitton, 1980) α-adrenergic stimulation may also enhance activation of phosphorylase. However, a similar mechanism probably did not account for the α-adrenergic enhancement of glycogen depletion in contracting skeletal muscle in the present study, since α-adrenergic blockade inhibited epinephrine-induced glycogen depletion without simultaneously inhibiting phosphorylase a activity (Figure 2). It has been reported that in perfused rat heart the activity of the enzyme phosphofructokinase is increased by α-adrenergic stimulation (Clark & Patten, 1981). If α-adrenergic stimulation also increases the activity of phosphofructokinase in contracting slow-twitch muscle this might provide a mechanism for increased glycolytic flux from glycogen and/or exogenous glucose.

Phosphorylase a activity in slow-twitch fibers was in control perfusions not higher after 10 min of contractions than at rest (Figure 2). However, when epinephrine was added to the perfusate, phosphorylase a activity was at the end of the 10 min of contractions higher than at rest (Figure 2). Thus, in agreement with findings in a recent study (Richter et al., 1982a) the enhancing effect of epinephrine on glycogenolysis during contractions may in part be ascribed to an epinephrine-induced sustained increase in phosphorylase a activity. When epinephrine was added to the perfusate, cAMP concentration in slow-twitch fibers was increased, whereas synthase activity ratio was decreased compared to control values (Figure 2). These effects of epinephrine were abolished by β-adrenergic but not by α-adrenergic blockade (Figure 2). In fact, epinephrine increased cAMP concentrations more during α-adrenergic blockade than when administered alone (Figure 2). This indicates that α-adrenergic stimulation inhibits the β-adrenergic increase of cAMP. Such an inhibitory effect of α-adrenergic stimulation on cAMP production has also been described in the heart (Keely et al., 1977).

In perfused skeletal muscle, epinephrine increased glucose uptake by α-adrenergic stimulation but this effect was counteracted by β-adrenergic stimulation (Figure 1). These findings are in accordance with previous studies showing an α-adrenergic increase and a β-adrenergic decrease in glucose transport in incubated resting muscle (Saitoh et al., 1974; Sloan et al., 1978). However, it has recently been shown in the perfused hindquarter that epinephrine increases basal glucose uptake but decreases glucose uptake when the latter is maximally stimulated by insulin (1mU/ml) (Chiasson et al., 1981). In the present study insulin was at 75 μU/ml, a concentration that is physiological and only stimulates glucose uptake 75% (Richter et al., 1982). Thus, in the perfused hindquarter, epinephrine seems to stimulate glucose uptake at physiological concentrations of insulin as well as in the absence of insulin but to inhibit maximally insulin-stimulated glucose uptake.

In conclusion, the present study suggests that in isolated, contracting, well-oxygenated muscle, epinephrine in a physiological concentration enhances muscle glycogen breakdown. In slow-twitch red muscle, in which this effect is largest, epinephrine seemingly enhances glycogenolysis by both α- and β-adrenergic mechanisms, the latter involving increased production of cAMP, phosphorylase activation and synthase inactivation. In contrast, in fast-twitch muscle only β-adrenergic mechanisms are involved in the glycogenolytic effect of epinephrine. The results also indicate that α-adrenergic stimulation increases the uptake of glucose and oxygen during electrical stimulation.

Acknowledgments

This study wsa supported in part by U.S.P.H.S. grants AM-19514 and T 32-AM-08201-06, by P. Carl Petersens Foundation, The Danish Medical Research Council, and by Idraettens Forskningsraad.

References

CHIASSON, J.-L., H. Shikama, D.T.W. Chu & J.H. Exton. Inhibitory effect of epinephrine on insulin-stimulated glucose uptake by rat skeletal muscle. J. Clin. Invest. 68:706–713, 1981.

CLARK, M.G. & G.S. Patten. Adrenaline activation of phosphofructokinase in rat heart mediated by α-receptor mechanism independent of cyclic AMP. Nature 292:461–463, 1981.

GALBO, H., E.A. Richter, J.J. Holst & N.J. Christensen. Diminished hormonal responses to exercise in trained rats. J. Appl. Physiol.: Respirat. Environ. Exercise Physiol. 42:953–958, 1977.

HEMS, D.A. & P.T. Whitton. Control of hepatic glycogenolysis. Physiol. Rev. 60:1–50, 1980.

KEELY, S.L., J.D. Corbin & T. Lincoln. Alpha adrenergic involvement in heart metabolism: Effects on adenosine cyclic 3′, 5′-monophosphate, adenosine cyclic 3′, 5′-monophosphate-dependent protein kinase, guanosine cyclic 3′, 5′-monophosphate, and glucose transport. Mol. Pharmacol. 13:965–975, 1977.

RICHTER, E.A., L.P. Garetto, M.N. Goodman & N.B. Ruderman. Muscle glucose metabolism following exercise in the rat: Increased sensitivity to insulin. J. Clin. Invest. 69:785–793, 1982.

RICHTER, E.A., N.B. Ruderman, H. Gavras, E. Belur & H. Galbo. Muscle glycogenolysis during exercise: Dual control by epinephrine and contractions. Am. J. Physiol. 242:E25–E32, 1982a.

SAITOH, Y., K. Itaya & M. Ui. Adrenergic α-receptor-mediated stimulation of the glucose utilization by isolated rat diaphragm. Biochem. Biophys. Acta 343:492–499, 1974.

SLOAN, I.G., P.C. Sawh & I. Bihler. Influence of adrenalin on sugar transport in soleus, a red skeletal muscle. Mol. Cell. Endocrinol. 10:3–12, 1978.

Nerve and Muscle

Differential Inter- and Intra-Muscular Responses to Exercise: Considerations in Use of the Biopsy Technique

R.B. Armstrong, M.H. Laughlin,
J.A. Schwane and C.R. Taylor
Oral Roberts University, Tulsa, Oklahoma, U.S.A.,
and Harvard University, Cambridge,
Massachusetts, U.S.A.

The biopsy technique has been widely used to study acute and chronic changes in human muscles with exercise. These studies have provided important information on human muscle function. However, there are two problems that require consideration in the interpretation of biopsy data. First, the different fiber types in the extensor muscles of both humans (Johnson et al., 1973) and animals (Armstrong, 1981) are stratified to varying extents both within and among muscles, as illustrated in Figure 1. Thus, a small muscle sample does not necessarily represent the fiber composition of the synergistic muscles or of other parts of the same muscle. Secondly, and perhaps more importantly in terms of human studies, muscle fibers of the same type in different muscles or parts of muscles may respond quite differently to a particular level of exercise (Sullivan & Armstrong, 1978). In this paper these differences within and among the muscles of the triceps surae group in rats will be used to illustrate these potential problems.

Methods and Results

Male Sprague-Dawley rats were used in all of the studies. The animals weighed 300 to 600 g, were housed under controlled environmental conditions at $23 \pm 2°$ C, and were provided food (commercial rat chow) and water ad libitum.

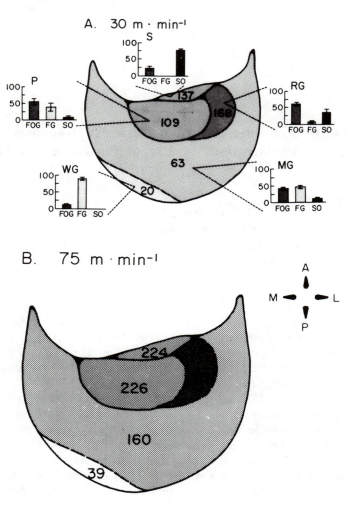

Figure 1(A,B)—Schematic illustrations of cross-sections of rat triceps surae muscles of the right hindlimb depicting average blood flows (ml 100 g^{-1} min^{-1}) during locomotion at two speeds (for each mean, n = 6 to 10). Relative shading intensities within the muscles emphasize the differences in blood flow. Blood flows in these muscles in control anesthetized rats were 6 to 9 ml 100 g^{-1} min^{-1}. Also included in A are the muscle names (S: soleus; P: plantaris; RG: red gastrocnemius; MG: middle gastrocnemius: WG: white gastrocnemius) and fiber type populations (histograms: $\bar{x} \pm$ SD). Fiber types are fast-twitch-oxidative-glycolytic (FOG), fast-twitch-glycolytic (FG), and slow-twitch-oxidative (SO). Compass directions are anterior (A), posterior (P), medial (M), and lateral (L).

Table 1

Percentages of FG Fiber Populations Showing Glycogen Loss

| | Treadmill exercise (m min^{-1} | | | |
Muscle	36	48	60	72
P	3	53	76	88
RG	25	70	62	81
WG	5	15	11	99

Data represent means with 5 to 9 per group for plantaris (P), red gastrocnemius (RG), and white gastrocnemius (WG) muscles. (From Sullivan & Armstrong, 1978).

Muscle Blood Flows (BF's) at Different Running Speeds

BF's within and among the muscles were determined after 1 min of running at 30 or 75 m min^{-1} using the radiolabeled microsphere technique (Laughlin et al., 1982). For controls, BF's were measured in anesthetized rats.

During running at 30 m min^{-1} there was a gradation of BF's from the deep muscles and parts of muscles to the more peripheral parts (Figure 1). Thus, deep S had about twice the BF as the more peripheral whole gastrocnemius muscle. RG had about 8X the BF as WG. During running at 75 m min^{-1} these relative differences were maintained at higher BF's.

Muscle Glycogen Loss Rates (Ġ's) at Different Running Speeds

Ġ's were determined within and among the muscles while rats ran on the treadmill at 36, 48, 60, or 72 m min^{-1} (Sullivan & Armstrong, 1978). At all running speeds the animals ran the same total distance (200 m). Ġ's were calculated from the differences between postexercise glycogen concentrations ([G]'s) and the [G]'s from control rats.

During trotting at 36 m min^{-1}, the highest Ġ's were observed in the deepest muscles (S) or the deepest portions within muscles (RG) (Figure 2). For example, RG lost glycogen (G) at 3X the rate of WG. On the other hand, during running at 72 m min^{-1} the more peripheral gastrocnemius muscle showed higher Ġ's than the deeper muscles (S and P). At this speed the whole gastrocnemius muscle lost G about 3X faster than S. Furthermore, marked differences in G loss occurred within the same types of fibers located within different muscles and parts of muscles at the various speeds (Table 1).

A. 36 m · min⁻¹   B. 72 m · min⁻¹

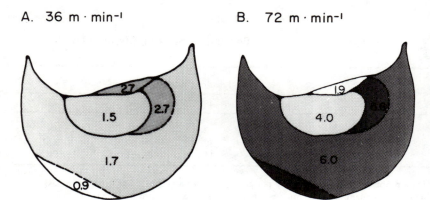

Figure 2—Schematic illustrations of triceps surae muscles depicting average rates of glycogen loss (mmoles glucosyl units kg^{-1} min^{-1}) during exercise (for each mean, n = 5 to 9). Relative shading intensities emphasize differences in rates of glycogen loss. Refer to legend for Figure 1 for further explanation of the illustration.

Muscle Ġ's in Rats Carrying Loads

Ġ's were determined during running on the treadmill for 10 min at 30 m min^{-1} with or without loads equaling 24% of body mass (Armstrong & Taylor, 1982). Ġ's were calculated from the postexercise [G]'s and [G]'s in control nonexercised animals.

Without loads at 30 m min^{-1}, the highest Ġ's were observed in the deepest muscle or the deepest parts of muscles (Figure 3). With loads, the increased Ġ's were observed in the more peripheral muscles or parts of muscles. S did not change its Ġ.

Muscle Glucose-6-Phosphate Dehydrogenase (G-6-PDase) Activity after Exercise

G-6-PDase activities were measured in the muscles 48 h following running for 90 min at 16 m min^{-1} down a 16° incline. Changes in G-6-PDase activity after exercise are related to accumulations of mononuclear cells in fibers undergoing necrosis or in connective tissue of the muscles (unpublished observations). Enzyme activities in the animals that exercised were compared with those in the muscles of control rats.

Forty-eight h following downhill running, the largest elevations in G-6-PDase activity occurred in the deepest muscle (S:100%) or in the deepest portions within muscles (RG:59%). The smallest changes in activity occurred in the most peripheral part of the muscle group, WG (25%).

A. 30 m · min⁻¹ : Unloaded   B. 30 m · min⁻¹ : Loaded

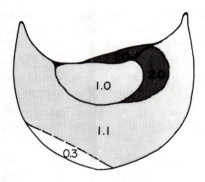 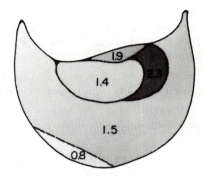

Figure 3—Schematic illustrations of triceps surae muscles depicting average rates of glycogen loss (mmoles glucosyl units kg⁻¹ min⁻¹) during running without or with a load equalling 24% of body mass (for each mean, n = 8 to 9). Relative shading intensities emphasize differences in rates of glycogen loss. Refer to the legend for Figure 1 for further explanation of the illustration.

Discussion

All of these experiments demonstrate that muscle does not behave as a unit, but that large differences in the response to exercise occur among and within muscles of extensor groups. The difficulties in using one biopsy sample to extrapolate to general muscular function during exercise are clear. Although rat muscle has a greater degree of stratification of the fiber types than that of most mammals (Armstrong, 1980) including humans (Johnson et al., 1973), the phenomenon is observed in all species. Also, fibers of the same type may respond differently at different levels of the muscle group during exercise. It is likely that peripherally-situated fibers of a given type will be recruited under different conditions, and display different metabolic patterns, than the same types of fibers located deeply within the muscle group. Although human studies employing the biopsy technique have provided valuable information, it is important to recognize these potential pitfalls in the interpretation of the data.

Acknowledgments

This work was supported by NIH Grants AM18123, AM18140 and AM25472.

References

ARMSTRONG, R.B., Properties and distribution of the fiber types in the locomotory muscles of mammals. In K. Schmidt-Neilsen, L. Bolis & C.R. Taylor (Eds.), Comparative Physiology: Primitive Mammals, pp. 243–254. Cambridge: Cambridge U. Press, 1980.

ARMSTRONG, R.B. & C.R. Taylor. Relationship between muscle force and muscle area showing glycogen loss during locomotion. J. Exp. Biol. (in press)

JOHNSON, M.A., J. Polgar, D. Weightman & D. Appleton. Data on the distribution of fiber types in thirty-six human muscles. J. Neurol. Sci. 18:111–129, 1973.

LAUGHLIN, M.H., R.B. Armstrong, J. White & K. Rouk. A method for using microspheres to measure muscle blood flow in exercising rats. J. Appl. Physiol.: Respirat. Environ. Exercise Physiol. (in press)

SULLIVAN, T.E. & R.B. Armstrong. Rat locomotory muscle fiber activity during trotting and galloping. J. Appl. Physiol.: Respirat. Environ. Exercise Physiol. 44:358–363, 1978.

Gas Tensions (O_2, CO_2, Ar and N_2) in Human Muscle during Static Exercise and Occlusion

F. Bonde-Petersen and J.S. Lundsgaard

University of Copenhagen, Copenhagen, Denmark,
and University of Odense, Odense, Denmark

With the introduction of mass spectrometers in the laboratories over recent years, new methods have been developed to measure gas tensions in blood and tissues (Lundsgaard et al., 1978 & 1980). We used this method in order to investigate the variation in gas tensions in human skeletal muscle during contraction. It was believed that such an experiment would throw some light on the relative importance of pO_2, pCO_2 and pH for metabolism and hyperemia of skeletal muscle during and after contraction.

Methods

A probe (1 mm stainless tube) fitted with a porous tip of sintered bronze supporting a 20 μ polyethylene membrane, was introduced into the vastus lateralis muscle in three subjects through a small incision in the skin and fascia. The probe was connected to the inlet of a mass spectrometer, and partial pressures for O_2, CO_2, Ar and N_2 were analyzed. Argon was used as an internal standard for pO_2. As the pressure in the spectrometer is 10^{-4} Torr, the gases dissolved in the muscle tissue will diffuse from the surroundings through the polyethylene membrane, to be analyzed in the mass spectrometer. The tensions of Ar, O_2, CO_2 and N_2 were recorded on a six-channel strip chart recorder (Figure 1). The gas consumption of this probe having a low resistance to diffusion is of significant magnitude relative to the concentration of the gases dissolved in the tissue. This influenced the recorded signal for Ar, O_2 and N_2 because of the existence of a gas depletion layer adjacent to the polyethylene membrane. In the case of CO_2 the gas concentration is so high in the tissue that the consumption of

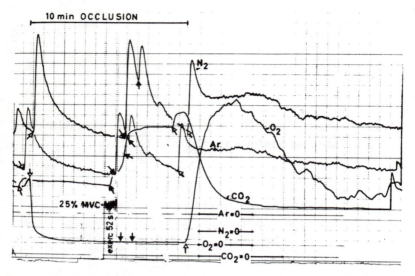

Figure 1—Original tracing from the mass spectrometer, demonstrating the signals measured for Ar, N_2, CO_2 and O_2. Also shown is the output from the static dynamometer. The individal curves are not aligned. Start and end of the 10 min occlusion is indicated by open arrows, and start and end of the occluded exercise by closed arrows on the individual curves. To obtain pO_2 the Ar curve was used as an internal standard (see text).

CO_2 by the probe was insignificant for the obtained signal. In vitro studies showed that the signals obtained for pAr and pO_2 were attenuated with the same factor of about four in an unstirred media. This attenuation of the Ar signal was used as an internal standard during the in vivo measurements (Lundsgaard & Degn, 1980) in order to calibrate the oxygen signal. The two molecules have the same low solubility in muscle tissue. Ar in tissue is in equilibrium with Ar in air, which contains about 7 Torr. Before the probe was introduced it was, therefore, calibrated against known gas tensions in water during stirring. Zero values and calibration values were shown to be stable within a few percent.

During the experiment the subjects (three males) were seated in a special chair (Tornvall, 1963) for the measurement of maximal voluntary contraction (MVC) during knee extension. The knee was in the position of 90° flexion, and the lower leg pressed against a spring steel bar mounted by strain guages. The signal was recorded on the strip chart recorder (Figure 1), which was placed in such a way that the subject by visual feedback was able to exert the desired percentage of MVC.

The subject was seated in the chair and given three trials to obtain the MVC. Local anesthetics were injected and a 4 mm opening was cut in the skin and fascia overlying the vastus lateralis muscle 6 to 8 cm proximal to

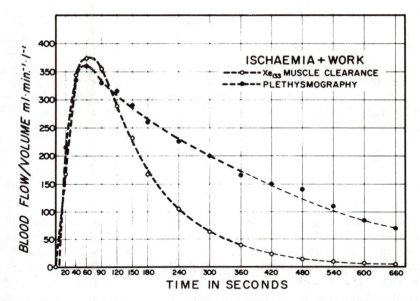

Figure 2—Post ischemic hyperemia in the tibialis anterior muscle as evaluated by 133-Xe clearance from a locally injected depot, and from venous occlusion plethysmography taking into account the difference between the post occlusion values after 5 min of occlusion with and without 2 min dynamic exercise. (Cited from Bonde-Petersen & Siggaard-Andersen, 1967).

the patella. Through this incision the catheter was inserted and gently pushed into position, dissecting its own way bluntly between the muscle fibers. The final position of the tip was 3 to 4 cm below the skin and 10 to 12 cm proximal to the incision. Then the muscle was allowed to remain undisturbed for about 10 min to obtain resting values. A series of experiments were then performed consisting of repeated sustained iso-metric contractions at different levels of MVC max. These experiments are described in detail elsewhere (Bonde-Petersen & Lundsgaard, 1981). Then a blood pressure cuff was placed around the proximal thigh and inflated to 300 Torr for 10 min. After 6 min of the occlusion period had elapsed, a 25% MVC contraction was performed until fatigue (about 1 to 2 min), but the cuff was kept inflated for 10 min. Recovery after the release of occlusion was followed for an additional 10 min period. The probe was then withdrawn and examined for blood clottings, which in no case were found on or near the tip (Figure 2).

The original tracings were measured at frequent intervals and tabulated for further calculation, where the attenuation of the Ar and oxygen signals were taken into account.

For the calculations the following equations were used:

$$pO_2 = Ar_{air}/Ar_{tissue} \times O_{2tissue}/O_{2air} = 0.2096(B - 47) \text{ Torr} \qquad (1)$$

$$pCO_2 = CO_{2tissue}/CO_{2air} \times 0.0388 (B - 47) \text{ Torr} \qquad (2)$$

where the signals measured on the recorder during calibration (air) and during the experiment (tissue) were used for Ar, O_2 and CO_2 and where 0.2096 and 0.0388 are the fractions of O_2 and CO_2 in the calibration gases, respectively.

Results

The mean resting pO_2 in the three subjects was 52 (31 to 79) Torr (mean and range) and for pCO_2 49 (43 to 57) Torr, respectively. In Figure 1 an example of an occlusion experiment is indicated. During the first min of the occlusion, there was a steep fall in pO_2, which was an indication of the resting oxygen consumption. After about 2 min pO_2 was zero and remained at this value during the occlusion period, and was not influenced by the sustained isometric contraction performed at 6 min of occlusion. When the 10 min occlusion period was terminated, the pO_2 signal rose rapidly to above resting values. When the values were corrected relative to the changes in the Ar signal, there was an overshoot in pO_2 already after 48 s of recovery to reach a maximum of 114 Torr after 160 s. Occlusion increased pCO_2 from a resting value of 47 Torr to 52 Torr. During occluded exercise pCO_2 further increased to 68 Torr at fatigue, which was reached after 52 s. During occluded recovery after exercise pCO_2 increased, leveling off at 90 Torr after 76 s. When occlusion was released there was a further increase to 97 Torr, reached at 12 s. After this there was a slow recovery, until 420 s, where the value of 35 Torr was read. The N_2 and Ar signals showed large variations at the beginning and the end of both exercise and contraction, respectively, related to the movement artifacts involved during inflation of the occlusion cuff and during contraction and relaxation. As mentioned above, this would seriously influence the signals from gases with low solubility like Ar, N_2 and O_2, but for O_2 these movement artifacts are cancelled out by the calibration procedure and only the influence of the metabolic events computed. The pCO_2 showed some variation in this subject, when the muscle was made to move but in the other two subjects pCO_2 was not affected by movement artifacts. All three subjects displayed the same pattern as in Figure 1 with very little variation: After occlusion pO_2 recovered within 87 (60 to 120) s, and a maximum pO_2 of 80 (63 to 114) Torr was reached after 193 (160 to 320) s while pCO_2 of 80

(63 to 114) Torr was reached after 193 (160 to 320) s while pCO_2 at the end of occlusion was 97 (90 to 101) Torr and recovered after 9 (7 to 10) min.

Discussion

The technique here presented for measurements of tissue gas tensions is promising for the further study of metabolic events in muscle tissue. However, it must be emphasized that this is not a micro- or intracellular method. The results can be considered as indicating mean values for interstitial gas tensions. These are indeed the tensions which will be of importance for the contraction state of the capillary and arteriolar sphincters as suggested by Krogh (1919). The content of myoglobin in the muscle has been shown to interfere with the diffusion transfer of oxygen (Hemmingsen, 1965), but will only be of importance below a pO_2 of 20 Torr. The present results were thus calculated on the basis of a constant diffusion rate of O_2 being equal to that of Ar. However, a correction for the error due to a pO_2 dependent diffusion coefficient will not change the trend of the presented results. The increase in pCO_2 during occlusion is related to O_2 metabolism and pH during the first min, but when pO_2 remains zero it must be considered exclusively related to pH variations. During contraction a decrease in pH will be related to the glycolytic production of lactic acid (HLa), but the simultaneous splitting of ATP and CP increases the amount of inorganic phosphate, which will tend to buffer the hydrogen ions in concert with the bicarbonate system. After contraction, but still under ischemia, ATP and CP will not regenerate but a continuous fall in pH will liberate CO_2, during this ischemic recovery phase possibly due to an outward diffusion of H^+ from the muscle cells. After occlusion there was a further increase in pCO_2. The recovery phase lasts as long as 7 to 10 min while pO_2 recovered already after 1.5 min. That pO_2 shows a longstanding overshoot must be related to a post-ischemic hyperemia, reinforced by the effects of exercise. It demonstrates an arterialization of the capillary blood, with the effect that the pO_2 in the intramuscular tissue space reached a mean value of 80 Torr. This demonstrates that the hyperemia is out of proportion to the need for oxygen.

The time course of muscle hyperemia expected after ischemic occluded exercise (Bonde-Petersen & Siggaard-Andersen, 1967) is demonstrated in Figure 2 (tibialis anterior muscle), and shows a close time correlation to the pCO_2 curve in the present experiment. It is, therefore, concluded that pO_2 cannot be responsible for reactive hyperemia in muscle. The pCO_2 (or a closely related factor as pH) is a much more likely candidate, like in the brain circulation. The peripheral cardiovascular system might be considered a sewer system, regulated to dispose of waste products, rather than a pipeline system, regulated to secure a normal pO_2 in the tissues.

References

BONDE-PETERSEN, F. & J.S. Lundsgaard. pO_2 and pCO_2 in human quadriceps muscle during exhaustive sustained isometric contraction. Adv. Physiol. Sci. 24:143–149, 1981.

BONDE-PETERSEN, F. & J. Siggaard-Anderson. Blood flow in skin and muscle, evaluated by simultaneous venous occlusion plethysmography and 133-Xe clearance. Scand. J. Clin. Invest. 19:113–119, 1967.

HEMMINGSEN, E., Accelerated transfer of oxygen through solutions of heme pigments. Acta Physiol. Scand. 64: Suppl 246, 1965.

KROGH, A., The supply of oxygen to the tissues and the regulation of the capillary circulation. J. Physiol. Lond. 52:457–474, 1919.

LUNDSGAARD, J.S. & H. Degn. The use of argon as an internal standard in mass spectrometric monitoring of oxygen in tissue. Proceedings of the International Conference of Blood Gas Monitoring, Blood Ion Concentrations, and Respiratory Gas Exchange, in View of Its Application during Extracorporeal Circulaton. 1980.

LUNDSGAARD, J.S., J. Grønlund & N. Einer-Jensen. In vivo calibration of flow-dependent blood gas catheters. J. Appl. Physiol. 44:124-128, 1978.

LUNDSGAARD, J.S., B. Jensen & J. Grønlund. Fast-responding flow-independent blood gas catheter for oxygen measurement. J. Appl. Physiol. 48:376–381, 1980.

TORNVALL, G., Assessment of physical capabilities with special reference to the evaluation of maximal voluntary isometric muscle strength and working capacity. Acta Physiol. Scand. 50 (Suppl. 201):21–36, 1963.

The Effects of Two Isokinetic Training Regimens on Muscle Strength and Fiber Composition

V.M. Ciriello, W.L. Holden and W.J. Evans
Liberty Mutual Research Center,
Hopkinton, Massachusetts, U.S.A., and Boston
University, Boston, Massachusetts, U.S.A.

The plasticity of human muscle fiber seems limited in comparison to animal studies. The experimental models used in animal studies induce a rapid and pronounced muscle hypertrophy. In human studies we have seen that pronounced strength increases can be elicited with muscle hypertrophy (Thortensson et al., 1976). Distribution of fiber types remains constant, independent of training regimens. However, the importance of muscle cross-sectional area for strength development has been shown (Ikai & Fukunaga, 1968). Top caliber bodybuilders train completely differently from Olympic lifters as far as the amount of sets and repetitions performed per body part to produce maximal hypertrophy. Their training stimulus (15 to 20 sets of 8 to 10 reps) may be more conducive in producing muscle hypertrophy than a typical strength training schedule (5 to 6 sets, 3 to 5 reps), but it should also be mentioned that Olympic lifters displayed an impressive musculature. This study investigated the effects of these two training schedules on percent of fiber types, FT/ST area ratios, specific hypertrophy of ST and FT muscle fibers, and maximal strength.

Methods

To study the effects of these two training regimens on skeletal muscle, nine men (mean [SD] age 25 years [2.7], wt 79.4 kg [6.5] and ht 179.3 cm [6.2]), with no previous exposure to strength or endurance training, trained each knee extensor three days per week for 16 weeks. Subjects trained one knee extensor with five sets of five repetitions (5-5), while the other knee extensor trained at 15 sets of 10 repetitions (15-10) using an Cybex II isokinetic dynamometer set at a velocity of 60°/sec. All subjects were tested pre- and post-training for maximal knee extension torque at the following speeds: 0°, 30°, 60°, 90°, 120°, 180°, 240° and 300°/sec. Maximal strength was determined by recording the highest torque produced during a series of three maximal voluntary knee extensions.

Skeletal muscle samples were obtained from the vastus lateralis muscle of both thighs using the needle biopsy technique. Muscle fiber classification was determined using myofibrillar ATPase. A projecting microscope was used to display the stained sample on a white surface. From this image, the number of fibers of each type was counted and the relative distribution of FT and ST fiber was determined (using approximately 500 fibers). Fiber areas were determined from samples stained with NADH-Diaphorase. Approximately 50 fibers of each type, identified by comparing serial sections stained with ATPase, were chosen from tightly packed cells which had a perpendicular orientation. Mechanical planimetry was then used to calculate the mean areas of the FT and ST fibers. The work output of each training session was quantified by integrating under the torque extension curve. Detailed methods and results of this study are reported elsewhere (Ciriello, 1982).

Results

Four months of training produced non-significant differences in fiber area and area ratios in both the 5-5 trained and 15-10 trained thighs. There were also no significant differences between the 5-5 and 15-10 thighs in fiber area and FT/ST area ratios either before or after 4 months of training. A significant decrease ($p < 0.05$) in the percent of FT (or increase in percent ST) did occur, however, in the 5-5 trained thighs. The percent of FT (and ST) of the 15-10 thigh was not significantly different after the four months of training (Table 1). In all cases peak torque significantly increased ($p < .01$) ranging 45% to 10% from low to high velocities (Figure 1). At the end of training there was no significant difference between the 5-5 and 15-10 thighs except for the 30°/sec speed test. In this case the 15-10 thigh was significantly stronger than the 5-5 thigh.

Discussion

The training regimens in this study did not significantly change the cross-sectional area of either the ST or the FT fibers. MacDougall (1979) using a total training stimulus of approximately 15 sets of 10 repetitions reported

Table 1

Muscle Fiber Composition and Muscle Fiber Area before and after 16 Weeks of Isokinetic Strength Training (N = 9)

	Before		After		% Difference
5-5, Regimen thigh	$\bar{X}$	SD	$\bar{X}$	SD	
FT, %	67	13	60	12	- 12 [a]
FT area, μm^2	6455	1775	6988	1727	8
ST area, μm^2	5408	156	6027	1270	11
FT/ST area ratio	1.19[b]	0.19	1.16[b]	0.17	- 2.5
15-10 Regimen thigh	$\bar{X}$	SD	$\bar{X}$	SD	
FT, %	65	15	64	15	- 1.5
FT area, μm^2	6462	1285	6800	1039	5
ST area, μm^2	5437	911	6252	1156	15
FT/ST area ratio	1.23[b]	0.34	1.11[b]	0.20	- 9.8

[a]$p < .05$
[b]Determined by averaging individual FT/ST ratios
$\bar{X}$ = Mean
SD = Standard deviation
FT = Fast-twitch
ST = Slow-twitch

increases in ST and FT cross-sectional areas of 27% and 33%, respectively. However, their training regimen involved the triceps, lasted six months, and both eccentric and concentric contractions were employed. The above differences were all significantly advantageous in producing a more pronounced hypertrophy than the present study. Tesch (1980) and Gollnick et al. (1972) sampled physically active, sprint, endurance and explosive trained individuals and found average fiber areas similar to the present study. These similarities in fiber areas indicate that the test subjects could be classified more as active than sedentary individuals. This may have had some relevance in their reaction to the training. Untrained or control groups used in other studies had fiber areas considerably smaller than the areas in this study (Larsson et al., 1979; Prince et al., 1976). The amount of hypertrophy may be related to the initial size and physical condition of the muscle. Moderately active individuals may need intense regimens over many months to achieve the same amount of hypertrophy that inactive individuals may achieve in a shorter period of time. This may explain why MacDougall et al. (1979) achieved considerable hypertrophy using the relatively inactive tricep muscle group.

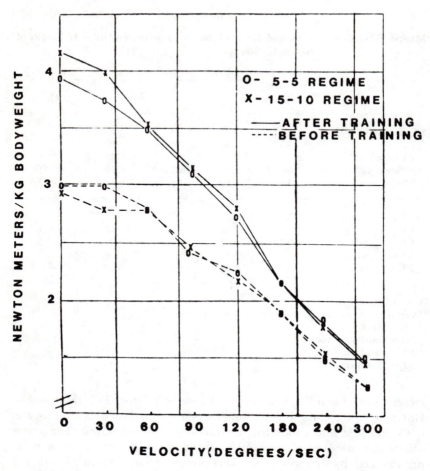

Figure 1—Average strength of knee extension torque before and after training. All post-training values are significantly greater (p < .01) than pre-training values. 15-10 regimen significantly greater (p < 0.05) than 5-5 regimen at 30°/sec (post training).

The FT/ST area ratio did not significantly change during the 4 months of training. However, the initial values of 1.19 and 1.23 did decrease to 1.16 and 1.11 for the 5-5 and 15-10 thigh, respectively. This indicates a trend toward selective hypertrophy of the ST fibers, especially in the 15-10 thigh. Isokinetic training at 60°/sec may be more conducive to ST hypertrophy than conventional weight training or strength building techniques. The significant decrease in percent of FT fibers in the 5-5 thigh may also be related to the slow isokinetic training. Several investigators have reported fiber type conversion from Fast to Slow in compensatory hypertrophy models (Baldwin et al., 1982; Ianuzzo et al., 1976) and this conversion might be a possible explanation for the percent decrease in FT fibers in the 5-5 trained thigh.

The significant gains in isometric and dynamic strength were not completely dependent on the quantity of training employed. The total ratio of work performed between the two regimens (15-10 vs 5-5) was 4.39/1. This extreme workload difference could only manifest a difference at one slow dynamic speed. Minimal regimens, such as the one employed on the 5-5 thigh, supply sufficient training stimulus to produce large increases in strength. Contralateral training effects were not tested in this experiment. Consequently, the cross-over effects were unknown.

The pretraining torque values are in close agreement with those of Thorstensson et al., (1977) and Larson et al. (1979). After 4 months of training, torque values increased to levels comparable to elite sprinters and jumpers (Thorstensson et al., 1977). These significant increases in maximal torque may have been the result of an increased number of myofibrils. Myofibrils have been shown to increase with training (MacDougall et al., 1976) and this increase in contracting elements, especially in the 15-10 thigh, might have been responsible for the significant improvements in maximal torque. Substantial increases in maximal torque after training may also have been the result of twitch potentiation. It has been documented (De Luca, 1979) that after repetitive stimulation, there is potentiation of motor unit twitch tension. This twitch potentiation or increase in twitch force of muscle fibers has been localized to muscle fibers having slow-twitch characteristics. This potentiation of muscle fibers may explain the increase in maximal torque without significant histometrical changes in muscle fibers.

Training may have also enhanced other neural mechanisms which increased maximal torque after training. The neural control of movement has been modeled (Houk & Henneman, 1974) and an important aspect of this control is inhibition of the antagonist muscle during movement. During the early phases of training, inhibition of the antagonist of knee extension, namely the knee flexors, may have increased, resulting in a greater net torque during extension. Further investigation is necessary to confirm these modifications of neural control or process of motor learning. Change in neural motor control probably does not include modification of motor unit control. Recent studies have demonstrated a highly ordered recruitment and decruitment scheme, based upon motoneuron excitability (De Luca et al., 1982).

References

BALDWIN, K.M., V. Valdez, R.E. Herrick, A.M. MacIntosh and & R.R. Roy. Biochemical properties of overloaded fast-twitch skeletal muscle. J. Appl. Physiol.: Respirat. Environ. Exercise Physiol. 52:467–472, 1982.

CIRIELLO, V.M., A Longitudinal Study of the Effects of Two Training Regimes on Muscle Strength and Hypertrophy of Fast-Twitch and Slow Twitch Fibers. (Doctoral Dissertation, Boston University) Boston, Mass.: University Microfilms, 1982.

DE LUCA, C.J., Physiology and mathematics of myoelectric signals. I.E.E.E. Transactions on Biomedical Engineering 26(6):313–325, 1979.

DE LUCA, C.J., R.S. Le Fever, M.P. McCue, & A.P. Xenakis. Behaviour of human motor units in different muscles during linearly-varying contractions. J. of Physiol. Accepted for publication, 1982.

GOLLNICK, P.D., R.B. Armstrong, C.W. Saubert IV, K. Piehl & B. Saltin. Enzyme activity and fiber composition in skeletal muscle of untrained and trained men. 33:312–319, 1972.

HOUK, J. & E. Henneman. Feedback control of muscle. In V.B. Mountcastle (Ed.), Medical Physiology. St. Louis: C.V. Mosby Co., 1974.

IANUZZO, C.D., P.D. Gollnick & R.B. Armstrong. Compensatory adaptations of skeletal muscle fiber types to a long-term functional overload. Life Sciences 19: 1517–1524, 1976.

IKAI, M. & T. Fukunaga. Calculation of muscle strength per unit cross-sectional area of human muscle by means of ultrasonic measurement. Int. Z. Angew Physiol. 26:26–32, 1968.

LARSSON, L., G. Grimby & J. Karlsson. Muscle strength and speed of movement in relation to age and muscle morphology. J. Appl. Physiol.: Respirat. Environ. Exercise Physiol. 46:451–456, 1979.

MACDOUGALL, J.D., G. Sale, G. Elder & J.R. Sutton. Ultrastructural properties of human skeletal muscle following heavy resistance training and immobilization (abstract). Medicine and Science in Sports 8:72, 1976.

MACDOUGALL, J.D., D.G. Sale, J.R. Moroz, G.C.B. Elder, J.R. Sutton & H. Howald. Mitochondrial volume density in human skeletal muscle following heavy resistance training. Medicine and Science in Sports 11:164–166, 1979.

PRINCE, F.P., R.S. Hikida & F.C. Hagerman. Human muscle fiber types in power lifters, distance runners and untrained subjects. Pflugers Arch. 363:19–26, 1976.

TESCH, P., Muscle fatigue in man with special reference to lactate accumulation during short term intense exercise. Acta Physiol. Scand. Supplement 480, 1980.

THORSTENSSON, A., B. Hulten, W. Von Dobeln & J. Karlsson. Effect of strength training on enzyme activities and fiber characteristics in human skeletal muscle. Acta Physiol. Scand. 96:392–398, 1976.

THORSTENSSON, A., L. Larsson, P. Tesch & J. Karlsson. Muscle strength and fiber composition in athletes and sedentary men. Medicine and Science in Sports 9(1):26–30, 1977.

Oxygen Consumption, Work and Efficiency as a Function of Contraction Velocity in Isolated Frog Sartorius Muscle

N.C. Heglund and G.A. Cavagna
Harvard University, Cambridge, Massachusetts,
U.S.A., and Instituto di Fisiologia Umana
dell'Universita' di Milano, Milano, Italy

The efficiency of transformation of chemical energy into mechanical work by the muscles (efficiency = work done/chemical energy consumed) has been most commonly measured in whole organisms (efficiency = work done/caloric equivalent of oxygen consumed) (Dickinson, 1929; Margaria, 1975), although a few measurements have been made in isolated muscles at $0°$ C (efficiency = work/work + heat) (Hill, 1939; Woledge, 1968). The peak efficiency determined in both cases, excluding tortoise muscle, was about 0.25. However recent results show that during some types of exercise, such as running in humans (Cavagna & Kaneko, 1977) and hopping in kangaroos (Cavagna et al., 1977), the peak efficiency of positive work production is much greater: 0.6 to 0.7. It has been assumed that these high efficiency values were due to the storage of elastic energy during a phase of negative work (when the subject lands) and a subsequent recovery of that energy during a phase of positive work (when the subject takes off). In this study the mechanical efficiency was determined in isolated frog sartorii at $12°$ C by measuring the recovery oxygen consumption (Kushmerick & Paul, 1976) simultaneously with the mechanical work done.

Two experimental procedures were followed: in the first the muscle was allowed to shorten from a state of isometric contraction, a situation where only chemical energy can be the origin of the work done (POS, the left half of Figure 1); and in the second the muscle was allowed to shorten immediately after being stretched 2.5 mm in 50 ms, a situation where part of the positive work done can be derived from mechanical energy stored in

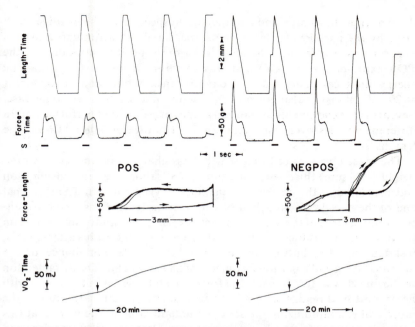

Figure 1—Experimental records showing the length changes imposed by the ergometer (Length-Time, top row) and the force developed simultaneously by the muscle (Force-Time, second row) as a function of time (in seconds) during four successive cycles; the time of stimulation is indicated by the horizontal bars (S, third row). The left column shows the muscle shortening from a state of isometric contraction (POS); the right column shows the muscle shortening immediately after having been stretched while active (NEGPOS). The force is plotted as a function of the length for the four cycles (Force-Length, fourth row). The spikes on the force records are artifacts due to the stirrer. The area below the force-length curve obtained during shortening represents the positive work done, the area below the curve obtained during stretching represents the negative work done. The oxygen recovery curve (VO$_2$-Time, fifth row) are recorded as a function of time on a much slower time base (in min) than the mechanical events described above. The arrows indicate the time of stimulation.

the muscle during the negative work phase (NEGPOS, the right half of Figure 1). In both cases the muscle was allowed to shorten 3.5 to 5.0 mm in 300 to 840 ms from the length at which the force exerted by the parallel elastic elements at equilibrium was about 2 g. Stimulation began 200 ms before shortening and continued for 60 ms after the beginning of shortening; this pattern was chosen in order to reach a substantial force during shortening while avoiding an appreciable redevelopment of isometric tension after the end of shortening. In both POS and NEGPOS the relaxed muscle was then lengthened to the original length and the cycle was repeated four times at an interval of about 1 s. The force-length diagrams obtained during the four cycles are superimposed in the middle part of Fig. 1. The area below these diagrams represents the positive and negative work done during shortening and lengthening, respectively. The positive work done by the parallel elastic elements, measured in unstimulated control experiments, was found to be small ($<3\%$ of the positive work done in the POS experiments, W^+_{POS}) and was therefore ignored. The metabolic energy input, calculated from the oxygen consumption (conversion factor $= 20.1$ J/ml O_2) is also given in the figure on a much slower time base. Seventy two experiments, organized as mirror pairs, were performed on 18 different muscles over a shortening velocity range of 0.08 to 1.6 muscle lengths/s.

The muscle was placed in a sealed glass chamber (diameter $= 8$ mm, height $= 80$ mm) filled with air equilibrated saline; oxygen consumption was measured as the decrease in chamber oxygen content. The proximal end of the muscle was attached to a force transducer by using a dovetail-type joint; a piece of the pelvic bone acted as the tenon and a slotted hole on the top of the transducer as the mortice. The transducer plus attachments, inserted into the bottom of the chamber, had a compliance of 3.5 microns/100 g and a natural frequency of 5600 Hz in air. The distal tendon of the muscle was folded back and forth through a series of knots tied with heavy cotton thread; the first knot was generally within 1.0 mm of the muscle fibers, minimizing the series compliance due to the tendon at this extremity. One end of a 1.5 mm diameter stainless steel rod was connected to a voice coil actuated ergometer, and the other end was bent into a small hook; the hook was placed around the remaining distal tendon and below the knots, forming a rigid connection between the ergometer and the muscle. The stainless steel rod passed through a diffusion barrier consisting of a glass pipe filled with saline inserted into the top of the chamber; the pipe was 65 mm long and had an internal diameter of 2 mm. The barrier's effectiveness was demonstrated in control experiments showing that O_2 was unable to diffuse into a N_2 equilibrated chamber. Platinum electrodes (30 x 6 mm) were fixed on opposite sides of the chamber for transverse stimulation; supramaximal stimulation was attained with 38 Hz alternating condenser discharges (16 V peak, time constant 60 microseconds). An oxygen electrode was placed between the stimulating electrodes on one side of the chamber; a Macor ceramic stirrer faced the oxygen electrode on the opposite side. The chamber was surrounded by a water jacket connected to a constant temperature bath controlling temperature to within 0.05° C. The chamber volume was adjusted from 2.5 to 3.5 ml, depending upon the length of the muscle, by changing the insertion depth of the force transducer and diffusion barrier. The movement of the ergometer was controlled by a microcomputer in discrete steps of less than 10 microns; compliance of the ergometer was less than 3 microns/100 g. As shown by control experiments without the muscle, the movement of the lever and the stimulation had no permanent effect on the chamber pO_2. Force and length changes were monitored by a digital oscilloscope and read into the computer in order to calculate the positive and negative work done by the muscle. The output of the oxygen electrode was fed into a digital picoammeter and read into the computer every 30 s in order to calculate the muscle oxygen consumption (using a solubility coefficient of 1.485 micromoles O_2/ml saline at 12° C and 1 atm of oxygen).

Glycolysis was not considered to be an important energy source for the following reasons: 1) Kushmerick and Paul (1976) measured insignificant amounts of lactate production during a 10 to 40 s tetanus obtained with 10 Hz stimulation at 0° C; in these experiments we used four tetani for a total of only 0.64 s of 38 Hz stimulation at 12° C; 2) despite an eight-fold weight range in the muscles used, our results do not show a size dependent variation in energy consumption (a regression of VO_2/g vs muscle weight yields a correlation coefficient r = 0.52 for POS, and r = 0.41 for NEGPOS, n = 36); and 3) the oxygen consumption of the muscles does not show a pO_2 dependence over the pO_2 range used in the experiments, 75 to 150 mm Hg (a regression of VO_2/g vs average pO_2 yields a correlation coefficient r = 0.23 for POS, and r = 0.16 for NEGPOS, n = 36).

The efficiency of transformation of chemical energy into mechanical work, W^+/kVO_2POS, attained a maximum at shortening speeds between 0.6 and 1.5 muscle l/s; the average value measured over this speed range was 0.26 ± 0.03 (mean and s.d., n = 18). This is in agreement with the maximum value predicted by Hill (1939) on the basis of heat and work measurements at 0° C. At shortening speeds greater than 1.5 l/s there was a large and unavoidable (at 12° C) redevelopment of tension, making measurements of the efficiency of positive mechanical work production pointless.

When the muscle shortens immediately after having been stretched while active, the positive work efficiency, $W^+/KVO_{2NEGPOS}$, can be greater than the efficiency of transformation of chemical energy into mechanical work. The ratio of NEGPOS efficiency to POS efficiency is 1.30 ± 0.24 (mean and s.d., n = 35). This increase in efficiency is due to an increase in the positive work done after stretching and not due to a decrease in the energy consumed. The additional amount of positive work done after stretching, $W^+_{NEGPOS} - W^+_{POS}$, is 5.0 ± 1.9 mJ/g (mean and s.d., n = 35). The additional amount of elastic energy released after prestretch, calculated on the basis of the elastic energy vs stress curves given by Cavagna et al. (1981) and the peak stress measured in the NEGPOS and POS experiments, is 4.0 mJ/g. Therefore the contractile component appears to be responsible for a non-elastic increase in positive work of 1.0 mJ/g, or 3.6% of the positive work done by the contractile component during shortening from a state of isometric contraction.

The peak efficiency attained after stretching in the present experiments

(0.38) is considerably less than the peak values attained during large animal locomotion (0.6 to 0.7), however these experiments were performed on a muscle which may have a relatively small capability for storing and utilizing elasic energy. In more recent experiments on mammalian fast (rat EDL) and slow (rat soleus) muscle we have found the peak efficiency without a prestretch is about 0.2, slightly less than the frog sartorius. however the peak efficiency after a prestretch was greater in the mammalian muscle, attaining values of 0.5 in the EDL and 0.4 in the soleus.

Acknowledgment

This study was supported by NIH post-doc#AM06022, NSF grant#PCM8119868 and Italian National Research Council contract #79.02366.65.

References

CAVAGNA, G.A., G. Citterio & P. Jacini. Effects of speed and extent of stretching on the elastic properties of active frog muscle. J. Exp. Biol. 91:131–143, 1981.

CAVAGNA, G.A., N.C. Heglund & C.R. Taylor. Mechanical work in terrestrial locomotion: two basic mechanisms for minimizing energy expenditure. Am. J. Physiol. 233(5):R243–R261, 1977.

CAVAGNA, G.A. & M. Kaneko. Mechanical work and efficiency in level walking and running. J. Physiol. 268:467–481, 1977.

DICKINSON, S., The efficiency of bicycle-pedalling, as affected by speed and load. J. Physiol. 67:242–255, 1929.

KUSHMERICK, M.J. & R.J. Paul. Aerobic recovery metabolism following a single isometric tetanus in frog sartorius muscle at 0° C. J. Physiol. 254:693–709, 1976.

HILL, A.V., The mechanical efficiency of frog's muscle. Proc. Roy. Soc(B) 127:434–451, 1939.

MARGARIA, R., Pisiologia Muscolare e Meccanica del Movimento. Milano: Mondadori, 1975.

WOLEDGE, R.C., The energetics of tortoise muscle. J. Physiol. 197:685–707, 1968.

Autonomic Nerve System and Metabolic Adaptations to Long-Term Exercise

M. Krotkiewski, P. Björntorp, U. Smith,
V. Marks, T. William-Olsson, A. Wirth,
and K. Mandroukas

University of Göteborg, Göteborg, Sweden,
University of Surrey, Gilford, Surrey, England,
and University of Heidelberg, Heidelberg,
Federal Republic of Germany

Physical training improves glucose tolerance after oral glucose administration (Krotkiewski et al., 1980). The known decrease (Björntorp et al., 1970) in insulin levels observed after training is partly dependent on increased peripheral uptake and to a large extent, on decreased insulin production by the pancreas (Wirth et al., 1980). Insulin production is known to be under autonomic control (Porte et al., 1969). Furthermore, the attenuating effect of anticholinergic blockade on insulin production is more pronounced after completion of long-term physical training (Krotkiewski et al., 1980). Thus, the decreased pancreatic insulin production may be due to either increased alfa-adrenergic inhibitory activity or decreased β_2-adrenergic stimulatory activity. The aim of the present study was to elucidate these possibilities and to evaluate the influence of chronic exercise on the autonomic nerve system.

Material and Methods

Two groups of 12 obese middle-aged women participated in the study (Table I). They were weight stable by history during the last 3 months preceding the start of the experiment. During a period of 4 days before and after the termination of the physical training the patients were admitted to the hospital. They were placed on a diet calculated to be sufficient to

Table 1

**Effect of Physical Training on Body Composition, Heart Rate (beats/min)
and Blood Pressure (mm Hg)**

	Before training	After training	p-level
Body weight			
(kg)	92.2 ± 4.1	94.4 ± 5.3	n.s.
Lean body mass			
(kg)	28.9 ± 1.2	29.2 ± 1.4	n.s.
Body fat			
(kg)	32.8 ± 2.5	33.5 ± 2.7	n.s.
Mean fat cell weight			
(μg)	0.60 ± 0.02	0.63 ± 0.02	n.s.
Fat cell number $\times 10^{10}$	6.2 ± 0.6	6.1 ± 0.5	n.s.
Plasma triglyceride			
(mmol/l)	1.8 ± 0.2	1.7 ± 0.2	n.s.
Plasma cholesterol			
(mmol/l)	5.7 ± 0.3	5.7 ± 0.4	n.s.
Work load			
At rest			
Heart rate (bpm)	78.8 ± 2.7	74.6 ± 2.2	< 0.05
Blood pressure			
syst	134.2 ± 2.8	128.4 ± 2.6	< 0.05
diast	88.0 ± 1.8	80.6 ± 2.0	< 0.01
100 watts			
Heart rate (bpm)	148.8 ± 2.9	136.7 ± 3.4	< 0.05
Blood pressure			
syst	186.0 ± 3.4	182.4 ± 4.8	< 0.05
diast	90.7 ± 2.2	84.2 ± 2.9	< 0.01
130 watts			
Heart rate (bpm)	172.0 ± 2.8	166.4 ± 2.6	< 0.01
Blood pressure			
syst	186.8 ± 3.6	183.0 ± 4.0	n.s.
diast	92.7 ± 2.3	90.3 ± 2.8	n.s.

Means ± SEM Group 1 and 2

maintain their caloric balance. The composition of the diet was identical before and after physical training. All blood samples and tests were performed during the hospitalization period.

The physical training consisted of three training sessions per week and followed a fixed schedule of dynamic interval exercise (Krotkiewski et al.,

Table 2

Effects of Physical Training and Atropine on Plasma Insulin (μU/ml), c-peptide (μmol/l), Pancreatic Polypeptide (PP) (pg/ml), Gastric Inhibitory Polypeptide (GIP) (pmol/l), and Glucose Values (mmol/l) after Oral Glucose Administration

	No atropine	With atropine	p-level atropine effect
Sum of insulin values before training	380 ± 20	284 ± 17	< 0.001
Sum of insulin values after training	344 ± 25	321 ± 29	n.s.
p-level	< 0.05	n.s.	
Sum of c-peptide values before training	3.15 ± 0.59	2.60 ± 0.38	< 0.05
Sum of c-peptide values after training	1.58 ± 0.34	1.40 ± 0.27	n.s.
p-level	< 0.01	< 0.02	
Sum of glucose values before training	38.2 ± 0.4	33.8 ± 0.5	< 0.05
Sum of glucose values after training	34.6 ± 0.3	32.8 ± 0.4	n.s.
p-level	n.s.	n.s.	
Sum of PP values before training	437.3 ± 89.9	150.8 ± 40.9	< 0.001
Sum of PP values after training	503.3 ± 95.2	147.2 ± 28.5	< 0.001
p-level	< 0.05	n.s.	
Sum of GIP values before training	7516 ± 777	7611 ± 534	n.s.
Sum of GIP values after training	6619 ± 851	8353 ± 825	< 0.05
p-level	< 0.05	n.s.	

Means ± SEM

1979). The training program was standardized for each subject in relation to the individual maximal working capacity. All participants were told not to adhere to any particular diet during the training period and to eat ad libitum.

Oral glucose tolerance tests were performed with and without the subcutaneous injection of atropine (0.1 mg/kg body weight) both before and after the training period.

The second group of patients were trained and followed in an identical fashion to the first group. Oral glucose tolerance tests (OGTT) were performed with and without the infusions of the beta-adrenergic agent isoprenaline, 0.015 μg/kg body weight/min during 100 min, and the alfa-adrenergic blocking agent phentolamine, 5 mg in a rapid intravenous injection followed by the infusion of 0.03 mg/min during 45 min. Both infusions started 10 min before the oral glucose administration. All studies were performed 4 days after the last bout of exercise. The patients were placed on the same diet as discussed above during the test period.

Results

Physical exercise three times/week for 3 months did not significantly change body weight, body composition, mean fat cell weight, fat cell number or plasma lipids in either group. However, heart rates tested on the ergonomic bicycle as well as systolic and diastolic blood pressures decreased at most of the submaximal work loads (Table 1).

Blood glucose values during OGTT did not change in either group while plasma insulin, c-peptide and gastric inhibitory polypeptide (GIP) decreased and pancreatic polypeptide increased after physical training. Insulin and c-peptide levels decreased after atropine administration before but not after the training period. Pancreatic polypeptide release significantly decreased after atropine, both before and after physical training. GIP levels increased after atropine administration only after physical training (Table 2).

Isoprenaline infusion significantly increased insulin release both before and after physical training. Isoprenaline also increased c-peptide levels, glucose concentration and pancreatic polypeptide concentrations but only after physical training. Isoprenaline infusion resulted in decreased glucagon and GIP levels but only before physical training (Table 3).

Phentolamine infusion increased c-peptide but not insulin levels both before and after physical training. Phentolamine infusion slightly decreased glucagon and increased pancreatic polypeptide levels but only after the termination of physical training (Table 4).

Table 3

Effects of Physical Training and Infusion of Isoprenaline on Plasma Insulin (μU/ml), c-peptide (μmol/l), Pancreatic Polypeptide (PP) (pg/ml), Glucose (mmol/l) and Gastric Inhibitory Polypeptide (GIP) (pmol/l) Values after Oral Glucose Administration

	No isoprenaline	With isoprenaline	p-level isoprenaline effect
Sum of insulin values before training	485.8 ± 72.9	638.1 ± 84.2	< 0.01
Sum of insulin values after training	386.9 ± 54.2	636.2 ± 83.6	< 0.001
p-level	< 0.05	n.s.	
Sum of c-peptide values before training	2.63 ± 0.37	3.09 ± 0.69	n.s.
Sum of c-peptide values after training	2.21 ± 0.38	4.70 ± 0.86	< 0.05
p-level	< 0.05	n.s.	
Sum of glucose values before training	55.3 ± 2.9	58.4 ± 2.8	n.s.
Sum of glucose values after training	53.1 ± 2.8	60.2 ± 2.8	< 0.01
p-level	n.s.	n.s.	
Sum of PP values before training	295.4 ± 115.8	348.0 ± 102.0	n.s.
Sum of PP values after training	251.6 ± 62.6	438.4 ± 105.3	< 0.01
p-level	n.s.	< 0.05	
Sum of GIP values before training	2894.4 ± 455.6	2576.5 ± 438.5	$0.05 < p < 0.1$
Sum of GIP values after training	2362.0 ± 405.8	2462.3 ± 364.6	n.s.
p-level	< 0.05	n.s.	
Sum of glucagon values before training	291.6 ± 27.2	240.7 ± 21.2	< 0.05
Sum of glucagon values after training	247.3 ± 30.1	225.5 ± 15.8	n.s.
p-level	$0.05 < p < 0.1$	n.s.	

Means ± SEM

Table 4

Effects of Physical Training and Infusion of Phentolamine on Plasma Insulin (μU/ml), c-peptide (μmol/l), Pancreatic Polypeptide (PP) (pg/ml) and Gastric Inhibitory Polypeptide (GIP) (pmol/l) Values after Oral Glucose Administration

	No phentolamine	With phentolamine	p-level phentolamine effects
Sum of insulin values before training	485.8 ± 72.9	445.9 ± 66.9	n.s.
Sum of insulin values after training	386.9 ± 54.2	425.2 ± 80.1	n.s.
p-level	< 0.05	n.s.	
Sum of c-peptide values before training	2.63 ± 0.37	4.28 ± 0.79	0.05 < p < 0.1
Sum of c-peptide values after training	2.21 ± 0.88	4.10 ± 0.89	0.05
p-level	< 0.05	n.s.	
Sum of glucose values before training	55.3 ± 2.9	58.1 ± 2.8	n.s.
Sum of glucose values after training	53.1 ± 2.8	56.1 ± 2.8	n.s.
p-level	n.s.	n.s.	
Sum of PP values before training	295.4 ± 115.8	326.5 ± 70.8	n.s.
Sum of PP values after training	251.6 ± 62.6	471.2 ± 156.9	< 0.05
p-level	n.s.	0.1 < p < 0.05	
Sum of GIP values before training	2894.4 ± 455.6	2569.0 ± 370.1	n.s.
Sum of GIP values after training	2362.0 ± 405.8	2711.2 ± 352.9	n.s.
p-level	< 0.05	n.s.	
Sum of glucagon values before training	291.6 ± 27.2	253.3 ± 25.6	n.s.
Sum of glucagon values after training	247.3 ± 30.1	206.0 ± 30.5	< 0.05
p-level	< 0.05	n.s.	

Means ±SEM

Discussion and Conclusion

Physical training seems to influence the autonomic nerve system and thereby the hormone release. This is documented by the following observations: 1) *Cholinergic blockade.* The increase in insulin and c-peptide levels as well as the increase in GIP concentration was significantly more pronounced after physical training. This finding is consistent with previous observations of increased vagal tonus after physical training; 2) *Beta-adrenergic stimulation.* The increased insulin, c-peptide, glucose and pancreatic polypeptide levels after isoprenaline infusion were, significantly higher after training, suggesting increased beta-adrenergic sensitivity; 3) *Alfa-adrenergic blockade.* Except for pancreatic polypeptide no evidence for an increased alfa-adrenergic tonus in the pancreas following physical training was found. Taken together, these results show alterations in the tonus and/or sensitivity of the autonomic nerve system following physical training. However, the changes so far noted cannot by themselves explain the lower plasma insulin levels.

References

BJÖRNTORP, P., K. de Jounge, L. Sjöström & L. Sullivan. The effect of physical training on insulin production in obesity. Metabolism 19:631, 1970.

KROTKIEWSKI, M., L. Sjöström & P. Björntorp. Physical training in hyperplastic obesity. V. Effects of atropine on plasma insulin. Inst. J. Obesity 4:49, 1980.

KROTKIEWSKI, M., L. Sjöström, K. Mandroukas, L. Sullivan, H. Wetterqvist, & P. Björntorp. Effects of long-term physical training on body fat, metabolism and blood pressure in obesity. Metabolism 28:1650, 1979.

PORTE, D., Sympathetic regulation of insulin section. Arch. Intern. Med. 123:252, 1969.

WIRTH, A., G. Holm, B. Nilsson, U. Smith & P. Björntorp. Insulin kinetics and insulin binding to adipocytes in physically trained and food-restricted rats. Am. J. Physiol. 238:E108, 1980.

The Effect of Training on Resting
Muscle Membrane Potentials

R.F. Moss, P.B. Raven, J.P. Knochel,
J.R. Peckham and J.D. Blachley
Texas College of Osteopathic Medicine, Fort Worth,
Texas, U.S.A., and University of Texas Health
Science Center, Dallas, Texas, U.S.A.

Known effects of endurance training in humans include changes in submaximum heart rate (Karvonen et al., 1957), maximal oxygen consumption (Ekblom, 1970), and cellular enzymes (Hollozy, 1967). Presently, very little work has been completed on membrane changes following endurance training. Training has been shown to selectively increase the cross sectional area occupied by subsarcolemmal mitochondria in rats (Muller, 1976) and to increase their density in humans and rats (Hoppeler et al., 1973; Krieger et al., 1980). The fact that these subsarcolemmal mitochondria, hypothesized to support membrane function, alter with training would indicate possible changes in the active transport mechanisms of the membrane. Knochel (1977), using mongrel dogs, showed a 10% increase in resting membrane potential following 6 weeks of treadmill running. The data suggest that the endurance type training affects the sarcolemma of muscle. The present study was designed to determine if the resting muscle membrane potential of endurance trained males was different than that of a sedentary population.

Methods

Sixteen healthy male subjects, age 15 to 35 years (mean = 24.4 ± 1.4 years), volunteered for study participation. A signed informed consent was obtained from each participant. On the basis of training status, eight were selected as trained (T) while the remaining subjects were classified as sedentary (S).

Maximum oxygen consumption ($\dot{V}o_2$max) during progressive treadmill exercise, using the open circuit technique, was determined for all subjects. $\dot{V}o_2$max was defined as the rate of oxygen consumption attained by an individual which failed to increase with an increase in work load.

Resting skeletal muscle membrane potential (Em) was measured by direct impalement with Ling electrodes of individual muscle fibers in the anterior tibialis muscle (Cunningham et al., 1971). Eight to twenty individual fiber potentials were measured and averaged for each subject. Potentials were obtained following a 48 h period during which no strenuous exercise occurred. At the time of Em measurement, venous blood was obtained for determination of serum potassium by flame photometry.

Statistical analysis of the data included unpaired t, Pearson product moment correlation, and linear regression analysis.

Results

Table 1 lists the composite data from these studies. There were significant differences (P $<$ 0.05) in weight and $\dot{V}o_2$max as would be expected when comparing trained and sedentary populations. No difference in serum potassium concentration was found between the two groups.

The Em of the two groups is found in Figure 1. The T group's resting skeletal muscle membrane potential was 8% higher than that of the S group (P $<$ .025).

Figure 2 shows the relationship of Em to $\dot{V}o_2$max in all subjects (T and S). There is a linear relationship described by the formula of Y = 0.21x +

Table 1

**Results of the T-Test for Compared Data on Age,
Weight, $\dot{V}o_2$max and Serum Potassium**

Group	N	Age (years)	Weight (kg)	$\dot{V}o_2$max (ml/kg/min)	Serum K (meq/l)
Trained	8	23.2 ($\pm$ 2.3)	63.2 ($\pm$ 3.9)	63.5 ($\pm$ 1.9)	4.1 ($\pm$ 0.25)
Untrained	8	24.6 ($\pm$ 2.2)	82.5 ($\pm$ 6.4)	39.8 ($\pm$ 2.9)	4.0 ($\pm$ 0.17)

(values are expressed as mean $\pm$ SEM, P $<$ 0.05)

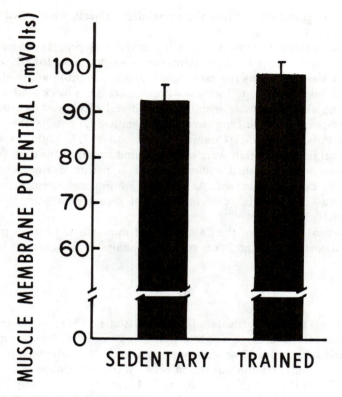

Figure 1—Resting membrane potential of sedentary and trained subjects. The Em of –98.8 mV for the Trained was more statistically significant (P < 0.05) than that of the –91.5 mV of the sedentary group.

84.3 with a correlation of +0.60 (P < .02). Intracorrelations were –0.12 (P > .05) and –0.23 (P > .05) for the S and T groups respectively. Although training improves both the $\dot{V}o_2$max and Em, it appears their relationship is contingent upon one or more other factors related to improved fitness.

Discussion

The data obtained in these preliminary studies seem to indicate that vigorous athletic training results not only in an increase in total body $\dot{V}o_2$ but also in skeletal muscle membrane hyperpolarization. A considerable degree of variability is apparent in the relationship between Em and $\dot{V}o_2$max. This may in part be explained by the observations of Klissouras

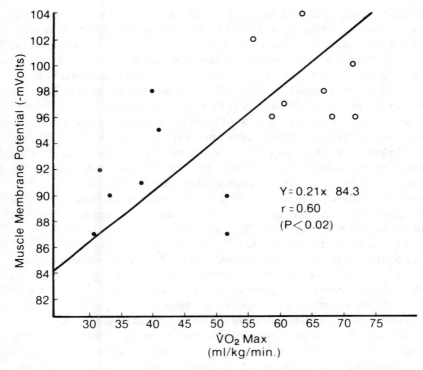

$Y = 0.21x$ 84.3
$r = 0.60$
$(P < 0.02)$

Figure 2—Relation of resting muscle membrane potential and $\dot{V}o_2max$ for both T and S groups, (• = S, o = T).

(1971) and Pollock (1973), who present evidence suggesting that the majority of the $\dot{V}o_2max$ is genetically determined and may increase as little as 10 to 20% with training. Possibly genetic differences between the groups can explain the higher $\dot{V}o_2max$ of the T group and bear somewhat on the resting skeletal muscle Em. Therefore, factors that allow for a higher $\dot{V}o_2max$ may also affect the resting skeletal Em without a direct cause and effect relation between the two variables.

The resting skeletal muscle Em of the sedentary group was -91.5 ± 1.2 mV which compares favorably with measurements previously reported for non-athletic normal subjects (Cunningham et al., 1971). The Em measured in our trained subjects, -98.9 ± 1.1 mV is the first reported example of membrane hyperpolarization in healthy subjects.

Muscle contraction results in a sharp increase in the potassium concentration in the venous effluent and muscle beginning uptake of this K^+ almost immediately (Saltin et al., 1981). These data indicate that skeletal muscle uptake of potassium may be an important mechanism to protect against exercise induced hyperkalemia. Because intracellular potassium concentration is 30 to 40 times that of the extracellular fluid, potassium uptake by muscle cells is the result of active transport. Active potassium influx is coupled to sodium efflux in an uneven and variable ratio, generally believed to be 3 Na^+ to 2 K^+. Such ionic exchange is electrogenic and would be expected to increase the membrane Em. The serum (extracellular) potassium concentrations measured in our trained subjects were essentially identical to those measured in sedentary subjects. This is not surprising in that these samples were obtained 48 h after exercise.

The potential differences (E) across the membrane are described by the Goldman equation (Casteels, 1970). However, this does not represent the Em measured by direct impalement. Included in this measure is that portion attributed to the electrogenic Na^+, K^+ pump (Flemming, 1980). The equation for Em, taking into consideration the electrogenic pump, becomes: $Em = E + \varphi pRm$. Where φp = the electrogenic contribution of sodium-potassium exchange, and Rm = membrane resistance.

Possibly training alters the sensitivity of the Na^+, K^+ pump to extracellular K^+ and/or intracellular Na^+, hence a greater activity of the pump mechanism. This would require increased amounts of ATP, fitting in nicely with the data reporting changes in the respiratory capacity of subsarcolemmal mitochondria following training (Krieger et al., 1980).

We postulate that the observed increase in skeletal muscle membrane potential is the result of either (1) increased intracellular potassium concentration, or (2) increased electrogenic potassium transport; i.e., increased sodium-potassium adenosine triphosphatase activity. This enzyme, the functional correlate of the Na^+, K^+ pump, has been shown to increase in activity under the influence of either increased intracellular sodium or extracellular potassium (Thomas, 1977).

In conclusion, we note that the ability of cells to adapt to changes in the extracellular ionic environment is not a new concept (Silva et al., 1977). However, data of this investigation is the first documenting that exercise may alter membrane transport characteristics, and perhaps intracellular ionic composition.

Acknowledgments

We appreciate the fine technical assistance of Julio Borroto and Kathleen Fouts Sims.

References

CASTEELS, R., The relation between the membrane potential and the ion distribution in smooth muscle cells. In E. Bülbring, A.F. Brading, A.W. Jones & T. Tomita (Eds.), Smooth Muscle, pp. 70–99. Baltimore: Williams and Williams, 1970.

CUNNINGHAM, J.N., N.W. Carter, F.C. Rector & D.W. Seldin. Resting transmembrane potential differences of skeletal muscle in normal subjects and severely ill patients. J. Clini. Invest. 50:49–59, 1971.

EKBLOM, B., Effects of physical training on circulation during prolonged severe exercise. Acta Physiol. Scand. 78:145–158, 1970.

FLEMMING, W.W., The electrogenic Na^+, K^+ pump in smooth muscle: physiologic and pharmacologic significance. Ann. Rev. Pharmacol. Toxicol. 20:129–49, 1980.

HOLLOZY, J., Effects of exercise on mitochondrial oxygen uptake and respiratory enzyme activity in skeletal muscle. J. Biol. Chem. 242:2278–2282, 1967.

HOPPELER, H., P. Luthi, H. Claassen, E.R. Weibel & H. Howald. The ultra structure of the normal human skeletal muscle. A morphometric analysis of untrained men, women and well trained volunteers. Pfluegers Arch. 344:217–232, 1973.

KARVONEN, M.J., E. Kentala & O. Mustala. The effect of training on heart rate. Annales Medicinae Experimentalis Et Biologiae Fenniae 35:307–315, 1957.

KLISSOURAS, V., Heredability of adaptive variation. J. Appl. Physiol. 31:338–344, 1971.

KNOCHEL, J.P., N. Carter & J.R. Cotton. Muscle hyperpolarization and potassium tolerance induced by exercise training. Clin. Res. 22:455, 1977.

KRIEGER, D.A., C.A. Tate, J. McMillin-Wood & F.W. Booth. Populations of rat skeletal muscle mitochondria after exercise and immobilization. J. Appl. Physiol. 48:23–28, 1980.

MULLER, W., Subsarcolemmal mitochondria and capillarization of soleus muscle fibers in young rats subjected to an endurance training program. Cell Tissue Research 174:367–389, 1976.

POLLOCK, M.L., Exercise and Sports Science Review. J.H. Wilmore (Ed.). New York, Academic Press, pp. 155–188, 1973.

SALTIN, B., G.S. Sjøgaard, F.A. Gaffney & L.B. Rowell, Potassium, lactate, and water fluxes in human quadriceps muscle during static contractions. Circ. Res. Supp I. 48:18–24, 1981.

SILVA, P., R.S. Brown & F.H. Epstein. Adaptation to potassium. Kidney Inter. 11:466–475, 1977.

THOMAS, R.C., Electrogenic sodium pump in nerve and muscle cells. Physiol. Rev. 52:563–594, 1977.

Myoglobin

Myoglobin Function in Exercising Skeletal Muscle in Hypoxia

R.P. Cole

Columbia University, New York, New York, U.S.A.

The concentration of myoglobin in skeletal muscle has been shown to increase under conditions of exercise training and chronic exposure to high altitudes (Pattengale & Holloszy, 1967; Reynafarje, 1962). In suspensions of pigeon breast muscle fiber bundles, conversion of myoglobin to forms incapable of reversible combination with oxygen led to a decline in steady state oxygen consumption (Wittenberg et al., 1975). Using mathematical models of muscle respiration, de Koning et al. (1981) and Fletcher (1980) demonstrated that the effect of myoglobin on muscle oxygen consumption should be dependent on the distribution and magnitude of tissue P_{O_2} levels. Conversion of intracellular myoglobin in vivo to non-functional forms has recently been shown to result in a decrease in oxygen consumption and tension generation in the exercising, normoxic dog gastrocnemius-plantaris muscle (Cole, 1982). In the present study, the effect of conversion of myoglobin to non-functional forms on the function of the same muscle under conditions of hypoxia was examined.

Methods

Detailed methods are presented elsewhere (Cole, paper submitted for publication). Briefly, mongrel dogs (n = 18) were anesthetized with pentobarbital and intubated. The left gastrocnemius-plantaris muscle was exposed and the arterial and venous circulation of the muscle was isolated. Muscle venous blood flow was determined volumetrically. The right femoral artery was cannulated to allow collection of arterial blood samples. The distal tendon of the muscle was cut and attached to a strain gauge. Conditions for isometric contraction were produced by fixing the proximal insertions of the muscle to the base plate with bone pins. Arterial blood gases were measured using standard electrodes, while mean arterial blood pressures and tension amplitudes were measured with precalibrated strain gauges. Blood oxygen content was determined with a galvanic oxygen cell. Muscle oxygen consumption was calculated from the product of muscle blood flow and arteriovenous O_2 content difference.

The animals were heparinized (2200 units/kg) and allowed to breathe 10% oxygen via a demand valve assembly. The muscle was then stretched to optimal length and stimulated via the cut sciatic nerve with 3 Hz supramaximal twitch stimuli for 20 min. In the control group (n = 9) the muscle was then set at resting length, blood flow was interrupted, and the muscle perfused with 400 ml of modified Ringer's lactate solution. The experimental group (n = 9) was treated similarly with the exception that the modified Ringer's lactate contained, in addition, 0.35 mM hydrogen peroxide.

Blood flow was then reestablished, and 3 Hz twitch stimuli continued for an additional 20 min. Twitch tension amplitude and mean perfusion pressure were recorded continuously while arterial blood gases, arterial blood hematocrit, muscle blood flow, and arterial and venous oxygen content were determined at 10 min intervals. The dogs were then sacrificed, and the left and right gastrocnemius-plantaris muscle were removed and weighed. Multiple cut sections of the muscles were compared for color changes. Comparisons between the control group and the hydrogen peroxide group were analyzed using the Student's unpaired t test while intra group comparisons were made with the paired t test. Results are reported as mean ± SD.

Results

The control and hydrogen peroxide groups were similar with respect to muscle weight, hematocrit, perfusate administration rate, and arterial P_{CO_2} and pH. The arterial PO_2 levels were lower in the hydrogen peroxide group after perfusate administration (32 ± 5 mm Hg versus 37 ± 6 mm Hg in the control group).

Mean arterial blood pressure, muscle blood flow, and muscle oxygen delivery were similar in both groups and were not influenced by perfusate administration. In the control group, muscle oxygen extraction was unchanged after Ringer's lactate perfusion while in the hydrogen peroxide group, oxygen extraction was significantly reduced from pre-perfusion values after hydrogen peroxide administration.

Muscle oxygen consumption remained relatively constant after perfusate administration in the control group. In the hydrogen peroxide group muscle oxygen consumption was initially similar to that of the control group and decreased after perfusate administration with the difference becoming significant at the 10 min period (Figure 1).

The time course of active twitch tension generation in response to 3 Hz stimulation for both groups is given in Figure 2. In the latter part of the post infusion period, tension generation was significantly reduced in the hydrogen peroxide group in comparison to the control group.

Cut sections of the muscle perfused with 0.35 mM hydrogen peroxide had the uniform brown appearance characteristic of higher oxidation states of in vivo myoglobin (Cole et al., 1978).

Discussion

In the present study, transient perfusion of the hypoxic gastrocnemius-plantaris muscle with hydrogen peroxide led to a reduction in muscle oxygen consumption and twitch tension generation. These results are thus similar to those reported previously in the isolated gastrocnemius-plantaris muscle with normal arterial P_{O_2} levels (Cole, 1982). In that study short term perfusion of skeletal muscle with hydrogen peroxide did not interfere with neuromuscular transmission, autoregulation of blood flow or the processes involved with excitation contraction coupling. Oxidative phosphorylation of isolated skeletal muscle mitochondria is also not adversely affected by hydrogen peroxide in the concentration used here (Cole, 1982). Hydrogen peroxide in these experiments thus appears to act specifically by forming higher oxidation states of intracellular myoglobin, thus rendering it incapable of reversible combination with oxygen. Changes in muscle function noted after hydrogen peroxide administration can therefore be attributed to loss of functional myoglobin. Spectroscopic evidence of alteration of intracellular myoglobin has been previously noted using sodium nitrite and ethyl hydrogen peroxide (Cole et al., 1978; Tamura et al., 1978).

In this study, post-perfusion P_{O_2} values were higher in the control group than in the hydrogen peroxide group. However, no dependence of oxygen consumption on the arterial P_{O_2} level was noted in these experiments, suggesting that the decline in muscle oxygen consumption after hydrogen peroxide administration could not be explained on the basis of a reduced P_{O_2}. Furthermore, P_{O_2} values in this range in exercising muscle have not been shown to produce appreciable decreases in muscle oxygen uptake (Stainsby & Otis, 1964).

In spite of similar muscle oxygen delivery in the two groups, muscle oxygen consumption declined in the muscles given hydrogen peroxide.

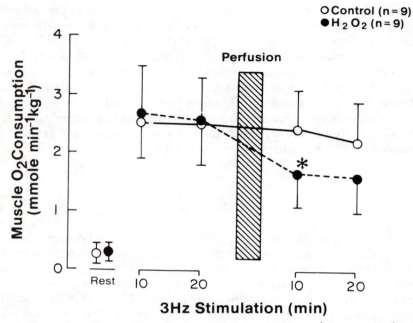

Figure 1—Relationship of muscle oxygen consumption (nmole $O_2 min^{-1}$ kg wet weight^{-1}) to the time course of 3 Hz twitch stimulation before and after perfusate administration. Means and standard deviations are illustrated. (o control, n = 9; • hydrogen peroxide treatment, n = 9). Asterisks indicate significance at p < 0.05 level. In the control group, muscle oxygen consumption did not change appreciably after perfusate administration. In the hydrogen peroxide group, muscle oxygen consumption declined after perfusate administration with the decrease becoming significant at the 10 min period.

This decline in oxygen consumption appears to result from a reduction in muscle oxygen extraction. As a result of this impaired oxygen extraction, muscle isometric tension generation also declines (Figure 2).

It therefore appears that intact functional myoglobin is important in the intracellular transport of oxygen to mitochondria. The mechanism of this effect is not well understood. Myoglobin does not enhance steady state oxygen consumption or ATP production in stirred suspension of skeletal muscle mitochondria, suggesting that it does not decrease the resistance to oxygen flux in unstirred layers present at the mitochondrial surface (Cole et al., 1979). In any event, these results suggest that in hypoxia induced by administration of 10% O_2, as in normoxia, formation of non-functional intracellular myoglobin resulted in a decline in oxygen consumption and tension generation in the isometrically contracting dog gastrocnemius-plantaris. Myoglobin thus appears to have an important role in oxygen transport and utilization in muscle function under hypoxemic conditions.

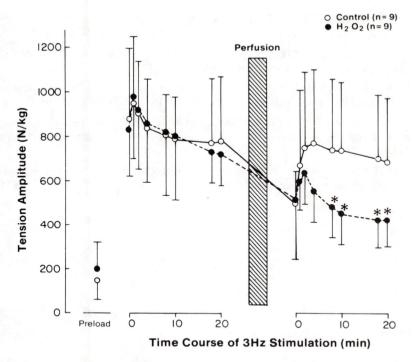

Figure 2—Relationship of muscle twitch tension amplitude (N kg^{-1} wet weight muscle) to the time course of 3Hz stimulation before and after perfusate administration. Symbols as in Figure 1. Both groups showed an initial decline in tension amplitude after perfusate administration. Tension generation then increased to near pre-perfusion values in the control group but decreased significantly after 8 min of stimulation in the hydrogen peroxide treated group.

References

COLE, R.P., Myoglobin function in exercising skeletal muscle. Science, 1982. (in press)

COLE, R.P., P.C. Sukanek, J.B. Wittenberg & B.A. Wittenberg. Mitochondrial function in the presence of myoglobin (Abstract). Physiologist 22:21, 1979.

COLE, R.P., B.A. Wittenberg & P.R.B. Caldwell. Myoglobin function in the isolated fluorocarbon perfused dog heart. Am. J. Physiol. 234:H567–H572, 1978.

DE KONING, J., L.J.C. Hoofd & F. Kreuzer. Oxygen transport and the function of myoglobin. Pflügers Arch. 389:211–217, 1981.

FLETCHER, J.E., On facilitated oxygen diffusion in muscle tissues. Biophys. J. 29:437–458, 1980.

PATTENGALE, P.K. & J.O. Holloszy. Augmentation of skeletal muscle myoglobin by a program of treadmill running. Am. J. Physiol. 213:783–785, 1967.

REYNAFARJE, B., Myoglobin content and enzymatic activity of muscle and altitude adaptation. J. Appl. Physiol. 17:301–305, 1962.

STAINSBY, W.N. & A.B. Otis. Blood flow, blood oxygen tension, oxygen uptake, and oxygen transport in skeletal muscle. Am. J. Physiol. 206:858–866, 1964.

TAMURA, M., N. Oshino, B. Chance & I.A. Silver. Optical measurements of intracellular oxygen concentration of rat heart *in vivo*. Arch. Biochem. Biophys. 191:8–22, 1978.

WITTENBERG, B.A., J.B. Wittenberg & P.R.B. Caldwell. Role of myoglobin in the oxygen supply to red skeletal muscle. J. Biol. Chem. 250:9038–9043, 1975.

Myoglobin Concentration and Training in Humans

E. Jansson, C. Sylvén and B. Sjödin
Karolinska sjukhuset and Laboratory for Human
Performance, Research Institute, Stockholm, Sweden

Based on animal experiments, physical training appears to increase the skeletal muscle myoglobin concentration and especially endurance training which increases the oxidative potential of the skeletal muscles involved (e.g., Lawrie, 1950). Treadmill running by rats resulted in a marked increase of both the myoglobin concentration and the activity of oxidative enzymes in the leg muscles (Baldwin et al., 1977; Hickson, 1981; Pattengale & Holloszy, 1967). In humans, endurance training increases the activity of oxidative enzymes. Long-distance runners, for example, have about twice as high succinate dehydrogenase activity as untrained subjects (e.g., Jansson & Kaijser, 1977). However, it is not known if the myoglobin concentration increases in parallel to the oxidative enzymes in humans subjected to endurance training. Thus, the aim of the present study was to compare thigh muscle myoglobin concentration of endurance trained and untrained subjects. To confirm that the trained subjects had a higher oxidative potential than the untrained subjects as well as for further characterization of the subjects, the muscle samples were also analyzed for the citrate synthase activity, capillarization, muscle fiber areas, and the relative number of type I (slow twitch) and type II (fast twitch) muscle fibers.

Methods

Mean and range of age, height and weight were for the six untrained subjects: 34 (27 to 42) years, 183 (178 to 192) cm and 79 (60 to 110) kg; for the five trained subjects (Swedish elite long-distance runners), 35 (24 to 43) years, 168 (158 to 176) cm and 56 (49 to 64) kg. The study was approved by the Ethical Committee of the Karolinska Hospital, Stockholm, Sweden.

Muscle biopsies were obtained at rest from the vastus lateralis muscle (Bergström, 1962). Each biopsy was divided into two pieces of which one was weighed and homogenized in an ice cooled 0.1 M phosphate buffer pH 7.7 with 0.5% bovine serum albumin (BSA). The myoglobin was analyzed by a radioimmunological assay ad modum, Möller and Sylvén (1981), the citrate synthase activity was determined according to Shepherd and Garland (1969) and protein according to Lowry (1951). The second piece was mounted in O.C.T. Compound embedding medium, frozen at $-150°$ C in isopentane. Transverse sections were cut at $-20°$ C and stained for myofibrillar ATPase at pH 9.4 after preincubation at pH 10.3 and for capillaries by the amylas-PAS method (Andersen, 1975). The classification of the muscle fibers into type I and type II was based on the ATPase stain. Stained sections were used for measuring the cross-sectional areas of the fibers by using a grid method.

Results

Muscle Characteristics: Trained vs Untrained (Table 1)

The citrate synthase (CS) activity, the percentage type I fibers and the numbers of capillaries per mm^2 (capillary density) was 67%, 48% and 37% higher in the trained than in the untrained subjects, respectively. In contrast, the mean fiber area tended to be smaller (19%) in the trained subjects. The myoglobin concentration was of similar magnitude in the two groups: 5.22 ± 0.6 mg $\times$ g^{-1} wet muscle in the trained and 5.30 ± 1.0 mg $\times$ g^{-1} wet muscle in the untrained subjects.

Single Correlations: All Subjects

No significant correlations were found between the myoglobin concentration and the percentage type I fibers, CS activity, mean fiber area, capillaries per mm^2 or capillaries per fiber. A direct correlation was found between the CS activity and the number of capillaries per mm^2 ($r = 0.91$; $p < 0.001$).

Untrained Subjects

A direct correlation was found between the myoglobin concentration and the height of the subjects ($r = 0.88$; $p < 0.05$).

Trained Subjects

A direct correlation was found between the myoglobin concentration and the type I fiber area ($r = 0.92$; $p < 0.05$).

Table 1

Mean ± SD for Myoglobin, Protein, Citrate Synthase, % Type I Fibers, Fiber Areas, Numbers of Capillaries per mm² and Capillaries per Fiber in Six Untrained (UT) and Five Trained (T, Elite Long-distance Runners) Subjects

Variables	Untrained	Trained	Ratio T/UT	Significance of the UT-T difference p
Myoglobin, mg × g⁻¹ wet muscle	5.30 ± 1.0	5.22 ± 0.6	0.98	n.s
Protein, g × 100 g⁻¹ wet muscle	0.161 ± 0.01	0.173 ± 0.01	1.07	n.s
Myoglobin, mg × g⁻¹ protein	33.2 ± 6.8	30.1 ± 2.7	0.91	n.s
Citrate synthase, μkat × g⁻¹ wet muscle	0.163 ± 0.057	0.272 ± 0.053	1.67	< 0.01
Type I fibers, %	44.3 ± 10.6	65.6 ± 6.6	1.48	< 0.01
Fiber area, type I, μm²	5540 ± 950	4750 ± 730	0.86	n.s.
Fiber area, type II, μm²	5130 ± 2360	3760 ± 1170	0.73	n.s.
Mean fiber area, μm²	5310 ± 1710	4370 ± 700	0.82	n.s.
Capillaries per mm²	412 ± 78	574 ± 56	1.39	< 0.01
Capillaries per fiber	2.69 ± 0.57	3.17 ± 0.44	1.18	n.s.

Discussion

The runners in the present study had a considerably higher citrate synthase (CS) activity than the untrained subjects, reflecting their higher capacity of aerobic energy turnover. Myoglobin is thought to serve the muscles by the facilitation of oxygen diffusion as well as by acting as an oxygen store (Wittenberg, 1970). It is thus tempting to suppose that the myoglobin concentration increases with training, in parallel to the increase in activity of oxidative enzymes. The finding of a basically similar myoglobin concentration in the thigh muscle of the trained and the untrained subjects was thus somewhat surprising. Earlier studies have shown that capillary density increases with endurance training. The present study confirmed a higher capillary density in the trained subjects. It is not possible to judge from the present study if the higher capillary density, expressed as capillaries per mm², is due to decreased fiber areas and/or to an absolute increase in the number of capillaries. Other studies, however, have stated that both these adaptive mechanisms seem to occur with endurance training (Schantz et al., 1982). Thus, the greater oxygen requirement of the trained muscles during exercise might to a large extent be provided for by an increased capillary density, as supported by the significant relationship between capillaries per mm^2 and the CS activity found in the present study. This finding of shorter oxygen diffusion distances in the trained subjects could explain why no difference was found in myoglobin concentration.

In an attempt to discover the factors governing myoglobin concentration in skeletal muscle, multiple linear regressions were calculated. A multiple linear regression analysis with the myoglobin concentration as the dependent variable and the CS activity, mean fiber area and the numbers of capillaries per mm^2 as the independent variables gave a multiple r value of 0.57 (p < 0.10). However, within the untrained group, body height and myoglobin concentration showed a strong direct correlation. Thus, a multiple linear regression analysis with the myoglobin concentration as the dependent and the CS activity, mean fiber area, capillaries per mm^2 and the body height as the independent variables was tested and gave a multiple r value of 0.84 (p < 0.01). In tall individuals in the upright position the leg blood flow and thereby the oxygen delivery to the muscles might be somewhat limited by venous pooling (Sjöstrand, 1953). It could be speculated that a higher myoglobin concentration could compensate for this by increasing the oxygen extraction from the blood to the tissue. In fact, Lawrie (1953), who demonstrated a direct correlation between myoglobin and the cytochrome oxidase activity in the psoas muscle from different animals, showed that there also was a direct correlation between myoglobin concentration and the size of the animal although he did not focus upon that finding. However, due to the low number of subjects in the present study, it is not possible to state which of the independent variables or which combination is the most important for determining the myoglobin concentration.

In conclusion, there seems not to be a simple correlation between oxidative potential and myoglobin concentration in the human skeletal muscle; the endurance trained subjects, with a higher oxidative potential than the untrained, showed a similar leg muscle myoglobin concentration as the untrained. It is suggested that 1) trained subjects with an increased aerobic capacity and thus an increased muscle oxygen requirement during exercise have adapted the peripheral oxygen transport system by increasing the capillary density, and 2) myoglobin might be of special importance when there is an imbalance between muscle oxygen requirement and supply.

Acknowledgments

This study was supported by grants from the Swedish Sports Research Council.

References

ANDERSEN, P., Capillary density in skeletal muscle of man. Acta Physiol. Scand. 95:203–205, 1975.

BALDWIN, K.M., D.A. Cooke & W.G. Cheadle. Time course adaptations in cardiac and skeletal muscle to different running programs. J. Appl. Physiol. 42:267–272, 1977.

BERGSTRÖM, J., Muscle electrolytes in man. Scand. J. Clin. Lab. Invest. Suppl. 68. 1962.

HICKSON, R.C., Skeletal muscle cytochrome c and myoglobin, endurance, frequency of training. J. Appl. Physiol. 51:746–749, 1981.

JANSSON, E. & L. Kaijser. Muscle adaptation to extreme endurance training. Acta Physiol. Scand. 100:315–324, 1977.

LAWRIE, R.A., Some observations on factors affecting myoglobin concentrations in muscle. J. Agric. Sci. 40:356–366, 1950.

LAWRIE, R.A., The activity of the cytochrome system in muscle and its relation to myoglobin. Biochem. J. 55:298–305, 1953.

LOWRY, O.H., H.J. Rosebrough, A.L. Farr & R.J. Randell. Protein measurement with the Folin Phenol reagent. J. Biol. Chem. 193:265–275, 1951.

MÖLLER, P. & C. Sylvén. Myoglobin in human skeletal muscle. Scand. J. Clin. Lab. Invest. 41:479–482, 1981.

PATTENGALE, P.K. & J.O. Holloszy. Augmentation of skeletal muscle myoglobin by program of treadmill running. Am. J. Physiol. 213:783–785, 1967.

SCHANTZ, P., J. Henriksson & E. Jansson. The adaptive response of human triceps brachii and quadriceps femoris muscle to endurance training of long duration. Clin. Physiol. (in press)

SHEPERD, D. & P.B. Garland. Citrate synthase from rat liver. Methods in Enzymology 13:11–16, 1969.

SJÖSTRAND, T., Volume and distribution of blood and their significance in regulating circulation. Physiol. Rev. 33:202–228, 1953.

WITTENBERG, J.B., Myoglobin-facilitated oxygen diffusion: Role of myoglobin in oxygen entry into muscle. Physiol. Rev. 50:559–636, 1970.

Myoglobin Content of Normal and Trained Human Muscle Fibers

P.M. Nemeth, M.M.-L. Chi,
C.S. Hintz and O.H. Lowry
Washington University School of Medicine,
St. Louis, Missouri, U.S.A.

Human muscle fibers can be clearly divided into two major groups, Type I (slow-red or slow-oxidative) and Type II (fast-white or fast-glycogenolytic). The nomenclature, derived from combined histological, biochemical and physiological studies, associates "color" with levels of myoglobin, enzymes of energy metabolism, and capillarization (Close, 1972). Microanalytical techniques now provide a means to quantitatively characterize individual muscle fibers on the basis of oxidative and glycogenolytic enzyme activities (Lowry et al., 1978). The present study extends the characterization to quantitative levels of myoglobin in single fibers with the development of a microanalytical radioimmunoassay.

Changes in functional demands of muscle are known to alter levels of energy related enzymes. Endurance exercise, for example, increases the capacity of muscle to oxidize carbohydrates and fats (Holloszy & Booth, 1976). This paper describes the effect of exercise on myoglobin content. Myoglobin was measured in muscle fibers, enzymatically identified, from a human subject after intense training and detraining programs.

Methods

A male cyclist engaged in intense uninterrupted training for 24 months. The training regimen incorporated continuous and high intensity inter- mittant exercise, and is described in detail elsewhere (Chi et al., in preparation). The training was followed by careful restriction of physical activity (detraining) for 83 days. Muscle samples were obtained by a needle biopsy of the vastus lateralis, at maximal training and detraining times. The vastus lateralis of a control male was also studied after a 400 day period of moderate inactivity.

Microanalytic procedures and the methods for the enzymatic assays for β-hydroxyacyl CoA dehydrogenase (βOAC), lactate dehydrogenase (LDH), citrate synthetase (CS), and adenylokinase (AK) are described by Lowry et al. (1972). Consecutive assays for each of these enzymes and for myoglobin were made on samples of the same fiber.

The radioimmunoassay was made with a commercial serum myoglobin kit (Nuclear Medical Systems, No. 1025) by reducing all volumes ten-fold to accomodate 20 to 40 ng of frozen dried muscle tissue (0.2 to 1 ng myoglobin). Incubation times followed the recommended procedure except that the antibody-antigen reaction was increased to 1 h. Bound ^{125}I-myoglobin was counted on a gamma counter (Beckman, Gamma 300 system). The amount of myoglobin was calculated from a standard curve obtained with each assay.

Tests were made to verify the uniformity of myoglobin levels along the fiber, the lack of day to day variations in levels, the optimal time of each reaction step, and the reliance of myoglobin measurements following extended times of exposure of the fibers to room temperature and atmosphere. Specificity for human myoglobin was confirmed by showing no reaction with red rat muscle.

Results

The enzymes LDH and AK are useful for enzymatically typing human muscle (Lowry et al., 1978). Fibers low in both have other enzyme characteristics of Type I fibers and fibers with relatively high levels of both conform to Type II characteristics. This is shown for the control individual in Figure 1, where LDH and AK activities are plotted on the horizontal axes. The vertical axis (length of each line) gives the level of myoglobin in each fiber. Type I fibers had an average of 29.9 mg myoglogin/g of muscle dry weight, while Type II fibers had 20.5 mg/g. Myoglobin was directly correlated to CS and βOAC among these fibers (data not shown).

The effects on muscle enzymes and myoglobin when the trained athlete stopped exercising for 83 days are summarized in Figure 2. Muscle fibers examined at the time of intense training separate distinctly into Types I and II. Myoglobin content was 20.4 mg/g in Type I and 12.8 mg/g in Type II. Detraining caused an increase in the activity of the two enzymes in both fiber groups. Myoglobin levels, however, were not significantly affected in either type by detraining. Type I fibers were 20.1 mg/g and Type II were 13.7 mg/g.

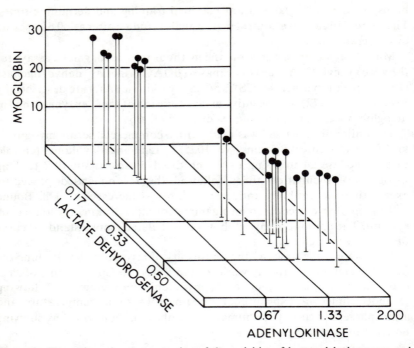

Figure 1—Three-dimensional presentation of the activities of lactate dehydrogenase and adenylokinase enzymes in Units/ g dry weight and the myoglobin content in mg/ g dry weight in individual fibers of the vastus lateralis muscle from a control human.

In this same individual there was a continuum of values for both βOAC and myoglobin across fiber types (Figure 3). The myoglobin content was correlated to βOAC, in both trained and detrained. Detraining led to a slight reduction in βOAC, but the value of myoglobin was not significantly changed.

Discussion

Rodent muscles that consist mainly of fast-twitch glycolytic fibers are very low in myoglobin and can be considered truly "white" (Peter et al., 1972). The association between myoglobin content and oxidative capacity in individual fibers is based on a histochemical stain for heme containing substances (Drews & Engel, 1961). In that study, considerably more myoglobin was estimated in the small Type I fibers than in Type II of rat, rabbit, cat and human muscle, most demonstrably in rat. Since the stain is somewhat non-specific for myoglobin, it seemed possible that myoglobin

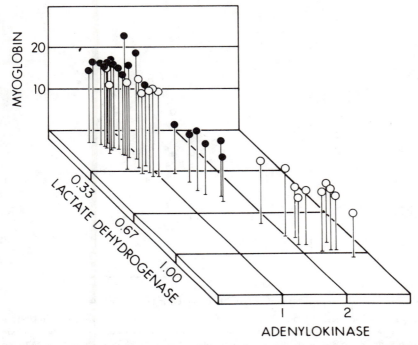

Figure 2—Three-dimensional presentation, as in Figure 1, of individual muscle fibers of the vastus lateralis following 24 months of endurance training (black circles) and following 83 days of detraining (open circles) in the same human athlete.

differences had been overestimated in human fibers, which do not show great differences in oxidative capacity. The present results with the specific quantitative immunoassay show that the fast glycolytic (Type II) fibers contain approximately two-thirds as much myoglobin as the slow oxidative (Type I) fibers. Clearly, in this case, the designation of "fast-twitch white" is not appropriate.

Myoglobin concentrations varied in the two individuals. In fact, the concentrations were higher in the control than the athlete (3% and 2% of muscle dry weight in Type I and II respectively in control and 2% and 1.4% in the athlete). Variations of this proportion were also reported for four control individuals by Jansson (1981) studying myoglobin levels in groups of fibers. This indicates that studies on myoglobin changes require that each individual serves as his own control or that large numbers of individuals be studied.

The myoglobin content was not affected in either fiber group by detraining despite changes in glycogenolytic and oxidative enzymes (Chi et al., in preparation). Animal studies have also shown that long-term

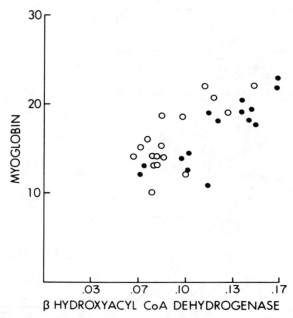

Figure 3—The relationship of the activity of β-hydroxyacyl CoA dehydrogenase in Units/g dry weight and myoglobin content in mg/g dry weight in individual human muscle fibers following 24 months of endurance training (black circles) and 83 days of detraining (open circles) in the same human athlete.

exercise increases mitochondrial proteins and oxidative enzymes while not affecting sarcotubular vesicles, Ca^{2+} capacity or Ca^{2+} affinity or myosin ATPase (Holloszy & Booth, 1976). Many of these properties are known to change with intense chronic stimulation. If myoglobin levels are also affected by more contraction, then one might conclude that the threshold for myoglobin modulation is above that for energy related enzymes.

Acknowledgments

This study was supported by grants from MDA and NIH NS-08862.

References

CHI, M.-Y., C.S. Hintz, E.F. Coyle, W.H. Martin III, J.L. Ivy, P.M. Nemeth, J.O. Holloszy & O.H. Lowry. Effects of detraining on enzymes of energy metabolism in individual human muscle fibers. (in preparation)

CLOSE, R.I., Dynamic properties of mammalian skeletal muscles. Phys. Revs. 52:129–197, 1972.

DREWS, G.A. & W.K. Engel. An attempt at histochemical localization of myoglobin in skeletal muscle by the benzidine-peroxidase reaction. J. Histochem. Cytochem. 9:206–207, 1961.

HOLLOSZY, J.O. & F.W. Booth. Biochemical adaptations to endurance exercise in muscle. Ann. Rev. Physiol. 38:273–291, 1976.

JANSSON, E., Myoglobin and fiber types in human skeletal muscle. Acta Physiol. Soc. (Abstract) 112(12):12A, 1981.

LOWRY, C.V., J.S. Kinney, S. Felder, M.M.-Y. Chi, K.K. Kaiser, P.N. Passonneau, K.A. Kirk & O.H. Lowry. Enzyme patterns in single human muscle fibers. J. Biol. Chem. 253:8269–8277, 1978.

LOWRY, O.H. & J.V. Passonneau. A Flexible System of Enzymatic Analysis. New York: Academic Press, 1972.

PETER, J.B., R.J. Barnard, V.R. Edgerton, C.A. Gillespie & K.E. Stemple. Metabolic profiles of three fiber types of skeletal muscle in guinea pigs and rabbits. Biochemistry 11:2627–2633, 1972.

Clinical Exercise

Rated Effort Angina, Perceived Leg Fatigue and Blood Lactate during Graded Exercise

H. Åström, A. Holmgren, J. Karlsson and E. Orinius

Karolinska Hospital, Stockholm, Sweden

Borg et al. (1981) described in patients with different severity of angina pectoris and verified coronary artery disease a curvilinear increase in perceived chest pain during exercise. This relation was achieved in each group of patients with low, intermediate and high exercise capacity respectively. If the highest tolerable work load was set up to 100% and the submaximal intensities as a fraction of that, it could be demonstrated that the progress in perceived angina followed approximately the same function indicating a common denominator. Karlsson et al. (1983) confirmed these results and were able to demonstrate that the relative work load concept could be substituted by blood lactate concentration even for patients with very low peak exercise capacity (approximately 50 W). The peak lactates amounted to 2.9, 3.1 and 4.3 mmol $\times$ 1^{-1} and corresponded to rated effort angina according to a 1–9 graded scale (Borg et al., 1981) 3.8, 4.5 and 4.8 for the three patient categories with 53, 90 and 125 W peak performance capacity before the angina pain got untolerable. Both the rated effort angina and blood lactates described curvilinear relationships to work load and for both variables inflexion points could be determined (expressed in W) for onset of angina pain (OAP) and onset of blood lactate accumulation (OBLA). Both the W_{OAP} and W_{OBLA} estimates disclosed a quantitative agreement. As a close relationship has been found between muscle and blood lactates on one hand and experience of local leg muscle and general fatigue as well on the other (Noble et al., 1982), it was thought of interest to include in the protocol rated perceived exertion (RPE) and compare RPE with effort angina and blood lactates.

Methods and Patients

A total of 44 patients, who were referred to the Thoracic Clinics for evaluation of coronary artery bypass surgery and with established coronary artery disease and angina pectoris, were investigated. Their mean age, height and weight were 54 (range 42 to 67) years, 172 (range 154 to 181) cm and 74 (range 63 to 91) kg, respectively. They performed graded exercise according to Åström and Jonsson (1976) with 10 W increments every min. A finger tip blood sample was drawn for subsequent blood lactate determination at the end of the work load (Karlsson et al., 1983; Rydevik et al., 1982) and the patient was questioned for perceived angina pain and leg muscle fatigue according to a one to nine graded scale (Borg et al., 1981).

Results

As can be seen in Figure 1 the rated angina pain and blood lactate concentration increased curvilinearly as earlier described by Borg et al. (1981) and Karlsson et al. (1982) for three groups of patients with low intermediate and high exercise capacity, respectively. When rated angina pain was plotted versus blood lactate an almost linear relationship was obtained irrespective of peak exercise performance.

The 16 patients, who were interviewed about their perceived leg muscle exertion, are described in Figure 2. Rated effort angina pain disclosed a tendency to larger individual variations vs blood lactate as compared to rated perceived exertion vs blood lactate. This variability might be related to the severity of the coronary artery disease and/or the extent of the ischemic heart muscle tissue. It was also possible to show that leg muscle exertion seemed to precede the effort angina (Figure 3).

Discussion

As earlier documented there was a relationship between effort angina and blood lactate concentration during graded exercise. Onset of blood lactate accumulation seemed to vary with onset of anginal pain.

It seems reasonable to suggest that the increased blood lactate concentrations originated in the exercising muscle and as suggested by Noble et al. (1983), the leg muscle fatigue reflected this phenomenon.

Mitchell and coworkers have suggested that the nerve fibers, which most probably are responsible for mediating the sensation of fatigue—the C-fibers—also are responsible for activation of central circulation (Shepherd et al., 1981). According to Shepherd and coauthors, there might

EFFORT ANGINA AND BLOOD LACTATE

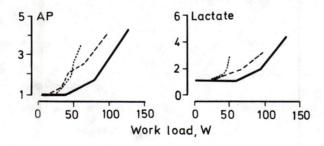

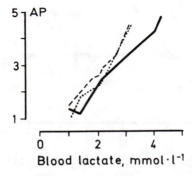

Figure 1—Upper panel. Rated effort anginal pain (AP) and blood lactate concentration vs work load during graded exercise.

Lower Panel. Rated effort anginal pain (AP) vs blood lactate concentration. Data from the upper panel.

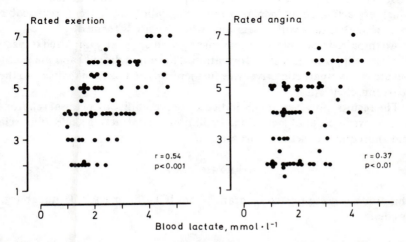

Figure 2—Individual data for rated effort anginal pain and rated perceived exertion in relation to blood lactate concentration.

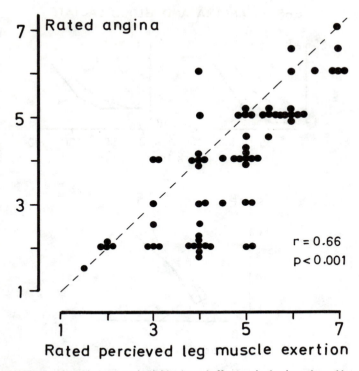

Figure 3—The relationship between individual rated effort anginal pain and rated leg muscle perceived exertion.

exist receptors referred to as ergoreceptors in the contracting muscles which are activated by increased muscle metabolism. It seems reasonable to assume that they will appear also when lactate is formed.

An impaired heart muscle function and a subsequently impaired oxygen delivery will promote lactate formation. This might directly and indirectly elevate the sympathetic drive with further impaired oxygen delivery to the heart muscle itself.

The results, however, might indicate the possibility of a causal relationship between onset of lactate accumulation in contracting skeletal muscle and subsequent onset of effort angina.

Acknowledgment

This study was supported by grants from ICI-Pharma AB, Gothenburg, Sweden.

References

ÅSTRÖM, H. & B. Jonsson. Design of exercise test with special reference to heart patients. Brit. Heart J. 38:289–296, 1976.

BORG, G., A. Holmgren & I. Lindblad. Quantitative evaluation of chest pain. Acta Med. Scand. Suppl. 644:43–45, 1981.

KARLSSON, J., A. Holmgren, H. Åström, E. Orinius & C. Kaijser. Angina pectoris and blood lactate concentration during graded exercise. (to be published)

KARLSSON, J., I. Jacobs, B. Sjödin, P. Tesch, P. Kaiser, O. Sahl & B. Karlberg. Semi-automatic blood lactate assay: Experiences from an exercise laboratory. Int. J. Sports Med. 4:45–48, 1983.

NOBLE, B., I. Jacobs, G. Borg, P. Kaijser & R. Ceci. Rated perceived exertion and muscle and blood lactate concentrations during graded exercise. (to be published)

RYDEVIK, U., L. Nord & F. Ingman. Automatic lactate determination by flow injection analysis. Int. J. Sports Med. 3:125–129, 1982.

SHEPHERD, J.T., C.G. Blomqvist, A.R. Lind, J.H. Mitchell & B. Saltin. Static (isometric) exercise. Retrospection and intraspection. Circ. Res. 48(Suppl. 1): 179–188, 1981.

Blood Lactate Threshold in Trained Ischemic Heart Disease Patients

E.F. Coyle, W.H. Martin, A.A. Ehsani, J.M. Hagberg
and J.O. Holloszy
Washington University School of Medicine,
St. Louis, Missouri, U.S.A.

The impetus for this study came from our clinical impression that some of the patients who had been training intensively in our coronary rehabilitation program for more than 1 year, had a remarkably high capacity for endurance exercise. While it is well documented that patients with ischemic heart disease can adapt to endurance exercise-training with large increases in $\dot{V}O_2$max and exercise capacity (Clausen, 1976; Mitchell, 1975), our patients were unusual in that they could maintain running speeds that seemed disproportionately fast relative to their $\dot{V}O_2$max. This paper describes a physiological evaluation of these patients. The results provide insights regarding the relative importance of blood lactate threshold and of $\dot{V}O_2$max as indicators of performance ability in endurance activities such as distance running.

Methods

Subjects and Training

Six patients who had previously suffered a myocardial infarction, participated in a 12 mo long exercise program (Ehsani et al., 1981). After completing the program these patients continued to train intensely and had been running 6 to 12 km/day, approximately 5 days/wk (average 37 km/wk) for the 6 mo prior to this study. The trained patients were compared to six healthy men (trained normals), who were individually matched to the trained patients with respect to age, distance run per wk and running pace during their training sessions. In addition, six untrained non-symptom limited patients with ischemic heart disease were studied to distinguish between the physiological effects of ischemic heart disease per se as opposed to the adaptation to training in patients with ischemic heart disease.

Measurement of Maximal O_2 Uptake

$\dot{V}O_2$max was measured using a continuous treadmill exercise test which resulted in a clear leveling off of $\dot{V}O_2$ in all subjects. Expired gases were collected in meteorological balloons and analyzed for O_2 and CO_2 using a mass spectrometer; volumes were measured in a Tissot spirometer.

Blood Lactate Threshold (LT)

Each subject's LT was determined from his blood lactate response to a series of 10 min long treadmill runs on the level at different speeds requiring between 70% and 100% of $\dot{V}O_2$max in the trained patients and trained normals and between 50% and 90% of $\dot{V}O_2$max in the untrained patients. Each subject was tested on three occasions over a 2 wk period, and performed two or three 10 min long exercise bouts of successively greater intensity during each laboratory visit. Blood samples were obtained before and immediately after exercise via a teflon catheter placed in an antecubital vein and assayed for lactate concentration.

The LT's reported in this paper represent the $\dot{V}O_2$ at which lactate increased 1 mM above the baseline lactates. This criterion was selected as opposed to previously reported methods which employ visual inspection of a graph of the relationship between blood lactate and $\dot{V}O_2$ or use linear regression to predict the breakpoint of this curvilinear relationship, because a 1 mM increase can be determined objectively in all subjects and in a standardized manner. Ventilation ($\dot{V}_E$), heart rate (HR), and $\dot{V}O_2$ were averaged during the 7th, 8th, 9th and 10th min of exercise. Running performance was evaluated during an 8 km road race.

Results

Trained Subjects

A close matching of the trained normals and trained patients was accomplished as evidenced by respective mean ($\pm$ SE) ages of 55.2 $\pm$ 3.5 yr vs 55.6 $\pm$ 2.7 yr, a similar distance run during training (41 $\pm$ 4 km/wk vs 37 $\pm$ 2 km/wk) and an identical training pace (173 $\pm$ 8 m/min vs 173 $\pm$ 9 m/min). Running efficiency was essentially the same in the two groups since both groups' $\dot{V}O_2$ averaged 28 ml/kg/min while running at 145 m/min on the level. Thus comparisons of the running performance of the two groups should not be confounded by differences in running skill.

Responses to Maximal Exercise

Despite a similar training stimulus, the trained patients had an 18% lower ($p < 0.02$) average $\dot{V}O_2$max than the trained normals (Table 1). Evidence that true $\dot{V}O_2$max was attained in all subjects includes 1) a leveling off of $\dot{V}O_2$ with increasing workrate, 2) hyperventilation and 3) R max values in excess of 1.15. Both maximum HR and HR at $\dot{V}O_2$max were significantly lower in the trained patients than in the trained normals.

Running Performance

Despite the trained patients' 18% lower $\dot{V}O_2$max, performance in an 8 km race was very similar in the two groups, with three of the patients outperforming the healthy runners with whom they were matched (Table 1).

Table 1

Comparison of the Mean ($\pm$ SE) Physiological Responses during Running at $\dot{V}O_2$max and the Blood Lactate Threshold (LT) Along with Running Performance

	Trained Normals	Trained Patients	% diff
Maximal responses			
$\dot{V}O_2$max (ml/kg/min)	45.0 ± 2.2	37.0 ± 1.9	$-18\%*$
HR at $\dot{V}O_2$max (bt/min)	175 ± 2	153 ± 7	$-13\%†$
HR max (bt/min)	182 ± 2	166 ± 5	$-9\%*$
Lactate threshold responses			
$\dot{V}O_2$ at LT (ml/kg/min)	37.8 ± 2.1	37.0 ± 1.9	NS
% $\dot{V}O_2$max at LT	83.3 ± 1.7	99.8 ± 1.2	$+16\%††$
HR at LT (bt/min)	155 ± 3	153 ± 7	NS
$\dot{V}_E$ at LT (l/kg/min)	0.91 ± 0.06	0.93 ± 0.05	NS
Running speed at LT (m/min)	178 ± 10	176 ± 10	NS
Running performance			
8 km Race Pace (m/min)	189 ± 12	185 ± 10	NS

% diff = (Trained normals – Trained patients)/Trained normals. * denotes $p < 0.02$; † denotes $p < 0.01$; †† denotes $p < 0.001$; NS denotes not statistically significant with $p > 0.05$; Student's t-test.

Blood Lactate Threshold (LT)

The LT was not significantly different in the two trained groups, occurring at a similar treadmill running speed and similar $\dot{V}O_2$ (Table 1). The lactate threshold occurred at 84% of $\dot{V}O_2$max in the trained normal subjects and 100% of $\dot{V}O_2$max in the trained patients. As shown in Figure 1B, the increases in lactate concentration in the two groups were indistinguishable from each other when plotted in terms of absolute oxygen uptake. The lactate responses of the two groups appear different only when expressed relative to $\dot{V}O_2$max (Figure 1A). The HR and $\dot{V}_E$ at LT were not significantly different in the two groups.

Untrained Patients

The untrained patients had an average $\dot{V}O_2$max of 26 ± 2 ml/kg/min, which is similar to that of the trained patients prior to training. The LT occurred at $67 \pm 3\%$ of $\dot{V}O_2$max in the untrained patients, which is significantly ($p < 0.01$) lower than the % of $\dot{V}O_2$max at which LT occurred in either of the trained groups (Table 1). This indicates that LT at $\dot{V}O_2$max is not a consequence of ischemic heart disease patients in general, but appears to result from the adaptation to training in patients with ischemic heart disease.

Discussion

The trained patients' ability to maintain metabolic steady state and run for prolonged periods at a speed that elicits $\dot{V}O_2$max seems incredible in

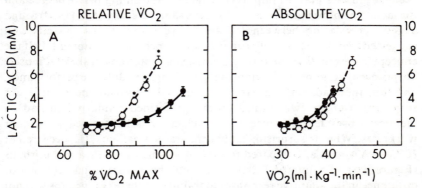

Figure 1—Relationship between blood lactate concentration following 10 min long treadmill runs at different speeds. A. percent of $\dot{V}O_2$max required by the exercise, and B. the oxygen uptake during the last 4 min of exercise, in the trained patients ●——●, and healthy subjects o——o. The highest two $\dot{V}O_2$ values shown for the trained patients are estimations of the energy cost of the exercise rather than measurements, as $\dot{V}O_2$max was exceeded at these two exercise intensities. *Normal subjects vs trained patients, $P < 0.05$.

context of the normal physiological response to exercise at $\dot{V}_{O_2}$max. These patients, however, are clearly not normal, and their adaptation to training and capacity for endurance exercise must be evaluated in the context of their heart disease. We have noted (Coyle et al., 1982 submitted), that the trained patients have a normal stroke volume and heart rate response to moderate exercise. However, on attainment of a higher heart rate, these individuals seem to develop some degree of left ventricular dysfunction which prevents them from increasing cardiac output. Maximal oxygen consumption (18% below trained normal) occurs at this point of apparent dysfunction. Despite having an impaired $\dot{V}_{O_2}$max, the trained patients performed 8 km at a similar pace as the trained normals, which clearly indicates that the capacity for prolonged aerobic performance is not entirely determined by $\dot{V}_{O_2}$max. $\dot{V}_{O_2}$max must, of course, set the upper limit for prolonged aerobic energy expenditure. However, the percentage of $\dot{V}_{O_2}$max that can be maintained during prolonged exercise or that intensity which elicits LT can vary widely as evidenced by values of 67% of $\dot{V}_{O_2}$max in untrained patients to the extreme case of 100% of $\dot{V}_{O_2}$max as observed in these trained patients.

The comparable 8 km performances of the two trained groups, although not significantly related to $\dot{V}_{O_2}$max, was highly related ($r = 0.93$) to LT running speed. As shown in Figure 1 and Table 1, the trained normals and trained patients demonstrated a virtually identical LT. As suggested by Farrell et al. (1979) and confirmed by Lafontaine et al. (1981) and Sjodin and Jacobs (1981), LT velocity is probably the best indicator of long distance running performance.

Holloszy (1973) has postulated that training induced adaptations in the skeletal muscles play an important role in determining performance and the lactate response to exercise. Along the same lines Ivy et al. (1980) found a good correlation between LT and muscle respiratory capacity and suggested that the latter is of primary importance in determining LT. If the concept is correct that the training induced adaptations in skeletal muscle are primarily responsible for improving prolonged endurance performance, and that these adaptations are reflected by LT measurements, it is not surprising that the two trained groups performed similarly and had an identical absolute LT, since their skeletal muscles trained at similar workrates. What is surprising is that the trained patients' LT occurred at 100% of $\dot{V}_{O_2}$max as compared to 84% of $\dot{V}_{O_2}$max in the trained normals (Figure 1A). To interpret this, it should be realized that the major difference in the adaptive responses to training of these two groups was not absolute LT (Figure 1B) or performance, which both appeared normal in the trained patients. Instead, the trained patients were unable to increase $\dot{V}_{O_2}$max to normal value for their level of training, because of ischemic heart disease. Therefore the trained patients had a remarkably high LT, when expressed relative to $\dot{V}O_2$max, simply because endurance training promoted disproportionate increases in LT relative to $\dot{V}O_2$max. This dissociation of increases in LT and $\dot{V}O_2$max provides strong evidence that $\dot{V}O_2$max and LT are determined, at least in part, by different factors. Studies showing that lactate starts to accumulate at a higher percent of $\dot{V}O_2$max in the trained than in the untrained state (Ekblom, 1969; MacDougall, 1977) provides additional evidence for such a dissociation. The present results support the concept that LT is more closely related with, and is a better predictor of, long distance running performance than $\dot{V}O_2$max.

References

CLAUSEN, J.P., Circulatory adjustments to dynamic exercise and effect of physical training in normal subjects and in patients with coronary artery disease. Prog. Cardiovas. Dis. 18:459–495, 1976.

EHSANI, A.A., G.W. Heath, J.M. Hagberg, B.E. Sobel & J.O. Holloszy. Effects of twelve months of intense exercise training on ischemic ST-segment depression in patients with coronary artery disease. Circulation 64:1116–1124, 1981.

EKBLOM, B., Effect of physical training on oxygen transport system in man. Acta Physiol. Scand. Suppl. 328:1–45, 1969.

FARRELL, P.A., J.H. Wilmore, E.F. Coyle, J.E. Billings & D.L. Costill. Plasma lactate accumulation and distance running performance. Med. Sci. Sports 11:338–344, 1979.

HOLLOSZY, J.O., Biochemical adaptation to exercise: aerobic metabolism. In J. Wilmore (Ed.), Exercise and Sport Sciences Reviews. New York: Academic Press, 1973.

IVY, J.L., R.T. Withers, P.J. Van Handle, D.H. Elger & D.L. Costill. Muscle respiratory capacity and fiber type as determinants of the lactate threshold. J. Appl. Physiol. 48:523–527, 1980.

LaFONTAINE, T.P., B.R. Londeree & W.K. Spath. The maximal steady state versus selected running events. Med. Sci. Sports Exercise 13:190–192, 1981.

MacDOUGALL, J.D., The anaerobic threshold: its significance for the endurance athlete. Can. J. Appl. Sports Sci. 2:137–140, 1977.

MITCHELL, J.H., Exercise training in the treatment of coronary heart disease. Adv. Intern. Med. 20:249–272, 1975.

SJODIN, B. & I. Jacobs. Onset of blood lactate accumulation and marathon running performance. Int. J. Sports Med. 2:23–26, 1981.

Peripheral Responses and Adaptation to Treadmill Exercise in Patients with Intermittent Claudication

U. Maass and K. Alexander
Medical School Hannover, Hannover,
Federal Republic of Germany

The effect of bicycle exercise on lactate and pyruvate as well as on blood gases has been of particular importance in patients with arterial occlusive disease. Until now, no data on the metabolism during treadmill exercise in patients with arterial occlusive disease were available.

Previously, blood specimens were taken more proximally from the femoral vein and not from the draining blood vessel of the exercising muscles. Consequently the analyzed blood samples contained a higher degree of blood from non-ischemic muscle. The methodological problem could be overcome by introducing a catheter into the popliteal vein. This regional catheterization technique permits the study of the spontaneous and reactive metabolic changes in the legs during and after treadmill exercise in patients with intermittent claudication. The present investigation was undertaken to study the kinetics of lactate, pyruvate and blood gases during and after treadmill exercise. For the purpose of comparison, the effects of treadmill exercise on metabolism were simultaneously investigated in the occluded and non occluded leg of the same patient.

Methods

Twelve men suffering from intermittent claudication participated in this study (ages 41 to 63 yr). The unilateral arterial occlusive disease was demonstrated by angiography. The occlusion was located within the femoral artery. All probands have been informed about the nature, purpose and possible risks involved in the study. The measurements were carried out after 12 h of bed rest. Catheters were introduced percutaneously into the popliteal veins and radial artery.

The exercise consisted of walking (3.7 km/h) on a 7% uphill treadmill. The maximal duration of exercise was 10 min. Blood was sampled 10 min before exercise, 4, 6 and 10 min after starting exercise and 5 min after exercise. Lactate and pyruvate concentrations were determined in duplicate by using an enzymatic method (Biochemical Test Combination, Boehringer, Mannheim, Germany). P_{O_2}, P_{CO_2} and pH were analyzed with radiometer equipment. Oxygen saturation, oxygen extraction and lactate-pyruvate ratio (L/P) were calculated. The t-test for paired data was used.

Results

P_{O_2}, P_{CO_2}, S_{O_2} and pH in the arterial and popliteal venous blood before, during and after treadmill exercise are shown in Figure 1. During treadmill exercise there was a significant increase in arterial P_{O_2} and a significant decrease in pH. The change of arterial P_{CO_2} was not significant. After treadmill exercise the arterial P_{O_2} was significantly different from the resting value. There was no difference in popliteal venous P_{O_2}, P_{CO_2}, S_{O_2} and pH at rest in the occluded and the control legs. During exercise there was a significant decrease in popliteal venous P_{O_2} from 18.5 mmHg to 15.3 mmHg in the control leg and from 18.3 mmHg to 12.6 mmHg in the

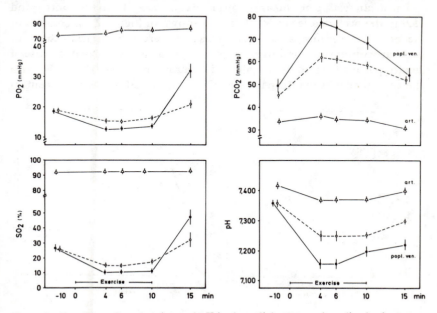

Figure 1—P_{O_2}, P_{CO_2}, O_2 saturation and pH in the radial artery and popliteal vein at rest, during treadmill exercise and after exercise in patients with occluded (•) and non occluded legs (o). Values are means ± SD.

occluded leg, respectively. We could also demonstrate a significant difference in the popliteal venous Pco_2, So_2 and pH between the occluded and the control leg. After exercise the increase of the popliteal venous Po_2 was higher in the occluded than in the control leg. No significant correlation existed between the popliteal venous Pco_2, pH and work load in the control legs, but there was a significant positive correlation between the work load and popliteal venous Pco_2 (r = 0.72, p < 0.05) and a negative correlation between the work load and popliteal venous pH (r = -0.76, p < 0.05) in the occluded legs. During exercise the oxygen extraction rose in the mean from 14.5 ml/100 ml to 16.2 ml/100 ml in the control legs and from 14.9 ml/100 ml to 18.0 ml/100 ml in the occluded legs, respectively (p < 0.001).

The arterial concentration of lactate rose to a maximum during exercise from 0.61 mmol/1 to 2.55 mmol/1 and then declined. Five min after the end of exercise the arterial concentration of lactate was still above normal. The arterial concentration of pyruvate also rose during exercise, but its maximal value occurred usually within the first 5 min after exercise. During exercise the mean arteriovenous differences of lactate increased considerably more in the occluded than in the control leg. In all subjects the arterial L/P ratio exhibited a considerable increase during exercise from 15.7 to 28.8.

The mean resting popliteal venous L/P ratio was 14.3 in the control and 15.9 in the occluded leg (p < 0.05). As shown in Figure 2 the changes of the popliteal venous lactate concentration were greater than those of pyruvate concentrations during exercise. Consequently, a sharp increase of lactate-pyruvate ratio was observed. During recovery the L/P ratio showed a tendency to return to normal with increased pyruvate concentra-

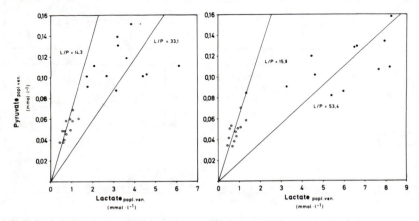

Figure 2—Popliteal venous concentration of lactate and pyruvate in relation to lactate-pyruvate ratio (L/P) at rest and during treadmill exercise in the non occluded legs (left figure) and in the occluded legs (right figure).

tions and less increased lactate concentrations in the control leg (Figure 3). However, the decrease of the popliteal venous L/P ratio was lower in the occluded than in the control legs.

Figure 4 shows the lactate-oxygen ratio before, during and after treadmill exercise. During exercise, there is an increasing production of lactate per unit of oxygen in the occluded leg. There was a significant positive correlation between the popliteal venous lactate-pyruvate ratio and the lactate-oxygen ratio in the control leg (r = 0.88) and in the occluded leg (r = 0.83) respectively.

Discussion

Doll et al. (1968) and Pirnay et al. (1972) found that the P_{O_2} in the femoral vein during maximal work never dropped to values near the calculated critical O_2 pressure. According to Stainsby (1964) the critical venous oxygen pressure for the resting skeletal muscle of the dogs is approximately 25 Torr and for the active muscle, 10 Torr. As our experiments have shown, the critical oxygen pressure was not reached in the control legs. However, during treadmill exercise the popliteal venous P_{O_2} dropped in some occluded legs to values near the calculated critical O_2 pressure. The unknown value of importance is the P_{O_2} in the region of the actively metabolizing sites in the cells.

A decrease of the popliteal venous blood P_{O_2} to values as low as 6.4 mmHg caused a high oxygen extraction in the occluded legs. During exercise, when the raised oxygen demand cannot be compensated by an

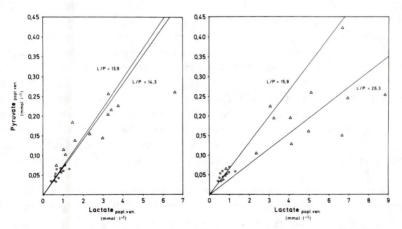

Figure 3—Popliteal venous concentration of lactate and pyruvate in relation to lactate-pyruvate ratio (L/P) at rest and after treadmill exercise in the non occluded legs (left figure) and in the occluded legs (right figure).

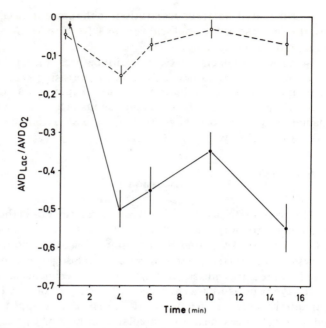

Figure 4—Changes in the lactate-oxygen ratio during and after treadmill exercise in the non occluded legs (o) and in the occluded legs (●).
AVD_{Lac} : arteriovenous lactate difference (mmol/ l)
AVD_{O2} : arteriovenous O_2 difference (mmol/ l)
Values are means ± SD.

increased regional blood flow the oxygen extraction is more increased in the occluded leg than in the control leg. The complete peripheral oxygen extraction observed towards the end of exercise in some patients reflects a severe metabolic acidosis in the blood, indicated by a lowering of the blood pH to low values (7.1) and high lactate-pyruvate ratio (74.0).

During exercise the decrease in the pH of the popliteal venous blood exceeds that of the arterial pH. On one hand, the changes in the arterial pH are mainly influenced by the enhanced concentration of lactate during and after treadmill exercise, since the arterial P_{CO2} shows only minor fluctuations. In the popliteal venous blood, on the other hand, the pH is influenced not only by the changes of the lactate level, but also by the remarkable alterations of P_{CO2}.

The increasingly negative arteriovenous difference in the lactate-pyruvate ratio which occurred in the occluded leg during exercise is consistent with the existence of local hypoxia and a low oxidation-reduction potential. The greatly diminished oxygen saturation of the popliteal venous blood is further evidence of this. The lactate-oxygen ratio indicates that there is an increase in the proportion of lactate produced to oxygen consumed during exercise in the occluded leg.

References

DOLL, E., J. Keul & C. Maiwald. Oxygen tension and acid-base equilibria in venous blood of working muscle. Am. J. Physiol. 215:23–29, 1968.

PIRNAY, F., M. Lamy, J. Dujardin, R. Deroanne & J.M. Petit. Analysis of femoral venous blood during maximum muscular exercise. J. Appl. Physiol. 33:289–292, 1972.

STAINSBY, W. & A.B. Otis. Blood flow, blood oxygen tension, oxygen uptake and oxygen transport in skeletal muscle. Am. J. Physiol. 206:858–866, 1964.

The Effect of Long-term Physical Training on the Relationship of Muscle Morphology to Metabolic State and Insulin Sensitivity in Normal and Hyperglycemic Obese and Diabetic Subjects

K. Mandroukas, M. Krotkiewski, P. Björntorp, G. Holm,
H. Lithell, U. Smith, P. Lönnroth and G. Strömblad
University of Göteborg, Göteborg, Sweden

Obese patients show high prevalence of glucose intolerance. Insulin sensitivity, a basic metabolic disturbance in obese patients, is improved after physical training (Björntorp, et al., 1970). The aim of the present study was to evaluate the effects of physical training in obesity with or without diabetes mellitus, and to evaluate its therapeutic value in both conditions.

Material and Methods

Fourteen middle-aged (38 ± 12 yr, mean $\pm$ SD) obese women (body weight 98.7 ± 3.9 kg, mean $\pm$ SD) and 12 middle-aged (40 ± 6 yr, mean $\pm$ SD) patients with diabetes mellitus type II (body weight 91.2 ± 17.2 kg, mean $\pm$ SD) as well as 10 patients (40.3 ± 8.3 yr) with diabetes mellitus type I and insulin treatment (with good metabolic control) were included in the study. Diabetics type II were treated with sulfonyl-urea tables. The weight stability in all groups, and the metabolic control of diabetics were ensured during the 2 months preceding the study. All measurements have been performed after 3 days of metabolic-wardlike conditions with standard diet (caloric requirements being calculated according to ideal body weight). All patients had sedentary occupations, and before training they were instructed to maintain their usual food intake and drinking habits. In all patients muscle biopsies were taken superficially from the middle portion of the right vastus lateralis muscle by surgical technique before and after training. Specimens were then taken for determination of muscle enzymes

(Aniansson et al., 1980; Lithell et al., 1981), glycogen (Aniansson et al., 1980), muscle fibers (Brooke et al., 1970) and capillaries (Andersen, 1975). All patients were tested on a Cybex II dynamometer for determination of muscle strength and dynamic and isometric endurance, and with an exercise test for determination of heart rate, blood pressure and oxygen uptake on submaximal and maximal work loads. The groups were trained for 3 mo three times/wk. The duration of the training program was of the interval type and lasted for 50 min with three heavy intervals each session. The intensity of the training was followed in each patient by telemetry and occasionally by oxygen uptake measurements with a portable oxygen flow meter. The euglycemic insulin clamp with different rates of infusions of insulin was performed before and after training. In surgical biopsy specimens from the hypogastric region of adipose tissue, lipoprotein lipase activity and insulin binding, as well as metabolic responsiveness of insulin in glucose metabolic pathways (Cigolini & Smith, 1979) were measured.

Results

Neither obese patients nor diabetics changed their body weight and body composition after training. Maximal oxygen uptake increased and pulse rate and blood pressure (on the submaximal work loads) significantly decreased in all groups. In comparison with women of normal weight, obese patients showed a significantly higher percent of type II$_B$ fibers. Physical training resulted in a decrease of the relative percent of this type of muscle fiber and a concomitant increase of type II$_A$ fibers. The number of capillaries around different fibers increased after training (mostly around the more oxidative type of muscle fibers, type I and type II$_A$). Diabetics showed a significantly higher percent of fast-twitch (type II) muscle fiber in comparison with other groups. A significant increase of enzyme activity has been observed in both obese and diabetic patients in comparison with the pre-physical training time. The increase (observed in all groups) concerned mostly oxidative enzymes (citrate-synthase and hydroxy-acyl-(CoA)-dehydrogenase. Glucose uptake, as judged from the euglycemic insulin clamp, was primarily lowest in the obese hyperglycemic patients, and increased significantly with physical training both in obese and diabetic patients. At low insulin concentrations the increase appeared to be highest in obese hyperglycemic patients where it was more than doubled. At high insulin concentrations, however, the highest increase was observed in patients with diabetes type II.

Insulin clearance increased significantly both in patients with diabetes type II, and at high insulin levels in obese hyperglycemic and normogly-cemic patients.

Lipoprotein lipase activity increased in adipose but not in muscle tissue in obese patients with normal glucose tolerance. In patients with diabetes type II, physical training resulted in an increase of lipoprotein lipase activity in both muscle and adipose tissue. In obese patients the number of capillaries around muscle fibers type I and type II$_A$ was inversely correlated with fasting insulin level ($r = 0.80$ and 0.62, respectively). The increase in the number of capillaries was significantly correlated with the decrease of fasting insulin and glucose after training. Diabetics type II, but not obese subjects, also showed a significant increase of the glycogen concentration in the vastus lateralis muscle. An increase of activity of hexokinase and trios-phosphate dehydrogenase was also found in the vastus lateralis muscle in diabetics. Diabetics, but not obese patients, showed slight improvement of peroral and intravenous glucose tolerance. Glucose excretion in urine decreased in diabetics during training but then increased again after termination of the training program.

Insulin binding to adipocytes did not change significantly, neither in obese nor in patients with diabetes type I and type II.

Discussion and Conclusion

These results seem to show that in obese hyperglycemic subjects, with a prevalent decrease of insulin sensitivity, the insulin sensitivity (measured with the euglycemic clamp technique) is increased by physical training. This was followed by adaptations in muscle fiber composition and enzyme activities, while insulin binding in adipose tissue was not changed. Patients with diabetes mellitus type I showed, at least partly, another result in these variables. These findings do not as yet allow a conclusion as far as the efficacy of physical training in improvement of diabetes mellitus, but rather points to a complex situation where probably physical training has a varying effect both in terms of the category of diabetic or obese patients tested, and in terms of adaptations in different tissues. Clearly, more studies are needed to clarify this important question and to sort out potentially beneficial effects of physical training for patients with diabetes mellitus.

References

ANDERSEN, P., Capillary density in skeletal muscle of man. Acta Physiol. Scand. 95:203–205, 1975.

ANIANSSON, A., G. Grimby, M. Hedberg & M. Krotkiewski. Muscle morphology, enzyme activity and muscle strength in elderly men and women (III page 1p19).

In A. Aniansson, Muscle Function in Old Age with Special Reference to Muscle Morphology, Effect of Training and Capacity in Activities of Daily Living. Thesis, Göteborg, 1980.

BROOKE, M.H. & K.K. Kaiser. The "myosin ATPase" systems: The nature of their pH lability and sulfhydryl dependence. J. Histochem. Cytochem. 18:670–672, 1970.

CIGOLONI, M. & U. Smith. Human adipose tissue in culture VIII. Studies on the insulin antagonistic effect of glucocorticoids. Metabolism 28:502–509, 1979.

LITHELL, H., M. Cedermark, J. Fröberh, P. Tesch & J. Karlsson. Increase of lipoprotein lipase activity in skeletal muscle during heavy exercise. Relation to epinephrine excretion. Metabolism 30:1130–1134, 1981.

Fatigue

Fatigue

Fatigue and Metabolic Patterns of Overloaded Fast-twitch Rodent Skeletal Muscle Contracting in Situ

K.M. Baldwin, S.L. Hillman and V. Valdez
University of California at Irvine, Irvine, California, U.S.A.

Slow-twitch skeletal muscle possesses a higher resistance to fatigue than fast-twitch skeletal muscle while performing a sustained isometric contraction in which the pathways for ATP synthesis are inhibited (Goldspink et al., 1970). This greater fatigue resistance has been attributed primarily to a slower rate of cross bridge cycling mediated by the low myosin ATPase isoenzyme present in slow-twitch skeletal muscle. The above conclusion assumes that the process of calcium cycling (sequestering and release) by the sarcoplasmic reticulum (SR), which is also energy dependent, is not a major influence during a sustained isometric contraction. Since SR functional capacity is also lower in slow-twitch as compared to fast-twitch skeletal muscle, the greater efficiency of the former would still be expected even if SR function were contributing to the energy cost of maintaining an isometric contraction.

Recently, we have observed that compensatory overloaded fast-twitch plantaris (OP) muscle contains a significantly greater amount of slow-myosin as compared to normal plantaris (NP) muscle (Baldwin et al., 1982). However, the capacity of isolated SR to sequester Ca^{++} (Baldwin et al., 1982) and the rate of tension development and relaxation were similar between NP and OP muscle (Roy et al., 1982). These findings led us to hypothesize that if myosin ATPase is primarily responsible for energy turnover during repeated isometric contractions (in which normal excitation-contraction-relaxation processes are operating), then OP muscle should have slower rates of fatigue and of glycogen and phosphagen depletion during repeated contractions performed with either an intact or occluded blood supply. Consequently, a series of experiments were conducted to test this possibility.

Methods

Animal Care and Experimental Design

Young female rats weighing approximately 150 g were obtained from Simonsen Laboratories and maintained in light and temperature controlled quarters. They were fed food and water ad libitum. The rats were randomly assigned to either a normal plantaris (NP) or overload plantaris (OP) group. Surgical overload was accomplished by bilaterally removing the gastrocnemius and soleus muscles (Baldwin et al., 1982). Following surgery, the rats were housed in groups of five and the muscles were studied approximately 12 wk later for fatigue and metabolic properties.

Stimulation Protocol

The rats were anesthetized with sodium pentobarbital (5 mg/100 g i.p.). The hindlimb was surgically isolated to expose the plantaris muscle and sciatic nerve on both sides. The left leg was pinned in a brace to anchor the lower portion at the patella, tibia, and ankle. The distal tendon of the plantaris muscle was cut and secured via wire to a Statham force transducer. The muscle was stimulated at L_O with trains (60 Hz, 150 ms duration, 60/min). Two stimulation protocols were used. The first (n = 7) involved contractions for 5 minutes with the blood supply intact. In the second series (n = 7), the blood supply was occluded (10 min prior to stimulation) and the muscle was stimulated as described above for 2 min. At the termination of each experiment, the tested muscle was clamp frozen with precooled aluminum tongs. Also, the contralateral control (resting) muscle was frozen, and the muscles were stored at $-80°$ C prior to biochemical analysis. Contractile force production was recorded on a Beckman R411 dynograph. All contractile data are expressed as a percent of the initial force generated.

Biochemical Analyses

The muscles were divided into three portions and weighed in the frozen state. One portion was extracted with 4.2% perchloric acid as described previously (Baldwin et al., 1977). A second portion was digested in boiling 30% KOH and processed for glycogen determination by the anthrone method (Hassid & Abraham, 1957). A third portion was processed for protein content (Gornall et al., 1949). Lactate (Hohorst, 1965), ATP (Lamprecht et al., 1974) and phosphocreatine (Ennor & Rosenberg, 1952). Tissue metabolites were expressed as μmol g^{-1} wet weight as the water content and protein concentration of the resting and stimulated muscles were not statistically significantly different.

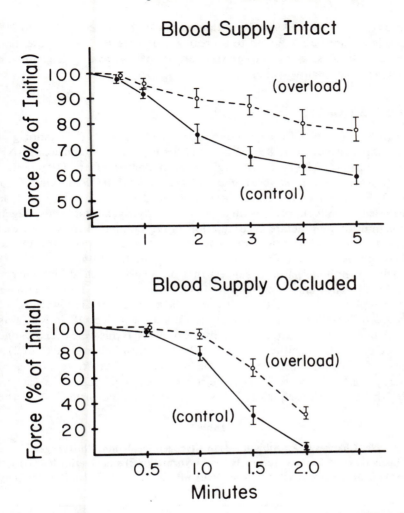

Figure 1—Relative force output of normal control and of overloaded plantaris muscles stimulated with trains at a rate of 60/min with either the blood supply intact or occluded. Each point represents an N of 7 muscles.

Results and Discussion

Fatigue Patterns

As shown in Figure 1, the OP muscle maintained approximately 20% greater relative force output than the NP muscle. This response was independent of blood supply. These data suggest that the fatigue resistance of the OP cannot be attributed to oxidative metabolic processes dependent on oxygen delivery. Also, the greater response of OP compared to NP muscle cannot be attributed to a greater anaerobic metabolic potential, because OP muscles have approximately 35 to 45% lower glycogenolytic enzyme levels (Baldwin et al., 1982).

Muscle Metabolites

Resting levels of ATP (NP 4.5 ± 0.56 vs OP 4.30 ± 0.45 μmol g^{-1}) phosphocreatine (NP 13.8 ± 1.5 vs OP 11.5 ± 0.9 μmol g^{-1}) and lactate (NP 1.61 ± 0.28 vs OP 1.15 ± 0.3 μmol g^{-1}) were similar for NP and OP muscles. Resting glycogen levels were higher in NP compared to OP muscle (28 ± 1 vs 23 ± 2 μmol g^{-1}). Contractions performed with either an intact or occluded blood supply resulted in lower glycogen and phosphagen consumption and lactate accumulation in OP compared to NP muscles (Table 1). Phosphocreatine breakdown accounted for over 80% of the phosphagen consumed for each muscle during contraction.

These metabolic patterns suggest that both substrate and high energy phosphate requirements were lower in OP compared to NP muscle under these experimental conditions. It cannot be argued that the lower rate of glycogen depletion and of lactate accumulation in the OP muscle was attributed to its lower capacity for glycogenolysis. If this were the case, one would expect to see either 1) a greater loss of phosphagen content or 2) a greater loss of force production in the OP muscles if substrate (glycogen) energy availability were limiting the contractile system. Clearly, such was not the case. Consequently, these findings are consistent with the

Table 1

Net Changes (Rest-Stimulated) in Glycogen Depletion, Phosphagen Depletion and Lactate Accumulation in Normal (NP) and Overloaded (OP) Plantaris Muscle Contracting Isometrically with Intact or Occluded Blood Supply

	(5 min stimulation) Blood supply intact		(2 min stimulation) Blood supply occluded	
	NP	OP	NP	OP
[a]Glycogen (7)	21 ± 2	12 ± 1*	20 ± 1	11 ± 2*
Lactate (7)	22 ± 3	10 ± 2*	26 ± 2	19 ± 2*
Phosphagen (7)	11.3 ± 0.78	9.3 ± 1.0	15.6 ± 1.4	11.4 ± 0.75*

Values are mean ± SEM; Number of measurements given in parentheses *p < .05 NP vs OP
a = μmol g^{-1} wet wt

hypothesis that muscles containing a greater content of the slow-myosin isoenzyme are more efficient in performing isometric contractions than muscles consisting predominantly of the fast-myosin isoenzyme. Interestingly, our previous findings also show that the isometric contractile response (time to peak tension and one-half relaxation time) of OP muscle remains similar to NP muscle (Roy et al., 1982). This was also the case for SR calcium uptake capacity (Baldwin et al., 1982). Consequently, these findings further suggest that the myosin enzymatic property may be the dominant factor in regulating isometric contractile efficiency in a given muscle.

References

BALDWIN, K.M., P.J. Campbell & D.A. Cooke. Glycogen, lactate and alanine changes in muscle fiber types during graded exercise. J. Appl. Physiol. 43:288–291, 1977.

BALDWIN, K.M., V. Valdez, R.E. Herrick, A.M. MacIntosh & R.R. Roy. Biochemical properties of overloaded skeletal muscle. J. Appl. Physiol. 52: 467–472, 1982.

ENNOR, A.H. & H. Rosenberg. The determination and distribution of phosphocreatine in animal tissues. Biochem. J. 51:606–613, 1952.

GOLDSPINK, G., R.E. Larson & R.E. Davies. The immediate energy supply and the cost maintenance of isometric tension for different muscles in the hamster. Z. Vergl. Physiologie 60:389–397, 1970.

GORNALL, A.G., C.J. Bardawill & M.M. David. Determination of serum proteins by means of the biuret reaction. J. Biol. Chem. 177:751–766, 1949.

HASSID, W.Z. & S. Abraham. Chemical procedures for analysis of polysaccharides. Methods Enzymol. 3:34, 1957.

HOHORST, H.J., Determination of L-lactate with LDH and DPN. In H.U. Bergmeyer (Ed.), Methods of Enzymatic Analysis, pp. 266–270. New York: Academic Press, 1965.

LAMPRECHT, W., P. Stein, P. Heinz & H. Weisser. Creatine phosphate. In H.U. Bergmeyer (Ed.), Methods of Enzymatic Analysis, pp. 1771–1781. New York: Academic Press, 1974.

ROY, R.R., I.D. Meadows, K.M. Baldwin & V.R. Edgerton. Functional significance of compensatory overloaded rat fast muscle. J. Appl. Physiol. 52:473–478, 1982.

Does a Reduction in Motor Drive Necessarily Result in Force Loss during Fatigue?

B. Bigland-Ritchie, R. Johansson and J.J. Woods
John B. Pierce Foundation and Quinnipiac College,
New Haven, Connecticut, U.S.A.

The causes of force loss during fatigue of human voluntary contractions have long been the subject of controversy. While it is largely due to changes in muscle biochemistry, both reduced motor drive ("central fatigue") and/or failure of neuromuscular transmission have also been implicated (Asmussen, 1980; Stephens & Taylor, 1972). Merton (1954) concluded that neuromuscular block was not involved, since the evoked muscle mass action potential (M wave) did not decline during sustained maximal voluntary contractions (MVC), despite near total loss of force. We have confirmed these observations (Bigland-Ritchie et al., 1982), and suggest explanations for the contrary conclusions of Stephens and Taylor (1972).

A sustained MVC is accompanied by a decline in muscle electrical activity (Bigland-Ritchie et al., 1979; Stephens & Taylor, 1972). In the absence of neuromuscular block this may be attributed to reduced motor drive from the central nervous system (CNS). However, if this fails to maintain full muscle activation much of the lost force should be restored by maximal tetanic motor nerve stimulation. No force increase has generally been observed (Bigland-Ritchie et al., 1978; Merton, 1954). This paradox may be resolved if the simultaneous slowing of muscle contractile speed equals or exceeds any decrease in motor neural firing rates.

In the current study, changes in the muscle contractile and electrical properties were compared during fatigue of sustained MVC's of the human adductor pollicis muscle. Measurements were made throughout each contraction of: 1) central versus peripheral fatigue; 2) changes in contractile speed (relaxation rate); 3) surface EMG; and 4) the discharge rates of single motor units.

Methods

Repeated experiments were performed on each of six subjects. The methods for recording force and surface EMG from the human adductor pollicis muscle, and for stimulating its motor nerve, have been described previously (Bigland-Ritchie et al., 1979 & 1982).

Relaxation Rate

The force signal was differentiated and two rate constants calculated: 1) the maximum rate of relaxation (MRR) expressed as the ratio of the differentiator output (df/dt), divided by the force at the onset of relaxation (Wiles et al., 1979); and 2) the inverse of the half relaxation time ($t_{1/2}^{-1}$), measured during the exponential phase of force decay, following its maximum rate.

Single Unit Firing Rates

Single unit potentials were recorded using tungsten microelectrodes. Once inserted, they were manipulated until regular spikes with a high signal-to-noise ratio were obtained. Brief trains of impulses were recorded and histograms constructed of the range of discharge rates of populations of motor units recorded during different time intervals. Changes in their mean rates could then be examined.

Protocol

Measurements of all parameters were made during a series of 10 s control MVC's, each separated by 3 min rest; and again while the maximum force was sustained for up to 90 s. When the sustained MVC was briefly interrupted for various test procedures the blood supply was occluded throughout by a pressure cuff to prevent any recovery during these times. All signals were recorded on FM tape.

Results

Central Fatigue

With practice subjects learned to generate voluntary force only in the direction produced by ulnar nerve stimulation. A good match was then obtained between the force from both maximal voluntary and stimulated (50 Hz) contractions. In an MVC sustained for 60 s the total force loss for

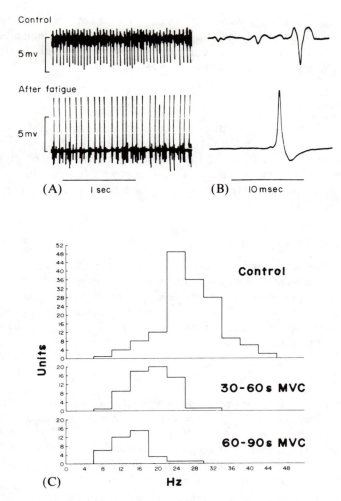

Figure 1—A = Trains of potentials from two single motor units recorded at the contraction onset and after 60 s MVC. B = Single spikes from each train. C = Histograms of motor unit firing rates recorded at different times during fatigue.

different subjects ranged from 30 to 50%. When these contractions were periodically interrupted by 2 to 3 s of maximal tetanic nerve stimulation no significant recovery of force was seen. The voluntary and stimulated forces declined in parallel, whether or not the blood supply to the arm was occluded.

Changes in EMG

Surface EMG. As reported earlier (Bigland-Ritchie et al., 1979; Stephens & Taylor, 1972) the force loss during a 60s sustained MVC was accompanied by a 50–70% decline in the surface smoothed, rectified EMG. This was similar to the decline in the rate of spike counts recorded under the same experimental conditions from groups of motor units using intramuscular fine wire electrodes (Bigland & Lippold, 1979).

Single Unit Firing Rates. Using tungsten microelectrodes, we recorded clearly identifiable single motor unit potentials with a high signal/noise ratio during fully maximal contractions. Their amplitude, duration and shape generally showed them to be potentials from single fibers (Figures 1A & B). In brief, 10 s maximal contractions executed once every 3 min the average mean firing rate of each of 200 units was 28.2 ± 6.4 Hz. During prolonged maximal effort both force and firing rates declined. Between 30 and 60 s and 60 and 90 s after the onset of the contractions the rates were 18.8 ± 4.6 Hz (n = 65) and 14.3 ± 4.4 Hz (n = 38), respectively (Figure 1C). As in previous experiments (Bigland-Ritchie et al., 1982) periodic monitoring of the evoked M wave showed no sign of neuromuscular block.

Changes in Relaxation Rate

The relaxation rate constants MRR and $t_{1/2}^{-1}$ were calculated: first, before and after uninterrupted 60 s MVC's both in the presence and absence of occluded blood supply; and second, with blood supply occluded, during sustained MVCs interrupted every 10 s by brief periods of relaxation, each sometimes followed by short bursts of nerve stimulation.

Before fatigue, control relaxation rates for both voluntary and stimulated contractions were about 10.1 ± 0.65 ms and 14.2 ± 1.25 ms for MRR and $t_{1/2}^{-1}$, respectively. During 60 s fatigue MRR and $t_{1/2}^{-1}$ declined by about 70% (Figure 2).

Discussion

The discharge rates of single motor neurons have been accurately measured during fully maximal voluntary contractions. In most previous experiments this was not possible because of interference from other surrounding active units. In brief MVCs of unfatigued muscle the rates ranged from about 15 to 50 Hz, with a mean value of 28.2 Hz. These rates declined by about 50% over the 60 s of fatigue. During this period there was 30 to 50% loss of force, suggesting that this may result partly from reduced muscle activation by the CNS, rather than from failure of the muscle cellular

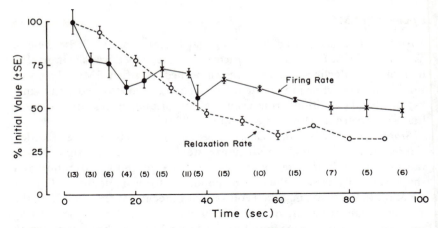

Figure 2—Changes in the mean rates of motor neuron discharge, and of muscle relaxation (average values for MRR and $t_{1/2}^{-1}$ combined) during sustained MVCs. The number of motor units recorded during each time bin is also shown.

contractile mechanisms. However, the force of a MVC was well matched by maximal nerve stimulation both before and throughout the fatigue process. Thus, voluntary effort can recruit all motor units, and excite them to discharge at rates sufficient to elicit full force at all times.

These two apparently conflicting results become compatible when changes of muscle contractile speed are considered. During a 60 s sustained MVC the slowing of contractile speed equals or exceeds the concomitant slowing of motor neuron discharge rates (Figure 2). Figure 3 illustrates the relation between muscle excitation frequency and the relative amount of force generated before and after fatigue. For the unfatigued adductor pollicis the minimum firing rate for full force production is probably 25 to 30 Hz. After fatigue and contractile slowing, this can occur at only 10 to 15 Hz. If the initial firing rates were maintained throughout fatigue these rates would become markedly supratetanic. The observed decline in neural discharge rates therefore results in no force loss that cannot be attributed to contractile failure. The reduction in neural firing rates has the functional advantage of minimizing any risk of neuromuscular block, and of optimizing motor control. Force regulation is most sensitive when the total range of motor neuron discharge rates is limited to the steepest parts of the force/frequency curve.

Thus the normal relationship between EMG and muscle force only applies for relatively brief contractions of unfatigued muscle. If changes in EMG are accompanied by a parallel slowing in the muscle contractile properties then no force loss need necessarily occur unless it is also accompanied by muscle fiber contractile failure.

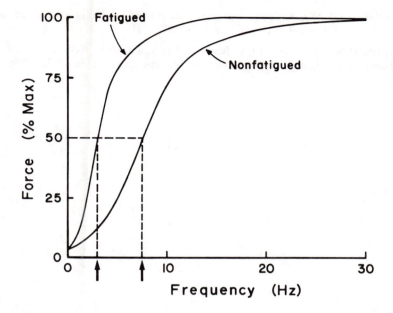

Figure 3—The percentage of the total available force generated by different excitation frequencies before and after contractile slowing.

References

ASMUSSEN, E., Muscle Fatigue. Med. Sci. Sports 11(4):313–321, 1979.

BIGLAND-RITCHIE, B., D.A. Jones, G.P. Hosking & R.H.T. Edwards. Central and peripheral fatigue in sustained maximum voluntary contractions of human quadriceps muscle. Clin. Sci. & Mol. Med. 54:609–614, 1978.

BIGLAND-RITCHIE, B., D.A. Jones & J.J. Woods. Excitation frequency and muscle fatigue: Electrical responses during human voluntary and stimulated contractions. Exp. Neurol. 64:414–427, 1979.

BIGLAND-RITCHIE, B., C.G. Kukulka, O.C.J. Lippold & J.J. Woods. The absence of neuromuscular transmission failure in sustained maximal voluntary contractions. J. Physiol. (London), 330:265–278, 1982.

BIGLAND-RITCHIE, B. & O.C.J. Lippold. Changes in muscle activation during prolonged maximal voluntary contractions. J. Physiol. (London) 292:14–15P, 1979.

MERTON, P.A., Voluntary strength and fatigue. J. Physiol. (London) 128: 553–564, 1954.

STEPHENS, J.A. & A. Taylor. Fatigue of maintained voluntary muscle contraction in man. J. Physiol. (London) 220:1–18, 1972.

WILES, C.M., A. Young, D.A. Jones & R.H.T. Edwards. Relaxation rate of constituent muscle fiber types in human quadriceps. Clin. Sci. 56:47–52, 1979.

Effects of Varied Dosages of Caffeine on Endurance Exercise to Fatigue

B.S. Cadarette, L. Levine, C.L. Berube, B.M. Posner and W.J. Evans

Boston University, Boston, Massachusetts, U.S.A., and U.S. Army Research Institute of Environmental Medicine, Natick, Massachusetts, U.S.A.

The purpose of this study was to investigate the effects of three dosages of caffeine on substrate levels and endurance performance during exercise. Previous studies (Costill et al., 1978; Ivy et al., 1979) indicated increased lipolysis with caffeine ingestion, and suggested that a glycogen sparing effect due to increased use of free fatty acids (FFA) may be the mechanism for increased performance during submaximal exercise. These studies were done with a single treatment of caffeine. The current study investigated the effects of three different caffeine dosages on levels of serum FFA, glycerol, glucose and lactate as well as time to fatigue in a submaximal endurance exercise.

Methods

Four men and four women were studied. After signing informed consent forms subjects were screened for level of aerobic fitness using a continuous treadmill $\dot{V}_{O_2max}$ test. Minimum requirements for acceptance were 50 ml/kg/min for males and 45 ml/kg/min for females. The $\dot{V}_{O_2max}$ was determined by a plateau in heart rate (HR), a plateau or decline in $\dot{V}_{O_2}$, and a respiratory exchange ratio (RER) of equal to or greater than 1.0.

Subjects maintained their normal exercise and dietary patterns during the study, but abstained from endurance exercise and caffeine products for 48 h before test runs. Four treadmill endurance runs to self determined fatigue at approximately 80% of $\dot{V}_{O_2max}$ were spaced at 1 wk intervals. For each test a pre-drink (pre-dr) venous blood sample was taken from the subject's antecubital vein to establish baseline values of FFA, glycerol, glucose, lactate and caffeine. A pre-exercise (pre-ex) sample was drawn 1 h after ingestion of the day's test dose, and a post exercise (post-ex) sample at completion of exercise. After the pre-dr sample subjects drank one of the four dosages. The placebo ($\emptyset$) consisted of 250 ml of saccharin-sweetened lemonade. Caffeine doses were 2.2 mg caffeine/kg body wt, light (L); 4.4 mg/kg, medium (M); and 8.8 mg/kg, heavy (H) dissolved in a drink otherwise like the placebo. Treatments were randomized and double blind. Expired gases were collected and heart rate was monitored throughout the run. Blood samples were centrifuged and the serum frozen for later analyses.

Time to fatigue, as well as the end point values of HR, percent $\dot{V}O_{2max}$, RER, and RPE were analyzed using a mixed factorial analysis of variance with each subject receiving all doses and the subjects divided into groups by sex. Analysis of the circulating metabolites was done similarly with the addition of repeated measures (pre-dr, pre-ex, post-ex) for each dose received. Significance was set at the 95% level of confidence.

Results

Time to fatigue was found to be significantly different between the placebo ($\emptyset$) and M dose of caffeine (53.4 < 73.4 min). No significant difference was found between the times to fatigue for the three caffeine doses, nor between $\emptyset$ and the L (67.8 min) and H (57.9 min) doses. Overall mean values taken at fatigue were 190 bt min^{-1} for heart rate, 83% for percent $\dot{V}O_{2max}$, .90 for RER and 18.5 for RPE. There were no significant differences between treatments.

Pre-ex serum caffeine levels increased significantly over the placebo (21.78 μM) levels with each of the caffeine treatments. Further, pre-ex serum caffeine increased significantly between each increasing caffeine treatment (L, 34.38 μM < M 48.75 μM < H, 74.84 μM). There was no significant change between pre-ex and post-ex serum caffeine with any of the treatments.

FFA showed significant increases after exercise regardless of dose. Neither males nor females showed significant increases from pre dr levels (males, .14 mM; females .12 mM) to pre-ex levels (males .25 mM; females .20 mM). The male post-ex values (.35 mM) increased enough to be significantly greater than the pre-dr value. However, the female post-ex value (.56 mM) proved to be significantly greater than both the pre-dr and pre-ex values (Figure 1).

Glycerol post-ex values were higher than pre-dr and pre-ex values which remained virtually the same regardless of dose. Glycerol post-ex $\emptyset$ (.248

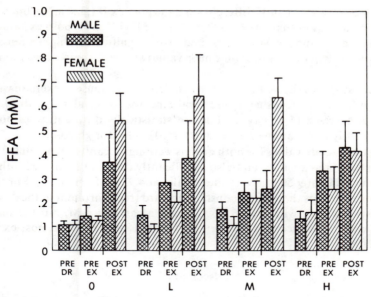

Figure 1—Mean (± SE) values of serum free fatty acids (FFA) concentration for males and females measured pre-drink, pre-exercise, and post-exercise during control (O) and light (L), medium (M) and heavy (H) caffeine trials.

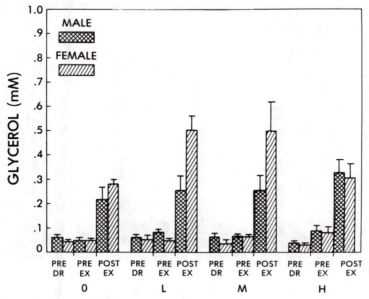

Figure 2—Mean (± SE) values of serum glycerol concentration for males and females measured pre-drink, pre-exercise and post-exercise during control (O), and light (L), medium (M) and heavy (H) caffeine trials.

mM) was not significantly different from the post-ex H (.320 mM), but was significantly lower than the L (.378 mM) and M (.377 mM) post-ex values. The higher L and M values resulted from significantly higher female glycerol levels at these doses while male values did not increase significantly (Figure 2).

Glucose values showed a significant increase in response to exercise with no significant changes due to either caffeine dose or gender. Pre dr (4.38 mM) and pre-ex (5.11 mM), while not statistically different, were both significantly less than the post-ex (7.18 mM) value for glucose.

Post-ex lactate values for both groups were significantly higher than the pre-dr and pre-ex values and also significantly different from each other with the males at 9.50 mM and the females at 4.93 mM (Figure 3). For the combined groups there was a constant upward trend throughout the doses in post-ex values with L at 6.81 mM and M at 7.48 mM, but the only significant difference was between post-ex ∅ (5.89 mM) and post-ex H (8.69 mM).

Discussion

The clear differences in serum caffeine levels with the varying doses of ingested caffeine did not appear to have a significant effect on performance

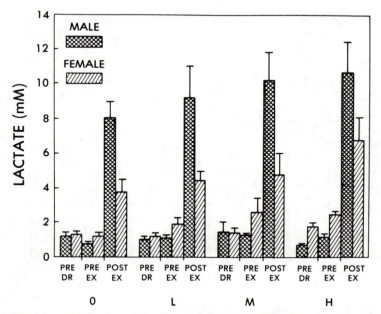

Figure 3—Mean (± SE) values of blood lactate concentration for males and females measured pre-drink, pre-exercise, and post-exercise during control (O), and light (L), medium (M) and heavy (H) caffeine trials.

parameters. While there were differences in post-exercise values of measured substrates between the placebo and caffeine trials there were no significant differences between the post-exercise values of the three doses. There was a significant difference in time to fatigue between the medium dose and the placebo. This is in agreement with Costill et al. (1978) who found increased exercise time to fatigue with caffeine, and Ivy et al. (1979) who found increased work production in a 2 h exercise bout with caffeine compared to placebo. However, in the current study this increase was almost entirely due to one subject who after receiving the medium dose ran 49 min longer than her second longest time.

The changes in the blood borne substrates FFA, glycerol, and glucose all appear to fall within reasonable expectations when compared to other studies on caffeine ingestion and exercise. FFA showed an increase with exercise but no difference in response between the placebo and any of the caffeine doses. This is in agreement with the studies of Costill et al. (1978) and Ivy et al. (1975). This would appear to indicate that the stress of exercise is a more important factor in FFA release than is serum caffeine level, at least during the time period of the runs in this study.

Glycerol values also followed a path similar to those in the Costill and Ivy studies cited above. On each run the post-ex values increased significantly over pre-ex values. Havel et al. (1964) have suggested that the ratio of glycerol to FFA may be indicative of FFA use by the cells with the ratio tending towards 1.00 as greater amounts of FFA are metabolized. While there were no significant changes in the FFA values seen post-ex, the ratio with glycerol levels would seem to indicate an increase in the use of FFA through β-oxidation.

Our study found no difference in the pre-ex blood glucose values regardless of the caffeine dosages, but significant increases post-exercise. This is substantiated by findings reported by Felig and Wahren (1975) of increases in blood glucose concentration during exercise of less than 90 min duration.

The lactate findings argue for possible increases in glycogenolysis relative to increasing serum caffeine levels. This is in apparent disagreement with earlier studies. However, the results should not be unexpected when considering the combination of caffeine with exercise. It is substantiated that caffeine increases the levels of circulating catecholamines (Levi, 1967). Additionally by working as an inhibitor of phosphodiesterase activity, caffeine allows the effect of the catecholamines, acting in the cells as cyclic-AMP, to continue for a longer than normal duration (Sutherland et al., 1968).

In adipose tissue triglyceride lipase activity is increased by cyclic-AMP activity while in muscles cyclic-AMP primarily increases glycogenolytic activity. While caffeine ingestion at rest yields increased lipolysis, no differences in FFA were seen between doses during exercise. It is possible that the increasing caffeine dosages had only a minor effect on adipose tissue due to an increased shunting of blood to the working muscles during exercise. On the other hand increased muscle blood flow would seem to potentiate the action of caffeine and the catecholamines in muscle tissue. Research by Ruderman et al. (1979) using mixed types I and II fiber muscle tissues indicates that increased use of Acetyl CoA as a substrate by exercising muscle reduces levels of active pyruvate dehydrogenase (PDH), therefore inhibiting oxidative use of pyruvate. However, it did not inhibit glucose uptake or metabolism in the muscle. If this theory is accepted it can be suggested that increasing levels of caffeine increase glycogenolysis within the muscle tissue which is metabolized to the level of pyruvate and then shunted to lactate as a result of increased Acetyl-CoA levels from β-oxidation inhibiting the active form of PDH.

This study indicates that a pharmacological dose of caffeine (200 to 300 mg) may increase the time of performance to fatigue in an endurance exercise when compared to a placebo. However, the mechanisms of this increase were not uncovered. Increased serum caffeine levels did not result in increased serum FFA, and if anything lactate results suggested a possible mechanism for an increase in glycogen utilization with increased serum caffeine levels.

References

COSTILL, D.L., G.P. Dalsky & W.J. Fink. Effects of caffeine ingestion on metabolism and exercise performance. Med. Sci. Sports 10(3):155–158, 1978.

FELIG, P. & J. Wahren. Fuel homeostatis in exercise. New England J. Med. 293:1078–1084, 1975.

HAVEL, R.J., L.A. Carlson, L.G. Ekelund & A. Holmgren. Turnover rate and oxidation of different free fatty acids in man during exercise. J. Appl. Physiol. 19(4):613–618, 1964.

IVY, J.L., D.L. Costill, W.J. Fink & R.W. Lower. Influence of caffeine and carbohydrate feedings on endurance performance. Med. Sci. Sports 11(1):6–11, 1979.

LEVI, L., The effect of coffee on the function of the sympathadrenomedullary system in man. Acta Medica Scand. 181(4):431–438, 1967.

RUDERMAN, N.B., M.N. Goodman, M. Berger & S. Hagg. Effect of starvation on muscle glucose metabolism: studies with the isolated perfused rat hindquarter. Fed. Proc. 36(2):171–176, 1977.

SUTHERLAND, E.W., G.A. Robinson & R.W. Butcher. Some aspects of the biological role of adenosine 3', 5' -monophosphate (cyclic AMP). Circulation 37:279–306, 1968.

Membrane Permeability Changes as a Fatigue Factor in Marathon Runners

H.M. Gunderson, J.A. Parliman, J.A. Parker and G. Bell

Northern Arizona University, Flagstaff, Arizona, U.S.A.

Our study of the biochemical causes of fatigue using volunteer marathon runners indicates that a change in the muscle cell membrane permeability is an important factor in the development of muscular fatigue. Although the popular marathon race of 42.2 km has many participants, its length causes muscular fatigue in nearly all runners. Costill et al. (1973) have suggested that selective depletion of muscle glycogen in slow-twitch fibers is the primary cause of muscular fatigue experienced by runners in marathon races. Other possible causes of fatigue have been reported in current biochemical literature (Fitts, 1977). These include lactic acid accumulation, hypoglycemia, dehydration, electrolyte losses and hyperthermia (Costill, 1974). This study was designed to determine if changes which occur in blood substituents can be related to fatigue in marathon runners. The findings of this investigation demonstrate a relationship between the permeability of the muscle cell membrane and fatigue in marathon runners.

Methods

Two dozen runners volunteered to take part in our study of biochemical causes of fatigue. To participate in the study, each volunteer signed an informed written consent. Venous blood samples were obtained from each volunteer within 1 h before the race and immediately after the race at 42.2 km. After allowing the collected blood samples to clot, they were spun down in a clinical centrifuge. Serum concentrations of 22 substances were assayed using a Technicon autoanalyzer. This report will discuss changes in five of these substances: lactate dehydrogenase, glutamate oxaloacetate transaminase, calcium ions, sodium ions and glucose. These were chosen because they relate to membrane permeability and energy supply. Significance of each change was appraised by an analysis of variance, in a randomized block design. The runners constitute the blocks and the single treatment considered is going from beginning to end of the race.

Table 1

Change in Concentration of Serum Constituents before and after a Marathon Run (N = 24)

Substance	Before	After	Change (significance)
lactate dehydrogenase (units/l)	202	365	81% (p < 0.001)
glutamate oxaloacetate transaminase (units/l)	22	41	86% (p < 0.001)
sodium, Na^{+1} (M)	0.139	0.143	2.9% (p < 0.001)
calcium, Ca^{+2} (mM)	2.50	2.65	6.0% (p < 0.001)
glucose (mM)	5.72	5.94	3.8% (p = 0.534)

Results

The concentration of the enzyme lactate dehydrogenase (LDH) increased in the blood serum in all the runners during the race (Table 1). Average concentrations in the samples at the start of the race were at the high end of the normal range. At the finish all runners had concentrations exceeding the normal range of 100 to 225 units/l. The smallest increase was from 198 to 253 units/l, or 28%, and the largest increase was from 166 to 467 units/l, or 181%.

Increases in glutamate oxaloacetate transaminase (GOT) parallel those of LDH. All volunteers registered an increase in GOT. Only six runners exceeded the normal serum values of 0 to 41 units/l.

The metal ions, sodium and calcium, registered slight increases during the race. Although the increase was small it was statistically significant (Table 1).

Glucose concentration in the blood averaged 5.72 mM prior to the start of the race and 5.94 mM at the finish. The small average increase was not statistically significant. Twelve runners had increased glucose from start to finish, and among these 12, six had increased more than 25%. Twelve also had decreased blood glucose concentration but of these 12, only two decreased more than 25%. These two runners had a glucose concentration below the normal values of 3.61 to 6.39 mM. Nine runners showed above normal glucose concentrations at the finish of the marathon.

Discussion

The results from our studies indicate that blood glucose concentration remained relatively constant and above the normal minimum value for greater than 90% of the volunteer marathon runners. Two runners had blood glucose concentrations of 3.39 mM, just below the normal value of 3.61 mM. Reports from several investigators (Bagby et al., 1978; Pirnay et al., 1977) suggest that muscle cells absorb large amounts of glucose from the blood. Pirnay et al. (1977) studied glucose utilization ingested during muscle exercise using [13-C] glucose. Following intake and with continued exercise the expired CO_2 became rapidly enriched in carbon-13. Bagby et al. (1978) reported a glycogen sparing effect when exercising rats were infused with glucose. These observations indicate that skeletal muscle transports large amounts of glucose out of the bloodstream. According to Wahren (1977) the dominant carbohydrate substrate is muscle glycogen during the initial phase of exercise. The utilization of blood glucose increases steadily as exercise continues. After 40 min of exercise, completely oxidized blood glucose can support 75 to 90% of the carbohydrate metabolism of the leg muscles. Muscle glycogen stores supply the remaining 10 to 25%. When muscle glycogen becomes depleted, a marked increase in the dependence by the muscle on blood glucose occurs. This dependence causes greater uptake of glucose from the blood and a sharp fall in its concentration in the blood. Clark and Conlee (1979) observed a marked decrease in blood glucose concentration in rats exercised to exhaustion. Exercise caused a virtual depletion (96 to 97%) of the liver glycogen.

Almost all our volunteers registered either an increase or a less than 25% decrease in blood glucose concentration. Depletion of liver or muscle glycogen will result in hypoglycemia (Segal & Brooks, 1979). Only two volunteers were hypoglycemic and the other 22 runners had normal serum glucose levels. Therefore, their glycogen reserves were not depleted and fatigue was not caused by a lack of carbohydrate substrate.

Two blood electrolytes, calcium and sodium, registered small but statistically significant increases from start to finish of the 42.2 km run. Both of these ions may increase because the cell membrane becomes more permeable. The increase in serum sodium concentration could be the result of increased activity of the Na^+ - K^+ pump which transports Na^+ ions from the inside to the outside of the cell. Increased activity of the Na^+ - K^+ pump places an increased demand on cellular ATP supply because of its use as the energy source for this transport system.

The two enzymes GOT and LDH were significantly elevated in every runner. Block et al. (1969) and others (Rose et al., 1970; Siegel et al., 1981) have studied the elevation of serum enzymes caused by exercise. These investigators have shown that the source of the additional enzyme is skeletal muscle or liver. Increase in creatine kinase (CK) isozyme MB and MM and LDH isozymes 3, 4, and especially 5 are evidence that the source of the serum enzyme increase is muscle or liver, not heart or kidney tissue. Direct evidence for tissue breakdown is not yet complete but these enzymes are used as biochemical indicators of tissue damage. Their dramatic increase in every volunteer provides indirect evidence that cell membrane damage had occurred. During the course of the marathon run a change occurs in the muscle or liver cell which allows these enzymes to leak out into the blood. The increase of CK indicates muscle tissue to be the probable source of these increased enzyme concentrations in the blood. Factors which influence their increased concentration include the number of cells damaged, type and degree of damage to the cells, and the permeability of the cell membrane. Magazanik et al. (1974) and others (Raven et al., 1970) have suggested that distance running may cause hyperpermeability of the cell membrane of skeletal muscle enabling a greater efflux of cytoplasmic substances. The change in permeability of the muscle cell membrane indicated by the increased serum concentration of GOT, LDH, and other enzymes indicates that other crucial cellular components may likewise leak out of the cell. When this leakage occurs the cell will become less able to function. The muscle tissue then will have a decreased ability to contract. Therefore this change in the permeability of the muscle cell may be an important contributing factor to fatigue in marathon runners.

References

BAGBY, G., H. Green, S. Katsuta & P. Gollnick. Glycogen depletion in exercising rats infused with glucose, lactate, or pyruvate. J. Appl. Physiol. 45:425–429, 1978.

BLOCK, P., M. Van Rumenant, R. Badjou, A.Y. Van Melsem & R. Vogeleer. The effects of exhaustive effort on serum enzymes in man. In J.R. Poortmans (ed.), Biochemistry of Exercise. New York: S. Karger Basel (Switzerland), 1969.

CLARK, J.H. & R.K. Conlee. Muscle and liver glycogen content: diurual variation and endurance. J. Appl. Physiol. 47:425–428, 1979.

COSTILL, D.L., Muscular exhaustion during distance running. Phys. Sportsmed. 36:41, 1974.

COSTILL, D.L., P.D. Gollnick, E. Jansson, B. Saltin & E. Stein. Glycogen depletion in human muscle fibers during distance running. Acta Physiol. Scand. 89:374–383, 1973.

FITTS, R.H., The effects of exercise training on the development of fatigue. Ann. N.Y. Aca. Sci. 301:424–430, 1977.

MAGAZANIK, A., Y. Shapiro, D. Meytes & I. Meytes. Enzyme blood levels and water balance during a marathon race. J. Appl. Physiol. 36:214–217, 1974.

PIRNAY, F., M. Lacroix, F. Mosora, A. Luycks & P. Lefebvre. Glucose oxidation during prolonged exercise evaluated with naturally labeled [^{13}C] glucose. J. Appl. Physiol. 43:258–261, 1977.

RAVEN, P.B., T.J. Conners & E. Evonuk. Effects of exercise on plasma lactic dehydrogenase isozymes and catecholamines. J. Appl. Physiol. 29:374–377, 1970.

ROSE, L.I., J.E. BOUSSER & K.H. Cooper. Serum enzymes after marathon running. J. Appl. Physiol. 29:355–357, 1970.

SEGAL, S.S. & G.A. Brooks. Effects of glycogen depletion and work load on post exercise O_2 consumption and blood lactate. J. Appl. Physiol. 47:514–521, 1979.

SIEGEL, A.J., L.M. Silverman & B.L. Holman. Elevated creatine kinase MB isozyme levels in marathon runners. J. Am. Med. Assoc. 246:2049–2051, 1981.

WAHREN, J., Glucose turnover during exercise in man. Ann. N.Y. Aca. Sci. 301:45–55, 1977.

Relationships between Fiber Type, Enzyme Activities, Anaerobic Capacity and Human Muscle Fatigue

G. Lortie, C. Bouchard, J.A. Simoneau, C. Leblanc
and G. Thériault
Laval University, Quebec, Canada

Several indices of muscular fatigue and anaerobic capacities have been studied in relationship with histochemical and biochemical properties of human muscle. However, it can be observed that some discrepancies exist in correlations reported which might be partially explained by the absence of control over age and activity levels of subjects.

The present study was conducted to 1) verify the relationship between the histochemical and biochemical properties of human muscle and muscle fatigue index (FI) and 2) examine the relationship between FI obtained during maximal isometric contractions and alactacid anaerobic capacity (AAC). Data were submitted to statistical control over age and current energy expenditure.

Methods

1) Subjects: Some of the physical characteristics of the 28 male subjects are presented in Table 1.

2) Muscle fatigue test: The fatigue test consisted of 25 maximal voluntary isometric contractions of leg extensors, each lasting 10 s, separated each by a 5 s interval rest period (Kroll et al., 1980). Fatigue indices were calculated from decreases in strength over 25 trials. From the output, the highest and the mean strengths exerted during each trial were used in the computation of FI-I and FI-II, respectively.

3) Alactacid anaerobic test: Subjects were submitted to a 10 s all-out ergocycle test and the total work (AAC) was expressed in Joules (J). AAC/kg of body weight (AAC/kg) and AAC/l of leg volume were also

Table 1

Descriptive Statistics for all Variables of the Study (N = 28)

Variable[a]	$\overline{X}$	SD	Min	Max
Age (yr)	25.1	4.2	16.3	34.5
Weight (kg)	68.1	9.9	52.3	100.0
Height (cm)	171.1	4.5	163.0	180.0
Leg volume (liter)	9.26	1.79	5.04	12.72
% ST	41.9	13.5	18.0	63.8
HK (μmol/g min)	1.102	0.253	0.642	1.763
LDH (μmol/g min)	200	59	59	343
PHOS (μmol/g min)	9.00	2.00	6.00	12.84
MDH (μmol/g min)	178	52	94	280
HADH (μmol/g min)	5.83	2.20	2.57	12.84
Energy expenditure[b]	5.5	3.6	7.4	14.3
Fatigue index I (%)	36.1	8.5	11.4	57.1
Fatigue index II (%)	32.3	8.7	10.9	53.6
AAC (J)	7616	1043	5107	9486
AAC/kg (J/kg)	112	15	75	143
AAC/l (J/l)	842	151	471	1153

[a]See text for description of variables
[b]Expressed as number of 15 min periods with energy expenditure $\geq$ 5.5 mets over 3 days.

obtained. Leg volume was computed according to the indirect method of Watson and O'Donovan (1978).

4) Muscle biopsy: Muscle biopsies were obtained from the vastus lateralis. Muscle fibers were classified on the basis of myofibrillar ATPase (Mabuchi & Sréter, 1980). Hexokinase (HK), lactate dehydrogenase (LDH), malate dehydrogenase (MDH), hydroxy-acyl-CoA-dehydrogenase (HADH) and total phosphorylase (PHOS) activities were determined by fluorometric methods (Lowry & Passonneau, 1972) and were expressed in μmol $\times$ (g of wet tissue $\times$ min)$^{-1}$

5) Current energy expenditure: Energy expenditure was computed from a 3 day energy expenditure record as described by Lortie et al. (in press).

Results

Descriptive data for all variables of the study are presented in Table 1. Mean percent ST fiber averaged 41.9 ± 13.5, while mean FI values ranged from 32% to 36%. AAC amounted to 7616 ± 1043 J, while AAC/kg/body weight and AAC/liter reached 112 ± 14 J/kg and 842 ± 151 J/liter, respectively. For all variables, the range of scores is rather high.

Table 2 presents partial correlation between FI and other variables of the study after statistical control over age and current energy expenditure. FI-I is significantly correlated with percent ST (r = -.41) and AAC/1 (r = .45), while LDH (r = .32) and HADH (r = -.29) activities are near the significant level. On the other hand, FI-II is well correlated with percent ST (r = -.56), MDH (r = -.40), HADH (r = -.46), LDH (r = .37), and AAC/1 (r = .54).

Discussion

The present results suggest that both FI-I and FI-II are moderately but significantly correlated with percent ST. It should be noted that FI-II accounts for total work during each trial, while FI-I is based on peak strength. Other studies have obtained similar findings. For example, Tesch et al. (1978) have reported significant correlations between the force decline due to repeated high velocity isometric contractions and percent FT. As FT fibers have higher glycolytic enzymes activities (Essén et al., 1975), predominantly fast muscle will produce more energy anaerobically, thus with higher fatigue level. Our results are generally in agreement with these observations, particularly for FI-II.

The relationships between FI and AAC measurements were also investigated. Results suggest that AAC/1 (in a test lasting 10 s) is the

Table 2

Partial Correlations between Fatigue Indices and Other Variables of the Study with Statistical Control over Age and Current Energy Expenditure

Variable	Fatigue index I	Fatigue index II
AAC	.10	.17
AAC/kg	.26	.33
AAC/1	.45*	.54**
% ST	-.41*	-.56**
HK	-.11	-.14
LDH	.32	.37*
PHOS	.21	.07
MDH	-.23	-.40*
HADH	-.29	-.46*

*for p ≤ .05
**for p ≤ .01

variable which is more highly correlated with FI of the leg muscles (in a test lasting 6 min). However, these correlations remain quite moderate, since the highest common variance reaches only 30%.

In summary, the present results are in agreement with those previously reported about the relationships between muscle histochemical and biochemical properties and human muscle fatigue. Furthermore, they show 1) that FI-II tends to be a better discriminant variable than FI-I and 2) that there exists a significant but moderate correlation between FI and AAC, although the degree of specificity remains high.

Acknowledgment

This study was supported by the Ministère de l'Education du Québec (EQ-1330 and CE-29) and NSERC (E6227).

References

ESSÉN, B., E. Jansson, J. Henriksson, A.W. Taylor & B. Saltin. Metabolic characteristics of fiber types in human skeletal muscle. Acta Physiol. Scand. 95:153–165, 1975.

KROLL, W., P.M. Clarkson, G. Kamen & J. Lambert. Muscle fiber type composition and knee extension isometric strength fatigue patterns in power and endurance trained males. Res. Quart. 51:323–333, 1980.

LORTIE, G., C. Bouchard, C. Leblanc, A. Tremblay, J.A. Simoneau, G. Thériault & J.P. Savoie. Familial similarity in aerobic power. Hum. Biol. (in press)

LOWRY, O.H. & J.V. Passonneau. A Flexible System of Enzymatic Analysis. New York: Academic Press, 1972.

MABUCHI, K. & F.A. Sréter. Actomyosin ATPase. II Fiber typing by histochemical ATPase reaction. Muscle & Nerve 3:233–239, 1980.

TESCH, P., B. Sjödin, A. Thorstensson & J. Karlsson. Muscle fatigue and its relation to lactate accumulation and LDH activity in man. Acta Physiol. Scand. 103:413–420, 1978.

WATSON, A.W.S. & D.J. O'Donovan. Factors relating strength of male adolescents. J. Appl. Physiol.: Repirat. Environ. Exercise Physiol. 43:834–838, 1977.

Delay of Fatigue Effects during Exhaustive Exercise by Aldosterone

W. Skipka and U. Schramm

Deutsche Sporthochschule Köln, Köln, Federal Republic of Germany

Earlier investigations demonstrated an increase of oxygen uptake caused by aldosterone during submaximal and maximal exercise (Skipka et al., 1978; Skipka & Stegemann, 1981). Experiments of Wong and Walsh (1971)—observing an increased oxygen uptake of rat diaphragm caused by aldosterone—indicate that this augmented oxygen uptake is due to an augmented aerobic metabolic rate of muscles. These findings are confirmed by investigations of Bedrak and Samoiloff (1967), which demonstrated a higher activity of muscular oxidative enzymes after aldosterone injection. The aim of the present investigations was to reveal whether the aldosterone-induced increase of the aerobic metabolism leads to a significant delay of fatigue, estimated from the slope of heart rate and acid base values.

Methods

Nine male students performed exercise on the cycle ergometer; the exercise intensity was increased stepwise by 40 W every 3 min until the level of exhaustion was reached. In the aldosterone experiments, Aldocorten (0.5 mg) was injected subcutaneously 5 and 3 h before starting exercise, while in the control experiments a placebo (isotonic solution of NaCl) was injected.

In a second study 10 experienced male swimmers were required to perform an 800 m crawl under competitive conditions. The procedure before the starting exercise intensity was equal to the study with cycle ergometer exercise.

In both studies half of the subjects performed control experiments 7 days before aldosterone experiments, the second half vice versa.

Results

Cycle Ergometer Exercise

In the experiments with aldosterone injections the period of the last exercise intensity was prolonged by approximately 0.5 min, resulting in a significant elevation of total exercise by 8.1 ± 4.2 kJ (X ± SE). The increase of heart rate was delayed in the aldosterone experiments (Figure 1); however, at the end of the exercise, heart rate values reached the level of control experiments. Acid base values of the blood, taken from the ear lobe before and after exercise did not differ between aldosterone and control experiments.

Swimming Exercise

Figure 2 demonstrates the swimming times of the 800 m crawl differentiated to 100 m distances. After aldosterone application the subjects swam significantly faster over the last 600 m. This resulted in a decrease of the total time of 6.5 ± 2.1 s (P < 0.01). Despite the fact that swimmers in the aldosterone treatment were faster over distances exceeding 200 m, they

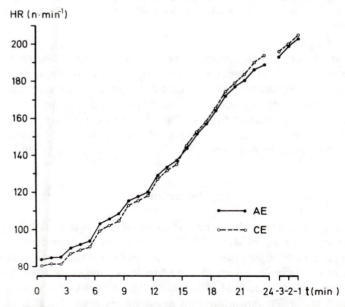

Figure 1—Behavior of heart rate during bicycle ergometer work load after aldosterone injection (AE) and during control experiments (CE). Until 24 min (= 280 W), values of AE and CE refer to the same power. Beyond this mean values of the last 3 min before ending work load are demonstrated.

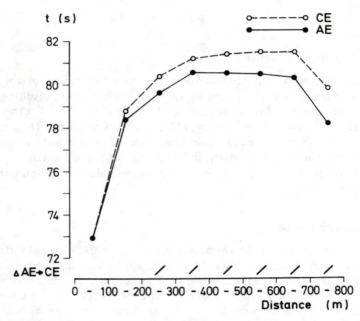

Figure 2—Mean values of the swimming times for each 100 m over the distance of 800 m crawl. Abbreviations as in Figure 1. Significant differences are indicated by symbol / (P < 0.05).

displayed diminished elevations in heart rate up to the 500 m distance (Figure 3). According to the results of the cycle ergometer experiments, the heart rate in aldosterone experiments did not reach those values attained under control conditions until the final stage of swimming. In a similar manner, acid-base values at the end of swimming were the same under both conditions. Observing the recovery period, a more pronounced decrement of heart rate could be observed after aldosterone injection (Figure 3).

Discussion

Our results concerning the behavior of heart rate and acid base values might be based on an influence of aldosterone on the aerobic metabolism. This seems mostly to be caused by an aldosterone-induced increase of the activity of oxidative enzymes observed in rat kidneys and toad bladders (Kirsten et al., 1968; Liu et al., 1972) as well as in skeletal muscles (Bedrak & Samoiloff, 1967). An additional argument is given by Pfaller et al. (1974), indicating a 200% increase in the density of cristae mitochondriales in proximal and distal tubules after aldosterone injection. Furthermore, an

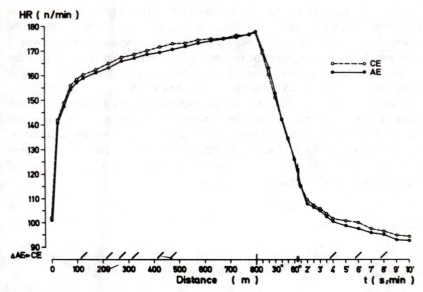

Figure 3—Mean values of heart rate before (0 m), during (25 to 800 m) and after (10 s to 10 min) swimming exercise. Abbreviations and symbols as in Figures 1 and 2.

increase of the fatty acid oxidation by aldosterone (Goodman et al., 1975; Kirsten et al., 1977) could be a reason for the enhanced oxygen uptake.

Under the present conditions the elevated activity of oxidative enzymes as well as the augmented fatty acid oxidation seems to induce a pronounced rate of turnover in tricarboxylic acid cycle. Therefore, production of lactate, which starts just in the middle ranges of exercise intensity, will be inhibited. Considering the investigations of Thimm et al. (1979), which described a dominant influence of lactate on regulation of heart rate, the slightly delayed increase of heart rate during exercise might express a delayed production of lactate under aldosterone conditions.

This suppression of lactate production caused by an augmented oxygen uptake after aldosterone injections is demonstrated by experiments with a constant, defined exercise intensity in aldosterone and control experiments (Schramm & Skipka, 1981). Aldosterone induced a less marked acidosis during exercise.

It seems to be evident that the anaerobic capacity is not influenced by aldosterone, because acid base values in the present investigations do not differ between aldosterone and control experiments. The increase of performance capacity observed during aldosterone experiments is almost caused by an increase of aerobic metabolism; meanwhile the anaerobic part of energy supply remains unchanged.

Since oxygen uptake was measured during the experiments with cycle ergometer exercise, we can criticize the behavior of oxygen pulse. The augmented oxygen uptake induced by aldosterone, especially during heavy exercise, leads to a significant increase of oxygen pulse (Figure 4). That means that, during intense exercise, aldosterone enhances the oxygen consumption referred to the same level of heart rate.

Our investigations indicate that aldosterone-inauced increases of the aerobic metabolic rate delay the process of fatigue and postpone the moment of exhaustion leading to an improvement in exercise capacity. Respecting the findings of Karasch and Müller (1951) the faster decrease of the heart rate after the end of swimming expresses an accelerated regeneration by aldosterone.

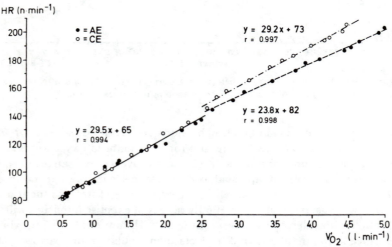

Figure 4—Heart rate of different levels of work load in relation to oxygen uptake. Abbreviations as in Figure 1.

References

BEDRAK, E. & V. Samoiloff. Aldosterone and oxidative phosphorylation in liver mitochondria. J. Endocrin. 36:63–71, 1966.

GOODMAN, D.B.P., M. Wong & H. Rasmussen. Aldosterone induced membrane phospholipid fatty acid metabolism in the toad urinary bladder. Biochemistry 14:2803–2809, 1975.

KARRASCH, K. & E.A. Müller. Das Verhalten der Pulsfrequenz in der Erholungsperiode nach körperlicher Arbeit. Arbeitsphysiol. 14:369–378, 1951.

KIRSTEN, E., R. Kirsten, A. Leaf & G.W.G. Sharp. Increased activity of enzymes of the tricarboxylic acid cycle in response to aldosterone in the toad bladder. Pflügers Arch ges Physiol. 300:213–225, 1968.

KIRSTEN, R., K. Nelson, U. Roschendorf, W. Seger, Th. Scholz & E. Kirsten. Effects of aldosterone on lipid metabolism and renal oxygen consumption in the rat. Pflügers Arch. 368:189–194, 1977.

LIU, D.R., C.C. Liew & A.G. Gornall. Effects of aldosterone on mitochondrial enzymes and cytochromes of rat tissues. Can. J. Biochem. 50:1219–1225, 1972.

SCHRAMM, U. & W. Skipka. The effects of aldosterone on aerobic and anaerobic metabolic rate during 5 min exercise. Pflügers Arch. 391:R54, 1981.

SKIPKA, W., L. Schöning, J. Stegemann & H. Vetter. Increase of $\dot{V}_{O_2max}$ induced by aldosterone. Pflügers Arch 377:R34, 1978.

SKIPKA, W. & J. Stegemann. Aldosterone-induced diminution of oxygen debt during submaximal exercise. In J. Poortmans & G. Niset (Eds.), International Series on Sports Sciences, Vol. 11-B: Biochemistry of Exercise IV-B, p. 255. Baltimore: University Park Press, 1981.

THIMM, F., M. Carvalho & M. Babka. Influence of different substances related to the exercise metabolism in an isolated muscle group on the heart rate of rat. Pflügers Arch. 382:R43, 1979.

Closing Session

Fatigue in Retrospect and Prospect: ^{31}P NMR Studies of Exercise Performance

Britton Chance, A. Sapega, D. Sokolow, S. Eleff, J.S. Leigh, T. Graham, J. Armstrong and R. Warnell
University of Pennsylvania, Philadelphia, Pennsylvania, U.S.A.

Continuous, noninvasive monitoring of muscle energy metabolism has been one of the goals of biophysical research for the past twenty years. Such an achievement would permit rapid changes in energy metabolism to be followed in undisturbed, functioning muscle. Figure 1 identifies ^{31}P nuclear magnetic resonance (NMR) and NADH fluorescence as two methods currently available for probing the bioenergetics of muscle energy metabolism.

A characteristic metabolic response of muscle during contraction is an increase in mitochondrial respiration rate concomitant with increasing tissue activity. This response is characterized by the transition from the resting state 4 to the active state 3, in which NADH and other components of the chain become more oxidized in response to "trigger" signals from ADP and inorganic phosphate. NADH oxidation can be readily measured at the surface of exposed tissue by spectrophotometry (Chance & Connelly, 1957; Chance & Weber, 1963; Jöbsis & Chance, 1957) and fluorometry as demonstrated in early experiments by Chance and Baltscheffsky (1958) and as applied to muscle by Chance and Jobsis (1959). Other methods of monitoring this metabolic activity depend on measurements of ATP. This product of oxidation re-phosphorylates creatine, replenishing phospho-creatine and reducing inorganic phosphate and NADH levels. It is noteworthy that Jacobus et al. (1982) have recently reaffirmed the essential role of ADP and presumably phosphate in the activation of mitochondrial respiration in skeletal muscle. They have found the ATP/ADP ratio to be less important in respiratory control than previously believed. Whatever the difference may ultimately be, the respiratory chain is "turned on" to a

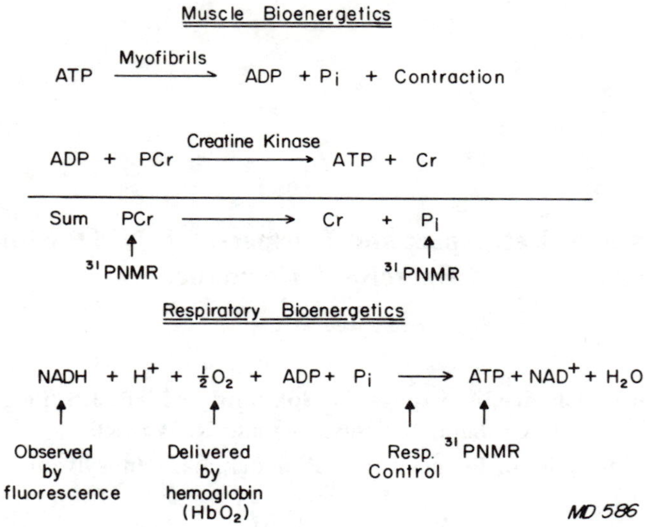

Muscle Bioenergetics

$$\text{ATP} \xrightarrow{\text{Myofibrils}} \text{ADP} + P_i + \text{Contraction}$$

$$\text{ADP} + \text{PCr} \xrightarrow{\text{Creatine Kinase}} \text{ATP} + \text{Cr}$$

$$\text{Sum} \quad \text{PCr} \longrightarrow \text{Cr} + P_i$$

$$\overset{\uparrow}{^{31}\text{PNMR}} \qquad\qquad \overset{\uparrow}{^{31}\text{PNMR}}$$

Respiratory Bioenergetics

$$\text{NADH} + \text{H}^+ + \tfrac{1}{2}O_2 + \text{ADP} + P_i \longrightarrow \text{ATP} + \text{NAD}^+ + H_2O$$

Observed	Delivered	Resp. ^{31}PNMR
by	by	Control
fluorescence	hemoglobin	
	(HbO_2)	*MD 586*

Figure 1—Chemical equations for the biochemistry of muscular contraction with labels identifying appropriate probes of noninvasive measurement by polargraphic, optical and NMR methods.

large extent by the simple addition of ADP and inorganic phosphate. ATP, as shown in these studies, remains constant, and is therefore not a highly significant factor in these metabolic control processes.

Since this talk was intended to be in part retrospective, I shall refer to my early experiments on skeletal muscle tissue and review its various states of functional metabolic control. Using NADH fluorescence, we measured the bioenergetic response of the frog sartorius to electrical stimulation (Chance & Jöbsis, 1959). As illustrated in Figure 2, electrical stimulation at 1 Hz transforms the resting state 4 muscle to the actively metabolizing and contracting state 3. Simultaneous with the initiation of contraction, ADP and P_i stimulate the oxidation of NADH, signifying the transition from state 4 to state 3. An increase in the stimulation rate from 1.0 to 1.5 Hz produces a slight increase in oxidation but is followed immediately thereafter by a precipitous shift from state 3 to state 5, as indicated by the downward sweep of the trace. Upon cessation of stimulation, O_2 better permeates the muscle and the NADH redox level is restored from that of state 5 to somewhat above that of state 3. At this point, the substrate supplies available to the muscle tissue are modestly depleted and a metabolic state between 2 and 3 is reached. From this point, there is a slow recovery of the oxygen debt, during which ATP and PCr return to resting values.

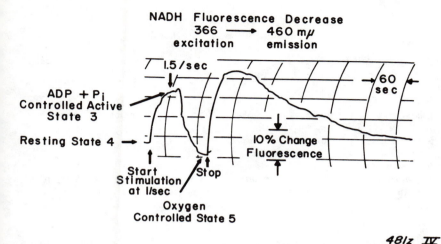

NADH Fluorescence Decrease
366 ⟶ 460 mμ
excitation   emission

1.5/sec

60 sec

ADP + Pi
Controlled Active
State 3

Resting State 4 ⟶

10% Change
Fluorescence

Start
Stimulation
at l/sec

Stop

Oxygen
Controlled State 5

481z IV

Figure 2—An identification of the metabolic states of mitochondrial activity with the contractile activity and the state of tissue oxygenation in frog sartorius muscle.

Five metabolic states have been assigned to mitochondrial function by Chance and Williams (1955) (Table 1). These states are designated on the basis of respiration rate as measured by platinum electrode and by optical signals measured spectrophotometrically and eventually fluorometrically. As Table 1 indicates, state 1 is the "endogenous" state of aerobic mitochondria where ADP and inorganic phosphate are minimal. In addition, state 1 is characterized by substrate levels insufficient to permit a true resting state 4, which has a well-supplied metabolic system. If ADP and inorganic phosphate are added to muscle in state 1, the small amount of substrate becomes exhausted and a transition to the substrate-depleted state 2 results. State 2 is characterized by a high oxidation of all the respiratory carriers. Upon addition of substrate to muscle in state 2, an abrupt transition to state 3 occurs. State 3 may also be achieved either through the excessive production of ADP and P_i or by the introduction of uncouplers. State 3 muscle contains ADP and inorganic phosphate levels sufficient to activate the respiratory chain, and enough substrate to cause a vigorous flow of electrons. This results in the phosphorylation of ADP and a reduction in P_i. The transition from the active state 3 back to the resting state 4 is characterized by a high reduction of the carriers and by a low electron transport rate. If the tissue is made anoxic at this point, a transition to the anaerobic state 5 ensues.

These characteristic states may be studied in a stimulated frog sartorius preparation where substrates are supplied via the glycolytic chain, ADP

Table 1

Metabolic States of Mitochondria

	State 1	State 2	State 3	State 4	State 5
Characteristics	Aerobic	Aerobic	Aerobic	Aerobic	Anaerobic
ADP level	Low	High	High	Low	High
Substrate level	Low-en-dogenous	Approach-ing 0	High	High	High
Respiration rate	Slow	Slow	Fast	Slow	0
Rate-limiting component	Phosphate acceptor	Substrate	Respiratory chain	Phosphate acceptor	Oxygen

and inorganic phosphate are produced from the breakdown of ATP during contraction, and where oxygen diffuses into the thin tissue specimen from the surrounding medium. These metabolic states, as identified in living muscle, serve as a useful model for understanding many of the results presented at this symposium, especially those dealing with substrate deprivation and hypoxia during exercise.

The possibility that nuclear magnetic resonance could be used to study the bioenergetics of muscular contraction was first exploited by the Oxford group (particularly Dawson et al., 1977 & 1978; also see Dawson, this volume). By simultaneously stimulating a group of amphibian muscles in a manner similar to that described in our single-muscle optical experiments, these investigators found similar relationships between muscle activity and the levels of ATP, PCr and P_i. These initial experiments were germinal to those which followed in which investigators used larger bore magnets and more sophisticated coils for detecting the weak signals.

Our approach to phosphorous NMR has been to emphasize the study of human energy metabolism. By employing an isokinetic ergometer to measure muscular work and power output, we have been able to directly compare work output and biochemical cost. As shown in Figure 3, the subject's forearm lies upon an NMR surface coil within the magnet bore while he performs wrist flexion exercise on a non-magnetic device coupled to a Cybex ergometer as shown in Figure 4. It is therefore possible to obtain a quantitative measurement of work output during submaximal, steady-state exercise, and to correlate this with the energy cost as determined by NMR.

During exercise, we see a rise in the muscular inorganic phosphate level and a fall in phosphocreatine, the sum of the two being constant for the exercise regimens we have employed. This bears out the assumption that

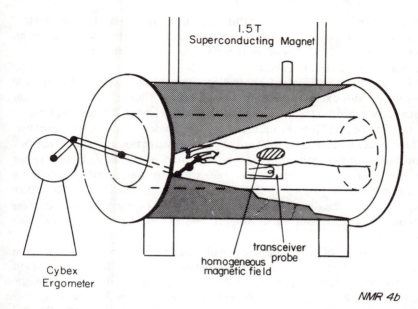

Figure 3—Coupling of ergometer measurement of muscular power output to NMR measurement of tissue metabolites.

Figure 4—An illustration of the mechanical linkage which couples the Cybex ergometer for wrist exercise within the NMR magnet.

the increase in phosphate level during exercise results almost completely from the breakdown of phosphocreatine. During contraction, there is no net depletion of ATP, nor is there significant utilization of P_i in the formation of sugar phosphates.

Under our conditions of controlled submaximal exercise, the muscle tissue operates exclusively on PCr as an energy source, with P_i as the net breakdown product. There is a constant level of ATP, and very little activation of glycolysis as indicated by the low levels of sugar phosphate. This forms an ideal situation for the precise evaluation of metabolic work and has afforded results of the type indicated in Figure 5. Here a linear relationship exists between the average external power output generated by 1 s contractions every 5 s, and the "energy cost index" (P_i/PCr ratio). Furthermore, the data are time-resolved to the extent that the data was accumulated only during a 100 ms interval at the very end of each contraction.

These data verify the suppositions that during intermittent submaximal exercise, ATP remains constant, and that there is a lack of significant glycolytic activity. The mitochondria are able to cyclically restore the deficit in the PCr/P_i level in between each contraction. We would expect that the rate of ATP synthesis is constant since the peak values of ADP and P_i are well over the K_m's for the mitochondria. The time interval between contractions over which the mitochondria operate "at full tilt" increases

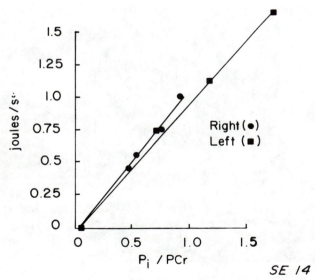

Figure 5—An illustration of muscle exercise efficiency as evaluated by the Cybex ergometer output and "biochemical cost" as indicated by P_i/PCr.

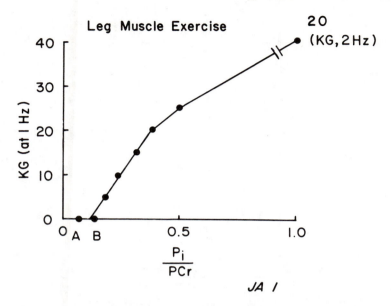

Figure 6—Performance of female distance runner in a steady state exercise efficiency test.

with the increase in mean work rate in a fashion appropriate to the negative feedback properties of the mitochondrial-myofibrillar system. At such low level work rates, a linearity will be obtained between P_i/PCr and work rate, but any condition which causes the mitochondria to operate at "full tilt" over the entire interval between contractions will cause a nonlinear response to further increases in work rate. This has been shown in experiments using the calf muscle exercise model (see Figure 6).

Maximal and supramaximal exercise regimens may also be followed with precision by the Johnson Foundation NMR system, as indicated in Figures 7 and 8. In Figure 7 the single F.I.D. measurements of metabolite levels were taken every 5 s. After 15 s, PCr/P_i falls below the level reached during maximally tolerable steady-state exercise. The PCr/P_i ratio soon reaches levels 1/10th of this, and is accompanied by ATP depletion, maximum glycolysis and formation of sugar phosphates.

Recovery from such a bout is extraordinarily prolonged (Figure 8) with a PCr/P_i recovery half-time of approximately 5 min. Presumably the deficits of phosphocreatine, ATP and oxygen incurred during exercise are very high. Furthermore, the possibility exists that oxidative phosphorylation may be inhibited here by extreme metabolic conditions, especially low pH. It is obvious that the submaximal exercise regimen affords a better evaluation of the physiological properties of mitochondrial oxidative

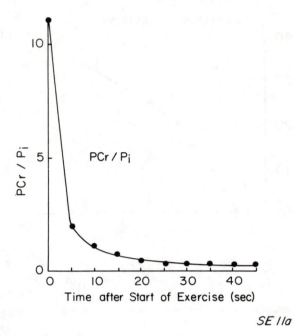

SE IIa

Figure 7—The kinetics of PCr/P$_i$ decrease during a maximal exercise bout.

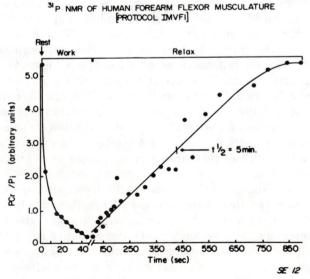

SE I2

Figure 8—Time course of PCr/P$_i$ decrease during a maximal exercise bout and subsequent recovery.

phosphorylation, while the maximal energy bout provides an index of mitochondrial function under extreme conditions.

The study of calf muscle exercise (Figure 9) is more difficult than that of the forearm because the clear bore of the magnet must be increased from the 6 inches appropriate for wrist movement to 10 to 12 inches, which is necessary for ankle motion. This is not possible with an Oxford TMR 32/200 System with its small-bore profiling shim coils installed. However, for studies of leg exercise, particularly the gastrocnemius muscle, it is not necessary to precisely define the tissue volume since cellular heterogeneity (of the fast- and slow-twitch fiber) is already present. As a result, the NMR data represents an average for different muscle fibers in varying states of activity. The average values we obtain appear to be appropriate for selective analysis and evaluation of muscle performance.

Figure 10 illustrates our calf exercise-NMR probe device. The gastrocnemius muscle rests upon the surface coil and the foot against the exercise pedal. This pedal is connected by cable to a weight-pulley system or Cybex ergometer, thereby allowing the subject to perform work in plantarflexion. Typical spectra for resting and contracting muscle are seen in Figure 11 which shows the triplicate of ATP peaks, the prominent phosphocreatine peak, and the significant increase in the inorganic phosphate peak from the resting state to a mildly exercised level as shown

Figure 9—Illustration of the insertion of the leg into the bore of the 10¼″ magnet for calf exercise studies.

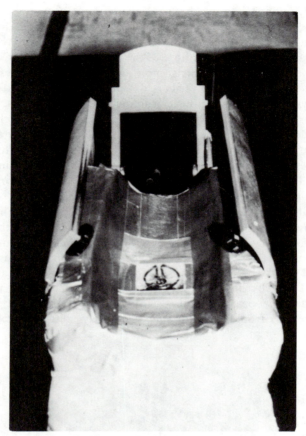

Figure 10—Photograph of Johnson Research Foundation surface coil and foot pedal appropriate for ankle exercise.

in the lower trace. A phosphodiester peak, labelled GPC, can frequently be observed in adult skeletal muscle.

Typical results obtained from a trained distance runner are indicated in Figure 6, where the changes in P_i/PCr are qualitatively similar to that of the forearm, being linear up to 20 kg weight lifted at 1 Hz. Thereafter, the curve takes a less linear shape as the mitochondria are no longer able to restore the energy levels in the interval between contractions. This break occurs when P_i/PCr equals approximately 0.4, significantly lower than the value of 1.0 obtained in our forearm exercise studies. The explanation for this is not clear, although the forearm is exercised isokinetically here, while the calf protocol involved lifting a fixed load. In the latter case, speed of

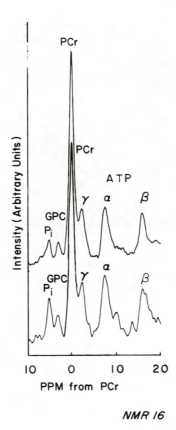

NMR 16

Figure 11—^{31}P NMR spectra of resting and exercising human calf muscle (Oxford Research Systems Spectrometer).

motion was not controlled and inertial effects may have artifactually affected the measured work rate.

It has been possible to compare normal and trained calf muscle; results are indicated in Figure 12. The ratio of slopes for the untrained subject to that of the athlete is nearly 3. Further sophistication by taking morphometric factors into account would be desirable in this type of comparison, since the subject's work output is a product of the full muscle volume available. This may differ from subject to subject.

Figure 13 compares the bioenergetic response of two members of the University of Pennsylvania Women's track team to submaximal, steady-state plantarflexion exercise. The two runners have distinctly different resting levels of P_i/PCr, both lying within the normal range. What is more

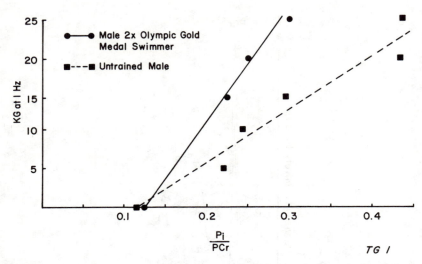

Figure 12—Comparison of trained and untrained calf muscle performance in the steady-state efficiency tests.

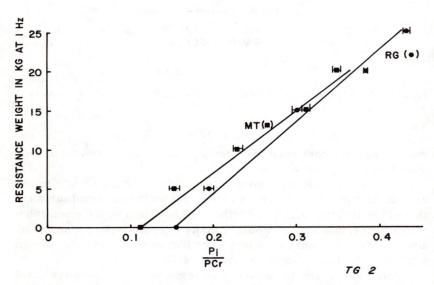

Figure 13—Comparison of work efficiencies of 2 female endurance runners of the University of Pennsylvania Track Team (courtesy of M.T. and R.G.).

PHOSPHOCREATINE DEPLETION DURING TOURNIQUET ISCHEMIA:
NMR vs. BIOCHEMICAL ASSAY

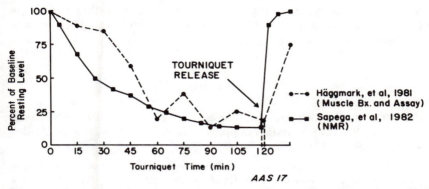

Figure 14—Comparison of analytic biochemical and NMR determinations of energy related metabolite levels in a beagle leg model.

significant is that the slope of R.G.'s trace appears to exceed that of M.T.'s trace. While these data are only preliminary, they suggest that differences in training levels and performance capability may be observed through use of this protocol. This study obviously needs a larger sample and a correlation between the metabolic performance evaluation and each athlete's abilities and state of training.

Concluding Remarks

NMR evaluation of muscular exercise performance is in preliminary stages of development but shows great promise for the future. The precision of NMR techniques compares favorably with current muscle biopsy and biochemical assay methods. This is exemplified in the comparison of the two methods shown in Figure 14. Phosphocreatine depletion in a canine and a human limb rendered ischemic by tourniquet as determined by each method is graphically depicted. The high time resolution capabilities of NMR are clearly demonstrated, particularly in the period following tourniquet release.

We hope to eventually use various forms of exercise testing in conjunction with ^{31}P NMR to characterize muscle bioenergetics in normal and diseased subjects (Chance et al., 1980), as well as further our study of trained and elite athletes.

Acknowledgment

The authors wish to acknowledge the great help in this research of Alan Bonner, John Sorge and others of the Johnson Foundation Shops; the Lumex Corporation; and the Department of Physical Therapy and Rehabilitation for the use of Cybex ergometers.

References

CHANCE, B. & H. Baltscheffsky. Respiratory enzymes in oxidative phosphorylation. VII. Binding of intramitochondrial reduced pyridine nucleotide. J. Biol. Chem. 233(3):736–739, 1958.

CHANCE, B. & C.M. Connelly. A method for the estimation of the increase in concentration of adenosine diphosphate in muscle sarcosomes following a contraction. Nature 179:1235, 1957.

CHANCE, B., S. Eleff, J.S. Leigh, Jr., D. Sokolow & A. Sapega. Noninvasive, nondestructive approaches to cell bioenergetics. Proc. Natl. Acad. Sci. USA 77(121):7430–7434, 1980.

CHANCE, B. & F. Jöbsis. Changes in fluorescence in a frog sartorius muscle following a twitch. Nature 184:195–196, 1959.

CHANCE, B. & A. Weber. The steady state of cytochrome b during rest and after contraction in frog sartorius. J. Physiol. 169:263–277, 1963.

CHANCE, B. & G.R. Williams. Respiratory enzymes in oxidative phosphorylation. III. The steady state. J. Biol. Chem. 217:409–427, 1955.

DAWSON, M., D. Gadian & D. Wilkie. Contraction and recovery of living muscles studied by [31]P Nuclear magnetic resonance. J. Physiol. (London) 267:703–705, 1977.

DAWSON, M.J., D. Gadian & D. Wilkie. Muscular fatigue investigated by phosphorus nuclear magnetic resonance. Nature 274:861–866, 1978.

JACOBUS, W.E., R.W. Moreadith & K.M. Vandegaer. Evidence against absolute control of respiration by extramitochondrial phosphorylation potentials or by ATP/ADP ratios. Biophys. J. 37:407a, 1982.

JÖBSIS, F.F. & B. Chance. Time relations between muscular contraction and response of cytochrome chain. Fed. Proc. 16(1):293a, 1957.

Fatigue in Retrospect and Prospect: Heritage, Present Status and Future

Philip D. Gollnick
Washington State University, Pullman, Washington, U.S.A.

As the final speaker in this the Fifth International Symposium for the Biochemistry of Exercise, I have been assigned the task of addressing the topic of fatigue in retrospect and prospect. This is a challenging assignment that necessitates a look backward and a look forward. In terms of the immediate symposium it could also be interpreted as a license to criticize or praise certain aspects of what has taken place over the last 4 days. I prefer not to view this assignment in the latter context since what has happened during these last 4 days only represents a microcosm in the continuing chain of events in a dynamic field. Rather, I will take a quick look to the past and consider a few points from the present meeting. Finally, I will speculate a bit about a possible mechanism by which endurance training may forestall the onset of fatigue during submaximal exercise by enhancing the use of fat and suppressing the depletion of muscle glycogen.

The Heritage for the Study of Exercise

In the opening address to this symposium Dr. Edwards indicated that the study of exercise has a rich culture. Unquestionably, it has a long history. I prefer to view this not as a culture, as I am uncomfortable with this distinction, but rather as a heritage or legacy left by those who have preceded us. This heritage of the biology of exercise comes from the important role that exercise has played in the study of responses to upsetting the equilibrium of the total body, of select systems, or of individual cells by which it has been possible to better understand how adjustments and adaptations are made to meet the challenge of altered metabolic states. One need only to read the older literature to experience the excitement of earlier investigations and to gain a foundation for additional studies. By reading the earlier works of such investigators as

Zuntz, Chauveau, Krogh, Lindhard, Hill, Meyerhof, Barcroft, Dill, Hohwü-Christensen, and Asmussen, to mention only a few, it is possible to develop an appreciation for the progress that has been made over the years that has led to a better understanding of the biology of exercise. In many instances this progress has come from the interdigitation of seemingly independent disciplines, into a multi-disciplinary approach to a general problem.

It is, of course, impossible to review the important contributions of all of the past workers. It is, at least in my mind, important for those entering the field of the biological aspects of exercise and training to realize that considerable effort has been expended in this field, that the work of the past forms the basis for what is being done today, and for work in the future. They should also realize that progress is a continual, steady process that requires years of effort. In some areas, the frontier of knowledge has had to await the development of new methodologies. We continue to see this in our current meeting. In most instances new information is added to the frontier in small increments or steps.

Our Immediate Past

As part of a retrospective view of the field, I thought it might be interesting to compare the types of problems that were studied and the progress that has been made in the 14 years since the convening of the First International Symposium on the Biochemistry of Exercise with some of those that occurred in the present, or Fifth Symposium. The first symposium was held in Brussels in 1968 and was organized by Dr. Jacques Poortmans (proceedings published as Poortmans, 1968). At the first symposium, one of the plenary papers was presented by Professor U.S. von Euler of Stockholm who subsequently was awarded a Nobel prize. He addressed the topic of the sympathoadrenal activity and physical exercise. The paper dealt with the urinary excretion of catecholamines following exercise. After his address there followed a series of papers dealing with the role of hormones in response to exercise and training. Clearly, this is a field where major advances have been made based on the development of methods to estimate the concentration of a variety of hormones, including the catecholamines, in small volumes of blood. Some papers in this, the fifth symposium, have dealt with this topic. I would point out that many studies of hormonal modification during exercise continue to be primarily the measurement of changes in the concentration of hormones in blood. It should be recognized that the blood only represents a transport system from the site of release of the hormone. In some cases this transport is to the site of action, whereas in other cases it is from the site of release, which may also be the site of action, to a disposal site. In these cases a gap still remains in establishing the exact role of the hormone(s) and identifying whether the response in the concentration in blood produces a response associated with exercise or is the result of the exercise.

Another speaker at the first meeting was Dr. Sven Fröberg, also of Stockholm, who addressed the question of lipid metabolism in blood and tissues during exercise. This problem still remains with us and has been the subject of several papers in the current meeting. Although the importance of fat as an energy source remains well known, the mechanisms controlling its mobilization and relative use continue as a fruitful area of investigation.

The problem of acid-base regulation and of electrolytes and their role during exercise was also considered in Brussels in 1968. Professor M. Scherrer, of Berne, delivered a plenary address on this topic. There were also numerous papers dealing with blood pH and the ionic composition of blood and sweat. Included were papers by Métivier and Saltin. This field remains one of major interest with attention now being focused on pH and electrolyte concentrations inside the muscle cells and the consequence of any changes that occur during exercise or fatigue. This topic has been addressed in this symposium by Drs. Donaldson, Kushmerick, Saltin and Sembrowich. Here is a vivid example of where technical advances such as application of methods such as nuclear magnetic resonance (NMR), the skinned muscle fiber, isolation of individual fibers, and the electron probe have been instrumental in pressing forward the leading edge of knowledge. Clearly, this is an area where we can expect to see rapid progress that will aid our understanding of what intracellular events occur during exercise and how they may be associated with or responsible for fatigue.

The role of lactate formation by muscle and its concentration in blood were also examined during the first Brussels symposium as part of the overall consideration of metabolism during exercise. This is an old problem (Hill & Lupton, 1923) and one that continues to haunt those interested in the metabolism of exercise. The relationship between exercise intensity and lactate formation was studied by Jervell (1928), Owles (1930), and Margaria, Edwards, and Dill (1933), and it has re-emerged and is presently being given extensive attention as the "anaerobic threshold." Reports in this symposium have dealt with the production and metabolism of lactate. I would once again caution, as above, that the concentration of lactate in blood only represents an instant in the dynamics of production in muscle or other tissues, its release into the blood, and its clearance. It may be possible that new information will be forthcoming on this rather old problem.

The 1968 symposium also included some of the first studies of human skeletal muscle from the standpoint of changes in enzymes with training and of metabolites in response to exercise. Clearly, this is an area where major progress was made following the application of the needle biopsy method to a variety of experimental models. It is presently possible to examine individual fibers for enzyme activities and substrate concentrations and to follow changes induced by exercise or training. In the present symposium considerable interest has been focused on the use of NMR for estimating the concentration of the high energy phosphates in the resting and active muscle. The application of this new and highly sensitive method gives promise of pushing the leading edge of knowledge still farther. Once again a precautionary comment seems appropriate. One must realize that, as is the case with the muscle biopsy, there are limits for interpreting data from studies with NMR. It does represent a sensitive, non-invasive method for instantaneously following changes in the phosphagens and pH of muscle. However, it is a method that scans areas of muscle that contain groups of fibers of different types—some of which may be active, whereas others are inactive during contractile activity—and by itself, it cannot give a clear indication of what conditions exist within active fibers or differentiate between types of motor units in muscle or identify active or inactive motor units. Thus, it gives an average value for a given unit volume of muscle. This is not intended to denigrate the method or to discourage its application to the study of exercise and fatigue. On the contrary, it is an exciting new area. It is only to caution that the need still exists to ask important, well-thought-out questions and to plan experiments in a manner that will lead to definitive answers and not to find new methods that are begging for questions.

This discourse on the past and present could continue for a considerable length of time. However, I would like to turn my attention to some other matters associated with this symposium. Before doing so, I would like to once again stress the importance of past experiments and to point out that much of what we are doing is only an extension of problems initially attacked by our predecessors and that we are only pushing the frontier of knowledge a little farther. We should also realize that knowledge is usually pushed forward in small steps. Some of the steps are backward steps and must be retraced. In all cases we should strive for good science that seeks the truth.

The Complexity of the Problem

The main focus of this symposium has been directed towards a consideration of factors relating to muscular fatigue. As such there have been very interesting and elegant presentations dealing with aspects of exercise on pH, ionic composition, contractile properties, substrate content and changes with exercise, hormonal influences, and others. When considered in toto it is little wonder that it is presently a conundrum. It is an extremely complex problem. Not only are there many factors involved but the permutations in the conditions such as the type of activity, type and number of motor units engaged, and state of physical training are extensive. Clearly the interactions among these factors may be such that we are faced with a complex system that will take a long time to unravel. Of particular interest and excitement in the present symposium were the probes made into the subcellular roles of pH and ionic composition on the contractile process. The studies reported with NMR also have demonstrated the limitation of the chemical methods for estimating intracellular compounds and how they change with exercise. We should be treated to some exciting new information in these areas in the near future. One area that has received relatively little attention in the present meeting was that of the motor end plate, the plasma membrane of the muscle, and the tubular systems. All these are involved in the chain of events associated with excitation-contraction coupling. There are only a few initial efforts having been reported in this area and it would appear such effort will increase in the near future.

A Common Ground for Fatigue

It does seem, however, that at present there is one common element associated with fatigue during prolonged exercise of moderate to high intensity. This appears to be closely related to a depletion of the intramuscular glycogen store. This depletion has been demonstrated by studies where biopsy samples were examined before, at selected intervals during, and after exercise (Bergström et al., 1967; Fitts et al., 1975; Hermansen et al., 1967). It occurs in a selective manner in given fiber types as a function of the intensity or duration of the exercise (Gollnick et al., 1973b & c). The effect of altering intramuscular glycogen stores and its influence on performance is well documented (Bergström et al., 1976). There is also considerable evidence demonstrating that endurance training increases endurance and is associated with a greater utilization of fats and a reduced lactate production and rate of muscle glycogen depletion (Costill et al., 1979; Davies et al., 1981; Fitts et al., 1975; Gollnick et al., 1973a; Holloszy & Booth, 1976). This increased use of fat during submaximal exercise after training is not due to a greater availability of fat since the plasma free fatty acid (FFA) concentration is similar before and after training (Karlsson et al., 1973, 1974). There are also observations to support the position that the capacity for lipid mobilization and the concentration of FFA in blood during exercise are not altered by endurance training (Gyntelberg et al., 1977). For subjects where only one leg is trained with an endurance program there is also a lower respiratory quotient, the depletion of glycogen and lactate production is less, and the use of fat is greater for the trained leg in spite of the fact that the power production, blood flow, oxygen uptake and availability of blood-borne substrates are the same to both legs (Henriksson, 1977; Saltin et al., 1976).

Adaptations to Training

A major question then is, "What are the effects of training on the body and how are these interrelated and expressed as a shift in the choice of fuel during prolonged submaximal exercise?" Training has been studied extensively. The response of the cardiovascular system and of the skeletal muscle to a variety of training regimens has been well documented. These changes can be summarized as: 1) an increase in the total body maximal oxygen uptake (Clausen et al., 1973; Davies et al., 1981; Henriksson & Reitman, 1977; Saltin et al., 1968); 2) an increase in maximal cardiac output (Clausen et al., 1973; Ekblom, 1969; Fredman et al., 1955; Rowell, 1974; Saltin et al., 1968); 3) an increase in the capillarization of skeletal muscle with endurance training (Andersen & Henriksson, 1977; Ingjer, 1979); 4) a widening of the $\alpha\text{-}\bar{V}o_2$ difference (Rowell, 1974); 5) increases in the activities of the enzymes of the citric acid cycle, electron transport system, and beta oxidation with endurance but not sprint training (Holloszy & Booth, 1976); 6) the magnitude of the increase in oxidative enzymes that occurs with training greatly exceeds that of the increase in total body maximal oxygen uptake (Davies et al., 1981; Gollnick et al., 1973a); and 7) a disparity exists between the time constant for the decay in muscle enzymes and the total body maximal oxygen uptake following cessation of training (Henriksson, 1977).

The role of the increase in oxidative potential and endurance capacity has also been illustrated in a recent study of Davies et al. (1981), where it was demonstrated that following endurance training of the rat there was an increase in maximal oxygen uptake, a greater use of fat during submaximal exercise, and an increased run time to exhaustion. An important aspect of these studies was the demonstration that the basic characteristics, that is activity per unit of protein and kinetic properties, of mitochondrial enzymes from the muscle of endurance trained animals were essentially the same as those from sedentary animals. The only major difference being that they were present in higher concentrations per unit of muscle. Conversely, sprint training increases maximal oxygen uptake (Davies et al., 1982; Saltin et al., 1976) whereas there is little or no change in the oxidative potential of muscle (Davies et al., 1982; Saltin et al., 1976; Saubert et al., 1973), in endurance capacity (Davies et al., 1982), or in the oxidation of fat during a standard submaximal exercise test (Davies et al., 1982).

It is also important to point out that most evidence to the present time indicates that at a given power production the oxygen consumption, cardiac output, and blood flow through working muscle are not altered by training (Clausen et al., 1973; Fredman et al., 1955; Rowell, 1974; Saltin et al., 1968). Since the same amount of ATP must be utilized to generate a given amount of muscular force, the availability of ADP must also be the same for both the nontrained and trained state during a standard exercise. This is important since ADP is a major regulator of oxygen consumption by mitochondria and its production is not altered at a given submaximal effort as a consequence of training.

Metabolic Regulation and Enzyme Concentration

As part of a look into the future, I would like to propose a hypothesis that may explain the role of the enhanced oxidative capacity of skeletal muscle to prolong exercise capacity at submaximal loads and to promote a greater use of fat. This hypothesis has been presented in greater detail elsewhere (Gollnick & Saltin, 1982). This hypothesis is centered around the general mechanisms of enzyme regulation. Basic to this is the principle that most enzyme (E) activity is the result of a random encounter (collision) of the substrate (S) with its enzyme or vice versa. Upon this encounter an enzyme-substrate interaction may occur with an enzyme-substrate complex (ES), being formed. This subsequently dissociates into the free enzyme E and the product P. This is summarized in the simple equation $E + S \rightleftharpoons ES \rightleftharpoons E + P$. At constant E concentration, the more substrate that is present the more likely an ES complex will be formed, and product formation increased. For many enzymes the substrate-reaction velocity relationships follow the form of a right rectangular hyperbola. The progressive decline (decrease in the increment of increase) at higher S concentration is due to more and more of E being present as ES complexes and many of the random collisions occur with these formed ES complexes and thus do not elevate substrate turnover. To saturate all of the active sites of the enzyme and produce maximal activity requires a near infinite substrate concentration, a condition which probably never exists in the intact cell. The relationship between substrate concentration and enzyme activity is described for many enzymes by standard Henri-Michaelis-Menten kinetics. This is described by the equation $v = [S] \times Vmax/[S] + Km$ where *v is the reaction velocity, [S] is the substrate concentration, Vmax is the maximal velocity at saturating S, and Km is that substrate concentration that results in a v of ½ Vmax.* It is important to realize that Km is a characteristic of the enzyme that is independent of its concentrations. However, Vmax is directly related to the total enzyme concentration. Thus, this latter characteristic indicates that in the random collision theory an increase in the number of enzyme molecules will also increase the likelihood of an E-S encounter and ES formation at constant [S] as would an increase in [S] at constant E concentration. The [S] needed to induce a given v can be estimated by rearranging the above equation to the form of [S] = Km/(Vmax/v) - 1. The effect of changing enzyme concentration on v at any substrate concentration on alpha-ketoglutarate activity of mitochondria from rat skeletal muscle is illustrated in Figure 1. The effect is similar to increasing the gain of an electronic system.

The importance of this enzyme concentration-activity relationship is, that a given v could be produced in a muscle with a lower [S] when E is increased. This would be particularly important at low substrate concentrations, that is, those below the Km where most intracellular regulation probably occurs. The implication of this to the muscle cell and to the regulation of substrate selection involves the fact that the cytosolic [ATP]/[ADP][Pi] ratio is important in the control of glycogenolysis at the level of phosphofructokinase (PFK) (Passonneau & Lowry, 1962). By increasing the concentration of mitochondria, the possibility for removing cytosolic ADP produced during muscular contraction by its translocation into the mitochondria will be enhanced. This is postulated as having two major effects. First, it will result in a more rapid stimulation of oxygen uptake and activity of the entire oxidative system in the mitochondria. Second, a more rapid removal the ADP from the cytosol when its production is unchanged, will keep the [ATP]/[ADP][Pi] ratio high and suppress glycogenolysis. Since lactate formation is the result of a competition between the pyruvate dehydrogenase system and lactate dehydrogenase there will be a reduced possibility for its production. Moreover, the Km of pyruvate dehydrogenase is lower than that of lactate dehydrogenase and at low pyruvate concentrations a majority of it will be shunted to the citric acid cycle. The maintenance of a higher cytosolic [ATP]/[ADP][Pi] ratio at a given oxygen tension by an increase in enzyme concentration can be likened to the induction of a local Pasteur effect. At a low mitochondrial concentration there will be a delay in the overall rate of ADP transport into the mitochondria. This may contribute to the delaying in the rate for the increased oxygen uptake that occurs in sedentary individuals as compared with trained persons at the onset of exercise. However, with time the [ATP]/[ADP][Pi] ratio in the cytosol will be lowered sufficiently to drive oxygen consumption at the same level in the muscle with low mitochondrial concentration as occurs in the muscle with high mitochondrial concentration. At this lower [ATP]/[ADP][Pi] ratio the inhibition of PFK will be relieved to a point where pyruvate production will be the dominate source of acetyl CoA and the excess will spill over to lactate formation.

The role of changing enzyme concentration on reaction rate is illustrated for alpha-ketoglutarate dehydrogenase in Figure 1 and isocitrate dehydrog-

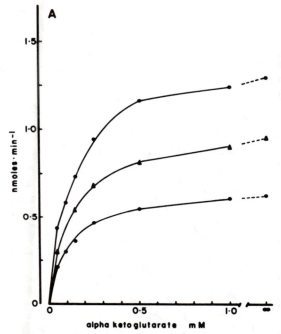

Figure 1—The relationship between enzyme concentration and the substrate-activity response for alpha-ketoglutarate dehydrogenase of mitochondria prepared from rat skeletal muscle. These data illustrate the influence of enzyme concentration on an enzyme that follows standard Henri-Michealis-Menten kinetics. Protein concentrations were 35 (●), 52 (▲), and 70 (o) $\mu g/$ ml.

enase in Figure 2. These data are from isolated mitochondria solubilized with triton. These data are presented to illustrate two types of enzymatic control. Alpha-ketoglutarate dehydrogenase is a non-equilibrium enzyme that obeys standard Henri-Michaelis-Menten kinetics in response to increasing substrate concentrations, whereas isocitrate dehydrogenase has a sigmoid curve response to increasing substrate concentrations and also is influenced allosterically by ADP. Figure 2 demonstrates that the response to ADP is also related to the concentration of enzyme in the reaction medium. This response is similar to the random collision response described above. Here the collision is with the allosteric site rather than the active site for substrate transformation. Results were similar with the intact mitochondria except that the activity per unit of protein was lower, probably due to a retarded transport of reducing equivalents across the membrane of isolated mitochondria. Although the mitochondria of the in vivo muscle fibers are intact, it is not unreasonable to expect that the overall system would operate in a similar manner where more mitochondrial surface area would be available after endurance training.

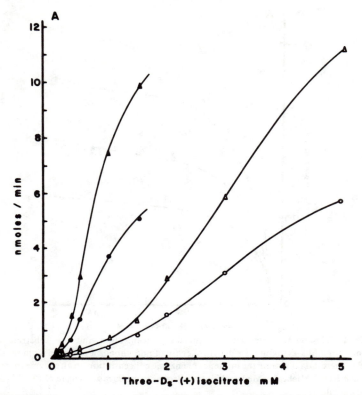

Figure 2—The influence of enzyme concentration on the substrate-activity relationship for isocitrate dehydrogenase of mitochondria from rat skeletal muscle. The substrate-reaction velocity relationship for isocitrate dehydrogenase follows a sigmoid curve and it is also allosterically modified by ADP, a non-reactive modulator. Activities were determined in the absence (open symbols) and presence (closed symbols) of ADP at protein concentrations of 120 (●), and 240 (▲) ug/ml.

Some evidence to support this concept comes from the studies of Molé et al. (1971) where it was demonstrated that the mitochondria from 1 g of muscle from trained rats oxidized palmitic acid faster at all substrate concentrations than mitochondria from 1 g of muscle from nontrained rats.

Summary

In this report, I have attempted to take a quick look at the overall study of the biology of exercise. This has included a glance at the past with some emphasis on how this has influenced and will influence the present and future. Clearly, this is an area where exciting frontiers still exist and where we can expect to see major advancements in coming years. I have also presented a scheme that attempts to explain how endurance training with its effect on muscle enzymes may influence substrate selection and work capacity during submaximal exercise. To some this may appear to be too speculative. It may be. It is, however, an attempt to stimulate thinking into this problem, as well as others. It is time to identify the roles of specific changes rather than to merely continue in their description. If there is a central theme in this presentation, it would be to build upon the past while working in the present and looking to the future.

References

ANDERSEN, P. & J. Henriksson. Capillary supply of the quadriceps femoris muscle of man: adaptive response to exercise. J. Physiol. 270:677–690, 1977.

BERGSTRÖM, J., L. Hermansen, E. Hultman & B. Saltin. Diet, muscle glycogen, and physical performance. Acta Physiol. Scand. 71:172–179, 1967.

CLAUSEN, J.P., K. Klausen, B. Rassmussen & J. Trap-Jensen. Central and peripheral circulatory changes after training of arms or legs. Am. J. Physiol. 225:675–682, 1973.

COSTILL, D.L., W.J. Fink, L.H. Getchell, J.L. Ivy & F.W. Witzmann. Lipid metabolism in skeletal muscle of endurance-trained males and females. J. Appl. Physiol.: Respirat. Environ. Exercise Physiol. 47:787–791, 1979.

DAVIES, K.J.A., L. Packer & G.A. Brooks. Biochemical adaptation of mitochondria, muscle, and whole-animal respiration to endurance training. Arch. Biochem. Biophys. 209:538–553, 1981.

DAVIES, K.J.A., L. Packer & G.A. Brooks. Exercise bioenergetics following sprint training. Arch. Biochem. Biophys. 215:260–265, 1982.

EKBLOM, B., Effect of physical training on oxygen transport system in man. Acta Physiol. Scand. Suppl. 328:1–45, 1969.

FREDMAN, M.E., G.L. Snider, P. Brostoff, S. Kimelblot & L.N. Katz. Effects of training on response of cardiac output to muscular exercise in athletes. J. Appl. Physiol. 8:37–47, 1955.

FITTS, R.H., F.W. Booth, W.W. Winder & J.L. Holloszy. Skeletal muscle respiratory capacity, endurance, and glycogen utilization. Am. J. Physiol. 228: 1029–1033, 1975.

GOLLNICK, P.D., R.B. Armstrong, B. Saltin, C.W. Saubert IV, W.L. Sembrowich & R.E. Shepherd. Effect of training on enzyme activity and fiber composition of human skeletal muscle. J. Appl. Physiol. 34:107–111, 1973a.

GOLLNICK, P.D., R.B. Armstrong, W.L. Sembrowich, R.E. Shepherd & B.

Saltin. Glycogen depletion pattern in human skeletal muscle fibers after heavy exercise. J. Appl. Physiol. 34:615-618, 1973b.

GOLLNICK, P.D., R.B. Armstrong, C.W. Saubert IV, W.L. Sembrowich, R.E. Shepherd & B. Saltin. Glycogen depletion pattern in human skeletal muscle fibers during prolonged work. Pfluegers Arch. 344:1-12, 1973c.

GOLLNICK, P.D. & B. Saltin. Significance of skeletal muscle oxidative enzyme enhancement with endurance training. Clin. Physiol. 2:1-12, 1982.

GYNTELBERG, F., M.J. Rennie, R.C. Hickson & J.O. Holloszy. Effect of training on the response of plasma glucagon to exercise. J. Appl. Physiol.: Respirat. Environ. Exercise Physiol. 43:302-308, 1977.

HENRIKSSON, J., Training adaptations of skeletal muscle and metabolism during submaximal exercise. J. Physiol. 270:661-675, 1977.

HENRIKSSON, J. & J.S. Reitman. Time course of changes in human skeletal muscle succinate dehydrogenase and cytochrome oxidase activities and maximal oxygen uptake with physical activity and inactivity. Acta Physiol. Scand. 99:91-97, 1977.

HERMANSEN, L., E. Hultman & B. Saltin. Muscle glycogen during prolonged severe exercise. Acta Physiol. Scand. 71:129-139, 1967.

HILL, A.V. & H. Lupton. Muscular exercise, lactic acid, and the supply and utilization of oxygen. Quart. J. Med. 16:135-171, 1923.

HOLLOSZY, J.O. & F.W. Booth. Biochemical adaptations to endurance exercise in muscle. Ann. Rev. Physiol. 38:273-291, 1976.

INGJER, F., Effects of endurance training on muscle fibre ATPase activity capillary supply, and mitochondrial content in man. J. Physiol. 294:419-422, 1979.

JERVELL, O., Investigation of the concentration of lactic acid in blood and urine. Acta Med. Scand. Suppl. 24:1-135, 1928.

KARLSSON, J., L.-O. Nordesjö & B. Saltin. Muscle lactate, ATP, and CP levels during exercise after physical training in man. J. Appl. Physiol. 33:199-203, 1973.

KARLSSON, J., L.-O. Nordesjö & B. Saltin. Muscle glycogen utilization during exercise after physical training. Acta Physiol. Scand. 90:210-217, 1974.

MARGARIA, R., H.T. Edwards & D.B. Dill. The possible mechanisms of contracting and paying the oxygen debt and the role of lactic acid in muscular contraction. Am. J. Physiol. 106:689-715, 1933.

MOLÉ, P.A., L.B. Oscai & J.O. Holloszy. Adaptation of muscle to exercise. Increases in the levels of palmityl CoA synthetase, carnitine palmityl transferase, and palmityl CoA dehydrogenase, and in the capacity to oxidize fatty acids. J. Clin. Invest. 50:2323-2330, 1971.

OWLES, W.H., Alterations in the lactic acid content of the blood as a result of light exercise, and associated changes in the CO_2-combining power of the blood and in the alveolar CO_2 pressure. J. Physiol. 69:214-237, 1930.

PASSONNEAU, J.M. & O.H. Lowry. Phosphofructokinase and the Pasteur effect. Biochem. Biophys. Res. Commun. 7:10–15, 1962.

POORTMANS, J.R. (Ed.). Biochemistry of Exercise, Medicine and Sport, Vol. 3, S. Karger, Basel, 1968.

ROWELL, L.B., Human cardiovascular adjustments to exercise and thermal stress. Physiol. Rev. 54:75–159, 1974.

SALTIN, B., G. Blomqvist, J.H. Mitchell, R.L. Johnson Jr., K. Wildenthal, & C.B. Chapman. Response to exercise after bed rest and after training. Circulation 38 (Suppl. VII): 1–78, 1968.

SALTIN, B., K. Nazar, D.L. Costill, E. Stein, B. Essén & P.D. Gollnick. The nature of the training response: peripheral and central adaptations to one-legged exercise. Acta Physiol. Scand. 96:289–305, 1976.

SAUBERT, C.W. IV, R.B. Armstrong, R.E. Shepherd & P.D. Gollnick. Anaerobic enzyme adaptations to sprint training in rats. Pfluegers Arch. 341:305–312, 1973.

Author Index

Adrian, T.E.	652	Brassard, L.	520
Alexander, K.	846	Brisson, G.R.	520, 631, 645
Allard, C.	219	Brooks, G.A.	397
Allenberg, K.	625	Brown, J.E.	252
Amery, A.	662	Bruno, J.F.	453
Angersbach, D.	264	Bukowiecki, L.	321, 326, 524
Armstrong, J.	895		
Armstrong, L.E.	312, 595	Cadarette, B.S.	871
Armstrong, R.B.	775	Cardinal, E.	321, 524
Åström, H.	835	Carlson, L.A.	502
		Cavagna, G.A.	794
Babij, P.	345	Cerny, F.J.	441
Baldwin, K.M.	859	Chance, B.	895
Barbee, R.W.	600	Chasiotis, D.	723
Barnett, A.	579	Chi, M.M.-L.	826
Bar-Or, O.	234	Christensen, N.J.	675
Belcastro, A.N.	545	Cintrón-Treviño, N.M.	351
Belda, M.C.R.	487	Ciriello, V.M.	787
Bell, G.	877	Cole, R.P.	815
Bennet, P.	625	Cook, J.D.	472
Berger, M.	97	Corbucci, G.G.	286
Bert, H.	385	Corey, P.	618
Berube, C.L.	871	Costill, D.L.	281, 312, 579, 595
Bianchi, C.P.	557	Côté, G.	321, 524
Bieger, W.P.	694, 715	Cousineau, D.	297
Biewener, A.A.	607	Coyle, E.F.	840
Bigland-Ritchie, B.	864	Craig, B.	281
Bilotto, G.	175	Cukes, V.G.	688
Bistrian, B.R.	497	Cutmore, C.M.M.	336
Björntorp, P.	799, 852		
Blachley, J.D.	806	Daniels, W.L.	258, 514
Blackburn, G.L.	497	Dawson, M.J.	116
Blomqvist, C.G.	472	DeCarufel, D.	645
Bolshakova, T.D.	688	DeGroof, R.C.	557
Bolton, T.	536	De Luca, C.J.	175
Bonde-Petersen, F.	781	Després, J.P.	326
Booth, F.W.	378	Ditunno, J.F., Jr.	557
Bouchard, C.	219, 326, 882	Divine-Spurgeon, L.	397
Braaten, J.T.	536	Dlin, R.	404

Dmi'el, R.	421	Hager, C.L.	31
Dobrovolsky, O.B.	688	Hagerman, F.C.	312
Doi, R.	479	Hall, G.M.	652
Dolny, D.G.	367	Hall, S.E.H.	536
Donaldson, S.K.B.	126	Haller, R.G.	472
Dotan, R.	234	Halliday, D.	345
Dreyer, A.C.	755	Haralambie, G.	441
Druckemiller, M.	373	Hashimoto, I.	657
Dufaux, B.	356	Heck, H.	239
Dulac, S.	631	Heglund, N.C.	794
		Henriksson, J.	447
Edgerton, V.R.	31	Herbison, G.J.	557
Edwards, R.H.T.	3, 286	Higuchi, M.	657
Ehsani, A.A.	87, 840	Hillman, S.L.	859
Eleff, S.	895	Hilsted, J.	675
Eller, A.K.	363	Himmelsbach, M.	738
Essen, B.	508	Hintz, C.S.	826
Essén-Gustavsson, B.	447	Hirata, F.	479
Evans, W.J.	225, 497, 514, 787, 871	Hochochka, P.W.	584, 590
		Höffken, K.	356
Fagard, R.	662	Holden, W.L.	787
Farrell, P.A.	637	Hollman, W.	239, 356
Farrell, S.	291	Holloszy, J.D.	87, 840
Fentem, P.H.	749	Holm, G.	852
Fink, W.J.	281, 312, 579	Holmgren, A.	835
Fisher, E.C.	225, 497	Holmquist, N.	625
Fixter, L.M.	336	Hultman, E.	63, 723
Fridén, J.	161	Hutchinson, T.E.	571
Frishberg, B.	291		
Fritz, J.T.	755	Imbach, A.	762
		Imelik, O.	550
Gad, P.	508	Ivy, J.L.	291
Gagnon, J.	645		
Galbo, H.	625, 675, 766	Jacobs, I.	234
Garetto, L.P.	681	Jacqmin, P.	428
Gates, W.K.	637	Jansson, E.	821
Gitel, E.P.	688	Jaweed, M.M.	557
Gleeson, T.T.	421	Jenkins, R.R.	467
Gohil, K.	286	Johansson, R.	864
Goldfarb, A.H.	453	Johnsen, S.G.	625
Goldspink, G.	607	Johnson, D.	571
Gollnick, P.D.	909	Jones, B.H.	514
Goodman, M.N.	681	Jones, D.A.	286
Gorski, J.	229		
Govindappa, S.	460	Kaijser, C.	404
Graham, T.	895	Kaijser, L.	502
Gregor, R.J.	31	Kaiser, P.	728
Gunderson, H.M.	877	Kapp, R.	738
		Karas, R.	607
Hagberg, J.M.	87, 840	Karlsson, J.	234, 404, 835
Hägele, H.	385	Katz, A.	281, 579

Kawarabayashi, T.	479	Martin, W.H.	840
Keilhoz, U.	385	Matthews, D.E.	345
Kennedy, J.M.	306	Matthews, S.M.	345
Kerr, M.G.	336, 564	Maughan, R.J.	433
Keul, J.	738	McKenzie, D.C.	584, 590
Kiens, B.	508	McLeod, A.A.	252
King, D.S.	595	McPhail, G.	286
Kjaer, M.	675	Mendez, J.	373
Knapik, J.J.	514	Menshikov, V.V.	688
Knochel, J.P.	806	Métivier, G.	667
Knuttgen, H.G.	225	Michel, G.	694
Komi, P.V.	197	Mikines, K.J.	675
Kondo, T.	479	Miller, W.	291
Krotkiewski, M.	799, 852	Millward, D.J.	345
Krywawych, S.	286	Mommsen, T.P.	584, 590
Kushmerick, M.J.	50	Mong, F.S.F.	302
		Montanari, G.	286
Labrie, A.	530, 733	Moore, R.	618
Lai, J.-S.	411	Morgan, W.P.	637
Laughlin, M.H.	775	Moss, R.F.	806
Lavoie, J.-M.	297	Muñoz, B.	749
Leach, C.S.	351	Murry, T.M.	312
Leblanc, C.	219, 882	Murthy, C.V.N.	460
Leblanc, J.	530, 702, 733		
Leclerc, S.	219	Nadeau, A.	702
Ledoux, M.	520, 631	Nadel, E.R.	134, 707
Lehmann, M.	738	Nemeth, P.M.	826
Leigh, J.S., Jr.	895	Newsholme, E.A.	144
Lemon, P.W.R.	367	Nielsen, J.	625
Levine, L.	871		
Lewis, S.F.	472	Ochlich, P.	264
Lien, I-N.	411	Oettinger, U.	385
Lijnen, P.	662	Ohno, H.	479
Lindström, L.	187	Orinius, E.	835
Lithell, H.	508, 852	Oscai, L.	331
Lönnroth, P.	852	Ovalle, W.K.	584, 590
Lortie, G.	882		
Low, M.P.	545	Parker, J.A.	877
Lowry, O.H.	274, 826	Parkhouse, W.S.	584, 590
Lucke, J.N.	652	Parliman, J.A.	877
Lundsgaard, J.S.	781	Patton, J.F.	225
Lupien, D.	733	Peckham, J.R.	806
Lupien, J.	321, 326, 524	Pedersen, P.K.	415
		Péronnet, F.	297, 520, 631, 762
Maass, U.	846	Pert, C.B.	637
Macdonald, I.A.	749	Petersén, I.	187
Mader, A.	239	Phinney, S.D.	497
Malan, D.D.J.	755	Poland, J.L.	302, 306
Mandroukas, K.	799, 852	Poland, J.W.	306
Marks, V.	799	Poortmans, J.R.	113

Posner, B.M.	871	Sturbois, X.	428
Provencher, P.J.	297	Sylvén, C.	821
Rambaut, P.C.	351	Talbot, J.	219
Raven, P.B.	806	Taniguchi, N.	479
Reddanna, P.	460	Taylor, C.R.	607, 775
Rennie, M.J.	345	Terayama, K.	479
Rettenmeier, A.	385	Tesch, P.A.	234, 258, 728
Rhodes, E.C.	584, 590	Thériault, G.	882
Richard, D.	530, 702, 733	Thoden, J.	536
Richter, E.A.	675, 681, 766	Toner, M.M.	514
Rössner, S.	502	Tremblay, A.	530, 702
Round, J.M.	286	Tremblay, R.R.	645
Roy, R.R.	31	Tronier, B.	675
Ruderman, N.B.	681, 766		
		Valdez, V.	859
Sabbahi, M.A.	175	Vallerand, A.L.	321, 524
Sahlin, K.	151, 723	Verde, T.	618
Sannerstedt, R.	404	Viru, A.A.	76, 363
Sapega, A.	895	Vocke, T.	694
Sargeant, A.J.	612	Voghel, L.	520
Savard, R.	326	Vollrath, W.	373
Schramm, U.	886	Vranic, M.	536
Schwane, J.A.	775		
Secher, N.H.	625	Wahlberg, F.	404
Seeherman, H.J.	421	Wahlqvist, M.L.	502
Sembrowich, W.L.	571	Wang, E.	571
Shand, D.G.	252	Warnell, R.	895
Sharp, R.L.	579, 595	Watson, P.A.	378
Shaw, J.H.	707	Weicker, H.	385
Shephard, R.J.	618	Wickiewicz, T.	31
Sherman, B.A.	367	Wilke, R.	264
Sherman, W.M.	291, 312	William-Olsson, T.	799
Shinn, S.L.	584, 590	Williams, C.	269, 749
Simoneau, J.	882	Williams, R.S.	252
Sjödin, B.	258, 821	Wirth, A.	799
Sjöholm, H.	63	Wolfe, R.R.	707
Sjöström, M.	161	Wolman, S.E.	345
Skipka, W.	886	Woods, J.J.	864
Smith, U.	799, 852	Wootton, S.A.	269, 749
Snow, D.H.	336, 564	Wright, J.E.	258
Sokolow, D.	895		
Sonne, B.	675	Yamakawa, K.	657
Sopper, M.M.	545	Young, D.A.	274
Srivastava, A.	378	Young, K.	433
Staessen, J.	662	Young, V.R.	497
Stainsby, W.N.	600		
Strömblad, G.	852	Zittel, R.	715
Strydom, G.L.	755	Zucas, S.M.	487
Stulen, F.B.	175		